Petrochemical Catalyst Materials, Processes, and Emerging Technologies

Hamid Al-Megren
King Abdulaziz City for Science and Technology, Saudi Arabia

Tiancun Xiao
Oxford University, UK

A volume in the Advances in Chemical and Materials Engineering (ACME) Book Series

ENGINEERING SCIENCE REFERENCE
An Imprint of IGI Global

Published in the United States of America by
Engineering Science Reference (an imprint of IGI Global)
701 E. Chocolate Avenue
Hershey PA, USA 17033
Tel: 717-533-8845
Fax: 717-533-8661
E-mail: cust@igi-global.com
Web site: http://www.igi-global.com

Library of Congress Cataloging-in-Publication Data

Names: Al-Megren, Hamid, 1962- editor. | Xiao, Tiancun, 1965- editor.
Title: Petrochemical catalyst materials, processes, and emerging technologies / Hamid Al-Megren and Tiancun Xiao, editors.
Description: Hershey, PA : Engineering Science Reference, [2016] | Includes bibliographical references and index.
Identifiers: LCCN 2015050629| ISBN 9781466699755 (hardcover) | ISBN 9781466699762 (ebook)
Subjects: LCSH: Catalytic reforming. | Catalysts.
Classification: LCC TP690.45 .P39 2016 | DDC 660/.2995--dc23 LC record available at http://lccn.loc.gov/2015050629

This book is published in the IGI Global book series Advances in Chemical and Materials Engineering (ACME) (ISSN: 2327-5448; eISSN: 2327-5456)

British Cataloguing in Publication Data
A Cataloguing in Publication record for this book is available from the British Library.

All work contributed to this book is new, previously-unpublished material. The views expressed in this book are those of the authors, but not necessarily of the publisher.

For electronic access to this publication, please contact: eresources@igi-global.com.

Advances in Chemical and Materials Engineering (ACME) Book Series

J. Paulo Davim
University of Aveiro, Portugal

ISSN: 2327-5448
EISSN: 2327-5456

Mission

The cross disciplinary approach of chemical and materials engineering is rapidly growing as it applies to the study of educational, scientific and industrial research activities by solving complex chemical problems using computational techniques and statistical methods.

The **Advances in Chemical and Materials Engineering (ACME) Book Series** provides research on the recent advances throughout computational and statistical methods of analysis and modeling. This series brings together collaboration between chemists, engineers, statisticians, and computer scientists and offers a wealth of knowledge and useful tools to academics, practitioners, and professionals through high quality publications.

Coverage

- Thermo-Chemical Treatments
- Numerical Techniques
- Artificial Intelligence Methods
- Industrial chemistry
- Ductility and Crack-Resistance
- Coatings and surface treatments
- Ceramics and glasses
- Fracture Mechanics
- Polymers Engineering
- Wear of Materials

IGI Global is currently accepting manuscripts for publication within this series. To submit a proposal for a volume in this series, please contact our Acquisition Editors at Acquisitions@igi-global.com or visit: http://www.igi-global.com/publish/.

The Advances in Chemical and Materials Engineering (ACME) Book Series (ISSN 2327-5448) is published by IGI Global, 701 E. Chocolate Avenue, Hershey, PA 17033-1240, USA, www.igi-global.com. This series is composed of titles available for purchase individually; each title is edited to be contextually exclusive from any other title within the series. For pricing and ordering information please visit http://www.igi-global.com/book-series/advances-chemical-materials-engineering/73687. Postmaster: Send all address changes to above address.

Titles in this Series

For a list of additional titles in this series, please visit: www.igi-global.com

Applying Nanotechnology to the Desulfurization Process in Petroleum Engineering
Tawfik A. Saleh (Department of Chemistry, King Fahd University of Petroleum and Minerals, Saudi Arabia)
Engineering Science Reference • copyright 2016 • 555pp • H/C (ISBN: 9781466695450) • US $235.00 (our price)

Quantitative Structure-Activity Relationships in Drug Design, Predictive Toxicology, and Risk Assessment
Kunal Roy (Jadavpur University, India)
Medical Information Science Reference • copyright 2015 • 703pp • H/C (ISBN: 9781466681361) • US $295.00 (our price)

Handbook of Research on Recent Developments in Materials Science and Corrosion Engineering Education
Hwee Ling Lim (The Petroleum Institute, UAE)
Engineering Science Reference • copyright 2015 • 493pp • H/C (ISBN: 9781466681835) • US $325.00 (our price)

Processing Techniques and Tribological Behavior of Composite Materials
Rajnesh Tyagi (Indian Institute of Technology (BHU), Varanasi, India) and J. Paulo Davim (University of Aveiro, Portugal)
Engineering Science Reference • copyright 2015 • 464pp • H/C (ISBN: 9781466675308) • US $225.00 (our price)

Handbook of Research on Diverse Applications of Nanotechnology in Biomedicine, Chemistry, and Engineering
Shivani Soni (Alabama State University, USA) Amandeep Salhotra (City of Hope National Medical Center, USA) and Mrutyunjay Suar (KIIT University, India)
Engineering Science Reference • copyright 2015 • 820pp • H/C (ISBN: 9781466663633) • US $395.00 (our price)

Innovative Applications and Developments of Micro-Pattern Gaseous Detectors
Tom Francke (Myon, Sweden) and Vladimir Peskov (CERN, Switzerland)
Engineering Science Reference • copyright 2014 • 303pp • H/C (ISBN: 9781466660144) • US $215.00 (our price)

Handbook of Research on Nanoscience, Nanotechnology, and Advanced Materials
Mohamed Bououdina (University of Bahrain, Bahrain) and J. Paulo Davim (University of Aveiro, Portugal)
Engineering Science Reference • copyright 2014 • 617pp • H/C (ISBN: 9781466658240) • US $365.00 (our price)

Quantum and Optical Dynamics of Matter for Nanotechnology
Mihai V. Putz (West University of Timisoara, Romania)
Engineering Science Reference • copyright 2014 • 527pp • H/C (ISBN: 9781466646872) • US $180.00 (our price)

IGI GLOBAL
DISSEMINATOR OF KNOWLEDGE
www.igi-global.com

701 E. Chocolate Ave., Hershey, PA 17033
Order online at www.igi-global.com or call 717-533-8845 x100
To place a standing order for titles released in this series, contact: cust@igi-global.com
Mon-Fri 8:00 am - 5:00 pm (est) or fax 24 hours a day 717-533-8661

Table of Contents

Detailed Table of Contents

Ethanol and higher alcohols (C2+OH) have attracted much attention owing to their wide applications. They can be produced from syngas using homogeneous and heterogeneous catalysts. Although homogeneous catalysts exhibit high productivity and selectivity of C2+OH, difficulties in separating and recycling homogeneous catalysts remain challenging. Among heterogeneous catalysts, Rh-based catalysts show promising higher selectivity of C2+OH. However, prohibitive cost of Rh metal hinders its large-scale application. Non-noble metal based heterogeneous catalysts include modified methanol synthesis catalysts, modified Fischer-Tropsch (F-T) synthesis catalysts, and Mo/MoS2-based catalysts. Compared with the modified F-T synthesis catalysts and Mo/MoS2-based catalysts, production of undesired byproducts on modified Cu-based catalysts can be well suppressed. Here, the influences of additives and supports on catalytic activity of modified Cu-based catalysts are discussed. Reaction mechanism and development of novel reactors are also included.

In this work, a series of supported CoMo or NiMo HDS catalysts have been prepared based on the organic-metal matrix decomposition method and tested for diesel deep HDS with minimum hydrogen consumption under relatively low hydrogen partial pressure conditions. The aim is to develop a HDS catalyst which can reduce sulphur in diesel fuel from 5000ppm down to 50 ppm in a single pass with minimum hydrogen consumption under the conditions of 340oC, 35 bar, LHSV 1.2 h-1 with low H2/ oil ratio. The catalysts preparation process was monitored and the resultant catalyst samples before and after the HDS performance test have been characterised, some interesting results have been obtained. The presence of citric acid as organic additive/dispersing agent/chelating agent in the impregnation solution improved HDS activity compared to the equivalent CoMo catalyst prepared without citric acid, The order of activity of the cobalt precursors is Co citrate > Co acetate > Co nitrate.

Hydrodesulphurization (HDS) is a standard process for removing sulphur compounds in fuel oils in industry. HDS is effective to remove simple aliphatic sulphur compounds while less effective to remove thiophenes, dibenzothiophenes, and their derivatives because of sterically hindered adsorption on catalyst surface. Application of ionic liquids (ILs, a new class of compounds) substituting for traditional volatile organic solvents in extractive desulphurization (EDS) or oxidative desulphurization (ODS), have been being studied intensively in the latest decades, and many very promising results have been obtained, showing a good prospect as complement method to HDS. In this chapter, these fresh research results of EDS and ODS using ILs are summarized along with comprehensive discussions on diversified desulphurization factors along with some potential problems. It can be inferred that ILs are a class of potential ideal solvents to realize clean fuel oil in future although some problems come too.

There has been still a growing interest in ionic liquids (ILs) for catalysis, in particular considering their tremendous success made over the past decades. Indeed, ILs for catalysis has been a main subject in chemistry and technology, and a rough estimate is that more than 6000 publications involving IL catalysis have been reported in the past fifteen years. Since there have been a large number of excellent reviews and books concerning the catalysis in ILs, in this chapter the authors mainly focus on the IL immobilized nano- or super-fine metal particles for catalysis, which could bridge or fill the gap between homogeneous and heterogeneous catalysis. Detailed IL-immobilized catalyst preparation, characterization and their application in hydrogenation, C-C coupling, oxidation, etc. will be discussed.

Kinetic models were developed to account for the partial contributions of intermediates in complex parallel–consecutive reactions. The models allow precise estimation of the apparent rate constants of all steps in such a reaction network. The hydrodesulfurization (HDS) of dibenzothiophene (DBT) over CoMo-based alumina and carbon catalysts, and over an unsupported molybdenum sulfide catalyst, were investigated in a batch reactor and used to represent this type of reaction. The HDS reactions proceeded

Preface

Dear Readers,

It is our great honor to present this IGI Global Series book on the "Petrochemical Catalyst Materials, Processes, and Emerging Technologies". As we know that Petrochemical Industry is a technology-intensive sector, where any small changes in the technology could lead to a big improvement. The development and understanding of chemical processes relies on knowledge of various disciplines, from raw materials, reaction mechanism, process parameters and the target materials. In the modern information age, lots of changes are happening every day in petrochemical industry, such as new materials discovery and application, process modification and automation, market and demand changes, thus leading to new and emerging technologies.

This IGI Global book is to address the latest development in the petrochemical technologies, especially focusing on the core technologies such as catalyst materials for fuel upgrading, especially on sulfur removal, syngas conversion, alkane and oxygen activation, functionization of polymer. Of course safety is always the top priority in petrochemical industry, so we also cover these topics for the expanding LNG industry. The chapters cover research from mechanism study to the industry catalysts development, from homogenous catalysis to the heterogeneous catalyst and from small molecules activation to the heavy oil upgrading. This book also includes both the conventional process like supported catalysts, and latest catalytic process and materials in ionic liquid, the chemicals resources include fossil fuel, syngas as well as the glycerol, the renewable resources. These will be helpful for the academic researchers and the industrial engineers for them to have a quick view on the latest development in these areas.

Compilation of a professional book is a trivial job, which takes us 15 months. However, we are encouraged by the very enthusiastic authors from various countries, they actively responded for the call for chapters and we initially have received more than 50 plans for chapters. We screened down to 26 chapters for full writing, and after submission and extensive reviewing, we are very pleased to include 15 chapters in this book.

Here we would like to express our sincere thanks to these contributors for their hard and serious working, also we like to thanks our reviewers for their constructive input. Last but not least, thanks are due to IGI Global for giving us the chance to edit the book. We hope readers would enjoy the reading and feel free to point any errors or mistakes which we may not be spotted. Your comments and suggestions are highly appreciated.

Yours sincerely,

Hamid Al-Megren
King Abdulaziz City for Science and Technology, Saudi Arabia

Tiancun Xiao
Oxford University, UK

Chapter 1
ADMET:
Functionalized Polyolefins

Taylor W Gaines
University of Florida, USA

Kathryn R. Williams
University of Florida, USA

Kenneth Boone Wagener
University of Florida, USA

ABSTRACT

The use of acyclic diene metathesis (ADMET) in the synthesis of functionalized polyolefins is discussed. The nature of polymerizations, catalysts, and techniques are briefly covered. An overview of recent functionalized ADMET polymers and major contributions to its methodology is given. This chapter will place special emphasis on the use of ADMET to synthesize increasingly complex and new morphologies, resulting in well-defined polyolefin structures and functionalized materials unlocked as a result of this powerful polymerization method. Presented are polyolefins, materials, and architectures not possible through conventional polymerization techniques.

1. INTRODUCTION

Functionalized polyolefins: incorporation of desired properties into the world's most common plastics. Currently, the production of all forms of polyethylene (PE) total above 80 million tons each year. (Piringer, Baner, & Editors, 2008) This tonnage is attributed to the availability of cheap starting materials derived from fossil fuels and now even agricultural products. Commercially produced functionalized polyolefins are synthesized by the copolymerization of ethylene and other vinyl monomers, such as vinyl acetate and acrylates. However, due to the many side-reaction products, commercially produced functionalized PE structures are not well defined and are limited by defects such as uncontrollable and random branching. ADMET is recognized as the premiere polymerization technique for synthesizing complex polyolefin structures requiring accurate placement of functionalities. (Schulz & Wagener, 2014)

DOI: 10.4018/978-1-4666-9975-5.ch001

Since the emergence of Acyclic Diene Metathesis (ADMET) polymerization in the early 1990s, many sophisticated structures and functionalized polyolefins have been produced. Moieties including alcohols, acids, ketones, halogens, aromatics and others have been incorporated into PE. (Opper & Wagener, 2011) Because of the mild reaction conditions and limited number of side reactions, ADMET provides a level of control not possible in conventional techniques. Thus, ADMET is the leading polymerization technique for synthesizing complex architectures requiring accurate and precise placement of functionalities, while limiting reaction defects. Imperfections such as crosslinking and uncontrollable branching make conventionally polymerized materials difficult to study fundamentally. Vasile & Editor, 2000) These side reactions are less desirable for advanced applications where precision is required.

Elimination of defects is made possible due to increasing availability of highly-selective olefin metathesis catalysts in recent years. Although metathesis polycondensation was first hypothesized as early as the 1950s, success was not achieved until 1991 with the successful polymerization of 1,9-decadiene and 1,5-hexadiene. (K. B. Wagener, Boncella, & Nel, 1991) The advent of ADMET coincides directly with the development of novel catalysts such as Schrock's and Grubbs' catalysts, which earned their developers (along with Yves Chauvin) a shared Nobel Prize in 2005. (Harries-Rees, Chauvin, Grubbs, & Schrock, 2005) The continuing development of new, more tolerant catalysts allows for more sophisticated and functionalized polyolefin structures to be produced.

To better understand the effects of defects in PE, precision models of PE functionalized with branches have been synthesized via ADMET. More recently, high-polarity, acidic, ionic, renewable chemicals and drug moieties have been employed as branches. These precision polymers have brought new properties to polyolefins and the potential for advanced applications, such as ion transport, renewable and biodegradable plastics, as well as drug therapeutics. This chapter describes ADMET, the reaction and techniques, followed by a discussion of precision functionalized polyethylenes, with an emphasis on major contributions as well as recent literature examples.

2. ADMET

Before discussing the details of various functionalized polyolefins synthesized by ADMET, it is first necessary to give the reader a brief description of the ADMET reaction and mechanism, including considerations and limitations. This section also provides a general procedure for performing a typical polymerization.

2.1 Polycondensation of α,ω-dienes

Although polyolefins are typically produced by chain-addition reactions, yielding molecular weights into the hundreds of thousands and even millions, the ADMET mechanism is based on stepwise-condensation. (Opper & Wagener, 2011) For every metathesis event between α,ω-diene monomers, a molecule of ethylene is evolved (Figure 1a). Removal of ethylene drives the reaction forward, slowly forming dimers, trimers, tetramers, and so on. All of these olefinic molecules may react at any time, continually increasing the molecular weight. Since, the reaction is step-growth in nature, the Carothers equation applies; i.e., only at near quantitative conversion of monomer is high molecular weight polymer formed (Figure 1b). Hence, monomer purity is absolutely crucial to prevent unintended chain capping and other side reactions.

Figure 1. a) ADMET – polycondensation of α,ω-diene monomers with one molecule of ethylene released per number of repeat units x. b) Carothers equation plot emphasizing that monomer conversion must approach 100% for high-molecular weight polymer via step polymerization

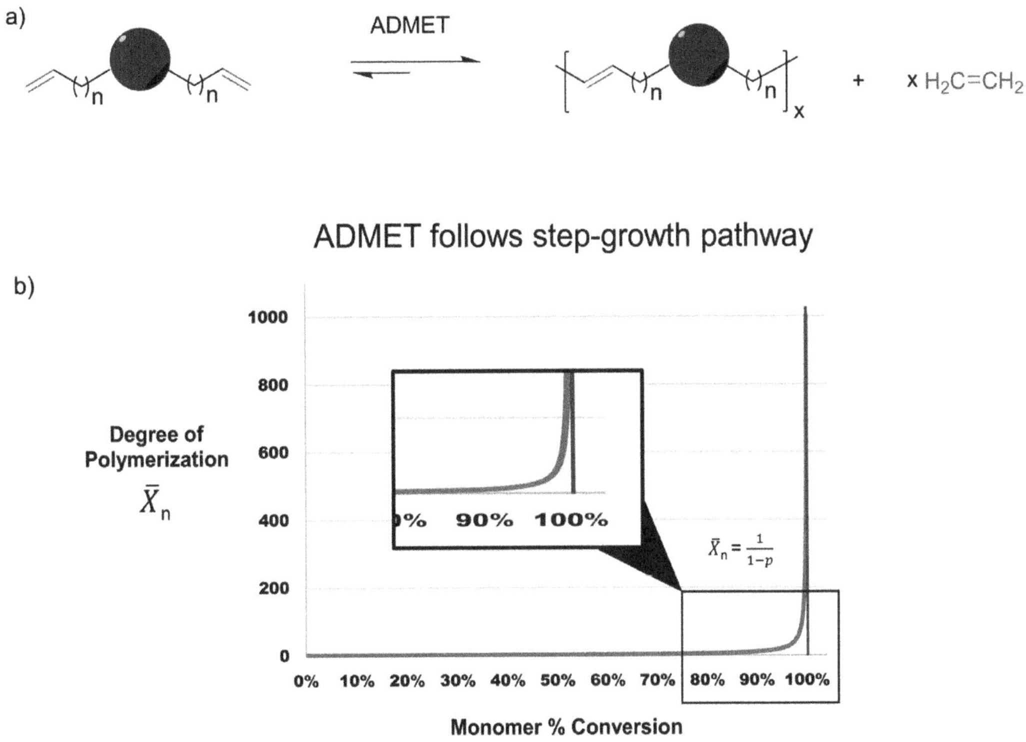

Even with these precautions, molecular weights of ADMET polymers are slightly lower than those prepared by radical, Zeigler/Natta and metallocene reactions. Also, dienes are quite sophisticated and expensive to produce, compared to ethylene and substituted vinyl monomers. Thus, commercial polymerization of dienes to produce materials capable of competing with commodity plastics in the world market is not feasible from a cost perspective. However, high-end engineering plastics requiring fine-tuned properties are easily synthesized via ADMET. As mentioned above, molecular weight is also a factor, and currently commercial polyolefins may be produced with much higher molecular weight (10^5 to 10^6 g/mol) than ADMET provides (ADMET upper limit ~100,000 M_w). On the other hand, functionalized polymers typically produced by step-growth conditions, while lower in molecular weight, have very strong intermolecular interactions (e.g., hydrogen bonding, pi stacking, ionic bonding), whereas PE depends on mechanical strength provided by molecular weight and van der Waals forces. For example, nylon at a number average molecular weight of 11,300 (g/mol) has sufficient mechanical integrity for certain applications. (Lewin & Editor, 2007) Strong physical interactions along with structural control can be exploited to provide advanced polyolefin materials capable of new applications, like ion transport, while maintaining the crystallinity of linear PE.

The advantage of ADMET is its utility in production of these precisely functionalized polyolefins, but the control arises from selective catalysts and symmetrical monomers, the key factors in ADMET sophistication. Design symmetry is built into the monomer and therefore transferred to the resulting polymer. This characteristic is particularly useful in precision systems and branch-functionalized models of PE varieties. While maintaining a linear, defect-free PE backbone, a functionality or purposely placed defect is installed at a known concentration with known frequency. Thus, an equal number of carbons can be placed between each functionality if desired. Such precision allows scrutinizing defects at the molecular level to answer questions such as: What is the effect of a specified defect placed on every 21st carbon versus every 75th carbon?(Inci & Wagener, 2011) The ability to provide such fundamental morphological insight is unique to ADMET.

2.2 ADMET Catalysts

ADMET's rise in the literature began shortly after efficient and robust olefin metathesis catalysts were introduced. However, olefin metathesis and metathesis polymerization were extensively theorized in the 1950s and 60s, building upon the polyolefin successes of Zeigler/Natta chemistry. In 1967 Nissim Calderon reported a successful polymerization of cyclic alkene monomers via metathesis using $MoCl_5$/$EtAlCl_2$/EtOH. (Calderon, Ofstead, & Judy, 1967) At that time, catalysis advancements were not sufficient for ADMET, but the efforts of Richard Schrock in the 1970s and Robert Grubbs in the 1980s led to catalysts later used to polymerize diene monomers via metathesis polycondensation.

Schrock-type catalysts feature an early 4d or 5d transition metal, such as tantalum, and positively charged ligands. Such a complex is termed a "Schrock carbene." More active Schrock catalysts feature tungsten and molybdenum metal centers (Figure 2). However, due to the highly electrophilic and oxophilic nature of early transition metals, these catalysts are not ideal for most functionalities. (Schrock & Hoveyda, 2003) Handling and storage also complicate their usage. Ruthenium catalysts are the most popular and practical for producing functionalized ADMET polymers because of their tolerance to many

Figure 2. A variety of olefin metathesis catalysts

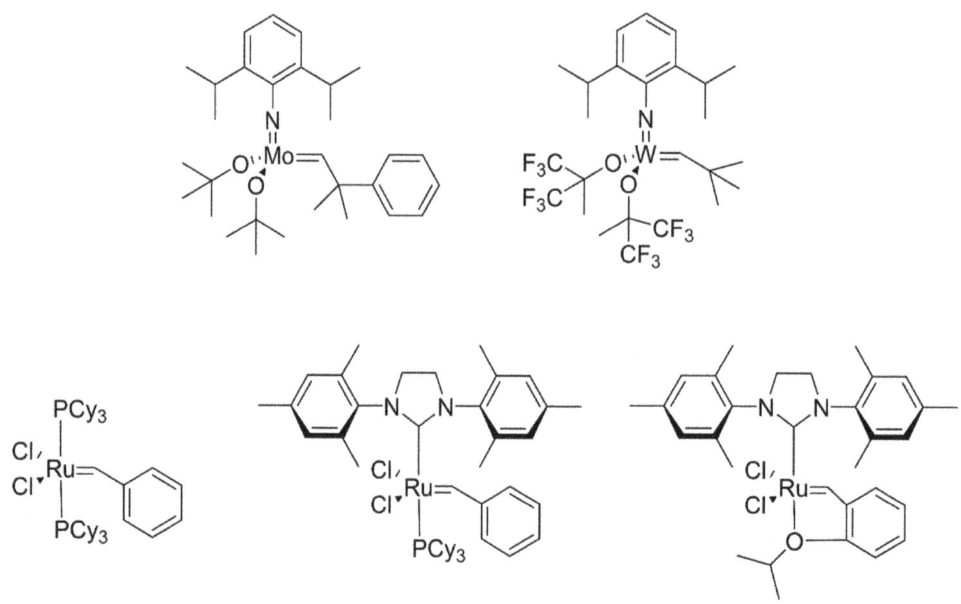

functionalities and polymerization conditions. (Anon, 2004) This versatility and robustness are attributed to the ruthenium metal center, which, unlike tungsten or molybdenum, is not as easily poisoned by oxygen-containing or other nucleophilic moieties.

To date, libraries of catalysts have been developed and are in use across the chemical industry. The chemistry of these catalysts will not be presented here, but it is worth noting that the catalyst should be selected with consideration of the desired functionalization and activity required.

2.3 Mechanism

The now widely accepted olefin metathesis mechanism was first proposed by Yves Chauvin in 1971. (Herisson & Chauvin, 1971) At the time, the mechanism was debated in the scientific community, but it was later substantiated by others in the field, including Grubbs. For his contribution to understanding the mechanistic details of olefin metathesis, Chauvin also shared the 2005 Nobel Prize with Grubbs and Schrock. Figure 3 displays the ADMET adaptation of Chauvin's proposed productive pathway mechanism. The catalysts shown in Figure 2 are actually considered precatalysts 1, which react to afford the

Figure 3. ADMET Mechanism

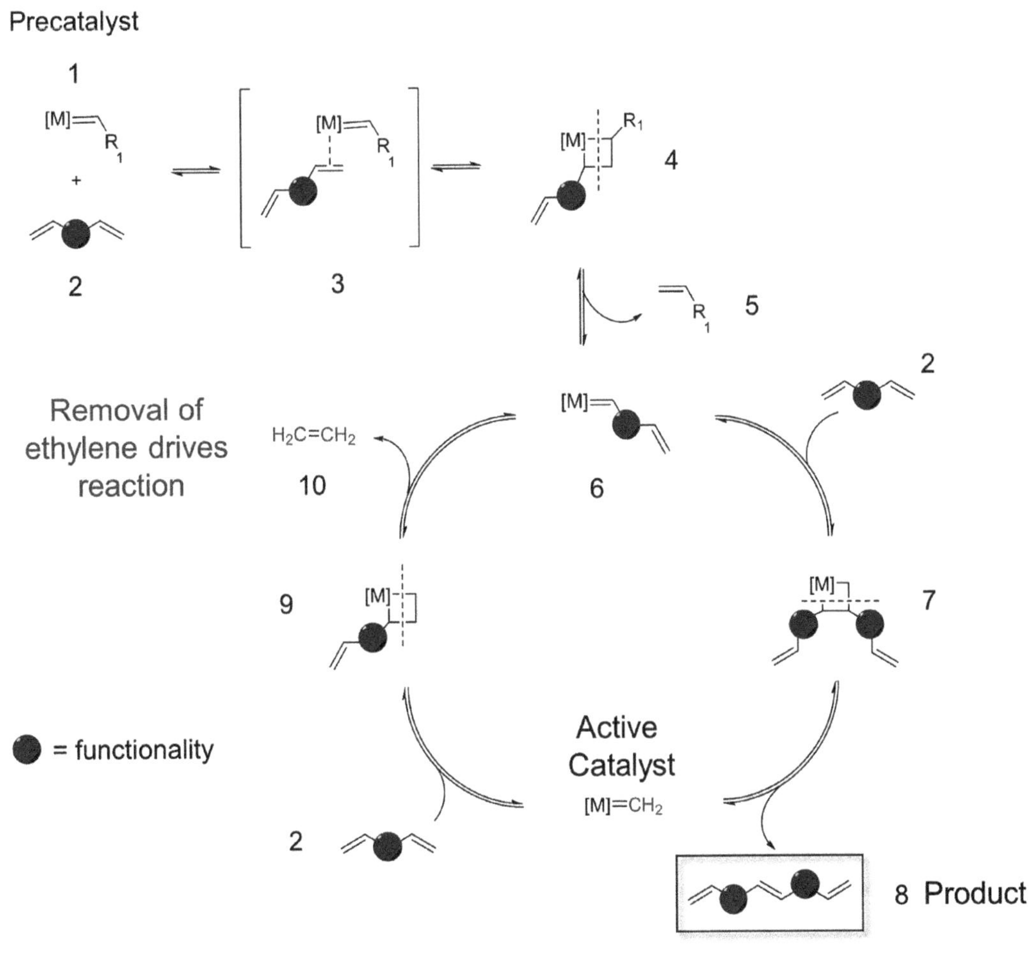

Figure 4. Typical ADMET apparatus a) bulk polymerization of a functionalized α,ω-diene monomer under high-vacuum. b) example of solution polymerization utilizing both high-vacuum and argon flow to maintain a partial pressure, preventing high-boiling solvent from evaporating

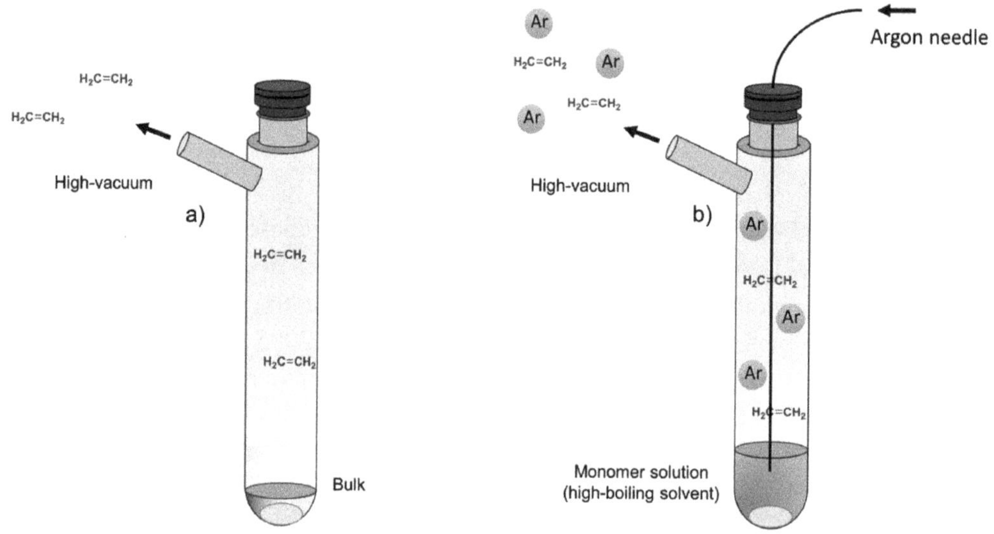

true catalyst, a methylidene carbene. An alkene from a functionalized α,ω-diene monomer 2 coordinates to the catalyst 3 allowing for a 2 + 2 cyclcoaddition and resulting in a metallocyclobutane intermediate 4. The metallocyclobutane then undergoes a 2 + 2 cycloreversion, decomposing into styrene 5 (in the case of Grubbs' first generation) and an α-substituted metal alkylidene 6. At this point, coordination of another alkene occurs followed by another cycloaddition 7. The newly coordinating alkene can be monomer, polymer, or any olefin species present, even ethylene if equilibrium is reversed. Assuming the reaction is driven forward towards polymer product, the final cycloreversion yields a larger olefin species 8 and the regenerated catalyst. Continuation will release ethylene gas 10 for terminal dienes. The catalyst remains active until poisoned or quenched (discussed in Section 2.4).

2.4 Performing an ADMET Polymerization

There exists a variety of polymerization setups possible for ADMET; this summary is in no way comprehensive of the techniques available, but is merely to serve as a starting point. Selection of an apparatus and conditions depends on the desired polymer product and monomer(s) to be polymerized. Typically, α,ω-diene monomers are viscous oils which can be polymerized in bulk. However, many monomers are solids even at high temperatures and require solution polymerization. The most crucial requirement to be met by any setup is the ability to remove the ethylene by-product, or adequate molecular weight is unachievable. Two effective methods for carrying out an ADMET polymerization of most monomers are: 1) bulk polymerization of liquid monomer under high vacuum (Figure 4a); and 2) solution polymerization in a high-boiling solvent (Figure 4b). Halogenated solvents prove to be the best for olefin metathesis reactions in general. (Schulz & Wagener, 2012) A method published by Plenio in 1,2-dichlorobenzene has been exploited extensively. Weychardt & Plenio, 2008) Opposed to the high-vacuum used in bulk conditions, a partial vacuum is used and is still adequate for removal of ethylene from solution, while maintaining the solvent. Another very crucial decision is the choice of catalyst. There are libraries of catalysts from which to choose, and each choice depends on a multitude of factors, including reaction temperature, substrate tolerance, and activity, to name a few.

After the successful synthesis and purification of a functionalized α,ω-diene, the dry monomer or monomer solution is degassed in a Schlenk tube either by freeze-vacuum-thaw cycles or by bubbling argon through the monomer while stirring. To induce an ADMET reaction the catalyst of choice is simply added to the Schlenk tube, after which the flask is purged and subjected to the necessary conditions to drive the reaction forward. If Schrock-type catalysts are being used, a glove box is most likely necessary – another disadvantage ruthenium catalysts overcome. Polymerization temperatures for ADMET reactions are usually kept at 50 °C or below, due to the propensity for alkene isomerization at higher temperatures. Isomerization of alkenes presents a few notable issues, the most serious being the polymer structure, which will no longer be exactly as intended with known run-lengths purposely installed in the monomer. Another result of isomerization is reduced reactivity due to higher molecular weight condensates which must be removed. Propylene, butylene, and pentene are more difficult to completely remove than ethylene from a viscous reaction solution, and their presence will inevitably inhibit the equilibrium shift to product. If precision spacing is not a requirement, nor is a sophisticated model of PE the desired product, temperature control is less important.

Polymerization times can vary from hours to days, again depending on a multitude of factors. Longer reaction times are known to result in more isomerization. Reaction time and temperature, mentioned above, result in ruthenium-hydride formation caused by catalyst decomposition. The Ru-hydride species is known to cause isomerization and will alter the location of alkenes. (Schmidt, 2004) However, all polymers will possess internal alkenes from cross-metathesis events. Because these alkenes are still active, to avoid further reaction of the polymer with the catalyst, ethyl vinyl ether is added to "quench" the reaction by rendering the metathesis catalysts virtually inactive. This prevents reversal of the equilibrium (depolymerization if ethylene is present), cross-metathesis of the polymer chain, etc.

For conventional polyolefins to be mimicked through ADMET, the resulting unsaturated polymers must be completely hydrogenated. Saturation can be achieved through a variety of methods, including catalyzed high-pressure hydrogenations or the use of quantitative amounts of hydrazide reducing agents. (Boz, Ghosal, Alamo, & Wagener, 2006; Opper & Wagener, 2011) The limitations of saturation methods also depend on structural tolerance to the conditions and catalyst tolerance to the functional groups. A common technique is the use of Wilkinson's catalyst, $RhCl(PPh_3)_3$, with a hydrogen pressure between 500 and 900 psi for periods of around 5 days. At these conditions, extreme caution should be exercised and choice of apparatus, as well as solvent, should be investigated prior to performing reactions in order to avoid mishaps. For example, ether solvents such as THF and dioxane (great solvents for some polymers) may form explosive peroxides under certain conditions required for alkene saturation.

3. FUNCTIONALIZED POLYOLEFINS BY ADMET

Now that the reader has some knowledge of how to make an ADMET polymer and the details of the reaction, the following section will provide a brief discussion of the wide array of possible functionalized polyethylenes successfully synthesized with ADMET since the technique's emergence. This is not meant to be a comprehensive list of every polyolefin derivative; rather, it is meant to give the reader a review of important structural advances and insight into the vast realm of possibilities achievable with this powerful, yet controllable, polymerization technique. The development of ADMET began in the early 1990s by Kenneth Wagener and his group, who have published a large number of functionalized polyolefin studies. However, researchers now use the method in many countries and institutions, attesting to the technique's power and versatility being realized world-wide.

Polyolefins display a range of properties, from those useful in commodity materials to those required of engineering plastics. These characteristics are suitable for many applications including cost-effective food storage, disposable bags, even artificial joints and bulletproof vests. Low cost, desirable properties, and practicality are the reasons why polyolefins are the most abundant plastics on Earth. However, alternative functionalization can improve and expand the use of polyolefins to new frontiers.

3.1 Branch Functionalization

Although hydrocarbon branches possess no heteroatoms nor are they particularly different from the backbone carbons, they do affect the properties of polyolefins. Branch sites act as impurities disrupting crystallinity, for example. Thus, branches are "functional" and serve purposes in regards to physical properties, and are therefore included as functionalities in this text. Commercial PE, especially low-density polyethylene (LDPE), contains random branching as a result of radical polymerization. Radical side reactions including chain transfer and back-biting result in LDPE with a mixture of long- and short-chain branches. Metallocene linear low-density polyethylene (LLDPE) has branches purposely installed by the copolymerization of ethylene with α-olefins, including 1-butene, 1-hexene, and 1-octene (Boz, E., Ghosal, A., Alamo, R. G., & Wagener, K. B., 2006). Using various lengths and differing concentrations of branches, properties such as tensile strength can be tailored to a range of applications. However, placement of intentional branches is still random and the fundamental understanding of their morphological effect remained a mystery for some time. An accurate and precise method of controlling branch frequency and concentration *exactly* was needed. This method is ADMET.

Perhaps one of the most significant contributions of ADMET to date is the ability to model these polyolefin systems by isolating the effect of branch defects. Utilizing the monomer symmetry mentioned previously, theoretically any desired length branch can be placed on a PE backbone at any desired frequency. This structural precision allows for isolation and study of concentration and branch identity effects on PE morphology. Such information provides basic fundamental knowledge which may be used to improve upon an industry supplying billions of dollars in plastics.

The branch identity has a significant effect on PE behavior. The degree of crystallinity in PE types is a consequence of branching types, frequency, and distribution. Multiple studies by Wagener and coworkers have provided PE functionalized with over a dozen different branches spanning methyl to pentadecyl on backbones where some groups were placed every 15th, 21st, and 39th carbon. (Rojas, Inci, Wei, & Wagener, 2009) One particular study, provided insight into the effect a butyl group would have if placed every 75th carbon, a synthetic challenge. Even a butyl every 75 carbons affects the properties of PE, lowering the melting point from 134 °C (linear PE) to 104 °C.[8] Because of these investigations, a better understanding of PE crystallization, lamellar thickness, and crystal type due to branching is available.

PE modeling studies began with the simplest branch: methyl groups. Symmetrical monomers were synthesized as described previously with an equal number of -CH_2- units between the methyl and terminal alkenes, translating to PE functionalized at precise intervals ranging from 8 to 20 carbons between methyls. (Smith, Brzezinska, Valenti, & Wagener, 2000) To serve as a comparison, random copolymerizations of 1,9-decadiene and methyl-substituted monomers were performed. (Sworen, Smith, Wagener, Baugh,

Figure 5. Comparison of ADMET precision methyl model versus random methyl model also synthesized by ADMET. Adapted with permission. (Sworen et al., 2003) Copyright 2003, American Chemical Society

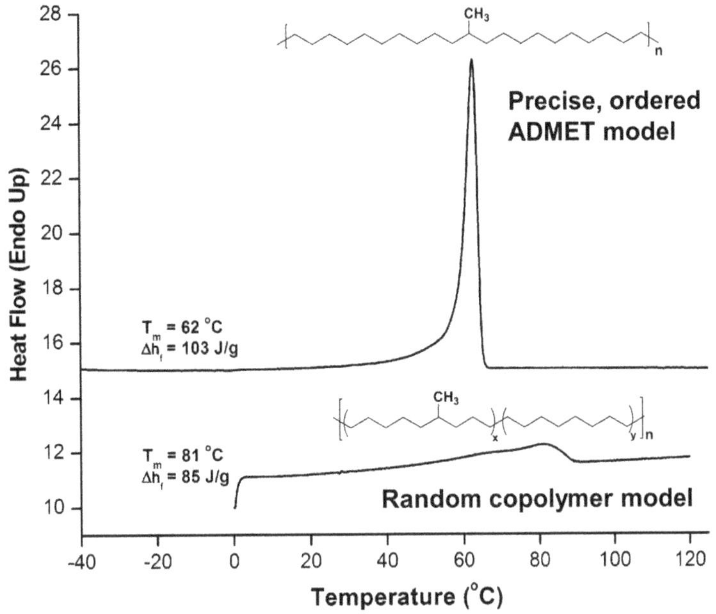

& Rucker, 2003) Very different behavior was observed for the precise and random methyl-containing polymers (Figure 5). Differential scanning calorimetry (DSC) data for PE with random methyl placement display a broad melting transition, in contrast to the sharp melting points of the precision methyl polymers. In other words, precision afforded by ADMET has a profound effect on polymer properties (an effect very important in other functionalized systems discussed below).

An ongoing synthetic challenge is the preparation of tactic branched polymers via ADMET. A regioregular, frequency-controlled polymer is one thing, but stereoregular polymers with long methylene run-lengths represent a supreme achievement, presently being addressed in the Wagener group. (Atallah, Wagener, & Schulz, 2013; Few, Wagener, & Thompson, 2014) To negate the atactic character of previous methyl studies, geminal dimethyl models were synthesized for comparison. (Schwendeman & Wagener, 2002) The atactic polymer containing a single methyl every 9th carbon is semicrystalline, whereas a gem-dimethyl group placed every 9th carbon leads to an amorphous polymer. This work exemplifies that size of functionality affects crystalline behavior, morphology, and in turn the properties of the resulting polymer.

Next, a systemic study of larger branches was undertaken. Ethyl branched polymers were included in this endeavor to gain information regarding crystallization of short-chain branched PE. Matsui et al., 2013) Utilizing small- and wide-angle X-ray scattering, PE functionalized with an ethyl group precisely located on every 21st carbon (ethyl21) was studied, with particular attention to the temperature dependence of crystallization for ethyl-fucntionalized PE. Samples were isothermally crystallized at different temperatures from 5 to 28 °C. Above 21 °C, the crystal structure is triclinic with two ethyl branches included per lamella stem. Crystallization at 17 °C results in only one ethyl per stem and a stable hexagonal mesophase. Lowering the crystallization temperature below 8 °C excludes all branches from the precise

Figure 6. PE branch inclusion and exclusion from the unit cell and effect on melting temperature. Reproduced with permission. (Rojas et al., 2009) Copyright 2009, American Chemical Society

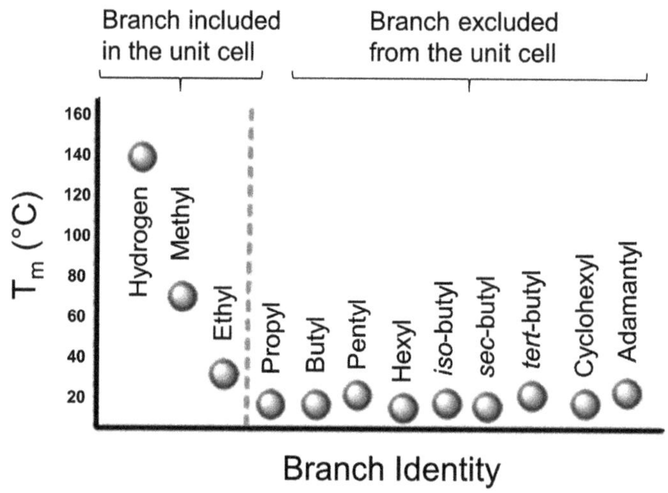

ethyl21 polymer lamellae. Therefore, lamellae undergo thickening and include more ethyl branches as the crystallization temperature increases. This is not observed in random versions of ethyl-fucntionalized PE, as the polymer rapidly crystallizes in a disordered manner to very small, orthorhombic phases.

In contrast to the ethyl precision polymers, hexyl versions exhibit branch exclusion from the lamellae under all crystallization conditions. When incorporated into the unit cell, small groups such as ethyls and methyls alter the crystal, resulting in a melting temperature change. Even in radical chain polymerization, short-branch inclusion and long-branch exclusions are known, but the effect on melting temperature was determined from the study of larger precisely placed branches. As shown in Figure 6, a branch size of propyl or larger on every 21st carbon is excluded from the unit cell, and these excluded longer chain branches have no further effect on the melting temperature. (Rojas et al., 2009)

These observations result directly from precision placement via ADMET functionalization. Although this discussion does not cover all PE modeling, it demonstrates some major findings. PE, although containing only carbon and hydrogen, is a treasure trove of fundamental polymer science knowledge still to be investigated.

3.2 Polar Moieties

Polar functionalization brings to polyolefins the properties attainable in all other walks of polymer science. Strong forces such as hydrogen bonding and interactions of carbonyl-containing functional groups introduce new structures, properties, and applications far beyond typical polyolefins.

Although, precision alcohol polymers were synthesized as early as 1998, (Valenti & Wagener, 1998) only recently has morphological characterization shed additional light on the effects of hydroxy groups placed every 21st carbon. (Thompson et al., 2015) X-ray scattering data confirmed hydroxy groups are expelled from the unit cell, much like the branches mentioned in Section 3.1. Orthorhombic reflection planes are observed in the X-ray data at angles of 21° and 23.5° identical to those of PE. Evidence of hydrogen bonding was confirmed by DSC, in which the hydroxy functionalized polymer exhibits a melting transition of 100 °C, in comparison to other excluded-branch precision PEs (propyl through adamantyl) that melt below 20 °C. Temperature dependent IR analysis (Figure 7) showed that between 100 and 150 °C the hydrogen bonds weaken, as indicated by OH group stretching shifts from 3429 to 3475 cm^{-1}. These results for precisely spaced alcohol functionalized PE are significant for multiple reasons: 1) OH exclusion from the unit cell may allow for easy PE surface modification; and 2) the stimulus-induced change in hydrogen bonding may make stimuli-responsive PE derivatives feasible.

Additional well-defined vinyl-alcohol copolymers were produced by Grubbs et al. via ADMET and ring-opening polymerization (ROMP). (Tuba, Al-Hashimi, Bazzi, & Grubbs, 2014) The work was motivated by the concept that more controlled and defined polyvinyl alcohol (PVA) copolymers could lead to better gas barriers and biomaterials for more advanced applications, other than coatings and simple commodity packaging. Other polar functionalities such as acetals,(Spring, Estrada, Vasilyeva, Rinzler, & Reynolds, 2014) carbonates,(Ortmann, Heckler, & Mecking, 2014) esters,(Unverferth & Meier, 2014) phosphoesters,(Ding, Qiu, Wei, & Zhu, 2014; Ding, Wei, Qiu, Wang, & Zhu, 2014) sulfonate esters,(Parkhurst, Balog, Weder, & Simon, 2014) and amides(Kreye, Trefzger, Sehlinger, & Meier, 2014) have been incorporated within the last 2 years. Additional examples of polar functionalities are discussed in Section 3.4 concerning renewable polyolefins via ADMET.

3.3 Acidic Moieties

Along with diverse polar functionalities (e.g., esters and alcohols), the ability to synthesize acid-functionalized PE is possible due to the development of robust, tolerant metathesis catalysts. With control and precision, shortfalls arising from commercially available materials may be overcome. For example,

Figure 7. IR spectra for PE with hydroxy groups precisely placed every 21 carbons. Reproduced with permission. (Thompson et al., 2015) Copyright, 2015 Wiley

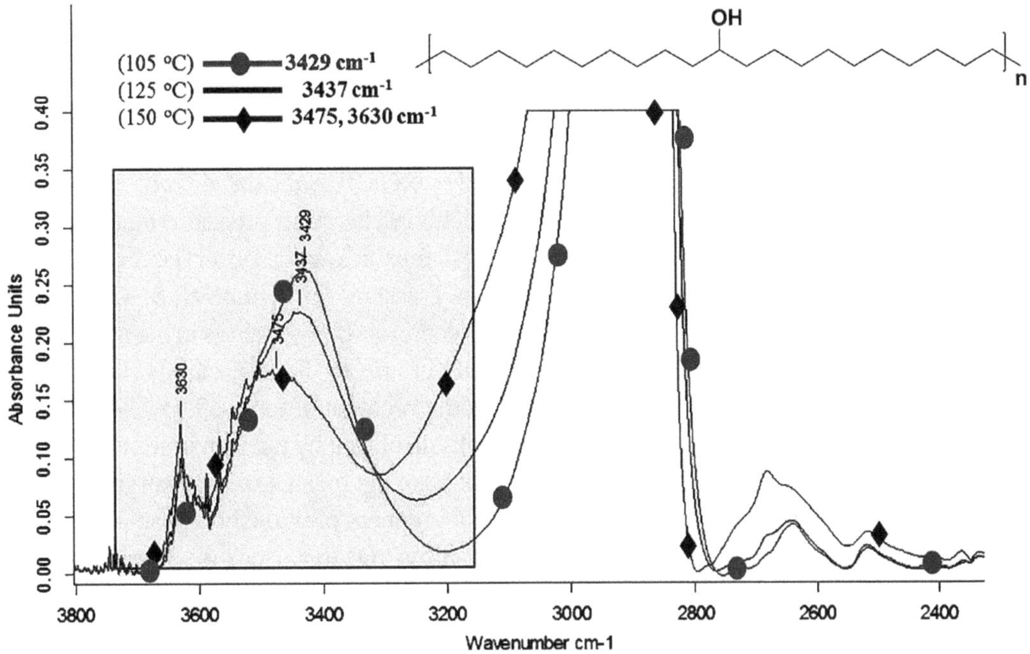

Figure 8. X-ray scattering for precise carboxylic acid functionalized PE displaying typical orthorhombic PE reflection and reflections pointing to the presence of acid layers. Reproduced with permission. (Baughman, Chan, Winey, & Wagener, 2007) Copyright, 2007 American Chemistry Society

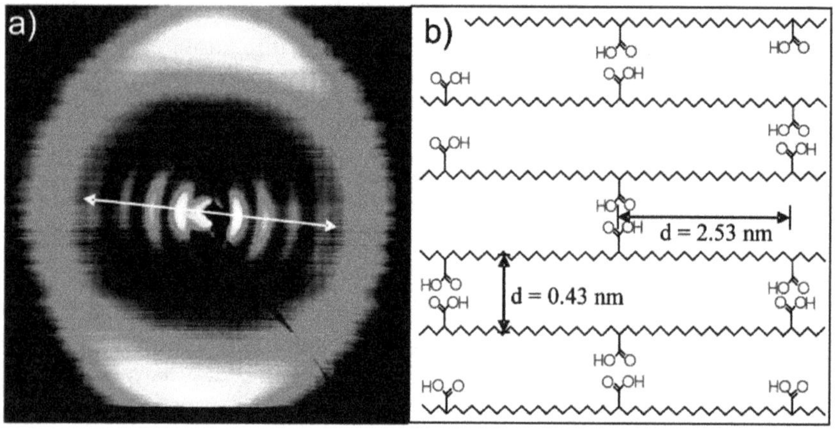

isolation of the acidic/ionic interactions from defects such as branching or random frequency/concentration of acid groups along a backbone leads to a better understanding of behavior. Acidic and ionic functionalized materials are presented in this section separate from other polar moieties because of their very different morphological trends. Neutral and ionic (acidic and basic) forms of PE functionalized with carboxylic acids exhibit pH responsive behaviors resulting in altered microstructures dependent on ionization. Phosphonic acids (-PO(OH)$_2$) are also presented, because they display drastically different properites, due to the presence of two acidic hydrogens per group. Thus, phosphonics have twice as many hydrogen bond donors compared to carboxylic acids.

According to X-ray diffraction data (Figure 8 for COOH every 21st carbon), precise carboxylic acid polymers containing -COOH at known frequency dimerize between lamellae, forming layers of acids between crystalline regions of PE backbones. (Rojas, G., Inci, B., Wei, Y., & Wagener, K. B., 2009) Random model polymers containing identical amounts of acids were synthesized for comparison via ROMP, a random polymerization method. Interestingly, random ROMP polymers, containing the same concentration of acid groups as precise versions, dimerize similar to ADMET precision polymers. However, ordered acid layers were not present in polymers prepared by ROMP. A recent study of similar precision carboxylic acids, found that the run-length of the PE backbone between precisely placed acid groups determines the crystal structure thickness, with 3 nm and 5-6 nm thicknesses observed for -COOH groups placed every 21st and 45th carbon, respectively. (Ortmann, Trzaskowski, Krumova, & Mecking, 2013)

Winey and coworkers presented an investigation of precision carboxylic acids in the anion-form (Figure 9). (Seitz et al., 2010) Upon neutralization of –COOH groups to the zinc carboxylates, aggregates form, in stark contrast to the layers which are present in the protonated forms. These aggregates show sharp X-ray scattering peaks, as opposed to the broad peaks displayed by random ionomers. With higher functionalization content (22 to 66 mol%), the zinc carboxylate aggregates are present in a cubic lattice. Such order has not been witnessed in carbon backbone monomers previously. These observations are no doubt by-products of the precision and control afforded by ADMET. Because other polymerization methods lack control over structure, most commercially produced ionomers are ill-defined, thus making

Figure 9. X-ray scattering of random and precise zinc carboxylates. High-concentration (22 mol% and 66 mol%) precision zinc carboxylate aggregates assume a cubic lattice. This is not observed in random ionomers with carbon backbones. Reprinted with permission. (Seitz et al., 2010) Copyright 2010, American Chemical Society

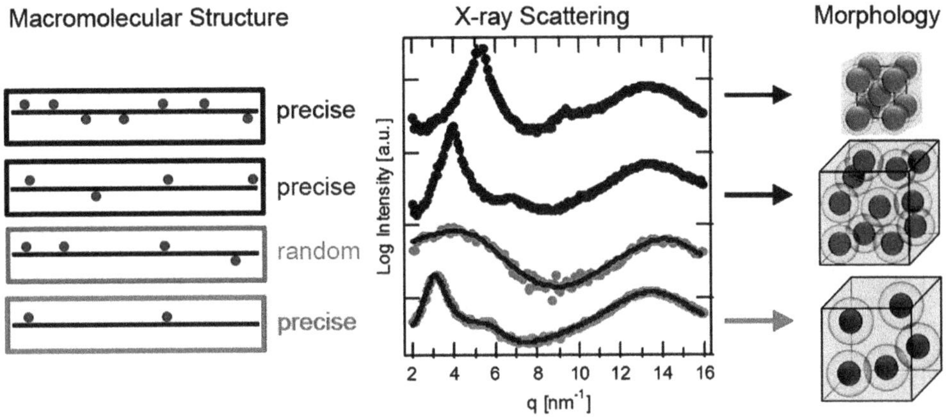

them difficult to understand fully. However, with precision ionomers and aggregates continuing to be studied in such detail, the ionic behavior of precision systems could provide insight concerning random ionomer morphologies and interactions.

Phosphoric acid-based materials are currently used in many applications, including high-temperature fuel cells. (Mader, Xiao, Schmidt, & Benicewicz, 2008) Studying phosphonic effects in defect-limited PE may allow better understanding of ion transport and conductivity in these systems. Opper et al. synthesized precision phosphonic-containing polymers via ADMET, including geminal phosphonic acid polymers having double the acid content on the same backbone. (Opper, Fassbender, Brunklaus, Spiess, & Wagener, 2009; Opper, Markova, Klapper, Mullen, & Wagener, 2010) These high-acid content polymers show very different properties from the carboxylics previously mentioned. Characterization of the geminal PA polymers by Winey and coworkers indicated the presence of a face-centered cubic lattice (Figure 10). Higher temperature X-ray scattering proved that the order of aggregates is virtually the same at 150 °C (high above the 47 °C T_m) as at room temperature. (Buitrago, Opper, Wagener, & Winey, 2012) This is not the case with carboxylic acids, for which the acid dimers between layers dissociate when the temperature is above 120 °C.[31] These investigations provide strong evidence that precision functionalization also leads to improved properties and that ADMET provides control of polymer architecture, much like other examples described in previous sections. Precisely engineered acid materials provided by ADMET may lead to improvements in ion transport and layered nanomaterials applications. ADMET itself may be used to make future conducting materials, as this behavior is not observed in random systems.

3.4 Renewable Polyolefins

Another group of functionalized polyolefins containing polar functionalities are presented as a separate category due to their synthesis from renewable materials, instead of fossil fuels.

Commercial polyurethanes are typically produced utilizing diols and diisocyanates. However, urethane functionalities have been installed into ADMET-capable α,ω-diene monomers using 10-undecenoic acid, a derivative of castor oil, as the starting material (Figure 11). (Lebarbe et al., 2014) These bio-based

Figure 10. a) Room temperature X-ray scattering for precision geminal phosphonic acids every 21st carbon (structure shown above). Peak values at arrows are evidence for FCC lattice. b) FCC lattice shown with Bragg distances acquired by X-ray scattering as a function of Miller indices. Reproduced with permission. (Buitrago et al., 2012) Copyright 2012, American Chemical Society.

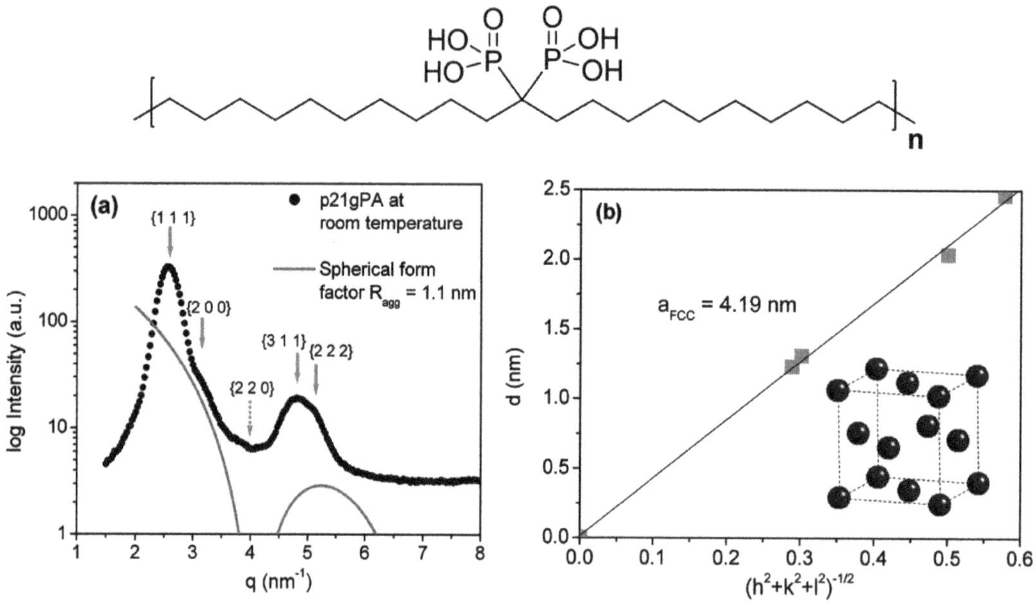

Figure 11. Recently published functionalized polyolefins synthesized from castor oil derivatives. Adapted from Seitz, Shearouse, and Sworen.

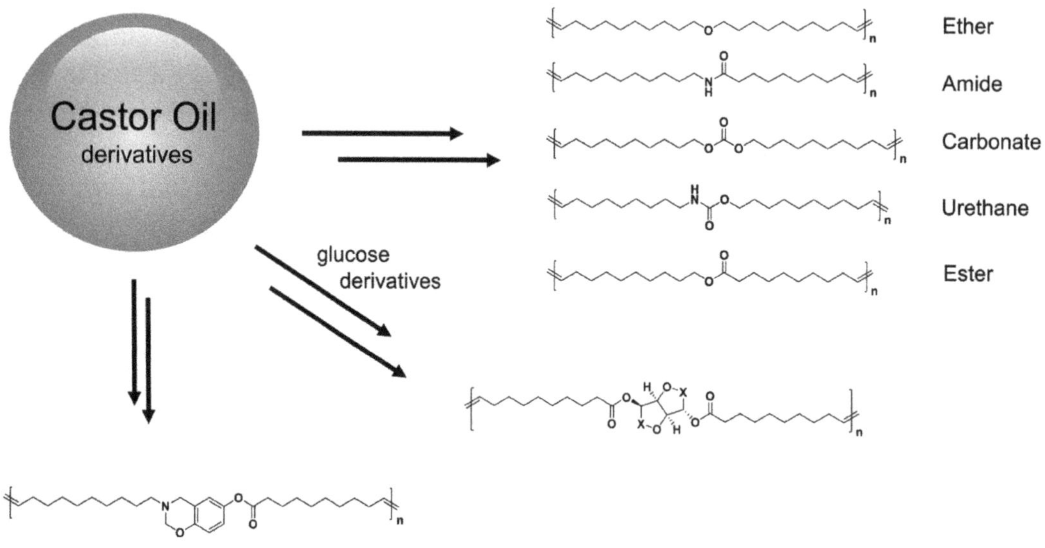

polyurethanes exhibit excellent thermal stability, decomposing well above 300 °C. Polyesters, polycarbonates, polyethers, and polyamides have also been prepared in a similar manner, with polyethers having the best thermal stability, degrading only 5% at 409 °C.

Another study was performed with polymers based upon the castor oil derivative 10-undecenoyl chloride. Two α,ω-diene monomers were synthesized by combining 10-undecenoyl chloride with either isosorbide or glucarodilactone, both derivatives of glucose, one of the most abundant and easily accessible renewable molecules. (Shearouse, Lillie, Reineke, & Tolman, 2015) Copolymers containing both isosorbide and glucarodilactone-based monomers were also prepared and characterized. They exhibit degradation temperatures ranging from 206 °C to 369 °C. An important aspect of this work was the study of hydrolytic degradations. Isosorbide-containing backbones provide stability under all aqueous conditions (neutral, acidic and basic), while lactone functionalization was found crucial to degradability. Hence, any polymer synthesized using glucarodilactone was readily broken down in solution. One copolymer composition even showed an elongation of 640% at the break – displaying the wide array of properties possible for functionalized polyolefins.

Additional examples are documented in the literature. In fact, a large percent of recently published ADMET literature is devoted to bio-based or green materials. Products from natural oils and acids comprise the majority of these studies. Polyolefins functionalized with ferulic acid,(Barbara, Flourat, & Allais, 2015) abietic acid,(Llevot, Grau, Carlotti, Grelier, & Cramail, 2014) and even benzoxazine(Tuzun, Lligadas, Ronda, Galia, & Cadiz, 2015) have recently been prepared through ADMET polymerization.

3.5 Biologically Active and Drug Functionalized Polyolefins

The controllable nature of ADMET is ideal for preparation of biologically active materials and drug delivery vehicles, where the minutest defect could have a negative health effect. Leonard et al. reported early uses of ADMET with amino acid and peptide branches; specifically, lysine functionalities were precisely placed along a PE backbone on every 21st carbon. (Leonard, Hopkins, Chaffin, & Wagener, 2008) Four distinctly different polymers were synthesized all differing by lysine's connectivity to the backbone. This project was synthetically challenging because free amines and carboxylic acids required protection due to their nucleophilicity and ability to poison metathesis catalysts. This highlights an issue faced with protection groups in precision systems: quantitative removal is essential to retain precision.

A similar study involved the precision placement of drugs on a PE backbone. (Leonard & Wagener, 2008) As proof of concept, diene monomers were synthesized with common drugs including ibuprofen and naproxen. Polymers with drugs attached directly to the backbone were prepared, as well as polymers containing hydrophobic (decanediol unit) or hydrophilic (tetraethylene glycol unit) spacers between the drug and backbone. Hydrolysis studies found that between 15 and 40% of drugs were hydrolyzed from the backbone over a 33 hour period, indicating a potential for timed release drug applications.

Knowing that large side chains and moieties are excluded from the crystal lattice of PE, one can imagine macromolecules similar to those described above to construct materials having bio-active surfaces with mechanical and thermal properties of PE. Considering the importance and interdisciplinary nature of biology and medicine, there is no doubt that a controllable synthetic method like ADMET will become a go-to technique for future bio-olefins.

3.6 Postpolymerization Modification of Functionalized Polyolefins

Post-polymerization chemistry has never been more popular, especially after the introduction of "click" chemistries that provide mild, efficient, and highly selective reactions. These are very useful in the preparation of some ADMET polymers, because the available catalysts do not allow installation of certain functionalities prior to polymerization. Also, the ADMET reaction alone is not as versatile in terms of accessing complex architectures directly, compared to the very popular ATRP and RAFT polymerizations. However, internal alkenes provided by the polymerization itself are now being exploited, as well as the reactivity of preinstalled compatible groups, to functionalize ADMET products further. Recently reported "click" reactions and other post-polymerization reactions are presented here.

Figure 12. Linear-dendron block polyphosphoesters synthesized by ADMET polymerization by Zhenshu and coworkers. Reproduced with permission. (Ding, Qiu, et al., 2014) Copyright, 2014 Royal Society of Chemistry

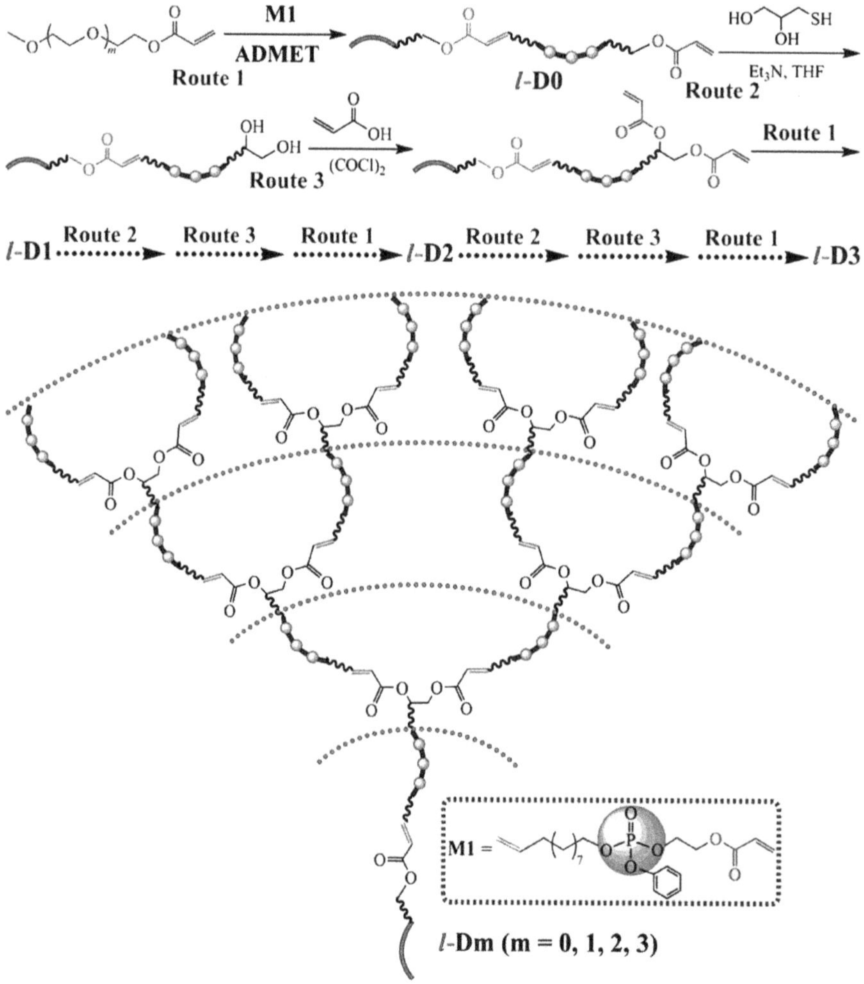

Unverferth et al. have produced star polymers via ADMET polymerization of 10-undecenyl acrylate with hexaacrylate and tetraacrylate monomers, resulting in 4 and 6-arm polyester stars, respectively. (Unverferth & Meier, 2014) Afterwards, the polymerization products were exposed to thiol-Michael conditions to functionalize the star-shaped systems with various mercaptoalcohols and methyl thioglycolates, as confirmed by ^{1}H NMR. The thiol-Michael reaction was also used by Zhenshu and coworkers to synthesize star and linear-dendron (Figure 12) polyphosphoester functionalized ADMET derivatives. (Ding, Qiu, et al., 2014; Ding, Wei, et al., 2014) The star polyphosphoesters were subsequently used to make gold nanoparticle-containing composite materials, which are potentially suitable for biomedical applications.

Thiol-Michael and thiol-ene reactions were utilized by Lowe and coworkers to synthesize a diversely functionalized group of materials containing alcohols, sugar derivatives, hydrocarbon tails, aromatics, fluorinated tails, silyls, and siloxanes. With an acrylate monomer, thiol-Michael reactions resulted in a series of drastically different linear functionalized polyolefins. (van Hensbergen, Gaines, Wagener, Burford, & Lowe, 2014) However, the divalent sulfur present after thiol-Michael addition was found to inhibit ADMET polymerization. Thus, sulfide bonds were oxidized selectively (no alkene oxidation) with hydrogen peroxide and a catalytic amount of triazotriphosphorine chloride, a process termed "protective oxidation" by the authors. (van Hensbergen et al., 2014) Post-polymerization, alkenes were subjected to thiol-ene conditions to yield further functionalization of the backbones.

Other post-ADMET reactions have been performed, including other types of polymerizations. ATRP was utilized in coordination with ADMET by O'Donnell et al. to create precision graft polymers containing phosphonic acids. (O'Donnell & Wagener, 2003) This is just one example of how combination of polymerization methods can produce new, complex precision architectures. Li et al. used post-polymerization lactonization to form γ-butyrolactones on ADMET polymer backbones (Figure 13). (Li, Lv, Du, & Li, 2014) The monomer design contained ethyl esters beta to the alkenes. After the initial ADMET polymerization, the alkenes were hydroxylated, and γ-lactonization was subsequently performed. The installed rings raised the T_g of one sample by 160 °C (from -33° to 127°C), dramatically changing the physical properties. These examples provide clear evidence that the emergence of click chemistry and efficient post-polymerization reactions has allowed even more functionalization of ADMET synthesized polymers.

Figure 13. Synthesis of lactone-containing functionalized polyolefins via ADMET. Reproduced with permission. (Li et al., 2014) Copyright, 2014 American Chemical Society

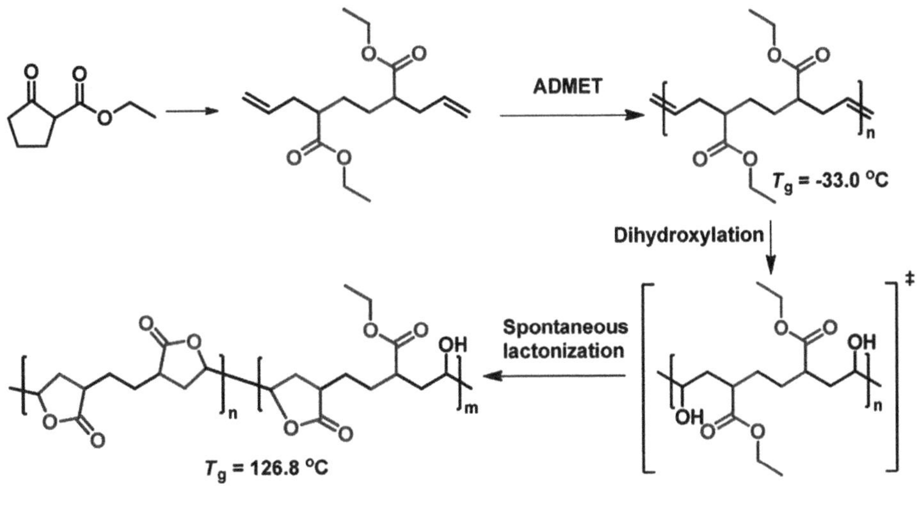

4. CONCLUSION

ADMET allows researchers to produce new structures, which alone may have applications, or whose resulting properties may be elucidated to better understand existing commercial materials. And, with the development of even more efficient, tolerant and active catalysts, increasingly sophisticated structures, properties and morphologies will be possible through ADMET. Whether it be a drug, ionic group, renewable feedstock, or post-polymerization functionalization, ADMET is becoming an increasingly published and important technique for polyolefin functionalization.

REFERENCES

Atallah, P., Wagener, K. B., & Schulz, M. D. (2013). ADMET: The Future Revealed. *Macromolecules*, *46*(12), 4735–4741. doi:10.1021/ma400067b

Barbara, I., Flourat, A. L., & Allais, F. (2015). Renewable polymers derived from ferulic acid and bio-based diols via ADMET. *European Polymer Journal*, *62*, 236–243. doi:10.1016/j.eurpolymj.2014.11.035

Baughman, T. W., Chan, C. D., Winey, K. I., & Wagener, K. B. (2007). Synthesis and Morphology of Well-Defined Poly(ethylene-co-acrylic acid) Copolymers. *Macromolecules (Washington, DC, U. S.)*, *40*(18), 6564–6571. doi:10.1021/ma070841r

Boz, E., Ghosal, A., Alamo, R. G., & Wagener, K. B. (2006). Polyolefins with precisely placed halogens. *Polym. Prepr. (Am. Chem. Soc., Div. Polym. Chem.)*, *47*(1), 219.

Buitrago, C. F., Opper, K. L., Wagener, K. B., & Winey, K. I. (2012). Precise Acid Copolymer Exhibits a Face-Centered Cubic Structure. *ACS Macro Lett.*, *1*(1), 71–74. doi:10.1021/mz2000237

Calderon, N., Ofstead, E. A., & Judy, W. A. (1967). Ring-opening polymerization of unsaturated alicyclic compounds. *J. Polym. Sci., Part A-1 Polym. Chem.*, *5*(9), 2209–2217. doi:10.1002/pol.1967.150050901

Ding, L., Qiu, J., Wei, J., & Zhu, Z. (2014). Convenient divergent synthesis of linear-dendron block polyphosphoesters via acyclic diene metathesis polymerization. *Polym. Chem.*, *5*(14), 4285–4292. doi:10.1039/c4py00327f

Ding, L., Wei, J., Qiu, J., Wang, J., & Zhu, Z. (2014). Star-shaped polyphosphoesters with reactive end groups synthesized via acyclic diene metathesis polymerization and their transformation to nanostructures. *RSC Advances*, *4*(43), 22342–22349. doi:10.1039/c4ra01546k

Durasul highly cross linked polyethylene. (n. d.). *Zimmer*. Retrieved from http://www.zimmernederland.nl/producten/heup/alternative-bearing-solutions/durasulr-highly-cross-linked-polyethylene/

Few, C. S., Wagener, K. B., & Thompson, D. L. (2014). Systematic Studies of Morphological Changes of Precision Polyethylene. *Macromolecular Rapid Communications*, *35*(2), 123–132. doi:10.1002/marc.201300672 PMID:24317938

Grubbs, R.H. (2004). Handbook of Metathesis. Wiley VCH: Weinheim.

Harries-Rees, K., Chauvin, Y., Grubbs, R., & Schrock, R. (2005). Three share prize for metathesis work. *Chem. World*, *2*(11), 42–44.

Herisson, J. L., & Chauvin, Y. (1971). Catalysis of olefin transformations by tungsten complexes. II. Telomerization of cyclic olefins in the presence of acyclic olefins. *Die Makromolekulare Chemie*, *141*(1), 161–176. doi:10.1002/macp.1971.021410112

Inci, B., & Wagener, K. B. (2011). Decreasing the Alkyl Branch Frequency in Precision Polyethylene: Pushing the Limits toward Longer Run Lengths. *Journal of the American Chemical Society*, *133*(31), 11872–11875. doi:10.1021/ja2040046 PMID:21766883

Kreye, O., Trefzger, C., Sehlinger, A., & Meier, M. A. R. (2014). Multicomponent Reactions with a Convertible Isocyanide: Efficient and Versatile Grafting of ADMET-Derived Polymers. *Macromolecular Chemistry and Physics*, *215*(22), 2207–2220. doi:10.1002/macp.201400187

Lebarbé, T., More, A. S., Sane, P. S., Grau, E., Alfos, C., & Cramail, H. (2014). Bio-Based Aliphatic Polyurethanes through ADMET Polymerization in Bulk and Green Solvent. *Macromolecular Rapid Communications*, *35*(4), 479–483. doi:10.1002/marc.201300695 PMID:24339353

Leonard, J. K., Hopkins, T. E., Chaffin, K., & Wagener, K. B. (2008). Semicrystalline lysine functionalized precision polyolefins. *Macromolecular Chemistry and Physics*, *209*(14), 1485–1494. doi:10.1002/macp.200700645

Leonard, J. K., & Wagener, K. B. (2008). Polyethylene prodrugs using precisely placed pharmaceutical agents. *Polym. Prepr. (Am. Chem. Soc., Div. Polym. Chem.)*, *49*(1), 874.

Lewin, M., (Ed.), (2007). *Handbook of Fiber Chemistry* (Third ed.). CRC Press.

Li, Z.-L., Lv, A., Du, F.-S., & Li, Z.-C. (2014). Intrachain Cyclization via Postmodification of the Internal Alkenes of Periodic ADMET Copolymers: The Sequence Matters. *Macromolecules (Washington, DC, U. S.)*, *47*(17), 5942–5951. doi:10.1021/ma5013732

Llevot, A., Grau, E., Carlotti, S., Grelier, S., & Cramail, H. (2014). Dimerization of abietic acid for the design of renewable polymers by ADMET. *Eur. Polym. J.* doi:10.1016/j.eurpolymj.2014.10.021

Mader, J., Xiao, L., Schmidt, T. J., & Benicewicz, B. C. (2008). Polybenzimidazole/acid complexes as high-temperature membranes. *Adv. Polym. Sci.*, *216*(Fuel Cells II), 63-124. doi:10.1007/12_2007_129

Matsui, K., Seno, S., Nozue, Y., Shinohara, Y., Amemiya, Y., Berda, E. B., & Wagener, K. B. et al. (2013). Influence of Branch Incorporation into the Lamella Crystal on the Crystallization Behavior of Polyethylene with Precisely Spaced Branches. *Macromolecules (Washington, DC, U. S.)*, *46*(11), 4438–4446. doi:10.1021/ma400608q

O'Donnell, P. M., & Wagener, K. B. (2003). Graft Copolymers by Acyclic Diene Metathesis and Atom Transfer Radical Polymerization Techniques. *Journal of Polymer Science. Part A, Polymer Chemistry*, *41*, 2816–2827.

Opper, K. L., Fassbender, B., Brunklaus, G., Spiess, H. W., & Wagener, K. B. (2009). Polyethylene Functionalized with Precisely Spaced Phosphonic Acid Groups. *Macromolecules (Washington, DC, U. S.)*, *42*(13), 4407–4409. doi:10.1021/ma900884f

Opper, K. L., Markova, D., Klapper, M., Mullen, K., & Wagener, K. B. (2010). Precision Phosphonic Acid Functionalized Polyolefin Architectures. *Macromolecules (Washington, DC, U. S.)*, *43*(8), 3690–3698. doi:10.1021/ma902659y

Opper, K. L., & Wagener, K. B. (2011). ADMET: Metathesis polycondensation. *Journal of Polymer Science. Part A, Polymer Chemistry*, *49*(4), 821–831. doi:10.1002/pola.24491

Ortmann, P., Heckler, I., & Mecking, S. (2014). Physical properties and hydrolytic degradability of polyethylene-like polyacetals and polycarbonates. *Green Chemistry*, *16*(4), 1816–1827. doi:10.1039/c3gc42592d

Ortmann, P., Trzaskowski, J., Krumova, M., & Mecking, S. (2013). Precise Microstructure Self-Stabilized Polymer Nanocrystals. *ACS Macro Lett.*, *2*(2), 125–127. doi:10.1021/mz300606n

Parkhurst, R. R., Balog, S., Weder, C., & Simon, Y. C. (2014). Synthesis of poly(sulfonate ester)s by ADMET polymerization. *RSC Advances*, *4*(96), 53967–53974. doi:10.1039/C4RA08788G

Piringer, O. G., Baner, A. L., (Eds.), (2008). *Plastic Packaging; Interactions with Food and Pharmaceuticals.*

Rojas, G., Inci, B., Wei, Y., & Wagener, K. B. (2009). Precision Polyethylene: Changes in Morphology as a Function of Alkyl Branch Size. *Journal of the American Chemical Society*, *131*(47), 17376–17386. doi:10.1021/ja907521p PMID:19891434

Schmidt, B. (2004). Catalysis at the interface of ruthenium carbene and ruthenium hydride chemistry: Organometallic aspects and applications to organic synthesis. *European Journal of Organic Chemistry*, *2004*(9), 1865–1880. doi:10.1002/ejoc.200300714

Schrock, R. R., & Hoveyda, A. H. (2003). Molybdenum and tungsten imido alkylidene complexes as efficient olefin-metathesis catalysts. *Angewandte Chemie International Edition*, *42*(38), 4592–4633. doi:10.1002/anie.200300576 PMID:14533149

Schulz, M. D., & Wagener, K. B. (2012, March 25-29). Exploring solvent effects with alternating acyclic diene metathesis (ADMET) polymerization. *Abstracts of Papers, Proceedings of the 243rd ACS National Meeting & Exposition, San Diego, CA, USA.*

Schulz, M. D., & Wagener, K. B. (2014). Precision Polymers through ADMET Polymerization. *Macromolecular Chemistry and Physics*, *215*(20), 1936–1945. doi:10.1002/macp.201400268

Schwendeman, J. E., & Wagener, K. B. (2002, April 7-11). Precisely placed gem-dimethyl branching in ADMET polyethylene. *Abstracts of Papers, Proceedings of the 223rd ACS National Meeting, Orlando, FL, USA.*

Seitz, M. E., Chan, C. D., Opper, K. L., Baughman, T. W., Wagener, K. B., & Winey, K. I. (2010). Nanoscale Morphology in Precisely Sequenced Poly(ethylene-co-acrylic acid) Zinc Ionomers. *Journal of the American Chemical Society*, *132*(23), 8165–8174. doi:10.1021/ja101991d PMID:20496894

Shearouse, W. C., Lillie, L. M., Reineke, T. M., & Tolman, W. B. (2015). Sustainable Polyesters Derived from Glucose and Castor Oil: Building Block Structure Impacts Properties. *ACS Macro Lett.*, Ahead of Print. doi:10.1021/acsmacrolett.5b00099

Smith, J. A., Brzezinska, K. R., Valenti, D. J., & Wagener, K. B. (2000). Precisely Controlled Methyl Branching in Polyethylene via Acyclic Diene Metathesis (ADMET) Polymerization. *Macromolecules*, *33*(10), 3781–3794. doi:10.1021/ma9920792

Spring, A. M., Estrada, L. A., Vasilyeva, S. V., Rinzler, A. G., & Reynolds, J. R. (2014). Carbon Nanotube Adsorptive Materials Derived from Acid Degradable Poly(acetals). *Macromolecules (Washington, DC, U. S.)*, *47*(8), 2556–2560. doi:10.1021/ma500009s

Sworen, J. C., Smith, J. A., Wagener, K. B., Baugh, L. S., & Rucker, S. P. (2003). Modeling Random Methyl Branching in Ethylene/ Propylene Copolymers Using Metathesis Chemistry: Synthesis and Thermal Behavior. *Journal of the American Chemical Society*, *125*(8), 2228–2240. doi:10.1021/ja020862v PMID:12590552

Thompson, D. L., Wagener, K. B., Schulze, U., Voit, B., Jehnichen, D., & Malanin, M. (2015). Spectroscopic Examinations of Hydrogen Bonding in Hydroxy-Functionalized ADMET Chemistry. *Macromolecular Rapid Communications*, *36*(1), 60–64. doi:10.1002/marc.201400545 PMID:25393938

Tuba, R., Al-Hashimi, M., Bazzi, H. S., & Grubbs, R. H. (2014). One-Pot Synthesis of Poly(vinyl alcohol) (PVA) Copolymers via Ruthenium Catalyzed Equilibrium Ring-Opening Metathesis Polymerization of Hydroxyl Functionalized Cyclopentene. *Macromolecules (Washington, DC, U. S.)*, *47*(23), 8190–8195. doi:10.1021/ma501976v

Tuzun, A., Lligadas, G., Ronda, J. C., Galia, M., & Cadiz, V. (2015). Integrating plant oils into thermally curable main-chain benzoxazine polymers via ADMET polymerization. *Eur. Polym. J.* doi: 10.1016/j.eurpolymj.2014.12.023

Unverferth, M., & Meier, M. A. R. (2014). Tuning the polarity of ADMET derived star-shaped polymers via thiol-ene chemistry. *Polymer*, *55*(22), 5571–5575. doi:10.1016/j.polymer.2014.06.035

Valenti, D. J., & Wagener, K. B. (1998). Direct Synthesis of Well-Defined Alcohol-Functionalized Polymers via Acyclic Diene Metathesis (ADMET) Polymerization. *Macromolecules*, *31*(9), 2764–2773. doi:10.1021/ma9714833

van Hensbergen, J. A., Gaines, T. W., Wagener, K. B., Burford, R. P., & Lowe, A. B. (2014). Functional α,ω-dienes via thiol-Michael chemistry: Synthesis, oxidative protection, acyclic diene metathesis (ADMET) polymerization and radical thiol-ene modification. *Polym. Chem.*, *5*(21), 6225–6235. doi:10.1039/C4PY00783B

Vasile, C., (Ed.), (2000). *Handbook of Polyolefins (Rev. Second ed.).*

Wagener, K. B., Boncella, J. M., & Nel, J. G. (1991). Acyclic diene metathesis (ADMET) polymerization. *Macromolecules*, *24*(10), 2649–2657. doi:10.1021/ma00010a001

Wagener, K. B., & O'Donnell, P. M. (2002). ATRP grafting on ADMET polyolefin backbones. *Polym. Prepr. (Am. Chem. Soc., Div. Polym. Chem.)*, *43*(2), 48.

Weychardt, H., & Plenio, H. (2008). Acyclic Diene Metathesis Polymerization of Divinylarenes and Divinylferrocenes with Grubbs-Type Olefin Metathesis Catalysts. *Organometallics*, *27*(7), 1479–1485. doi:10.1021/om701277p

Chapter 2
Advances in Catalytic Technologies for Selective Oxidation of Lower Alkanes

Srikant Gopal
SABIC Technology Center, Saudi Arabia

Mohammed H Al-Hazmi
SABIC Technology Center, Saudi Arabia

ABSTRACT

The main challenge for selective oxidation catalysis has always been achieving an economically viable selectivity. In this chapter, the advances made around the last decade in catalyst development for the selective oxidation of C2-C4 alkanes to carboxylic acids and olefins are discussed. To assess the progress made and to study the trends, particular attention has been given to the results reported in patents. A review of the developments reported in the research literature and new knowledge about the catalysts and their functioning have also been summarized. A comparison with existing processes is provided to obtain an idea of which selective oxidation routes are approaching closer to commercial implementation. Finally, some of the challenges that the selective oxidation routes must overcome for widespread commercial adoption are discussed and suggestions for future research are provided.

INTRODUCTION

This chapter discusses the recent advancements made in heterogeneous catalysts for the selective oxidation of light alkanes to obtain oxygenated products such as carboxylic acids. Since oxidative dehydrogenation carried out a low temperatures, is closely related, the production of olefins using this route is also discussed. In the review of the literature, particular attention has been given to the patent literature, especially patents published since 2000. Based on the assessment of the published literature, the current status, opportunities for further advancement of the technologies and their commercial implementation are discussed.

DOI: 10.4018/978-1-4666-9975-5.ch002

The continued sustainability of the chemical industry is dependent on moving to lower cost feedstocks and obtaining products with fewer processing steps. In this context, the selective oxidation of alkanes offers an attractive route to directly convert the alkanes into value added oxygenated products. Selective oxidation of alkanes offers the possibility of a lower cost as well as a lower environmental impact process.

Activation of lower alkanes such as ethane, propane and butanes through an oxidative dehydrogenation process offers advantages such as no equilibrium limitations, operation at a lower temperature, no coke formation on catalyst and control on downstream product impurity or selectivity (Karim, Al-Hazmi, & Al-Andis, 1999). Controlling selectivity to the desired product and developing catalyst with the required activity remains the key challenge in selective oxidations.

The replacement of existing routes by selective alkane oxidation can bring about huge environmental and economic benefits. As an example, the commercial process for methacrylic acid production is based on the acetone-cyanohydrin route involving three process steps, use of hydrogen cyanide as a raw material and production of ammonium bisulfate as a co-product (Jing, Katryniok, Dumeignil, Bordes-Richard, & Paul, 2014). Development of a selective isobutane oxidation process would provide a one-step route to methacrylic acid, using a lower cost raw material and with lower negative impact on the environment (Jing, Katryniok, Dumeignil, Bordes-Richard, & Paul, 2014). This illustrates the need for the chemical industry to pursue research towards the development of selective oxidation catalysts that would make these routes competitive with the existing technology.

Although offering several advantages, partial oxidation of alkanes does have its own set of limitations. The oxygen concentration fed to the reactor has to be limited to be outside the explosive limit, which puts limitations on the amount of hydrocarbon converted per pass. Loss of selectivity due to further oxidation of the desired products, removal of reaction heat, possibility of runaway (Grabowski, 2006) and avoidance of hotspots in the reactor are other potential issues in selective oxidation processes.

Oxidation catalysis has the greatest potential for improvement in the chemical industry, and substantial progress has been made. However, selective oxidation routes need to offer a substantial benefit over the existing routes to be successful. The yield and selectivity targets needed to make many of the processes economically viable have not yet been achieved (Cavani F., 2010). The best results reported in the literature indicate that some of the processes such as ethane oxidative dehydrogenation to ethylene and propane to acrylic acid are quite promising. Further activity and selectivity improvement and process intensification approaches such as combining separation and reaction steps could make these processes attractive for commercialization.

In this chapter, the goal has been to review the literature on selected routes, focusing especially on advances made over the last decade. The trends in the patent as well as research literature have been assessed and any new understanding obtained has been summarized. Another objective was to identify what research approaches have been successful to make advances in catalyst performance. The key challenges faced in each of the new routes are discussed and suggestions are provided on what is needed to make the new process technology based on selective oxidation replace an existing process.

The scope of this chapter has been limited to selective oxidation and oxidative dehydrogenation carried out at low temperatures, using molecular oxygen as the oxidant. Therefore, the ammoxidation of propane to acrylonitrile, although closely related to propane oxidation, is not discussed. The use of other oxidants such as N_2O and CO_2 and oxidative dehydrogenation carried out at high temperatures are not covered in this chapter. Oxidation of *n*-butane maleic anhydride is a well-established commercial process, and is excluded from the discussion in this chapter to focus more on the emerging technologies.

BACKGROUND

Selective oxidation of lower alkanes for the production of oxygenates and olefins must compete against existing commercial processes, many of which have been operational for decades and has seen continuous improvement. To obtain a better picture of where the selective oxidation processes stand in terms of commercialization and widespread commercial adoption, it is necessary to be fully aware of the existing processes and upcoming alternative processes to understand the targets that need to be met.

For ethylene production, ethane ODH competes against the well-known ethane steam cracking process. Although steam cracking is said to be highly energy and capital intensive, it remains an established and highly optimized process at very large scales. The selectivity to ethylene in this process is around 80%, and several valuable by-products are produced. The direct oxidation of ethane to acetic acid has been commercialized by SABIC in a 30,000 tons per annum scale. It is reported that the ethane oxidation route is competitive with the methanol carbonylation route. Further improvements in the selectivity of the ethane route and availability of lower cost ethane are likely to lead to a wider adoption.

Oxidative dehydrogenation of propane to propylene would have to compete against on-purpose propylene processes, such as propane dehydrogenation (PDH). Based on the information available, PDH provides a selectivity of 90-92%. Therefore, ODH of propane needs to exceed these selectivity targets in order to displace this route to propylene, for large scale production.

The direct oxidation of propane to acrylic acid is a very interesting alternative to the two step propylene to acrylic acid route. The existing route based on propylene to acrolein and then to acrylic acid provides a yield of 86% (European Patent No. EP2179790, 2010). Reports in the literature have frequently targeted a yield of 60% for the viability of the propane to acrylic acid process. There are other routes to acrylic acid under development; the interesting ones are shown along with the propylene oxidation route in Figure 1.

Figure 1. Possible routes for Acrylic Acid synthesis from lower alkane & alkenes

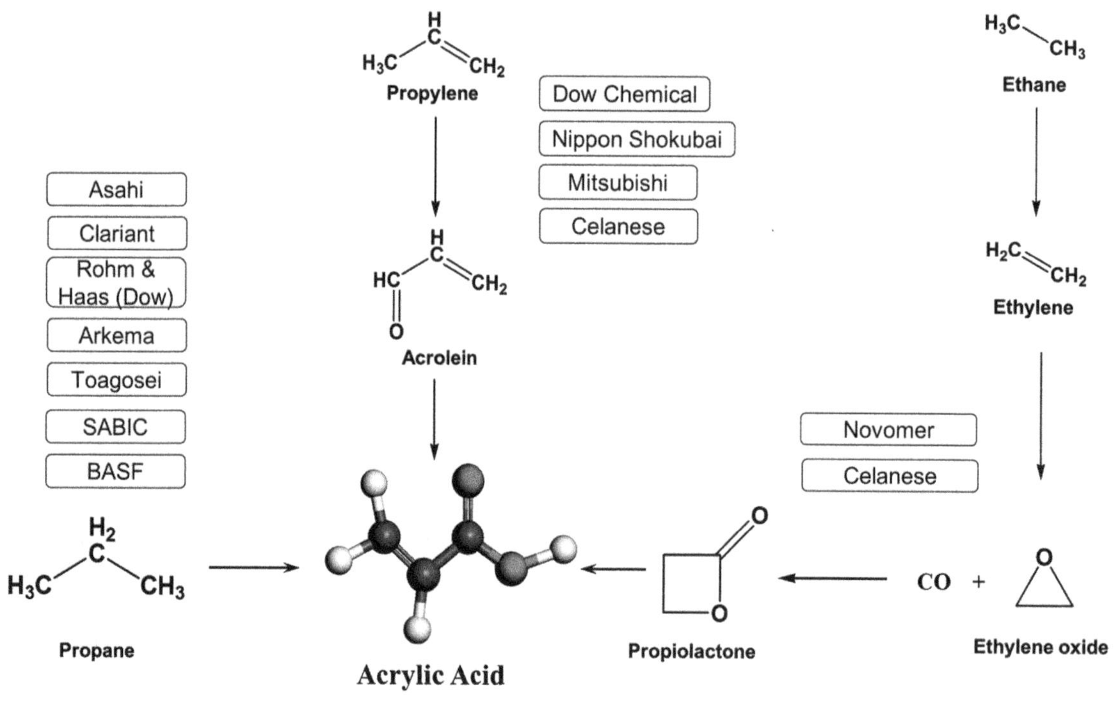

A new route developed by Novomer appears interesting, which is a two stage process for the synthesis of acrylic acid from ethylene oxide and carbon monoxide. The core of the technology is basically developing a novel catalytic system for reaction of ethylene oxide with carbon monoxide to produce propiolactone selectively which then can be converted to acrylic acid in the second step with ultra-high purity glacial Acrylic acid. A homogenous based catalytic system was developed and optimized to obtain more than 99% selectivity with almost no raw material loss at relatively low temperature (30-50 °C). These technological features allow a lower carbon and energy footprint as well as low capital investment (Novomer, 2014).

Several routes are available for the production of methacrylic acid and methylmethacrylate, some them are shown in Figure 2. However, if a highly selective catalyst is developed for direct oxidation of isobutane to methacrylic acid, it is likely to be the most attractive route.

In the following section, the advancements made in the selective alkane oxidation route to each of the above products is reviewed. The expectation is to identify the oxidation routes that have reached an advanced stage closer to commercialization and present the further challenges that need to be resolved.

SELECTIVE CATAYTIC OXIDATION OF LOWER ALKANES

Progress and Current Challenges in Selective Oxidation of Lower Alkanes

The progress in catalyst technology and current challenges for selective oxidations of lower alkanes are discussed in greater detail in the following sections. The patenting trends and the best results reported in patents and research literature have been examined.

Figure 2. Routes to MMA from lower alkanes and alkenes

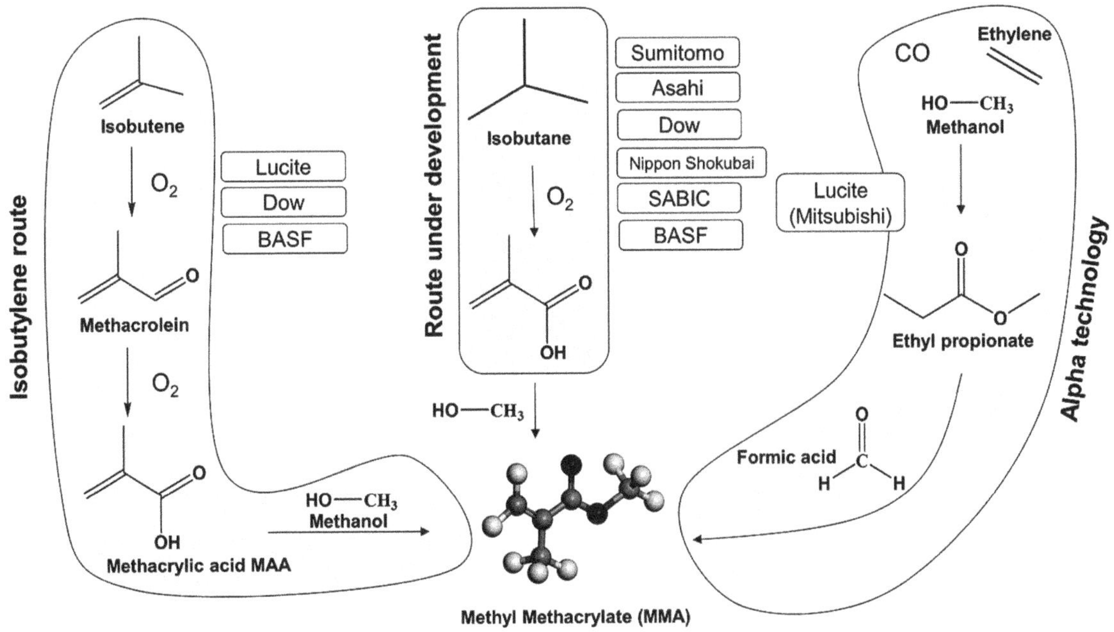

Ethane Oxidation to Acetic Acid

As in any selective oxidation reaction, the challenge in oxidation of ethane to acetic acid is to maximize the yield of acetic acid while minimizing the carbon loss in the form of CO_2. Patent data shows that industry research on the oxidation of ethane to acetic acid is dominated by three companies, BP, Celanese and SABIC. Most of the patent filings have been made in the period from 1996-2004 as shown in Figure 3. Earlier patents in this area were filed in the late 1980s by Union Carbide and Standard Oil/BP. Although a combined selectivity to ethylene and acetic acid of greater than 90% were obtained in these early patents, the product was predominantly ethylene and these selectivities were generally achieved at ethane conversions below 10%. An effort to improve the selectivity and yield of acetic acid can be observed in the subsequent patent literature, with acetic acid selectivity of 70-90% being achieved at ethane conversions around 10%. The best results from the patent literature are summarized in Table 1.

The catalyst compositions used for the oxidation are based on mixed oxides of molybdenum and vanadium. In addition most of the compositions contain niobium, as well as a metal such as palladium for further oxidation of the ethylene. The most successful catalysts are complex multimetallic oxides containing several metals. The highest reported selectivity to acetic acid with minimal CO_2 formation has been reported by Celanese using a catalyst consisting of MoPdVNbSbCaOx (European Patent No. EP1286942B1, 2001); selectivity to acetic acid of 91% at ethane conversion greater than 10%, with less than 8% CO_2 formation was achieved.

The research efforts on the catalyst side appear to have centered on identifying novel catalyst compositions providing improved performance, through the replacement of certain metals on the catalyst or by addition of new metals to the MoV oxide. In the catalyst disclosed in the Celanese patent, Sb and Ca are present in the catalyst composition, in addition to Nb and Pd (European Patent No. EP1286942B1, 2001). Addition of Sb and Ca to the catalyst is found in the 1987 patent of Union Carbide (US Patent

Figure 3. Number of patents (patent families) filed vs. year of filing for ethane oxidation to acetic acid

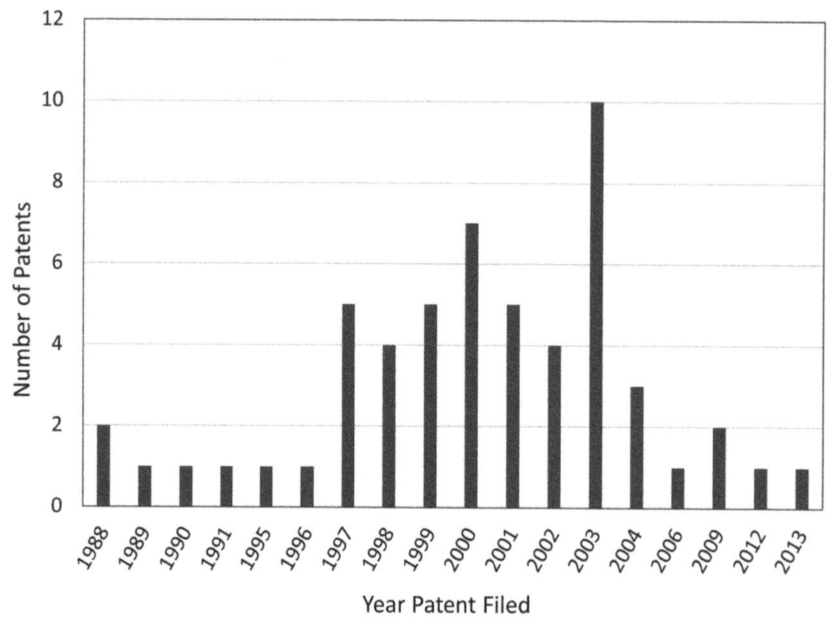

Table 1. Best results reported in patents for ethane oxidation to acetic acid

Catalyst	Ethane Conversion, %	Ethylene Selectivity, %	Acetic Acid Selectivity, %	Patent Reference	Year	Assignee
MoVNbSbCa + Mol sieve	3	56	39	EP294845(A1), US5162578(A)	1987	Union Carbide
MoVNbSbCa	7	65	28	EP294845(A1), US5162578(A)	1987	Union Carbide
MoReVNbSbCa	14.3	12.1	78	EP407091(A1), US5210293(A)	1989	BP
MoReVNbSbCa	4	55	39	EP915821(B1), US6194610(B1)	1996	Celanese
MoReVNbSbCaPd	8	2	90	EP901454(B1), US6034270(A)	1997	Celanese
MoVNbPd	9.6	6.03	73.7	US6030920	1997	SABIC
MoVNbP	33.6	14.7	50.2	EP0938378	1997	SABIC
	20.7	60.03	31.25		1997	SABIC
MoPdVNbSbCa	15	1	91	EP1025075(B1), US6399816(B1)	1998	Celanese
WPdVNb	9	0	81	US6274765(B1)	1998	Celanese
MoReVNbSbCaAg	5.7	19.3	76.2	WO1999051339(A1), US6350716(B1)	1998	BP
MoVNbB	64.4	38.2	26.32	US5907056(A), US6013597(A)	1998	SABIC
MoVPdLa	86		80	US6274764B1	1999	SABIC
MoVNbPd	9.62	6	73.7	US6310241B1	1999	SABIC
MoReVNbSbCaIr	2.6	33.9	63	EP1069945(B1)	1999	BP
MoVLaPdNbAl	17	41	45	WO2001021570(A1)	1999	SABIC
	20.7	60.03	31.25	US6383977(B1), US20030100794(A1)	1999	SABIC
MoPdVNbSbCa	13.3	0.7	91.5	EP1286942(B1), WO1999020592(A1)	2000	Celanese
MoVLaPdNbAl	10		85	US6130356A, EP1058579(B1), US6060421(A)	2000	SABIC
MoVPdLa	26.85	0	71.65	US6156928A, US6087297(A)	2000	SABIC
MoVAlMnSbCaPd	8	22	71	US6531631B1, EP1276558(B1)	2000	SABIC
MoVNbAu	11.8	46.5	40.1	EP1201631(B1)	2000	BP
MoVNbAu	11.3	55.8	39.3	EP1043064(A2,A3), US6333444(B1,X601)	2000	BP
MoVAlMnSbCa	21.5	62	30	US6531631B1	2000	SABIC
MoPdVNbSbCa	10.5	3.5	90.4	EP1286942(B1), US20030158440(A1)	2001	Celanese
MoPdVNb (and optional Sb, Ca)	3.1	0.19	81.8	US6906221(B2)	2001	Celanese
	8.9	45.1	48.8	US7078563(B2)	2001	BP

continued on following page

Table 1. Continued

Catalyst	Ethane Conversion, %	Ethylene Selectivity, %	Acetic Acid Selectivity, %	Patent Reference	Year	Assignee
MoVNbAu	7.8	34.4	47	WO2003033138(A1), US7368599(B2)	2001	BP
	10.5	50.1	44.3	US7078563(B2)	2001	
MoVNbAu	4.6	9	74.5	EP1549432(B1)	2002	BP
MoVNbAuSb	6.3	24.7	56.9	US7368599(B2)	2002	BP
MoVNbAu	7.2	38.2	48.8	WO2004033090(A1), US20070185349(A1), EP1549432(B1)	2002	BP
MoVNbAu	7.2	38.2	48.8	WO2004033090(A1)	2002	BP
MoWAuVNbY	4.1	0	80.8	US20070185349(A1)	2003	
MoVNbTi, MoWVNbTi		5	62	US8084388(B2)	2003	BP
MoVNbWZ	3	18	61	US20060217264(A1)	2003	BP
		51.8	43	US7262322(B2)	2003	BP
Not specified		51.8	43	EP1638915(B1)	2003	BP
MoVNbAu	5	59	33	US8242048(B2), EP1587617(B1)	2003	BP
	9	60	29	US20060217264(A1), US7727928(B2)	2003	BP
MoVNbAu spray dried on alpha alumina support	9	64	28	WO2004067167(A1)	2003	BP
MoVNbAu		63	28	EP1587617(B1), US8242048(B2)	2003	BP
		51.8	43	US7262322(B2)	2004	BP
MoVNb	10	58	32	US20060128988(A1), US20100094047(A1), US8084388(B2)	2004	BP
MoPdVNbSbCa	14.3	69	22.3	US20080132723(A1)	2006	Celanese
MoVNbW	4	41	43	US7727928(B2)	2009	BP
MoVNb		58	32	US20100094047(A1), WO2004108277(A1)	2009	BP
MoVTeNbPd	12.6	65.8	33.8	WO2013148006(A1)	2012	Rohm and Haas
MoVNbPPd	10.66	0.82	85.85	US20140235893(A1)	2013	King Saud U

No. US5162578(A), 1987) and the patent of BP (US Patent No. US5210293(A), 1993); these elements appear to promote the selectivity further to acetic acid and suppress CO_2 formation. In fact, the highest selectivity to acetic acid in the BP patents is obtained with catalyst compositions incorporating Sb and Ca (US Patent No. US5210293(A), 1993), (US Patent No. US6350716(B1)). SABIC has also explored catalyst compositions containing Sb and Ca (US Patent No. US6531631B1), these are in Nb free formulations with Al and Mn present in the catalyst. Interestingly these Sb and Ca containing catalysts are among those that produce the lowest amount of CO+, a trend observed across patents from various companies (Figure 4).

Figure 4. Best results from patents showing acetic acid and ethylene selectivity plotted against ethane conversion

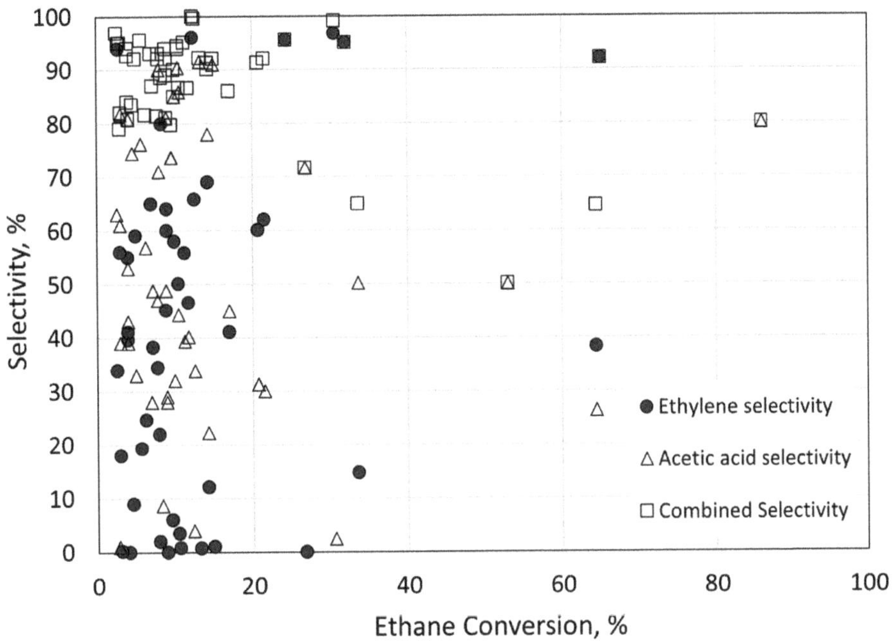

Addition of other metals to boost oxidation of the intermediate ethylene to acetic acid is observed in catalysts for the direct oxidation of ethane to acetic acid. As per the data disclosed in the patents, SABIC and Celanese have relied on Pd to perform this function (US Patent No. US6310241B1), (US Patent No. US6906221(B2)). The addition of Pd to the MoVNb oxide catalyst composition helped increase the selectivity to acetic acid to about 80% and completely oxidize CO to CO_2 (Mamedov & Karim, 2014), (Al-Zeghayer, Al-Awadi, Y, A, & A, 2013). This led to led to the commercialization of a 30,000 tons/year direct ethane to acetic acid plant by SABIC. The amount of Pd in the catalyst composition is crucial, excessive amount of Pd leads to CO_2 formation. Palladium is not observed in the best catalyst compositions disclosed in BP patents, instead, the catalysts mainly feature Re or Au in the composition. A recent patent application from King Saud University (US Patent No. US2014235893A1, 2013) claims the use of Pd in nanometallic form to boost the selectivity to acetic acid. Niobium is present in most of the catalyst compositions reported. A couple of SABIC patents (US Patent No. US6531631B1), (US Patent No. US6274764B1, 1999) disclose compositions without niobium; these catalysts contain La or Al and Mn in the composition.

A promising recent development from the patent literature for production of acetic acid as well as ethylene from ethane are catalyst compositions reported in the patent application from Rohm and Hass (Patent No. WO2013148006(A1), 2012). A combined selectivity to ethylene and acetic acid of over 99% at an ethane conversion of 12.6% has been reported for a MoVTeNbPd catalyst system. Although the ethylene to acetic acid ratio is approximately 2, this direction shows promise for further reducing the hydrocarbon loss in the form of CO_2.

Similar to the trend observed in the patents, a major portion of the work on ethane to acetic acid is found in the years from 1996-2005 in the academic literature. These studies have primarily focused on developing an understanding of the functioning of the MoVNb based catalyst, reaction mechanisms and reaction kinetics.

Among the attempts to improve catalyst performance, Iglesia and co-workers (Li & Iglesia, 2008) reported that the activity of $Mo_{0.61}V_{0.31}Nb_{0.08}O_x$ was enhanced by dispersion of the active oxides on TiO_2. The colloidal TiO_2 was introduced during precipitation, which led to a higher surface area for the supported material and gave higher acetic acid synthesis rates compared to the unsupported (bulk) material. Al-Zeghayer (Al-Zeghayer Y., 2014), (Al-Zeghayer, Al-Awadi, Y, A, & A, 2013) explored this idea further using different grades of titania as supports and found that TiO_2 grades having higher surface area provided improved performance. Both the above studies were able to shift the selectivity towards acetic acid by adding a small amount of Pd to the catalyst. The Pd addition results in a mutually synergistic effect of the Pd and Mo-V-Nb oxide functions. A catalyst composition not containing Nb, reported in the research literature is a MoVAlOx composition yielding approximately equimolar amounts of ethylene and acetic acid (Sankaranarayanan, et al., 2008).

Several studies have focused on the kinetics and mechanism of ethane oxidation to acetic acid over MoVNb based catalysts. Prior literature has also been reviewed in these references. Linke et al. (Linke, et al., 2002), concluded that over a MoVNbPdOx catalyst, the reaction mechanism is temperature dependent. At low temperatures, a consecutive reaction scheme with ethylene as intermediate dominates, subsequently forming acetic acid; while at high temperatures ethylene and acetic acid are formed in parallel. The proposed initial step for both pathways was the reaction of ethane with lattice oxygen with vanadium acting as the redox center; this was based on the observation that a change in oxidation state was observed for V after treatment with an ethylene oxygen mixture, but not for Mo.

The C-H bond activation in ethane is said to take place over surfaces essentially saturated with lattice oxygens (Li & Iglesia, 2008). Pd was found to increase the rate of ethylene oxidation to acetaldehyde which is rapidly converted to acetic acid on the MoVNbOx sites. Iglesia et al. argue that all products including acetic acid and CO oxides can form directly from ethane based on the extrapolation of the selectivities to zero conversion (Li & Iglesia, 2008) (Li & Iglesia, 2008).

The primary product of ethane oxidation is ethylene (Rahman, et al., 2010), which through consecutive reactions transforms to acetic acid and products of deep oxidation such as CO_x. They also postulate that acetic acid is formed from ethylene via both an oxygen route and a hydroxyl route, based on the shift in catalyst behavior without water and with addition of water in the feed.

The influence of Nb and/or Pd dopants on the $MoV_{0.4}O_x$ system has been studied by (Roussel, Bouchard, Karim, Al-Sayari, & Bordes-Richard, 2006). The Nb free catalysts were characterized by the presence of crystalline oxides; however, the Nb containing catalysts were amorphous materials that the authors hypothesize to be made up of nanocrystalline oxides. The Nb containing catalysts were found to be slightly enriched with vanadium on the surface (Roussel, et al., 2009). An Mo free VNbPd catalyst was also found to be selective to acetic acid and CO_x, indicating the role of vanadium in the catalyst (Roussel, Bouchard, Karim, Al-Sayari, & Bordes-Richard, 2006). The promotion of MoVNb by Pd did not strongly modify the structural characteristics of MoVNb (Roussel, et al., 2009); Pd was believed to act as a catalyst for ethylene oxidation.

The economics of direct oxidation of ethane to acetic acid process has been studied and compared with methanol carbonylation (Smejkal, Linke, & Baerns, 2005), (Soliman, Al-Zeghayer, Al-Awadi, & Al-Mayman, 2012). The studies found that although addition of water has a positive effect on the acetic acid selectivity, the separation effort greatly increases, therefore from an energy and economic point of view the water free reaction feed was the most attractive. In their evaluation, Smejkal et al. (Smejkal, Linke, & Baerns, 2005) found that the acetic acid price was slightly lower in methanol carbonylation; however, the investment costs are higher due to the requirement of special materials of construction. The cost of compressor for ethane oxidation process increased dramatically with production amount and for larger plants can be the major contributor to equipment costs. They concluded that selective oxidation can compete with the existing methanol carbonylation process for 10-200 KTA plants, for a new plant located in Europe. If the acetic acid selectivity could be improved to 90%, the ethane oxidation process becomes even more attractive. Soliman et al. (Soliman, Al-Zeghayer, Al-Awadi, & Al-Mayman, 2012) also conclude that acetic acid production from ethane is very competitive with methanol carbonylation and the economics can be greatly improved by further catalyst development and process optimization. Based on the above analyses it is clear that for acetic acid production from ethane, further improvements in the selectivity to acetic acid and acetic acid production rate per kg catalyst at a higher per pass conversion, would make the economics of the process more attractive.

BP has recently announced a new acetic acid process starting from syngas, not much is known about the economics of the process, but it is expected to be an improvement over the methanol carbonylation process. BP's announcement states that iodides are not used in this process, which would lead to lower capital costs and eliminate corrosion issues.

Ethane Oxidative Dehydrogenation to Ethylene

Unlike ethane oxidation to acetic acid, there is more recent activity in the patent literature on ethane oxidative dehydrogenation to ethylene. As seen in Figure 5, there is an increase in the number of patents filed covering oxidative dehydrogenation from 2007 onwards. The companies appearing more frequently in the recent patents are Nova Chemicals, Solvay, Dow Chemical (Rohm and Haas) and Lummus Technology.

In the patent applications filed since 2010, significant improvements in catalyst performance can be observed compared to the earlier literature as shown in Figure 6. Ethylene selectivity greater than 95% at ethane conversions in the range of 40-50% have been reported (US Patent No. US8642825(B2), 2012), (US Patent No. US8105972(B2), 2009), (WO Patent No. WO2013164418(A1), 2012), (WO Patent No. WO2013148006(A1), 2012). Even at ethane conversions of 60-70%, ethylene selectivity of greater than 90% has been achieved (US Patent No. US8642825(B2), 2012), (US Patent No. US8105971(B2), 2009), (US Patent No. US20140114109(A1), 2012). The catalyst composition in almost all the recent patent literature on low temperature ethane oxidative dehydrogenation is based on the MoVTeNb system. A summary of the best catalysts from each patent is presented in Table 2.

Different approaches have been used to achieve improved performance by modifying the MoVTeNb catalyst. In the Rohm and Haas patent application (Patent No. WO2013148006(A1), 2012), excellent selectivity to ethylene is achieved with an un-promoted $Mo_{1.0}V_{0.3}Te_{0.23}Nb_{0.17}O_x$ composition. The catalyst is subjected to a grinding step in a freezer/mill after calcination. Using this catalyst, at low ethane conversions around 12%, only formation of ethylene and acetic acid is reported. There appears to be no yield loss through production of carbon oxides at lower conversion. Addition of 0.01 Pd to the above catalyst composition increases the catalyst activity as well as results in increased selectivity to acetic acid. Even with the addition of Pd, a combined selectivity of 99% to ethylene and acetic acid is achieved with ethylene to acetic acid produced approximately in a 2:1 ratio.

Figure 5. Number of patents focusing on production of ethylene by ethane oxidative dehydrogenation plotted against year of patent filing

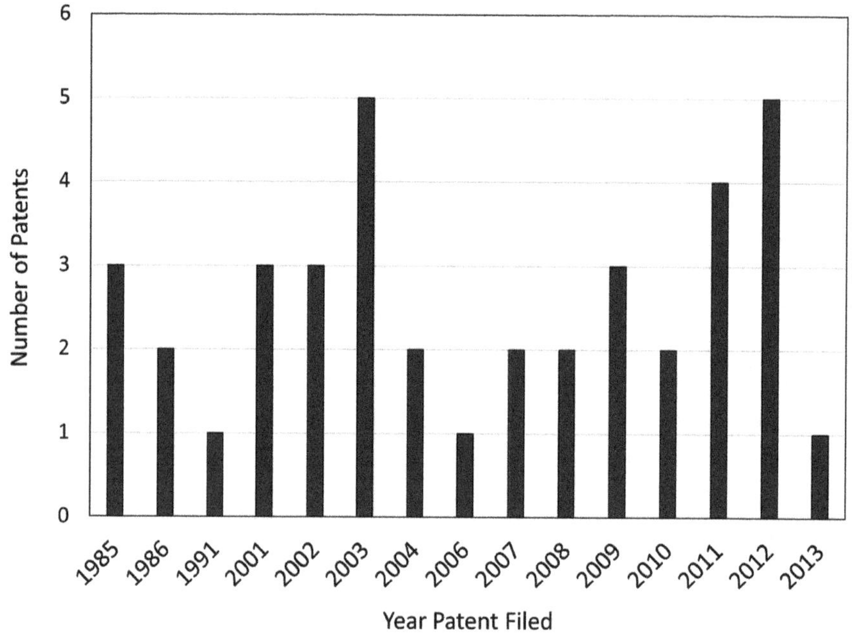

Figure 6. Best results from patent literature showing the trend in ethane conversion and ethylene selectivity obtained over the years

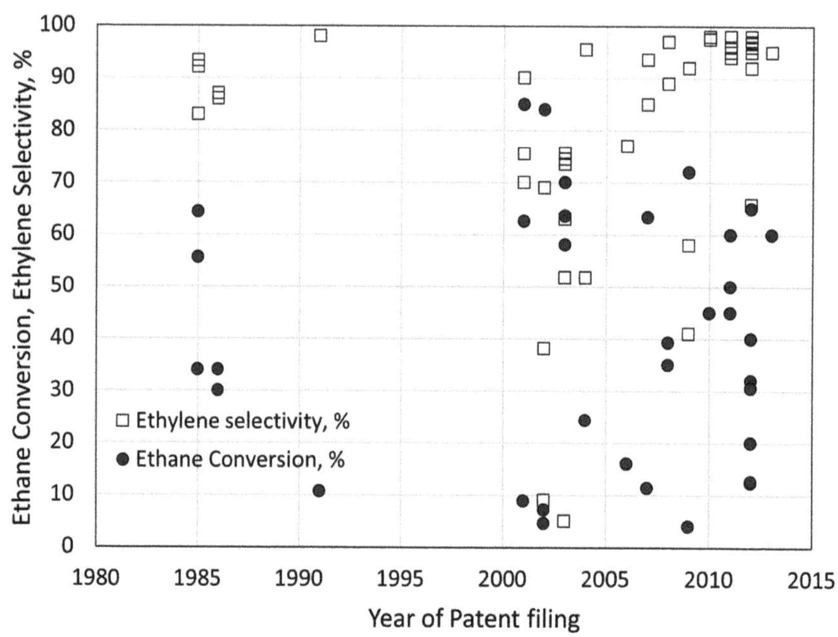

Table 2. Best results reported in each patent for the oxidative dehydrogenation of ethane to ethylene

Catalyst	Ethane Conversion, %	Ethylene Selectivity, %	Patent Reference	Year	Assignee
Sn/P oxides	64.3	93.3	EP189282(B1)	1985	National Research Development, London
Li/MgO	55.6	92.1	EP205765(B1)	1985	Phillips petroleum
MoVNbSb	34	83	EP166438(B1)	1985	Union Carbide
MoVNbSbMg	30	87	EP167109(B1)	1986	Union Carbide
MoVNbSbCa	34	86	EP167109(B1)	1986	Union Carbide
KPMoSbFeCeCr	10.7	98	EP544372(A1)	1991	Eni
NiNbTa	8.9	90.1	US6891075(B2)	2001	Symyx
Rh-Cu/Al2O3	62.6	75.5	EP1263699(B1)	2001	Ineos
Cr/Al2O3	85	70	WO2003026787(A2,A3)	2001	Conoco Phillips
Sm/FeCrAlY	84	69	US6831204(B2)	2002	Conoco Phillips
Pd-Cu/Al2O3	58.1	75.6	US7067455(B2)	2003	Conoco Phillips
Pd-Cu/Mn/ZrO2	63.6	74.5	US7067455(B2)	2003	Conoco Phillips
Ru-Cn/Mn/ZrO2	70	73.5	US7067455(B2)	2003	Conoco Phillips
MoVTeNbA (Ex 16)	24.4	95.5	US7319179(B2), EP1479438(A1)	2004	CSIC-UPV
MoVTaTe	16	77	US20100222623(A1)	2006	Celanese
NiNbTa	11.4	93.6	US7674944(B2)	2007	Celanese
	63.4	85	EP2093205(B1)	2007	Rohm and Haas
MoVNbSbTe (cryogenic grinding, oxalic acid treatment, annealing)	35	97	US8105971(B2)	2008	Lummus
PtSn	39.3	89	US7829753(B2)	2008	Saipem SpA
MoVNbSbTe (cryogenic grinding, oxalic acid treatment, annealing)	72	92	US8105971(B2)	2009	Lummus
MoVNbTe (grinding, oxalic acid treatment)	45	98	US8105972(B2)	2010	Lummus
MoVNbTe ball milled with 50% TiO_2		97.5	US20120016171(A1)	2010	Nova Chemicals
Mg-La	45	96	EP2606963(A1)	2011	Solvay
Chlorides	50	95	WO2013092179(A1)	2011	Solvay
LiRbCl/MgDyO	60	94	EP2606963(A1)	2011	Solvay
MoVNbTe (membrane supported)	40	98	US8642825(B2)	2012	Nova Chemicals
MoVNbTe (calcined in 500 ppm O2)	20	97	WO2013164418(A1)	2012	Shell
MoVTeNb	30.6	96.7	WO2013148006(A1)	2012	Rohm and Haas
MoVTeNbPd	30.6	96.7	WO2013148006(A1)	2012	Rohm and Haas
MoVTeNb	12.4	96	WO2013148006(A1)	2012	Rohm and Haas
MoVSbNb	32	95	US20140114109(A1)	2012	Mexican Petroleum Institute
MoVSbK	65	92	US20140114109(A1)	2012	Mexican Petroleum Institute
MoVNbTe (membrane supported)	60	95	US8642825(B2)	2013	Nova Chemicals

Use of supports such as TiO_2, alpha-alumina or coating the active catalyst composition on a ceramic membrane has been disclosed by Nova Chemicals (US Patent No. US8642825(B2), 2012), (US Patent No. US20120016171(A1), 2010). With the MoVNbTe composition supported on a ceramic membrane, a selectivity to ethylene of 98% at 40% ethane conversion and 95% selectivity at 60% conversion are reported (US Patent No. US8642825(B2), 2012). In another patent, the MoVNbTe catalyst is ball milled along with an equal amount of a support such as alpha-alumina or TiO_2. Excellent catalyst performance with an ethylene selectivity of 97.5% is reported for the TiO_2 supported composition (US Patent No. US20120016171(A1), 2010).

Patents assigned to Lummus Technology, Inc. also disclose various post-synthesis treatments of the MoVNbTe catalyst. After formation from a solution, the MoVNbTe oxide particles are subjected to treatments such as cryogenic grinding, oxalic acid treatment and annealing (US Patent No. US8105971(B2), 2009). A $Mo_{1.0}V_{0.29}Nb_{0.17}Te_{0.21}$ oxide catalyst composition subjected to grinding and an oxalic acid treatment is reported to provide an ethylene selectivity of 98% at 45% ethane conversion (US Patent No. US8105972(B2), 2009). The addition of Sb to the catalyst composition is reported in a Lummus patent (US Patent No. US8105971(B2), 2009); the Sb incorporated MoVNbSbTe oxide catalyst subjected to cryogenic grinding, oxalic acid treatment and annealing is reported to provide a high ethylene selectivity of 92% at ethane conversion as high as 72% at 340 °C.

The method of calcination of the MoVNbTe catalyst appears to be crucial to achieve a good performance. Typically these catalysts are calcined in air at temperatures below 300 °C and the atmosphere is then switched to an inert gas such as Argon for calcination at 600 °C. Loss of Te from the catalyst composition is reported when calcined in an oxidizing atmosphere at higher temperatures (European Patent No. EP2179790, 2010). A patent application from Shell claims the calcination of a catalyst in a gas containing oxygen concentrations of 10 to 10,000 ppm O2 (WO Patent No. WO2013164418(A1), 2012). The best selectivity is reported for a MoVNbTe oxide catalyst calcined at 500 ppm O2, yielding 97% ethylene selectivity at 20% ethane conversion.

A tellurium free catalyst composition providing a high ethylene selectivity has been disclosed in a recent patent application (US Patent No. US20140114109(A1), 2012). Ethylene selectivity of 95% is achieved at 32% ethane conversion with a MoVSbNb oxide composition. With a MoVSbK catalyst, ethylene selectivity of 92% is obtained at a high ethane conversion of 65%. Nickel based catalyst compositions also appear to be interesting for low temperature ethane ODH as disclosed in Symyx and Celanese patents (US Patent No. US6891075(B2), 2001), (US Patent No. US7674944(B2), 2007). The best combination appears to be a NiNbTa oxide providing 93.6% ethylene selectivity at 11.4% ethane conversion at 300 °C.

A summary of the best results from patent literature, including those from patents not discussed in the above paragraphs is given in Table 2. As can be seen in the table, there has been a clear advancement in catalyst performance in recent years.

The oxidative dehydrogenation (ODH) of ethane to ethylene continues to be an active area in the research literature. Several reviews on ethane ODH are available in the literature, including a recent one by Gartner et al. (Gartner, van Veen, & Lercher, 2013) where the mechanism of ethane activation and ODH over various catalysts, including new reactor concepts, have been reviewed. The authors summarize that the best performance in ethane ODH is not restricted to one class of catalyst or reactor concept, therefore multiple possible solutions exist.

The MoVNbTe catalyst system happens to be the most active and selective for ethane oxidative dehydrogenation as well as the oxidation of propane to acrylic acid. Therefore, a lot of attention has been provided to this catalyst system in the recent research literature with a focus mainly on developing an understanding of the structure and functioning of the MoVTeNb oxide system.

The MoVTeNb mixed oxides are mainly composed of two crystalline phases named as M1 and M2 phases (Grasselli, et al., 2004), (Valente, et al., 2014). The M1 phase is believed to contain the most active and selective sites for ODH of ethane. Ueda (Ueda, 2013) found that pure, well-crystallized Mo-$_3VO_x$ materials having the orthorhombic structure showed higher activity than the MoVTeNb catalyst. They attribute the high activity obtained with both catalysts to the structure and suggest that Te and Nb are not necessary for catalytic activity. Trigonal, tetragonal and amorphous phases of the Mo_3VO_x materials with similar compositions were found to have poor activity (Ueda, 2013). The microporous heptagonal channel of the orthorhombic structure is believed to contribute to activation of ethane as well as in attaining high ethylene selectivity (Ishikawa, Yi, Murayama, & Ueda, 2014), (Ishikawa, Yi, Murayama, & Ueda, 2014).

The effect of external surface area of orthorhombic Mo_3VO_x oxide was studied by Ueda and co-workers. Ethane conversion activity was found to be independent of the external surface area of, while a slight effect was found for propane (Ishikawa, Yi, Murayama, & Ueda, 2014). Schlögl and co-workers (Celaya Sanfiz, et al., 2008) found that for propane oxidation to acrylic acid the orthorhombic structure was the one that showed good activity. Ueda and co-workers attribute the activity of M1 phase to the reaction taking place in the heptagonal channel, whereas Schlögl and co-workers suggest formation of specific surface structures in the M1 phase.

The loss of Te from the catalyst during activation as well as operation appears to be a restriction on the scale-up of the MoVTeNb catalyst to industrial levels (Valente, et al., 2014). Therefore, Valente et al. (Valente, et al., 2014) studied the changes occurring over a MoVTeNb catalyst after exposure to

Figure 7. Ethylene selectivity vs. ethane conversion; the best results (highest selectivity data) from each patent have been used.

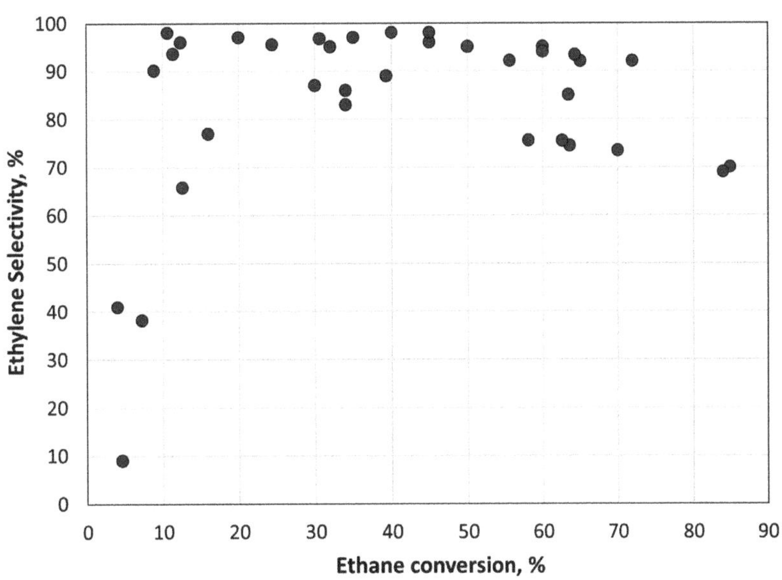

reaction conditions and evaluate their impact on catalytic performance. They concluded that the catalyst is stable for operation at temperatures below 500 °C, although the catalyst showed reduction in activity and ethylene selectivity even after operation at 440 °C. The loss of Te was found to preferentially occur from destruction of the M1 phase with simultaneous formation of MoO_2, thereby decreasing the catalyst activity but not impacting the selectivity much. The detachment of tellurium occurs by reduction of Te^{4+} to Te^{0}; the authors therefore recommend minimizing reactor hot spots.

As can be seen from the literature, substantial variation in catalyst performance can be observed depending on the preparation and treatment methods, physical properties, slight variation in compositions and catalyst calcination/activation conditions. These issues could pose additional challenges during the preparation of the catalyst on a larger scale.

The oxidative dehydrogenation of ethane to ethylene has to compete with steam cracking of ethane. Although ethylene selectivity in steam cracking is around 80%, other useful hydrocarbon products and hydrogen are produced. Therefore, the overall selectivity of the process is very high as there no yield loss in the form of CO or CO_2.

Ethane ODH to ethylene has its own set of drawbacks and challenges. In ODH, although the equilibrium restriction is removed, operating at a high ethane conversion and achieving high selectivity to the olefin is still a challenge due to further oxidation of the olefin to CO and CO_2. CO is not desired in the product mix and is difficult to separate from the olefin; CO impurity in ethylene can affect further processing of ethylene as it acts as a poison for many catalysts. In oxydehydrogenation, the hydrogen is removed as water, therefore there is a loss of hydrogen production, which can have a very significant economic impact (Cavani & Trifiro, 2004). The ethylene produced in ODH is accompanied by a small amount of corrosive acetic acid by-product. The safety issues of handling the hydrocarbon oxygen mixture in a very large scale also needs to be considered. The oxygen concentration in the feed to the reactor would be limited below the explosive limit; therefore, in an ethane rich feed the conversion would be limited. Although there are no thermodynamic limitations, the safety aspect would limit the amount of ethane per pass conversion achievable.

According to Cavani and Trifiro (Cavani & Trifiro, 2004), ODH of ethane to ethylene will be economically competitive with steam cracking only when CO formation is avoided. Another major problem is the formation of oxygenated products such as acetic acid. A thorough economic analysis comparing the detailed ODH process considering all aspects with the steam cracking of ethane would provide clarity on the catalyst activity and selectivity targets needed for commercial viability of the ODH process.

Propane Oxidation to Acrylic Acid

In the oxidation of ethane to acetic acid and propane ammoxidation to acrylonitrile, the products obtained are stable. However, in the case of propane and isobutane oxidation to acrylic and methacrylic acid, the products have a high reactivity under conditions necessary for alkane activation and easily undergo further oxidation (Cavani & Trifiro, 2004), (Lopez Nieto, et al., 2010). Therefore, the oxidation of propane to acrylic acid is a lot more challenging.

The patent filing trends for the direct single step oxidation of propane to acrylic acid indicate a sustained research effort since 1996. Although in Figure 8, no patent filings are indicated for 2009 and 2010, a few continuations were filed in these years for earlier patent applications. Several companies continue to remain interested in this technology as indicated by the number of patent applications filed

Figure 8. Number of patents on propane oxidation to acrylic acid vs. the year filed

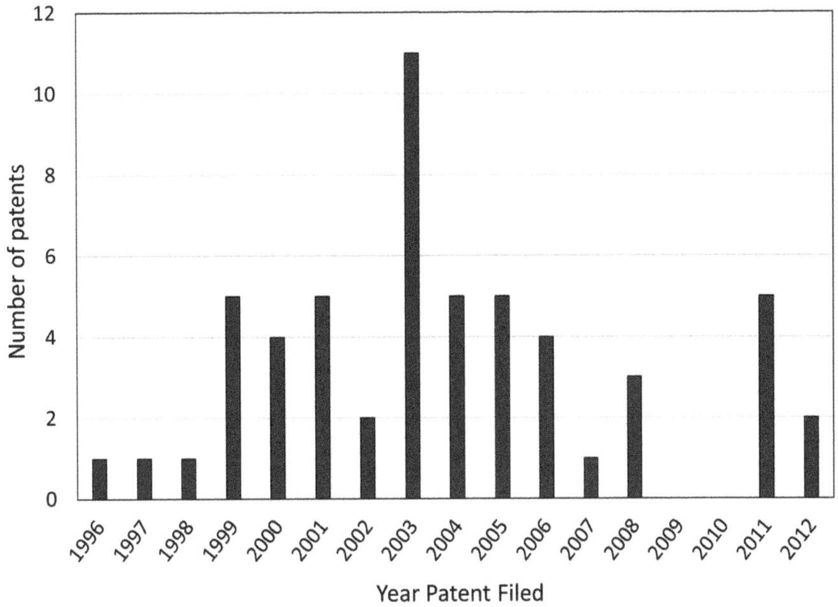

since 2011. Also, compared to the acetic acid and ethane ODH technologies, the number of companies interested in this route to acrylic acid is larger and consists of Clariant, Rohm and Haas, Arkema, SABIC, LG Chem, Evonik, Toagosei, BASF.

A plot of how the best reported propane conversion and acrylic acid selectivity has changed over the years is shown in Figure 9; propylene selectivity has been plotted wherever data was reported. A continuous increase in the selectivity to acrylic acid can be observed. Many patents report achieved a maximum selectivity around 80%. In patents filed by Rohm and Haas, acrylic acid selectivity greater than 90% at propane conversions around 40% have been reported (US Patent No. US7396955(B2), 2007), (US Patent No. US7718568(B2), 2006).

The catalyst system of choice in most of the patent applications is the MoVTeNb oxide. These four elements have remained a part of the catalyst composition; however, large performance improvements have been achieved in recent years through modifications to the catalyst preparation method and addition of other elements to the composition. After 2002, a jump in acrylic acid selectivity to around 80% can be observed in Figure 7; this has been achieved by variations in the catalyst preparation such as addition of surfactant during catalyst preparation (WO Patent No. WO2008068332(A1), 2006), inert carrier (US Patent No. US20070149390(A1), 2003), cryo-grinding (US Patent No. US7718568(B2), 2006), addition of Te in the form of fine particles (US Patent No. US7375052(B2), 2004). Addition of other elements such as Si (US Patent No. US7645897(B2), 2005), Pd (US Patent No. US7304014B2, 2004), In (US Patent No. US7718568(B2), 2006), Sm (US Patent No. US7396955(B2), 2007) to the catalyst composition has also resulted in improved acrylic acid selectivity around 80%. These catalysts also produce some propylene, including the propylene selectivity, the combined selectivity achieved over many of these catalysts is in the range of 85-90%. However, full details of the selectivity to propylene and other by-products are not available in most of the patents. A summary of best results from patent literature is given in Table 3.

Figure 9. Trend in propane conversion and acrylic acid and propylene yields over the years, obtained from the best results reported in the patent literature

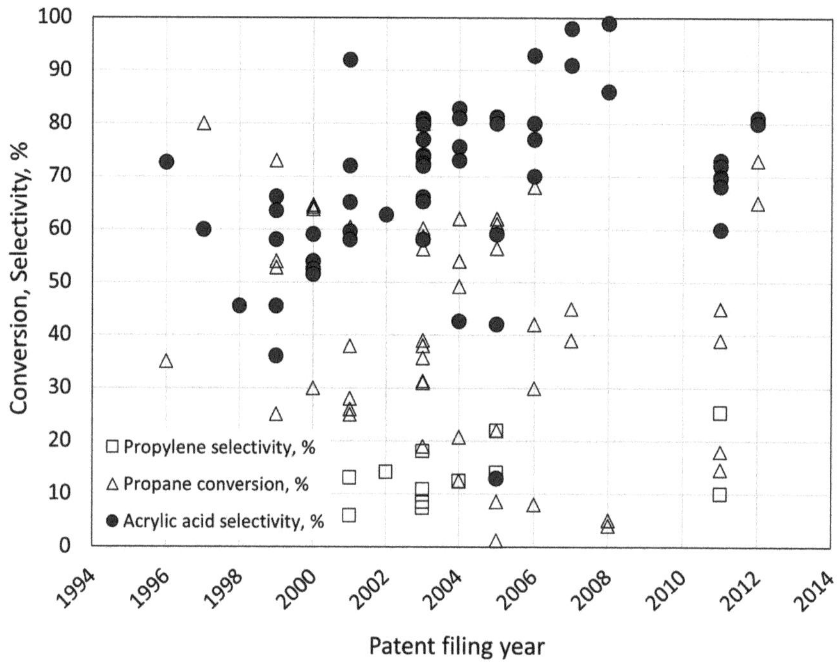

Figure 10. Acrylic acid selectivity vs. propane conversion, based on the best results from the patent literature

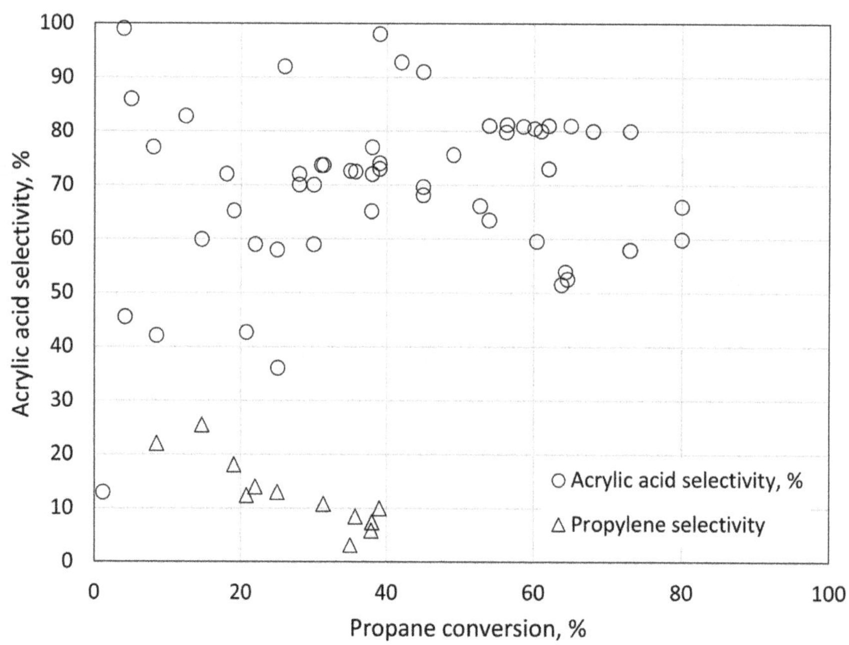

Table 3. Oxidation of propane to acrylic acid; summary of best result in each patent

Catalyst	Propane Conversion, %	Acrylic Acid Selectivity, %	Patent Reference	Year	Assignee
MoSbVNb	35	72.6	US5994580(A), US6060422(A)	1996	Toagosei
MoVNbTe	80	59.9	US6036880(A), EP895809(B1)	1997	Asahi
MoVSbNb	52.8	66.1	US6291393(B1), US20010029234(A1), US6346647(B2)	1999	Toagosei
MoVSbNbTl	54	63.5	US6432870(B1), US20020072628(A1), US6512141(B2)	1999	Toagosei
Mo1V0.3Te0.23Nb0.10	73	58	EP1090684(A1)	1999	Rohm and Haas
MoVGaPdNbTe		45.5	US 6646158	1999	SABIC
MoVTeNb supported on polyoxometallate salts	25.1	36	EP1078687(A1)	1999	Rohm and Haas
MoVTiSbK	30	59	EP1346766(B1), US20040054221(A1), US7642214(B2)	2000	Nippon Kayaku
MoVTeNb	64.3	53.9	US6610629(B2), US20030017944(A1)	2000	Asahi
MoVSbNbTi/SiO2	64.6	52.5	US20030088118(A1)	2000	Asahi
MoVSbNb	63.8	51.5	US20080234513(A1)	2000	Asahi
MoVTeNb	26	92	US6841699(B2)	2001	Rohm and Haas
MoVTeNb (coated)	28	72	US6867328(B2,X601), US20030187298(A1), EP1301457(B1)	2001	BASF
MoTeVNb	37.9	65.1	WO2003008096(A1), US20040230070(A1), EP1473081(A1)	2001	UPV
MoVTeNb	60.4	59.6	EP1253132(A1)	2001	Nippon Shokubai
MoVTeNb (coated)	25	58	US7005403(B2), US7214822(B2), US20040102648(A1)	2001	BASF
MoVSbNbSi		62.7	US20060183941(A1)	2002	Arkema
MoVNbTe (fine particle metallic Te)	58.6	80.9	US20070161767(A1), US7375052(B2), EP1618952(A1)	2003	Toagosei
MoVTeNb	60.1	80.5	US20060047137(A1), WO2004065004(A1)	2003	Toagosei
	56.3	79.8	WO2004065004(A1), US20060047137(A1)	2003	Toagosei
MoVTeNb with inert carrier	38	77	US20070149390(A1)	2003	Avantium, U Malaya
Coated catalyst	39	74	US7019169(B2)	2003	BASF
Mix of phases TeMo and Vmo	31.3	73.7	US7683213(B2), WO2004105938(A2,A3)	2003	Atofina
MoVTe + MoVNbTe mixture	31	73.7	US7683213(B2), WO2004105938(A2,A3)	2003	Arkema
MoVTe+MoVTeNb	35.7	72.5	US20060293538(A1)	2003	Arkema

continued on following page

Table 3. Continued

Catalyst	Propane Conversion, %	Acrylic Acid Selectivity, %	Patent Reference	Year	Assignee
MoVNbTe	38	72	US7091377(B2), WO 2010/121006, US20090287019(A1)	2003	SABIC
MoVNbTe		72	US8193387(B2), US20090287019(A1)	2003	SABIC
MoVTeNbNi	80	66	US7091377(B2)	2003	BASF
MoVNbTe	19	65.2	WO2005063672(A1,R6A1), WO2010121006(A1), EP1694625(B1), US8193387(B2)	2003	SABIC
MoVTeNb	73	58	US6812366(B2), US20040054223(A1)	2003	Rohm and Haas
VWSbBiTiMo	12.4	82.8	US7087782(B2), EP1593663(A1)	2004	Haldor Topsoe
MoVTeNbPd	54	81	US20050202964(A1), EP1574253(A2,A3), US7304014B2	2004	Rohm and Haas
	54	81	US7304014B2	2004	Rohm and Haas
MoVTeNb	49.2	75.6	US7304180(B2), EP1591161(A1)	2004	Toagosei
MoVTeNb	62	73	WO2006008177(A1)	2004	Fritz Haber
MoVTeNb (treatment)	62	73	US20070203022(A1)	2004	Fritz Haber
	20.8	42.6	WO2006072682(A1)	2004	Arkema
MoVTeNbSi	56.4	81.2	US20090030224(A1), EP1808227(A1), US7645897(B2)	2005	Toagosei
MoVTeNbPd	62	81	US20050202964(A1), EP1574253(A2,A3), EP1574255(A2,A3)	2005	Rohm & Haas
	61	80	US20050202964(A1), EP1574253(A2,A3), EP1574255(A2,A3)	2005	Rohm and Haas
MoVNbTe (Nb-S)	22	59	WO2006100128(A1)	2005	Arkema
Nb doped MoVSbO	8.5	42	EP1704919(A1)	2005	Arkema
Polyoxometallate	1.2	12.9	US7019165(B2), US20050192461(A1)	2005	Caltech
InMoVTeNb	42	92.8	US7718568(B2)	2006	Rohm and Haas
MoVTeNbPd with cryo grinding and extraction using OA-MeOH	68	80	US7718568(B2)	2006	Rohm and Haas
	68	80	US7718568(B2)	2006	Rohm and Haas
MoVTeNb (with nonionic surfactant)	8	77	WO2008068332(A1), EP1930074(A1)	2006	Fritz Haber
MoCrTeCo	30	70	US8222461(B2)	2006	Evonik
MoVTeNbSmO	39	98	US7396955(B2)	2007	Rohm and Haas
MoVTeNbSmO	45	91	US7396955(B2)	2007	Rohm and Haas

continued on following page

Table 3. Continued

Catalyst	Propane Conversion, %	Acrylic Acid Selectivity, %	Patent Reference	Year	Assignee
MoVPNbO diluted with SiO2	4	99	EP2179793(A1)	2008	Süd-Chemie
MoVBiNb	5	86	EP2179790(A1)	2008	Süd-Chemie
MoVNbPtMZOx	39	73	WO 2013/082514	2011	SABIC
MoCrTeCo	18	72	US20110263415(A1), US20100286450(A1)	2011	Evonik
MoVNbSbTePt		70	US20130144085(A1)	2011	SABIC
MoVTeNbPd	45	69.6	US20110178333(A1)	2011	LG Chem
MoVTeNbW	45	68.1	US20110178332(A1), WO2009151254(A2,A3)	2011	LG Chem
MoVNbSbTePt	14.6	59.9	US20130144085(A1)	2011	SABIC
MoVNbTeMnWNi	65	81	WO2013021020(A1)	2012	Clariant
MoVNbTeMnCoNi	73	80	WO2013021034(A1)	2012	Clariant

Reports of acrylic acid selectivity greater than 90% can be found in patents from Rohm and Haas (Dow Chemical) and Süd-Chemie (Clariant). A very high selectivity of 98% has been achieved with a Samarium containing MoVTeNbSm composition, with 39% propane conversion (US Patent No. US7396955(B2), 2007). A selectivity of 99% to acrylic acid with a Te free composition, although at low conversion, has been achieved with phosphorus containing MoVPNbO composition (EP Patent No. EP 2179793, 2008).

High per pass yields of acrylic acid approaching 60% have been reported in the 2012 patent applications from Clariant (WO Patent No. WO2013021020(A1), 2012), (WO Patent No. WO2013021034(A1), 2012). In addition to MoVNbTe, the above catalyst compositions contain Mn, Co and Ni, and Mn, W and Ni, providing selectivity of 80% at around 70% conversion. High acrylic acid yields while maintaining good selectivity has also be achieved through the addition of Pd to the MoVNbTe composition (US Patent No. US7304014B2, 2004) and cryo-grinding the material. The cluster of data points between 60-70% propane conversion in Figure 10, indicate the high yields achieved.

Mo-V oxides as selective oxidation catalysts has been reviewed by Ueda (Ueda, 2013). The research literature has also mainly focused on the MoVTeNb oxide system for propane oxidation to acrylic acid, with many research groups extensively studying the synthesis and characterization of these catalysts. The progress in mixed metal oxide catalysts for propane oxidation to acrylic acid was reviewed by Lintz and Mueller (Lintz & Muller, 2009). Most of the research has focused on catalysts containing Mo, V, Te and Nb as they showed a far superior performance and acrylic acid yield than other catalyst systems.

In general, the MoVTeNb oxide catalysts obtained are found to consist of two main phases, named M1and M2. The M1 phase has been known to be the most active and selective. The preparation of the catalyst involves a calcination under inert atmosphere at 600 °C; M1 phase is not formed in an oxidative atmosphere (Lintz & Muller, 2009). Due to concern about the volatility of Te, catalysts containing Sb instead of Te, leading to similar M1 and M2 phases have been studied. However, the performance of antimony containing catalysts is considerably lower than the Te containing phases (Lintz & Muller, 2009).

Based on assessment of the reaction network of propane oxidation over MoVTeNb oxide catalysts reported in the literature Lintz et al. (Lintz & Muller, 2009) pointed out two main drawbacks, limited selectivity of propylene oxidation to acrylic acid and further oxidation of acrylic acid. Several studies state that the oxidative dehydrogenation of propane to propylene is the rate determining step. Trunschke and co-workers (Hävecker, et al., 2012) studied the surface chemistry and reaction network in propane oxidation over a pure M1 phase MoVTeNb oxide. The initial formation of propane followed by its oxidation directly to acrylic acid without desorption of the intermediates was found to be dominant route (d'Alnoncourt, et al., 2014). However, at higher conversions, the acrylic acid selectivity was restricted due to its degradation to CO_2 and ethylene, which was further oxidized to acetic acid then to CO and CO_2. The catalyst was also found to possess high stability in the selective oxidation of propane under reaction conditions. In a study of the surface chemistry of the pure M1 phase, Havecker et al. (Hävecker, et al., 2012) found that acrylic acid formation correlates with enrichment of the surface with V^{5+} sites, in close vicinity to Te^{4+} sites. The surface composition of M1 was also different from the bulk, implying that the active sites are formed in a thin layer on the surface of the catalyst.

The influence of composition of the synthesis mixture on the phases obtained and their catalytic performance has been studied. Lercher and co-workers (Naraschewski, Praveen Kumar, Jentys, & Lercher, 2011), found that V, Te, Nb, all were essential for formation of the M1 phase, while too high Te content led to the M2 phase. Higher concentrations of Nb increasingly led to more amorphous material, which was also observed by Ivars et al. (Ivars, Solsona, Hernandez, & Lopez Nieto, 2010). The M1 phase was the most catalytically efficient and activity was proportional to the concentration of V incorporated. Lopez Nieto and co-workers (Lopez Nieto, et al., 2010) made similar observations, finding that catalysts with exclusively M1 phase to be the most selective. The M2 phase was not able to activate propane. A loss of Te was observed during the heat treatment step, which was attributed to the presence of reducing species such as oxalate ions, due to the correlation between Te loss and oxalate content.

As noted in the patent literature on propane to acrylic acid as well as ethane oxidative dehydrogenation, the MoVTeNb oxide catalyst is sensitive to the preparation procedure, especially the calcination and heat treatment steps leading to evolution of the active crystalline phases. By carrying out an in-situ X-ray diffraction study Girgsdies et al. (Girgsdies, Schlögl, & Trunschke, 2012) found that formation of the M2 phase starts at significantly lower temperatures than M1, occurring simultaneously with evaporation of Te. Other phases, M_5O_{14} and MoO_3 formed competitively with M1, the availability of residual Te playing a role in distribution of the phases, and explaining the dependence on the method of preparation and conditions used.

Recently, doping of the catalysts with various elements to modify the catalyst surface has been reported as a successful way to improve the catalyst performance. Lopez Nieto and co-workers (Ivars, Solsona, Soriano, & Lopez Nieto, 2008) studied the performance of alkali doped MoVSbO catalysts, where the alkali was incorporated by hydrothermal synthesis. On this catalyst, Na, K, Rb incorporation was found to improve acrylic acid selectivity and lower the CO_2 and acetic acid formation. The authors suggest that in addition to eliminating acidic sites on the surface, the alkali metal is also incorporated in the catalyst framework.

Grasselli et al. (Grasselli, Lugmair, & Volpe Jr, 2008) studied the doping on separate M1 and M2 phases of MoVNbTe oxide composition. The doping was carried out by incipient wetness impregnation of low levels of selected elements having acidic (P, B, W), basic (Cs) and redox (Cu) properties. Addition of acidic elements, especially P, was found to give a good acrylic acid yield enhancement, which is attributed to more efficient conversion of intermediate acrolein to acrylic acid. Similar to the observation of Lopez Nieto and co-workers (Ivars, Solsona, Soriano, & Lopez Nieto, 2008), addition of Cs was found to suppress the catalytic activity.

Xu et al. (Xu, et al., 2013) added a small amount of Cr to the raw material mixture during synthesis to obtain a MoVTeNbCr oxide catalyst. With the Cr doping a large increase in the activity MoVTeNbOx catalyst was observed with a simultaneous reduction in CO_2 formation. A selectivity of 78.3% to acetic acid at 64.8% conversion could be obtained. It is suggested that Cr doping increases the number of surface sites that activate propane leading to activity improvement, and reduces the surface acidity which reduces acrylic acid oxidation.

Mestl et al. (Mestl, Margitfalvi, Végvári, Szijjártó, & Tompos, 2014) used combinatorial and high throughput tools to identify additional promoters Mn, Ni, W and In to boost the yield of acrylic acid obtained with the M1 phase of the MoVTeNb oxide catalyst system. A maximum yield of 59% at 76% propane conversion was achieved. Patent applications filed by Clariant related to these catalyst compositions have been discussed above.

Doping by in-situ synthesis as well as impregnation is successful; therefore, it holds promise as a versatile approach for improving alkane oxidation catalysts. The number of variations in the synthesis methods, catalyst compositions, dopants and levels of dopants can lead to a very large number of combinations; high throughput experimentation is a very useful tool to tackle such a large number of catalyst preparations and testing.

Although showing a good selectivity for production of acrylic acid directly from propane, the MoVTeNb catalyst composition appears to have some challenges. The long term stability of this catalyst is not clear, the main reason for this being the possibility of Te loss from the catalyst due to volatilization.

The existing process starting from propylene is very efficient; although consisting of multiple steps this process is more cost effective with both lower capital and operating costs. Therefore, it appears that further improvement in catalyst performance in all aspects, selectivity, per-pass yield, productivity (activity) and proven lifetime are necessary to commercially realize the full potential of the direct propane to acrylic acid process.

Propane Oxidative Dehydrogenation to Propylene

The recent patent literature on the oxidative dehydrogenation of propane to propylene is sparse. The best results reported are found in patents granted to academic institutions; a recent patent from King Fahd University of Petroleum and minerals mentions achieving 100% propylene selectivity at 4.8% propane conversion (US Patent No. US8623781(B2), 2011) using a MoV oxide prepared by a sol-gel technique. A patent assigned to Ohio State University mentions a selectivity of 96.5% at 5% conversion over a K/Mo catalyst supported on a 1:1 silica-titania mixed oxide (US Patent No. US6521808(B1), 2000). Other patents reporting selectivity above 80% are assigned to SABIC and Union Carbide (US Patent No. US6235678(B1), 1999), (WO Patent No. WO2003028881(A1), 2001); however, the propane conversion is around 5%.

In contrast to the patents, a substantial amount of research literature is found on the topic of oxidative dehydrogenation of propane. A review of propane oxidative literature is provided by Cavani et al. (Cavani, Ballarini, & Cericola, 2007). Although propylene selectivity above 90% have been reported recently for low temperature propane oxidative dehydrogenation, the propane conversions remain low at 5% or less (De Leon, et al., 2014), (Taylor, Carley, Davies, & Taylor, 2009), (Ruettinger, et al., 2011). One of the more interesting reports on low temperature propane ODH is with a Ce–Ni–O mixed oxide catalyst, with a selectivity of 82.7% obtained at 12.8% at 275 °C (Liu, et al., 2009). A catalyst system receiving a lot of attention is vanadium dispersed on high surface area oxides such as mesoporous silica, with the goal of achieving highly isolated VOx sites.

Compared to the commercial dehydrogenation processes where the selectivity obtained is over 90%, there are only a few reports where propane ODH provides such selectivity, and at very low conversions. Until such high selectivity are achieved it appears unlikely that low temperature propane ODH would find a place among propylene production technologies in the near future. The conclusions reached by Cavani et al. (Cavani, Ballarini, & Cericola, 2007) and Bhasin et al. (Bhasin, McCain, Vora, Imai, & Pujado, 2001) that the conversions and selectivity in propane ODH are still far from being commercially attractive are still valid.

Isobutane Oxidation to Methacrylic Acid

A patent search for selective oxidation of isobutane to methacrylic acid yields a large set of results. However, in most of these patents, examples are typically found for propane oxidation or ammoxidation, only a few patent examples with experimental data on isobutane oxidation to methacrylic acid are available. The most promising selectivity to methacrylic acid has been reported in a patent granted to CNRS, France. A selectivity of 77% to methacrylic acid and 8% to methacrolein at 10% conversion was obtained using a mixture of two crystal phases, one of which was a phosphomolybdic acid (US Patent No. US8207080(B2), 2007). A similar high combined selectivity of 81.5% (50.5% methacrylic acid and 31% methacrolein) at 4.4% conversion was reported in an older patent using a molybdenum heteropoly acid catalyst (US Patent No. US5380932, 1993). Phosphomolybdic acids containing other elements such V, Cu, have been reported to show a high selectivity; a composition consisting of PMoVCuAsB provided a selectivity of 56.2% to methacrylic acid and 19.2% to methacrolein at 8.8% isobutane conversion with the reaction carried out at 340 °C (US Patent No. US5191116(A), 1989).

Phosphomolybdate based catalysts are one of the major areas of focus in the research literature on selective oxidation of isobutane to methacrylic acid, with the best recent results also being reported with these materials. Millet and co-workers provide further details of their patented catalyst composition in their publications (Huynh, Selmi, Corbel, Lacorre, & Millet, 2009), (Huynh, Schuurman, Delichere, Loridant, & Millet, 2009). A phosphomolybdic polyoxometallate with Cs, Te, V, counter cations in a mechanical mixture with a lanthanum molybdate (either pure or with Mo partially substituted with V) was found to have a synergistic effect, leading to very efficient catalysts for isobutane oxidation. Similar CsTeHPMoV heteropoly compounds have been reported others as well to show high selectivity to methacrylic acid (Ding, Kan, Liu, Liu, & Guan, 2012). In an evaluation of MMA process options considering economic and environmental aspects, Sugiyama et al. (Sugiyama, et al., 2009) found that the isobutane oxidation route would have a better economic performance at a selectivity of above 70% to methacrylic acid.

Solutions and Recommendations

Great progress has been made in selective oxidation reactions over the past decade. Oxidation of ethane to acetic acid was commercialized by SABIC; this route has been found to be competitive with methanol carbonylation. Propane ammoxidation to acrylonitrile has also been commercialized by Asahi. For widespread commercial adoption the new processes need to provide a substantial improvement over the existing processes. Without economic benefits that are proven, companies will not move away from well-established technologies carried out in highly reliable and efficient plants (Cavani, Ballarini, & Cericola, 2007).

For the selective oxidation of lower alkanes, catalyst selectivity continues to be the main issue for commercial viability of the process. There appears to be a need to reassess the target selectivity and catalyst productivity required for the selective oxidation routes discussed. For example, for ethane ODH it would be interesting to know the selectivity/yield and catalyst productivity required to compete with the existing process. Would the ethane ODH be competitive at the 95+% selectivity reported? For the oxidation of propane to acrylic acid, a 60% yield of acrylic acid is frequently mentioned as the criteria for commercial viability; it would be useful to revisit this number and also determine the catalyst productivity required. Such studies comparing the selective oxidation process with the established manufacturing routes, using the latest published data would be helpful.

To advance the development of selective oxidation processes, catalyst research to identify and develop more active and selective catalyst compositions should continue. Understanding of the catalyst structure, nature of active sites, the role of different elements in the composition has helped lead to advancements; more knowledge in these areas would lead to further refinement of the catalyst compositions. Several publications have shown the value of high throughput experimentation in the identification of new catalyst compositions; wider use of these tools would further speed up the catalyst discovery efforts.

Apart from achieving the selectivity, productivity/yield targets, a few other aspects also need to be considered for commercial implementation of a process. Factors like by-product formation, separation and purification steps needed, product quality and catalyst life would also impact the feasibility of the process. Resolving these issues and building confidence would be necessary to proceed with process and catalyst scale-up.

Production of olefins by oxidative dehydrogenation offers the potential advantages of improved energy efficiency and lower operating temperature; however, ODH offers no advantage in terms of feedstock cost and reduction in number of steps. Other issues in ODH are separating out the olefin from a dilute stream, large recycle, and by-products like CO and acids. Development of process intensification solutions is likely to greatly benefit the ODH routes to produce olefins, especially ethane ODH to ethylene. At present the selectivity to propylene in the oxidative dehydrogenation of propane is well below that obtained in propane dehydrogenation processes.

Among the emerging selective oxidation routes, the major focus in the literature has rightly been on propane oxidation to acrylic acid and ethane ODH to ethylene. Although single step, the direct propane oxidation route to acrylic acid is not yet attractive, compared to the propylene route. Further improvements to catalyst selectivity, productivity and eliminating by-product and impurity formation is likely to make this route very attractive.

FUTURE RESEARCH DIRECTIONS

Looking at the progress that has been made in recent years in ethane ODH to ethylene and propane oxidation to acrylic acid, it is clear that this trend is likely to continue. It is likely that serious consideration would be given to these processes as viable alternatives to the current routes. The economics of these routes would need to be reevaluated considering the latest reported results and developments. For commercial implementation other questions that would need to be answered include those related to product quality, by-products, operational safety and reliability, and long term stability of the catalyst. Other routes that are emerging such as BP's new route to acetic acid from syngas and Novomer's EO carbonylation route to acrylic acid would provide challenges to the selective oxidation routes.

In propane oxidative dehydrogenation to propylene, the selectivity currently achievable is far away from that in the current dehydrogenation processes.

Catalyst compositions based on the MoVTeNb oxide have provided the best performance; these catalysts are becoming increasingly complex with more elements being incorporated during synthesis of the mixed oxide as well as post-synthesis addition of promoters and dopants. The quest to develop more selective catalysts that avoid by-product and impurity formation should continue. The issue of Te loss from the catalyst during preparation as well as reaction has been raised and alternatives to Te have also been studied. Further work is required to answer these questions related to Te or develop Te free compositions with equivalent or better performance. The commercialization of the propane ammoxidation to acrylonitrile using a MoVTeNbO catalyst provides confidence in the feasibility of commercially using the MoVTeNbO catalyst in other processes.

Future research opportunities on the fundamental side would include continuing to develop a better understanding of the role of different metals in the mixed metal oxide catalyst, especially considering the increase in complexity of the catalyst composition. This would help in the design and optimization of catalyst formulations with improved performance. Generating knowledge on how to produce the desired catalyst phases and surface characteristics in easier ways would be very useful for the preparation and scale-up of the catalyst. In process research the opportunities are in new reactor designs, better separation processes, eliminating separation and purification steps. It would be helpful to revisit the economic assessments and determine the selectivity, yield and productivity targets that need to be achieved in the selective oxidation processes to make the process economically viable.

CONCLUSION

Selective oxidation of lower alkanes remains a very active area of research, with a tremendous amount of work done in the past decade. Mixed metal oxide catalysts consisting of MoVTeNbO have emerged as the best performing catalysts for propane oxidation to acrylic acid and oxidative dehydrogenation of ethane to ethylene. Knowledge generated about the catalyst structure, characteristics and reaction mechanisms have helped in advancing the performance of the catalysts.

Very high selectivity above 95% have been achieved in ethane ODH to ethylene. Similarly big improvements in the selectivity and yield of acrylic acid from propane have been made with selectivities around 90% reported. Clearly, these routes have moved a step closer to commercialization. It would be interesting to see detailed economic comparisons of these processes with the existing routes, using the latest results. Such studies would also provide a better picture of the targets that need to be achieved to make the new routes economically favorable than the current dominant route. For both ethane ODH and propane oxidation to acetic acid the questions about Te loss and the long term stability of MoVTeNbO catalysts also need to be answered.

The oxidation of ethane to acetic acid was shown to be competitive with the methanol carbonylation process. In the last decade there has not been much focus on ethane oxidation to acetic acid in the literature and no new plants based on this route have come up. However, with reports of very high selectivity being achieved in ODH of ethane to ethylene, it appears possible to further improve catalyst performance for ethane to acetic acid.

Selective oxidation catalyst research must continue to focus on improving the selectivity and activity of the catalysts. Approaches to increase catalyst surface area, optimizing the catalyst composition, incorporation/doping with promoters have all been successful approaches to improve the catalyst performance. Combinatorial and high throughput testing tools have enabled catalyst preparation and testing to be carried out faster and screen a larger catalyst composition space, leading to successful results. Further work in these directions offers the promise of more success in improving catalyst selectivity and activity. From a fundamental point of view, a better understanding of the role of each metal/element in the catalyst composition would help in the preparation and optimization of catalyst compositions with improved performance.

The production of oxygenated products such as acrylic acid and acetic acid in a single step from an alkane appears more feasible since they compete with multi-step processes. In addition to use of a lower cost feedstock, these routes can also provide sustainability benefits. However for the oxidative dehydrogenation routes to produce olefins, the challenge is much higher as they have to compete with highly mature steam cracking and dehydrogenation processes that make use of the same alkane feedstock. The ODH routes must therefore provide a big advantage in energy and capital intensity; in addition to catalyst developments, process intensification opportunities could help provide this advantage.

REFERENCES

Ahmed, S., Alasiri, H. S., Rahman, F., & Al-Amer, A. M. (2011). *US Patent No. US8623781(B2).*

Al-Zeghayer, Y. (2014). Partial Oxidation of Ethane to Acetic Acid Catalyzed by MoVNbPd Catalyst Supported on Titania. *International Journal of Chemical Engineering and Applications*, *5*(1), 50–57. doi:10.7763/IJCEA.2014.V5.350

Al-Zeghayer, Y. S., & Al-Awadi, A. S. (2013). Partial Oxidation of Ethane to Acetic Acid on Titania Supported MoVNbPd Catalyst. *Asian Journal of Chemistry*, *25*(14), 7979–7985. doi:10.14233/ajchem.2013.14886

Al-Zeghayer, Y. S., Soliman, M. A., & Al-Mayman, S. & S, A.-A. A. (2013). *US Patent No. US2014235893A1.*

Bardin, B. B., & Hammond, R. R. (2001). *WO Patent No. WO2003028881(A1).*

Bhasin, M., McCain, J., Vora, B., Imai, T., & Pujado, P. (2001). Dehydrogenation and oxydehydrogenation of paraffins to olefins. *Applied Catalysis A, General*, *221*(1-2), 397–419. doi:10.1016/S0926-860X(01)00816-X

Bielmeier, E., Haeberle, T., Siegert, H.-J., & Gruber, W. (1993). *US Patent No. US5380932.*

Cavalcanti, F. A., Chaturvedi, S., Gaffney, A. M., Han, S., Song, R., & Vickery, E. M. (2004). *US Patent No. US7304014B2.*

Cavani, F. (2010). Catalytic selective oxidation: The forefront in the challenge for a more sustainable chemical industry. *Catalysis Today*, *157*(1-4), 8–15. doi:10.1016/j.cattod.2010.02.072

Cavani, F., Ballarini, N., & Cericola, A. (2007). Oxidative dehydrogenation of ethane and propane: How far from commercial implementation? *Catalysis Today*, *127*(1-4), 113–131. doi:10.1016/j.cattod.2007.05.009

Cavani, F., & Trifiro, F. (2004). Partial Oxidation of C2 to C4 Paraffins. In M. Baerns, Basic Principles in Applied Catalysis (pp. 19-84). doi:10.1007/978-3-662-05981-4_2

Celaya Sanfiz, A., Hansen, T. W., Sakthivel, A., Trunschke, A., Schlogl, R., Knoester, A., & Hamid, S. B. et al. (2008). How important is the (001) plane of M1 for selective oxidation of propane to acrylic acid? *Journal of Catalysis*, *258*(1), 35–43. doi:10.1016/j.jcat.2008.05.028

Celaya Sanfiz, A., Timpe, O., Trunschke, A., & Schlogl, R. (2008). *EP Patent No. EP 2179793.*

Celaya Sanfiz, A., Timpe, O., Trunschke, A., & Schlogl, R. (2010). *European Patent No. EP2179790.*

Cook, J., Ellis, B., Howard, P., Jones, M. D., & Kitchen, S. J. (n.d.). *US Patent No. US6350716(B1).*

De León, M. A., Santos, C., Latronica, L., Cesio, A. M., Volzone, C., Castiglioni, J., & Sergio, M. (2014). High catalytic activity at low temperature in oxidative dehydrogenation of propane with Cr–Al pillared clay. *Chemical Engineering Journal*, *241*, 336–343. doi:10.1016/j.cej.2013.10.086

de Rooij, F., & Schoonebeek, R. J. J, S. J., & Verhaak, M. J. (2012). *WO Patent No. WO2013164418(A1).*

Ding, W., Kan, Q., Liu, H., Liu, L., & Guan, J. (2012). Catalytic activity of V-substituted Cs2Te0.2H0.6 + x PMo12 − x V x O n heteropoly compounds in selective oxidation of isobutane. *Kinetics and Catalysis*, *53*(3), 404–408. doi:10.1134/S0023158412030068

Gaffney, A. M., Ghosh, R., Song, R., Yeh, C. C., & Langner, T. (2009). *US Patent No. US8105972(B2).*

Gaffney, A. M., Ghosh, R., Song, R., Yeh, C. Y., & Langner, T. (2009). *US Patent No. US8105971(B2).*

Gaffney, A. M., Heffner, M. D., & Song, R. (2007). *US Patent No. US7396955(B2).*

Gaffney, A. M., & Song, R. (2006). *US Patent No. US7718568(B2).*

Gartner, C. A., van Veen, A. C., & Lercher, J. A. (2013). Oxidative Dehydrogenation of Ethane: Common Principles and Mechanistic Aspects. *ChemCatChem*, *5*(11), 3196–3217. doi:10.1002/cctc.201200966

Girgsdies, F., Schlögl, R., & Trunschke, A. (2012). In-situ X-ray diffraction study of phase crystallization from an amorphous MoVTeNb oxide catalyst precursor. *Catalysis Communications*, *18*, 60–62. doi:10.1016/j.catcom.2011.11.009

Grabowski, R. (2006). Kinetics of Oxidative Dehydrogenation of C2-C3 Alkanes on Oxide Catalysts. *Catalysis Reviews*, *48*(2), 199–268. doi:10.1080/01614940600631413

Grasselli, R., Buttrey, D., DeSanto, P. Jr, Burrington, J., Lugmair, C., Volpe, A. Jr, & Weingand, T. (2004). Active centers in Mo-V-Nb-Te-Ox (amm)oxidation catalysts. *Catalysis Today*, *91-92*, 251–258. doi:10.1016/j.cattod.2004.03.060

Grasselli, R. K., Lugmair, C. G., & Volpe, A. F. Jr. (2008). Doping of MoVNbTeO (M1) and MoVTeO (M2) Phases for Selective Oxidation of Propane and Propylene to Acrylic Acid. *Topics in Catalysis*, *50*(1-4), 66–73. doi:10.1007/s11244-008-9097-9

Guerrero-Perez, O. M., Lopez-Medina Ricardo, B. M., M, R. J., & Rodriguez-Mirasol Jose, C. T. (2010). A New and Highly Selective Vanadium Carbon-based catalysts. *American Carbon Society Proceedings*.

H, M. J., W, K. S., & L, O. G. (1987). *US Patent No. US5162578(A)*.

Hagemeyer, A., Mestl, G., Neumann, S., Margitfalvi, J., Tompos, A., & Vegvari, I. L. (2012). *WO Patent No. WO2013021020(A1)*.

Hagemeyer, A., Mestl, G., Neumann, S., Tompos, A., Margitfalvi, J., & Vegvari, I. L. (2012). *WO Patent No. WO2013021034(A1)*.

Han, S., & Frick, C. D. & J, M. D. (2012). *WO Patent No. WO2013148006(A1)*.

Han, S., & Frick, C. D. & J, M. D. (2012). *Patent No. WO2013148006(A1)*.

Hävecker, M., Wrabetz, S., Kröhnert, J., Csepei, L.-I., d'Alnoncourt, R. N., Kolen'ko, Y. V., & Trunschke, A. et al. (2012). Surface chemistry of phase-pure M1 MoVTeNb oxide during operation in selective oxidation of propane to acrylic acid. *Journal of Catalysis*, *285*, 48–60.

Huynh, Q., Schuurman, Y., Delichere, P., Loridant, S., & Millet, J. (2009). Study of Te and V as counter-cations in Keggin type phosphomolybdic polyoxometalate catalysts for isobutane oxidation. *Journal of Catalysis*, *261*(2), 166–176. doi:10.1016/j.jcat.2008.11.013

Huynh, Q., Selmi, A., Corbel, G., Lacorre, P., & Millet, J. (2009). Atypical synergetic effect between Te- and V-substituted phosphomolybdic cesium salt and LAMOX-type phases for the oxidation of isobutane into methacrylic acid. *Journal of Catalysis 266 (2009) 64–70*, *266*, 64-70.

Ishikawa, S., Yi, X., Murayama, T., & Ueda, W. (2014). Catalysis field in orthorhombic Mo3VOx oxide catalyst for the selective oxidation of ethane, propane and acrolein. *Catalysis Today*, *238*, 35–40. doi:10.1016/j.cattod.2013.12.054

Ishikawa, S., Yi, X., Murayama, T., & Ueda, W. (2014). Heptagonal channel micropore of orthorhombic Mo3VOx as catalysis field for the selective oxidation of ethane. *Applied Catalysis A, General*, *474*, 10–17. doi:10.1016/j.apcata.2013.07.050

Ivars, F., Solsona, B., Hernandez, S., & Lopez Nieto, J. (2010). Influence of gel composition in the synthesis of MoVTeNb catalysts over their catalytic performance in partial propane and propylene oxidation. *Catalysis Today*, *149*(3-4), 260–266. doi:10.1016/j.cattod.2009.09.018

Ivars, F., Solsona, B., Soriano, M., & Lopez Nieto, J. (2008). Selective Oxidation of Propane Over AMoVSbO Catalysts (A = Li, Na, K, Rb or Cs). *Topics in Catalysis*, *50*(1-4), 74–81. doi:10.1007/s11244-008-9103-2

Jing, F., Katryniok, B., Dumeignil, F., Bordes-Richard, E., & Paul, S. (2014). Catalytic selective oxidation of isobutane to methacrylic acid on supported (NH4)3HPMo11VO40 catalysts. *Journal of Catalysis*, *309*, 121–135. doi:10.1016/j.jcat.2013.09.014

Karim, K., Al-Hazmi, M. H., Khan, A. A., & Zaheer, S. I. (n.d.). *US Patent No. US6531631B1*.

Karim, K., Mamedov, E., Al-Hazmi, M. H., Fakeeha, A. H., Soliman, M. A., Al-Zeghayer, Y. S., . . . Al-Arify, A. A. (n.d.). *US Patent No. US6310241B1*.

Karim, K., & Sheikh, K. (1999). *US Patent No. US6274764B1.*

Karim, K. S., Al-Hazmi, M. H., & Al-Andis, N. (1999). Catalysts for Oxidative Dehydrogenation of Ethane. *The Arabian Journal for Science and Engineering*, 41-48.

Kitson, M. (1993). *US Patent No. US5210293(A).*

Kustov, L. M., Kucherov, A. V., Finashina, E. D., Kucherova, T. N., Isaeva, V. I., Krzywicki, A., & Cai, H. (2010). *US Patent No. US20120016171(A1).*

Kustov, L. M., Kucherov, A. V., Finashina, E. D., Shimanzhenkov, V., & Krzywicki, A. (2012). *US Patent No. US8642825(B2).*

Li, X., & Iglesia, E. (2008). Kinetics and Mechanism of Ethane Oxidation to Acetic Acid on Catalysts Based on Mo-V-Nb Oxides. *The Journal of Physical Chemistry C, 112*(38), 15001–15008. doi:10.1021/jp801488y

Li, X., & Iglesia, E. (2008). Support and promoter effects in the selective oxidation of ethane to acetic acid catalyzed by Mo-V-Nb oxides. *Applied Catalysis A, General, 334*(1-2), 339–347. doi:10.1016/j.apcata.2007.10.021

Linke, D., Wolf, D., Baerns, M., Timpe, O., Schlogl, R., Zeyb, S., & Dingerdissen, U. (2002). Catalytic Partial Oxidation of Ethane to Acetic Acid over Mo1V0.25Nb0.12Pd0.0005Ox. I. Catalyst Performance and Reaction Mechanism. *Journal of Catalysis, 205*, 16–31. doi:10.1006/jcat.2001.3367

Lintz, H.-G., & Muller, S. P. (2009). The partial oxidation of propane on mixed metal oxides - A short overview. *Applied Catalysis A, General, 357*(2), 178–183. doi:10.1016/j.apcata.2009.01.018

Liu, Y. (2001). *US Patent No. US6891075(B2).*

Liu, Y. (2007). *US Patent No. US7674944(B2).*

Liu, Y.-M., Wang, L.-C., Chen, M., Xu, J., Cao, Y., He, Y.-H., & Fan, K.-N. (2009). Highly Selective Ce–Ni–O Catalysts for Efficient Low Temperature Oxidative Dehydrogenation of Propane. *Catalysis Letters, 130*(3-4), 350–354. doi:10.1007/s10562-009-9977-z

Lopez Nieto, J. M., Solsona, B., Concepcion, P., Ivars, F., Dejoz, A., & Vazquez, M. I. (2010). Reaction products and pathways in the selective oxidation of C2–C4 alkanes on MoVTeNb mixed oxide catalysts. *Catalysis Today, 157*(1-4), 291–296. doi:10.1016/j.cattod.2010.01.046

Mamedov, E., & Karim, K. (2014). Selective Oxidation at SABIC: Innovative Catalysts and Technologies. In D. Duprez, & F. Cavani (Eds.), Handbook of Advanced Methods and Processes in Oxidation Catalysis: From Laboratory to Industry (pp. 291-301). Imperial College Press.

Mamedov, E. A., & Shaikh, S. N. (1999). *US Patent No. US6235678(B1).*

Mestl, G., Margitfalvi, J. L., Végvári, L., Szijjártó, G. P., & Tompos, A. (2014). Combinatorial design and preparation of transition metal dopedMoVTe catalysts for oxidation of propane to acrylic acid. *Applied Catalysis A, General, 474*, 3–9. doi:10.1016/j.apcata.2013.08.022

Millet, J.-M., Lacorre, P., & Huynh, Q. (2007). *US Patent No. US8207080(B2).*

Naraschewski, F. N., Praveen Kumar, C., Jentys, A., & Lercher, J. A. (2011). Phase formation and selective oxidation of propane over MoVTeNbOx catalysts with varying compositions. *Applied Catalysis A, General*, *391*(1-2), 63–69. doi:10.1016/j.apcata.2010.07.005

Naumann d'Alnoncourt, R., Csepei, L.-I., Hävecker, M., Girgsdies, F., Schuster, M. E., Schlögl, R., & Trunschke, A. (2014). The reaction network in propane oxidation over phase-pure MoVTeNb M1 oxide catalysts. *Journal of Catalysis*, *311*, 369–385. doi:10.1016/j.jcat.2013.12.008

Novomer. (2014). Retrieved from http://www.novomer.com/novomer-co-process

Ozkan, U. S., & Watson, R. B. (2000). *US Patent No. US6521808(B1).*

Rahman, F., Loughlin, K. F., Al-Saleh, M. A., Saeed, M. R., Tukur, N. M., Hossain, M. M., & Mamedov, A. et al. (2010). Kinetics and mechanism of partial oxidation of ethane to ethylene and acetic acid over MoV type catalysts. *Applied Catalysis A, General*, *375*(1), 17–25. doi:10.1016/j.apcata.2009.11.026

Roussel, M., Barama, S., Lofberg, A., Al-Sayari, S., Karim, K., & Bordes-Richard, E. (2009). MoV-based catalysts in ethane oxidation to acetic acid: Influence of additives on redox chemistry. *Catalysis Today*, *141*(3-4), 288–293. doi:10.1016/j.cattod.2008.04.046

Roussel, M., Bouchard, M., Karim, K., Al-Sayari, S., & Bordes-Richard, E. (2006). MoVO-based catalysts for the oxidation of ethane to ethylene and acetic acid. Influence of niobium and/or palladium on physicochemical and catalytic properties. *Applied Catalysis A, General*, *308*, 62–74. doi:10.1016/j.apcata.2006.04.017

Ruettinger, W., Benderly, A., Han, S., Shen, X., Ding, Y., & Suib, S. L. (2011). Influence of Support Surface Area and Niobium Addition on the Reactivity of Vanadium Catalysts for Propane Oxidative Dehydrogenation. *Catalysis Letters*, *141*(1), 15–21. doi:10.1007/s10562-010-0459-0

Sankaranarayanan, T. M., Ingle, R. H., Gaikwad, T. B., Lokhande, S. K., Raja, T., Devi, R. N., & Manikandan, P. et al. (2008). Selective Oxidation of Ethane Over Mo–V–Al–O Oxide Catalysts: Insight to the Factors Affecting the Selectivity of Ethylene and Acetic Acid and Structure-activity Correlation Studies. *Catalysis Letters*, *121*(1-2), 39–51. doi:10.1007/s10562-007-9289-0

Schlogl, R., & Trunschke, A. (2006). *WO Patent No. WO2008068332(A1).*

Sijpkes, A. H., Wilhelmus, R. H., van der Puil, N., & Abdul Hamid, S. B. (2003). *US Patent No. US20070149390(A1).*

Smejkal, Q., Linke, D., & Baerns, M. (2005). Energetic and economic evaluation of the production of acetic acid via ethane oxidation. *Chemical Engineering and Processing*, *44*(4), 421–428. doi:10.1016/S0255-2701(04)00146-1

Soliman, M., Al-Zeghayer, Y., Al-Awadi, A. S., & Al-Mayman, S. (2012). Economics of Acetic Acid Production by Partial Oxidation of Ethane. *APCBEE Procedia*, *3*, 200–208. doi:10.1016/j.apcbee.2012.06.070

Sugiyama, H., Fischer, U., Antonijuan, E., Hoffmann, V. H., Hirao, M., & Hungerbühler, K. (2009). How do different process options and evaluation settings affect economic and environmental assessments? A case study on methyl methacrylate (MMA) production processes. *Process Safety and Environmental Protection*, *87*(6), 361–370. doi:10.1016/j.psep.2009.08.002

Taylor, M. N., Carley, A. F., Davies, T. E., & Taylor, S. H. (2009). The Oxidative Dehydrogenation of Propane Using Vanadium Oxide Supported on Nanocrystalline Ceria. *Topics in Catalysis*, *52*(12), 1660–1668. doi:10.1007/s11244-009-9307-0

Tu, X., Furuta, N., & Sumida, Y. (2005). *US Patent No. US7645897(B2).*

Tu, X., Sumida, Y., Takahashi, M., & Niizuma, H. (2004). *US Patent No. US7375052(B2).*

Ueda, W. (2013). Establishment of Crystalline Complex Mo-V-Oxides as Selective Oxidation Catalysts. *Journal of the Japan Petroleum Institute*, *56*(3), 122–132. doi:10.1627/jpi.56.122

Valente, J. S., Armendáriz-Herrera, H., Quintana-Solórzano, R., del Ángel, P., Nava, N., Masso, A., & López Nieto, J. (2014). Chemical, Structural, and Morphological Changes of a MoVTeNb Catalyst during Oxidative Dehydrogenation of Ethane. *ACS Catalysis*, *4*(5), 1292–1301. doi:10.1021/cs500143j

Valente, J. S., Lopez-Nieto, J. M., Herrera, H. A., Ramirez, A. M., Barcelo, F. I., Castillo, M., . . . Flores, E. M. (2012). *US Patent No. US20140114109(A1).*

Xu, A., Wang, Y., Ge, H., Chen, S., Li, Y., & Lu, W. (2013). An outstanding Cr-doped catalyst for selective oxidation of propane to acrylic acid. *Chinese Journal of Catalysis*, *34*(12), 2183–2191. doi:10.1016/S1872-2067(12)60671-8

Yamamatsu, S., & Yamaguchi, T. (1989). *US Patent No. US5191116(A).*

Zeiss, S., Dingerdissen, U., Baerns, M., Wolf, D., & David, L. (n. d.). *US Patent No. US6906221(B2).*

Zeyss, S., & Dingerdissen, U. (2001). *European Patent No. EP1286942B1.*

Chapter 3
Palladium in Heterogeneous Oxidation Catalysis

Andreas Martin
Leibniz-Institute for Catalysis, Germany

Venkata Narayana Kalevaru
Leibniz-Institute for Catalysis, Germany

Jörg Radnik
Leibniz-Institute for Catalysis, Germany

ABSTRACT

Palladium is one of the precious group metals mainly used in automobile catalytic converters. Besides, it has an importance in various catalytic processes. Although it is well known for hydrogenation reactions, various oxidations can also be catalyzed by palladium. This chapter gives an overview on the most common application of palladium catalysts in heterogeneously catalyzed acetoxylation, i.e. the acetoxylation of ethylene to vinyl acetate. Derived from this knowledge, the authors summarize in detail recently accumulated research results in acetoxylation of toluene to benzyl acetate that can be easily converted to benzyl alcohol. The chapter includes a detailed description of catalyst syntheses, gas phase oxidation runs, comprehensive characterizations and a deep understanding in catalyst-feed interaction. This development can turn away the manufacture of important petrochemicals from chlorine chemistry to oxidations using molecular oxygen.

BACKGROUND

Palladium (Pd) is a silvery-shining metal discovered in 1803 by W.H. Wollastone. It is a member of the nickel group standing in group 10 (Ni, Pd, Pt) of the periodic table of elements and it is also a part of the platinum group metals (PGM) together with platinum, rhodium, ruthenium, iridium and osmium.

In general, metallic Pd and Pd ores are rare and often found together with other PGMs (e.g. alloyed with gold or platinum). However, the most commercial sources are deposits containing nickel and copper. The total Pd supply in 2012 was ca. 6.5 million oz.; 44% were extracted from Russian mines,

DOI: 10.4018/978-1-4666-9975-5.ch003

Figure 1. Palladium supply (%) by regions (Johnson Matthey, 2014)

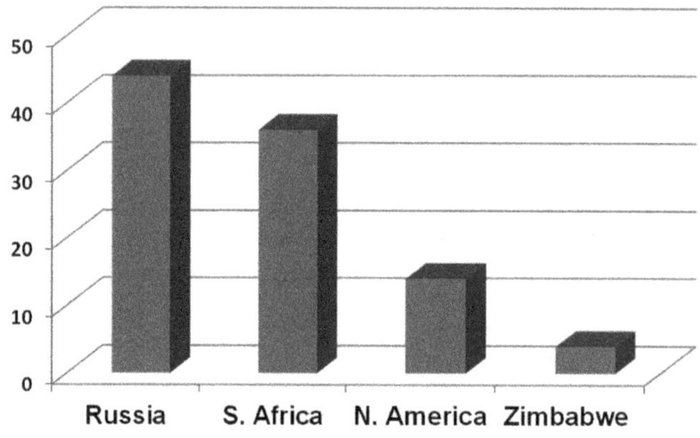

36% in South Africa (Johnson Matthey, 2014; Stillwater Palladium, 2012). Other important producing regions are located in North America and Africa; Figure 1 depicts the shares of the four largest producing countries. In addition, Pd is also obtained from recycling of noble metal containing scrap, mainly catalytic converters from automobiles.

Most of the Pd is used for automotive catalysts (65%), electronic, dental and jewelry application consume ca. 20%, the remaining part is for investment and other applications including its use as catalysts or dopants in catalyst formulations (Figure 2) (Manhattan Gold & Silver, 2011). In automotive catalysts, it is combined with platinum or rhodium and removes over 90% of harmful carbon monoxide, unconverted hydrocarbons and nitrogen oxides into water vapor, nitrogen and carbon dioxide. Pd plays also a key role in fuel cell technology to produce electricity and heat.

Besides its use in automotive catalysts and fuel cell technology, palladium has a large number of applications in homogeneous and heterogeneous catalysis (Ugo, 1980). For example, finely dispersed Pd on several supports, in particular carbon is well known as hydrogenation or dehydrogenation catalyst in chemical industries (Coq, & Figueras, 2001; Mahata, &Vishwanathan, 2000) as well as petroleum cracking (Gallezot, 1979; Bradow, Grenoble, Foley, Murray, Winquist, & Milham, 2001). Pd is also widely

Figure 2. Palladium demand (%) by application fields; AC = Automotive catalysts, E = Electronics, D = Dental, I = Investment and J = Jewelry (Johnson Matthey, 2014)

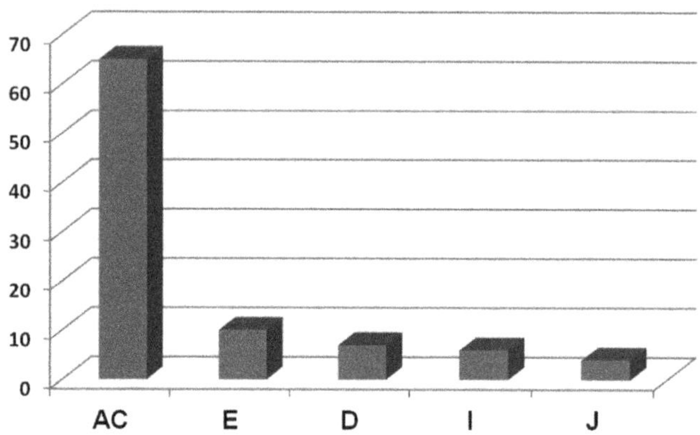

applied for manufacture of membranes for hydrogen separation (Yun, &Oyama, 2011). Moreover, Pd is also an important component of Wacker-catalysts used in the well-known Wacker-process for the synthesis of various aldehydes from olefins (Keith, &Henry, 2009; Jira, 2009). In addition, the Wacker-process is similarly applied for the synthesis of acetates, in particular vinyl acetate by acetoxylation of ethylene in the presence of acetic acid and molecular oxygen (Hagemeyer, Dingerdissen, Kühlein, Manz,, & Fischer, 2003). It was also shown that Pd can be used as an electro catalyst when finely dispersed on conductive materials; such catalysts can be used for a selective oxidation of primary alcohols (Tsuji, 2004). In homogeneous catalysis and organic chemistry, a large number of C-C coupling reactions are facilitated in the presence of Pd-containing catalysts, e.g. Heck reaction (Heck, 1982; Beletskaya, &Cheprakov, 2000) and Suzuki coupling (Old, Wolfe, & Buchwald, 1998). Richard F. Heck was awarded in 2010 with the Nobel Prize in chemistry for the development of the reaction enabling access to substituted alkenes. In addition, Pd is also found in the Lindlar catalyst used for example in hydrogenation of alkynes to alkenes (Lindlar, &Dubuis, 1973) and various fine chemical syntheses (Blaser, Indolese, Schnyder, Steiner, & Studer, 2001). Besides, Pd is also applied in a wide variety of homogeneous catalysts in combination with various ligands for selective transformations. Over the past few decades the Pd catalyzed oxidation of alpha-olefins and alkyl aromatics has become an increasingly attractive field for the selective synthesis of vinyl, allyl, benzyl alcohols and esters. Pd is also used as catalytic material in cluster from and such Pd clusters were applied for acetoxylation of olefins and alcohols (Vargaftik et al., 1989) or toluene (Choi, Mizugaki, Ebitani, & Kaneda, 2002). In the latter case, the authors claimed that the 8-shell Pd-cluster of about 38 Å in diameter showing high activity in the named reaction.

Thus, the development of more effective and economic synthesis routes (e.g. heterogeneously catalyzed oxidation and acetoxylation of methyl aromatics) to benzylic esters and/or alcohols and even aldehydes, as intermediates will be certainly much interesting for further synthesis of various industrially important chemicals. For example, benzyl alcohol being an important solvent and an intermediate in various organic reactions (Bauer, Garbe, & Surburg, 1988), is produced in some thousand tons/y. It is mainly manufactured in a multistep process by toluene chlorination and subsequent hydrolysis leading to larger amounts of HCl. Therefore, this process is not an environmentally benign synthesis. Alternatively, an oxidative coupling reaction of toluene and acetic acid to benzyl acetate may open a new chlorine-free and, thus, environmental friendly route to benzyl alcohol production through hydrolysis of the target product. In addition, several Pd-catalyzed reactions in the area of organic reactions are still under study, such as cross-coupling reactions, CC bond formation or allylations etc. (Yin, &Liebscher, 2007). New catalysts could consist of single-phase intermetallic compounds, e.g. PdBi on silica (Komatsu, &Onda, 2008). In addition, multi-component Pd-Sn-K on silica based catalysts were proposed (Shu, Wang, Ren, & Shi, 2005).

PALLADIUM IN OXIDATION CATALYSIS

A direct and controlled partial oxidation of various hydrocarbons to obtain desired oxygenates is one of the greatest challenges in the field of heterogeneous catalysis (Centi, Cavani, & Trifirò, 2000). The major problem in such oxidation reactions is to prevent the further oxidation of the desired products into the unwanted by-products, e.g. the thermodynamically favored carbon oxides. This difficulty can however be overcome to a certain extent by carrying out such oxidations in presence of suitable catalyst compositions, defined catalytic reaction engineering actions and last but not least using auxiliary

Figure 3. General reaction equation of the acetoxylation reaction of ethylene to vinyl acetate

$$C_2H_4 + CH_3COOH + 0.5\,O_2 \longrightarrow CH_2CHOCOCH_3 + H_2O$$

reagents to convert intermediates to stable compounds. For example, the partially oxidized intermediate in ammoxidation of aromatics is quenched by ammonia to form a nitrile (Martin, & Kalevaru, 2011). In a similar manner, in acetoxylation reactions, such oxidation is also carried out over Pd-containing solids but in presence of acetic acid as an auxiliary agent, which prevents the further oxidation of target products (Madaan et al., 2011a).

Acetoxylation reaction is a reaction where the hydrocarbon moiety is directly converted into an ester moiety in a single step in presence of acetic acid (source of acetyl group) and in an oxidizing atmosphere (O_2 or air). The reaction is applicable to the conversion of vinyl, allyl, benzyl and aryl groups into their corresponding esters. The most extensively studied reaction is the acetoxylation of ethylene to vinyl acetate in the presence of oxygen and acetic acid over a palladium-based catalyst system (Figure 3). The commercial production of vinyl acetate monomer (VAM) involves a heterogeneous process where the reaction is catalyzed by metallic palladium (Krekel, & Kronig, 1969). The overall reaction is the same as for the homogeneous process. The initial work on acetoxylation of ethylene and acetic acid to produce vinyl acetate was studied by Moiseev during the 1960's (Moiseev, Vargaftic, & Syrkin, 1960). Moiseev's results claimed that the Wacker process for the manufacture of acetaldehyde could be altered in favor of the formation of vinyl acetate by using $PdCl_2$ as a catalyst. The technical acetoxylation process was developed in cooperation with National distillers and Bayer (Schwerdtel, 1968). Since the discovery of this process, extensive studies have been focused on improving the catalyst (e.g. Benkalowycz, Blum, Cirjak,, Lemanski,, Paparizos, Pepera, & Wagner, 1996; Nicolau, Broussard, & Colling, 1999). All catalysts employed are palladium based (e.g. Kragten, van Santen, Neurock, & Lerou, 1999). Further catalyst compositions with the addition of gold as second metal (e.g., Pd: Au: K at 0.56: 0.25: 2.9 wt%) (Crathorne, MacGowan, Morris, & Rawlinson, 1994) or cadmium (e.g. Pd: Cd: K at 2.1: 2.0: 2.4 wt%) were also developed in the last years (Hornig, Fernholz, Schmidt, Wunder, & Quadflieg, 1972; Kroenig, & Scharfe, 1974). Silica is generally used as a catalyst carrier material. This process is usually operated at 130 to about 190 °C and 5 to 10 bar pressure (Miller, 1969). The space-time-yield of vinyl acetate is increased with both the total pressure and the partial pressure of oxygen. Considerable work regarding the reaction mechanism and the nature of the active sites and intermediates in VAM synthesis including the industrial application has been reported elsewhere (Schunk, &de Oliveira, 2008; Voskanyan, 2010; Han, Kumar, & Goodman, 2005; Kumar, Chen, & Goodman, 2007; Macleod, Keel, & Lambert, 2004). Among different mechanistic proposals, Samanos proposed mechanism is widely accepted for the formation of vinyl acetate from ethylene (Samanos, Boutry, & Montarnal, 1971). The mechanistic discussions are however described below in a separate section.

VINYL ACETATE AND BENZYL ACETATE MANUFACTURE

The introduction of an acetoxy group (CH_3-C(=O)-O) into olefins or aromatics by reaction with acetic acid lead to the formation of acetates. An outstanding example for acetoxylation is the reaction of ethylene with acetic acid in the presence of molecular oxygen leading to vinyl acetate (VAM) (Hagemeyer, & Jahn, 2012). The reaction is carried out in industry in a global production of ca. 5,000,000 t in 2011 to reach ca. 7,100,000 t in 2020 (Vertical Edge Limited, 2013). The largest production capacity is concentrated in China followed by United States, Western Europe, Taiwan and Japan. Worldwide annual average consumption growth rate of about 3% is expected for the next five years. The highest growth rate of ~5% is expected in China in the coming future. The production of VAM in China accounts to more than 30% of global production followed by United States (ca. 20%) and W. Europe (~10%). Celanese is the largest producer of VAM (~25% of the worldwide production), other leading players are SINOPEC, Dairen Chemical Corp., Dow Chemical Co and some other big chemical companies (PRWeb, 2014.).

The oldest synthesis process is a gas-phase acetoxylation of acetylene over a zinc acetate on carbon catalyst. In the early 1960's several processes appeared using ethylene such as the homogeneously catalyzed acetoxylation using Pd and Cu chloride catalysts. The reaction running in the presence of such a catalyst is similar to the Wacker process being introduced in the 1960's to convert olefins to aldehydes by molecular oxygen; the prominent example is the manufacture of acetaldehyde (Keith, &Henry, 2009). Later, the process was again replaced by gas phase acetoxylation of ethylene using supported Pd-Au egg-shell catalysts. Various aluminas, silica and silicates are used as supports. In addition, acetylene-acetic acid process is rejuvenated and mainly used by Chinese companies due to the easy availability of cheap acetylene in China.

In the ethylene based technology, ethylene, acetic acid and O_2 are used as reactants to produce vinyl acetate as the desired product of the reaction under gas phase conditions by a process called acetoxylation. The typical industrially applied catalysts consist of bimetallic catalysts supported on silica, e.g. Pd-Au/SiO_2 promoted generally with potassium acetate. Typical commercial catalyst consists of 1.5%Pd-0.75%Au-5%KOAc/SiO_2 and the lifetime of catalyst is reported to be around three years (Voskanyan, 2013). The selectivity of ethylene up to 95% and space-time-yields of 700 g VAM/(l cat • h) could be achieved. The reaction conditions usually applied are: T = 130 to 200 °C and p = 5-10 bar. The industrial heterogeneous catalysts based on silica supported Pd-Au alloys with excess of alkali acetate gave ca. 96% yield of VAM (Bissot, 1977).

In addition, research efforts were also made to develop a process for acetoxylation of olefins using two successively arranged catalyst layers. In this approach, the reactant feed mixture (olefin, acetic acid, O_2) was passed over at least two successively arranged catalyst layers of varying reactivity, i.e. the catalyst layers are arranged in one or more adjacent reaction tubes (Eberle, Jung, Heidenreich, & Roscher, 2013). In late 70s, some efforts were also made to apply heteropoly acids as catalysts for acetoxylation of ethylene. For instance, catalytic acetoxylation of ethylene and propylene (Kozhevnikov, Taraban'ko, Matveev, & Vardanyan, 1977) in presence of oxygen and by using Pd(II) complex of the heteropoly acid ($H_9PMo_6V_6O_{40}$) gave not only the products of acetoxylation (alkenyl and allyl acetates) but also some additional oxidation products (e.g. CH_3CHO and $(CH_3)_2CO$). However, the selectivity for acetoxylation products increases upon the addition of NaOAc to heteropoly acid catalyst.

The VAM manufacture is a very exothermic process; therefore heat management is crucial because total oxidation towards carbon oxides occurs as consecutive reactions from reactants and main product. In fact, the VAM demand is gradually shifting to China and other Asian countries.

VAM is extensively used for the manufacturing various polymers employed as paints, coatings, adhesives, fibers etc. Furthermore, VAM is also used in textile industries, for making safety glass sheets, in automobile industry and also for different architectural applications. The majority of VAM produced is used for poly vinyl acetate (PVA) manufacture. VAM is the intermediate for the synthesis of various polymers such as polyvinylacetate or polyvinylalcohol. It is also used for the synthesis of various co-polymers like ethylene-vinyl acetate (EVA).

Another important acetate compound is benzyl acetate, which is presently manufactured by esterification of benzyl alcohol using acetic acid anhydride and sodium acetate; benzyl alcohol is manufactured by chlorination of toluene and subsequent hydrolysis of the formed benzyl chloride. Benzyl acetate can also be obtained by direct conversion of benzyl chloride and alkali acetate. The global benzyl acetate market is rather small with only 4 kt/a (Hartmann-Schreier, 2012) but the benzyl alcohol demand is much bigger (roughly 130-150 kt/a). Whereas benzyl acetate is used mainly in flavor and fragrance industry, the much more demanded benzyl alcohol is used as general solvent for inks, paints, lacquers, coatings and a widely used intermediate in chemical industry.

BENZYL ACETATE SYNTHESIS BY HETEROGENEOUSLY CATALYZED GAS-PHASE ACETOXYLATION OVER Pd-CONTAINING CATALYSTS

In the last years some efforts were made to develop new Pd based catalysts for the gas phase acetoxylation of toluene to benzyl acetate, which is an environmental benign route for the production of benzyl alcohol than the traditional chlorination process. This reaction is a good example for making use of beneficial effects of more than one component in a catalytic system. In fact, supported one component system (e.g. Pd, Sb) showed only a poor activity, i.e. less than 5% toluene conversion, on the other hand, bi- and/or three component catalysts displayed at least 16 times higher activity (Benhmid, Kalevaru, Martin, & Lücke, 2004).The most promising catalytic system was found to be anatase supported Pd with Sb as a co-component. These catalysts were prepared by two-step impregnation method. In a first step, the impregnation of $SbCl_3$ on anatase is done, followed by oven drying and calcination at 400 °C, while in the second step $PdCl_2$ was impregnated on the above solid followed by oven drying at 120 °C for 16 h. Before catalytic testing the solids obtained with this procedure were activated for 2 h at 300 °C in the reactor under air. For a catalyst with 5 wt% Pd and 8 wt% Sb, a selectivity of >85% to benzyl acetate at a toluene conversion of nearly 40% was reached (Figure 4a). The drawback of this catalyst was the long activation period of 11 h and the deactivation probably due to coke deposition which spontaneously began at maximum activity. For an application of this catalyst these problems, especially the second one, had to be solved. In the subsequent years, systematic investigations were carried out to overcome these drawbacks. Different co-components, supports, pretreatment and regeneration procedures were attempted to achieve these goals. Additionally, it was tried to optimize the process parameters, e.g. temperature and pressure. After observing this interesting result on the synergistic effect between Pd and Sb, the individual effects of Pd and Sb loadings on the catalytic performance were also carried out separately in subsequent experiments. As an example, the time-on-stream behavior of 20%Pd-8%Sb/TiO_2 (anatase) catalyst is shown in Figure 4b.

Influence of the Co-Components

For a deeper insight into the beneficial influence of the co-components on the catalytic performance some additional systematic investigations were performed with different co-components with varying standard reduction potentials (Mn^{2+}/Mn: E^0 = -1.18 eV; Co^{2+}/Co: E^0 = -0.28 eV; Sb^{3+}/Sb: E^0 = +0.2 eV; Au^{3+}/Au: E^0 = +1.52 eV) (Gatla et al., 2011a). Although the benefit of co-components is well known, the way they influence the catalytic performance is still in discussion in many cases. In our case, the influence of the co-component on the electronic nature of the active Pd species is of interest. In some

Figure 4a. Synergetic effect between Pd and Sb for catalysts with 5 wt% Pd and 8 wt% Sb on TiO_2 (X-Tol = conversion of toluene, Y-BA = yield of benzyl acetate, S-BA = selectivity of benzyl acetate)

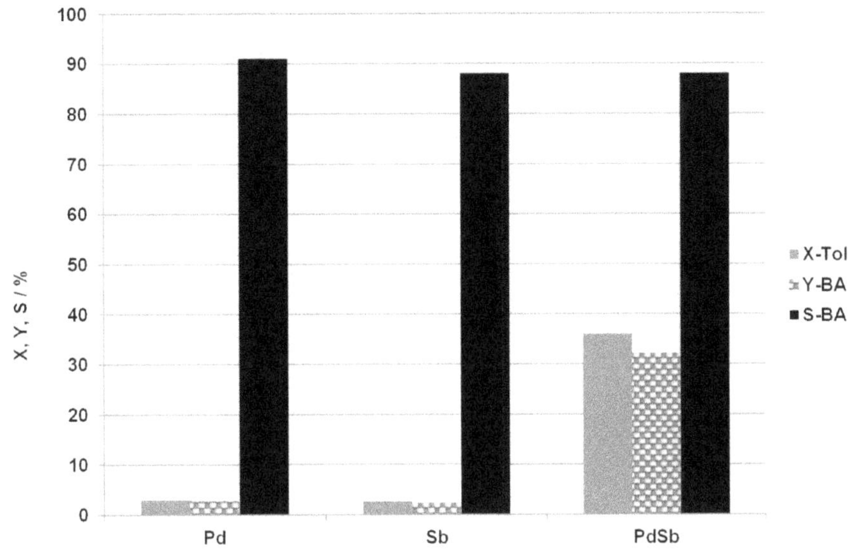

Figure 4b. Conversion of toluene and yield of benzyl acetate with time-on-stream over 20%Pd-8%Sb/ TiO_2 catalyst (X-Tol = conversion of toluene, Y-BA = yield of benzyl acetate, S-BA = selectivity of benzyl acetate)

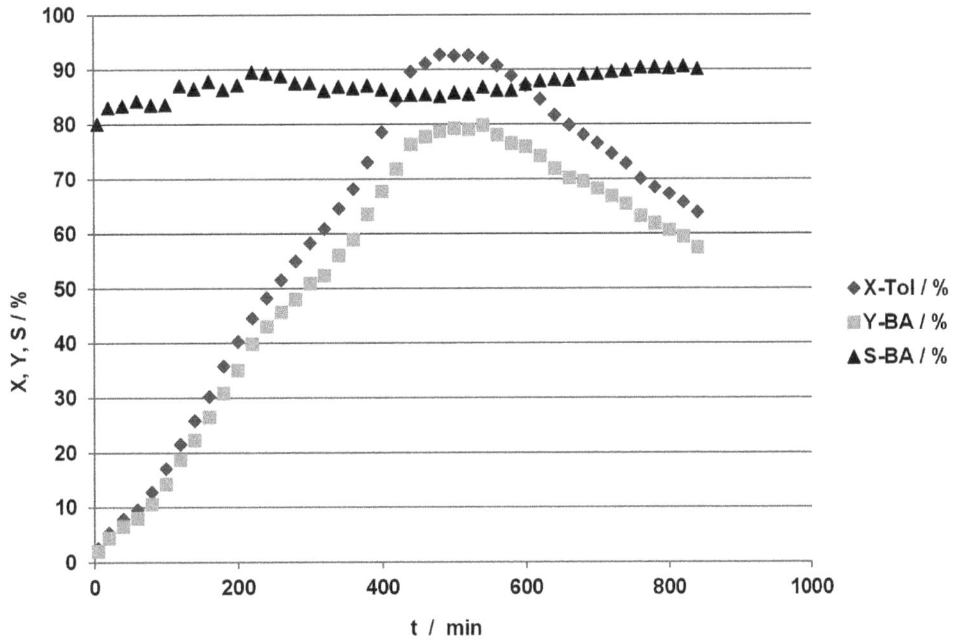

Figure 5a. Effect of nature of co-component on the catalytic performance (X-Tol = conversion of toluene, Y-BA = yield of benzyl acetate, S-BA = selectivity of benzyl acetate)

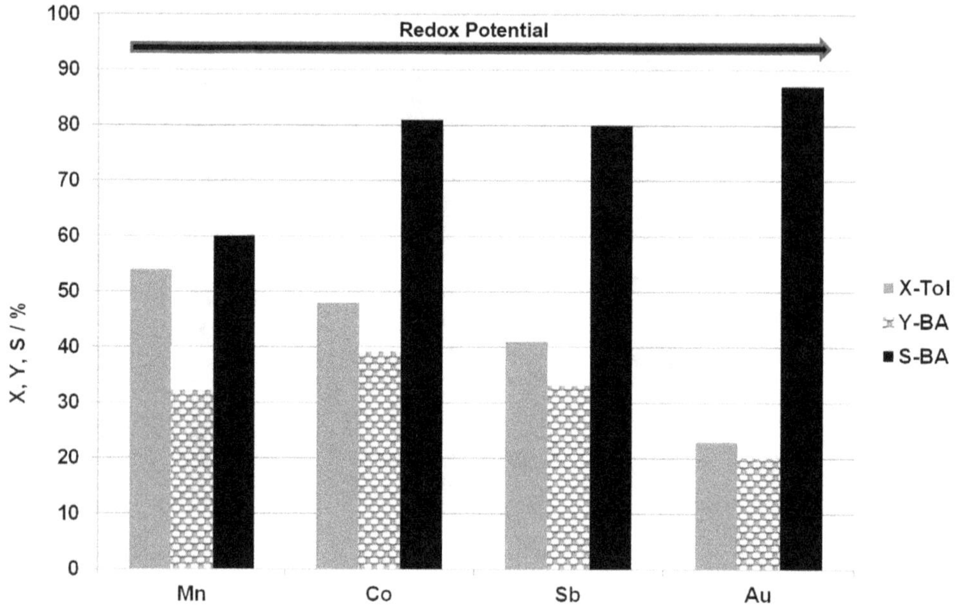

oxidation reactions, non-noble metals, e.g. Bi or Mn located in the vicinity of the Pd species are preferentially oxidized and, therefore, can stabilize metallic Pd (Mondelli et al., 2007). Otherwise, it was reported that alloy formation between Au and Pd can promote the decomposition of PdO, and, thereby the formation of metallic Pd (Qian, &Huang, 2011).

For the gas phase acetoxylation a clear correlation between the catalytic performance and the standard reduction potential was observed. Non-noble metals, e.g. Mn, led to a high activity, whereas with noble metal co-components, e.g. Au, a high selectivity was achieved (Figure 5a). Co and Sb as co-components were found to be a good compromise between these two extremes, i.e. with an acceptable activity and a high selectivity. For the determination of the oxidation state of Pd at the outermost surface investigations of CO adsorption were performed with Infrared Spectroscopy. For non-noble co-components a relatively high amount of oxidized Pd was detected, whereas for Au only a very small amount of oxidized Pd could be found (Figure 5b). Obviously, oxidized Pd enhanced the activity of the catalysts, while, metallic Pd was beneficial for the selectivity. Recent investigations performed with X-ray Absorption Spectroscopy (XAS) showed that the components except Au are mainly highly dispersed on the support. Surprisingly, only for Sb a stronger interaction between the co-component and the Pd particles were observed leading to an enrichment of Sb in the surrounding of the Pd particles and to an intermixing of a minor part of the Sb with the Pd (Gatla et al., 2015). This system is a good example that co-components allow the tuning of the catalytic properties.

Another important development of further investigations was the clear improvement in the long-term stability of Pd based catalysts. In fact, the majority of catalysts usually exhibited a very low initial activity (<5% conversion), which was observed to increase with time (≥10 h), and then remained stable

Figure 5b. Influence of the co-component on the amount of oxidized ($Pd^{\delta+}$) and metallic Pd (Pd^0) on the outermost surface determined by FTIR using CO as a probe molecule

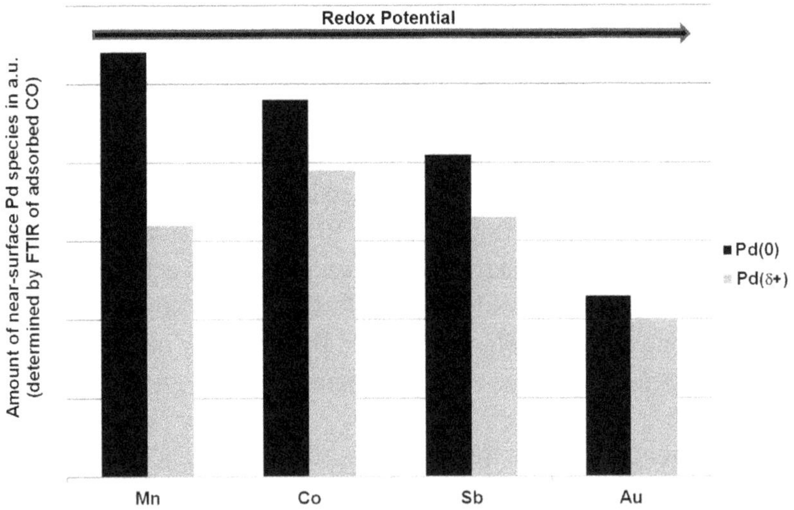

for few more hours. Afterwards, the catalysts deactivated due to coke deposition. Therefore, one aim was to enhance the long-term stability of the catalysts. To reach this aim, the influence of a further co-component on the coke deposition was investigated. A clear enhancement in the long-time stability could be achieved with Bi as a co-component, but, unfortunately the activation period to reach steady-state conditions has been increased dramatically from 11 h to 50 h (Radnik et al., 2005).Presence of Bi weakened the interaction between surface C and Pd and prevented the carbon deposition. Such results revealed that in the X-ray Photoelectron spectra (XPS) a new Pd state appeared with a lower electron binding energy than metallic Pd for the deactivated catalyst, the so-called $Pd^{\delta-}$ state. This state could be explained by an electron transfer from the carbon deposit to metallic Pd and was characteristic for the deactivated catalysts. Next to the prevention of the carbon deposition another possibility to avoid the deactivation is the modification of the nature of the deposited carbon in a less detrimental form. In subsequent studies, Cu was also used as a component. Interestingly, it has a similar standard reduction potential like Sb (Cu^{2+}/Cu: $E^0 = + 0.34$ eV) and comparable catalytic results were obtained with the Pd,Cu/TiO_2 catalyst as that of Pd,Sb/TiO_2 concerning the activity and selectivity. But, in contrast, the deactivation of this Sb-containing catalyst occurred only after a few hours while the Cu containing catalysts displayed stable behavior even after 160 h on-stream. However, this amazing result could not be explained simply by a lower carbon deposition, but with the formation of an onion-like carbon deposit which was not so deactivating like the closed carbide-like deposit formed on the Pd,Sb catalyst (Radnik, Pohl, Narayana, & Martin, 2007). Another less deactivating kind of carbon was observed for the Mn and Co containing catalysts. In contrast to Pd,Sb at which carbon was deposited on the surface of the Pd particles, for the catalysts with Mn or Co as co-components carbon is incorporated into the bulk of the particles, which was proved with X-ray Diffraction (XRD), Thermogravimetric and Differential Thermal Analysis (TG-DTA) and Extended X-Ray Absorption Fine Structure (EXAFS) investigations of the spent catalysts (Gatla et al. 2011, 2015).

Catalyst Preparation, Pretreatment and Regeneration Procedures

The long activation period of several hours (~10 to 50 h) was explained with forming Pd particles of more than 50 nm which was proposed to be active. Smaller particles seemed to be inactive in the gas phase acetoxylation of toluene (Benhmid et al., 2004; Radnik et al., 2005). To check this surprising result and to shorten the activation phase, variations in the preparation method were applied, which included variations in the precursors, varying ammonia additives and thermal pretreatment procedures. XRD measurements revealed that the decomposition of the metal salts used as educts are the critical step for the formation of the desired Pd particles (Gatla et al., 2010). The best suitable precursors for obtaining better performance were the chlorides. Alternatively usage of nitrate precursors led to the formation of undesired PdO. In situ XRD measurements showed that $PdCl_2$ was decomposed via an intermediate Pd-complex, e.g. Pd-chloroamine complex, which explained the necessity of ammonium salts addition to remove the chlorides during the preparation and conditioning. Among all the applied ammonium salts, $(NH_4)_2SO_4$ was the most effective one.

Additionally, three different thermal pretreatment procedures were applied such as i) oxidizing (air), ii) inert (He) and iii) reducing (10% H_2 in He) atmospheres in the temperature range from 400 °C to 650 °C (Gatla et al., 2011b). A major problem after the pretreatment in inert and reducing atmosphere was the formation of sulfides from $(NH_4)_2SO_4$, which ultimately poisoned the catalyst. This problem could be circumvented by using Sb_2O_3 as a chloride free precursor instead of $SbCl_3$ precursor so that the addition of the ammonium compound was no longer required. With this solid obtained after the two-step impregnation (firstly Sb_2O_3 on anatase, and secondly $PdCl_2$ on Sb_2O_3/anatase) different pretreatment procedures were applied. As expected, increasing the calcination temperature led to the formation of the desired larger Pd containing crystallites, which are beneficial for catalysis. Treatments at 600 °C were found to be optimum. After the thermal treatment in air, PdO particles with a size of 5 to 10 nm were formed, whereas such treatment in He led to the formation of larger Pd particles in the µm range. Surprisingly, for both materials after these different pretreatments, the same activation behavior was found. Fortunately, the activation period should be shortened to 6 h in contrast to the former 10 h. The sample calcined in the reducing atmosphere consisted of a stable Pd_8Sb_3 alloy, which was surprisingly catalytically inactive. These results are however summarized in Figure 6.

Although no significant differences were found during the activation phase for the air- and He-pretreated, afterwards, during time on-stream the pretreatment influenced the performance of the catalysts. The air-pretreated samples deactivated much faster than the other one. To obtain detailed insights into this unexpected behavior, some additional analytical investigations with XRD, X-ray Photoelectron Spectroscopy (XPS) and Transmission Electron Microscopy (TEM) were performed for the most active catalysts. The most remarkable differences between the Pd containing particles formed from PdO (after air treatment) and pure Pd (after He-treatment) was the incorporation of Sb into the Pd particles. PdO particles were transformed to nearly pure Pd particles, whereas the large pure Pd particles were decomposed to smaller ones With a Pd: Sb ratio of 5. This result was obtained with TEM elemental mapping using Energy-Dispersive X-ray (EDX) analysis for qualitative and quantitative analysis of single particles in the nm range. XRD confirmed this observation showing a significant shift of the Pd(111) reflections to lower 2Θ values typical for the expansion of the Pd lattice due to incorporation of other elements. It must be noted that such small incorporation of Sb seems to be beneficial for the catalytic performance, whereas the higher amounts >20% of Sb in the Pd particles lead to inactive catalysts as observed for the hydrogen-pretreated sample. XPS measurements provided hints for this positive influ-

Figure 6. TEM images of the catalysts subjected to different thermal pretreatments (after the pretreatment in a) air, b) helium and c) 10% hydrogen in helium), at maximum activity (images d) and e)) and after deactivation (image f)

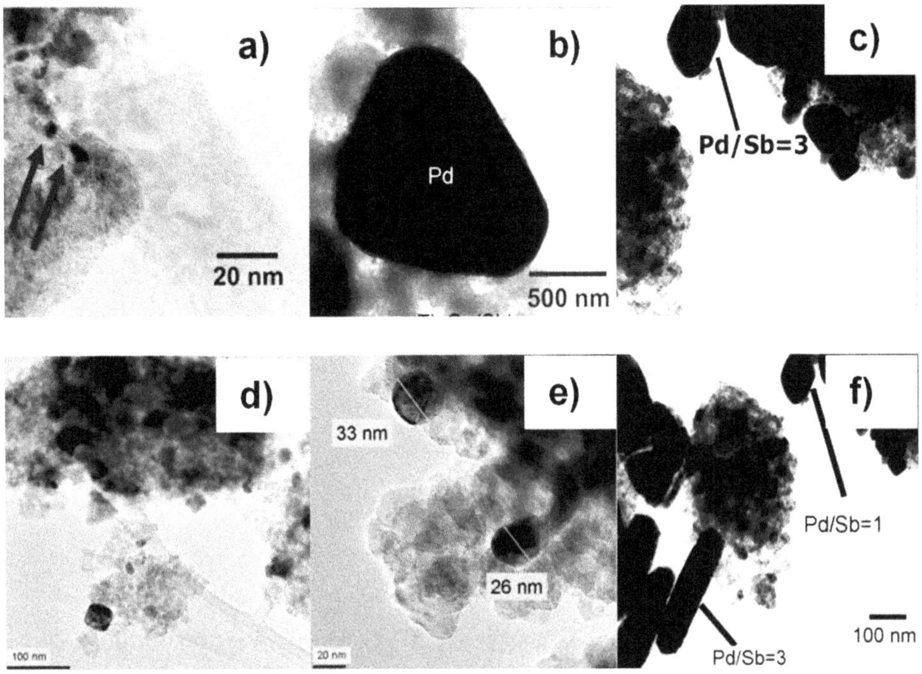

ence of the incorporated Sb. PdO is stabilized at such kind of particles, whereas on pure Pd particles no oxidized Pd could be detected after several hours on-stream. This finding confirmed the importance of the coexistence of Pd and oxidized Pd for an efficient catalyst described above.

The stability of the Pd-PdO_x pair had an important influence for the optimization of the regeneration procedure. It is well known, that deposited coke can be burnt off by air at adequate temperatures. To optimize this regeneration procedure, the spent catalysts obtained by the He-pretreatment procedure at 600 °C and used for 42 h on-stream were treated with air at temperatures in the range from 250 °C and 400 °C for two to six hours (Madaan et al., 2011b). At all these temperatures the coke deposits could be burned out effectively. Nevertheless, the catalysts could be regenerated at temperatures between 250 and 300 °C, but not at higher temperatures. TEM and XRD revealed no differences concerning the particle size and phase composition, but XPS showed clearly the existence of metallic and oxidized Pd only for the samples regenerated at the lower temperatures, whereas for the sample regenerated above 300 °C only metallic Pd was found. This observation explains the poor performance of this regenerated sample and emphasizes the importance of the coexistence of both, Pd and PdO_x, in the near-surface-region for a well-working catalyst.

Effect of the Reaction Pressure

The influence of the reaction pressure on the catalytic performance is well investigated, but the correlation between structural rearrangement, catalytic performance and reaction pressure is rather unknown. To close this knowledge gap, some additional catalytic tests at varying pressures (p = 1 to 10 bar) were

Figure 7a. Catalytic performance in dependence of the reaction pressure

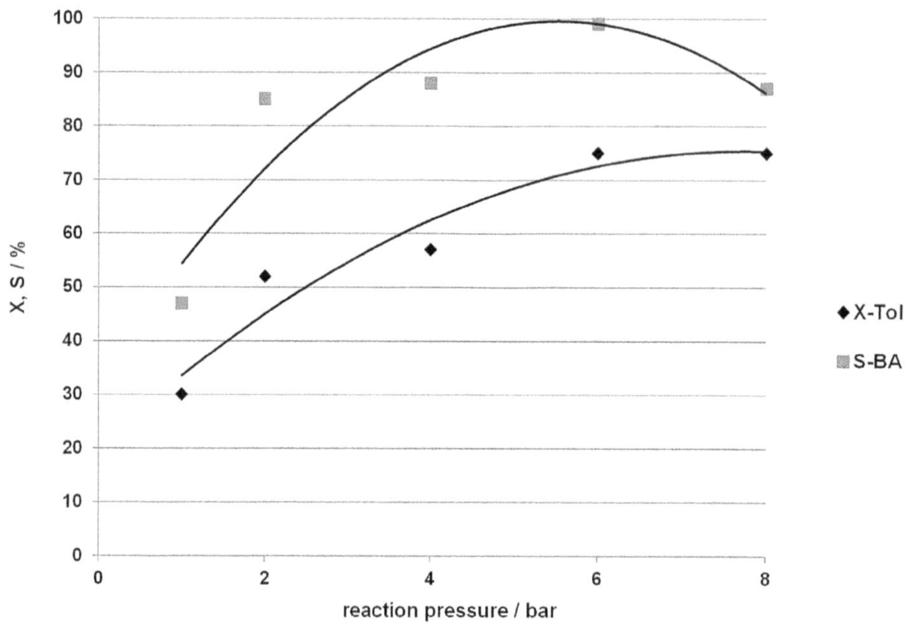

combined with structural investigations (Madaan et al., 2013). The conversion of toluene increased from 34% to 75% on increasing the reaction pressure from 1 bar to 6 bar, and then remained constant up to 10 bar (Figure 7a). In a similar way, the selectivity increased from 60% at 1 bar to nearly 100% at 6 bar, but decreased slightly to 86% beyond this pressure. Simultaneously, the pressure shortened the activation period from 11 h (at 1 bar) to only 3 h (at 6 bar), but, unfortunately, the long-term stability was found to be shortened from 40 h (at 1 bar) to only 12 h (at 10 bar). This latter observation could be explained by a more pronounced coke deposition at higher pressure. Interestingly, the extent of the carbon deposition increased marginally up to 4 bar, but at higher pressure it rose up significantly hinting a self-promoting of the coke deposition at these pressures (Figure 7b). From all these findings, it could be concluded, that the reaction pressure clearly influences the formation of the catalytically active Pd,Sb particles. To verify this hypothesis extended analytical investigations with XRD, XPS and TEM were performed.

In situ XRD investigations showed the restructuring of the Pd particles on-stream. The large Pd particles obtained after the He pretreatment at 600 °C dissolved and gradually disappeared with time-on-stream under reaction conditions applied and after certain reaction time small Pd,Sb particles of a few nanometers were formed. This is indicated by the Pd(111) reflections loosing intensity and a new reflection at lower 2 theta values growing with time-on-stream, which is typical for Pd with incorporated Sb. However, the generation of the active particles depended on the reaction pressure. At the low pressure of 1 bar particles between 10 and 25 nm were formed, at 6 bar the most active particles between 40 and 100 nm particles were generated (Figure 8). Higher pressures did not influence the size of the particles, but considerably affected the composition of the particles. At 6 bar and less, the optimum particle composition of a Pd: Sb ratio of 5: 1 were found, higher pressure led to a higher amount of Pd in the particle which deactivates faster, as described above. One of the most exciting results of this pressure study was achieving the selectivity of nearly 100%, which was reached at 6 bar. TEM showed some well-facetted

Figure 7b. Effect of the reaction pressure on the coke formation

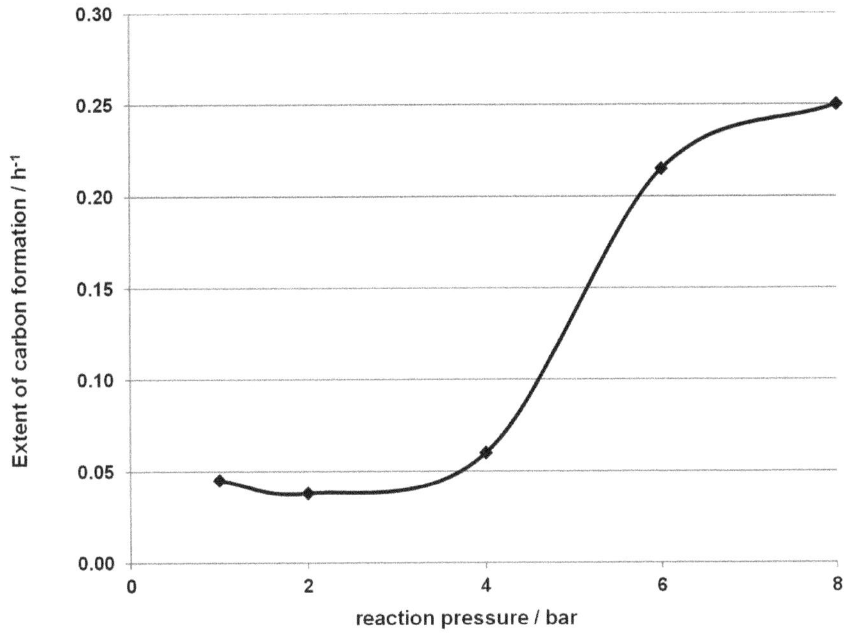

Pd,Sb particles formed only at this pressure. It seemed that particles with such morphology seemed to be beneficial for the selectivity. One aim of our study was also to obtain such well-facetted particles, but with an improved long-term stability.

Effect of the Support

In subsequent systematic studies, varying catalyst supports with different acid-base and redox properties were also investigated in order to find out a potential carrier for the desired Pd,Sb particles (Kalevaru et al., 2007). From such studies, anatase was identified as the best suitable support for the gas phase acetoxylation of toluene. Recently, the efficiency of the two most abundant modifications of TiO_2, i.e. anatase and rutile, were compared for this reaction (Gatla et al., 2013). Anatase is the common support in catalysis, because of its high surface area and its stronger interaction with metal particles deposited on it. This well-known Strong Metal Support Interaction (SMSI) effect is often observed for anatase, but it is less pronounced for rutile (Li et al., 2004; Iddir, Skavysh, Öğüt, Browning, & Disko, 2006).

Surprisingly, we found that the catalyst with rutile as support have a better performance in terms of selectivity and long-term stability than the system with anatase. As expected, the rutile catalyst was slightly less active, but highly selective (~100%) and stable (Figure 9a). TEM investigations revealed that on rutile the desired uniform, well-facetted Pd,Sb particles with a Pd: Sb ratio of 5 and a diameter of ca. 80 nm were formed (Figure 9b). Additionally, no coke deposit was found on these particles; instead, a carbon layer was detected on the surface of the rutile support. Besides, XRD and TG-DTA investigations indicated the incorporation of C into the rutile supported Pd,Sb particles. XPS showed, that in the near-surface region of these particles metallic and oxidized Pd coexists, in contrast to the anatase supported particle, over which the formation of metallic Pd and the detrimental $Pd^{\delta-}$ state was observed under the similar reaction conditions.

Figure 8. TEM images of Pd-Sb/TiO_2 catalysts tested at different reaction pressures of 1 bar (a), 2 bar (b), 6 bar (c) and 8 bar (d)

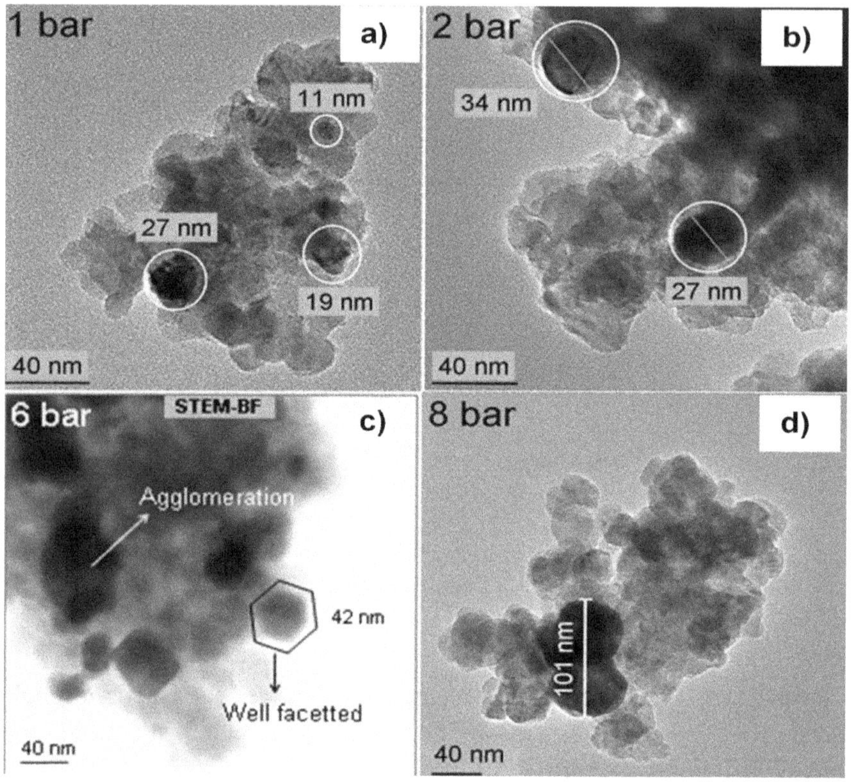

Figure 9a. Catalytic performance of the anatase (A) and rutile (R) supported catalysts with 10 wt% Pd and 16 wt% Sb (X-Tol = conversion of toluene, S-BA = selectivity of benzyl acetate)

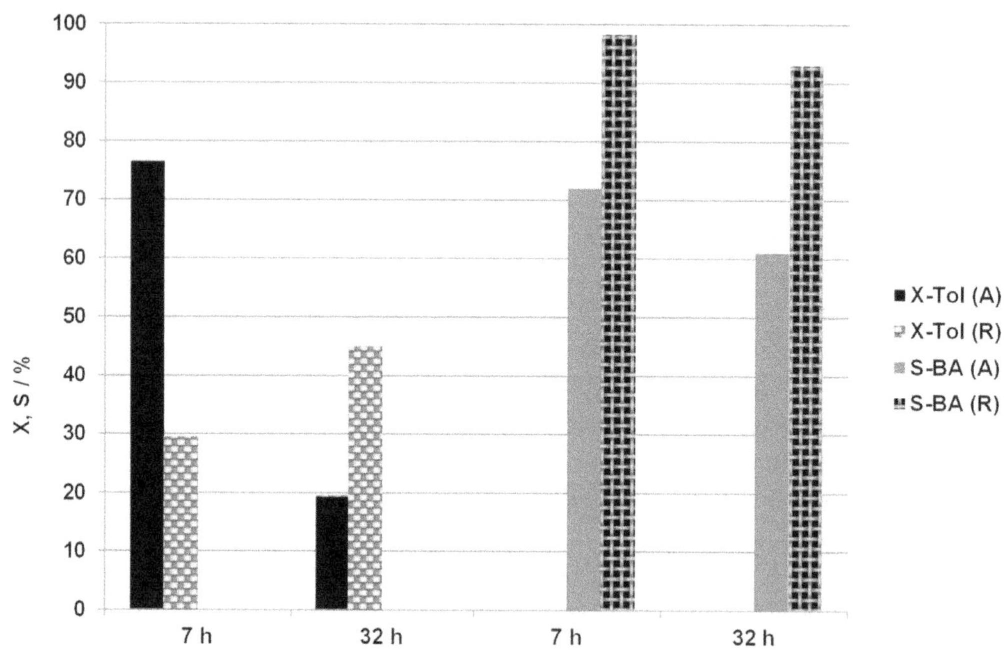

Figure 9b. TEM images of the anatase (upper row) and rutile (bottom row) supported catalysts after 32 h on-stream

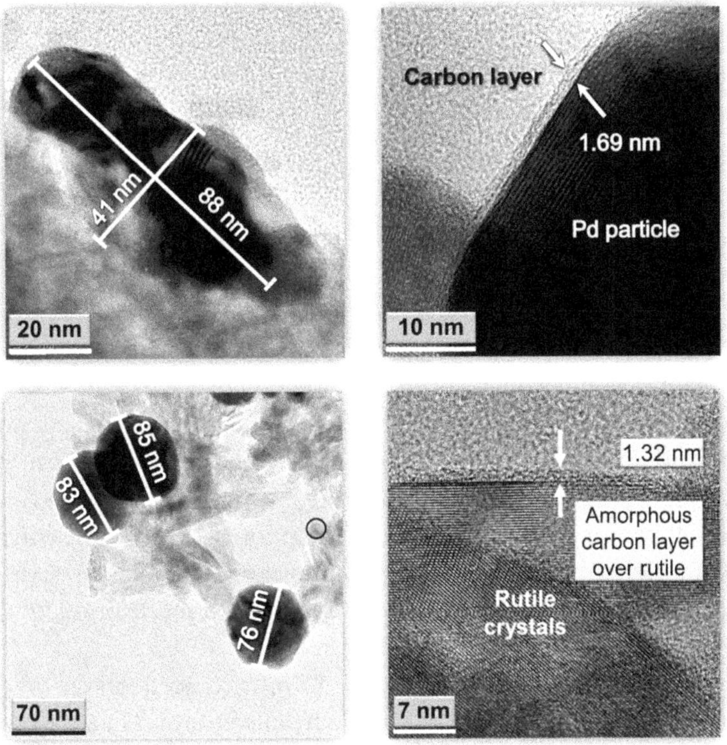

The reason for the formation of the well-facetted Pd,Sb particles is probably the weak interaction between the Pd particles and the rutile compared to anatase. Such less pronounced interaction between the active metal particles and the support favors the formation of larger, and well-facetted particles, whereas on supports interacting strongly with the particles, smaller particles or agglomerates of smaller particles with an irregular shape are preferentially formed (Lai, Clair, Valden, & Goodman, 1998). More remarkable is the high selectivity for benzyl acetate of these well-facetted particles. This observation shows that an ensemble of a number of Pd atoms on the outermost surface of the particles is necessary for the reaction, from which a part of the Pd atoms must be oxidized. Obviously, such cluster of atoms could be easily accommodated on well-facetted particles with large smooth crystal planes observed on rutile than on particles with rough surfaces and a lot of irregularities formed on anatase. This last finding enables us to obtain deeper insights into the mechanism of the acetoxylation of toluene on this catalytic system.

MECHANISTIC DISCUSSIONS

Vinyl acetate synthesis via acetoxylation is a well-known process, which is already running on industrial scale since the last few decades. The literature available on this topic is more abundant. On the other hand, the synthesis of benzyl acetate by means of gas phase acetoxylation of toluene is relatively a new approach and is still under developmental stage. Therefore, extensive studies particularly dealing with

Figure 10. Possible mechanistic route for the synthesis of vinyl acetate by Samanos et al. (Samanos, Boutry, & Montarnal, 1971)

$$Pd + 0.5\,O_2 + 2\,AcOH \longrightarrow Pd(OAc)_2 + H_2O$$

$$Pd(OAc)_2 + AcO^- \longrightarrow Pd(OAc)_3^-$$

$$Pd(OAc)_3^- + C_2H_4 \longrightarrow C_2H_4Pd(OAc)_3^-$$

$$C_2H_4Pd(OAc)_3^- \longrightarrow AcO\text{-}CH{=}CH_2 + AcOH + AcO^- + Pd$$

kinetic and mechanistic aspects of this reaction are still missing in the literature. To the best of our knowledge, there is not even single source of publication on the acetoxylation of toluene particularly related to the reaction mechanism and identifying reaction intermediates, respectively, under working conditions or dealing with the kinetic investigations. Even though, we have been intensively engaged with this interesting reaction for the past several years, the majority of our past efforts were focused on the catalyst development part and understanding the nature of active sites, deactivation and regeneration issues etc. using a variety of Pd-based catalyst compositions. Herein, we made attempts to give an overview on the similarities and differences between the reaction mechanisms of these two reactions, i.e. VAM synthesis vs. BA synthesis via acetoxylation route.

The reaction of ethylene, acetic acid and oxygen to form vinyl acetate over Pd-containing catalysts is actually initiated on palladium surface by means of ethylene insertion into the oxygen-metal bond of an adsorbed acetate species to form an acetoxyethyl-palladium intermediate followed by a β-hydride elimination to yield vinyl acetate. This reaction pathway was proposed by Samanos in early 70s (Samanos, Boutry, & Montarnal, 1971). However, a decade earlier (in 60s) Moiseev also proposed another alternative reaction mechanism for homogeneous process, where the reaction is initiated by the activation of ethylene to form a surface vinyl intermediate, which will be inserted into the adsorbed acetate to form vinyl acetate in subsequent steps (Moiseev, Vargaftic, & Syrkin, 1960). In fact, van Helden et al. (van Helden, Kohll, Medema, Verberg, & Honkhoff, 1968) have given the first proposal of the reaction mechanism several years before Samanos proposal and claimed that the elimination of the β-hydrogen atom was the rate limiting step. The ethylene-acetate coupling involves the insertion of the ethylene into one of the Pd-O bonds. During the initial structure, ethylene and acetate are adsorbed on two different Pd atoms on the catalyst surface. Among different mechanistic proposals, Samanos proposed mechanism is widely accepted for the production of vinyl acetate from ethylene (Samanos, Boutry, & Montarnal, 1971). This mechanism involves the coupling of ethylene directly with chemisorbed acetate as shown below in Figure 10. The acetoxy ethyl-Pd intermediate is formed first, which is then undergoes β-hydride elimination to form VAM.

Another mechanism proposed by Nakamura et al. is given below (Nakamura, &Yasui, 1970). They also reported that the elimination of β-hydride from acetoxy ethyl-Pd intermediate is crucial. All the steps involved during the course of the reaction are described below in detail (Figure 11).

Plata et al. (Plata, García-Mota, Braga, López, & Maseras, 2009) based on their DFT studies further suggest that the β-hydrogen elimination can take place from two different situations: i) in the case of homogeneous process, this step consists of the transfer of the β-hydrogen to the metal in the intermedi-

Figure 11. Possible mechanistic route for the synthesis of vinyl acetate proposed by Nakamura et al. (Nakamura, &Yasui, 1970)

$$2\,Pd + C_2H_4 \longrightarrow CH_2\text{-}CH\text{-}Pd + PdH$$

$$2\,Pd + O_2 \longrightarrow 2\,PdO$$

$$CH_3COOH + Pd \longrightarrow Pd\text{-}CH_3COOH$$

$$Pd\text{-}CH_3COOH + PdO \longrightarrow Pd\text{-}O\text{-}COCH_3 + Pd\text{-}OH$$

$$Pd\text{-}O\text{-}COCH_3 + CH_2CH\text{-}Pd \longrightarrow Pd\text{-}CH_2CHOCOCH_3 + Pd$$

$$Pd\text{-}CH_2CHOCOCH_3 \longrightarrow Pd + CH_2CHOCOCH_3$$

$$Pd\text{-}OH + PdH \longrightarrow 2\,Pd + H_2O$$

ate resulting from the C-O bond formation step, and ii) in the case of heterogeneous situation, either the oxygen or the hydroxyl groups on the surface act as acceptors for the β-hydrogen atom. At the transition state, the proton is close to the hydroxyl group, (i.e. 1.307 Å), which then evolves in the form of a water molecule. The basicity of such hydroxyl groups on Au centers is known to be large and hence the desorption process takes place easily (López, García-Mota, & Gómez-Diaz, 2008). The main difference in this step is thus the nature of the proton acceptor. In the homogeneous case, the metal center acts as an acceptor, while in the heterogeneous process the availability of oxygen or hydroxyl species facilitates this particular reaction step. The oxygen employed in the reaction is to drive the acetic acid dehydrogenation, and hence it is evolved in the form of water molecules. In the reaction mechanism, the oxygen undergoes dissociative adsorption on the catalyst surface, and then one of the oxygen atoms initiates the dehydrogenation process of acetate. Ethylene can be adsorbed in a neighboring Pd center and then the C-O coupling takes place; finally an oxygen atom or a hydroxyl group can provoke the β-hydrogen elimination leading to the formation of the desired VAM product (Plata, García-Mota, Braga, López, & Maseras, 2009).

According to Calaza et al. (Calaza, Stacchiola, Neurock, & Tysoe, 2010) the acetic acid adsorbed on Pd(111) surface easily dissociates to form acetate species at saturation coverage. The activation barrier for this step was reported to be <30 kJ/mol and <5 kJ/mol when assisted by surface oxygen or hydroxyl intermediates. In the subsequent reaction steps between the adsorbed acetate and the co-adsorbed ethylene leads to the formation of the acetoxyethyl intermediate. The transition state for this reaction involves the formation of a surface oxymetallocycle in which C-Pd and O-Pd atoms of the adsorbed ethylene and acetate are broken in order to form the C-O bond of the adsorbed acetoxyethyl intermediate. The reaction proceeds via the transformation of ethylene from a di-σ to a π-bound intermediate and then onto the acetate to form the acetoxyethyl intermediate. Both acetate and ethylene are nucleophiles; thus the changes in acetate coverages alter the activation barriers. The adsorbed acetate intermediates, however, compete with one another for electrons from the surface at higher coverages. As a result, there is a decrease in the charge transfer from the metal to the oxygen atoms on the acetate. This leads to a more positive charge on the oxygen atoms of the acetate which is then reduce their nucleophilicity. The results from a Hirshfeld charge analysis on the transition states for both low and high coverages of acetate indicate that the charge on the O atom of acetate that reacts with the ethylene changes from low to high acetate coverages. The positive charges on the Pd atoms also increase at higher acetate coverage, which is however not enough to meet the demand of the oxygen atoms on the acetate (Calaza, Stacchiola, Neurock, & Tysoe, 2010). The authors further investigated the effect of the surface coverage on the kinetics of Pd-catalyzed reactions over well-defined Pd(111) surfaces and also explored the influence of surface coverage on the formation of vinyl acetate. Such results were compared between theoretical DFT calculations and experimental observations. It is stated that high adsorbate coverages lower the activation energies for reactions that involve the coupling of two adsorbed intermediates. The activation energies calculated for elementary steps in the Samanos and Moiseev pathways for vinyl acetate synthesis performed on acetate-saturated palladium surfaces. Such results reveal that the reaction proceeds via the Samanos mechanism, which is consistent with experimental results carried out on acetate-saturated Pd(111) surfaces. The activation energy for the rate-limiting β-hydride elimination reaction from the acetoxyethyl intermediate calculated at saturation coverage is compared with experimental measurements for a saturated acetate overlayer as a function of temperature. The acetate coverage is measured using infrared spectroscopy. The authors claim that the rate-limiting step involves a β-hydride elimination from the adsorbed acetoxyethyl intermediate that proceeds with a calculated activation barrier of 53 kJ/mol, which is in good agreement with the experimental barrier of 55 ± 4 kJ/mol determined from kinetic measurements. The authors concluded that the changes in the atomic charges and the occupation of the frontier orbitals indicate that higher acetate coverages lower the electron density on the oxygen, making it less nucleophilic, and decrease the occupation of the antibonding C-O σ* state, which lowers the activation energy (Calaza, Stacchiola, Neurock, & Tysoe, 2010). The acetate removal kinetics was measured by monitoring the acetate asymmetric OCO vibrational mode at 1414 cm^{-1}. Of course, the intense acetate OCO mode can however be distinguished from other modes such as vinyl acetate (1788 cm^{-1}), ethylidyne (1330, 1090 cm^{-1}), and the acetoxyethyl intermediate (1718 cm^{-1}) (Stacchiola, Calaza, Burkholder, & Tysoe, 2004).

Rivalta et al. (Rivalta, Mazzone, Russo, & Sicilia, 2010) explored the influence of surface oxygen and hydroxyl groups on the dehydrogenation activity of ethylene on Pd second-neighbor ensembles of both PdAu(100) and PdAu(111) surface alloys with a cluster model approach. The calculated activation barriers for the dehydrogenation of ethylene to form vinyl species on O_s- and OH_s-covered PdAu surfaces are significantly lower than those on clean surface alloys. They suggest that a Langmuir-Hinshelwood-type mechanism involving a vinyl surface species cannot be excluded during the course of ethylene acetoxylation reaction on PdAu catalysts. Additionally, Stacchiola et al. also described the elementary steps during the acetoxylation of ethylene over the Pd-catalyzed solids, which involve the reaction between surface acetate species and oxygen species to form vinyl acetate and water (Stacchiola, Calaza, Burkholder, & Tysoe, 2004; Stacchiola et al., 2005). Other studies (e.g. Chen, Kumar, Yi, & Goodman, 2005; Han, Axnanda, Lyubinetsky, & Goodman, 2007; Kumar, Chen, & Goodman, 2007) demonstrate that this reaction is catalyzed by unique non-contiguous Pd sites isolated within a Au(100) surface. Recent theoretical calculations also help to support such ideas (García-Mota, & López, 2008; Mazzone, Rivalta, Russo, & Sicilia, 2009).

Furthermore, Hanrieder et al. (e.g. Hanrieder, Jentys, & Lercher, 2014) claim the formation of a condensed acetic acid film during the course of the reaction and such phenomenon leads to the restructuring of the Pd and Au particles towards a more active and selective surface that consists of Pd-Au particles in the ratio of 1: 1 along with some separated and well dispersed Pd particles. The addition of alkaline acetates to Pd-Au/SiO_2 is beneficial to enhance the overall activity, selectivity as well as the long-term stability. In fact, the presence of potassium acetate is required to control the equilibrium between Pd in the Pd_1Au_1 particle and the dispersed Pd. On the other hand, the absence of KOAc during the reaction leads to Au-rich particles. Moreover, Pd can be incorporated into monometallic Au particles in the presence of KOAc by forming a reaction intermediate $K_2Pd_2(OAc)_6$, while the alkaline metal could be incorporated in the SiO_2 support and on the surface of the PdAu particles (Hanrieder, Jentys, & Lercher, 2014). Chen et al. (Chen, &Goodman, 2008) investigated the role of Au in Pd-Au/SiO_2 catalyst and reported that the presence of Au atoms can break contiguous Pd atom ensembles at the surface into isolated Pd monomers. The absence of contiguous Pd sites considerably reduces the formation of combustion by-products, thus suppresses the poison effects of CO, and thereby enhances the selectivity of VAM.

Plata et al. (Plata, García-Mota, Braga, López, & Maseras, 2009) determined the reaction mechanism of ethylene acetoxylation by means of density functional theory for both homogeneous and heterogeneous processes involving Pd atoms as reaction centers. Their studies unraveled the similarities and differences in the reaction networks (see Figure 12) and claimed that the reaction network is similar for both homogeneous and heterogeneous processes but the rate determining step is different. The calculated barriers for the rate determining step are significantly higher (127 KJ/mol) for the heterogeneous process of the catalyst compared to homogeneous process (96 KJ/mol). In fact, the thermochemistry for the complete process depends on the nature of the oxidant employed, and hence the rate determining steps in both cases are different. In the case of homogeneous catalysts, the most demanding step is C-H cleavage

Figure 12. Reaction mechanism for vinyl acetate synthesis from ethylene and acetic acid over Pd-based catalysts (Homogeneous vs. Heterogeneous) (VA=vinyl acetate, BQ=benzoquinone, HQ=hydroquinone) (Plata, García-Mota, Braga, López, & Maseras, 2009)

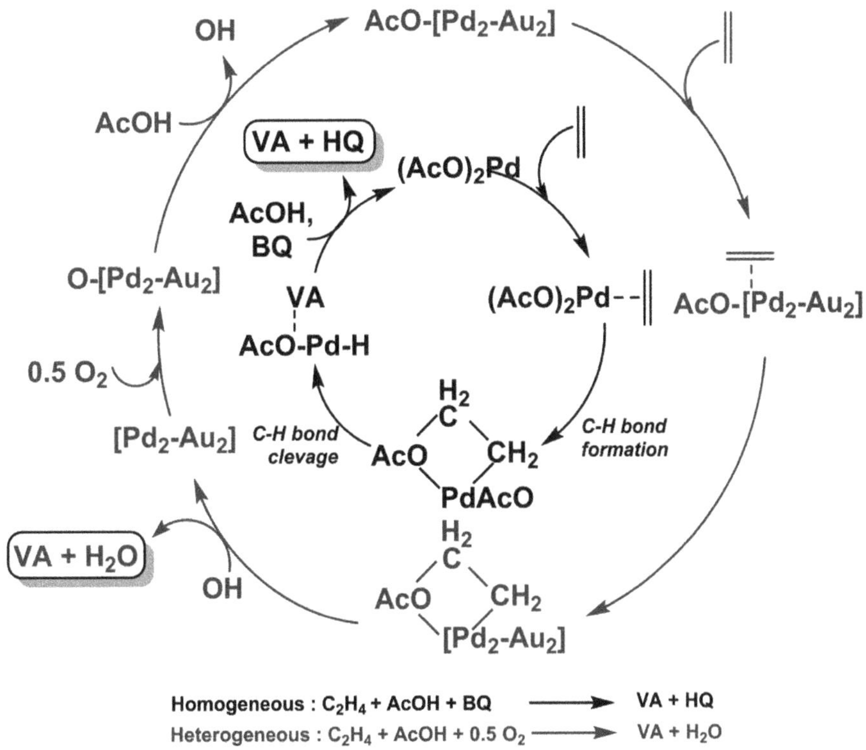

and for the heterogeneous catalysts, the C-O coupling is the rate determining step, which is of course in agreement with the experiments. The reoxidation process is far simpler in the heterogeneous situation, but it is not the rate-determining step. Another difference is that the reaction takes place at lower temperatures (50-60 °C) in the homogeneous case while at relatively higher temperatures (130-200 °C) in case of heterogeneous process.

Zhang et al. (Zhang, Gao, Li, Tao, & Yu, 2014) report the effect of CO on the acetoxylation activity of ethylene to vinyl acetate over a model surface of Pd-Au (100) by a DFT study. They claim that the effect of CO on the acetoxylation of ethylene to vinyl acetate was reflected in two ways: i) the adsorption of CO on a Pd atom is more stable than that of ethylene, as a result the adsorption of ethylene is blocked and the catalytic activity of the Pd-Au alloy decreased, and ii) the Pd-Au alloy surface catalyzes the CO oxidation to CO_2. Thus the catalytic activity will be influenced by the reaction rate and the content of CO in the reactant feed mixture. Such DFT calculations also indicate that the adsorption energy of CO is far higher than that of ethylene. Additionally, ensemble effects of Pd-Au and Pd-Sn bimetallic catalysts were also checked for vinyl acetate synthesis and elucidated the surface concentration, nature of adsorption sites surface ensembles in these two systems (Chen, Gath, Wei, Kumar, & Goodman, 2007). The authors report that the isolated Pd monomers formed on both the systems, (e.g. $PdAu_4$, $PdAu_6$, $PdSn_4$ and $PdSn_6$) are the active sites in VAM synthesis. They also suggest that two non-contiguous, suitably spaced Pd monomers are required for VAM synthesis. The promotional effects of Au and Sn in Pd-Au and Pd-Sn are similar, i.e. to isolate single Pd sites which then significantly weaken undesirable surface adsorption species and suppress the formation of reaction by-products. Chen et al. (Chen, Kumar, Yi, & Goodman, 2005) also demonstrated that the critical reaction site for VA synthesis consists of two non-contiguous, suitably spaced, Pd monomers in Pd-Au alloy catalyst. According to Macleod et al. (Macleod, Keel, & Lambert, 2004) the fresh Pd-Au-K/SiO_2 catalysts for VAM synthesis contain Pd in two forms such as i) Pd-Au alloy particles (4-5 nm in diameter) and ii) highly dispersed Pd metal particles.

In case of toluene acetoxylation, we do not have any β-hydrogen and hence this reaction involves somewhat different reaction mechanism, which is however not known yet due to lack of extensive scientific studies on this topic. We have recently investigated the reaction network of oxidative conversion of toluene with acetic acid and found that benzaldehyde and benzyl acetate are formed through two independent reaction pathways (Reining, Kondratenko, Kalevaru, & Martin, 2014). The most probable reaction mechanism for this reaction is that the toluene molecule gets adsorbed on the Pd surface by the oxidative addition of one of the methyl C-H bonds forming surface Pd-π-aromatic complex (i.e. Pd_x-π-$C_6H_5CH_3$ and/or Pd_x-π-$C_6H_5CH_2$-Pd species) followed by π-benzyl delocalisation, and then nucleophilic attack by acetate group that leads to the formation of BA. Similar investigations on mechanism of acetoxylation in liquid phase were carried out on Pd-nanoclusters (Ebitani, Choi, Mizugaki, & Kaneda, 2002). Oxygen might serve to clean the surface Pd-H via the formation of surface super oxides and/or peroxides and finally to produce water and oxygen. Some of the major differences between these two processes (VAM vs. BA by acetoxylation) are described below.

Similarities and Differences Between VAM and BA Synthesis

First of all, both the reactions run with palladium as an active component and at least one or more co-components (second metal) are required for obtaining enhanced performance. For VAM synthesis, the best catalyst is Pd-Au based while for benzyl acetate synthesis (from toluene acetoxylation) supported Pd-Sb catalysts are the best. In case of VAM, the role of second metal (e.g. Au, Sn etc.) is to isolate single Pd atoms, while in the case of BA, the second metal (e.g. Sb, Au, Mn, Co etc.) is needed to regulate the growth of Pd particles and also controls the distribution of Pd and PdO species on the surface, which in turn play a key role on the catalytic performance. Even though both the reactions run well over bi-metallic catalysts, VAM synthesis requires the smaller Pd particles, while the bigger Pd particles are necessary for BA synthesis. β-hydrogen elimination (from acetoxyethyl reaction intermediate) is the rate determining step for VAM, while no such β-hydrogen is present in BA production. Therefore, the mechanistic steps for both the reactions are different. Coking and deactivation is a major constraint in BA case (deactivate within 30 hours at lab scale), while the catalysts for VAM run for longer time, i.e. around three years on industrial scale. Frequent make up of KOAc is needed (due to its leaching during the course of the reaction) for VAM case to maintain high catalyst performance while no such problem exists with BA case. The presence of KOAc is required to control the equilibrium between Pd in the Pd_1Au_1 particle and the dispersed Pd for VAM synthesis while no such problem persists with BA.

CONCLUDING REMARKS

Oxidation reactions in general constitute an important class of heterogeneous catalytic processes with high commercial significance. Palladium catalyzed oxidation of alpha-olefins and alkyl aromatics in particular has become an increasingly attractive route for the past few decades and in some instances, the only route for the selective synthesis of vinyl, allyl, benzyl alcohols and esters. In this context, using more effective and economic synthesis (e.g. heterogeneously catalyzed acetoxylation of olefins or alkyl aromatics) of different esters and alcohols and even aldehydes, as intermediates will be of much interesting for further synthesis of various industrially important chemicals.

Besides, catalyst deactivation, the loss of catalytic activity and selectivity, respectively, is indeed a problem of great concern in the practice of industrial catalytic processes. Catalyst replacement costs and process shut down operations account to several billions of dollars across the globe. However, time scales for catalyst deactivation vary considerably from catalyst to catalyst and reaction to reaction, which is in the order of few seconds to several years. Thus the deactivation issues greatly impact research, development, design and operation of various commercial processes. Hence there is a great stimulus to understand and treat catalyst deactivation problems. On the other hand, prevention of catalyst deactivation poses substantial challenges in the design and operation of large scale catalytic processes. At this juncture, application of efficient and economic processes for producing olefinic and benzylic esters in a single step from their corresponding olefins and alkyl aromatics, respectively, particularly represents an attractive synthetic pathway for further preparing various commercially important chemicals. It is always desirable to find a better method for increasing the lifetime of the catalyst.

Vinyl acetate is one of the large scale chemicals with high industrial importance across the globe. It is possible to synthesize vinyl acetate from both homogeneous and heterogeneous processes involving Pd atoms as active centers. Supported Pd-Au catalysts promoted by alkali acetate are the best known catalysts for VAM synthesis. The heterogeneous gas phase acetoxylation process of ethylene is already running on industrial scale and the Pd based catalysts applied gave $\geq 95\%$ yield of vinyl acetate. In these catalysts, isolated Pd atoms are the active sites, and the role of Au is to isolate the Pd atoms. Reaction mechanisms for the synthesis of vinyl acetate from ethylene from both homogeneous and heterogeneous processes are closely linked up and follow the similar reaction steps. However, there also exist some differences between these two approaches particularly in the rate determining steps, nature of oxidant and reoxidation process of Pd species. Even though, the acetoxylation of ethylene to vinyl acetate over Pd-Au-K/SiO_2 catalysts is almost a matured commercial process, certain issues related to the surface reaction sites and reaction mechanism are still under debate. Although great deal has already been achieved in the field of VAM synthesis, new approaches and new attempts are certainly necessary to discover novel catalyst compositions to make the process much more green and economical.

On the other hand, the gas phase acetoxylation of toluene to benzyl acetate route is relatively a new approach, a green process and also possesses many superior qualities compared to the classical methods. Furthermore, a considerable progress has been made through this approach in improving the efficiency of the current method and the yield of benzyl acetate compared to the existing literature reports. Pd-containing catalysts are widely used for this reaction and the catalytic performance is strongly dependent on the Pd particle size; bigger Pd particles (50-100 nm) are more active and selective compared to smaller Pd particles (≤10 nm). Besides, the reaction variables such as temperature, pressure, toluene: AcOH ratio, toluene: O_2 ratio, Gas Hourly Space Velocity (GHSV) etc. are some of the key parameters that can affect the catalytic performance considerably. Among them, the reaction pressure is very critical for catalytic performance. Both the activity and selectivity parameters strongly depend on the reaction pressure. In the case of toluene acetoxylation, an extremely highly selectivity of (~100%) at relatively high conversion of toluene (>75%) could be successfully achieved at 6 bar pressure. In addition, long-term stability also depends on the pressure. Therefore, one needs to find a balance between activity and selectivity and stability in the process of selection of reaction pressure. The stability of the catalyst is also related to the amount of carbon deposition during the course of the reaction. At the same time, surface composition, reducibility of Pd species, Pd oxidation states and their distribution in the near-surface region also play a crucial role on the catalytic performance. Furthermore, nature of catalyst carrier, active phase-support interaction as well as acidity characteristics of the catalysts are some additional issues that need to be considered in the process of catalyst development. Even though significantly higher conversion of toluene and higher yield of benzyl acetate could be achieved through the current acetoxylation process, the content of Pd used in this approach is still considerably high. Therefore, the future investigations should aim at reducing the content of noble metal (e.g. Pd) in the catalyst composition in such a way that the process becomes more economical.

Nonetheless, there are bright chances and ample scope for this acetoxylation route to open the new way for the selective oxidation of various other alkyl aromatics and hetero aromatics to produce a variety of commercially useful chemical products. In parallel to catalyst developments made by the chemists, chemical engineers must also aim at developing new concepts, new reactor designs etc. It is hoped that these results may further stimulate innovative ideas on understanding and tailoring the formation of desired Pd particles with suitable composition and stability under the conditions applied. New approaches for the preparation of Pd catalysts are the usage of metal nanoparticles pre-formed by colloidal chemistry and their deposition on suitable supports. This method allows new possibilities in designing the particles in terms of shape, size and size distribution which was already shown recently. Due to the higher effort needed for the synthesis this approach is rather promising for fine chemical synthesis than for bulk chemicals. In addition, the development of microstructured reactors and application of suitable membranes could be further helpful to improve catalyst stability. Moreover, rational approaches still need to be worked out in the coming future to sequence various functionalities in the direction of developing novel, attractive and eco-friendly catalytic processes.

REFERENCES

Bauer, K., Garbe, D., & Surburg, H. . (1988). Common Fragrance and Flavor Materials. In Ullmann's Encyclopedia of Industrial Chemistry (Vol. 5, 92-117). Weinheim: VCH.

Beletskaya, I. P., & Cheprakov, A. V. (2000). The Heck Reaction as a Sharpening Stone of Palladium Catalysis. *Chemical Reviews*, *100*(8), 3009–3066. doi:10.1021/cr9903048 PMID:11749313

Benhmid, A., Kalevaru, V. N., Martin, A., & Lücke, B. (2004). One-step synthesis of benzyl acetate by gas phase acetoxylation of toluene over highly active and selective Pd-Sb-TiO_2 catalysts. *Chemical Communications*, 2004, 2118–2119. doi:10.1039/b406396a PMID:15368003

Benkalowycz, N. C., Blum, P. R., Cirjak, L. M., Lemanski, M. F., Paparizos, C., Pepera, M. A., & Wagner, D. R. (1996). *U.S. Patent No. 5,550,281*. Washington, DC: U.S. Patent and Trademark Office.

Bissot, T. C. (1977). *U.S. Patent No. 4,048,096.* Washington, DC: U.S. Patent and Trademark Office.

Blaser, H.-U., Indolese, A., Schnyder, A., Steiner, H., & Studer, M. (2001). Supported palladium catalysts for fine chemicals synthesis. *Journal of Molecular Catalysis A Chemical*, *173*(1-2), 3–18. doi:10.1016/S1381-1169(01)00143-1

Bradow, C. W., Grenoble, D. C., Foley, R. M., Murray, B. D., Winquist, B. H. C., & Milham, S. N. (2001). *U.S. Patent No. 6,210,561*. Washington, DC: U.S. Patent and Trademark Office.

Calaza, F., Stacchiola, D., Neurock, M., & Tysoe, W. T. (2010). Coverage Effects on the Palladium-Catalyzed Synthesis of Vinyl Acetate: Comparison between Theory and Experiment. *Journal of the American Chemical Society*, *132*(7), 2202–2207. doi:10.1021/ja907061m PMID:20121234

Centi, G., Cavani, F., & Trifirò, F. (2001). *Selective Oxidation by Heterogeneous Catalysis*. New York, NY: Kluwer Academic Press/Plenum Publishers. doi:10.1007/978-1-4615-4175-2

Chen, M., Gath, K. K., Wei, T., Kumar, D., & Goodman, D. W. (2007, August). Ensemble effects of Pd-Au and Pd-Sn bimetals for vinyl acetate and benzene synthesis. *Proceedings of the 234th ACS National Meeting*, Boston, MA, United States.

Chen, M., & Goodman, D. W. (2008). Promotional effects of Au in Pd-Au catalysts for vinyl acetate synthesis. *Cuihua Xuebao*, *29*, 1178–1186.

Chen, M., Kumar, D., Yi, C. W., & Goodman, D. W. (2005). The Promotional Effect of Gold in Catalysis by Palladium-Gold. *Science*, *310*(5746), 291–293. doi:10.1126/science.1115800 PMID:16224016

Choi, K.-M., Mizugaki, T., Ebitani, K., & Kaneda, K. (2002, April 7-11). Novel preparation of monodispersed palladium nanoclusters and their catalysis for oxidative acetoxylation of toluene using molecular oxygen. *Proceedings of the 223rd ACS National Meeting*, Orlando, FL, USA.

Coq, B., & Figueras, F. (2001). Bimetallic palladium catalysts: Influence of the co-metal on the catalyst performance. *Journal of Molecular Catalysis A Chemical*, *173*(1-2), 117–134. doi:10.1016/S1381-1169(01)00148-0

Crathorne, E., MacGowan, D., Morris, S. R., & Rawlinson, A. P. (1994). Application of Isotopic Transient Kinetics to Vinyl Acetate Catalysis. *Journal of Catalysis*, *149*(2), 254–267. doi:10.1006/jcat.1994.1294

García-Mota, M., & Lopez, N. J. (2008). Template Effects in Vinyl Acetate Synthesis on PdAu Surface Alloys: A Density Functional Theory Study. *Journal of the American Chemical Society*, *130*(44), 14406–14407. doi:10.1021/ja805519v PMID:18839957

Eberle, J., Jung, H., Heidenreich, R., & Roscher, A. (2013). *DE-OS 102011081786*. München, Germany: Deutsches Patent und Markenamt.

Ebitani, K., Choi, K.-M., Mizugaki, T., & Kaneda, K. (2002). Novel Preparation of Palladium Nanoclusters Using Metal Nitrates and Their Catalysis for Oxidative Acetoxylation of Toluene in the Presence of Molecular Oxygen. *Langmuir*, *18*(5), 1849–1855. doi:10.1021/la011359j

Gallezot, P. (1979). The State and Catalytic Properties of Platinum and Palladium in Faujasite-type Zeolites. *Catalysis Reviews. Science and Engineering*, *20*(1), 121–154. doi:10.1080/03602457908065108

Suresh, G., Radnik, J., Kalevaru, V. N., Pohl, M.-M., Schneider, M., Lücke, B., & Brückner, A. et al. (2010). Tailoring the synthesis of supported Pd catalysts towards desired structure and size of metal particles. *Physical Chemistry Chemical Physics*, *12*(18), 4833–4842. doi:10.1039/b920464d PMID:20428566

Gatla, S., Madaan, N., Radnik, J., Kalevaru, V. N., Lücke, B., Martin, A., & Brückner, A. et al. (2011a). Impact of co-components on the state of Pd and the performance of supported Pd/TiO_2 catalysts in the gas-phase acetoxylation of toluene. *ChemCatChem*, *3*(12), 1893–1901. doi:10.1002/cctc.201100225

Gatla, S., Madaan, N., Radnik, J., Kalevaru, V. N., Pohl, M.-M., Lücke, B., & Brückner, A. et al.(2013). Rutile – a superior support for highly selective and stable Pd-based catalysts in the gas phase acetoxylation of toluene. *Journal of Catalysis*, *297*, 256–263. doi:10.1016/j.jcat.2012.10.016

Gatla, S., Madaan, N., Radnik, J., Kalevaru, V. N., Pohl, M.-M., Lücke, B., & Brückner, A. et al. (2011b). Key properties promoting high activity and stability of supported PdSb/TiO_2 catalysts in the acetoxylation of toluene to benzyl acetate. *Applied Catalysis A, General*, *398*(1-2), 104–112. doi:10.1016/j.apcata.2011.03.025

Gatla, S., Radnik, J., Madaan, N., Pohl, M.-M., Mathon, O., Rogalev, A., & Brückner, A. et al. (2015). New insights into the nature of co-components and their impact on Pd structure: X-ray Absorption studies on toluene acetoxylation catalysts. *Chemistry (Weinheim an der Bergstrasse, Germany)*, *21*(43), 15280–15289. doi:10.1002/chem.201502162 PMID:26471442

Hagemeyer, A., & Jahn, U. (2012). *Vinylacetate*. Retrieved from https://roempp.thieme.de/roempp4.0/do/data/RD-22-00794

Hagemeyer, A., Dingerdissen, U., Kühlein, K., Manz, A., & Fischer, R. (2003). *U.S. Patent No. 6,603,038*. Washington, DC: U.S. Patent and Trademark Office.

Han, P., Axnanda, S., Lyubinetsky, I., & Goodman, D. W. (2007). Atomic-Scale Assembly of a Heterogeneous Catalytic Site. *Journal of the American Chemical Society*, *129*(46), 14355–14361. doi:10.1021/ja074891n PMID:17967014

Han, Y. F., Kumar, D., & Goodman, D. W. (2005). Particle size effects in vinyl acetate synthesis over Pd/SiO_2. *Journal of Catalysis*, *230*(2), 353–358. doi:10.1016/j.jcat.2004.12.018

Hanrieder, E.-K., Jentys, A., & Lercher, J.-A. (2014, August 10-14). Interaction of the alkaline acetate promoter with bimetallic PdAu catalysts for the gas phase vinyl acetate synthesis. *Proceedings of the 248th ACS National Meeting & Exposition*, San Francisco, CA, USA.

Hartmann-Schreier, J. (2012). *Benzylacetate*. Römpp Online. Retrieved from https://roempp.thieme.de/roempp4.0/do/data/RD-02-00911

Heck, R. F. (1982). Palladium-catalyzed vinylation of organic halides. *Organic Reactions*, *27*, 345–390.

Hornig, L., Fernholz, H., Schmidt, H.-J., Wunder, F., & Quadflieg, T. (1972). *U.S. Patent No. 3,658,888*. Washington, DC: U.S. Patent and Trademark Office.

Iddir, H., Skavysh, V., Öğüt, S., Browning, N. D., & Disko, M. M. (2006). Preferential growth of Pt on rutile TiO_2. *Physical Review B: Condensed Matter and Materials Physics*, *73*(4), 041403. doi:10.1103/PhysRevB.73.041403

Jira, R. (2009). Acetaldehyde from Ethylene - A Retrospective on the Discovery of the Wacker Process. *Angewandte Chemie International Edition*, *48*(48), 9034–9037. doi:10.1002/anie.200903992 PMID:19834923

Johnson Matthey plc, (2014). *Palladium-Market Data*, Retrieved from http://www.platinum.matthey.com/services/market-research/market-data-charts/palladium

Kalevaru, V. N., Benhmid, A., Radnik, J., Pohl, M.-M., Bentrup, U., & Martin, A. (2007). Marked influence of support on the catalytic performance of PdSb acetoxylation catalysts: Effects of Pd particle size, valence states, and acidity characteristics. *Journal of Catalysis*, *246*(2), 399–412. doi:10.1016/j.jcat.2007.01.002

Keith, J. A., & Henry, P. M. (2009). The Mechanism of the Wacker Reaction: A Tale of Two Hydroxypalladations. *Angewandte Chemie International Edition*, *48*(48), 9038–9049. doi:10.1002/anie.200902194 PMID:19834921

Komatsu, T., & Onda, A. (2008). Catalytic properties of single phase intermetallic compounds. *Catalysis Surveys from Asia*, *12*(1), 6–15. doi:10.1007/s10563-007-9031-3

Kozhevnikov, I. V., Taraban'ko, V. E., Matveev, K. I., & Vardanyan, V. D. (1977). Acetoxylation of olefins catalyzed by a palladium(II)-heteropolyacid system. *Reaction Kinetics and Catalysis Letters*, *7*(3), 297–302. doi:10.1007/BF02062459

Kragten, D. D., van Santen, R. A., Neurock, M., & Lerou, J. J. (1999). A Density Functional Study of the Acetoxylation of Ethylene to Vinyl Acetate Catalyzed by Palladium Acetate. *The Journal of Physical Chemistry A*, *103*(15), 2756–2765. doi:10.1021/jp982956q

Krekel, H., & Kroenig, W. (1969). Ethylene and its Industrial Derivatives. In S. A. Miller (Ed.), *Proc. 7th World Petrol. Congr.*, London, (pp. 24-108). Ernest Benn Limited.

Kroenig, W., & Scharfe, G. (1974). Process for the production of vinyl esters. *U.S. Patent No. 3,822,308.* Washington, DC: U.S. Patent and Trademark Office.

Kumar, D., Chen, M. S., & Goodman, D. W. (2007). Synthesis of vinyl acetate on Pd-based catalysts. *Catalysis Today*, *123*(1-4), 77–85. doi:10.1016/j.cattod.2007.01.050

Lai, X., Clair, T. P. S., Valden, M., & Goodman, D. W. (1998). Scanning Tunneling Microscopy studies of metal clusters supported on TiO_2 (110): Morphology and electronic structure. *Progress in Surface Science*, *59*(1-4), 25–52. doi:10.1016/S0079-6816(98)00034-3

Li, X., Xu, B., Feng, N., Qiu, A., He, J. M. J., Yang, H., & Chen, Y. (2004). The effect of titania polymorph on the strong metal support interaction of Pd/TiO_2 catalysts and their application in the liquid phase selective hydrogenation of long chain alkadienes. *Journal of Molecular Catalysis A Chemical*, *216*(1), 107–114. doi:10.1016/j.molcata.2004.02.007

Lindlar, H., & Dubuis, R. (1973). Palladium catalyst for partial reduction of acetylenes. *Organic Syntheses. Collect.*, *5*, 880–883.

López, N., García-Mota, M., & Gómez-Diaz, J. (2008). NH_3 Oxidation on Oxygen-Precovered Au(111): A Density Functional Theory Study on Selectivity. *The Journal of Physical Chemistry C*, *112*(1), 247–252. doi:10.1021/jp077205f

Macleod, N., Keel, J. M., & Lambert, R. M. (2004). The effects of ageing a bimetallic catalyst under industrial conditions: A study of fresh and used Pd-Au-K/silica vinyl acetate synthesis catalysts. *Applied Catalysis A, General*, *261*(1), 37–46. doi:10.1016/j.apcata.2003.10.032

Madaan, N., Gatla, S., Kalevaru, V. N., Radnik, J., Lücke, B., Brückner, A., & Martin, A. (2011a). Optimization of reaction conditions and regeneration procedure of the PdSb/TiO_2 catalyst for acetoxylation of toluene. *Topics in Catalysis*, *54*(16-18), 1197–1205. doi:10.1007/s11244-011-9743-5

Madaan, N., Gatla, S., Kalevaru, V. N., Radnik, J., Lücke, B., Brückner, A., & Martin, A. (2011b). Deactivation and regeneration studies of a PdSb/TiO_2 catalyst used in the gas phase acetoxylation of toluene. *Journal of Catalysis*, *282*(1), 103–111. doi:10.1016/j.jcat.2011.06.005

Madaan, N., Gatla, S., Kalevaru, V. N., Radnik, J., Pohl, M.-M., Lücke, B., & Martin, A. et al. (2013). The Impact of Reaction Pressure on the Catalytic Performance of the Pd-Sb/TiO2 Catalyst in the Acetoxylation of Toluene into Benzyl Acetate. *ChemCatChem*, *5*(1), 185–191. doi:10.1002/cctc.201200522

Mahata, N., & Vishwanathan, V. (2000). Influence of Palladium Precursors on Structural Properties and Phenol Hydrogenation Characteristics of Supported Palladium Catalysts. *Journal of Catalysis*, *196*(2), 262–270. doi:10.1006/jcat.2000.3041

The Industrial Uses of Palladium. (2011). Manhattan Gold & Silver. Retrieved from http://www.mgsrefining.com/blog/post/2011/09/13/The-Industrial-Uses-for-Palladium.aspx

Martin, A., & Kalevaru, V. N. (2010). Heterogeneously Catalyzed Ammoxidation: A Valuable Tool for One-Step Synthesis of Nitriles. *ChemCatChem*, *2*(12), 1504–1522. doi:10.1002/cctc.201000173

Mazzone, G., Rivalta, I., Russo, N., & Sicilia, E. (2009). The geometric effect in palladium-gold catalysis. Is the coupling the rate-determining step in the vinyl-acetate synthesis? *Chemical Communications*, *14*(14), 1852–1854. doi:10.1039/b823207e PMID:19319423

Miller, S. A. (1969). *Ethylene and its industrial derivatives* (pp. 24–108). Ernest Benn Limited London.

Moiseev, I. I., Vargaftic, M. N., & Syrkin, Y. K. (1960). Mechanism of the reaction of palladium salts with olefins in hydroxyl-containing solvents. *Doklady Akademii Nauk SSSR*, *133*, 377–380.

Mondelli, C., Ferri, D., Grunwaldt, J.-D., Krumeich, F., Mangold, S., Psaro, R., & Baiker, A. (2007). Combined liquid-phase ATR-IR and XAS study of the Bi-promotion in the aerobic oxidation of benzyl alcohol over Pd/Al_2O_3. *Journal of Catalysis*, *252*(1), 77–87. doi:10.1016/j.jcat.2007.09.013

Nakamura, S., & Yasui, T. (1970). Mechanism of the palladium-catalyzed synthesis of vinyl acetate from ethylene in a heterogeneous gas reaction. *Journal of Catalysis*, *17*(3), 366–374. doi:10.1016/0021-9517(70)90113-2

Nicolau, I., Broussard, J. A., & Colling, P. M. (1999). *U.S. Patent No. 5,948,724*. Washington, DC: U.S. Patent and Trademark Office.

Old, D. W., Wolfe, J. P., & Buchwald, S. L. (1998). Highly Active Catalyst for Palladium-Catalyzed Cross-Coupling Reactions: Room-Temperature Suzuki Couplings and Amination of Unactivated Aryl Chlorides. *Journal of the American Chemical Society*, *120*(37), 9722–9723. doi:10.1021/ja982250+

Plata, J. J., García-Mota, M., Braga, A. A. C., López, N., & Maseras, F. (2009). Vinyl Acetate Synthesis on Homogeneous and Heterogeneous Pd-Based Catalysts: A Theoretical Analysis on the Reaction Mechanisms. *The Journal of Physical Chemistry A*, *113*(43), 11758–11762. doi:10.1021/jp902929x PMID:19769362

Global Vinyl Acetate Production to Post 4.5% Annual Growth Through 2017, According to In-demand Report by Merchant Research & Consulting. (2014). *PRWeb*. Retrieved from http://www.prweb.com/releases/2014/06/prweb11958764.htm

Qian, K., & Huang, W. (2011). Au–Pd alloying-promoted thermal decomposition of PdO supported on SiO_2 and its effect on the catalytic performance in CO oxidation. *Catalysis Today*, *164*(1), 320–324. doi:10.1016/j.cattod.2010.10.018

Radnik, J., Benhmid, A., Kalevaru, V. N., Pohl, M.-M., Martin, A., Lücke, B., & Dingerdissen, U. (2005). Deactivation of Pd acetoxylation catalysts: Direct observations by XPS investigations. *Angewandte Chemie International Edition*, *44*(41), 6771–6774. doi:10.1002/anie.200501650 PMID:16187392

Radnik, J., Pohl, M.-M., Kalevaru, V. N., & Martin, A. (2007). First knowledge on the formation of novel core-shell structures in PdCu catalysts and their Influence on the prevention of catalyst deactivation. *The Journal of Physical Chemistry C*, *111*(28), 10166–10169. doi:10.1021/jp073184d

Reining, S., Kondratenko, E. V., Kalevaru, V. N., & Martin, A. (2014). Reaction pathways in the palladium-catalyzed acetoxylation of toluene. DGMK Tagungsbericht, 3, 51-56.

Rivalta, I., Mazzone, G., Russo, N., & Sicilia, E. (2010). The influence of surface oxygen and hydroxyl groups on the dehydrogenation of ethylene on PdAu surface alloys. A theoretical cluster model study. *Chemical Physics Letters*, *493*(1-3), 87–93. doi:10.1016/j.cplett.2010.05.033

Samanos, B., Boutry, P., & Montarnal, R. (1971). The mechanism of vinyl acetate formation by gas-phase catalytic ethylene acetoxidation. *Journal of Catalysis*, *23*(1), 19–30. doi:10.1016/0021-9517(71)90019-4

Schwerdtel, W. (1968). Vinyl acetate via ethylene. *Hydrocarbon Processing*, *47*, 187–191.

Schunk, S. A., & de Oliveira, A.L. (2008). Acetoxylation of Ethylene. In G. Ertl, H. Knözinger, F. Schüth, J. Weitkamp (Eds.), Handbook of Heterogeneous Catalysis (pp. 3464-3479). Weinheim, Germany: Wiley-VCH Verlag GmbH & Co. KGaA. doi:10.1002/9783527610044.hetcat0176

Shu, Q., Wang, X., Ren, S., & Shi, W. (2005). Gas-phase oxidative acetoxylation of toluene over Pd-Sn-K/SiO_2 catalyst. *Cuihua Xuebao*, *26*, 869–873.

Siyo, B., Schneider, M., Radnik, J., Pohl, M.-M., Langer, P., & Steinfeldt, N. (2014). Influence of support on the aerobic oxidation of HMF into FDCA over preformed Pd nanoparticle based materials. *Applied Catalysis A, General*, *478*, 107–116. doi:10.1016/j.apcata.2014.03.020

Stacchiola, D., Calaza, F., Burkholder, L., Schwabacher, A. W., Neurock, M., & Tysoe, W. T. (2005). Elucidation of the reaction mechanism for the palladium-catalyzed synthesis of vinyl acetate. *Angewandte Chemie International Edition*, *44*(29), 4572–4574. doi:10.1002/anie.200500782 PMID:15988776

Stacchiola, D., Calaza, F., Burkholder, L., & Tysoe, W. T. (2004). Vinyl Acetate Formation by the Reaction of Ethylene with Acetate Species on Oxygen-Covered Pd(111). *Journal of the American Chemical Society*, *126*(47), 15384–15385. doi:10.1021/ja044641w PMID:15563157

Palladium Alliance International, (2012). *Stillwater Palladium*. Retrieved from http://stillwaterpalladium.com/

Tsuji, J. (2004). *Palladium reagents and catalysts: new perspectives for the 21st century* (p. 90). John Wiley and Sons. doi:10.1002/0470021209

Ugo, R. (Ed.), (1980). *Palladium catalyzed oxidation of hydrocarbons*. Dordrecht, Netherlands: D. Reidel Publishing Company.

van Helden, R., Kohll, C. F., Medema, D., Verberg, G., & Jonkhoff, T. (1968). Palladium-catalyzed reactions of unsaturated compounds in nonaqueous solvents. Synthesis of vinyl acetate via palladium salt/ethylene complexes. *Recueil des Travaux Chimiques des Pays-Bas*, *87*(8), 961–991. doi:10.1002/recl.19680870813

Vargaftik, M. N., Zagorodnikov, V. P., Stolyarov, I. P., Moiseev, I. I., Kochubey, D. I., Likholobov, V. A., & Zamaraev, K. I. et al. (1989). Giant palladium clusters as catalysts of oxidative reactions of olefins and alcohols. *Journal of Molecular Catalysis*, *53*(3), 315–348. doi:10.1016/0304-5102(89)80066-5

Vinyl Acetate Monomer (VAM) Global Market to 2020. (2013). *Vertical Edge Limited*. Retrieved from http://www.companiesandmarkets.com/Market/Chemicals/Market-Research/Vinyl-Acetate-Monomer-VAM-Global-Market-to-2020/RPT1134065

Voskanyan, P. (2010). Effect of the amounts and proportions of active components on the activity and selectivity of a catalyst for vinyl acetate synthesis by gas-phase ethylene acetoxylation. *Catalysis in Industry*, *2*(2), 167–172. doi:10.1134/S2070050410020121

Voskanyan, P. S. (2013). Effect of the nature of a support on the catalytic activity of a palladium catalyst in the synthesis of vinyl acetate by gas-phase ethylene acetoxylation. *Catalysis in Industry*, *5*(1), 90–97. doi:10.1134/S2070050412040198

Yin, L., & Liebscher, J. (2007). Carbon-Carbon Coupling Reactions Catalyzed by Heterogeneous Palladium Catalysts. *Chemical Reviews*, *107*(1), 133–173. doi:10.1021/cr0505674 PMID:17212474

Yun, S., & Oyama, S. T. (2011). Correlations in palladium membranes for hydrogen separation: A review. *Journal of Membrane Science*, *375*(1-2), 28–45. doi:10.1016/j.memsci.2011.03.057

Zhang, M., Gao, M., Li, Q., Tao, M., & Yu, Y. (2014). DFT study on effect of CO on the system of acetoxylation of ethylene to vinyl acetate. *RSC Advances*, *4*(34), 17709–17716. doi:10.1039/c3ra47794k

Chapter 4
Production of Ethylene and its Commercial Importance in the Global Market

Ahmad Alshammari
King Abdulaziz City for Science and Technology, Saudi Arabia

Abdulaziz Bagabas
King Abdulaziz City for Science and Technology, Saudi Arabia

Venkata Narayana Kalevaru
Leibniz Institute for Catalysis, Germany

Andreas Martin
Leibniz Institute for Catalysis, Germany

ABSTRACT

Ethylene is the largest of the olefin markets and is also one of the most important petrochemically derived monomers that are used as a feedstock for the production of various commercially useful chemical products (e.g. polyethylene, polymers, fibers etc.). The primary objective of this chapter is to provide a comprehensive overview about olefins particularly ethylene production technologies and its commercial significance in the world market. The content of this chapter is presented as follows: a general overview about olefins production is given. This is followed by introducing the reader to ethylene including its properties importance/applications. The next section describes the production technologies of ethylene and some of its selected derivatives, followed by an overview of the technology, market, costs, capacity, global demand and supply of ethylene technology. Finally, main points and outlook of this highly industrially important commodity chemical are summarized.

OLEFIN PRODUCTION

General Introduction (Solomons, 1996)

Olefins or alkenes are hydrocarbons structurally distinguished by their carbon-carbon double bond. They are generally represented by the molecular structure shown in Box 1.

Ethene or ethylene is the simplest olefin because both of its double bond carbon atoms are not substituted (i.e. they are only attached to hydrogen atoms). Propene or propylene is a gaseous mono-substituted olefin where a methyl group ($-CH_3$) is attached to one of the double bond carbon atoms, thereby extend-

DOI: 10.4018/978-1-4666-9975-5.ch004

Box 1. General molecular structure of olefins

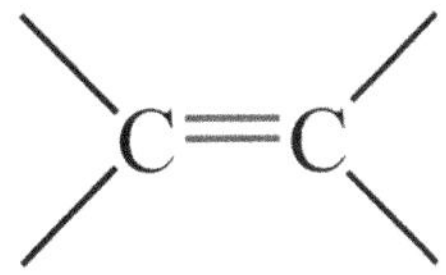

Box 2. Molecular structure of cis- and trans-2-butene

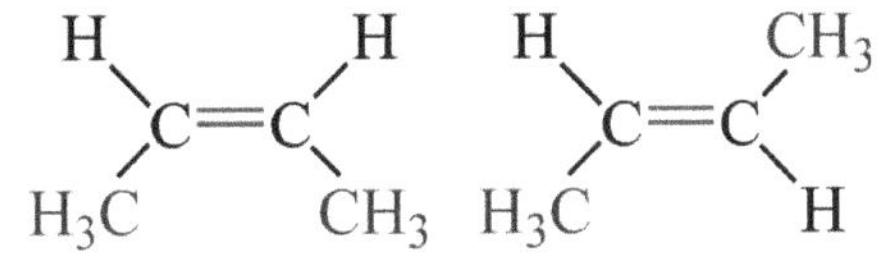

Box 3. Molecular structure of 2-methylpropene

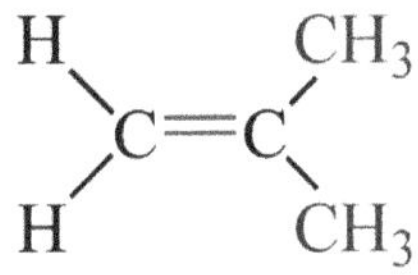

ing the hydrocarbon chain by one carbon atom. Disubstitution of both of the double bond carbon atoms leads to the emergence of two diastereomers: *cis* when both of the substituent groups are on the same side and *trans* when they are on opposite sides. Box 2 shows an example of the diastereomers *(cis* and *trans*) of 2-butene.

However, diastereomerism disappears when both of the substituent groups are attached to one carbon atom of the double bond, as in the case of isobutene or 2-methylpropene (Box 3).

For trisubstituted and tetrasubstituted alkenes, alkene stereochemistry cannot be described anymore in terms of *cis* and *trans*; it is conveniently described by the Cahn-Ingold-Prelog convention, known as the (*E*)–(*Z*) system, on the basis of group priority. Both E and Z symbols come from German Language, i.e. the E comes from a German word "entgegen", which means opposite, and also the symbol Z comes from a German word "zusammen", which means together. Alkene stereochemistry is designated (*E*) when groups of higher priority are on opposite sides of the carbon atoms of the double bond. On the other hand, alkene stereochemistry is designated (*Z*) when groups of higher priority are on the same side of the carbon atoms of the double bond. Box 4 shows an example of the two diastereomers (*E*) and (*Z*) of 1-bromo-1-chloro-1-pentene.

Box 4. Molecular structure of (E)- and (Z)-1-bromo-1-chloro-1-pentene

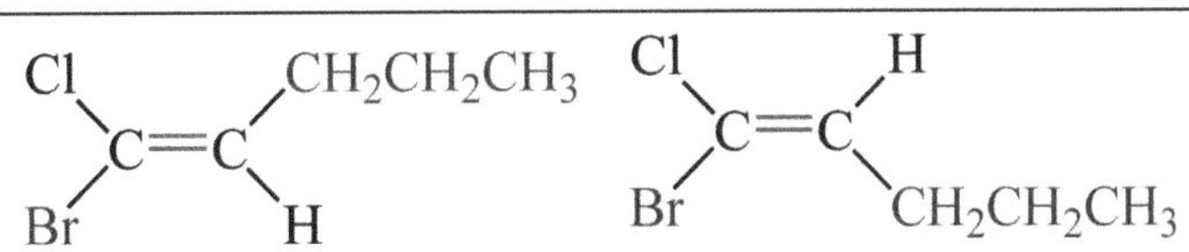

Box 5. Comparison of the relative stabilities of 2-methyl-2-butene, 2-methyl-1-butene, and 3-methyl-1-butene

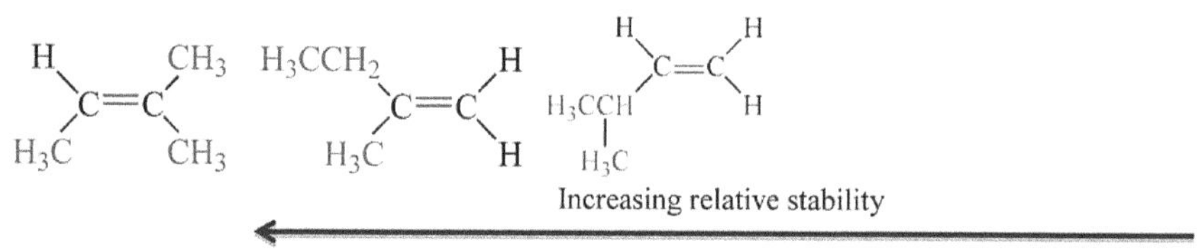

The number of alkyl group substituents, which are connected to the carbon atoms of the double bond, affects the relative stabilities of alkenes. The greater the number of alkyl substituents on the carbon atoms of the double bond is, the greater the stability of alkene is owing to the electron-donating effect of alkyl substituents. Hence, 2-methyl-2-butene, a five-carbon trisubstituted alkene, is more stable than 2-methyl-1-butene, a five-carbon disubstituted alkene, which in turn is more stable than 3-methyl-1-butene, a five-carbon monosubstituted alkene (Box 5).

Furthermore, the trans isomer of an alkene is more stable than the cis isomer because of the elimination of the steric effect in the trans isomer. The relative stabilities of alkenes can be evaluated by the heat of hydrogenation or combustion.

Alkenes occur naturally, e.g., β-pinene, a component of turpentine, and ethylene, a plant hormone for fruit ripening.

Ethylene, propylene, and butadiene are among the most important alkenes for polymer synthesis. Ethylene is also a raw material for making important chemicals such as ethanol and acetaldehyde. Propylene is used to synthesize acetone and cumene.

HISTORICAL DEVELOPMENT (Weissermel & Arpe, 2003)

Olefins importance has been increased remarkably since the Second World War owing to the enormous growth in petrochemistry and the usage of petrochemical products. Specifically, two main factors were behind the development of olefins and their chemistry. The first factor was the increased demand for gasoline, which had its quality improved by reaction with olefins. Thus, olefins demand was increased and crude oil cracking processes were in turn developed. The second factor was the development of methods for the conversion of olefins into useful chemicals, for the preparation of monomers and for their subsequent polymerization.

The crude oil cracking processes can be achieved through catalytic, hydrocatalytic, or thermal cracking. The catalytic cracking produces a little bit of olefins and a large amounts of saturated branched paraffins, cycloparaffins, and aromatics using aluminum silicates catalyst doped with chromium(III) oxide or manganese(II) oxide in a fluidized bed or a riser reactor at 450-500 °C. Mixture of crystalline zeolites with amorphous, synthetic, or naturally-occurring aluminum silicates are currently being used as catalysts in their acidic form or doped with lanthanides. The catalyst contains small amounts of platinum, needed for the catalytic oxidation of coke and carbon monoxide to carbon dioxide. The catalytic cracking is advantageous for the production of gasoline and motor fuels. However, the catalyst must be reactivated under oxygen atmosphere for the removal of coke layer on the catalyst surface.

On the other hand, hydrocracking process is performed using hydrogen to convert higher boiling distillation fractions into lower boiling products such as isobutane, gasoline, naphtha, and fuel oil using bifunctional catalyst system, made up of hydrogenation-dehydrogenation and acidic cracking components. This process, however, does not produce olefins at all. Such process is also relatively expensive due to its high consumption of hydrogen gas and its operation at high temperatures and pressures.

Thermal cracking is an uncatalyzed, radical process involving homolysis of C–C bonds of hydrocarbons under pressure at 400-500 °C. Each produced alkyl radical is able to abstract a hydrogen atom from another alkane molecule, resulting in a shorter alkane and an alkyl radical, which is then undergoes a β-cleavage to form ethylene or propylene and a shorter alkyl radical. Therefore, thermal cracking process involves dehydrogenation and hydrogen transfer from hydrogen-rich hydrocarbon fractions, classified as hydrogen-content change reactions, and cracking of hydrogen-deficient higher molecular fractions, classified as carbon skeleton change reactions. Both of these reactions, called primary reactions, lead to chain shortening, isomerization, and cyclization. However, olefin polymerization, alkylation, and condensation of aromatics, may take place. These latter unfavorable reactions are called secondary reaction. The reaction conditions (temperature, residence time, and partial pressure) are the keys to optimize thermal cracking for the production of olefins, i.e. favoring primary reactions over secondary reactions. Low molecular weight olefins such as ethylene and propylene are preferably produced under high temperature, short residence time, and low partial pressure conditions. The yield of low molecular weight olefins can be increased by the addition of steam (water vapor), a process called steam cracking. The higher the content of steam is, the higher the yield of olefins is. In addition, the high amount of steam eliminates the carbon deposition on the catalyst surface. The steam cracking can be either low-severity cracking (<800 °C and one second residence time) or high-severity cracking (~900 °C and ~0.5 second residence time).

METHODS FOR OLEFIN SYNTHESIS (Solomons, 1996)

Alkenes are widely prepared by elimination reactions. The dehydrohalogenation of alkyl halides, dehydration of alcohols, debromination of vicinal dibromides, and the Cope elimination are among the important elimination reactions.

Dehydrohalogenation is the elimination of a hydrogen halide (HCl, HBr, or HI) from a haloalkane using a strong base at high temperature; as shown in Box 6 (equation 1).

Dehydrohalogenation is also called β-elimination, caused by the abstraction of a hydrogen atom from the β-carbon atom. For synthesizing alkenes, dehydrohalogenation preferentially occurs through a bimolecular mechanism, E2 reaction, to avoid the rearrangement of the carbon skeleton of the generated

Box 6. Dehydrohalogenation

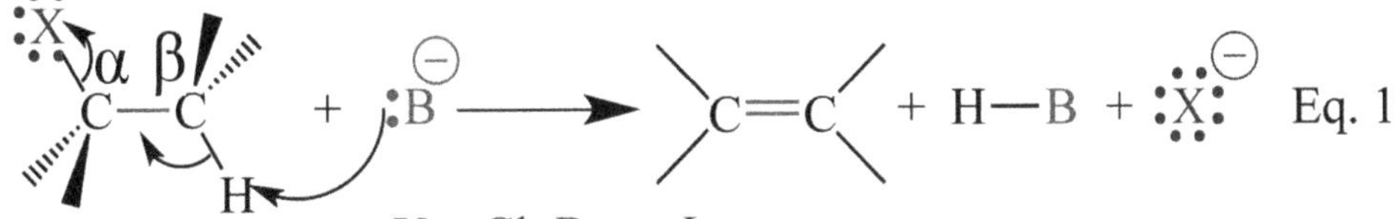

X = Cl, Br, or I

Box 7. Dehydrobromination of 2-bromo-2-methylbutane

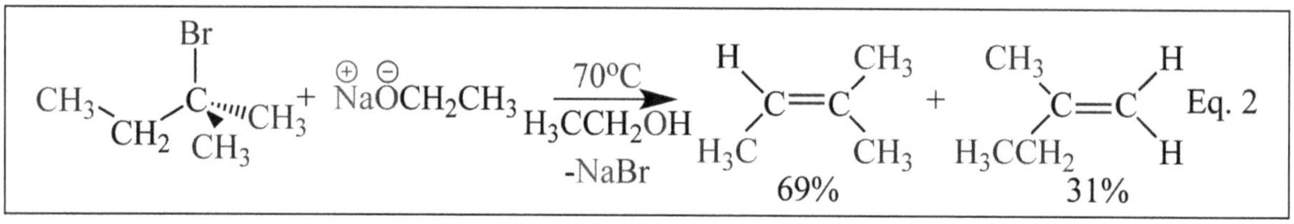

carbocation, which might occur through a unimolecular mechanism, E1 reaction. Thus, for the E2 reaction, secondary or tertiary alkyl halides are used. In addition, the E2 reaction is facilitated using a high concentration of a strong, relatively nonpolarizable base in a relatively nonpolar solvent, at a relatively high temperature. A bulky base is used when a primary alkyl halide is used as the starting material for the desired alkene.

Dehydrohalogenation often results in more than one product. The major product is the most stable, most highly substituted alkene when using a small, unhindered base such as ethoxide ion ($H_3CCH_2O^-$) or hydroxide ion (OH^-) in ethyl alcohol. On the other hand, the minor product is the least stable, least substituted alkene. In such case, elimination is described by following Zaitsev's rule. Box 7 (equation 2) shows the dehydrobromination of 2-bromo-2-methylbutane under conditions favoring Zaitsev's rule as an example.

However, dehydrohalogenation primarily forms the less substituted alkene by using a bulky base such as *tert*-butoxide ion [$(H_3C)_3CO^-$] in *tert*-butyl alcohol. Such elimination is prescribed by following the Hofmann rule. The formation of the least substituted alkene (Hofmann yield) is partially attributed to the bulkiness of the used base and to the association of solvent molecules with the base. Therefore, the bulky base cannot easily access the internal secondary hydrogen atoms to abstract them; however, it can readily abstract one of the readily accessible primary hydrogen atoms of the methyl group. Box 8 (equation 3) shows the favored formation of 2-methyl-1-butene on the dehydrobromination of 2-bromo-2-methylbutane under conditions favoring the Hofmann rule as an example.

The Hofmann elimination for the synthesis of alkene through dehydrohalogenation involves the use of an electrically neutral alkyl halide, as discussed above. However, the Hofmann elimination extends to positively charged substrates such as quaternary ammonium hydroxide and sulfonium when they are heated. The preference of the Hofmann yield in such case is due to the bulkiness of the leaving group and the higher acidity of the hydrogen atoms of the less substituted carbon atom. The Hofmann elimination of 2-trimethylammoniumbutane and 2-dimethylsulfoniumbutane is shown in Box 9. (equations 4 and 5), respectively.

Box 8. Favored formation of 2-methyl-1-butene on the dehydrobromination of 2-bromo-2-methylbutane

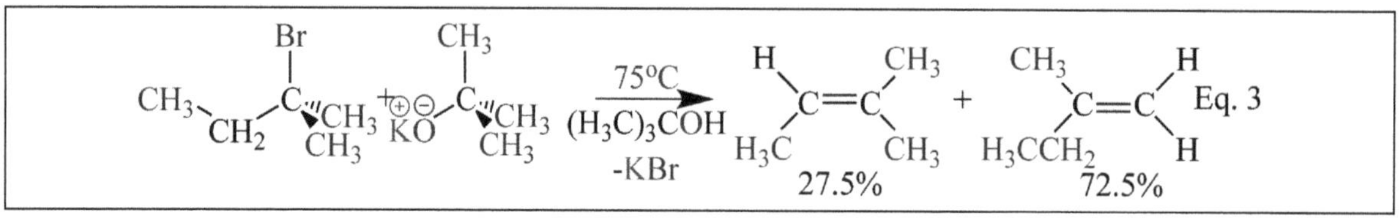

Box 9. Hofmann elimination of 2-trimethylammoniumbutane and 2-dimethylsulfoniumbutane

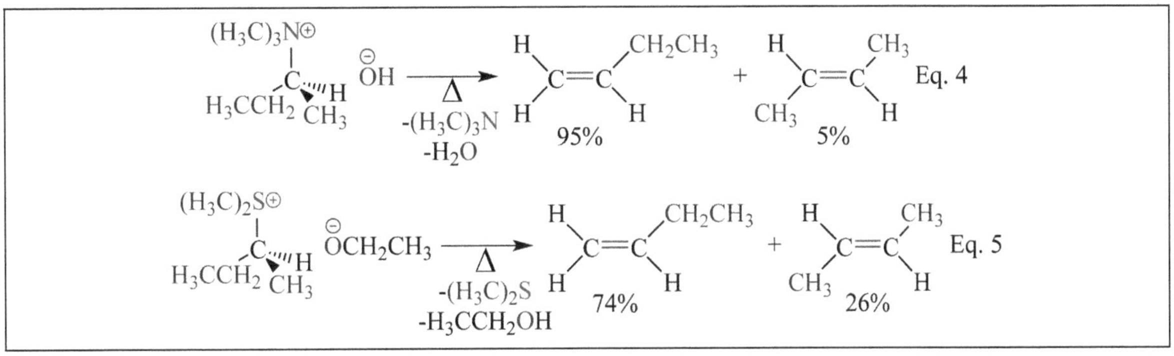

Box 10. Formation of alkenes and water from β-elimination

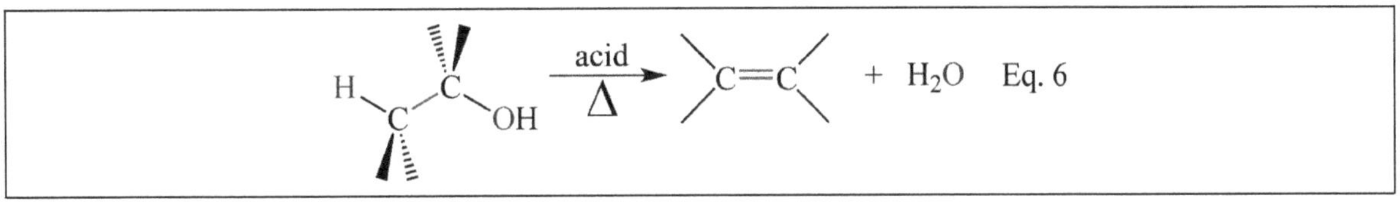

β-elimination involves the dehydration of alcohols at high temperatures using a strong acid; it leads to the formation of alkenes and water, as shown in Box 10 (equation 6).

Brønsted acids such as sulfuric acid and phosphoric acid are usually used for alkenes synthesis by the dehydration of alcohols in the laboratory, while Lewis acids such as aluminum oxide are used for the heterogeneous, gas-phase dehydration in industry.

The structure of the starting alcohol strongly impacts the required temperature and acid concentration for its dehydration to an alkene. The ease of dehydration in alcohols increasing in the following order: Tertiary alcohols > secondary alcohols > primary alcohols. In Box 11 (equations 7, 8, and 9), we can

Box 11. Dehydration of tert-butyl compared to ethanol

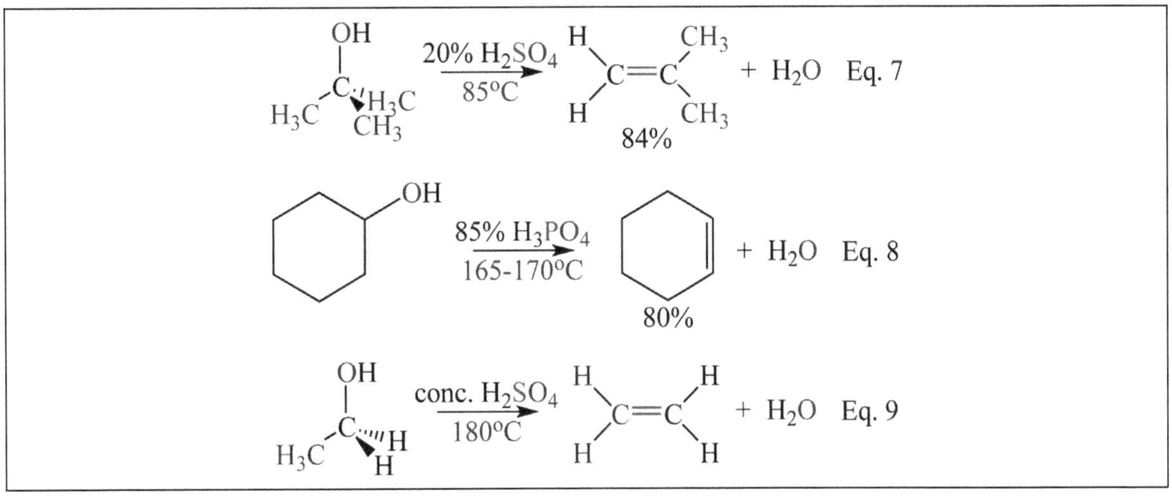

Box 12. Alcohol is protonated to form an alkyloxonium ion by the transfer of a proton from the Brønsted acid catalyst

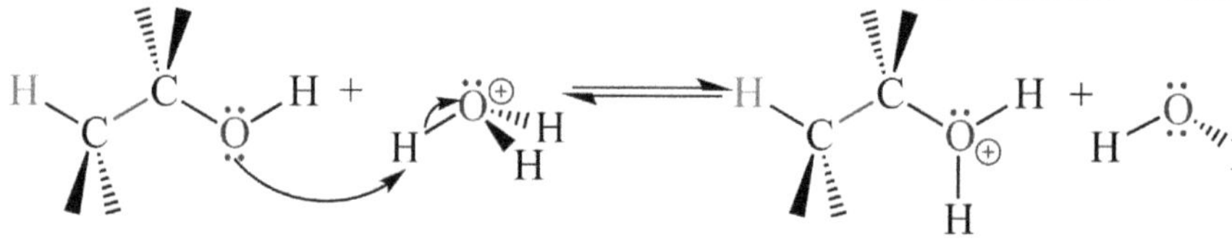

Eq. 10

Box 13. Carbocation is formed by heterolytic cleavage

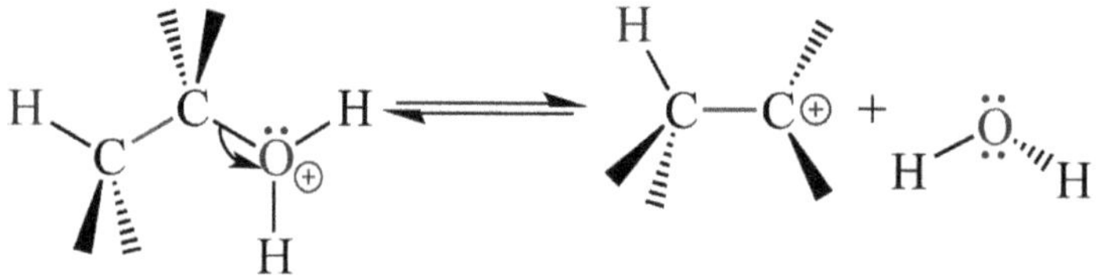

Eq. 11

Box 14. Alkene is formed by the abstraction of a β-proton from the carbocation

Eq. 12

observe that the dehydration of *tert*-butyl alcohol requires the least concentration of the Brønsted acid catalyst and temperature, while ethanol requires the highest concentration of the Brønsted acid catalyst and temperature.

The dehydration of alcohols to alkenes is classified as the E1 reaction depending on the Whitmore mechanism, which consists of three consecutive steps. First, the alcohol is protonated to form an alkyloxonium ion by the transfer of a proton from the Brønsted acid catalyst, as shown in Box 12 (equation 10).

Second, a carbocation is formed by the heterolytic cleavage of the C–O bond owing to the positive charge on the oxygen of the alkyloxonium ion, as shown in Box 13 (equation 11).

Lastly, an alkene is formed by the abstraction of a β-proton from the carbocation and the regeneration of the Brønsted acid catalyst, as shown in Box 14 (equation 12).

Box 15. Products obtained from the dehydration of 3,3-dimethyl-2-butanol

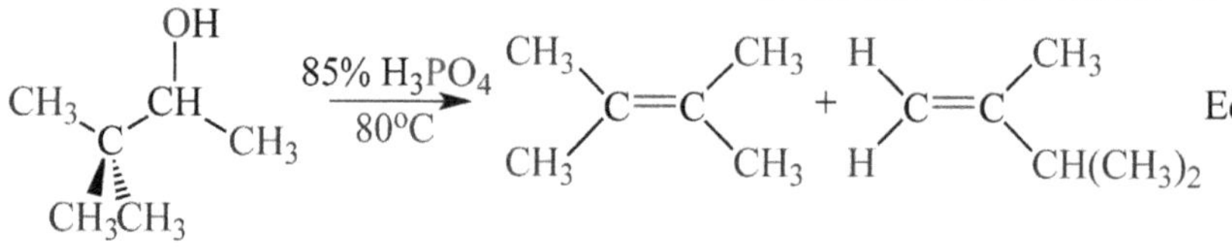

Eq. 13

Box 16. Representation of the carbon skeleton of the alkene products resulting from the dehydration of 3,3-dimethyl-2-butanol

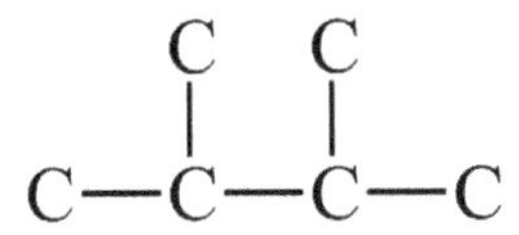

Box 17. Representation of the carbon skeleton of 3,3-dimethyl-2-butanol

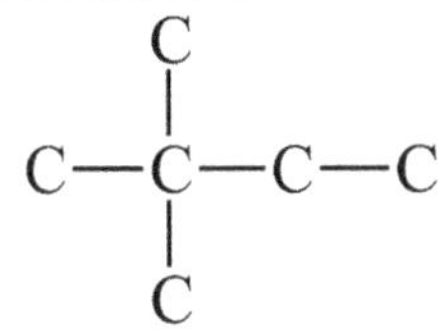

The Whitmore mechanism along with the stability of the generated carbocation (tertiary > secondary > primary > methyl) can explain the observed dehydration of alcohols, formation of more than one product, and carbon skeleton rearrangement. The dehydration of 3,3-dimethyl-2-butanol, a secondary alcohol, is an example for carbon skeleton rearrangement, as it can deduced from the obtained products in Box 15 (equation 13).

The carbon skeleton of the products is shown in Box 16.

However, the carbon skeleton of the starting alcohol 3,3-dimethyl-2-butanol is shown in Box 17.

This carbon skeleton rearrangement is viable by the migration of a methanide ion (1,2-methanide shift) for the conversion of the less stable, secondary carbocation to a more stable tertiary carbocation, as shown in Box 18 (equation 14).

The stable tertiary carbocation loses a proton to primarily form the Zaitsev product, the more stable higher substituted alkene (2,3-dimethyl-2-butene) according to Zaitsev's rule.

Alkenes can be prepared by the debromination of vicinal dibromides (*vic*-dibromides). This reaction is not extensively employed for the synthesis of alkenes because *vic*-dibromides are prepared by the addition of bromine (Br_2) to an alkene. However, bromination followed by debromination is a good method to purify alkenes and to protect their double bond.

The debromination of vicinal dibromide occurs via the E2 mechanism with the preferred anti-periplanar conformation for the transition state. This type of a reaction is accomplished using either a solution of sodium iodide in acetone or a mixture of zinc dust in acetic acid or ethanol. Box 19 (equations 15 and 16) show the debromination of 2,3-dibromo-2,3-dimethylbutane.

Alkenes can also be prepared by the Cope elimination, a syn elimination that involves the heating a tertiary amine oxide. Box 20 (equation 17) shows the synthesis of methylenecyclohexane by the Cope elimination.

Box 18. The conversion of the less stable, secondary carbocation to a more stable tertiary carbocation

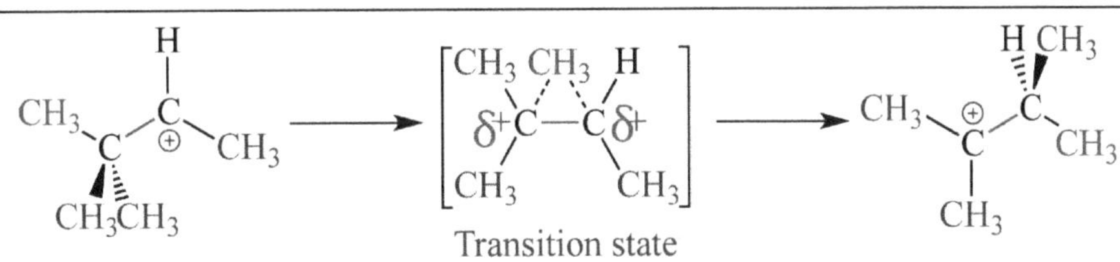

Eq. 14

Box 19. The debromination of 2,3-dibromo-2,3-dimethylbutane

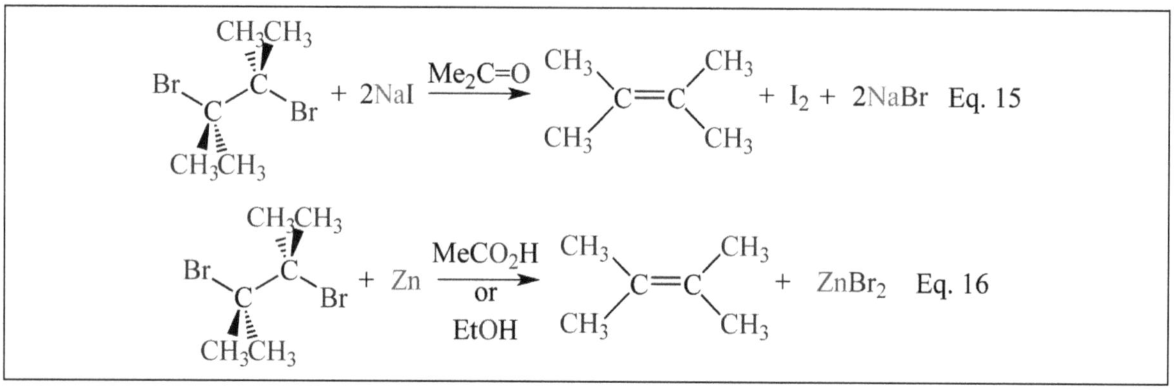

Box 20. The synthesis of methylenecyclohexane by the Cope elimination

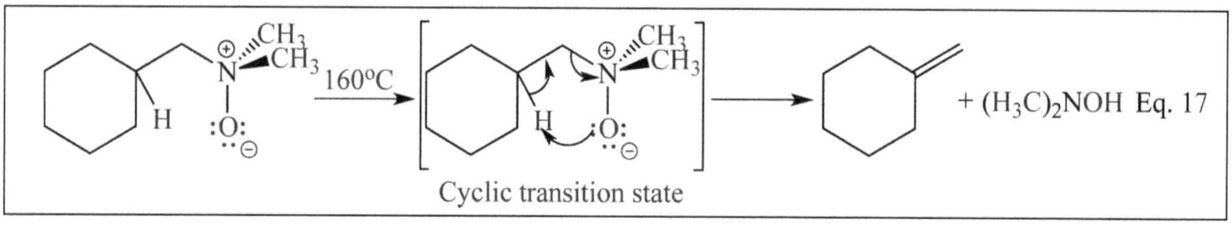

ETHYLENE

Properties of Ethylene

Ethylene or ethene (C_2H_4) is considered to be one of the most important raw materials in the chemical industry. Its significance is driven by its molecular structure, i.e., carbon-carbon double bonds (C=C). This π-bond is responsible for its chemical reactivity. The double bond is also a place of high electron density; therefore it is susceptible to attack by electrophiles. It is classified as the second simplest unsaturated hydrocarbon because it has only two H atoms. It is a volatile substance, colorless at room temperature, noncorrosive, nontoxic, flammable gas, slightly soluble in water, and soluble in most organic solvents. It has boiling and melting points of -104 and -169.2 °C, respectively, at a pressure of 1 atm. Ethylene is a very active chemical; as exemplified by the reaction between ethylene and water to produce ethyl alcohol. Most of the ethylene reactions are catalyzed by transition metals, which bind transiently to the ethylene using both the π and π* orbitals. Ethylene is also an active alkylating agent, which can be used for the production of important monomers, such as ethyl benzene (EB), which is dehydrogenated to styrene.

Importance and Applications of Ethylene

Undoubtedly, ethylene is the largest contributor to the olefin market and is also one of the most important petrochemical derived monomer used as a feedstock for the production of various commercially useful chemical products such as plastics (e.g., polyethylene), polymers, (e.g., poly vinyl chloride, polyester, and polystyrene), fibers, resins, and packaging materials. Figure 1 presents the most key chemicals based

Figure 1. Key chemicals based on ethylene

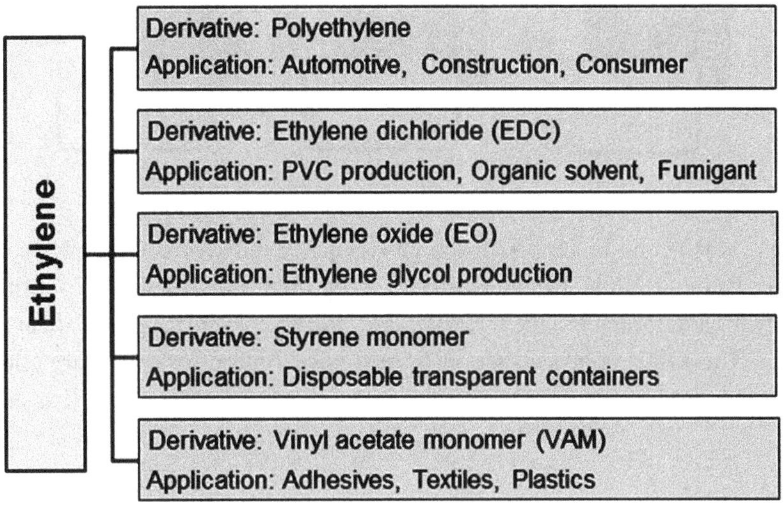

on ethylene. In addition, ethylene is also used to produce various valuable chemicals, e.g., ethyl benzene (EB), styrene, ethylene oxide (EO), ethylene glycols, synthetic lubricants, surfactants, plasticizers, and detergents. Moreover, ethylene is also used to produce plastics, fibers, and other organic chemicals in packaging, transportation, construction, and other industries.

Ethylene is also consumed in the production of ethylene dichloride (EDC), which can be directly applied to synthesize PVC. This type of polymer is currently used for a wide variety of applications such as production of pipes and rigid and flexible foams. EO is another important type of an intermediate chemical, which is obtained from ethylene. Majority of EO is used for the production of ethylene glycols, including diethylene glycol and triethylene glycol, which account for the majority of the global consumption. A small amount of EO is used in the production of a wide range of detergents and solvents including ethylene glycol ethers, ethanolamines, and ethoxylates. Furthermore, styrene monomer is used in a broad range of products, mainly styrene-based polymers. The majority of styrene monomer is used to produce polystyrene, expandable polystyrene, and acrylonitrile. Vinyl acetate monomer (VAM) is also used to produce a wide variety of polymers such as polyvinyl acetate, insulation for magnetic wires, inter-layers for safety glass, and polyvinyl alcohol is used to produce adhesives, coatings, and water-soluble packaging films. Finally, small amounts of ethylene are applied as a co-monomer for the production of polypropylene and ethylene propylene diamine monomer (EPDM), as well as in the synthesis of acetaldehyde and ethanol.

PRODUCTION OF ETHYLENE

Generally, the most common feedstock for the production of ethylene is ethane caused by its high ethylene yield. Ethylene is currently generated commercially by the thermal cracking of ethane, propane, or naphtha with steam at high temperatures (Albright, Crynes, & Nowak, 1992). Other production routes, besides conventional steam cracking, are fluidized-bed cracking and catalytic pyrolysis. (Chalavekilian, & Dumon, 1976; Stratton, Hemming, & Teper, 1983). In addition to these routes, new methods such as

Figure 2. Typical flow diagram for producing ethylene using steam cracking

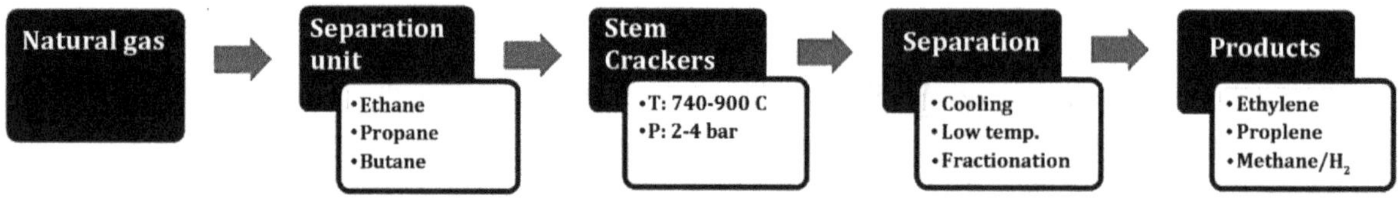

methanol to olefins (MTO), the Fischer-Tropsch process, and refinery off-gas recovery are also being developed (Worrell, Faaij, & Blok, 1994). Quite recently, other new methods for ethylene production (e.g., oxidative dehydrogenation of ethane (ODHE)) are being explored (e.g. Brazdil et al., 2006, Gaspar, & Pasternak, 1971). This ODH approach seems to be a good option for the future ethylene production at relatively low temperature, pressure, and also by using cheaper feedstock such as ethane.

STEAM CRACKING

Steam cracking (IEA, 2004) typically refers to the production of valuable chemical products in which saturated hydrocarbons are broken down into smaller hydrocarbons. Steam cracking technology is considered to be the backbone for many industrial processes, despite some problems associated with these processes, for instance i) easiness of catalyst deactivation caused by carbon deposition, ii) occurrence of a significant amount of side reactions and iii) high operating cost caused by high energy requirement (e.g., temperatures above 800°C are usually essential). Figure 1 shows a simplified process flow diagram for producing ethylene using steam cracking (e.g., naphtha or gas oil). Technically, natural gas first enters the gas separation unit to obtain different types of hydrocarbons (e.g., ethane). Next, feedstock gases are thermally cracked (up to 1100°C) in the presence of steam using pyrolysis furnaces to produce lighter hydrocarbons. At this stage, two chemical reactions occur: splitting of C-H and C-C single bonds. The products obtained in this step depend on feedstock feed composition, hydrocarbon-to-steam ratio, and cracking temperature. After reaching the cracking temperature, the hot gas mixture is quickly quenched in the transfer line exchangers (TLE) to 550-650 °C. TLE will be subsequently cooled down to 300°C to avoid degradation by secondary reactions and generate high-pressure steam for driving compressors.

FLUID CATALYTIC CRACKING

Fluid catalytic cracking (FCC) (Gary, Handwerk, & Kaiser, 2007) is important for converting heavy feedstock to light olefinic hydrocarbons such as gasoline and distillate fuels. Historically, FCC was discovered in 1942, and it is still the major secondary conversion process in petroleum refinery. It uses an alumina catalyst in a fluidized bed reactor, which is usually prepared by crushing pumice stones; these stones mainly contain small particles of aluminum oxide and silica. FCC uses solid materials and heat to break large molecules of gas oil into smaller molecules, which form gasoline, distillate, and other higher-value products such as butane and propane. These materials are further treated to separate them into fuel gas, propane, propylene, butane, and butane, which can be used for different applications. Technically, three basic steps are involved in catalytic cracking. First step involves the reaction between the feedstock and catalyst and cracking into different hydrocarbons. Second step involves catalyst regeneration by burning off coke and being recirculated into the reactor. Third step involves the separation of the cracked hydrocarbon stream into several products such as gasoline. Such products usually have a very low concentration of sulfur and other contaminants, which need to be desulfurized and reformed before they can be blended into finished products (e.g., gasoline). On the other hand, hydrocracking is the main source for the production of jet fuel, kerosene, and diesel. Such products contain a very low sulfur concentration. In addition, a low concentration of ethylene obtained from FFC can be also used as a feedstock for some particular processes such as the production of ethylbenzene (EB) and metathesis. Further, FCC product streams in the C_4-C_8 range can also be utilized to produce ethylene, propylene, and butylene.

OXIDATIVE DEHYDROGENATION OF ETHANE (ODHE)

Light olefins (e.g., ethylene) are mainly obtained by steam cracking hydrocarbon feedstocks (e.g. Matar & Hatch, 2001) and catalytic dehydrogenation of light alkanes (e.g., CA Patent 2701089 A1, 2011). These endothermic processes are typically conducted at high temperatures (usually >800 °C), and thus become energy intensive. Moreover, significant side reactions and catalyst deactivation are possible because of coke deposition. However, an extensive research activity is being undertaken to overcome some of these drawbacks as to make the reaction exothermic and to operate it at relatively lower temperatures and at atmospheric pressure. In view of such activities, oxidative dehydrogenation of ethane (ODHE) is found to be the most attractive alternative for the production of ethylene, which does not need internal heat input. In the last two decades, it has been reported that ODHE has been studied using several catalysts systems, including alkali, alkaline-earth, rare-earth-metal oxides, and transition metal oxides (Solsona, López Nieto, Concepción, Dejoz, Ivars, & Vázquez, 2011; Frank, Morassutto, Schomäcker, Schlögl, & Su, 2010). Heracleous and Lemonido obtained an ethylene yield of 46% with high selectivity (ca. 90%) using a NiBbOx mixed-oxide catalyst (Heracleous, & Lemonidou, 2005). In addition, some promising catalysts based on multicomponent MoVTeNbO mixed-oxide catalysts have been reported and exhibited good catalytic activity towards ODHE with high selectivity to ethylene (≥80%) at high ethane conversion (>80%) (López Nieto, Botella, Vázquez, & Dejoz, 2002; Botella, García-González, Dejoz, López Nieto, Vázquez, & González-Calbett, 2004). It has also been reported elsewhere (Gaab, Machli, Find, Grasselli, & Lercher, 2003) that ethylene with a yield of almost 75% has been achieved using molten alkali chlorides such as LiCl, KCl, NaCl, Li-K-Cl, Li-Na-Cl, Li-Sr-Cl, and Li-Ba-Cl, which are supported by Dy_2O_3/MgO. However, the use of these chloride-containing materials can generate different Cl-containing byproducts, which are considered to be environmentally harmful. Such harmful by-products are a big concern on the industrial scale since their separation requires additional costs, thereby making such a process expensive and unattractive. Interestingly, it has been reported that the melting point of chloride inversely controls the selectivity toward ethane. As the temperature increases, the selectivity has been found to linearly increase below the melting point. However, a further increase in the reaction temperature above the melting point does not affect the selectivity, which remains constant. Among these catalysts, transition-metal oxide systems (e.g., Mo-based catalysts) are gaining importance caused by their selective nature. Tsilomelekis and Boghosian (Christodoulakis, Heracleous, Lemonidou, & Boghosian, 2006) have reported broad remarks after an operando Raman study of molecular structure and reactivity of molybdenum(VI) oxide supported on TiO_2 (anatase) for ODHE. The structural and catalytic properties of MoO_3 catalysts supported on ZrO_2, Al_2O_3, TiO_2, and SiO_2 with Mo surface densities in the range of 0.5-18.5 Mo/nm^2 have also been investigated for ODHE using *in situ* Raman spectroscopy and catalytic activity measurements at 400-540 °C (Tsilomelekis, G., Christodoulakis, A., & Boghosian, S., (2007). In addition, different research groups have also investigated the structural and catalytic properties of Al-supported Mo systems by *in situ* Raman spectroscopy at 400-550 °C (Christodoulakis, Heracleous, Lemonidou, & Boghosian, 2006). They have suggested that the anchoring of Mo-O support bonds is significant for ODH reactivity, consequently suggesting that both the support and Mo species play a key role on the catalytic performance.

OTHER ETHYLENE PRODUCTION PROCESSES

In addition to above-described ethylene production technology, new processes for ethylene production have also been reported. Such processes include MTO, the Fischer-Tropsch route, and off-gas recovery. These routes are also briefly described below.

METHANOL TO OLEFINS ROUTE

The production of hydrocarbons (e.g. ethylene) from methanol (Figure 3) was discovered at Mobil Oil (now ExxonMobil) in 1977. This process utilizes the ZSM-5 catalyst to convert methanol to products such as olefins and gasoline (Chang, et al., 1983; Yurchak, et al.,1988; Chang, et al. 1992). However, the selectivity of methanol to ethylene was low, and the process was not attractive for industrial scale application. During the 1980s, UOP and Norsk Hydro found a new type of material based on a highly selective silicoaluminophosphate (SAPO-34) catalyst using a fluidized bed reactor. The discovered SAPO-34 catalyst was observed to be a breakthrough technology for the production of ethylene from methanol caused by its unique pore size, acidity, and geometry. It is reported elsewhere that (Dahl & Kolboe, 1993; Dahl & Kolboe, 1994) according to the hydrocarbon pool mechanism, the reaction steps included the methylation of "hydrocarbon pool intermediates," confined in the cages of SAPO-34, by methanol and subsequent elimination of ethene, propene, and butenes from the intermediates. Other investigations (Song, Haw, Nicholas, & Heneghan, 2000; Arstad, & Kolboe, S., 2001) have currently shown that polymethylbenzenes comprises the largest part of the materials retained in the catalyst and that hexamethylbenzene is found to be the most active species for the production of olefin from methanol. Overall, the MTO process has some limitations, namely, the applied catalyst needs to be normally

Figure 3. Reaction figure for the MTO process

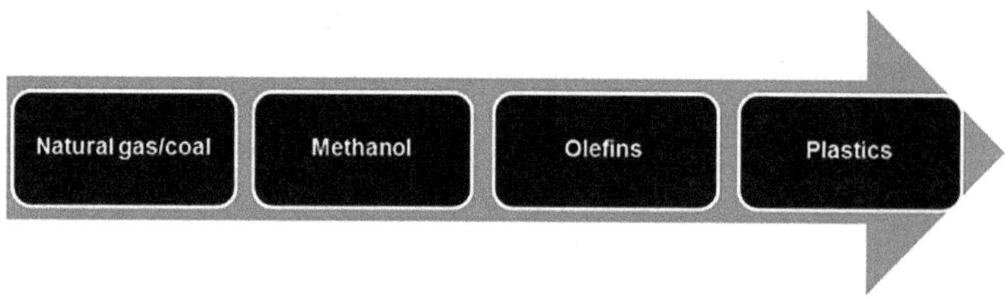

regenerated because of significant coke formation, and strong temperature control is needed because of the highly exothermic nature of the process. Therefore, it has been suggested elsewhere (Chang, et. al. 1992) that two fluidized bed reactors, one for reaction and one for catalyst regeneration, are highly recommended to overcome these limitations.

FISCHER-TROPSCH ROUTE

The (Synthol) Fischer–Tropsch process (Schulz, et al., 1999) converts coal catalytically to synthetic fuel as well as a wide range of hydrocarbons and oxygenates:

Coal $\rightarrow$ CO/H2 $\rightarrow$ hydrocarbons + alcohols

This method has mostly been used to produce liquid fuels such as gasoline, diesel, and jet fuel. This process was developed by Sasol, South Africa, during the fuel crisis in the 1970s. However, it has never been commercialized because of several reasons such as high cost of investment for building a production line and petrochemicals. The Fischer-Tropsch route can be modified for the production of light olefins using different types of catalysts. However, low selectivity toward ethylene, high methane formation, and insufficient catalyst activity and stability are challenging issues to utilize this process on an industrial scale.

TYPE OF CATALYSTS USED

The first publications on this topic of oxidative dehydrogenation of ethane (ODHE) to ethylene began to appear in early 70s. Since then, various catalysts have been used to achieve higher yields of ethylene as well as improve the long-term stability of the catalysts for using them commercially. In the early 70s, H_2S promoted ODH of ethane using Cd/Alundum catalysts were applied to investigate the influence of process conditions (Gaspar, & Pasternak, 1971). However such a composition has never been utilized again in recent times probably due to the toxic nature of cadmium. In the middle of the 80s, Mo-V-Nb-Sb-O catalysts were investigated, which yielded a 22-57% conversion of ethane and 70-80% selectivity of ethylene at 350-425°C [McCain, 1986]. In the 90s, this topic gained significant interest from the scientific community, followed by several publications using various types of catalyst systems. Gradually, Sr-Ce-Yb-O [Velle, Andersen, & Jens, 1990], Mo-V-Nb-O [Burch, & Swarnakar, 1991], $(VO)_2P_2O_7$ [Merzouki, Taouk, Monceaux, Bordes, & Courtine, 1992], SrO/La_2O_3 [Choudhary, Uphade, & Mulla, 1995], LaF_3-CeO_2 [Luo, & Wan, 1997], BaF_2-promoted Sm_2O_3-LaF_3 [Luo, Zhou, Chao, & Wan, 1997], Li-Mn-W-O [Liu, Zhang, Liu, Xue, & Li, 1998], Mo_2C/SiO_2 [Solymosi, & Nemeth, 1999],Cr_2O_3/SiO_2 [Wang, Murata, Hayakawa, Hamakawa, & Suzuki, 1999], NiO/Al_2O_3 [Zhang, Yu, Gong, Jiang, & Xie, 2000], and Cr/H-ZSM-5 [Mimura, Takahara, Inaba, Okamoto, & Murata, 2002] catalysts were used. Nevertheless, to the best of our knowledge, no such efforts have been successful in commercializing this ethane ODH technology.

Literature survey reveals that ODHE can be catalyzed by a wide variety of catalysts (Solsona, López Nieto, Concepción, Dejoz, Ivars, & Vázquez, 2011; Håkonsen, Walmsley, & Holmen, 2010). Konya et al. reasonably classified them into three groups (Konya, Katou, Murayama, Ishikawa, Sadakana, Buttrey,

& Ueda, 2013),: First group represents MgO- or CaO-based mixed metal oxides and rare earth metal oxides. These catalysts need high reaction temperatures (>500 up to 650°C) as they does not exhibit high intrinsic oxidation ability. Interestingly, chloride-mediated ODHE by using oxyhalides (-chlorides) was intensively studied and found to be very active materials. Second group represents V-containing solids, e.g., MoV oxides and Te, Nb, Sb-doped MoV oxides. These solids can catalyze the ODHE at 350-400°C. Third group contains transition metal oxides (group 8 elements) being active at lower temperatures (c.a. 400°C). Prominent members belong to NiNb oxides. Gärtner et al. presented a similar arrangement (Gärtner, van Veen, & Lercher, (2013). They grouped the catalysts in (i) transition-metal oxides containing all vanadium, molybdenum, and nickel oxide systems, (ii) rare-earth-metal oxides, (ii) supported alkali oxides, and (iv) supported alkali chlorides.

Recently, the research attention has been focussed not only on developing catalysts but also looking for novel approaches, reactor concepts/designs (e.g., microstructured reactors, membrane reactors, and monoliths), and change of oxidants to realize such results on commercial scale in the coming future. As mentioned above, this ODH approach has also been extended toward the application of microstructured reactors; e.g., a Pt-Sn catalyst (2.4:1 atomic ratio) coated on a coupon surface was used in microstructured reactors (Yang, Yuschak, Mazanec, Tonkovich, & Perry, 2008, Watson, Daly, Mazanec, & Yang, 2005) at 950°C and gave a considerably high ethane conversion (>75%) and high selectivity of ethylene (84%). Furthermore, different other catalysts were also used in membrane reactors to achieve higher ethylene selectivity. Some examples used in membrane reactors have been described below. Hamel et al. (Hamel, Wolff, & Seidel-Morgenstern, 2011) used three differently doped $VO_x/\gamma\text{-}Al_2O_3$ catalysts for ODHE and reported the compatibility of transport and reaction in membrane reactors. They stated that metal membranes possess favourable mechanical stability, relatively low production costs, and the possibility to control mass transfer if the rate of reaction and mass transfer in the membrane is compatible. Rodriguez et al. used a Ni-Nb-O mixed-oxide catalyst (Rodriguez et al., 2010, Rodriguez et al., 2009) in a multitubular packed-bed membrane reactor for ODHE and reported that the introduction of a membrane leads to lower oxygen partial pressure inside the catalyst tube, which results in an improved ethylene selectivity and high effective heat transfer area per unit volume. They also suggested that the multi-tubular membrane reactor significantly improves ethylene production per tube and gives milder temperature profiles than in a conventional wall-cooled fixed-bed reactor. Efforts were also made to overcome equilibrium conversion by using a multistep permeating Ba-Co-Fe-Zr-O perovskite hollow-fiber membrane reactor for ODHE with successive parts of permeable and passivated surface segments (Czuprat, Werth, Schirrmeister, Schiestel, & Caro, 2009). This study reports that this geometry allows for controlled oxygen insertion into the reactor over an extended period to overcome equilibrium conversion and also provides lower oxygen concentration to improve ethylene selectivity. Furthermore, Chalakov et al. used electrochemical packed-bed membrane reactors for ODHE using $VO_x/\gamma\text{-}Al_2O_3$ catalyst in the temperature range of 500–620°C (Chalakov, Rihko-Struckmann, Munder, & Sundmacher, 2009, Chalakov, Rihko-Struckmann, Munder, & Sundmacher, 2007). Their model includes the kinetics of oxygen reduction on the cathode, transfer of oxygen ions from the cathode to the anode, and formation of gaseous dioxygen on the anode. Besides membrane reactors, some monoliths were also used for ODHE. For instance, NiNbO/AISI 430 ferritic stainless steel monoliths were prepared by dip coating and tested towards ODHE (Santander, Lopez, Tonetto, & Pedernera, 2014). These investigations indicated that the structured catalysts are active, highly selective, and even show a constant catalytic performance for 170 hours on stream at 400°C. Another such approach is the application of Cr-based monolithic catalysts (e.g., Cr/SBA-15/Al_2O_3/FeCrAl), which gave 66.5% ethane conversion and extremely high ethylene selectivity (99.5%) at 750°C (Shi, Ji, Wang, & Li, 2008). The authors claimed that the hexagonal mesoporous structure of SBA-15 is still present after 1130 hours on stream and the pore walls of SBA-15 prevent the aggregation of Cr species. They also proposed that Cr^{6+} species are responsible for higher activity, and the redox cycle operates between Cr^{6+} and Cr^{3+} species. However, interactions among Cr, SBA-15, and the Al_2O_3/FeCrAl support modify the redox properties of the Cr/SBA-15/Al_2O_3/FeCrAl catalysts.

Apart from the conventional approaches, attempts were also made by various researchers to change the nature of the oxidant from O_2 to CO_2 (sometimes co-feeding of CO_2 as a soft oxidant) or N_2O in ODHE to improve the selectivity of the target product ethylene. The application of CO_2, a global warming gas, in the ODH of ethane is expected to offer some advantages of moderating the heat inside the reactor and decreasing coke formation. In addition, the replacement of O_2 by CO_2 as an oxidant can also give H_2 (a valuable byproduct) in the product stream instead of H_2O (a waste byproduct) in the conventional ODH of ethane, where O_2 is used as an oxidant. The presence of CO_2 in the feed mixture can also alleviate thermodynamic equilibrium limitations (Mimura, Takahara, Inaba, Okamoto, & Murata, 2002). Mimura et al. stated that the higher oxidation state of chromium (Cr^{+6}) is key for the improved activity in Cr-based catalysts and the presence of CO_2 in the feed maintains the activity of the catalyst by removing coke from the catalyst surface (Mimura, Takahara, Inaba, Okamoto, & Murata, 2002). Significant improvements in olefin selectivity were also observed on replacing O_2 with N_2O, caused by the suppression of CO_x formation over highly dispersed VO_x species supported on an MCM-41 catalyst (Ovsitser, & Kondratenko, 2009).

Recently, V_2O_5/Al_2O_3 (with enhanced pentacoordinated Al sites) (Qiao et al. 2014), modified NiO (Zhu et al. 2014), V_2O_5/Nb_2O_5 (Qiao et al. 2013a), heteropoly acids (Sri Hari Kumar et al. 2013), Ni-Me-O (Me=Li, Mg, Al, Ga, Ti, Nb, Ta) (Heracleous, & Lemonidou, 2010), Fe-Cr/ZrO_2 (Deng, Li, S., Li, H., & Zhang, 2009), and Co-$BaCO_3$ [Zhang, Ye, Xu, & He, 2007] catalysts have also been used. Moreover, Cr-Mg-Al and Cr-Mg mixed-oxide catalysts prepared from layered double hydroxides have been reported sporadically (Tsyganok, Green, Giorgi, & Sayari, 2007). Zeolite-based catalysts (e.g., Cr/H-ZSM-5 at 650°C) (Mimura, Takahara, Inaba, Okamoto, & Murata, 2002) and Cr/TS-1 (Zhao, & Wang, 2006) have also been rarely used and have demonstrated over 90% ethylene selectivity. Among various catalysts, V_2O_5- and MoO_3-loaded catalysts have been widely used because of their unique potential and active/selective nature toward ethylene; these materials generally operate well via the redox cycle (Mars and Krevelen mechanism) and acid–base properties (Argyle, Chen, Bell, & Iglesia, 2002a).

On the other hand, heteropolyoxometallates have also been successfully used for different partial oxidation reactions. The interest in such materials is related to both their acidic and redox properties, which could be controlled either by substituting the constituent elements and/or by partially exchanging the protons by redox metal cations. Many articles that demonstrate the catalytic potentiality of such compounds for the partial oxidation of alkanes have been published (Mizuno, Suh, Han, & Kudo, 1996). Niobium- and pyridine-exchanged salts of phosphomolybdic acid ($NbPMo_{12}Pyr$) and phosphovanadomolybdic acid, $NbPMo_{12}Pyr$ and $NbPMo_{11}VPyr$, respectively, have been studied as precursors for synthesizing catalytic materials for the ODHE to ethylene (Galownia, Wight, Blanc, Labinger, & Davis, 2005). It has been shown that the catalytic performance and thermal stability are enhanced by incorporating V^{5+} in the $H_3PMo_{12}O_{40}$ Keggin structure and by partial exchanging protons with Cs^+ cations. Redox elements are known to play an important role in the redox processes as reservoirs for electrons and active sites for the activation of hydrocarbons and molecular oxygen (Lin, 2001). Some rare examples are the usage of foams, e.g., Pt/SiO_2 foams (Gudgila, & Leclerc, 2011) and co-feeding of H_2 to the reactant feed in the ODH of ethane, and both approaches claim an improved ethylene selectivity over Pt catalysts (e.g. Hkonsen, & Holmen, 2007).

Among several catalyst systems, MoVTeNbO seems to be one of the best catalyst compositions, which gave an ethane conversion of 88% with an ethylene selectivity of over 80% at relatively low temperature (400°C) (Nieto, Botella, Vázquez, & Dejoz, 2002). However, catalyst synthesis is complex, and its composition is also guarded by a good number of patents caused by its unique potential applications even for other oxidation and ammoxidation reactions. Very recently, the kinetic modelling of ODHE using MoVTeNbO catalysts was investigated by the Araiza research group (Che-Galicia, Quintana-Solorzano, Ruiz-Martinez, Valente, & Castillo-Araiza, 2014). Ethylene selectivity of 76-96% for an ethane conversion of 17-85% was reported at 440°C. The authors claimed that ethylene formation requires the lowest activation energy, while the total oxidation of ethane needs the highest activation energy; reaction rates including that of catalyst oxidation are weakly affected by changes in the oxygen partial pressure, which explains the high ethylene selectivity over the MoVTeNbO catalyst. Similarly, Lercher et al. (Gaertner, van Veen, & Lercher, 2013) reported that it is possible to obtain ethylene yields up to 80% on MgO-supported mixed alkali chloride catalysts. However, the use of such chloride-containing materials can also produce different Cl-containing by-products in the product stream, whose separation on a commercial scale makes the process more expensive and complex.

Although several articles have appeared in the literature on the ODH of ethane, deactivation is still one of the major constraints. Thus, it is imperative to develop highly active, selective, and economically attractive catalysts with improved long-term stability. Nonetheless, there is still ample scope for further improvements in the performance, economic viability, as well as understanding the nature of active sites.

STRUCTURE-ACTIVITY RELATIONSHIPS/MECHANISTIC INVESTIGATIONS

Very recently, an extensive and very informative summary from the Lercher's group discussed common principles and mechanistic aspects in ODHE (Gärtner, van Veen, & Lercher, 2013). It combines the groups of catalysts for ODHE, mechanistic considerations, thoughts on advanced reactor concepts, and main results with respect to ethane conversion and ethylene selectivity.

As shown in Figure 4, ODHE mainly involves three partial reactions: selective reaction of ethane to ethylene, i.e., the formation of the desired olefin (R1), the total oxidation of ethane to carbon dioxide (CO_2, R2) running in parallel to R1, and the consecutive oxidation of ethylene to CO_2 (R3). This picture

Figure 4. Main reaction paths running in the ODHE reaction

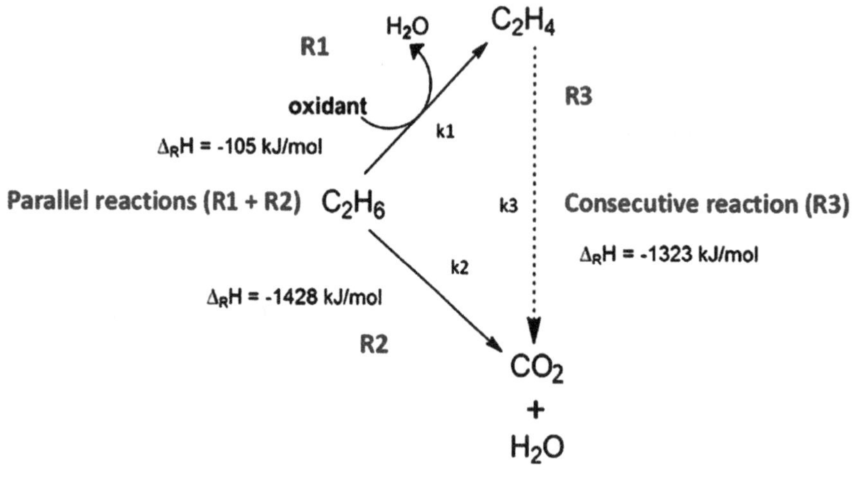

can be completed by two more general reactions: intermediate formation of CO during the nonselective reactions R2 and R3 and its further oxidation to CO_2. Klose et al. satisfactorily summarized their investigations on the ODHE mechanism over alumina-supported vanadia catalysts (Klose, Joshi, Hamel, & Seidel-Morgenstern, 2004). For the network of these five main reactions, kinetic parameters were estimated. The selective formation of ethylene from ethane (R1) was described by the Mars–van Krevelen mechanism, the total oxidative reactions by Langmuir–Hinshelwood equations (Klose, Joshi, Hamel, & Seidel-Morgenstern, 2004).

However, the formation of significant amounts of thermodynamically favoured but undesired products (e.g., CO_x and H_2O) cannot be excluded; hence, ethylene yield obtained using most catalysts is still not very satisfactory for industrial scale processing. Therefore, the key issue of ethylene production by ODHE is linked to catalytic system development, which is capable of selectively converting ethane to ethylene, and meanwhile preventing the total oxidation of ethane and consecutive ethylene oxidation to CO_2 at reasonable conversions. Therefore, numerous studies over the past 25 years dealt with trials to uncover the mechanistic problems depending on catalytic systems used and reaction parameters such as feed, oxygen ratio, and reaction temperature to understand the overall reaction figure and to discover the catalytic sites for single reactions with respect to different catalytically active solids (e.g., Cavani, Ballarini, & Cericola, 2007; Savova, Loridant, Filkova, Millet, 2010; Mamedov, & Cortés Corberán, 1995; Grabowski et. al. 2006; Solsona, López Nieto, Concepción, Dejoz, Ivars, & Vázquez, 2011; Solsona, Concepción, Hcrnández, Demicol, López Nieto, 2012).

In the previous chapter, the catalysts used primarily were introduced. In contrast, mechanistic features in this chapter will be discussed regarding i) catalysts containing either non-redox-active metals or ii) redox-active metal oxides. Besides this classification, mechanistic insights can also be grouped by the homolytic- or heterolytic-initiated cleavage of the first C-H bond, leading to a single-electron process to the radical chain mechanism or in a paired-electron process to redox cycles. Often, mechanistic research is also simultaneously conducted for ethane and propane ODH.

The first group contains rare-earth-metal oxides, alkaline-doped alkaline-earth-metal oxide-based catalysts, and supported alkali chlorides or oxychlorides (e.g., $SrNdO_x$, $SrLaNdO_x$, Li-Dy-doped MgO, Li/MgO, LiCl/ZrON, or SmOF) showing very good ethylene yields of 50-75% at elevated temperatures up to 700°C (Gaertner, van Veen, & Lercher, 2013). For alkaline (Li)-doped solids, Li appears as an essential component of these materials to create active sites. Kinetic and mechanistic effects were enlightened by Swaan et al. (Swaan, Toebes, Seshan, van Ommen, Ross, 1992). Measurements at 600°C revealed that the Li/MgO catalyst produces C_2H_4, CO_2, CO, and H_2 by parallel reactions, whereas the Na-promoted catalyst forms only C_2H_4 and CO_2 again by parallel reactions. The groups of Lefferts and Lercher studied alkane ODH mechanism in detail (Leveles, Seshan, Lercher, & Lefferts, 2003). Alkane activation might occur on the $[Li^+O^-]$ site acting as an initiator. Consequently, the increase of the number of such sites may increase the reaction rate. Alkane C-H bond scission forms $[Li^+OH^-]$ and an alkyl radical. Later, the radical undergoes chain propagation in the gas phase to form the desired olefin (Leveles, Seshan, Lercher, & Lefferts, 2003, Morales, & Lunsford, 1989).

Note that there is a significant influence of the oxygen concentration in the reaction mixture because it considerably affects the radical gas-phase chemistry. A similar mechanism for the ODH of propane and n-butane has also been elucidated for a Co-nitride/Al_2O_3 catalyst (Kaliya, Kogan, Froumin, & Herskowitz, 2002).

Further catalytic systems exhibiting excellent performance are those based on rare-earth-metal oxides and oxychlorides, often in combination with alkali or alkaline-earth-metal oxides. The activation of the first C-H bond using rare-earth-metal oxides as catalysts might occur in a single-electron process, leading to ethyl radicals. These catalysts are active between 600 and 900°C, and the reaction mechanism might include the relevant contribution of homogeneous reactions (Cavani, & Trifirò, 1995).

Various studies on the recognition of the ODHE mechanism using Pt-loaded honeycomb catalyst short-contact time reactors at 800-1000°C were conducted by Schmidt's group (e.g., Huff, & Schmidt, 1993). At the beginning, ODHE was believed to occur via pure heterogeneous catalysis on the basis of various observations made from the partial oxidation of methane (Torniainen, Chu, & Schmidt, 1994). Consequently, a heterogeneous model was developed to describe ODHE over a Pt catalyst in more than 24 steps of adsorption, surface reaction, and desorption (Huff, & Schmidt, 1996). Later experiments revealed a significant contribution of homogeneous reactions occurring at high temperatures. Recently, the same group presented an improved kinetic model for ODHE on Pt- and PtSn-coated monoliths combining the gas-phase mechanism (Donsı`, Williams, & Schmidt, 2005). Experimental data obtained over Pt foam are in good agreement with simulations.

The second group of catalysts contain vanadia, mixed molybdate, or nickel oxide species as active sites. However, the most prominent catalysts always contain vanadium oxides (Gaertner, van Veen, & Lercher, 2013). Surface metal-oxygen species, -OH groups, and oxygen vacancies are the most abundant reactive intermediates during ODH on active VO_x domains (Gaertner, van Veen, & Lercher, 2013, Argyle, Chen, Bell, & Iglesia, 2002b). Most of these vanadium-oxide-containing catalysts reveal kinetically relevant steps involving the dissociation of C-H bonds (e.g., methylenic C atom in propane) by interaction with vanadyl sites and supply of nucleophilic oxygen according to the Mars–van Krevelen redox mechanism (Mars, & van Krevelen, 1954) by using lattice oxygen (O^{2-}) for C-H bond activation. The resulting alkyl species desorb as the olefin from the solid and an -OH group remains on the surface. Subsequently, the -OH groups recombine with neighbouring -OH groups to form water and reduce V sites. The latter are re-oxidized through the oxidic lattice; defect sites are compensated by its interaction with gas-phase oxygen. In contrast, CO_x formation seems to be caused by the adsorbed electrophilic oxygen species, i.e. nonselective reactions might occur via the Langmuir-Hinshelwood mechanism (Klose, Joshi, Hamel, & Seidel-Morgenstern, 2004).

López-Nieto's group studied for the first time the effects of reaction temperature, oxygen and ethane concentrations, and feed space velocity on the catalytic performance of a MoVTeNb mixed-oxide catalyst used for ODHE. Their kinetic modelling indicates that CO_2 is mainly generated from ethane, while CO is formed from both ethane and ethylene (Valente, Quintana-Solórzano, Armendáriz-Herrera,, Barragán-Rodríguez, López-Nieto, 2014). In analogy to vanadia-based catalysts, doped MoO_x are also active catalysts, and the molybdenyl group initiates the first step of C-H bond activation. Another interesting catalyst in this group is a microporous $MoVO_x$ solid recently introduced by the group of Ueda (Konya, Katou, Murayama, Ishikawa, Sadakana, Buttrey, & Ueda, 2013) also acting in ODHE according to the Mars-van Krevelen mechanism.

A further redox-active metal oxide group is based on nickel oxide promoted with a wide variety of dopants, in particular Nb (up to 15 wt%) seems to be very useful. The catalysts are very active at low reaction temperatures (300-400°C) with ethylene yields of ca. 50% (Heracleous, & Lemonidou, 2006). ODHE over these catalysts follows the Mars-van Krevelen mechanism. Pure NiO reveals electrophilic oxygen species and minor nucleophilic O_2^- (Iwamoto, Yoda, Egashira, & Seiyama, 1976). Nb addition suppresses the appearance of electrophilic oxygen, thereby increasing ethylene selectivity.

OWN RESULTS

For the past few years, we have been engaged in the oxidative dehydrogenation of ethane to ethylene process and applied a variety of catalyst compositions. The results achieved from those attempts are briefly described below. In the beginning, we have applied Nb_2O_5 as a novel support for vanadia catalysts. In this case, we have prepared V_2O_5/Nb_2O5 catalysts by wet impregnation method (excess solvent) and varied the vanadia contents in the range from 5 to 20 wt%. Those catalysts were characterized by various available techniques and tested their potentiality towards the oxidative dehydrogenation of ethane in a fixed bed quartz reactor in the temperature range from 500 to 600 °C (Qiao et al. 2013a). UV-vis diffuse reflectance spectroscopy revealed the nature of VOx species formed. XPS analysis indicated a clear enrichment of vanadium in the near-surface-region. As expected (Figure 5), pure Nb_2O_5 exhibited poor performance (Conversion (X) = 12% and Yield (Y) = 4% at 600 °C), while the pure V_2O_5 also displayed relatively low conversion of ethane (22% at 600 °C) and low selectivity (S) of ethylene (27%).

In subsequent studies, we have applied vanadia based catalysts further and explored the influence of nature of alumina (as a support material) on the catalytic properties in ODHE reaction. A series of V_2O_5/Al_2O_3 catalysts with varying the source of Al_2O_3 supports (at fixed V_2O_5 content (10 wt %)) were prepared and tested towards the ODHE again in the temperature range from 500 to 600 °C. The nature of alumina support applied had a remarkable influence on the catalytic properties. Among different aluminas applied, the high surface area γ-Al_2O_3 (200-300 m^2/g) supported vanadia catalyst showed the best performance. Such improved performance of this solid can be attributed to the high dispersion of VO_x species on the support surface., However, other aluminas used exhibited higher ethane conversions under the same reaction conditions but low ethylene selectivity due to the existence of larger V_2O_5 crystallites in those solids. Quite interestingly, a structurally disordered γ-Al_2O_3 (ca. 6 m^2/g) containing a high proportion

Figure 5. Effect of V_2O_5 loading on the conversion of ethane at varying temperatures (Reaction conditions: mole ratios of $C_2H_6:O_2:N_2=1:0.7:2.6$, GHSV=3250 h^{-1}, τ=1.1 s, catalyst amount: 1 g) (Qiao et al. 2013a)

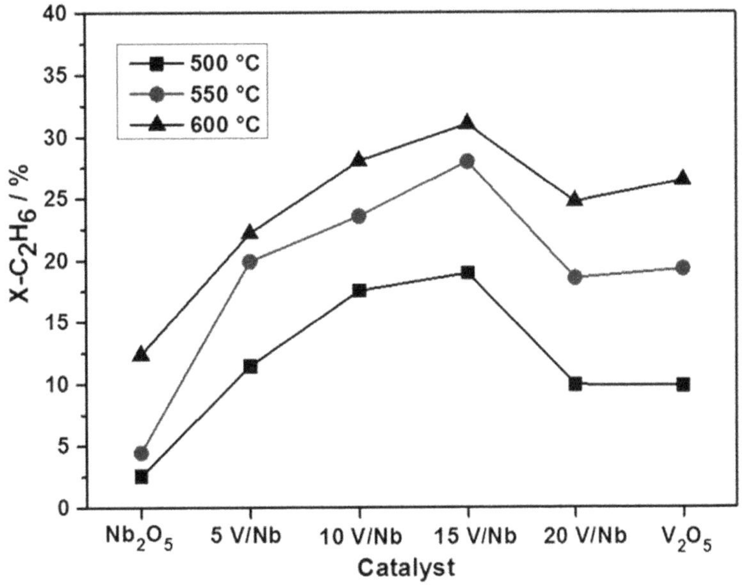

Figure 6. Effect of V_2O_5 loading on the yield of ethylene at varying temperatures (Reaction conditions: mole ratios of C_2H_6:O_2:N_2=1:0.7:2.6, GHSV=3250 h^{-1}, τ=1.1 s, catalyst amount: 1 g) (Qiao et al. 2013a)

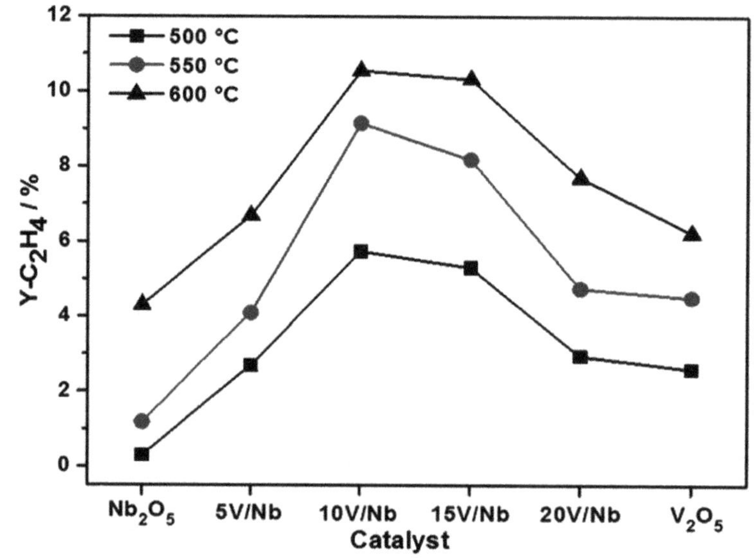

Figure 7.Effect of temperature on the ethane conversion in ODHE over V/Al-x catalysts (Reaction conditions: C_2H_6/O_2/N_2, 1:0.7:2.7; GHSV, ~3250 - 3270 h−1; τ= 1.1 s; V/Al-1: high-energy ball milled γ-Al_2O_3 with high amount of penta-coordinated Al sites (BET-SA, 5.8 m^2/g); V/Al-2: α-Al_2O_3 (SASOL, Germany; BETSA, 5.2 m^2/g); V/Al-3: γ-Al_2O_3 (Engelhard Italiana, S.p.A.; BET-SA, 102 m^2/g); V/Al-4: γ-Al_2O_3 (CONDEA Chemie GmbH, Germany; BET-SA, 201 m^2/g); and V/Al-5: γ-Al_2O_3 (CONDEA Chemie GmbH, Germany; BET-SA, 294 m^2/g)). (Qiao et. al. 2014)

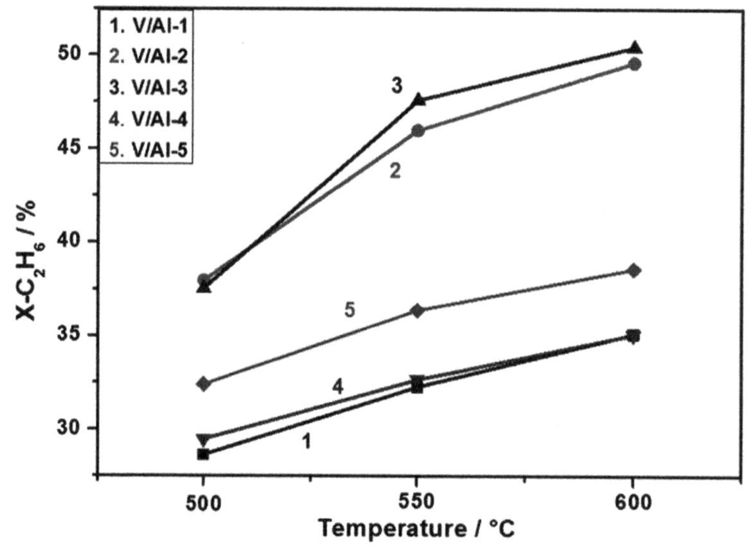

Figure 8. Effect of temperature on the ethylene selectivity in ODHE over V/Al-x catalysts (Reaction conditions applied and the sample denotation are the same as given in Figure 7) (Qiao et al. 2014)

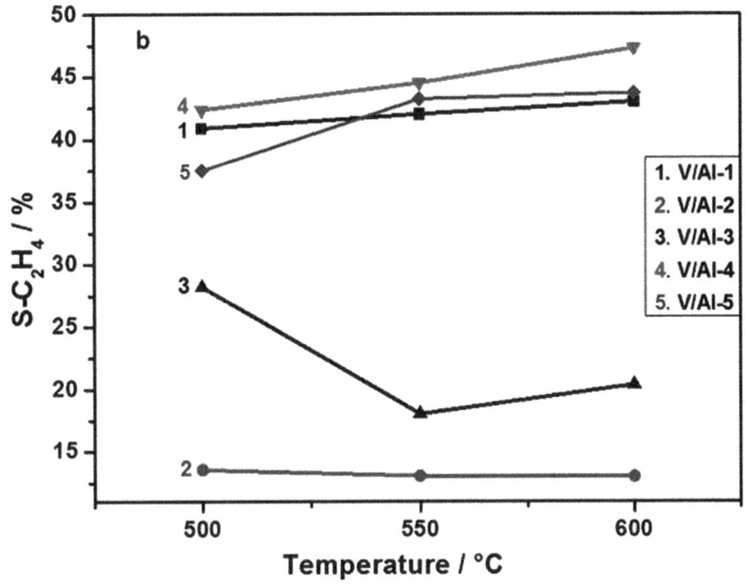

of penta-coordinated Al-sites were also included in this series, which has displayed a good performance in ODHE that is comparable to the high surface areas alumina sample. Such amazing performance of this structurally disorder alumina is attributed to the presence of the penta-coordinated Al surface sites acting as anchors for preferentially monomeric and oligomeric VO_x species. The impact of the nature of alumina on the catalytic performance is depicted in Figures. 7 and 8. Ultimately, it was concluded from these studies that the nature of alumina support exhibits remarkable influence on the nature of VOx species formed and their dispersion on the support surface. Moreover, the acidity characteristics, morphology and surface composition are also significantly altered by changing the nature of alumina.

In addition, we have explored the potentiality of Ni-Nb-O catalysts with an intention to reduce the reaction temperature so that the selectivity of target product ethylene can be enhanced. With such intention, Ni-Nb-M-O catalysts with different promoters (M = Cr, Mo, W) were applied in the ODH of ethane to ethylene and additionally the effect of CO_2 admixture on the catalytic performance was also investigated. The results showed in Figure 9 compares the results in absence and presence of CO_2 on the catalytic performance of NI-Nb-M-O catalysts. It is obvious that the CO_2-admixture to the reactant feed caused a slight decrease in the conversion of ethane but considerably improved the selectivity of ethylene. In case of the best catalyst (i.e. Ni-Nb-Cr-O), the CO_2 admixture improved the selectivity from over 60% to ca. 85% while the conversion of ethane is slightly reduced from 25% to 20%. Similar such tendency is also observed with the other two promoters. Among the three promoters applied, Cr displayed superior performance compared to other two components.

Besides the above mentioned catalysts, we have also investigated another class of catalysts such as different heteropoly acids based materials for ODHE reaction. One such example is the application of alumina supported molybdophosphoric acid (MPA) catalysts (Sri Hari Kumar et al. 2013). In this study, we describe the preparation, characterization and application of Mo based catalysts derived from MPA for ODHE reaction. It is evident from Figure 10 that the catalytic activity and selectivity were found

Figure 9. Effect of CO_2 admixture on the activity & selectivity of Ni-Nb-M-O catalysts (Qiao et al. 2012)

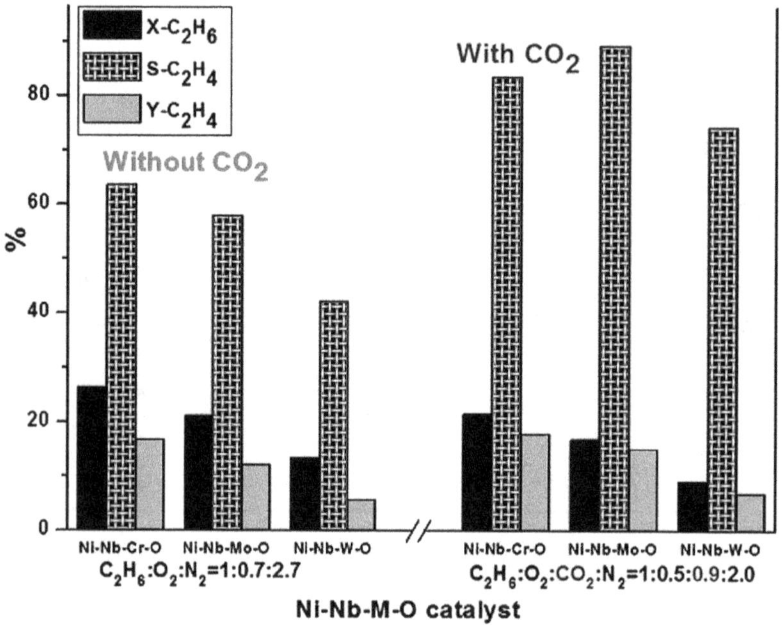

to depend on MPA loading. The conversion of ethane is varied in the range from 12 - 24% while the selectivity is varied from ca. 30-70%, respectively. The conversion of ethane was observed to increase progressively up to 20 wt % MPA loading and then decreased, while the selectivity of ethylene increased continuously (30 to 70%) at the expense of CO_2 selectivity (reduced from 65 to 30%). Among all load-

Figure 10. Effect of MPA loading on the conversion of ethane and selectivity of products: (Reaction conditions: mole ratio of $C_2H_6/O_2/He$ = 1/0.7/2.3; T = 550°C; catalyst weight of 1 g; GHSV = 3600 h^{-1}; τ = 1 s). (Sri Hari Kumar et al. 2013)

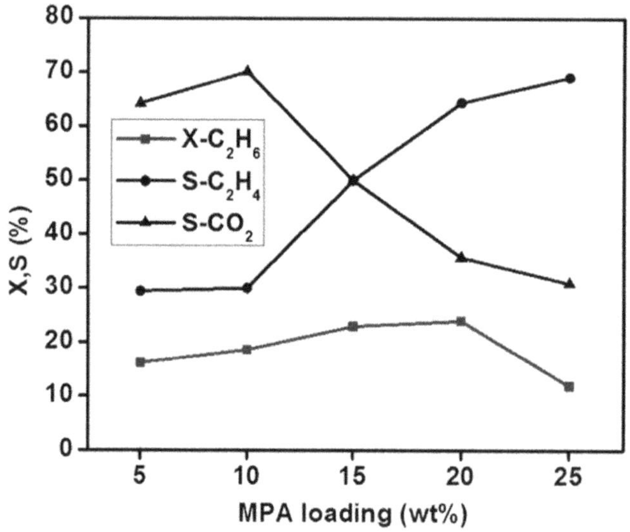

ings, the 20 wt % MPA/Al_2O_3 catalyst (at 550°C) displayed the superior activity and selectivity. We have observed through these investigations that the MPA was transformed into surface Mo oxides on Al_2O_3 when subjected to calcination at 600 °C. The results showed that MPA loading and the source of Mo precursor had a clear influence on the catalytic performance. The evaluation of the catalysts for ODHE at temperatures between 450 and 550°C revealed an ethane conversion of ~25% and ethylene selectivity of ca. 65% over 20 wt % MPA/Al_2O_3 catalyst (Figure 10). The transformation of MPA into finely dispersed Mo oxides on Al_2O_3 appeared to be responsible for this improved performance.

ECONOMIC ANALYSIS OF ETHYLENE

Global Market

Ethylene is an important building block for the chemical industry. In fact, it is one of the largest-volume petrochemical products with a diverse derivative portfolio. It is also a key raw material for the production of surfactants, plasticisers, detergents, ethylene glycols, EB, synthetic lubricants, among others. Among all products derived from ethylene, polyethylene dominates the world market, and this particular product alone constitutes approximately 60% of the total ethylene consumed. After polyethylene, EO and PVC are the two other derivatives of high commercial significance. In view of its enormous consumption and wide range applications, ethylene can certainly dominate the whole petrochemical industry to a large extent in the coming future. As expected, its applications will continue to increase daily. As the gap between demand and supply increases, scientists are trying to develop alternative processes for manufacturing ethylene. Ethane is currently the most preferred feedstock; it is the second major component of shale gas, making it a potential source of various chemicals including ethylene. Nearly 30% of ethylene produced by steam cracking is from ethane. It should be noted that the trading of ethylene across international borders is in the form of derivative chemicals (e.g., polyethylene, styrene, and ethylene glycol), not as a pure monomer.

The global economic crisis (between late 2007 and 2009) has demonstrated a clear impact both on the world GDP and ethylene demand/supply scenario. Even though an unprecedented decline of ethylene sales had been registered during the severe economic recession from 2008 to 2009, the world ethylene consumption has somehow managed to increase again. Interestingly, ethylene is the only petrochemical product that is less affected in comparison with all other petrochemicals during the economic crisis. This fact clearly demonstrates the high commercial importance of ethylene in the world market.

North America and Asia are the world's two largest ethylene producing regions. Recent investments (i.e., new capacity additions) in the Middle East boosted this region to fall into the group of the largest ethylene producing regions. In Asia, the major ethylene exporters are Japan, Malaysia, South Korea, and Singapore. In 2013, the worldwide production of ethylene accounted to ca. 135 Mio. t. Northeast Asia produces the highest (~25%) amount of ethylene and more or less the same amount comes from N. America caused by shale gas revolution. The global demand for ethylene as a whole is expected to grow by ~3% per year over a period of next 5 years. Ethane is the second major component of shale gas; hence, there will be abundant ethane supply in the US for producing ethylene by cracking. Besides these two regions, Middle East (20%) and W. Europe (15%), are the two other major ethylene producing regions.

In other words, more than 80% of world ethylene production comes from these four regions. Among them, the Middle East is the most suitable location for new capacity additions based on ethane feedstock. On the other hand, new mixed feed crackers are expected to face more challenges in the coming future owing to the shortage of ethylene supply in some countries. Nonetheless, no such problem exists in the regions where shale gas exploration continues.

Currently, the US is the largest ethylene exporter, while China is its largest importer worldwide. The fastest consumption is expected in the Asiatic region, particularly in China and India. Equally high consumption rates are also expected in Africa and the Middle East. In fact, the fastest growing regions concerning ethylene capacity build up would be the Middle East (e.g., Saudi Arabia and Iran) followed by China, where an ~8% growth rate is expected. Across the globe, the Asia Pacific region has been the largest consumer of ethylene, which accounts to more than 30% of worldwide production.

Historically and even now, the major feedstock worldwide for producing ethylene is naphtha. However, this situation may change in the coming future. In other words, ethylene producers must consider the issue of global feedstock competitiveness before investing in new capacities. In the early 2000, heavy liquid feedstocks were found to be more competitive for ethylene production than natural gas feedstock, while from 2008 onwards, the situation changed considerably, and then ethane became the most favoured feedstock for ethylene production. Consequently, the amount of ethylene produced from ethane feedstock has increased dramatically (i.e., from <40% to more than 70%). The low price of ethane (e.g., from shale gas) in the US further boosts the ethylene margins and also creates a profitable atmosphere for producers. As long as the shale gas reserves are present and shale gas exploration continues, the production of high amounts of natural gas is certain, thereby maintaining a high ethane supply; as a result, ethane prices are low compared to other feedstocks. Currently, ethylene produced from naphtha cracking is much more expensive (ca. 1000 US$ per metric ton) compared to that produced from ethane cracking (300-400 US$ per metric ton). This means that a change in the feedstock remarkably affects the price of ethylene. In view of extensive shale gas explorations in the US, ethane supply is high; hence, ethylene production from ethane cracking is much cheaper. Unlike the US, the ethylene prices in W. Europe will be relatively higher because of restricted shale gas exploration, and hence restricted supply of ethane. Generally, the ethylene price in Europe strongly depends on the changes in naphtha prices. In fact, the cost of ethylene produced from naphtha in Europe will be around 1000 US$ per metric ton, which is much higher (ca. 600–700 US$ more per one ton of ethylene) than that produced from ethane cracking in the US. In Asia, the majority of ethylene (>60%) is produced from naphtha feedstock, while the remaining is from ethane and liquid petroleum gas (LPG). Thus, the cost of ethylene in Asia is mainly related to the cost of naphtha.

Ethylene investments will be particularly high in the Asia Pacific region, Middle East, as well as in USA in the coming years. However, the investments for ethylene in Europe will dramatically decrease in the near future. Indeed, the scale of ethylene production is enormous where the size of a typical ethylene unit ranges from a production capacity of 700,000 to 1.7 million tons per annum. Some examples of such big units are ExxonMobil's Baytown, INEOS Americas, Shell's Deer Park (all in the US), Sinopec and PetroChina in China, SABIC, and the new Saudi Arabian units. The joint venture between NOVA and DOW (Canada) is one of the largest single units for ethylene production with an annual capacity of 1.27 Mio. t. in North America. The world's largest single new units in Saudi Arabia such as Petro-Rabigh (1.5 Mio. t.) and YANSAB (1.7 Mio. t.) for ethylene production are also known. Among all, SABIC has now become the largest ethylene producer in the world.

FUTURE DEVELOPMENTS

Even though the oxidative dehydrogenation of ethane (ODHE) to produce ethylene is still under developmental stage, it is indeed an attractive alternative process of the future and is more likely to compete with the existing cracker technologies of ethylene manufacture on commercial scale. However, the requirement of an air separation unit for getting pure oxygen could be possibly one of the drawbacks of this approach and hence it may be cost-wise prohibitive. To the best of our knowledge, the ODHE process has not been commercialised yet due to various reasons, for instance the development highly potential catalyst composition with high yields of ethylene as well as good mechanical and long-term stability is yet to be achieved. In addition, lot many things have yet to be worked out. Although ODHE has not been commercialized, considerable progress has been made recently, and a significant number of patents (Bos, Schoonebeek, Spies, & Verhaak, 2015; Kumar & Schoonebeek, 2015; Tarasov, Kustov, Kirichenko, & Kucherov, 2014; Han, Martenak, & Suib, 2010) and articles (e.g. Che-Galicia, Ruiz-Martínez López-Isunza & Castillo-Araiza, 2015) have been published concerning the process and catalyst development. Critical analysis of voluminous literature concerning ODHE reveals that the most active and selective catalyst systems that are generally used are based on multicomponent systems. Among several catalysts reported in the literature, vanadia-based catalysts form an important class of oxidation catalysts. However, pure vanadium-oxide-containing catalysts demonstrate lower selectivity. For solving this problem, more complicated and multicomponent oxide catalysts have also been recently proposed. Nevertheless, most of them are again based on vanadium, albeit in combination with various other elements such as Cr, Co, Sb, Sn, Mo, Bi, and alkali and alkali earth metals.

It seems that China will continue to dominate ethylene derivative imports in the near future despite several new investments and new ethylene capacities. Moreover, the new capacity additions also offer the Middle Eastern producers to increase their polyethylene and ethylene glycol exports into Asia. North American exports are also expected to considerably grow in the coming period caused by the availability of low cost ethane feedstock from shale wells. Such feedstock cost advantages in turn cause a decline in ethylene exports from other parts of Asia (e.g., from Japan and South Korea) caused by their less competitiveness compared to the producers in other regions. Economic trends can shift demand to developing nations at a faster rate than expected. At the same time, unconventional feedstocks also play a key role in shaping the ethylene industry. The change of feedstock and its availability can also impact the global ethylene market. Ultimately, feedstock prices and co-product values determine ethylene production costs. New investments are expected to shift toward regions that offer certain advantages either in terms of demand growth (e.g., China) or cheaper feedstock availability (e.g., N. America and Middle East). Shale gas fields in the US will supply considerably low-cost feedstocks in the next years; hence, global producers will consider North America as a suitable place and stable investment region.

The development of new technologies for alternatives of ethylene and its derivatives will gain momentum. Such alternatives could be the usage of coal or renewable feedstocks (e.g., sugarcane and corn). China is already producing olefins from coal; hence, this could also be an option for the future. In addition, the production of coal-based ethylene glycol has been known since 2009. For the past few years, Brazil has already started producing polyethylene from sugarcane. Changes in the price of natural gas relative to crude oil determine the competitiveness of regions, countries, and producers of ethylene.

Polyethylene remains the key derivative and main driver of ethylene in the future. New ethylene capacity additions will be concentrated in regions where there is sufficient supply of cheaper feedstock and an increasing growth in demand. There is a possibility that new capacity additions for ethylene production can shift back to N. America in the very near future. Simultaneously, alternative ethylene derivative technologies are also growing currently, albeit at a slower pace. Regional competitiveness and production costs are also important parameters to be considered.

SUMMARY AND OUTLOOK

Ethylene is indeed one of the most important raw materials for the production of numerous chemicals and polymers. It is occasionally referred to as the "king of petrochemicals". It has abundant commercial significant among other chemicals and, therefore, developments should be constantly made in its industrial production to increase the manufacturing processes for other types of important petrochemicals. Considering its enormous consumption and wide range applications, ethylene can certainly dominate the whole petrochemical industry to a large extent in the coming future.

The global changes of the ethylene markets are expected to happen in the years to come mostly driven by the cost of feedstock and sharply diverging consumption growth between geographic areas. The global demand for ethylene as a whole is expected to grow by ~3% per year over a period of next 5 years. Ethane is the second major component of shale gas; hence, there will be abundant ethane supply in the US for producing ethylene by cracking. Besides these two regions, Middle East (20%) and W. Europe (15%), are the two other major regions for ethylene production; roughly 80% of world ethylene production comes from these four regions. In 2013, the worldwide production of ethylene accounted to be almost 135 Mio. t. North America and Asia are the world's two largest ethylene producing regions. However, Middle East recently boosted this region to fall into the group of the largest ethylene producing regions.

Currently, the US is the largest ethylene exporter, while China is its largest importer worldwide. The fastest consumption is expected in the Asiatic region, particularly in China and India. Equally high consumption rates are also expected in Africa and the Middle East. Across the globe, the Asia Pacific region has been the largest consumer of ethylene, which accounts to more than 30% of worldwide production.

A number of novel technologies for the production of ethylene have been briefly reviewed. It has been shown that the ethylene worldwide production is based on thermal cracking with steam. Other routes for the production of ethylene besides conventional thermal cracking have been also discussed which include fluidized-bed cracking, catalytic pyrolysis, methanol to olefins, the Fischer–Tropsch route, and off-gas recovery. Among these routes, steam cracking will remain for the years to come as the only industrial method for the production of ethylene in large-scale level. Nearly 30% of ethylene produced by steam cracking is from ethane. Quite recently, oxidative dehydrogenation of ethane (ODHE) is being explored. This ODHE approach seems to be a good option for the future ethylene production at relatively low temperature, pressure, and also by using cheaper feedstock such as ethane. However, ODHE is still under development and more active catalyst systems and optimized reaction conditions need to be implemented for its commercial implementation.

Overall, future and current developments technologies for the production of ethylene will take their drive from both market forces and technological improvements. Therefore, the production methods of ethylene will be based on the cheapest raw material, lowest energy consumption, market forces, and lowest investment.

REFRENCES

Albright, L. F., Crynes, B. L., & Nowak, S. (1992). *Novel production methods for ethylene, light hydrocarbons, and aromatics*. New York, NY: Marcel Dekker Inc.

Argyle, M. D., Chen, K., Bell, A. T., & Iglesia, E. (2002a). Effect of catalyst structure on oxidative dehydrogenation of ethane and propane on alumina-supported vanadia. *Journal of Catalysis*, *208*(1), 139–149. doi:10.1006/jcat.2002.3570

Argyle, M. D., Chen, K., Bell, A. T., & Iglesia, E. (2002b). Ethane oxidative dehydrogenation pathways on vanadium oxide catalysts. *The Journal of Physical Chemistry B*, *106*(21), 5421–5427. doi:10.1021/jp0144552

Arstad, B., & Kolboe, S. (2001). The Reactivity of Molecules Trapped within the SAPO-34 Cavities in the Methanol-to-Hydrocarbons Reaction. *Journal of the American Chemical Society*, *123*(33), 8137–8138. doi:10.1021/ja010668t PMID:11506579

Bos, A. N., Schoonebeek, R., Spies, F., & Verhaak, M. J. (2015). Alkane oxidative dehydrogenation and/or alkene oxidation using mixed metal oxide catalyst (WO 2015082602 A1).

Botella, P., García-González, E., Dejoz, A., López Nieto, J. M., Vázquez, M. I., & González-Calbett, J. (2004). Selective oxidative dehydrogenation of ethane on MoVTeNbO mixed metal oxide catalysts. *Journal of Catalysis*, *225*(2), 428–438. doi:10.1016/j.jcat.2004.04.024

Brazdil, J. F. (2006). Strategies for the selective catalytic oxidation of alkanes. *Topics in Catalysis*, *38*(4), 289–294. doi:10.1007/s11244-006-0027-4

Burch, R., & Swarnakar, R. (1991). Oxidative dehydrogenation of ethane on vanadium-molybdenum oxide and vanadium-niobium-molybdenum oxide catalysts. *Applied Catalysis*, *70*(1), 129–148. doi:10.1016/S0166-9834(00)84159-7

Cavani, F., Ballarini, N., & Cericola, A. (2007). Oxidative dehydrogenation of ethane and propane: How far from commercial implementation? *Catalysis Today*, *127*(1-4), 113–131. doi:10.1016/j.cattod.2007.05.009

Cavani, F., & Trifirò, F. (1995). The oxidative dehydrogenation of ethane and propane as an alternative way for the production of light olefins. *Catalysis Today*, *24*(3), 307–313. doi:10.1016/0920-5861(95)00051-G

Chalakov, L., Rihko-Struckmann, L. K., Munder, B., & Sundmacher, K. (2007). Feasibility study of the oxidative dehydrogenation of ethane in an electrochemical packed-bed membrane reactor. *Industrial & Engineering Chemistry Research*, *46*(25), 8665–8673. doi:10.1021/ie070089i

Chalakov, L., Rihko-Struckmann, L. K., Munder, B., & Sundmacher, K. (2009). Oxidative dehydrogenation of ethane in an electrochemical packed-bed membrane reactor: Model and experimental validation. *Chemical Engineering Journal*, *145*(3), 385–392. doi:10.1016/j.cej.2008.08.017

Chalavekilian, B., & Dumon, R. (1976). New route to ethylene-pyrolysis. *Chemical and Engineering News*, 1976, 176–178.

Chang, C. D. (1983). Hydrocarbons from methanol. *Catalysis Reviews. Science and Engineering*, *25*(1), 1–118. doi:10.1080/01614948308078874

Chang, C. D. (1992). The New Zealand gas-to-gasoline plant: An engineering tour de force. *Catalysis Today*, *13*(1), 103–111. doi:10.1016/0920-5861(92)80190-X

Che-Galicia, G., Quintana-Solorzano, R., Ruiz-Martinez, R. S., Valente, J. S., & Castillo-Araiza, C. O. (2014). Kinetic modelling of the oxidative dehydrogenation of ethane to ethylene over a MoVTeNbO catalytic system. *Chemical Engineering Journal*, *252*, 75–88. doi:10.1016/j.cej.2014.04.042

Che-Galicia, G., Ruiz-Martínez, R. S., López-Isunza, F., & Castillo-Araiza, C. O. (2015). Modeling of oxidative dehydrogenation of ethane to ethylene on a MoVTeNbO/TiO2 catalyst in an industrial-scale packed bed catalytic reactor. *Chemical Engineering Journal*, *280*, 682–694.

Christodoulakis, A., Heracleous, E., Lemonidou, A. A., & Boghosian, S. (2006). An operando Raman study of molecular structure and reactivity of MoO_3/Al_2O_3 catalysts for ethane ODH. *Journal of Catalysis*, *242*, 16–25. doi:10.1016/j.jcat.2006.05.024

Czuprat, O., Werth, S., Schirrmeister, S., Schiestel, T., & Caro, J. (2009). Olefin production by a multistep oxidative dehydrogenation in a perovskite hollow-fiber membrane reactor. *ChemCatChem*, *1*(3), 401–405. doi:10.1002/cctc.200900176

Dahl, I. M., & Kolboe, S. (1993). On the reaction mechanism for propene formation in the MTO reaction over SAPO-34. *Catalysis Letters*, *20*(3-4), 329–336. doi:10.1007/BF00769305

Dahl, I. M., & Kolboe, S. (1994). On the Reaction mechanism for hydrocarbon formation from methanol over SAPO-34: I. Isotopic labeling studies of the co-reaction of ethene and methanol. *Journal of Catalysis*, *149*(2), 458–464. doi:10.1006/jcat.1994.1312

Deng, S., Li, S., Li, H., & Zhang, Y. (2009). Oxidative dehydrogenation of ethane to ethylene with CO2 over Fe-Cr/ZrO_2 catalysts. *Industrial & Engineering Chemistry Research*, *48*(16), 7561–7566. doi:10.1021/ie9007387

Donsı`, F., Williams, K. A., & Schmidt, L. D. (2005). Multistep mechanism for ethane oxidative dehydrogenation on Pt and Pt-Sn coated monoliths. *Industrial & Engineering Chemistry Research*, *44*(10), 3453–3470. doi:10.1021/ie0493356

Frank, B., Morassutto, M., Schomäcker, R., Schlögl, R., & Su, D. S. (2010). Oxidative dehydrogenation of ethane over multiwalled carbon nanotubes. *ChemCatChem*, *2*(6), 644–648. doi:10.1002/cctc.201000035

Gaab, S., Machli, M., Find, J., Grasselli, R. K., & Lercher, J. A. (2003). Oxidative dehydrogenation of ethane over novel Li/Dy/Mg mixed oxides: Structure–activity study. *Topics in Catalysis*, *23*(1/4), 151–158. doi:10.1023/A:1024836707308

Galownia, J. M., Wight, A. P., Blanc, A., Labinger, J. A., & Davis, M. E. (2005). Partially reduced heteropolyanions for the oxidative dehydrogenation of ethane to ethylene and acetic acid at atmospheric pressure. *Journal of Catalysis*, *236*(2), 356–365. doi:10.1016/j.jcat.2005.10.010

Gärtner, A., van Veen, A. C., & Lercher, J. A. (2013). Oxidative dehydrogenation of ethane: Common principles and Mechanistic aspects. *ChemCatChem*, *5*(11), 3196–3217. doi:10.1002/cctc.201200966

Gary, J. H., Handwerk, G. E., & Kaiser, M. J. (2007). *Petroleum refining technology and economics* (5th ed.). Boca Raton, FL: CRC Press.

Gaspar, N. J., & Pasternak, I. S. (1971). H_2S promoted oxidative dehydrogenation of ethane. *Canadian Journal of Chemical Engineering*, *49*(2), 248–251. doi:10.1002/cjce.5450490213

Grabowski, R. (2006). Kinetics of oxidative dehydrogenation of C_2-C_3 alkanes on oxide catalysts. *Catalysis Reviews. Science and Engineering*, *48*(2), 199–268. doi:10.1080/01614940600631413

Gudgila, R., & Leclerc, C. A. (2011). Support effects on the oxidative dehydrogenation of ethane to ethylene over platinum catalysts. *Industrial & Engineering Chemistry Research*, *50*(14), 8438–8443. doi:10.1021/ie200035q

Håkonsen, S. F., Walmsley, J. C., & Holmen, A. (2010). Ethene production by oxidative dehydrogenation of ethane at short contact times over Pt-Sn coated monoliths. *Applied Catalysis A, General*, *378*(1), 1–10. doi:10.1016/j.apcata.2010.01.032

Hamel, C., Wolff, T., & Seidel-Morgenstern, A. (2011). Compatibility of transport and reaction in membrane reactors used for the oxidative dehydrogenation of short-chain hydrocarbons. *International Journal of Chemical Reactor Engineering,* 9(1), 1542-6580.

Han, S., Jin, L.; Martenak, D. J., Suib, S. L., (2010). Improved process for oxidative dehydrogenation of ethane and related reactions (EP 2165997 A1).

Heracleous, E., & Lemonidou, A. A. (2006). Ni–Nb–O mixed oxides as highly active and selective catalysts for ethene production via ethane oxidative dehydrogenation. Part I: Characterization and catalytic performance. *Journal of Catalysis*, *237*(1), 162–174. doi:10.1016/j.jcat.2005.11.002

Heracleous, E., & Lemonidou, A. A. (2006). Ni-Nb-O mixed oxides as highly active and selective catalysts for ethene production via ethane oxidative dehydrogenation. II. Mechanistic aspects and kinetic modelling. *Journal of Catalysis*, *237*(1), 175–189. doi:10.1016/j.jcat.2005.11.003

Heracleous, E., & Lemonidou, A. A. (2010). Ni-Me-O mixed metal oxides for the effective oxidative dehydrogenation of ethane to ethylene - Effect of promoting metal Me. *Journal of Catalysis*, *270*(1), 67–75. doi:10.1016/j.jcat.2009.12.004

Hkonsen, S. F., & Holmen, A. (2007). Oxidative dehydrogenation of ethane at short contact times. *Studies in Surface Science and Catalysis*, *167*, 337–342. doi:10.1016/S0167-2991(07)80154-6

Huff, M. C., & Schmidt, L. D. (1993). Ethylene formation by oxidative dehydrogenation of ethane over monoliths at very short contact times. *Journal of Physical Chemistry*, *97*(45), 11815–11815. doi:10.1021/j100147a040

Huff, M. C., & Schmidt, L. D. (1996). Elementary step model of ethane oxidative dehydrogenation on Pt-coated monoliths. *AIChE Journal. American Institute of Chemical Engineers*, *42*(12), 3484–3497. doi:10.1002/aic.690421218

IEA. (2004). *Energy statistics of OECD countries 2001/2002 and energy statistics of non-OECD countries 2001/2002*. Paris: International Energy Agency.

Iwamoto, M., Yoda, Y., Egashira, M., & Seiyama, T. (1976). Study of metal oxide catalysts by temperature programmed desorption. 1. Chemisorption of oxygen on nickel oxide. *Journal of Physical Chemistry*, *80*(18), 1989–1994. doi:10.1021/j100559a008

Kaliya, M. L., Kogan, S. B., Froumin, N., & Herskowitz, M. (2002). Novel nitrogen containing heterogeneous catalysts for oxidative dehydrogenation of light paraffins. *Catalysis Communications*, *3*(8), 327–333. doi:10.1016/S1566-7367(02)00141-3

Klose, F., Joshi, M., Hamel, C., & Seidel-Morgenstern, A. (2004). Selective oxidation of ethane over a VOx/g-Al_2O_3 catalyst-investigation of the reaction network. *Applied Catalysis A, General*, *260*(1), 101–110. doi:10.1016/j.apcata.2003.10.005

Konya, T., Katou, T., Murayama, T., Ishikawa, S., Sadakane, M., Buttrey, D., & Ueda, W. (2013). An orthorhombic Mo_3VOx catalyst most active for oxidative dehydrogenation of ethane among related complex metal oxides. *Catalysis Science & Technology*, *3*(2), 380–387. doi:10.1039/C2CY20444D

Kumar, D., & Schoonebeek, R. J. (2015). Process for the oxidative dehydrogenation of ethane to ethylene (WO 2015059275 A1).

Kustov, L. M., Kucherov, A. V., Finashina, E. D., Stakheev, A. Y., Sinev, I. M., & Krzywicki, A. (2011). CA Patent 2701089 A1.

Leveles, L., Seshan, K., Lercher, J., & Lefferts, A. L. (2003). Oxidative conversion of propane over lithium promoted magnesia catalyst I. Kinetics and mechanism. *Journal of Catalysis*, *218*(2), 296–306. doi:10.1016/S0021-9517(03)00112-X

Lin, M. M. (2001). Selective oxidation of propane to acrylic acid with molecular oxygen. *Applied Catalysis A, General*, *207*(1-2), 1–16. doi:10.1016/S0926-860X(00)00609-8

Liu, Y., Zhang, Y., Liu, X., Xue, J., & Li, S. (1998). The excellent performance of TiO2-supported Li-Mn-W oxide catalyst for oxidative dehydrogenation of ethane to ethylene. *Chemistry Letters*, 1998(10), 1057–1058. doi:10.1246/cl.1998.1057

Luo, J. Z., & Wan, H. L. (1997). Oxidative dehydrogenation of ethane over LaF_3-CeO_2 catalysts. *Applied Catalysis A, General*, *158*(1-2), 137–144. doi:10.1016/S0926-860X(97)00039-2

Luo, J. Z., Zhou, X. P., Chao, Z. S., & Wan, H. L. (1997). Oxidative dehydrogenation of ethane over BaF_2 promoted SmO_3-LaF_3 catalysts. *Applied Catalysis A, General*, *159*(1-2), 9–19. doi:10.1016/S0926-860X(97)00046-X

Mamedov, E. A., & Cortés Corberán, V. (1995). Oxidative dehydrogenation of lower alkanes on vanadium oxide-based catalysts. The present state of the art and outlooks. *Applied Catalysis A, General*, *127*(1-2), 1–40. doi:10.1016/0926-860X(95)00056-9

Mars, P., & van Krevelen, D. W. (1954). Oxidations carried out by means of vanadium oxide catalysts. *Chemical Engineering Science*, *3*(1), 41–59. doi:10.1016/S0009-2509(54)80005-4

Matar, S., & Hatch, L. F. (2001). *Chemistry of petrochemical processes* (2nd ed.). Woburn: Gulf Professional Publishing.

McCain, J. H. (1986). Oxydehydrogenation of ethane to ethylene. EP 166438 A2.

Merzouki, M., Taouk, B., Monceaux, L., Bordes, E., & Courtine, P. (1992). Catalytic properties of promoted vanadium oxide in the oxidation of ethane to acetic acid. *Studies in Surface Science and Catalysis*, *72*, 165–179. doi:10.1016/S0167-2991(08)61669-9

Mimura, N., Takahara, I., Inaba, M., Okamoto, M., & Murata, K. (2002). High-performance Cr/H-ZSM-5 catalysts for oxidative dehydrogenation of ethane to ethylene with CO_2 as an oxidant. *Catalysis Communications*, *3*(6), 257–262. doi:10.1016/S1566-7367(02)00117-6

Mizuno, N., Suh, D.-J., Han, W., & Kudo, T. (1996). Catalytic performance of $Cs_{2.5}Fe_{0.08}H_{1.26}PVMo_{11}O_{40}$ for direct oxidation of lower alkanes. *Journal of Molecular Catalysis A Chemical*, *114*(1-3), 309–317. doi:10.1016/S1381-1169(96)00332-9

Morales, E., & Lunsford, J. H. (1989). Oxidative dehydrogenation of ethane over a lithium-promoted magnesium oxide catalyst. *Journal of Catalysis*, *118*(1), 255–265. doi:10.1016/0021-9517(89)90315-1

Nieto, J. M. L., Botella, P., Vázquezb, M. I., & Dejoz, A. (2002). The selective oxidative dehydrogenation of ethane over hydrothermally synthesized MoVTeNb catalysts. *Chemical Communications*, 2002, 1906–1907. doi:10.1039/b204037a

Ovsitser, O., & Kondratenko, E. V. (2009). Similarity and differences in the oxidative dehydrogenation of C2-C4 alkanes over nano-sized VOx species using N_2O and O_2. *Catalysis Today*, *142*(3-4), 138–142. doi:10.1016/j.cattod.2008.09.012

Qiao, A., Kalevaru, V. N., Radnik, J., Duevel, A., Heitjans, P., Kumar, A. S. H., & Martin, A. et al. (2014). Oxidative Dehydrogenation of Ethane to Ethylene over V_2O_5/Al_2O_3 Catalysts: Effect of Source of Alumina on the Catalytic Performance. *Industrial & Engineering Chemistry Research*, *53*(49), 18711–18721. doi:10.1021/ie5008344

Qiao, A., Kalevaru, V. N., Radnik, J., Srihari Kumar, A., Lingaiah, N., Sai Prasad, P. S., & Martin, A. (2013a). Oxidative dehydrogenation of ethane to ethylene over V_2O_5/Nb_2O_5 catalysts. *Catalysis Communications*, *30*, 45–50. doi:10.1016/j.catcom.2012.10.018

Qiao, A., Kalevaru, V. N., Sri Hari Kumar, A., Lingaiah, N., Sai Prasad, P. S., & Martin, A. (2012). Effect of CO2-admixture on the catalytic performance of Ni-Nb-M-O catalysts in oxidative dehydrogenation of ethane to ethylene. DGMK Tagungsbericht, 3, 97-103.

Rodriguez, M. L., Ardissone, D. E., Heracleous, E., Lemonidou, A. A., Lopez, E., Pedernera, M. N., & Borio, D. O. (2010). Oxidative dehydrogenation of ethane to ethylene in a membrane reactor: A theoretical study. *Catalysis Today*, *157*(1-4), 303–309. doi:10.1016/j.cattod.2010.01.053

Rodriguez, M. L., Ardissone, D. E., Lemonidou, A. A., Heracleous, E., Lopez, E., Pedernera, M. N., & Borio, D. O. (2009). Simulation of a membrane reactor for the catalytic oxydehydrogenation of ethane. *Industrial & Engineering Chemistry Research*, *48*(3), 1090–1095. doi:10.1021/ie800564v

Santander, J. A., Lopez, E., Tonetto, G. M., & Pedernera, M. N. (2014). Preparation of NiNbO/AISI 430 ferritic stainless steel monoliths for catalytic applications. *Industrial & Engineering Chemistry Research*, *53*(28), 11312–11319. doi:10.1021/ie501884g

Savova, B., Loridant, S., Filkova, D., & Millet, J. M. M. (2010). Ni-Nb-O catalysts for ethane oxidative dehydrogenation. *Applied Catalysis A, General*, *390*(1-2), 148–157. doi:10.1016/j.apcata.2010.10.004

Schulz, H. (1999). Short history and present trends of Fischer–Tropsch synthesis. *Applied Catalysis A, General*, *186*(1-2), 3–12. doi:10.1016/S0926-860X(99)00160-X

Shi, X., Ji, S., Wang, K., & Li, C. (2008). Oxidative dehydrogenation of ethane with CO_2 over novel Cr/SBA-15/Al_2O_3/FeCrAl monolithic catalysts. *Energy & Fuels*, *22*(6), 3631–3638. doi:10.1021/ef800567v

Solomons, T. W. G. (1996). *Organic Chemistry* (6th ed., pp. 260–268, 283–316, 940–941). New York, USA: John Wiley & Sons, Inc.

Solsona, B., Concepción, P., Hernández, S., Demicol, B., & Nieto, J.M.L. (2012). Oxidative dehydrogenation of ethane over NiO–CeO_2 mixed oxides catalysts. *Catalysis Today*, *180*(1), 51–58. doi:10.1016/j.cattod.2011.03.056

Solsona, B., López Nieto, J. M., Concepción, P., Dejoz, A., Ivars, F., & Vázquez, M. I. (2011). Oxidative dehydrogenation of ethane over Ni–W–O mixed metal oxide catalysts. *Journal of Catalysis*, *280*(1), 28–39. doi:10.1016/j.jcat.2011.02.010

Solymosi, F., & Nemeth, R. (1999). The oxidative dehydrogenation of ethane with CO_2 over Mo_2C/SiO_2 catalyst. *Catalysis Letters*, *62*(2/4), 197–200. doi:10.1023/A:1019027912597

Song, W. G., Haw, J. F., Nicholas, J. B., & Heneghan, C. S. (2000). Methylbenzenes are the organic reaction centers for methanol-to-olefin catalysis on HSAPO-34. *Journal of the American Chemical Society*, *122*(43), 10726–10727. doi:10.1021/ja002195g

Sri Hari Kumar, A., Kalevaru, V. N., Qiao, A., Alshammari, A., Lingaiah, N., Sailu, C., & Martin, A. et al. (2013b). Catalytic behavior of decomposed molybdophosphoric acid supported on alumina for oxidative dehydrogenation of ethane to ethylene. *Kinetics and Catalysis*, *54*(5), 615–619. doi:10.1134/S002315841305008X

Stratton, A., Hemming, D. F., & Teper, M. (1983). *Ethylene production from oil, gas and coal-derived feedstock*. London, UK: IEA Coal Research.

Swaan, H. M., Toebes, A., Seshan, K., van Ommen, J. G., & Ross, J. R. H. (1992). The kinetic and mechanistic aspects of the oxidative dehydrogenation of ethane over Li/Na/MgO catalysts. *Catalysis Today*, *13*(2-3), 201–208. doi:10.1016/0920-5861(92)80143-B

Tarasov, A. L., Kustov, L. M., Kirichenko, O. A., & Kucherov, A. V. (2014). Method for production of ethylene RU 2528830.

Torniainen, P. M., Chu, X., & Schmidt, L. D. (1994). Comparison of monolith-supported metals for the direct oxidation of methane to syngas. *Journal of Catalysis*, *146*(1), 1–10. doi:10.1016/0021-9517(94)90002-7

Tsilomelekis, G., Christodoulakis, A., & Boghosian, S. (2007). Support effects on structure and activity of molybdenum oxide catalysts for the oxidative dehydrogenation of ethane. *Catalysis Today*, *127*(1-4), 139–147. doi:10.1016/j.cattod.2007.03.026

Tsyganok, A., Green, R. G., Giorgi, J. B., & Sayari, A. (2007). Non-oxidative dehydrogenation of ethane to ethylene over chromium catalysts prepared from layered double hydroxide precursors. *Catalysis Communications*, *8*(12), 2186–2193. doi:10.1016/j.catcom.2007.04.031

Valente, J. S., Quintana-Solórzano, R., Armendáriz-Herrera, H., Barragán-Rodríguez, G., & López-Nieto, J. M. (2014). Kinetic study of oxidative dehydrogenation of ethane over MoVTeNb mixed-oxide catalyst. *Industrial & Engineering Chemistry Research*, *53*(5), 1775–1786. doi:10.1021/ie402447h

Velle, O. J., Andersen, A., & Jens, K. J. (1990). Oxidative dehydrogenation of ethane by perovskite type catalysts containing oxides of strontium, cerium and ytterbium. *Catalysis Today*, *6*(4), 567–574. doi:10.1016/0920-5861(90)85053-Q

Wang, S., Murata, K., Hayakawa, T., Hamakawa, S., & Suzuki, K. (1999). Oxidative dehydrogenation of ethane by carbon dioxide over sulfate-modified Cr2O3/SiO2 catalysts. *Catalysis Letters*, *63*(1/2), 59–64. doi:10.1023/A:1019084030344

Watson, J. M., Daly, F. P., Mazanec, T. J., & Yang, B. (2005). Catalysts having catalytic material applied directly to thermally grown alumina and catalytic methods using same, improved methods of oxidative dehydrogenation (US 2005/0272965).

Weissermel, K., & Arpe, H.-J. (Fourth Ed.) (2003). Industrial Organic Chemistry. (pp. 59-63). Weinheim, Germany: Wiley-VCH GmbH & Co. KGaA.

Worrell, E., De Beer, J. G., Faaij, A. P. C., & Blok, K. (1994). Potential energy savings in the production route for plastics. *Energy Conservation and Management*, *35*(12), 1073–1085. doi:10.1016/0196-8904(94)90011-6

Yang, B., Yuschak, T., Mazanec, T., Tonkovich, A. L., & Perry, S. (2008). Multi-scale modeling of microstructured reactors for the oxidative dehydrogenation of ethane to ethylene. *Chemical Engineering Journal*, *135*(1), S147–S152. doi:10.1016/j.cej.2007.07.050

Yurchak, S. (1988). Development of Mobil's fixed-bed methanul-to-gasoline (MTG) process. *Studies in Surface Science and Catalysis*, *36*, 251–272. doi:10.1016/S0167-2991(09)60521-8

Zhang, X., Ye, Q., Xu, B., & He, D. (2007). Oxidative dehydrogenation of ethane over Co-BaCO3 catalysts using CO2 as oxidant: Effects of Co-promoter. *Catalysis Letters*, *117*(3-4), 140–145. doi:10.1007/s10562-007-9122-9

Zhang, X., Yu, G., Gong, Y., Jiang, D., & Xie, Y. (2000). Oxidative dehydrogenation of ethane to ethylene over NiO/Al2O3 catalyst. *Studies in Surface Science and Catalysis*, *130*, 1835–1840. doi:10.1016/S0167-2991(00)80468-1

Zhao, X., & Wang, X. (2006). Oxidative dehydrogenation of ethane to ethylene by carbon dioxide over Cr/TS-1 catalysts. *Catalysis Communications*, *7*(9), 633–638. doi:10.1016/j.catcom.2006.02.005

Zhu, H., Dong, H., Laveille, P., Saih, Y., Caps, V., & Basset, J.-M. (2014). Metal oxides modified NiO catalysts for oxidative dehydrogenation of ethane to ethylene. *Catalysis Today*, *228*, 58–64.

Chapter 5
Membrane Engineering and its Role in Oil Refining and Petrochemical Industry

Adele Brunetti
Institute on Membrane Technology, ITM-CNR, Italy

Enrico Drioli
Institute on Membrane Technology, ITM-CNR, Italy

Miriam Sellaro
Institute on Membrane Technology, ITM-CNR, Italy

Giuseppe Barbieri
Institute on Membrane Technology, ITM-CNR, Italy

ABSTRACT

Nowadays, an important contribution to the sustainable industrial growth might come not only from the development of new materials but also from the re-design of process engineering, which leads to the development and the re-designing of more compact and efficient new processes in many areas of applications. As already done in large part of processes in Nature, membrane operations can serve in molecular separations, chemical transformations, water and energy transfer between different phases. Indeed, these technologies well fit the principles of this design philosophy and this is testified by the various applications of membranes in chemical and petrochemical industry and many are the successful examples of processes that have moved from lab-scale to pilot plant up to a demonstration unit in refinery. The main aim of this chapter is to show the main applications of membrane technology in the petrochemical industry and the related current research trends, focusing on the impact that membrane technology can have on the process sustainability.

INTRODUCTION

The ever more pressing necessity for sustainability is inducing to substitute traditional industrial growth with a sustainable one. In the recent Research Agenda published as part of the Delft Skyline Debates (Gorak & Stankiewicz, 2011), the development of the industrial system in the year 2050 has been well defined. In many sectors such as water, energy, food, health, etc. the necessary transformations will take

DOI: 10.4018/978-1-4666-9975-5.ch005

place consistently with process intensification principles. These are based on innovative equipment, design and process development methods that can bring important improvements in petrochemical, chemical and any other manufacturing and processing procedures, such as decreasing production costs, equipment size, energy consumption, waste generation, and improving remote control, information fluxes and process flexibility (Drioli, Brunetti, Di Profio, & Barbieri, 2012). *Green process engineering* can give a great contribution to the attainment of a sustainable industrial development.

Green process engineering is based on the principles of Process Intensification Strategy, which can lead to the development and the re-designing of more compact and efficient new processes allowing better exploitation of the raw material, lower energy consumption and a reduced plant volume. Membrane engineering can play a crucial role in the implementation of this strategy and, in the last few years, membrane technology has grown so much that various large unit operations of process engineering have already been redesigned as membrane units.

Membrane operations are an interesting solution for the rationalization of chemical productions, thanks to their high selectivity and permeability in the transport of specific components, efficiency and operational simplicity, low energetic requirement, good stability under operating conditions, compatibility between different operations in integrated systems, environment compatibility, easy control and scale-up, and large operational flexibility.

There are already interesting cases where the potentialities of membrane technologies, such as membrane reactors, membrane gas separation, pervaporation, membrane emulsifiers, membrane distillation, membrane crystallizers, membrane contactors, membrane strippers and scrubbers, in their various configurations and functionalities, were explored. The most successful case is that of pressure driven membrane operations in water and brackish water desalination and in water reuse, where reverse osmosis (RO), nanofiltration (NF), ultrafiltration (UF) and microfiltration (MF) are the most important technologies. The development of immersed bioreactors in the treatment of municipal wastewater is another significant example.

There are various applications of membranes in the petrochemical industry, the most important of which is membrane gas separation.

This type of application is particularly relevant for hydrogen recovery and it is also considered as a suitable technology for the recovery of olefins, hydrocarbon separations, natural gas processing, etc. Membrane reactors, specifically usable for dehydrogenation or hydrogenations of hydrocarbons and for hydrogen production in pre-combustion capture, are other notable potential applications.

Pervaporation processes are generally used for the separation of liquid organic mixtures, the desulfurization of gasoline as well as for isomers separations.

Another way to separate organic-organic mixtures is by organic solvent nanofiltration (OSN), usually used to separate aromatic and aliphatic hydrocarbons, alcohols, ketones and esters. Dewax process is a notable new application of this kind of technology in the refining of lubricants.

The wastewater produced in the petrochemical industry can be advantageously purified by using pressure driven membrane operations such as MF, UF, RO and by membrane bioreactors also in combined configurations. Recently, membrane contactors have attracted great interest as an alternative technology to the other membrane technologies for their application in CO_2 separation from the flue gas stream as well as for air purification.

The most important applications of membrane technology in the petrochemical industry are described in Table 1 where the status of development and the main limitations for scaling up are also summarized.

Table 1. Main industrial applications of membrane technology in petrochemical industry

Process		Separation/Reaction	Development Status	Hurdles Limiting Scaling Up
Hydrogen recovery		See Table 2		Membrane plasticization caused by strongly adsorbing component and condensation of light hydrocarbons on its surface
Enhanced oil recovery		CO_2 from hydrocarbon gaseous phase	Industrial scale (Natco)	-
Olefins/paraffins separation		Ethylene/ethane; propylene/propane	Lab scale	Difficult components separation (similar molecular size and condensability); membrane performance drop in hydrocarbon-rich environment under pressure
Natural gas processing		CO_2/CH_4	Offshore platforms (Cynara-NATCO in Thailand gulf; Grace-Separex)	Low CO_2/CH_4 commercial membrane selectivity and consequent extra cost related to compression units to limit methane loss
		H_2O/CH_4	Field testing at a processing plant (Towne Exploration Company in Isleton, CA)	High membrane area requirement and high methane loss owing to low feed-to-permeate pressure
Vapour/gas separation		Vinyl chloride monomer/air	Industrial scale (MTR)	-
		Ethylene/argon, CO_2, ethane, methane, O_2 (ethylene recovery)	Lab scale	Membrane stability (plasticization; pre-treatment required)
		Gasoline vapour recovery	Industrial scale (GKSS; MTR & OPW)	-
		Ethylene or propylene/N_2 (hydrocarbon monomers recovery from polyolefins production plants)	Industrial scale	-
Air separation		O_2/N_2	See Table 2	-
		H_2O/air (air drying)	Commercial scale (Air Products; Air Liquide; Ube; Asahi Glass)	-
CO_2 capture	Post-combustion capture	CO_2/N_2 (CO_2 from flue gas)	Pilot plant (NTNU Norway at Sines power plant; ITM-CNR at Italcementi factory)	Low CO_2 concentration and pressure of flue gas associated to low CO_2/CH_4 commercial membrane selectivity
	Pre-combustion capture	CO_2/H_2	Lab scale	Pd costs, membrane sealing and polarization phenomena
	Oxy-fuel combustion	O_2/N_2 to produce CO_2/H_2O mixture	Lab scale	Membrane sealing and durability, low permeability and high costs
Water capture with membrane condensers		H_2O from waste gaseous streams	Lab scale	Low hydrophobic membrane resistance and stability towards contaminants (SO_x, NO_x, NH_3, HF, HCl)

continued on following page

Table 1. Continued

<table>
<tr><th colspan="2">Process</th><th>Separation/Reaction</th><th>Development Status</th><th>Hurdles Limiting Scaling Up</th></tr>
<tr><td colspan="2" rowspan="4">Pervaporation</td><td>Sulphur compounds/ gasoline</td><td>Pilot plant (Grace Davison S-Brane plant);
Demo plant (S-Brane plant at Bayway Refinery)</td><td>Refrigeration costs and membrane life-time</td></tr>
<tr><td>VOCs/H_2O</td><td>Commercial scale (MTR)</td><td>-</td></tr>
<tr><td>Ethanol from biomass</td><td>Industrial scale (BioSep by MTR & USEPA)</td><td>-</td></tr>
<tr><td>Organics/organics</td><td>Lab scale;
Pilot plant (Separex; Exxon)</td><td>Difficult components separation (similar molecular size and boiling points)</td></tr>
<tr><td colspan="2" rowspan="2">Organic solvent nanofiltration</td><td>Methylethylketone and toluene/lube oil</td><td>Industrial scale (MAX-DEWAX process)</td><td>-</td></tr>
<tr><td>Solvent deoiling; catalyst recovery; separation of phase-transfer agents; solvent exchange</td><td>Lab scale;
Pilot plant</td><td>Membrane stability</td></tr>
<tr><td colspan="2" rowspan="3">Membrane reactors</td><td>Water gas shift</td><td>Pilot plant (HYGEAR)</td><td>Membrane life-time, sealing and Pd high costs</td></tr>
<tr><td>Benzene hydroxylation to phenol</td><td rowspan="2">Lab scale</td><td>Oxidants high costs, limited benzene conversion and by-products formation</td></tr>
<tr><td>Dehydroisomerization of butane to isobutene</td><td>Low isobutene selectivity and membrane life-time</td></tr>
<tr><td rowspan="2">Membrane systems for water recovery</td><td>Pressure driven processes (UF, MF, RO, MD)</td><td>Oils/H_2O</td><td>Pilot plant (North Sea oilfield-produced water treatment; API separator wastewater treatment in Saudi Arabia; NATCO);
Lab scale</td><td>Operational costs for RO units because of polarization</td></tr>
<tr><td>Membrane bioreactors for water treatment</td><td>Biochemical conversion and physical separation</td><td>Industrial scale (Clereflo by Conder Products, UK; ZeeMOD® by General Electric; MBR based plant in Oman by Kubota and in Italy by ENI and General Electric)</td><td>-</td></tr>
</table>

Along in the chapter a detailed analysis of each of these technologies, the related current research trends and the impact that membrane technology can have on the processes are described. Moreover, a case study is provided to illustrate the analysis of a membrane process, also in terms of Process Intensification.

BACKGROUND AND MAIN FOCUS

Membrane Gas Separation

In 1950, Weller and Steiner (1950) considered membrane processes as usable for the separation of hydrogen from hydrogenation tail gas, air separation and enrichment of refinery gas. However, only in 1970 commercial membranes capable of performing these separations became available. In 1980, Permea launched the first large industrial application of gas separation (GS) membranes (Baker, 2002; Spillman, 1989), proposing the hydrogen separating Prism-membrane. Since then, membrane-based

Table 2. – Main industrial applications of membrane technology for GS. Published from Brunetti, A., Bernardo, P., Drioli, E., & Barbieri, G. (2011). Membrane engineering progresses and potentialities in gas separations. In Y. Yampolskii & B. Freeman (Eds.), Membrane Gas Separation (pp. 281- 312) with permission of Wiley & Sons

Separation	Process	Traditional Technology	Membrane Technology	
			Materials	Status
H_2/N_2	Ammonia purge gas	PSA	Polysulfone	Plant installed (Prism by Permea)
			Pd-based	Lab scale
H_2/CO	Adjustment of H_2/CO ratio in syngas plant	PSA	Silicon rubber	Lab scale
			Polyimide	Plant installed (Separex)
			Pd-based	Lab scale
H_2/hydrocarbons	Hydrogen recovery in refineries	PSA	Silicon rubber Polyimide	Plant installed (Sinopec Zhenhai (china) Prism by Permea Du Pont)
			Pd-based	Lab scale
H_2/light hydrocarbons	Ethylene cracker old trains	Cryogenic distillation	PTMSP; PMP	Pilot plant
			Pd-based	Lab scale
O_2/N_2	Air separation	Cryogenic distillation	Silicon rubber	Plant installed (Cynara; Separex; GMS; Air Products)
			Polysulfone	Plant installed (Permea)
			Polyimide	Plant installed (Medal; Dow-Generan; UBE)
			Polyphenylene	Plant installed (Aquilo)
			Ethyl Cellulose	Plant installed (Air Liquide)
			Ion transport (perovskite)	Lab scale
			Pd-based	Lab scale
N_2/CH_4	Nitrogen removal	Cryogenic distillation	Silicon rubber PMP Parel	Pilot plant
			PEBAX	Lab scale
H_2O/CH_4	Dehydration	Glycol absorption	Cellulose acetate Polyimide Polyaramide	Plant installed
CO_2/CH_4	Sweetening of natural gas	Amine absorption	Cellulose acetate	Offshore platforms In Thailand gulf. (Cynara-NATCO); Grace-Separex

operations have had a rapid growth, substituting or being integrated with the traditional ones. Many companies, such as Cynara-Natco, Separex-UOP, GMS, Generon, Praxair, AirProducts and UBE have been involved in this field (Sanders et al., 1988; Ekiner, Hayes, & Manos, 1992).

A list of the main industrial applications of membrane technology in GS, along with the membrane materials used and the status of membrane technology is available in Table 2 (Brunetti, Bernardo, Drioli, & Barbieri, 2011a).

Hydrogen Recovery

Over 50 million tonnes of hydrogen is produced annually for use in synthesis gas, ammonia and methanol production. Steam methane reforming, where H_2 has to be separated from CO_2, is the most important technology of producing large scale hydrogen. Other sources of low cost hydrogen are refinery and chemical industry waste gases.

The un-reacted hydrogen, widely used in hydrotreating (to remove sulphur from gasoline and diesel fuels to the new lower standards) and in hydrocracking (conversion of heavy hydrocarbons to light ones), is generally at a high pressure and membrane processes are ideal for its recovery. The first widespread commercial application of membranes in GS was the separation of the hydrogen in the ammonia purge stream, by using Permea Prism™ systems. Nowadays, hydrogen recovery is applicable to many processes and it is divided into three main categories:

- Hydrogen recovery in refineries,
- Hydrogen recovery from ammonia purge streams,
- Syngas ratio adjustment.

The demand for hydrogen recovery in refineries is constantly increasing also because of the environmental regulations. The hydrogen content in the refinery purges and off-gases ranges between 30-80%$_{molar}$ mixed with light hydrocarbons (C_1-C_5). Since 90-95% hydrogen purity is required to recycle it to a process unit, the separation of the hydrogen contained in the stream coming out from the hydrocracker is a typical operation carried out in a refinery. Membranes can be used for this purpose alone or together with an absorber system, at a reduced capital cost and better process efficiency. The Prism™ system (using polysulfone hollow fibres with a thin silicone film on it) is dominant on the market for this type of separation, showing interesting selectivities (Baudot, 2011). More than 400 hydrogen permeators were installed worldwide whereas approximately one hundred were operated in refineries in 2003.

The main problems, limiting the application of membrane technology, are membrane plasticization caused by strongly adsorbing components, e.g. CO_2, and the condensation of light hydrocarbons on its surface, which strongly influence the membrane performance and durability. The development of new membrane materials more resistant to a high hydrocarbon partial pressure and the introduction of a more efficient pre-treatment stage capable to reduce pollutant content, made this technology able to rapidly expand (Baker, 2002).

In the ammonia process, the purge stream consists of a mixture of hydrogen, nitrogen, methane and argon, delivered at a high pressure (136 bar). Therefore, because hydrogen has a higher permeability with respect to the other gases and stream already provides the necessary driving force for promoting the permeation, this type of process can be considered the ideal application for membrane technology. Actually, Permea, Inc. (now owned by Air Products and Chemicals, Inc.) designed a two-step membrane process for this separation.

Membrane systems are already applied also for syngas ratio adjustment (H_2/CO ratio) of the streams coming out of reformers. In this case, membranes are used for stripping hydrogen out of the syngas in order to reduce the H_2/CO ratio. Several hundred hydrogen separation plants were installed in 2007 (Ockwig & Nenoff, 2007).

Recently many processes for syngas clean up and hydrogen recovery were developed. For example, Ku et al. (2013) proposed a syngas clean-up section, including a water-gas shift reactor associated with two operation units. The first of these units comprises a high permeance membrane with high H_2/CO_2 selectivity in flow communication with the water-gas shift reactor to furnish a H_2-rich permeate stream and an H_2-poor retentate stream. The second unit recovers H_2 and CO from the retentate to produce a CO_2-rich product stream, by means of a membrane with selectivity for permeating H_2, CO and CO_2. The permeate streams are then combined to produce a H_2 and CO-rich fuel stream.

Enhanced Oil Recovery (EOR)

The aim of EOR is to increase the oil fields recovery yield thanks to a high pressure injection of gaseous carbon dioxide, leading to maintain high pressures in the reservoir and to improve oil displacement. When the oil reaches the surface, carbon dioxide is purged with the associated gas.

Because of the large volumes of carbon dioxide concerned (from 140 to 280 Nm^3, 0°C and 1 atm, per extracted barrel) (Piroux, Espuche, Mercier, Pinéeri, & Gebel, 2002), it is necessary to separate carbon dioxide from the hydrocarbon gaseous phase in order to have it recycled to the reservoir, after pressurization. A typical EOR process starts with >50%$_{molar}$ CO_2 and a high pressure (up to 140 bar) and then the CO_2 content increases significantly over time. The CO_2 from the CO_2 enriched natural gas is recovered using membrane systems and then it is pressurized and re-injected in the EOR wells.

Natco (former Cynara) was the first to install a membrane plant for such an application. The plant was designed to reduce CO_2 concentration from 45%$_{molar}$ down to 28%$_{molar}$, for processing 60,000 $Nm^3\ h^{-1}$ of gas. This plant was expanded and, at the moment, it is processing 120,000 $Nm^3\ h^{-1}$ of gas, decreasing CO_2 concentration from 80%$_{molar}$ to less than 10%$_{molar}$.

Olefin/Paraffin Separation

Olefins such as ethylene and propylene are produced in the petrochemical industry by steam cracking of paraffins, followed by repeated compression and distillation to separate the complex vapour mixtures.

The common technology used to obtain the separation of olefins is a highly energy-intensive cryogenic distillation, at temperatures between 233 and 183 K, in a single or double column process with 150-200 trays at pressures ranging from 16 to 20 bar (Faiz & Li, 2012). Because of the harsh conditions necessary for this type of operation, several efforts were made to find an alternative solution, with the extractive distillation, pressure swing adsorption and thermal swing adsorption being the most investigated technologies. Membrane technologies are an attractive alternative to reduce by four times the capital and operating costs of cryogenic distillation for olefin/paraffin separation.

Facilitated transport membranes are the most important type for this separation, even though their use on large scale is limited by the harsh conditions of hydrocarbon-rich environment under pressure which strongly reduce membrane lifetime..

Recently Imtex Membranes Corporation has developed Chitosan based membranes with silver nitrate as a facilitation agent, which have shown a good performance stability over thousands of hours of operation. Performance evaluation of such membranes in the olefin/paraffin separation, in particular in C_2 and C_3 splitter applications, was carried out in collaboration with LyondellBasell. The results were encouraging as over 99.5% purity and 90% recovery were obtained in each case, also rendering this type of membrane promising for butane/butene separations (Hamza, Towe, & Keyvani, 2013).

Natural Gas Membrane Processing

Natural gas is mainly composed of methane, from 75 to 90$\%_{molar}$. It also contains inert gases, such as nitrogen and CO_2, which lower the calorific value of natural gas; condensable compounds (higher C_{2+} hydrocarbons and water) that can cause condensation problems, hydrate formation or corrosion when flowing natural gas in pipelines; undesired components such as acid gaseous impurities, like carbon dioxide or hydrogen sulphide that should be removed to prevent pipeline corrosion (Baudot, 2011).

Removal of carbon dioxide (*natural gas sweetening*) increases the calorific value and transportability of the natural gas stream. Carbon dioxide content in the natural gas of gas or oil well can vary from 4 to 50$\%_{molar}$ and it has to be reduced down to ca. 2-5$\%_{molar}$. This aim is generally achieved by means of absorption with an aqueous alkanolamine solution, having the tendency to corrode equipment and to lose amine properties by increases in degradation. Membrane GS systems are an alternative technology for the separation of carbon dioxide from natural gas, particularly for offshore applications (Datta & Sen, 2006). They offer several strong advantages with respect to conventional solvent columns, such as compactness, low investment and maintenance costs, especially with single-stage configurations that do not operate compressors. Currently CO_2/CH_4 selectivity of commercial membranes is not so high (12-25 in field conditions), so partial losses of treated methane in the low-pressure permeate can occur. Methane losses are in the range of 10-25% on a single-stage configuration and can be from 33 to 80% of the whole operating cost of a membrane unit (Baudot, 2011). The introduction of two-stage configurations leads to an overall methane recovery higher than 95% even though an extra cost owing to the stage compression unit is introduced (Baudot, 2011).

Baker proposed two carbon dioxide membrane systems to treat natural gas with low CO_2 concentration (Baker & Lokhandwala, 2008). Both of them allow to treat a feed stream with 10$\%_{molar}$ carbon dioxide. One-stage systems are preferred for very small gas flows and in this case methane loss in the permeate is often 10-15%.

Membrane systems can also be integrated into traditional units. The design of a hybrid membrane separation system is determined by different parameters, such as membrane permeance and selectivity, CO_2 concentration of the inlet gas and the target required, the gas value and the location of the plant (on an offshore platform, the weight, footprint, and simplicity of operation are critical; onshore, the total cost is more significant) (Baker, Hofmann, Kaschemekat, & Lokhandwala, 2001).

Several natural gas reserves are considered sub-quality because of the high nitrogen content, with gas pipeline specifications fixing a 4% limit (Tannerhill, 1999). At the moment, cryogenic distillation is used to separate N_2 from natural gas but membrane technology could be used in combination with a cryogenic plant (Figure 1) (Baker, 2002). The feed gas, containing 15$\%_{molar}$ nitrogen, is separated by a membrane in two streams: a retentate stream with a 30$\%_{molar}$ nitrogen content to be sent to the cryogenic plant and a permeate stream containing 6$\%_{molar}$ nitrogen to be sent to the product pipeline gas. The use of the membrane can reduce the volume of the gas to be treated by the cryogenic unit by more than a half.

At the same time, water, C_{3+} hydrocarbons and carbon dioxide preferentially permeate through the membrane and so their concentration is significantly lowered prior to cryogenic condensation, avoiding or reducing freezing problems in the plant.

The savings produced by using a smaller, simpler cryogenic plant more than offset the cost of the membrane unit (Lokhandwala, 1997).

Figure 1. Scheme of a hybrid membrane/cryogenic distillation plant for removal of nitrogen from natural gas. Reprinted from Baker, R.W. (2002). Future directions of membrane gas separation technology. Industrial & Engineering Chemistry Research, 41, 1393-1404 with permission from American Chemical Society

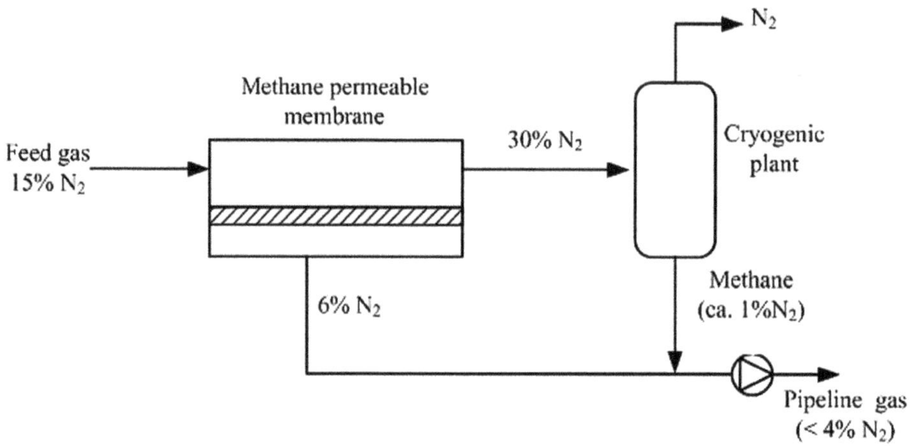

Natural gas can be processed by means of membranes also to obtain its dehydration. Actually, membrane technology is an interesting alternative to conventional glycol dehydrators for natural gas. However, membrane systems separation performance is limited by the low feed-to-permeate pressure ratio, resulting in high membrane area requirement and high methane loss.

In order to overcome such problems, a design with dry gas on the permeate side as a countercurrent sweep was developed. This type of design was evaluated by field testing a spiral-wound module, with Pebax membranes with promising H_2O/CH_4 separation properties, at a natural gas processing plant in Isleton, CA, by Towne Exploration Company. The use of a dry sweep gas on the permeate side significantly increased water vapour flux through the membrane. As sweep/feed flow rate ratio increased from 0 to $5.9\%_{molar}$ at a permeate pressure of 4.8 bar, water flux increased by 2.5 times, confirming the suitability of this system for natural gas dehydration (Lin et al., 2013).

Vapour/Gas Separation

The removal of organic solvents from gas streams was widely developed at the industrial level. The principal industrial applications of vapour/gas membrane separation are (Baker, 2006):

- Polyvinyl chloride manufacturing vent gas,
- Ethylene recovery,
- Gasoline vapour recovery systems at large terminals,
- Polyolefin plant resin degassing,
- Natural gas processing/fuel gas conditioning.

Figure 2. Membrane recovery of VCM monomer in a polyvinyl chloride plant. Adapted from Baker, R.W. (2006). Membranes for vapour/gas separation. From http://www.mtrinc.com/publications/MT01%20 Fane%20Memb%20for%20VaporGas_Sep%202006%20Book%20Ch.pdf

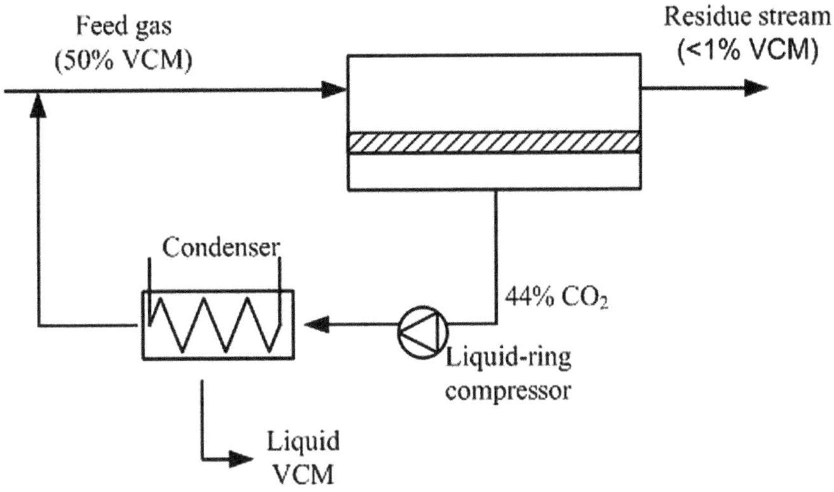

In the process of *polymerization of polyvinyl chloride*, side reactions generate unwanted gases into the reactors. These inerts must be removed from the process but the vented gas stream may contain several hundred thousand dollars of monomer values, owing to the high volatility of the vinyl chloride monomer. Feed gas containing vinyl chloride monomer and air is sent to the membrane system. Vinyl chloride monomer-enriched permeate from the membrane system is compressed in a liquid-ring compressor and cooled to liquefy the vinyl chloride monomer, whose recovery is more than 99% (Figure 2). The non-condensable gases are mixed with the feed gas and returned to the membrane section. In 1992, MTR installed the first unit of this type and since then about 40 of such systems have been installed (Lahiere, Hellums, Wijmans, & Kaschemekat, 1993).

From the catalytic oxidation reaction of ethylene with 99.6% pure oxygen, e*thylene oxide* is produced, with carbon dioxide and water being by-products. The purge gas for a typical ethylene oxide plant contains about 20-30$\%_{molar}$ ethylene, 10-12$\%_{molar}$ argon, 1-10$\%_{molar}$ carbon dioxide, 1-3$\%_{molar}$ ethane, 50$\%_{mol}$ methane and 4-5$\%_{molar}$ oxygen and a similar vent gas mixture is obtained during the production of vinyl acetate. This purge gas can be treated in a membrane-based recovery unit: ethylene permeates the membrane, producing an ethylene-enriched permeate stream and an argon-enriched retentate stream (Brunetti, Barbieri, & Drioli, 2015).

Membranes have been widely applied in the last few years for g*asoline vapour recovery*, especially for the recovery of the hydrocarbon vapours during the transfer of hydrocarbons from tankers to holding tanks and then to trucks. GKSS licences have installed about 30 gasoline vapour recovery systems at fuel transfer terminals, mostly in Europe (www.gkss.de), whereas MTR and OPW Fuelling Components have developed a membrane vapour recovery system for the fuel storage tanks of retail gasoline stations. The OPW Vaporsaver™ system, fitted with MTR membranes, reduces hydrocarbon emissions by 95-99% and pays for itself with the value of the recovered gasoline (www.mtrinc.com) (Brunetti et al., 2015).

The *recovery of hydrocarbon monomers from ethylene in polyethylene and polypropylene* plants is the most important application of vapour separation membranes. Actually, during the last 10 years, almost 50 of membrane based systems were installed around the world.

Air Separation

Air separation is generally obtained by cryogenic distillation or by adsorption processes.

Membrane air separation is used exclusively to separate nitrogen and to generate nitrogen stream mixture of low purity (even if with a N_2 content higher than in air) on a small scale and in remote locations.

Most of the membranes in use are made of polymeric material and are oxygen selective, so the nitrogen-rich stream is recovered in the high-pressure side, whereas O_2 enriched stream is obtained as permeate at a low pressure. Thousands of compact on–site membrane systems producing nitrogen gas are installed in the offshore and petrochemical industry. Air Products Norway has delivered more than 670 PRISM® systems to produce N_2 for different ship applications and more than 160 PRISM® systems for offshore installations. In December 2006, Air Products started with another PRISM® production plant in Missouri (USA) (Brunetti et al., 2015).

Air Liquide installed a new air separation unit with a capacity of 550 ton day^{-1} of oxygen in Dalian (China) (Hunter, 2007). In Japan, Ube Industries is increasing the production of polyimide hollow fibres for nitrogen separation to introduce a number of ethanol refining plants, mainly in the USA and in Europe, because of the increasing demand for bio-ethanol as an additive for oil products (Brunetti et al., 2015).

Rarely also *oxygen enriched air* with oxygen contents less than $30\%_{molar}$ is produced with membrane separation systems. In this case modest selectivity (3-5) and higher oxygen flux are preferred. Depending on the quality of the oxygen to be obtained, the separation can be developed in a one or two stages process, even though many studies are focused on new membrane materials that can allow to have both high selectivity and high fluxes with the aim to produce pure oxygen in a single stage system.

Air Drying

Adsorption or cryogenic processes were the only air drying processes carried out, even though such operations are not energy suitable because much energy is spent to condense water vapour to liquid. Moreover, adsorbents need to be regenerated by heat, vacuum or sweep in order to remove the adsorbed water.

Because the great part of polymers have higher water permeability with respect to air permeability, polymeric membrane based systems seem to be very suitable for the air drying applications, even if a selectivity of over 1000 would be required to reduce air loss during the process.

Various commercial products are based on either glassy polymers like polysulfone, polyimide, etc. or ionomers which are fluorine-containing polymers. In such systems water vapour is the permeate whereas dry air is recovered as the retentate.

All of the major membrane manufacturing companies such as Air Products, Air Liquide, Ube and Asahi Glass produce membrane modules for air drying applications. However, most of the business in this area is handled by other equipment manufacturers.

Membrane systems can supply oil and particulate free, dry compressed air to a dew point as low as -40°C at air pressure of 275 to 2050 kPa (Pushpinder, 2011), handling flows from 0.1 to 2.3 m^3 min^{-1}.

CO_2 Capture

Because of the stringent regulation in carbon dioxide emissions, efficient CO_2 capture systems have to be developed in order to be applied to already existing power plants, steel and cement factories and coal gasification plants as well as to be designed into new plants. The aim of such technologies is to allow to capture at least 90% of the CO_2 emissions, limiting the increase of the electricity cost (Ciferno, Fout, Jones, & Murphy, 2009). Therefore, the recovery of CO_2 from large emission sources has been a challenge attracting incredible attention for several years. At the moment, the principal technologies for carbon dioxide capture in a fossil fuel combustion process are:

- **Pre-Combustion Capture:** This strategy consists of two steps. The first one is the conversion of the fuel in a mixture of H_2 and CO (syngas mixture) by means, *e.g.*, of partial oxidation, steam reforming or auto-thermal reforming of hydrocarbons, followed by water-gas shift. The second one is the separation of CO_2 (at 30-35%) from the H_2, which is then used as clean chemical in ammonia process and fuel for turbines. In these cases, the CO_2 separation conditions could be very harsh: pressure difference can vary up to 80 bar, whereas temperatures can range from 300 to 700°C.

Membrane reactors could be used in the pre-combustion capture process in order to redesign the conventional units as more compact and efficient ones, associating the reaction to the separation in the same unit. In this case, a membrane reactor is used to carry out the hydrocarbons reforming and a second one for the water gas shift reaction, without the need for further separation steps.

- **Post-Combustion Capture:** In this case, the CO_2 is separated from N_2 in the flue gas emitted after the combustion of fossil fuels and separation conditions are milder. Streams with low CO_2 concentration are treated and SO_2, NO_2 and O_2 may also be present in small amounts.
- **Oxy-Fuel Combustion:** This procedure consists in performing the oxygen/nitrogen separation on the oxidant stream, so that a CO_2/H_2O mixture is produced through the combustion process.

The post-combustion process is the most common procedure adopted in the industrial areas (power, kiln and steel production), also because it is suitable to be retrofitted to already existing plants (Favre, 2007).

Nevertheless, it is the most difficult strategy to be applied because diluted, wet and too hot for polymeric membranes CO_2/N_2 mixture streams at low pressures have to be treated.

The conventional processes for CO_2 separation are absorption (with amines), adsorption (with highly adsorbing porous solids such as zeolite or active carbon) and cryogenic separation.

The first one of these three options is the most commonly adopted even though it has a very unfavourable energetic cost, especially owing to the regeneration step. Moreover, its main drawback is the need of chemical handling and large-scale equipment. About 3 kg of monoethanolamine per ton of CO_2 are released to carry out the absorption process.

Membranes are the most interesting alternative to the classical CO_2 post-combustion capture strategies but their application is limited by the low CO_2 concentration and pressure of the flue gas. These conditions require the use of membranes with selectivity of about 100, in order to allow a CO_2 recovery of 80%, with purity of at least 80%, as established by the *International Energy Agency* (Bounaceur, Lape, Roizard, Vallieres, & Favre, 2006). As CO_2/CH_4 selectivity of current commercial membranes, used at high pressures, is of ca. 50, they cannot be applied for one-stage operations, because of the large membrane area and high compression costs implied.

Within Nanoglowa EU project, polymeric polyvinylamine fixed site membranes were developed at NTNU Norway for CO_2 removal from flue gas. Subsequently, a pilot scale testing using real flue gas was performed at Sines power plant of EDP in Portugal. The pilot scale module used was built in a "plate and frame" configuration, consisting of 24 membrane sheets (25 cm x 25 cm) for a total membrane area of 1.5 m^2. Gas permeation experiments and material analysis showed that membranes did not lose their separation performances during more than six months operation in very harsh and challenging conditions, such as high concentration of NOx and SO_2. CO_2 permeances between 0.2 and 0.6 m^3 (STP) m^{-2} bar^{-1} h^{-1} and CO_2/N_2 selectivity between 80 and 300 were measured (Sandru, Kim, Capala, Huijbers, & Hägg, 2013).

Within an Italian MIUR co-funded project so-called FotoRiduCO$_2$ (PON-Ricerca e Competitività 2007-2013 – 01_02257) polyimide based membranes modules produced at ITM-CNR were installed in a pilot plant built in a cement factory (Italcementi, Calusco D'Adda (BG)-Italy) and are under testing, feeding them with real flue gas stream coming out from the chimney of the plant.

In 2010 Brunetti et al. (2010) introduced some general guidelines, useful to engineers, for the application of membrane gas separation technology for CO_2 capture from flue gas emissions. These general principles, introduced as maps, allow to evaluate the separation performance in terms of, e.g., CO_2 recovery and concentration on permeate side.

Membrane Condensers for Water Capture from Waste Gaseous Streams

Industrial water withdrawals account for around 22% of global water consumption. Its largest single use is in power generation which needs water for boilers, cooling and cleaning purposes. A coal-fired power station requires 1.6 liters of water per kWh of energy produced, that is a 500 MW thermoelectric power plant utilizes 45,000 m^3 h^{-1} of water (Brunetti et al., 2013). As a consequence, large amounts of evaporated water are emitted each day by the chimneys or cooling towers of power plants, paper and mills industry, etc. The new stringent rules on emissions in the atmosphere limit the amount of vapor that can be discharged, owing to the environmental concerns of these emissions in terms of greenhouse effects. Therefore, the separation and recovery of the water contained in the waste gaseous streams is a challenge to be addressed, but at the same time it can be considered as a new water source (Judd & Jefferson, 2003).

A new membrane technology called membrane condenser, with microporous hydrophobic membranes for the recovery of evaporated waste water from industrial gases, could be used to separate water vapor from a vapor/gas mixture.

Figure 3 schematizes the membrane condenser principle. The gaseous stream, e.g. flue gas, from power plants, cooling towers, stacks, etc. at a certain temperature and generally water saturated, is fed to the membrane condenser kept at a lower temperature, in order to cool the gas down to a super-saturation state. The water condenses in the membrane module and when this stream is in contact with the retentate side of the microporous membranes, the membrane hydrophobic nature prevents the penetration of the liquid into the pores, whereas the dehydrated gases pass through the membrane. Therefore, the liquid water is recovered at the retentate side and the other gases at the permeate side of the membrane unit.

Membrane condensers give higher water recovery with respect to all of the other technologies and they can be considered as clean unit operations. In addition, they do not undergo corrosion phenomena that usually are the problems related to traditional condensers or desiccant units. However, whether the gaseous stream contains hashes or particles, a pre-treatment stage could be required before feeding the membrane condenser.

In most of the cases, the waste gaseous streams coming out from chimneys of industrial plants contain also contaminants such as SO_x, NO_x, NH_3, HF, HCl, etc. that on a long term basis could affect the performance of the membrane condenser, depleting the hydrophobic character of the membrane. Therefore,

Figure 3. Scheme of a membrane condenser process for evaporated "waste" water recovery from a gaseous stream

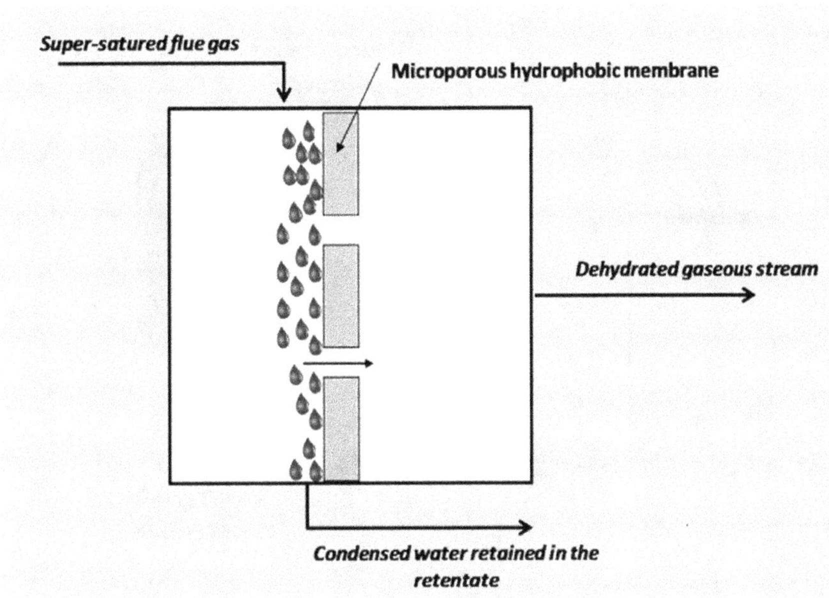

the challenge is to find out materials to make membranes exhibiting excellent resistance and stability toward these components. At the moment, such materials are PVDF or other super hydrophobic materials (e.g. Hyflon, Teflon, etc.), whose the only disadvantage is to be quite expensive. A reasonable cost as well as a long term stability of the unit could be obtained by means of the synthesis of new composite membranes, where the hydrophobic layer is deposited on a cheap support.

Pervaporation Processes

Removal of Sulphur Compounds from Gasoline

Pervaporation (PV) is today a promising operation for the separation of organic liquid mixtures such as aromatic/alicyclic mixtures, aromatic/aliphatic mixtures, polar/non-polar mixtures and aromatic isomers, both in chemical and petrochemical industries (Smitha, Suhanya, Sridhar, & Ramakrishna, 2004; Lin, Zhang, & Kong, 2009), thanks to its efficiency in separating azeotropic and close-boiling mixtures, isomers and heat-sensitive compounds.

PV processes have been recently investigated for the application in the removal of sulphur compounds of gasoline, from which they can be removed by means of a membrane because of their higher affinity with, and/or quicker diffusivity in the membrane itself.

A PV membrane module consists of two chambers divided by a non-porous polymeric membrane. One side of the membrane is in contact with the gasoline feed, whereas the other side is under vacuum or sweep gas to maintain a very low absolute partial pressure at the downstream side. Sulphur components permeate through the membrane and evaporate into the vapour phase that is then condensed.

Figure 4. Integration of S-Brane with HDS unit to improve clean gasoline output. Adapted from White, L.S. (2006). Development of large-scale applications in organic solvent nanofiltration and pervaporation for chemical and refining processes. Journal of Membrane Science, 286, 26-35

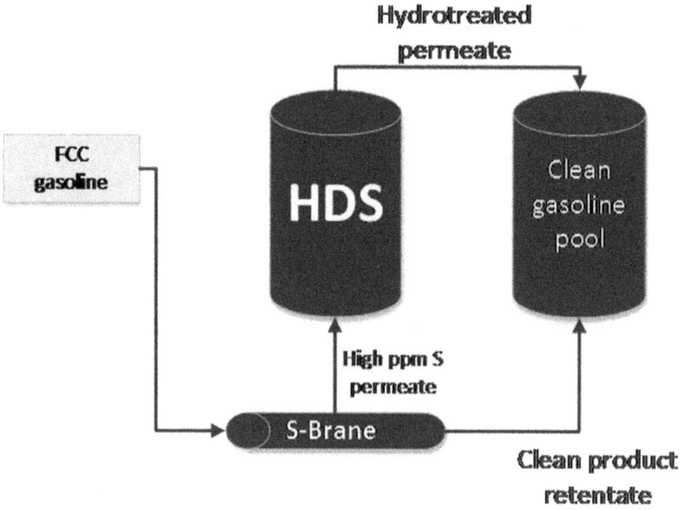

Figure 4 shows the integration of S-Brane with an HDS (hydrodesulphurization) unit to improve clean gasoline output, producing a sulphur-depleted retentate and a sulphur-enriched permeate (White, 2006). In such a system, the HDS unit uses hydrogen under high pressure and temperature to convert sulphur of the sulphur compounds to hydrogen sulphide, whereas the membrane unit allows to increase the total quantity of naphtha that can be treated, bypassing a large fraction of the feed around the HDS unit itself. In 2001, Grace Davison achieved the first scaled up installation of this technology with the S-Brane pilot plant, which had a capacity of 1 barrel day^{-1} treating streams with less than 30 ppm of sulphur.

Afterwards, in 2003 an S-Brane Demonstration Plant (300 barrels day^{-1} capacity) was installed at the Conoco Phillips, Bayway Refinery (Lin et al., 2009).

The actual membrane area required by these large-scale systems is of about 10,000 m^2 and it is determined by the feed stream composition, volume, target purities, and capabilities of available splitters and hydrogenation units. Given the successful scale-up, commercial installations should expect similar performance, with additional cost savings. Actually, the capital costs can be expected as 20-25% of that for other gasoline sulphur compounds removal technologies (White, 2006).

Trans Ionics Corporation developed another PV process, called TranSep™, for sulphur removal from gasoline (Brunetti et al., 2015). In this case a Venturi nozzle and a special operating fluid are used to create the vacuum necessary for the recovery of the permeate from the downstream side of the membrane, allowing the elimination of expensive refrigeration condensation systems.

Other Applications of PV

MTR Inc. has proposed another commercial pervaporation application for the separation of VOCs (volatile organic compound) from water, obtaining very high separation factors for toluene, benzene, chlorinated solvents, esters and ethers by means of silicon rubber coated on microporous polyimide support membranes.

MTR together with USEPA has also developed a new technology called BioSep to recover ethanol, associating pervaporation to an innovative condensation technology (dephlegmation). A dephlegmation process, consisting of partial condensation with a counter current flow of rising vapour and falling condensate, allows better separation and only vapour condensing at the top of the column has to be cooled to the lowest temperature. The BioSep process revealed to be cost-effective and energy efficient in recovering ethanol from biomass and currently is addressing new issues in bio-butanol production.

Separation of organics from water is one of the main challenges in the refineries and PV processes could be applied to solve this type of problem. Nonetheless, only few cases exist at industrial level (Smitha et al., 2004; Ravanchi, Kaghazchi, & Kargari, 2009).

Recently, Exxon installed a PV pilot plant for the separation of aromatic/aliphatic mixtures, using polyimide/polyurethane multiblock copolycondensates comprising alternate flexible (soft) and rigid (hard) sequences membranes.

However, the first pilot plant for organic-organic separation (methanol from methyl-*t*-butyl ether/isobutene mixtures) was reported in 1988 by Separex.

Xylene production and recovery has recently become an important industrial process, especially in petrochemicals. This substance presents three different isomers (ortho-, meta- and para-xylene) which are often used as chemical intermediates or solvents. Their separation is, however, very difficult to achieve because of their similar molecular structures and boiling points, requiring complex separation processes such as cryogenic crystallization or selective adsorption. If a membrane with ideal separation performance could be obtained (Qu, Dong, Zhou, Zhang, & Chen, 2010; Aliabadi, Aroujalian, & Raisi, 2012), PV could be considered as a simpler and efficient alternative to such technologies.

Qu et al. (2010) demonstrated that these isomers can be separated, at laboratory scale, by using hybrid membranes made up of poly-(acrylic acid) sodium (PAAS) membranes loaded with surface-modified zeolites.

Smitha et al. (2004) proposed an interesting analysis about the other potential PV applications in petrochemical sector, such as the recovery of aromatics like toluene and styrene contained in heavy, intermediate and light catalytic naphtha streams, the benzene content reduction in the C_6 reformates used in the production of high-grade leadless gasoline, the fractionation of alcohols/alkanes and alcohols/ether mixtures, related to the production of methyl-*tert*-butyl ether and ethyl-*tert*-butyl ether and the removal of aromatics from the feedstock of ethylene plants in order to enhance their production capability.

Organic Solvent Nanofiltration

The organic solvent nanofiltration (OSN) is a recent membrane technology, which together with pervaporation presents itself as an alternative to classical processes for the separation of organic mixtures.

Recent studies have demonstrated that its energy duty is only 10% with respect to the traditional distillation columns (Boam & Nozari, 2006). The first large-scale application of this technology was in solvent recovery from the dewaxing operation in lubes processing (Bhore et al., 1999; Gould, White, & Wildemuth, 2001). MAX-DEWAX is a large-scale OSN application used in refineries (White, 2006). In conventional operations, waxy feed is mixed with a mixture of volatile solvents like toluene and methylethylketone to precipitate the wax components as crystals after successive cooling and refrigeration. After crystal filtration, the resultant mixture of lube oil with methylethylketone and toluene is treated by multi-stage flash and successive distillations to recover the solvents. By OSN, the high costs of this particularly energy intensive step can be reduced.

Lubes and solvents mixture is fed to the membrane modules: the retentate goes to the conventional recovery section plant, whereas the permeate stream of clean and cold solvent is integrated into the chilled solvent recycle loop. One main feature of this type of system is that both the membranes and modules have to be solvent resistant. The feed in MAX-DEWAX at the refinery in Beaumont, TX takes in a volume of 36,000 barrels day^{-1}.

OSN was also applied in pilot plant trial or in laboratory experiments in solvent deoiling, homogeneous catalyst recovery, separation of phase-transfer agents and solvent exchange (White, 2006).

White (2006) provides a very interesting review about the large-scale applications of OSN in the petrochemical industry, presenting some guidelines on the convenience of using OSN or pervaporation (PV). In general, the choice between these two technologies depends on the trade-off between selectivity and costs. Aromatics-selective membranes can be used in either pervaporation or OSN processes. The lower capital and operation cost of OSN suggests that it should be considered when streams aromatics concentration is high (>80% content). On the contrary, when aromatics contents are lower, PV allows better aromatics enrichment. For reduction to very low levels of aromatics, such as reducing benzene to less than 1% in gasoline, PV is the preferred process owing to its higher selectivity.

Membrane Reactors

Membrane reactors are units in which chemical reactions are carried out assisted by a membrane separation process. The role of the membrane can be to separate one or more of the end products, shifting the reaction toward further conversion, especially for equilibrium-limited reactions, or to execute a catalytic function. Generally, the result is in both cases an improvement in productivity.

Some promising applications using inorganic membranes include dehydrogenation, hydrogenation and oxidation reactions such as dehydrogenation of butane to butadiene or of ethane to ethene, styrene production from dehydrogenation of ethyl benzene, steam reforming of hydrocarbons, water-gas shift reaction and oxidative coupling of methane, etc.

Pre-Combustion Capture by Means of Inorganic Membrane Reactors

The pre-combustion process is one of the technological solutions for allowing CO_2 capture from power plant emissions (see section on CO_2 capture).

Traditional technologies can be substituted by alternative solutions such as membrane reactors (MRs), which can reduce the number of units necessary for the fuel conversion and the successive hydrogen separation, by the association of the reaction to the separation in the same unit, once integrated in the power plant.

Global hydrogen production is mainly based on processes extracting hydrogen from fossil fuel feedstock: 96% hydrogen is directly produced from fossil fuels and about 4% is produced indirectly by using electricity generated through them. Therefore, membrane reactor technology can have a great impact at an industrial level.

As an example of MR application, it can be considered the case of fuel upgrading. The stream coming out from a reformer or a coal gasification plant contains around $50\%_{molar}$ hydrogen (on a dry basis) that must be recovered and $40\text{-}45\%_{molar}$ CO that is reduced in an upgrading stage, producing more hydrogen at the same time. In traditional applications (Figure 5), the upgrading stage is a multi-stage CO-shift process based on a series of catalytic reactors followed by a unit for the separation of the produced hydrogen from the rest of the stream (Brunetti, Barbieri, & Drioli, 2011b).

Figure 5. Traditional plant for H_2 production from natural gas. Reproduced from Brunetti, A., Barbieri, G., Drioli, E. (2011). Membrane reactors in hydrogen production. In E. Drioli & G. Barbieri (Eds.), Membrane engineering for the treatment of gases: volume 2: gas-separation problems combined with membrane reactors, 2 (pp. 87-109) with permission from The Royal Society of Chemistry

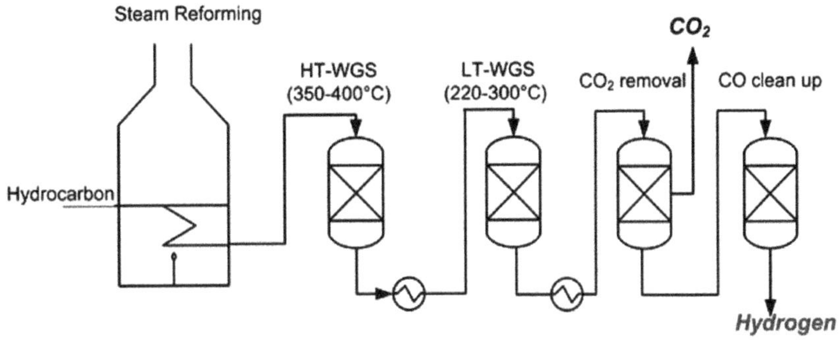

MR use allows to reduce reaction and separation stages also obtaining better performances with respect to the traditional operations. Moreover, the reaction volume required by the MR is significantly lower than the one required by the whole traditional system (Barbieri, Brunetti, Caravella, & Drioli, 2011).

The traditional process can thus be redesigned as more compact and efficient (Figure 6). A first MR is used for carrying out the reforming of the light hydrocarbons and another one for the water gas shift (WGS) reaction. The separation step is avoided, since the hydrogen stream coming out of the MR has already a high purity level, which is strictly determined by the membrane type used in both MRs.

Also in selective CO oxidation for CO removal from H_2-rich gas streams, MRs can be applied in order to maximize process efficiency and plant size reduction. By use of membrane reactors, the CO conversion as well as the purity of the hydrogen stream can be increased.

Figure 6. Integrated membrane plant for hydrogen production from natural gas. Reproduced from Brunetti, A., Barbieri, G., Drioli, E. (2011). Membrane reactors in hydrogen production. In E. Drioli & G. Barbieri (Eds.), Membrane engineering for the treatment of gases: volume 2: gas-separation problems combined with membrane reactors, 2 (pp. 87-109) with permission from The Royal Society of Chemistry

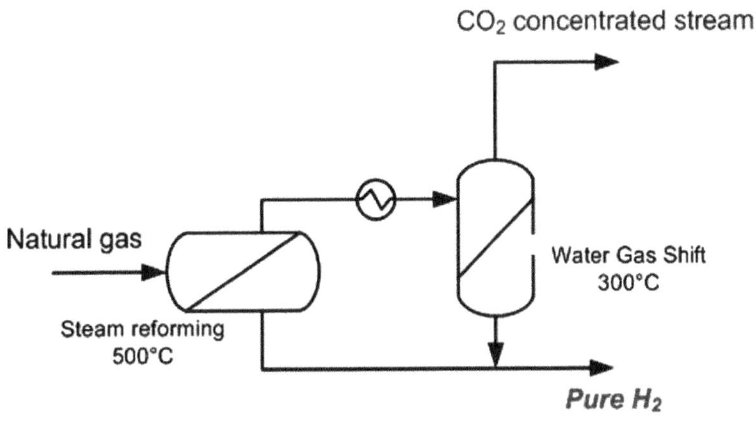

The synergic effects given by MRs through the combination of the reaction and the separation in the same unit, their simplicity and the possibility of advanced levels of automation and control, offer an interesting possibility to redesign industrial processes.

Membrane Reactors for Benzene Hydroxylation to Phenol

One of the most important petrochemical industry processes is production of phenol, which is used as raw material for synthesis of petrochemicals, agrochemicals and plastics.

A minor part of phenol is obtained by toluene oxidation, whereas the most of it is produced from benzene by the common process called Hock or cumene process. However, it requires high-energy consumption and an equimolecular quantity of acetone is produced as a by-product.

As an alternative, hydroxylation of benzene to phenol using H_2O_2 and N_2O as oxidants was studied, but in both cases neither benzene conversion nor phenol yield resulted as suitable for industrial applications. Another route is the production of *in situ* H_2O_2 by a mixture of O_2 and H_2. An example of such a procedure is given by Niwa et al. (2002), who used a shell-and-tube reactor, in which a gaseous mixture of benzene and oxygen was fed into a porous alumina tube coated with a Pd layer, whereas hydrogen was fed into the shell. Hydrogen dissociated on the Pd surface permeated onto the back and reacted with oxygen, to give active oxygen species responsible of benzene conversion to phenol. However, despite lower investments costs, the poor yield made this technology more expensive in terms of variable costs.

Another example is given by Bortolotto and Dittmeyer (2010), who proposed a novel reactor with separate membranes for distributed dosage of H_2 ($Pd_{60}Cu_{40}$, 50 μm) and O_2 (Ag 15 μm). At a concentration ratio $H_2/O_2 = 1.4$, phenol selectivity reached its highest value (9.6%).

Sato et al. (2010) studied the effect of metal particles loaded into the pores of an α-Al_2O_3 substrate support tube of the Pd-MR. Noble metal presence caused an enhancement of side reactions, whereas Cu loading promoted hydroxylation activity.

Molinari et al. (2006) carried out direct benzene hydroxylation with H_2O_2 as oxidant, using a biphasic system separated by a flat-sheet membrane: on one side of the membrane there was an aqueous phase, containing an iron catalyst and the oxidant, whereas on the other side there was an organic phase, only containing benzene. The latter was used as reactant, permeating through the membrane and reacting at the aqueous interface, and as solvent, as the produced phenol permeated back through the membrane and dissolved into benzene, overcoming over-oxidations. Nevertheless, the low permeation of phenol caused its further oxidation to other by-products and tar.

In order to overcome tar formation, in 2013, Al-Megren et al. (Al-Megren et al., 2013a) proposed a new continuous membrane reactor, in which the produced phenol was recovered using water on the stripping side of the membrane unit. Both hydrophilic and hydrophobic membrane materials were tested as well as different feed and stripping phases flow rates. 25% and 94% of phenol recovery and selectivity, respectively, were obtained by using a hydrophilic membrane, with a feed flow rate of 2 mL min^{-1} and stripping flow rate of 1 mL min^{-1}.

High costs of oxidants and the limited conversion of benzene, with formation of by-products, are the limits to membrane reactors for such a reaction.

Membrane Reactors for Dehydroisomerization of Butane to Isobutene

Isobutene, owing to its reactive double bond, can take part to several chemical reactions, such as oxidation, hydrogenation and other additions, resulting in a wide range of products. So it can be considered an important intermediate in the petrochemical industry.

On a large scale, isobutene is produced from crude oil by petrochemical cracking and from butanes, supplied from natural gas reserves and refinery streams (Romanow-Garcia & Hoffman, 2007). The most important industrial method for the production of isobutene from butanes is a two-step process (dehydroisomerization), based on *n*-butane catalytic dehydrogenation and successive isomerization to isobutene (Buonomo, Sanfilippo, & Trifirò, 1997; Pirngruber, Seshan, & Lercher, 2000). An alternative is a direct one-step process, based on the direct conversion of n-butane to isobutene (Li & Iglesia, 2008; Scirè, Burgio, Crisafulli, & Minicò, 2004).

Isobutane dehydrogenation using catalytic MRs was widely studied (Van der Bergh, Gücüyener, Gascon, & Kapteijn, 2011; Yanglong, Guanzhong, Yunsong, & Ren, 2003; Babak, Babak, Zakiev, & Kholpanov, 2009) and many membranes, such as γ-alumina, zeolite MFI, Pd/Ag and Pd, dense silica and carbon molecular sieves (Casanave, Giroir-Fendler, Sanchez, Loutaty, & Dalmon, 1995; Van Dyk et al., 2003; Liang & Hughes, 2005; Sznejer & Shientuch, 2004) were tested for this type of application.

In 2013, Al-Megren et al. (2013b) analyzed by simulation a Pd-based membrane reactor utilization in the direct conversion of n-butane to isobutene. The selective removal of hydrogen allowed the exploitation of the positive effect of the reaction pressure on the hydrogen permeation and consequentially on the conversion. Investigating temperature and pressure ranges, resulted that operating at 500°C, 20 and 0.1 bar as reaction and equilibrium hydrogen partial pressures, respectively, it was possible to obtain a conversion seven times higher with respect to that of a traditional reactor.

Membrane Systems in Ethylene Cycle

An interesting analysis in the ethylene cycle has been proposed by Bernardo et al. (2002) replacing some traditional units with membrane operations (Figure 7).

The author compared the separation and recovery of the H_2, required in the feed for improving the reactor performance, conventionally performed by a cryogenic system with the one carried out with membranes. The H_2 removal by means of membranes resulted in a reduction of energetic duties related to the refrigeration units, owing to a very different value of the operating temperature and to the successive fractionations operated at lower pressure.. Moving with the same approach, in a second study (Bernardo et al., 2004) the same authors considered ethane oxidative dehydrogenation, replacing the traditional reactor for ethylene production with a membrane reactor containing a dense perovskite oxygen-conducting membrane allowing only oxygen to permeate the membrane towards the reaction side. Comparing this improved system with the one where the conventional reactor was followed by a GS unit, it was observed that the membrane reactor showed better energy efficiency, 1.3-fold more than the traditional reactor+gas separation system.

Membrane Systems for Water Recovery

Refineries use of water is very large, with a production of wastewater being of approximate 3.5-5 m^3 per ton of crude.

Figure 7. Block flow diagram of a plant redesigned with the integration of some membrane unit operations (developed and innovative). Reproduced from Bernardo, P., Barbieri, G., & Drioli, E. (2006). An exergetic analysis of membrane unit operations integrated in the ethylene production cycle. Chemical Engineering Research and Design, 84(A5), 405-411, with permission from Elsevier

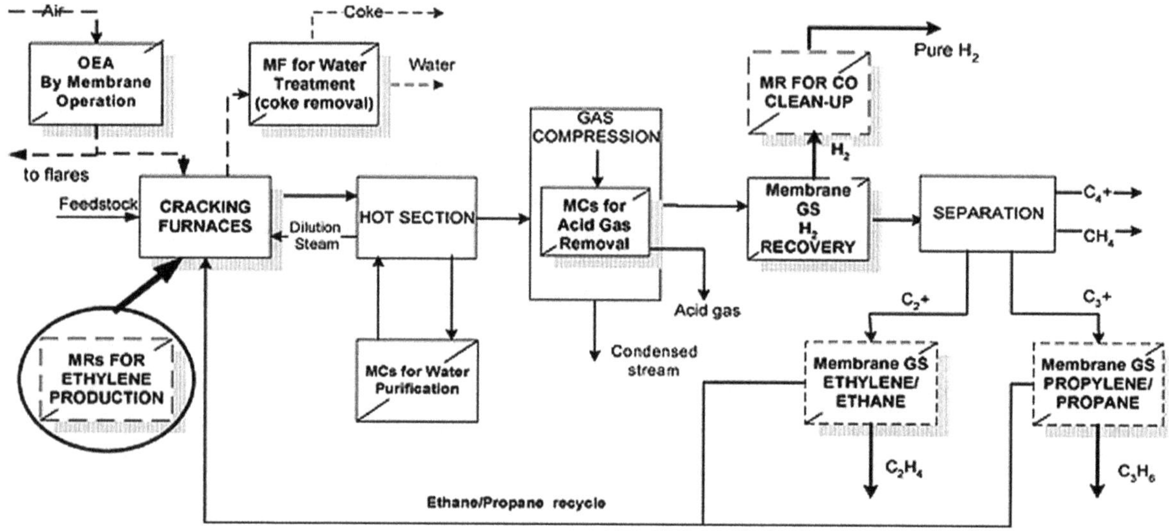

Typically, according to the use, the wastewater can be characterized by significant concentrations of suspended solids, chemical and biochemical oxygen demand (COD, BOD), oil and grease, sulphide, ammonia, phenols, hydrocarbons, benzene, toluene, ethylbenzene, xylene, polycyclic aromatic hydrocarbons (PAHs) and heavy metals (Al Zarooni & Elshorbagy, 2006; Diya'uddeen, Wan Daud, & Abdul Aziz, 2011). The typical treatment of petrochemical wastewater is constituted by (Di Fabio et al., 2013):

- Physical processes for the removal of oil, grease and heavy hydrocarbons,
- Physicochemical processes for the removal of macromolecular and colloidal substances,
- Biological wastewater processes with the aim to remove organics, nutrients and particularly lighter hydrocarbons.

In addition, refineries use water for cooling, boilers and domestic purposes and the make-up water has to be treated with specific processes such as sand-filtration, iron removal, (partial) softening, desalination, dosing of chemicals to control corrosion and bio-fouling, according to the source (surface, ground or reused water).

Pressure Driven Processes for the Purification of Water from Oil

One of the technologies for enhanced oil recovery is the injection of steam into the oil well. The injected water produced with the oil is obviously contaminated: in most of the cases, it is separated from the oil and only slightly treated and then sent back to the formation of a poor quality stream (Bernardo & Drioli, 2010). After some cycles of reuse, the water produced has to be disposed of and this operation is about the 10% of the final cost of the oil production. The disposed water contains large amounts of solutes and a pre-treatment before discharging underground at a remote site or on the surface is needed.

The possibility of treating this water, in order to reuse it, would imply significant advantages in the reduction of water consumption, particularly in the so-called water-stressed regions of the world. The conventional technologies for water treatment are gravity separation, air flotation, coagulation, flocculation, etc. but they have several disadvantages such as high operating costs, low efficiency (especially in the removal of low concentration residual oils), use of chemicals, corrosion.

Microfiltration (MF) or Ultrafiltration (UF) or combined solutions are suitable technologies to separate oils from water. Bilstad and Espedal (1996) compared MF and UF membranes in a pilot trial to treat the North Sea oilfield-produced water. The total hydrocarbon concentration can be reduced from 50 mg L^{-1} down to 2 mg L^{-1}, by applying UF membrane treatment. Benzene, toluene, and xylene (BTX) were reduced by 54%, and heavy metals like Cu and Zn were reduced to the extent of 95%. Lee and Frankiewicz (2005) of NATCO group tested a hydrophilic UF membrane to treat oilfield-produced water, placing the membrane units after a hydrocyclone used to desand and de-oil the wastewater, with a final reduction of oil and gas concentration to less than 2 mg L^{-1}.

In the Kingdom of Saudi Arabia two pilot-scale membrane processes were installed in a Riyadh petroleum refinery in order to treat wastewater discharged after an API separator. One of the processes was performed in a cross flow mode, whereas the other was performed immersing the membrane. Results on the permeate flux showed that the second type of process produced a stable flux value for more than 800 hours, whereas the flux for the cross flow membrane process declined after 600 hours. However, both membrane processes produced permeates containing less than 1.4 mg L^{-1} oil and grease (Al-Malack, 2013).

Combined membrane pre-treatment and reverse osmosis (RO) technology are effective techniques for producing water that can be used for irrigation or domestic purposes. Doran et al. (Bernardo & Drioli, 2010) present a pilot process to treat refinery water aiming to make it usable for irrigation. The system effectively produced irrigation water allowing the reduction of treatment costs of produced water, by combining precipitative softening, biological oxidation of organics, a filtration step, ion exchange softening and an RO unit.

The main problem limiting the diffusion of this technology is related to the polarization occurring often during the operation, so that RO units needs to be accompanied by pre-treatment stages and periodical chemical cleanings, increasing the operational costs and limiting the economic efficiency of this process.

Membrane distillation (MD) is another innovative membrane based process allowing to treat highly concentrated solutions and that can be used for the treatment of oilfield-produced water. MD is a temperature gradient driven process in which two phases at different partial pressure are separated through a hydrophobic microporous membrane. The vapours from the side with the highest partial pressure diffuse through the membrane and condense on the permeate side.

Macedonio et al. (2014) investigated a direct contact membrane distillation process to treat oil produced water. Four different membranes (two PVDF-based and two PP-based) were tested, varying the feed temperature and flow rate. Moreover, also the influence of different variables, such as membrane type, driving force, feed temperature, MD recovery and membrane life-time, on the process economics was investigated. In all the experimentation, the polymeric membranes showed reliable and stable performances, with rejection factors higher than 99% with respect to total dissolved solids and higher than 90% with respect to total carbon. Data from cost analysis indicated that MD can be considered a competitive technology as a produced water treatment process because the obtained water cost is intermediate between those for thermal processes operated at higher temperatures.

Membrane Bioreactors (MBR) for Water Treatment

Activated sludge processes (ASPs) were widely used for biological wastewater and sewage treatment. In order to overcome the difficulty of solid-liquid separation in biological wastewater treatment, MBR were proposed and since the 1960's many researches have been carried out on using MBR for water treatment.

MBRs are combined unit operations in which a biochemical conversion (by means of a catalyst of biological origin, i.e. an enzyme) is associated to a physical separation process.

For many applications, biological catalysts such as enzymes demonstrate higher reaction rates, greater stereo specificity and milder reaction conditions required, with respect to chemical catalysts.

A membrane bioreactor is operated similarly to a conventional activated sludge process, but without the need for secondary clarification and tertiary steps like sand filtration, thanks to the combination of biological treatment with membrane separation (Melin et al., 2006; Zhang et al., 2003). The water to be treated is separated from the purifying bacteria (active sludge) by a process of membrane filtration (Mazzei, Drioli, & Giorno, 2010). The purified effluent passes through the membrane and it is then pumped out, the sludge being recovered. It is worth of note that a MBR is up to five times more compact than a conventional activated sludge plant and capable of produce significantly less excess sludge.

The membrane filtration can occur either externally through recirculation (external loop) or inside the bioreactor (immersed configuration). In immersed membrane bioreactors, the membranes are directly immersed in the tanks containing the biological sludge and the treated permeate is extracted. This type of configuration was developed out because of the need to simplify the use of such systems and to operate more cost effectively with respect to the external loops, in terms of both energy consumption and cleaning requirements. This operating mode limits the energy consumption associated with the filtration from 0.2 to 0.4 kWh m^{-3}.

The MBR also provides a considerable level of physical disinfection, because the membrane acts as a barrier to bacteria and viruses (in UF case). The resulting high quality and disinfected effluent implies that MBRs can be especially suitable for reuse and recycling of wastewater. Since 1990's, when the early MBR systems were installed, their number has considerably grown. Commercially available systems include the package treatment plant Clereflo MBR (Conder Products, UK), designed to service populations up to 5,000 and the ZeeMOD® (Zenon Environmental Inc., today General Electric, GE), treating flow rates of up to 7,500 m^3 day^{-1}.

In refineries, the highly treated wastewater can directly be used in cooling tower make up, boilers and utilities. In 2007, Kubota installed in Oman an MBR based plant producing 78,000 m^3 day^{-1} of water reused for irrigation (Bernardo & Drioli, 2010).

ENI installed the world largest centralized MBR based plant treating petrochemical wastewater, located in the industrial area of Porto-Marghera, Venice, Italy (Figure 8). The plant receives wastewater from different petrochemical and chemical industries and it was upgraded to an MBR based plant in order to meet the strict legislation governing the effluents discharged into the Lagoon of Venice (Cattaneo et al., 2008).

Five runs were carried out, in order to reproduce the operating conditions of a full-scale MBR, and then alternations were introduced, including the addition of more external carbon source, the reduction of the anoxic compartment volume, changes in configuration and an increase of influent load. The MBR demonstrated to safeguard the effluent quality even under sudden and drastic transient conditions (Malamis et al., 2015).

Figure 8. ENI membrane bioreactor plant treating petrochemical wastewater, located in the industrial area of Porto Marghera, Venice, Italy. http://www.emag.suez-environnement.com/en/venice-water-under-protection-2558

Another Italian town, Loano, upgraded a wastewater treatment plant with General Electric (GE) advanced technology making it the second largest municipal MBR based plant in Italy (Brunetti et al.,2015).

Borghetto Santo Spirito is an Italian town located at about 60 km from Genoa: here a membrane plant was built initially to treat 7,000 m^3 day^{-1} of wastewater flow. With the new expansion, by using GE ZeeWeed technology, the treatment plant has quadrupled the wastewater flow rate and serves approximately 140,000 residents in the towns of Borghetto Santo Spirito, Toirano, Boissano, Balestrino and Ceriale. The expanded plant handles a maximum wastewater flow rate of about 35,000 m^3 day^{-1} .

An integrated membrane pilot plant for oil refinery wastewater reuse in a refinery in Indonesia was tested. This pilot plant consisted of a multimedia filter, an ultrafiltration (UF) unit, a reverse osmosis (RO) unit and an MBR unit. The UF system was used as a pretreatment prior to RO. The RO unit produced cooling water make-up and the MBR was used for treating RO concentrate. The MBR treatment in association with adsorption using granular activated carbon (GAC) led to a sensitive COD level reduction, so that the permeate quality was compliant with the wastewater quality standards for oil refinery industry in Indonesia (Widiasa, Susanto, & Susanto, 2014).

Metrics and Process Intensification

The choice of the technology suitable for the specific separation is related to different parameters like economics, stream conditions, product target and, also, to design considerations. In this logic, new design parameters have been introduced by (Miller & Stöcker, 1989) used for H_2 separation technologies, however they can be considered valid, in general, for gas separation. In addition to these latter, a new way to identify the level of progress towards sustainability was defined by introducing some industrial process indicators (Criscuoli & Drioli, 2007; Sikdar, 2003). These are usually defined as ratios, giving the impact that an operation can have independently of the scale and the balance between costs and benefits and they can be used to compare different operations (Brunetti et al., 2015). To compare membrane operation performances with those of the conventional ones, new metrics have been introduced (Criscuoli & Drioli, 2007), as to take into account the size, the weight, the flexibility, the yield and the modularity of the plants.

All of these design parameters and metrics can be used to immediately identify the benefits of a membrane operation over a traditional one and to understand in particular which kind of membrane or traditional operation should be used for a specific process. The case study reported in the next section intends to highlight these aspects and evaluations.

Case Study: Hydrogen Separation in Refineries

Recovery of hydrogen is one of the most common operations in refineries and pressure swing adsorption (PSA) or cryogenic separation are the means generally used to carry out such an operation.

Thanks to low capital costs, low energy required and modularity involved in the use of membrane systems, they are becoming more and more usual in the application of hydrogen separation.

Table 3 summarizes the comparison among the project parameters described above for the three operations considered in the case of H_2 separation.

Membrane systems can maintain product purity even at reduced capacity (down to 10% of the original design). This can be done by reducing feed pressure, increasing permeate pressure or by isolating modules from the system. They are quite able to operate under variable feed quality conditions, either on a short- or long-term basis. When membranes are used, the increase of impurity concentrations in the feed can cause a lowering in the level of product purity. However, when small changes are involved, by varying the ratio between the feed and permeate pressures it is possible to maintain the desired product purity, with the response time of membrane systems being instantaneous.

Table 3. Comparison among some important project parameters

		Membrane System	**PSA**	**Cryogenic**
Operating flexibility	Ability to operate under variable feed quality conditions, either on a short- or long-term basis.	Moderate	High	Low
Response to variations	Time to restore the separation after an unscheduled disturbance or flux variations	Instantaneous	Rapid (5-15 minutes)	Slow
Start up after the variations	Capability to restore the separation after an unscheduled disturbance	Extremely short (10 minutes)	1 h	8-24 h
Turndown	Capability to operate with under-dimensioned flow rates	Down to 10%	Down to 30%	Down to 30-50%
Reliability	It takes into account the on-stream factors, which can cause unscheduled shutdowns. It is an important project consideration, particularly if the process is a primary source of make-up hydrogen to a hydro processor or other mainstream refinery processes.	100%	95%	Limited
Control requirement	Equipment required for control	Low	High	High
By-product value	Capability of producing a high-value tail gas	Moderate	Not economical	High

Membrane systems can be considered very reliable with respect to the on-stream factor, as the membrane separation process is continuous and has few control components, which can cause a shutdown. Typically, the response to unscheduled shutdowns is rapid.

As already introduced, the choice of the appropriate hydrogen recovery system depends not only from economic considerations, but also on the kind of process from which hydrogen is produced and on its final use. Table 4 reports some selection guidelines of the most important processes of hydrogen production in refineries.

The *high pressure and low pressure purge gases from high pressure hydrocrackers and hydrotreaters* are good candidates for hydrogen upgrading. The recovered hydrogen is returned to the hydro processor with the make-up hydrogen.

Table 4. Selection guidelines for specific application processes. Published from Brunetti, A., Bernardo, P., Drioli, E., & Barbieri, G. (2011). Membrane engineering progresses and potentialities in gas separations. In Y. Yampolskii & B. Freeman (Eds.), Membrane Gas Separation (pp. 281- 312) with permission of Wiley & Sons

Process Application	Feed Conditions	Membrane System	PSA	Cryogenic
Catalytic reformer off-gas	• 70-90%$_{molar}$ H_2 • 30-10%$_{molar}$ C_1 to C_{6+} • ppv of aromatics and HCl • Pressure: 15-30 bar • Room Temperature	• H2 purity: 98%. • Final destination: uses such as catalyst regeneration	• H2 purity: 99+% • H2 recovery: 82-90% • Final destination: primary source of make-up hydrogen for hydrocracker	Not used owing to the necessity of pre-treatment for aromatic removals
Hydro processor purge gases	• High pressure purge gas: • 75-90%molar H2 • Hydrocarbon balance • Pressure: 50-200 bar • Room Temperature	• H2 purity: 92-98% • H2 pressure: 20-40 bar • H2 recovery: 85-95% • Final destination: make-up hydrogen compressor	Not used because of the high feed pressure	Not used owing to the low flow rates
	• Low pressure purge gas: • 50-75%molar H2 • Hydrocarbon balance • Pressure: 5-20 bar • Room Temperature	Not used because of the low feed pressure	• H2 purity: 99% • H2 pressure: 1 bar • H2 recovery: 80-90% • Final destination: primary source of make-up hydrogen for hydrocracker	
FCC off-gas and other refinery purge streams	• 15-50%molar H2 • Hydrocarbon or olefins balance • Pressure: 5-20 bar • Room Temperature	• Feed compression: > 25 bar • H2 purity: 80-90% • H2 pressure: 5-20 bar • H2 recovery: 50-85% • Final destination: low pressure hydrotreaters • The tail gas is sent to downstream hydrogen recovery units	Not used because of the low H_2 concentration in the feed	• H2 purity: >95% • H2 pressure: 15-30 bar • Recovery of a mixed stream containing >99+% of C2+ components
Ethylene off-gas	• 80-90%molar H2 • CO, CO2, CH4 balance • Pressure: 15-30 bar	• H2 purity: 95-97% • H2 recovery: 80-90% • Final destination: make-up hydrogen compressor	• H2 purity: 99.5% • H2 recovery: 80-90% • Final destination: primary source of make-up hydrogen for hydrocracker	• H2 purity: 95% • H2 pressure: 15-30 bar • Final destination: make-up hydrogen compressor

The high-pressure purge streams are available at 50-200 bar and contain 75-90%$_{molar}$ hydrogen, with the balance being hydrocarbons. Low pressure purge streams are available at much lower pressures (around 5-20 bar) and have hydrogen contents of about 50-75%$_{molar}$.

The membrane process is the most economical process for high pressure purge gas upgrading, whereas low pressure purge gases are usually upgraded by the PSA process.

The *ethylene production process* produces large amounts of hydrogen, which can be easily upgraded for refinery use if the ethylene plant is in close proximity to the refinery. The hydrogen-rich ethylene off-gas normally contains 80-90%$_{molar}$ hydrogen, with CO, CH_4, ethylene and nitrogen as impurities. The membrane systems are the most suitable for upgrading. The pressure and the high H_2 concentration, in fact, allow a high purity (95-97%$_{molar}$) product to be produced, even though lower than for the PSA system. Furthermore, the retentate gas stream is already compressed, on the contrary to the PSA systems that require tail gas compression, from 0.1-0.3 bar to fuel system pressure. Hydrogen can be recovered from *FCC off-gas and other low pressure refinery purge streams* of low hydrogen content. According to flow rate, feed composition, pressure and variability, and required hydrogen product purity, either the cryogenic or the membrane process can be used. If there are valuable hydrocarbons, particularly olefins, which can be recovered in addition to hydrogen, or if a hydrogen product purity in excess of 90%$_{molar}$ is required, the cryogenic process is normally used. However, feed quality variations and contaminant levels are important considerations in determining whether the cryogenic process is appropriate. The membrane process can recover hydrogen efficiently from these streams at a hydrogen purity 80-90%$_{molar}$.

The largest source of easily recovered hydrogen in a refinery is the *off-gas from catalytic reforming*. This off-gas typically contains from 70-90+%$_{molar}$ hydrogen with the balance being C_1 to C_{6+} hydrocarbons. If large quantities of reformer off-gas are to be upgraded for use as a primary source of make-up hydrogen for a hydrocracker or hydrotreater, the PSA process is normally used. The high hydrogen content makes the adsorbent requirements low, and the hydrocarbons in the tail gas sent to the fuel are relatively small with respect to the feed gas. Membrane systems using feed compression may be most economical for these applications requiring less than about 1000 $m^3\ h^{-1}$ of hydrogen product, owing to low capital costs. For large flows, cryogenic upgrading has sometimes been used. However, the feed hydrogen purity is often too high, and external refrigeration is required. Aromatic removal in the pre-treatment section is also an important design consideration. Unless recovered hydrocarbons are of high value, the cryogenic process is not normally used to process reformer off-gas alone.

As it could be seen, in the logic of a process intensification, membrane systems, owing to their high flexibility, reliability, modularity and ease of control, can be a competitive alternative to the other two classical technologies, in particular for some specific applications typical of refinery hydrogen upgrading.

ACKNOWLEDGMENT

King Abdulaziz City for Science & Technology is gratefully acknowledged for funding the project "Membrane reactor for petrochemical processes."

REFERENCES

Al-Malack, M. H. (2013). Treatment of petroleum refinery wastewater using crossflow and immersed membrane processes. *Desalination and Water Treatment*, *51*(37-39), 6985–6993. doi:10.1080/19443994.2013.778798

Al-Megren, H. A., Barbieri, G., Mirabelli, I., Brunetti, A., Drioli, E., & Al-Kinany, M. C. (2013b). Direct conversion of *n*-butane to isobutene in a membrane reactor: A thermodynamic analysis. *Industrial & Engineering Chemistry Research*, *52*(31), 10380–10386. doi:10.1021/ie400006c

Al-Megren, H. A., Poerio, T., Brunetti, A., Barbieri, G., Drioli, E., Al-Hebaid, B. S. A., & Al-Kinany, M. C. et al. (2013a). Liquid phase benzene hydroxylation to phenol using semi batch and continuous membrane reactors. *Separation and Purification Technology*, *107*, 195–203. doi:10.1016/j.seppur.2013.01.019

Al Zarooni, M., & Elshorbagy, W. (2006). Characterization and assessment of Al Ruwais refinery wastewater. *Journal of Hazardous Materials*, *136*(3), 398–405. doi:10.1016/j.jhazmat.2005.09.060 PMID:16859828

Aliabadi, M., Aroujalian, A., & Raisi, A. (2012). Removal of styrene from petrochemical wastewater using pervaporation process. *Desalination*, *284*, 116–121. doi:10.1016/j.desal.2011.08.044

Babak, V. N., Babak, T. B., Zakiev, S. E., & Kholpanov, L. P. (2009). Theoretical study of hydrocarbon dehydrogenation at high temperatures. *Theoretical Foundations of Chemical Engineering*, *43*(1), 74–87. doi:10.1134/S0040579509010102

Baker, R. W. (2002). Future directions of membrane gas separation technology. *Industrial & Engineering Chemistry Research*, *41*(6), 1393–1404. doi:10.1021/ie0108088

Baker, R. W. (2006). *Membranes for vapor/gas separation*. Retrieved from http://www.mtrinc.com/publications/MT01%20Fane%20Memb%20for%20VaporGas_Sep%202006%20Book%20Ch.pdf

Baker, R. W., Hofmann, T., Kaschemekat, J., & Lokhandwala, K. A. (2001). *Field Demonstration of a Membrane Process to Recover Heavy Hydrocarbons and to Remove Water from Natural Gas* (Annual Report). Contract Number DE-FC26-99FT40723.

Baker, R. W., & Lokhandwala, K. (2008). Natural Gas Processing with Membranes: An Overview. *Industrial & Engineering Chemistry Research*, *47*(7), 2109–2121. doi:10.1021/ie071083w

Barbieri, G., Brunetti, A., Caravella, A., & Drioli, E. (2011). Pd-based Membrane Reactors for one-stage Process of Water Gas Shift. *RSC Advances*, *1*(4), 651–661. doi:10.1039/c1ra00375e

Baudot, A. (2011). Applications of membrane gas separation in hydrocarbon industries. In E. Drioli & G. Barbieri (Eds.), *Membrane Engineering for the treatment of gases* (pp. 150–195). Cambridge, The United Kingdom: The Royal Society of Chemistry.

Bernardo, P., Barbieri, G., & Drioli, E. (2002). An exergetic analysis of membrane unit operations integrated in the ethylene production cycle. *Chemical Engineering Research & Design*, *84*(5), 405–411. doi:10.1205/cherd05014

Bernardo, P., Criscuoli, A., Clarizia, G., Barbieri, G., Drioli, E., Fleres, G., & Picciotti, M. (2004). Applications of membrane unit operations in ethylene process. *Clean Technologies and Environmental Policy*, *6*(2), 78–95. doi:10.1007/s10098-003-0225-8

Bernardo, P., & Drioli, E. (2010). Membrane technology: latest applications in the refinery and petrochemical field. In E. Drioli & L. Giorno (Eds.), Comprehensive Membrane Science and Engineering - Volume 4: Membrane Contactors and Integrated Membrane Operations (pp. 211-239). Kidlington, United Kingdom: Elsevier B.V. doi:10.1016/B978-0-08-093250-7.00054-2

Bhore, N. A., Gould, R. M., Jacob, S. M., Staffeld, P. O., McNally, D., Smiley, P. H., & Wildemuth, C. R. (1999). New membrane process debottlenecks solvent dewaxing unit. *Oil & Gas Journal*, *97*(46), 67–74.

Bilstad, T., & Espedal, E. (1996). Membrane separation of produced water. *Water Science and Technology*, *34*(9), 239–246. doi:10.1016/S0273-1223(96)00810-4

Boam, A., & Nozari, A. (2006). Fine chemical: OSN – a lower energy alternative. *Filtration & Separation*, *43*(3), 46–48. doi:10.1016/S0015-1882(06)70819-9

Bortolotto, L., & Dittmeyer, R. (2010). Direct hydroxylation of benzene to phenol in a novel microstructured membrane reactor with distributed dosing of hydrogen and oxygen. *Separation and Purification Technology*, *73*(1), 51–58. doi:10.1016/j.seppur.2009.10.019

Bounaceur, R., Lape, N., Roizard, D., Vallieres, C., & Favre, E. (2006). Membrane processes for post-combustion carbon dioxide capture: A parametric study. *Energy*, *31*(14), 2556–2570. doi:10.1016/j.energy.2005.10.038

Brunetti, A., Barbieri, G., & Drioli, E. (2011b). Membrane reactors in hydrogen production. In E. Drioli & G. Barbieri (Eds.), Membrane engineering for the treatment of gases (Vol. 2, pp. 87-109). Cambridge, The United Kingdom: The Royal Society of Chemistry.

Brunetti, A., Barbieri, G., & Drioli, E. (2015). Membranes applications in oil refining and petrochemical industry. In A. K. Pabby, S. S. H. Rizvi, & A. M. Sastre (Eds.), *Handbook of Membrane Separations: Chemical, Pharmaceutical, Food and Biotechnological Applications* (2nd ed., pp. 77–99). London, United Kingdom: CRC – press, Taylor and Francis Group. doi:10.1201/b18319-7

Brunetti, A., Bernardo, P., Drioli, E., & Barbieri, G. (2011a). Membrane engineering progresses and potentialities in gas separations. In Y. Yampolskii & B. Freeman (Eds.), *Membrane Gas Separation* (pp. 281–312). Chichester, United Kingdom: Wiley & Sons.

Brunetti, A., Santoro, S., Macedonio, F., Figoli, A., Drioli, E., & Barbieri, G. (2013). Waste Gaseous Streams: From Environmental Issue to Source of Water by Using Membrane Condensers. *Clean – Soil, Air. Water (Basel)*, *41*(9999), 1–9.

Brunetti, A., Scura, F., Barbieri, G., & Drioli, E. (2010). Membrane technologies for CO_2 separation. *Journal of Membrane Science*, *359*(1-2), 115–125. doi:10.1016/j.memsci.2009.11.040

Buonomo, F., Sanfilippo, D., & Trifirò, F. (1997). Dehydrogenation reactions. In G. Ertl, H. Knozinger, & J. Weitkamp (Eds.), *Handbook of Heterogeneous Catalysis* (Vol. 5, pp. 2140–2151). Weinheim, Germany: Wiley-VCH.

Casanave, D., Giroir-Fendler, A., Sanchez, J., Loutaty, R., & Dalmon, J. A. (1995). Control of transport properties with a microporous membrane reactor to enhance yields in dehydrogenation reactions. *Catalysis Today*, *25*(3-4), 309–314. doi:10.1016/0920-5861(95)00129-4

Cattaneo, S., Marciano, F., Masotti, L., Vecchiato, G., Verlicchi, P., & Zaffaroni, C. (2008). Improvement in the removal of micropollutants at Porto Marghera industrial wastewater treatment plant by MBR technology. *Water Science and Technology*, *58*(9), 1789–1796. doi:10.2166/wst.2008.531 PMID:19029720

Ciferno, J. P., Fout, T. E., Jones, A. P., & Murphy, J. T. (2009). Capturing Carbon from Existing Coal-Fired Power Plants. *Chemical Engineering Progress*, *105*(4), 33–41.

Criscuoli, A., & Drioli, E. (2007). New metrics for evaluating the performance of membrane operations in the logic of process intensification. *Industrial & Engineering Chemistry Research*, *46*(8), 2268–2271. doi:10.1021/ie0610952

Datta, A. K., & Sen, P. K. (2006). Optimization of membrane unit for removing carbon dioxide from natural gas. *Journal of Membrane Science*, *283*(1-2), 291–300. doi:10.1016/j.memsci.2006.06.043

Di Fabio, S., Malamis, S., Katsou, E., Vecchiato, G., Cecchi, F., & Fatone, F. (2013). Are centralized MBRs coping with the current transition of large petrochemical areas? A pilot study in Porto-Marghera (Venice). *Chemical Engineering Journal*, *214*, 68–77. doi:10.1016/j.cej.2012.10.015

Diya'uddeen, B. H., Wan Daud, W. M. A., & Abdul Aziz, A. R. (2011). Treatment technologies for petroleum refinery effluents: A review. *Process Safety and Environmental Protection*, *89*(2), 95–105. doi:10.1016/j.psep.2010.11.003

Drioli, E., Brunetti, A., Di Profio, G., & Barbieri, G. (2012). Process intensification strategies and membrane engineering. *Green Chemistry*, *14*(6), 1561–1572. doi:10.1039/c2gc16668b

Ekiner, O. M., Hayes, R. A., & Manos, P. (1992). Novel multicomponent fluid separation membranes, *U.S. Patent 5,085,676*.

Faiz, R., & Li, K. (2012). Polymeric membranes for light olefin/paraffin separation. *Desalination*, *287*, 82–97. doi:10.1016/j.desal.2011.11.019

Favre, E. (2007). Carbon dioxide recovery from post-combustion processes: Can gas permeation membranes compete with absorption? *Journal of Membrane Science*, *294*(1-2), 50–59. doi:10.1016/j.memsci.2007.02.007

Górak, A., & Stankiewicz, A. (Eds.). (2011). *Research Agenda for Process Intensification: Towards a Sustainable World of 2050. Creative Energy - Energy Transition*. Retrieved from http://3me.tudelft.nl/fileadmin/Faculteit/3mE/Actueel/Nieuws/2011/docs/DSD_Research_Agenda.pdf

Gould, R. M., White, L. S., & Wildemuth, C. R. (2001). Membrane separation in solvent lube dewaxing. *Environment and Progress*, *20*(1), 12–16. doi:10.1002/ep.670200110

Hamza, A., Towe, G., & Keyvani, M. (2013). Stable Facilitated Transport Membrane for Olefin/Paraffin Separation. *Presented at the Fuel and Petrochemical Division AIChE Spring Meeting and Global Congress on Process Safety*. Retrieved from http://www.cepmagazine.net/resources/chemeondemand/conference-presentations/stable-facilitated-transport-membrane-olefinparaffin-separation

Hunter, D. (2007). Air Liquide to install ASU for Dongbei Special Steel Group. *China Chemical Reporter, 18*(14), 10.

Judd, S., & Jefferson, B. (Eds.). (2003). *Membrane for Industrial Wastewater Recovery and Re-Use.* Oxford, United Kingdom: Elsevier B.V.

Ku, A. Y. C., Acharya, H. R., Kulkarni, P. P., Miller, S. M., & Cook Livingood, W., III. (2013). Syngas cleanup section with carbon capture and hydrogen-selective membrane. *U.S. Patent 8,495,882 B2.*

Lahiere, R. J., Hellums, M. W., Wijmans, J. G., & Kaschemekat, J. (1993). Membrane Vapor Separation: Recovery of Vinyl Chloride Monomer from PVC Reactor Vents. *Industrial & Engineering Chemistry Research, 32*(10), 2236–2241. doi:10.1021/ie00022a006

Lee, J. M., & Frankiewicz, T. (2005). Treatment of Produced Water with an Ultrafiltration (UF) Membrane - A Field Trial. *Presented at the SPE Annual Technical Conference and Exhibition*, Dallas, Texas. doi:10.2118/95735-MS

Li, X., & Iglesia, E. (2008). Catalytic dehydroisomerization of *n*-alkanes to isoalkenes. *Journal of Catalysis, 255*(1), 134–137. doi:10.1016/j.jcat.2008.01.021

Liang, W. Q., & Hughes, R. (2005). The catalytic dehydrogenation of isobutane to isobutene in a palladium/silver composite membrane reactor. *Catalysis Today, 104*(2-4), 238–243. doi:10.1016/j.cattod.2005.03.045

Lin, H., Thompson, S. M., Serbanescu-Martin, A., Wijmans, J. G., Amo, K. D., Lokhandwala, K. A., & Merkel, T. C. et al. (2013). Dehydration of natural gas using membranes. Part II: Sweep/countercurrent design and field test. *Journal of Membrane Science, 432*, 106–114. doi:10.1016/j.memsci.2012.12.049

Lin, L., Zhang, Y., & Kong, Y. (2009). Recent advances in sulfur removal from gasoline by pervaporation. *Fuel, 88*(10), 1799–1809. doi:10.1016/j.fuel.2009.03.031

Lokhandwala, K. (1997). Membrane-Augmented Cryogenic Methane/Nitrogen Separation. *U.S. Patent 5,647,227.*

Luis, P., Van der Bruggen, B., & Van der Gerven, T. (2011). Non-dispersive absorption for CO_2 capture: From the laboratory to industry. *Journal of Chemical Technology and Biotechnology (Oxford, Oxfordshire), 86*(6), 769–775. doi:10.1002/jctb.2614

Macedonio, F., Ali, A., Poerio, T., El-Sayed, E., Drioli, E., & Abdel-Jawad, M. (2014). Direct contact membrane distillation for treatment of oilfield produced water. *Separation and Purification Technology, 126*, 69–81. doi:10.1016/j.seppur.2014.02.004

Malamis, S., Katsou, E., Di Fabio, S., Frison, N., Cecchi, F., & Fatone, F. (2015). Treatment of petrochemical wastewater by employing membrane bioreactors: A case study of effluents discharged to a sensitive water recipient. *Desalination and Water Treatment, 53*(12), 3397–3406. doi:10.1080/19443994.2014.934112

Mazzei, R., Drioli, E., & Giorno, L. (2010). Biocatalytic membranes and membrane bioreactors. In E. Drioli & L. Giorno (Eds.), Comprehensive Membrane Science and Engineering (Vol 3., pp. 195-212). Oxford, United Kingdom: Elsevier B.V. doi:10.1016/B978-0-08-093250-7.00058-X

Melin, T., Jefferson, B., Bixio, D., Thoeye, C., De Wilde, W., De Koning, J., & Wintgens, T. et al. (2006). Membrane bioreactor technology for wastewater treatment and reuse. *Desalination*, *187*(1-3), 271–282. doi:10.1016/j.desal.2005.04.086

Miller, G. Q., & Stöcker, J. (1989). Selection of a Hydrogen Separation Process. *Presented at the 1989 NPRA Annual Meeting*, San Francisco, California (USA).

Molinari, R., Poerio, T., & Argurio, P. (2006). One-step production of phenol by selective oxidation of benzene in a biphasic system. *Catalysis Today*, *118*(1-2), 52–56. doi:10.1016/j.cattod.2005.11.089

Niwa, S., Eswaramoorthy, M., Jalajakumari, N., Anuj, R., Naotsugu, I., Hiroshi, S., & Fuji, M. et al. (2002). A one-step conversion of benzene to phenol with a palladium membrane. *Science*, *295*(5552), 105–107. doi:10.1126/science.1066527 PMID:11778042

Ockwig, N. W., & Nenoff, T. M. (2007). Membranes for hydrogen separation. *Chemical Reviews*, *107*(10), 4078–4110. doi:10.1021/cr0501792 PMID:17927157

Pirngruber, G. D., Seshan, K., & Lercher, J. A. (2000). Dehydroisomerization of *n*-butane over Pt-ZSM5: II. Kinetic and Thermodynamic Aspects. *Journal of Catalysis*, *190*(2), 338–351. doi:10.1006/jcat.1999.2753

Piroux, F., Espuche, E., Mercier, R., Pinéri, M., & Gebel, G. (2002). Gas transport mechanism in sulfonated polyimides. Consequences on gas selectivity. *Journal of Membrane Science*, *209*(1), 241–253. doi:10.1016/S0376-7388(02)00350-2

Pushpinder, P. (2011). Commercial Applications of Membranes in Gas Separations. In E. Drioli & G. Barbieri (Eds.), *Membrane Engineering for the treatment of gases* (pp. 215–244). Cambridge, United Kingdom: The Royal Society of Chemistry.

Qu, X. Y., Dong, H., Zhou, Z. J., Zhang, L., & Chen, H. L. (2010). Pervaporation Separation of Xylene Isomers by Hybrid Membranes of PAAS Filled with Silane-Modified Zeolite. *Industrial & Engineering Chemistry Research*, *49*(16), 7504–7514. doi:10.1021/ie100256u

Ravanchi, M. T., Kaghazchi, T., & Kargari, A. (2009). Application of membrane separation processes in petrochemical industry: A review. *Desalination*, *235*(1-3), 199–244. doi:10.1016/j.desal.2007.10.042

Romanow-Garcia, S., & Hoffman, H. L. (2007). Petroleum and its products. In J.A. Kent & Riegel (Eds.), Handbook of industrial chemistry and biotechnology (pp. 801-842). New York: Springer. doi:10.1007/978-0-387-27843-8_18

Sanders, E., Clark, D. O., Jensvold, J. A., Beck, H. N., Lipscomb, G. G., & Coan, F. L. (1988). Process for preparing powadir membranes from Tetrahalobisphenol A polycarbonate, *U.S. Patent 4,772,392*.

Sandru, M., Kim, T. J., Capala, W., Huijbers, M., & Hägg, M. B. (2013). Pilot scale testing of polymeric membranes for CO_2 capture from coal fired power plants. *Energy Procedia*, *37*, 6473–6480. doi:10.1016/j.egypro.2013.06.577

Sato, K., Hamakawa, S., Natsui, M., Nishioka, M., Inoue, T., & Mizukami, F. (2010). Palladium-based bifunctional membrane reactor for one-step conversion of benzene to phenol and cyclohexanone. *Catalysis Today*, *156*(3-4), 276–281. doi:10.1016/j.cattod.2010.04.045

Scirè, S., Burgio, G., Crisafulli, C., & Minicò, S. (2004). Dehydroisomerization of *n*-butane over H-Y zeolite supported Pt and Pt,Sn catalysts. *Applied Catalysis A, General*, *274*(1-2), 151–157. doi:10.1016/j.apcata.2004.06.019

Sikdar, S. K. (2003). Sustainable development and sustainability metrics. *AIChE Journal. American Institute of Chemical Engineers*, *49*(8), 1928–1932. doi:10.1002/aic.690490802

Smitha, B., Suhanya, D., Sridhar, S., & Ramakrishna, M. (2004). Separation of organic-organic mixtures by pervaporation – a review. *Journal of Membrane Science*, *241*(1), 1–21. doi:10.1016/j.memsci.2004.03.042

Spillman, R. W. (1989). Economic of gas separation by membranes. *Chemical Engineering Progress*, *85*(1), 41–49.

Sznejer, G., & Sheintuch, M. (2004). Application of a carbon membrane reactor for dehydrogenation reactions. *Chemical Engineering Science*, *59*(10), 2013–2021. doi:10.1016/j.ces.2004.01.051

Tannerhill, C. C. (1999). Nitrogen Removal Requirement from Natural Gas. *Gas Research Institute Topical Report GRI-99/0080, Washington, D.C.*

van den Bergh, J., Gücüyener, C., Gascon, J., & Kapteijn, F. (2011). Isobutane dehydrogenation in a DD3R zeolite membrane reactor. *Chemical Engineering Journal*, *166*(1), 368–377. doi:10.1016/j.cej.2010.11.021

van Dyk, L., Miachon, S., Lorenzen, L., Torres, M., Fiaty, K., & Dalmon, J. A. (2003). Comparison of microporous MFI and dense Pd membrane performances in an extractor-type CMR. *Catalysis Today*, *82*(1-4), 167–177. doi:10.1016/S0920-5861(03)00229-3

Weller, S. M., & Steiner, W. A. (1950). Separation of gases by fractional permeation through membranes. *Journal of Applied Physics*, *21*(4), 279. doi:10.1063/1.1699653

White, L. S. (2006). Development of large-scale applications in organic solvent nanofiltration and pervaporation for chemical and refining processes. *Journal of Membrane Science*, *286*(1-2), 26–35. doi:10.1016/j.memsci.2006.09.006

Widiasa, I. N., Susanto, A. A., & Susanto, H. (2014). Performance of an integrated membrane pilot plant for wastewater reuse: Case study of oil refinery plant in Indonesia. *Desalination and Water Treatment*, *52*(40-42), 7443–7449. doi:10.1080/19443994.2013.830759

Yanglong, G., Guanzhong, L., Yunsong, W., & Ren, W. (2003). Preparation and characterization of Pd–Ag/ceramic composite membrane and application to enhancement of catalytic dehydrogenation of isobutene. *Separation and Purification Technology*, *32*(1-3), 271–279.

Zhang, S., van Houten, R., Eikelboom, D. H., Doddema, H., Jiang, Z., Fan, Y., & Wang, J. (2003). Sewage treatment by a low energy membrane bioreactor. *Bioresource Technology*, *90*(2), 185–192. doi:10.1016/S0960-8524(03)00115-9 PMID:12895562

APPENDIX: LIST OF SYMBOLS AND ACRONYMS

API: American Petroleum Institute
BOD: Biochemical oxygen demand
BTX: Benzene, toluene and xylene
COD: Chemical oxygen demand
EDP: Energias de Portugal
EOR: Enhanced oil recovery
FCC: Fluid catalytic cracking
GAC: Granular activated carbon
GE: General Electric
GS: Gas separation
HDS: Hydrodesulphurization
MBR: Membrane bioreactor
MD: Membrane distillation
MF: Microfiltration
MR: Membrane reactor
MTR: Membrane Technology & Research
NF: Nanofiltration
NTNU: Norges teknisk-naturvitenskapelige universitet
OSN: Organic solvent nanofiltration
PAAS: Poly-(acrylic acid) sodium
PMP: Polymethylpentene
PSA: Pressure swing adsorption
PTMSP: Polymer poly [1-(trimethylsilyl)-1-propyne]
PV: Pervaporation
RO: Reverse osmosis
STP: Standard temperature and pressure
UF: Ultrafiltration
USEPA: US Environmental Protection Agency
VCM: Vinyl chloride monomer
VOC: Volatile organic compounds
WGS: Water gas shift

Chapter 6
Sustainable Process Integration in the Petrochemical Industries

Hisham S Bamufleh
King Abdulaziz University, Saudi Arabia

Mohamed M.B. Noureldin
Texas A&M University, USA

Mahmoud M. El-Halwagi
Texas A&M University, USA

ABSTRACT

The petrochemical industry has a substantial impact on the environment through the use of large quantities of natural resources (mass and energy) and discharge of wastes into the environment. Process integration offers a powerful framework for the sustainable development of the petrochemical through conservation of mass and energy and synergism among the different building blocks in the supply chain. The chapter provides an overview of process integration basics and tools that can be used to enhance the performance of the petrochemical industry. The chapter also describes the key building blocks of the petrochemical supply chain and how they may be integrated. Finally, the chapter describes key drivers and opportunities of process integration within individual petrochemical plants, among various petrochemical facilities, and the relationship to the existing refining infrastructure.

1. INTRODUCTION TO SUSTAINABLE DESIGN THROUGH PROCESS INTEGRATION

The petrochemical industry involves the usage of substantial quantities of natural resources in the form of raw materials, solvents, water, and energy. Furthermore, the impact of this industry on the environment is substantial. Industry is now launching various initiatives towards sustainable development. One definition of sustainable development emerged from the "Brundtland Report" of the 1987 World Commission on Environment and Development (WCED, 1987) which characterized sustainable development as "meeting the needs of the present without compromising the ability of future generations to

DOI: 10.4018/978-1-4666-9975-5.ch006

meet their own needs." Another definition was provided by Sikdar (2003): "sustainability occurs when we maintain or improve the material and social conditions for human health and the environment over time without exceeding the ecological capabilities that support them". For the petrochemical industry, sustainable development includes enhancing mass and energy efficiency, mitigating pollution, and increasing profitability. El-Halwagi (2012) defines sustainable design as: "the design activities that lead to economic growth, environmental protection, and social progress for the current generation without compromising the potential of future generations to have an ecosystem which meets their needs. The following are the principal objectives of a sustainable design:

- Resource (mass and energy) conservation;
- Recycle/reuse;
- Pollution prevention;
- Profitability enhancement;
- Yield improvement;
- Capital-productivity increase and debottlenecking;
- Quality control, assurance, and enhancement;
- Process safety.

As a result of the complex interaction among the different building blocks in the petrochemical industry and the need to reduce the associated usage of mass and energy resources and environmental impact, there is a need to develop and apply a holistic approach to improving the design and operation, enhancing mass and energy efficiency, and mitigating negative impact on the environment. Process integration offers a definitive and powerful framework for the sustainable design and optimization of the petrochemical industry and supply chains. Process integration is defined as "a holistic approach to process design, retrofitting, and operation which emphasizes the unity of the process" (El-Halwagi, 1997). Process integration may be classified into three branches: mass integration for the tracking and optimization of chemical species, energy integration for the transfer, transformation, and optimization of different forms of energy, and property integration for the functionality-based design of industrial processes. For a detailed description of the basics, methodologies, tools, and applications of process integration, the reader is referred to literature (e.g., El-Halwagi and Foo, 2014; El-Halwagi, 2012; Foo et al., 2011; Noureldin, 2011; Majozi, 2010; Rossiter, 2010; Foo, 2009; Kemp, 2007; El-Halwagi, 2006; Smith, 2005; El-Halwagi et al., 2004; Dunn and El-Halwagi, 2003; Hallale, 2001; El-Halwagi and Spriggs, 1998; El-Halwagi, 1997; and Shenoy, 1995).

2. OVERVIEW OF THE PETROCHEMICAL INDUSTRY

The chemical industry and in particular, petrochemicals, continues to play an important role in the advancement of other major industries such as construction, automotive, packaging, and consumer goods. The chemical industry primarily provides raw materials for other industrials and not directly to consumers, and thus is sometimes referred to as the hidden industry. This means that the chemical business is strongly influenced by weakness or strength in those industries. The rise of the petrochemical industry in the 20th century was strongly driven by the availability of feedstock, advances in chemistry and technology, the demand for new products and alternative processing pathways.

For example, the discovery of crude oil in the 19th century led to a growing industry centered on the utilization (refining) of this new feedstock. Technological advances led to improved productivity and process economics leading to further investment. This includes the production of methanol from coal at the start of the 20th century which was in the past produced through the inefficient distillation of wood. In many cases new demand was a result of disruption to existing supplies. Supply disruptions mainly in times of war (e.g. World War II) led to the rise of new processing pathways including Fischer-Tropsch synthesis, synthetic rubbers production, and nylon. These major drivers continue to strongly influence the evolution of the chemical industry.

The chemical industry has six major building blocks divided into two groups: olefins (ethylene, propylene, butadiene) and aromatics (benzene, toluene, xylene). These building blocks are primarily produced through steam cracking (e.g. naphtha, ethane, propane) and refining processes (e.g. fluid catalytic cracking, catalytic reforming). The steam cracker is the single most important process in the petrochemical industry. The major product of steam cracking is ethylene but depending on the feed composition the product distribution can vary. Ethane cracking primarily produces ethylene, while naphtha cracking produces a wide slate of products including ethylene, propylene, butadiene, and aromatics. This variability in product distribution means that different countries can vary greatly in the product distribution based on the feedstock available locally. In the Middle East, ethane is an abundant and cheap feedstock and thus mostly ethylene is produced while Europe relies more heavily on naphtha cracking given the lack of local supply of natural gas (i.e. ethane).

Recently discoveries of shale gas reserves in the United States have led to a considerable focus on ethane cracking, at a time when naphtha cracking is decreasing due to higher crude oil prices. This has significant implications for the chemical industry. A shift in steam cracker product mix impacts the availability of the building blocks for the downstream derivatives. Given the wide range of products produced by naphtha cracking, reduced capacity will mean that demand for the other building blocks not produced by ethane cracking will need to be met in new ways. In these market conditions, new innovative processing pathways and technology can take advantage of these changes.

The fluid catalytic cracking (FCC) and catalytic reforming are the major sources of building blocks for the petrochemical industry in refineries. These building blocks are produced as by-products to gasoline and diesel production. The FCC is an important source of propylene while catalytic reforming is a major source of aromatics. Changes in refinery operation to adjust the gasoline to diesel production ratio can influence the availability of these building blocks and again influence the downstream derivatives. The interaction between the natural/shale gas supply chain and the petroleum-based supply chain can change over a long period of time as resources become scarce or new discoveries are made. Figure 1 represents the key derivatives in the petrochemical supply chain. The following sections detail the different product-chains for the six building blocks, derivative uses, and important interaction between the different supply-chains.

3. OLEFINS

Olefins are typically produced by the steam cracker and used in a variety of applications including: packaging, consumer goods, and fibers.

Figure 1. Major petrochemical derivatives and supply-chains

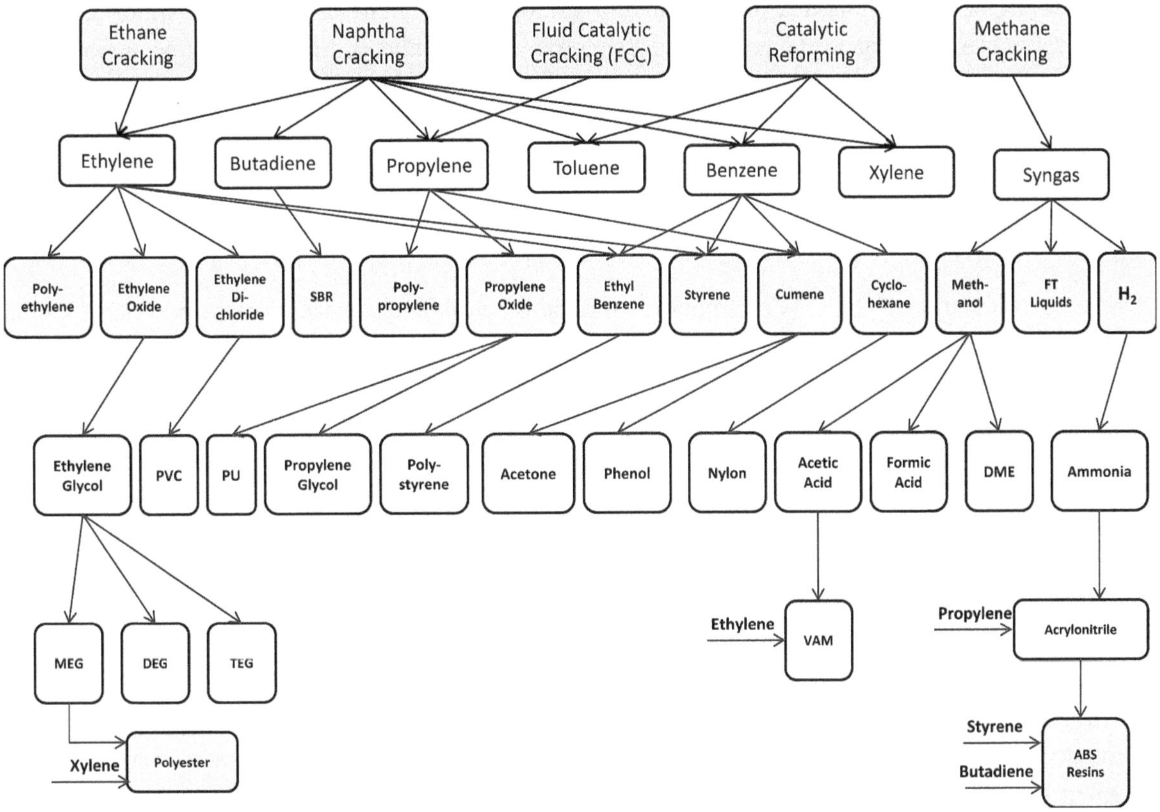

3.1 Ethylene

Ethylene is the most consumed of the building blocks with three main derivatives: polyethylene, ethylene oxide, and ethylene dichloride. Polyethylene (PE) is produced by the polymerization of ethylene. It is also the single biggest derivative of ethylene and mostly used for packaging and film applications including: food, carry-out bags, injection molded household goods, bottles, and pipes. Ethylene oxide (EO) is through the oxidation of ethylene in the presence of air or oxygen. Ethylene oxide, unlike PE, is not a final product but primarily converted to ethylene glycol (EG). Other EO derivatives include consumer chemical products used in toothpaste, shampoo, kitchen cleaners and other consumer products.

The three chemically different EGs produced from EO are: monoethylene (MEG), diethylene (DEG), and triethylene glycol (TEG). MEG is mainly converted to polyester fibers used primarily in clothing while DEG and TEG are mainly used in natural gas dehydration. The third major derivative of ethylene is ethylene dichloride (EDC) which is produced through the chlorination of ethylene. This EDC is subsequently cracked to make vinyl chloride monomer (VCM) which is almost exclusively used for polyvinyl chloride (PVC) production. This PVC is a major component in the construction industry for a wide range of applications.

With respect to the derivatives mentioned above, it is mostly a vertical chain in which ethylene is converted to a product that is subsequently converted into another value-added product. This mostly results in vertical supply-chains dependent mostly on the availability of the necessary precursor. Another example of multiple supply chains combining to produce a product is vinyl acetate monomer (VAM) an important component in paints. The production of VAM requires ethylene and acetic acid which originate from the same feedstock (natural gas) but enter two processing pathways. Ethylene produced via ethane cracking and acetic acid via methane reforming, methanol synthesis and acetic acid synthesis. This creates opportunity for integration across the different processing pathways that share a similar feedstock and produce products which combine to form value-added products.

3.2 Propylene

As previously mentioned, FCC is an important source of propylene. In fact, propylene from FCC is linked to the production of gasoline and as gasoline demand increases (e.g. summer in United States) the supply of propylene increases. With the global shift in liquid fuel consumption from gasoline to diesel this will lead to a decrease in propylene output from refineries, and while steam crackers represent the other major source of propylene, interest has recently focused on the design of processes specifically for propylene production. Another important factor is that with the substantial discoveries of shale gas in the US, many ethylene crackers are replacing the naphtha feedstock with ethane from shale gas. Since most of the propylene in the US has been produced as a co-product of naphtha cracking to ethylene, the switch from naphtha to ethane will lead to significant reduction in propylene availability (Jasper and El-Halwagi, 2015). This reduction in availability reduction is spurring to the move to alternate pathways. For example, dehydrogenation of propane (PDH) is an opportunity to utilize the propane fraction in shale gas. Other efforts have centered on the conversion of methanol olefins (MTO) and methanol to propylene (MTP) (e.g. BASF) as an opportunity to monetize the methane fraction in shale gas. The propylene supply chain has two key derivatives similar to the ethylene supply chain (polypropylene (PP), propylene oxide (PO)). Figure 2 is a representation of the propylene value chain.

Figure 2. Propylene Value Chain (Jasper and El-Halwagi, 2015)

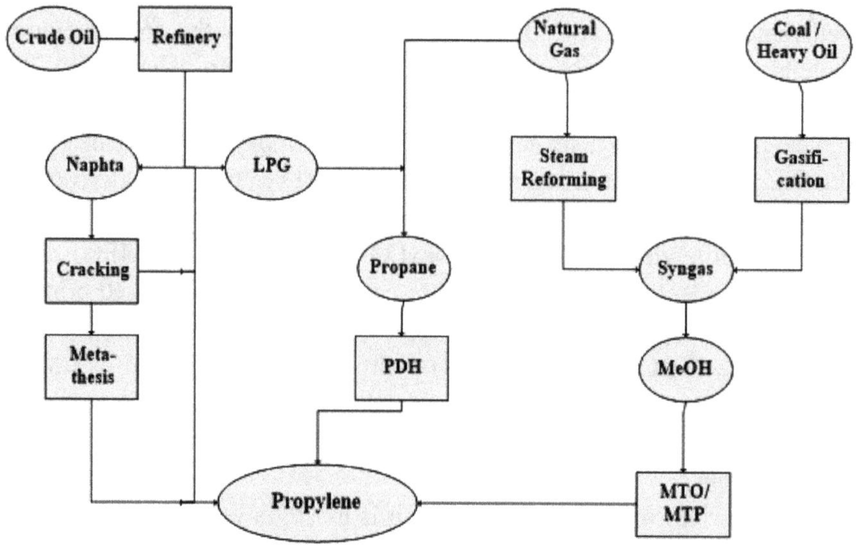

Polypropylene has a higher melting point than polyethylene making it more suited for applications in the automotive industry (e.g. dashboards) and electronic films. Propylene glycol (PG) is one of the major derivatives of PO similar to EG from EO. While EGs are typically used as antifreeze in automotive applications, a major use of PGs is in aircraft de-icing. In the United States, this creates a unique supply-demand relationship for PG based on the severity of weather and air travel. Another major use of PO is in the production of polyurethanes (PU) which are crucial in the construction industry for insulation, furniture, bedding, and a variety of foam applications.

Another important derivative of propylene is acrylonitrile which is produced through the reaction of propylene, ammonia and air. Acrylonitrile gives rise to acrylic fibers which are important in the textile industry (sweaters, blankets, socks) and acrylonitrile butadiene styrene (ABS) resins which are a class of resins important in the electronics industry (computers, TVs, telephones). From a process integration perspective, ABS resins are very unique because they combine two building blocks (propylene, butadiene) with styrene which is produced from two other building blocks (ethylene, benzene). This again provides opportunities for integration across the various processing pathways which combine to produce ABS resins.

3.3 Butadiene

Butadiene is a C4 olefin produced in FCCs and in small quantities by steam crackers. The major use of butadiene is in the production of synthetic rubbers (e.g. styrene butadiene rubber (SBR)) mostly for tire production.

4. AROMATICS

Benzene, toluene, and xylenes (BTX) are mostly produced as a by-product in refineries (catalytic reforming), from the pyrolysis gasoline (pygas) of steam crackers, and as a by-product of steel manufacture. These aromatics are important precursors for a variety of applications including home appliances, electronics, construction and automotive.

4.1 Benzene

Two of the most important derivatives of benzene, ethyl benzene and cumene, are a result of combining benzene with ethylene and propylene respectively. This creates opportunities for integration between the various supply chains and inter-process integration. Ethyl benzene is the key intermediate for styrene and its main derivative polystyrene. Polystyrene plays an important role in electric appliances most notably refrigerators but also as insulation in packaging. Cumene is the precursor for phenol production which produces acetone as a by-product.

Current processing technology produces approximately 60 tonnes of acetone for every 100 tonnes of phenol. The market demands for acetone and phenol are driven by different product supply chains. Phenol demand is driven by polycarbonate (PC), a major derivative of phenol and important in numerous applications including: glazing, bottles, housings for appliances, and automotive applications. While surplus acetone can be used in solvent applications the unmatched demand for phenol and acetone requires innovative solutions. Opportunities to convert acetone to high value products, or recycle back to ethyl benzene directly or indirectly through producing one of the original building blocks may create competitive advantages. From chemistry perspective, new paths to phenol may also be beneficial to avoid the acetone generation.

The other important benzene derivative is cyclohexane a key intermediate for the nylon supply chain. Nylon, an important polymer discovered in Dupont, was used during World War II as a replacement for silk and is now used extensively in the textile industry.

4.2 Toluene

The importance of toluene lies in its use as a source of the other two important building benzene and xylene. The limited industrial application of toluene means that significant amounts need to be converted to benzene and xylenes. Toluene conversion includes processes to produce benzene (e.g. toluene hydrodealkylation), to produce xylenes (e.g. transalkylation), and to produce both (e.g. toluene disproportionation). Other typical toluene applications include use as a solvent, as an octane booster in gasoline, and in foam applications.

4.3 Xylene

While xylenes exist in three different states (para-xylene, ortho-xylene, meta-xylene), by far the most industrially important is para-xylene. Para-xylene is a key intermediate in the production of polyester (PET) an important fiber in the clothing, bedding, and curtains industries. Polyester is also important for the bottling of carbonated drinks, food products, and water.

5. METHANE MONETIZATION

Depending on the natural/shale gas well, the composition can vary significantly. The six building blocks in the petrochemical industry that have been discussed originate as a by-product of petroleum refining for fuel production and natural gas liquids (NGLs) which includes ethane, propane, butane. Methane or pipeline natural gas is used in the electric power generation industry, residential heating, and industrial heat generation. However, methane can also play a critical role in providing important petrochemicals. Direct methane conversion is difficult due to its chemical stability. A key intermediate for methane conversion is synthesis gas (syngas), a mixture of hydrogen and carbon monoxide. Natural gas reforming is the typical method for syngas generation. This includes the addition of an oxidant (e.g. steam, oxygen) or combining multiple oxidants such is the case in autothermal reforming (e.g. steam and oxygen). Each reforming approach produces syngas with a different H_2: CO ratio.

Syngas is converted to a wide variety of chemicals, fuels, and polymers. Methane can be converted via syngas to methanol, dimethyl ether (DME), and Fischer-Tropsch (FT) liquids. Each conversion technology has specific syngas quality requirements this includes: H_2: CO ratio, pressure, CO_2 composition and acceptable impurities. Methanol is an important chemical and key feed in the production of acetic acid, formaldehyde, and formic acid. Methanol can also be converted to gasoline (MTG) and olefins (MTO). Fischer-Tropsch synthesis converts syngas into a mixture of hydrocarbons ranging from C_1 to C_{100+} which can yield once separated liquid fuel fractions including gasoline, diesel, and kerosene. The hydrogen portion of the syngas can be separated using pressure-swing absorption (PSA) or membrane technology. One of the major uses for hydrogen produced from methane is ammonia production, the majority of which is used in fertilizer and urea production. These processing pathways provide an opportunity to utilize the methane fraction of natural gas in chemical industry.

6. PETROCHEMICAL INDUSTRY DRIVERS

The petrochemical industry provides the important intermediates for a wide variety of industries and products. This includes large industries such as housing, construction, automotive, textile, and electronics. Demand in these industries is a major driver for the petrochemical industry. As discussed earlier, the petrochemical industry is also driven by feedstock cost and availability, technology, markets, logistics, and integration opportunities.

6.1 Feedstock

Feedstock availability and pricing has a tremendous impact on the petrochemical industry. Discovery of local supply of hydrocarbon feedstock provides an opportunity for countries to produce domestic liquid fuels and petrochemicals. The chemical industry is not a labor intensive industry with the major cost being associated with feedstock cost and energy, both of which are typically hydrocarbons. The chemical industry landscape can be dramatically altered by hydrocarbon feedstock price changes. Currently, crude oil is an internationally traded commodity in the world markets while gas is typically a more local price. For example, natural gas prices can range from \$3-5 per MMBTU in the US gulf coast region while being close to \$8-10 per MMBTU in Europe or Asia. Historically crude oil and natural gas prices tended to increase and decrease together. More recently, the shrinkage in the supply-demand gap for crude oil and the risk of supply disruptions has led to increased crude oil prices while the discovery of abundant supplies of shale gas has led to a decrease in natural gas prices.

The fact that natural gas prices vary from country to country creates feedstock advantage. This feedstock advantage helps the petrochemical to grow and thrive. In the United States during the late 1990s and early 2000s the high price of natural gas led petrochemical industries reliant on natural gas (e.g. methanol, ammonia) to move overseas to countries with lower feedstock costs. The current vast discoveries of shale gas reserves have once again given the US a local supply and revitalized the petrochemical industry. This has led to significant investment in additional capacity and new processing pathways. The variability natural gas prices worldwide leads to local dilemmas. Cheap gas produced in the US can be exported to Europe or Asia for a premium price while the local petrochemical anticipates vast supplies of cheap feedstock. Currently there is a moratorium on the construction of new LNG export terminals. Long term major decisions will be required to resolve such conflicts between exportation and utilization.

6.2 Technology

Technological progress plays an important role in advancing the petrochemical industry. Advanced reservoir engineering and predication tools help to find new feedstock reserves for utilization. In addition, innovation in drilling techniques can recover hydrocarbons that were not possible in the past (e.g.

horizontal drilling) and make it economic to drill wells previously not economically viable. These advances increase the availability of feedstock for the petrochemical industry and a reduction in feedstock prices typically follows.

Technological advances also play a crucial role in the advancement of new processing pathways which are more economically viable or safer than previous processing options. Finally, it is important to recognize the important impact of technological advances in other industries on the petrochemical industry. For example, a new television technology creates a demand for new chemicals and thus can benefit companies able to quickly meet this demand. These technological changes can also negatively impact the petrochemical industry, such as the reduction in the optical media (e.g. CDs) demand and subsequently the associated chemicals.

Countries which lack the feedstock advantage require greater investment in innovation and technological advances in order to compete in the global markets. Advances in catalyst design, material synthesis, and process integration can all help companies compete. In addition, a shift from low value commodity chemicals (e.g. ethylene, propylene) to more high value products such as specialty chemicals and polymers can help to alleviate higher feedstock costs.

6.3 Markets and Logistics

Market demands can influence the choice of feedstock, processing pathway, and the products produced. For example, in Europe greater demand for diesel over gasoline and the lack of natural gas production leads to greater Naphtha cracking compared to the US. In the Middle East, cheap ethane and demand for maximum crude oil export means that ethane cracking is the choice for the petrochemical industry. In turn, this influences the distribution of products produced and the types of industries that can flourish locally.

In some instances, products beyond local demand can be exported to external markets. Such export requires technological and logistic advances to ensure product is delivered safely to final destination. For example in Qatar, natural gas is physically transformed into liquefied natural gas (LNG) and exported to Asian and European markets (e.g. Japan, India). In other situations, pipelines can be used to transport the resource (e.g. Russia's natural gas transport to Western Europe). This ability to ship feedstock, intermediates, and products, gives rise to important trade-off decisions. For instance, a country may export crude oil, or convert the crude oil to an intermediate and ship the intermediate, or produce and ship the final product.

Depending on the local feedstock cost, access to market, logistics and available work force different circumstances can lead to unique supply-chain models. The fact that the petrochemical industrial is heavily reliant on end-user demand means that it is strongly influenced by global macroeconomic patterns. In good economic times, increased car sales, electronic sales and housing construction significantly improve the prospects for the petrochemical industry. Many labor intensive industries (e.g. textile) rely on intermediates from the petrochemical industry. In countries such as China, the profitability of the local petrochemical industry becomes secondary to providing a steady supply of intermediates for other major industries. This is to ensure the competitiveness of labor intensive industries and thus employment for local populations.

7. EXAMPLES OF PROCESS INTEGRATION

Process integration can be used to enhance the efficiency of individual petrochemical plants as well as petrochemical complexes. The following are two examples: one using mass integration and the other using energy integration. The mass integration example shows how multiple plants can be integrated through the concept of an eco-industrial park (EIP) where materials may be exchanged among the different plants. Figure 3 shows an EIP that integrates several plants such as gas-to-liquid (GTL) process, ethane cracking to ethylene, dimethyl ether (DME) process, biodiesel production, and acetic acid manufacture. Sources are streams from the individual plants and sinks are the manufacturing facilities in the EIP. Interceptors are units to be added to handle the various streams and to produce chemicals that are to be used in the individual plants of the EIP. The numbers in Figure 3 represent flows in kmol/hr.

Figure 3. An Eco-Industrial Park for Integrating Mass in an EIP (Noureldin and El-Halwagi, 2015)

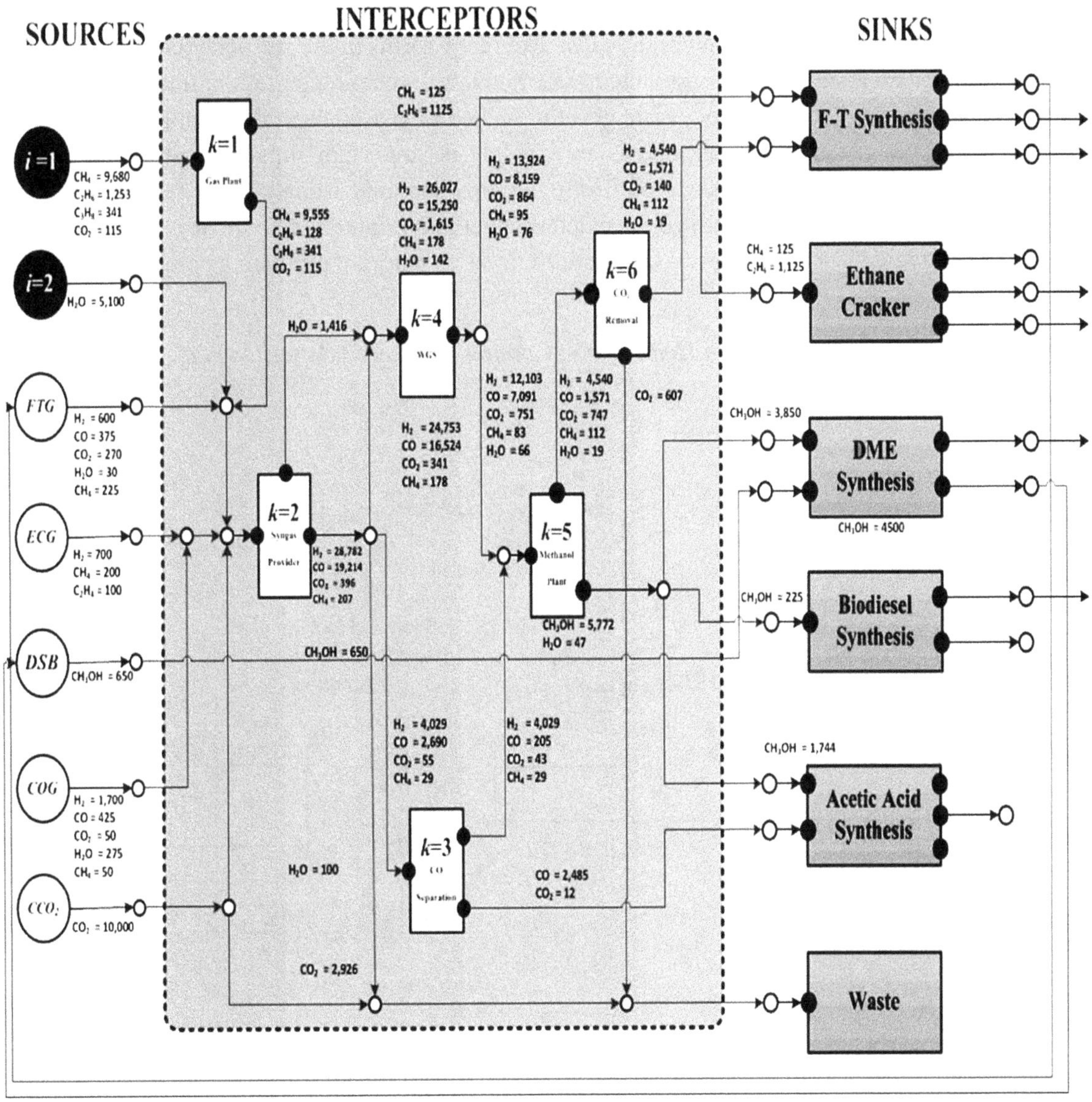

Next, an energy integration example is given. Consider the petrochemical process described shown in Figure 4.

Since there are hot streams to be cooled and cold streams to be heated, the process hot and cold streams may exchange heat to reduce the external heating and cooling utilities. Thermal pinch analysis and heat integration methods may be used to identify these targets and to synthesize a network of heat exchangers. For details on the heat integration methods, the reader is referred to literature (e.g., Kemp, 2009; Noureldin, 2011; El-Halwagi, 2012). Figure 5 shows the use of thermal pinch analysis to generate a grand composite curve that determines the targets for minimum heating and cooling utilities for the petrochemical plant of Figure 4.

8. INTEGRATION OPPORTUNITIES AND RECOMMENDATIONS

Integration can lead to significant improvement in the petrochemical industry. In particular, integration allows regions with a feedstock disadvantage an opportunity to compete and regions with a feedstock advantage greater competitiveness. Integration can take many forms in the petrochemical industry; this includes supply-chain integration, process integration and inter-process integration. Integration leads to improved resource utilization, reduced waste generation, and improved capital utilization.

Refinery and petrochemical integration is an example of important supply-chain integration. As discussed, refineries are major sources of olefins (e.g. propylene) and aromatics. By refining the crude oil and converting the by-products into value-added product in our integrated complex, a petrochemical company would avoid the feedstock price fluctuations. In fact, major oil producing companies also have

Figure 4. Petrochemical Process with Heating and Cooling Demands (Bamufleh et al., 2013)

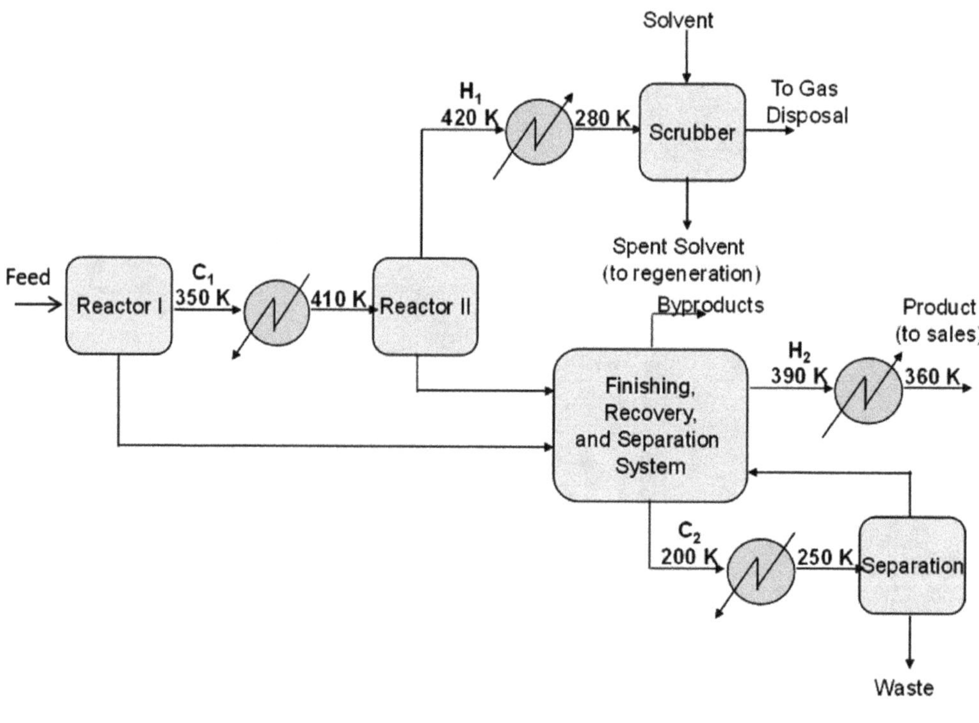

Figure 5. Determining the Minimum Heating and Cooling Utilities for the Petrochemical Plant of Fig. 4 (Abdelhady et al., 2015)

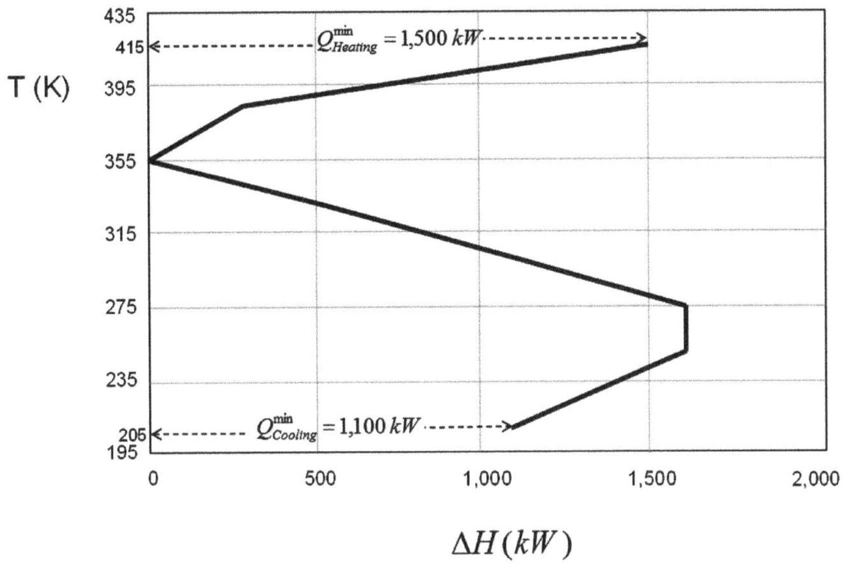

some of the biggest chemicals production (e.g. ExxonMobil). This provides a win-win opportunity, in which, if crude oil prices are low then a drop in the liquid transportation fuel profit may be somewhat offset by a benefit in the chemical business. If crude oil prices are high, then high liquid transportation fuel profits offset a loss in revenue in the chemical business.

Some companies also have strong supply-chain integration within the petrochemical business. For example, by cracking ethane to ethylene a chemical company would not be subjected to fluctuations in the ethylene market price (i.e. due to supply disruptions) which would hinder ethylene derivatives (e.g. Dow). In other situations it can be beneficial to co-locate different chemical plants to take advantage of feedstock synergies. For example, ethylene oxide production requires the presence of an air separation unit (ASU) to provide the required oxygen. On the other hand, an ammonia plant requires an ASU for the production of nitrogen. Both processes require the same unit for different portions of the feedstock. By co-locating the plants, each plant can utilize the portion of the feedstock necessary for operation while also sharing the capital investment of the ASU.

Finally, feedstock integration can also be very beneficial for petrochemicals that utilize multiple processing pathways. For example, a company producing multiple products such as ethylene, propylene and methanol would benefit from controlling more of the supply-chain by using shale gas to produce the three precursors instead of purchasing the precursors individually. The global nature of the petrochemical industry requires companies to pursue opportunities to improve. Process and inter-process integration can provide tremendous opportunities to improve process economics and reduce the environmental impact of numerous process systems.

REFERENCES

Abdelhady, F., Bamufleh, H., El-Halwagi, N. M., & Ponce-Ortega, J. M. (2015). Optimal Design and Integration of Solar Thermal Collection, Storage, and Dispatch with Process Cogeneration Systems. *Chemical Engineering Science*, *136*, 158–167. doi:10.1016/j.ces.2015.04.043

Bamufleh, H. S., Ponce-Ortega, J. M., & El-Halwagi, M. M. (2013). Multi-Objective Optimization of Process Cogeneration Systems with Economic, Environmental, and Social Tradeoffs. *Clean Technologies and Environmental Policy*, *15*(1), 185–197. doi:10.1007/s10098-012-0497-y

Dunn, R. F., & El-Halwagi, M. M. (2003). Process Integration Technology Review: Background and Applications in the Chemical Process Industry. *Journal of Chemical Technology and Biotechnology (Oxford, Oxfordshire)*, *78*(9), 1011–1021. doi:10.1002/jctb.738

El-Halwagi, M. M. (1997). *Pollution Prevention through Process Integration: Systematic Design Tools*. San Diego: Academic Press.

El-Halwagi, M. M. (2006). *Process Integration*. Amsterdam: Elsevier.

El-Halwagi, M. M. (2012). *Sustainable Design through Process Integration: Fundamentals and Applications to Industrial Pollution Prevention, Resource Conservation, and Profitability Enhancement*. Butterworth-Heinemann/Elsevier.

El-Halwagi, M. M., & Spriggs, H. D. (1998). Solve design Puzzles with Mass Integration. *Chemical Engineering Progress*, *94*(August), 25–44.

El-Halwagi, M. M., & D. C. Y. Foo. (2012). Process Synthesis and Integration. In Kirk and Othmer Encyclopedia of Chemical Technology. Wiley.

Foo, D. C. Y. (2009). State-of-the-Art Review of Pinch Analysis Techniques for Water Network Synthesis. *Industrial & Engineering Chemistry Research*, *48*(11), 5125–5159. doi:10.1021/ie801264c

Foo, D. C. Y., El-Halwagi, M. M., & Tan, R. R. (Eds.). (2014). *Recent Advances in Sustainable Process Design and Optimization. Series on Advances in Process Systems Engineering*. World Scientific Publishing Company.

Hallale, N. (2001). Burning Bright: Trends in Process Integration. *Chemical Engineering Progress*, *97*(7), 30–41.

Jasper, S., & El-Halwagi, M. M. (2015). A Techno-Economic Comparison of Two Methanol-to-Propylene Processes. *Processes*, *3*(3), 684–698. doi:10.3390/pr3030684

Kemp, I. (2009). *Pinch Analysis and Process Integration - A User Guide on Process Integration for the Efficient Use of Energy* (2nd ed.). Butterworth-Heinemann.

Majozi, T. (2010). *Batch Chemical Process Integration: Analysis, Synthesis, and Optimization*. Heidelberg: Springer. doi:10.1007/978-90-481-2588-3

Noureldin, M. B. (2011). *Pinch Technology and Beyond: New Vistas on Energy Efficiency Optimization*. Nova Science Publishers.

Noureldin, M. M. B., & El-Halwagi, M. M. (2015). Synthesis of C-H-O Symbiosis Networks. *AIChE Journal. American Institute of Chemical Engineers*, *61*(4), 1242–1262. doi:10.1002/aic.14714

Rossiter, A. P. (2010). Improve Energy Efficiency via Heat Integration. *Chemical Engineering Progress*, *106*(12), 33–42.

Sikdar, S. K. (2003). Sustainable Development and Sustainability Metrics. *AIChE Journal*, *49*(8), 1928–1932. doi:10.1002/aic.690490802

Smith, R. (2005). *Chemical Process Design and Integration*. New York: Wiley.

WCED (World Commission on Environment and Development). (1987). *"Our Common Future", (The Brundtland Report)*. New York: Oxford University Press.

Chapter 7
Safety and Efficiency Enhancement in LNG Terminals

Ravinder Singh
Lamar University, USA

Helen Huiru Lou
Lamar University, USA

ABSTRACT

Liquefaction of natural gas helps in transporting it over long distances by sea vessels. It is then regasified and transported through pipelines to the consumer. Due to large energy density of Liquefied Natural Gas (LNG), and associated flammability issues, the LNG terminal involves high risk. Consequently, safety is an important factor in the operation of LNG terminals. Although a substantial amount of time money and effort has been put in this area, there is always some possibility of improving the process so that less risk is involved. Rapid advancement in process simulation software like Aspen Plus and Aspen HYSYS, has led to the convenience of experimenting the various control methodologies on the computer offline from the actual plant operation, before they are implemented in real time. In this chapter, main hazards associated with LNG terminal operation will be highlighted. Further, recent advancements in research for safety enhancement and efficiency enhancement in the liquefaction and regasification processes will also be included.

INTRODUCTION

Natural gas is widely used as gaseous fuel in heating, transportation and power generation. Due to recent advancements in the fracking technology, it is now possible to economically produce natural gas from large shale formations (U.S Energy Information Administration, 2014). It is liquefied at the Liquefied Natural Gas (LNG) shipping terminals which are usually located along the coast and transported by LNG tankers over the sea. It is stored and regasified at the LNG receiving terminal and is transported through a network of pipelines to the end consumer.

DOI: 10.4018/978-1-4666-9975-5.ch007

Natural Gas Imports and Exports in US

In the regional U.S markets, natural gas imports play an important role. In 2012, 4 percent of natural gas consumed in US was imported. Ninety eight percent of imported natural gas came from Canada through integrated pipe line network, while 2 percent was imported as LNG. However, imports from Canada are declining since the increase in U.S shale production. Net U.S imports fell from 3,785 Bcf in 2007 to 1,181 Bcf in 2014. As the shale gas production increased further, it is estimated the imports will continue to decrease. LNG imports accounted for 1-3 percent of total natural gas demand between 2003 and 2008 according to EIA. In summer 2007, LNG imports reached maximum at about 100 BcF per month (Federal Energy Regulatory Commission, 2015).

United States exports natural gas to Mexico, Canada and Japan. Growth in shale gas production in U.S between 2011 and 2014 led to proposals to export large volumes of LNG. Consequently, many LNG export facilities have been approved, as of January 2015, but none have begun operations (Federal Energy Regulatory Commission, 2015).

Safety in LNG Terminals

Due to the hazardous nature of LNG, it is important to reduce the risk it poses to the plant personnel and the community. Risk in the process is defined as a product of probability that an undesired event will occur and the consequences that occur as a result of the undesired event. (Cramer, 1991)

i.e. Risk = Probability * Consequences

The calculation and mitigation of risk is governed by a Risk Management Plan (RMP), which a facility must maintain as a management tool in order to comply with Section 112(r) of the Clean Air Act of the Environment Protection Agency (EPA). A typical RMP includes

- Hazard identification,
- Consequence analysis,
- Hazard control and mitigation,
- Procedures for:
  - Operation of the plant,
  - Maintenance,
  - Testing and inspection
  - Management of change.
- Training,
- Emergency planning,
- Accident investigation,
- Audits.

Effciency in LNG Terminals

Operating the terminal in safe and efficient manner is a challenging task. Process simulation is an important tool in the hands of process engineers because they can experiment with various production routes on a computer before actually implementing their ideas in the plant. The process under observation can be optimized for better performance by tweaking the process variables, thereby increasing its efficiency. Process simulation is basically of two types: steady state simulation and unsteady state simulation. Although there are a number of process simulators available in the market, Aspen Plus for steady state simulation and Aspen Hysys for unsteady state simulation, may be used for simulation and efficiency enhancement of LNG terminals.Major LNG emission sources are (ConocoPhillips and EPA's Natural Gas STAR Program, 2006)

- Ship in transit,
- Ship unloading,
- Ship loading,
- Terminal operations,
- Storage tanks,
- Boil-off gas recovery system,
- Send-out (pumps, vaporizers),
- Liquefaction,
- Vehicle fuel (trucking).

The efficiency of the operations in the LNG terminals can be enhanced by

- Improving connecting and disconnecting practices,
- Improving tank pressure management,
- Improved vapor recovery system maintenance and availability,
- Better enforcement of vessel venting rule,
- Improving the liquefaction cycle efficiency,
- Improving the compressor and driver efficiency,
- Utilizing the waste heat generated in the plant.

This chapter discusses various options to identify hazards associated with the LNG terminals and describes how those hazards may be mitigated for enhancing their safety and efficiency.

BACKGROUND

After natural gas is rid of water, condensates, acid gases (e.g. CO_2), mercury and other impurities, its methane content varies between 80 to 99% by volume. Apart from methane, natural gas also contains ethane, propane, butane and few other higher hydrocarbons in small quantities. Its typical composition is shown in Table 1.

LNG is non-toxic, non-corrosive, colorless and odorless. In order to detect its leakage, t-butyl mercaptan is added as an odorant to it, which gives it its characteristic smell of rotten eggs. It is a flammable compound and its vapor flammability limit is between 5 - 15% by volume. The typical physical properties of natural gas are shown in Table 2.

Table 1. Natural Gas Composition

Methane concentration, volume %	80-99
Ethane concentration, volume %	2.7-4.6
Nitrogen concentration, volume %	0.1-15
Carbon dioxide, volume %	1-5
Sulfur concentration, weight % ppm	<5

Source: Handbook of Natural Gas Transmission and Processing by Mokhatab, Saeid

Table 2. Physical properties

Property	Value
Boiling point, °C	-162
Flash point, °C	-188
Autoignition temperature, °C	540-560
Vapor flammability limits, volume %	5-15
Flammability limits	0.7-2.1

Source: Handbook of Natural Gas Transmission and Processing by Mokhatab, Saeid

In a closed room, it can cause asphyxiation if it is present in high concentration. The auto ignition temperature of 10% LNG vapors is 540 °C, which is high, but if LNG does get ignited, the resulting "pool fire" burns much hotter and faster than gasoline fuel. LNG is a cryogenic fuel, so it can cause frostbite when it comes in contact with skin tissue, so protective clothing is required to handle it.

Liquefied natural gas is lighter than water, due to which its spills over water cause it to form a layer on the top and its vapors rise very rapidly. If there is thin spill on soil, it evaporates very fast, leaving behind no residues. On the other hand, if there is large spill on the ground, it will cause the ground to cool rapidly as the LNG evaporates. Consequently, evaporation of LNG will slow down with time. Depending upon the spill's volume and shape, it may take hours before LNG evaporates completely. The primary risk scenarios for LNG spill over water are:

- Rate of release of LNG from a ship or facility,
- Spread of unconfined pool on water,
- Vapor generation from unconfined spills on water,
- Thermal radiation from pool fires on water,
- Distance for flammable vapor dispersion following spills,
- Rapid phase transitions,
- Effects of thermal radiation on people and structures (ABSG, 2004),
- Environmental.

Decision making related to scenario risks should be considered in context of both potential consequence and expected likelihood (or frequency). LNG vessel and associated facility operations are highly regulated and closely monitored/ controlled by authorities; so many layers of protection exist against losses (ABSG, 2004).

Liquefaction Processes

Pure and mixed refrigerants are used to bring down the temperature of natural gas to liquefy it at -162 ℃ and 1 atm. Important liquefaction processes are:

- **Single mixed refrigerant (SMR):** This process is used when the size of the facility is small and capital costs have to be low. The disadvantage is that the efficiency of the plant is not very high. The refrigerant is a mixture of hydrocarbons and nitrogen, and their composition is optimized to match the cooling curve of the natural gas.
- **Propane Precooled Mixed Refrigerant (C3MR) Process:** This process was developed by Air Products and Chemicals, Inc., and is used for large scale liquefaction of natural gas. The feed natural gas is precooled by propane to -35 ℃ and then further cooled down to -162 ℃ by mixed refrigerants.
- **Cascade Process:** This process was developed by Conoco Phillips. It is highly efficient process which uses several different components arranged in separate loops to closely match the cooling curve of natural gas. Due to its complexity, the associated cost is high.
- **Double Mixed Refrigerant (DMR) Process:** This process was developed by Shell. It has fewer number of equipment as compared to C3MR, is more flexible and offers wider operating ranges (Nibbelke, Kauffman, & Pek, 2002).

Hazards in LNG Terminals

During the unloading of LNG cargo from the sea vessel into the LNG storage tank, there might exist difference in densities of LNG inside the tank due to tank temperature profile. This may cause different unstable layers of liquid to form inside the tank. Due to heat losses, the lower portion of the tank might get heated faster than the upper portion of the tank, and the density of liquid in the lower layer will decrease causing sudden vaporization of LNG called "rollover". If the rate of vaporization is high, and excess pressure cannot be released by pressure safety valves, rollover can cause cracks and structural failures in the tank. To prevent rollover and stratification, the density of the LNG is constantly monitored during unloading process. The rollover protection system includes pump-around mixing systems and temperature sensors distributed inside the tank.

When LNG is spilled over sea water, it may forcibly be mixed with water due to the turbulent nature of the sea. If the rate of heat transfer from the water to LNG is quite high, it might cause Rapid Phase Transition (RPT), due to which LNG instantly undergoes phase change and vaporizes, expanding up to six hundred times its original volume. The RPT results in either an overpressure or dispersion of "puff" of LNG expelled into the atmosphere. Although combustion is not involved in this transition, is very similar to explosion, and may be a hazard to the surrounding community if the rate of release is high

To minimize the risk involved in operation of large scale LNG terminals, industry follows a protection scheme which includes

- Primary containment,
- Secondary containment,
- Safeguards systems, and
- Separation distance.

The major objective of primary containment is to contain the LNG and prevent it from forming a flammable mixture with air in the ratio between 5% - 15% by weight. This is achieved by constructing a double walled storage tank with wrap around insulation and reinforced concrete roof. Many built-in safety mechanisms along with seismic isolators and foam glass insulation floor of the tank are standard features of modern onshore tanks. Dykes and berms form the secondary containment layer to isolate LNG spills, in case the primary containment fails. It is designed to exceed the volume of the primary containment system.

Various process control systems including Safety Instrumented Systems (SIS) form the tertiary layer of protection to mitigate the effects of LNG spills or leaks. SIS consists of automated leak detection systems and remote shutoff systems. In order to ensure their effectiveness and reliability, it is essential to conduct periodic safety audits including but not limited to HAZOP studies, what-if analysis, checklists studies, layers of protection analysis etc.

The surrounding community and LNG terminal are separated by federally-mandated minimum distance. This minimum distance varies depending upon the location of the site, LNG vapor dispersion data and thermal radiation contours. Detailed provisions of various different government regulations control the siting of the terminal. The thermal radiation exclusion zones are made large enough so that heat from LNG fire does exceed specified limits and cause danger to the community. Safety zones are identified for LNG cargo ships to avoid accidents while the ships are at sea or moored.

Risk Analysis

There are a number of risk analysis methodologies which have been developed for identification and evaluation of risk in process plants. The usual criterion is the combination of qualitative, quantitative, deterministic and probabilistic methodologies. Almost all of them go through three main phases i.e.

- Hazard identification,
- Evaluation of risk for its quantification,
- Ranking the risk.

During the conceptual design phase of the project, preliminary hazard identification technique is used. This helps in the elimination of time-consuming and costly delays due to the changes made in the later stage of the project (Cramer, 1991). Consequently, by including safety factors in the design criteria, potential hazards may be taken care of. This method is qualitative in nature and the methodology is quite general. To start with, the conceptual design is thoroughly assessed for identifying hazards related to raw materials in the feedstock, unit operations, utilities, facilities and safeguards.

When the process plant is under normal operation, safety audits are performed. They are performed periodically in order to certify that the plant complies with its safety program. The main feature of the audit is that a team thoroughly inspects critical safety variables and makes sure that the risk management plan is effectively implemented. The plant personnel are interviewed to improve upon any lapses in the safety procedures. These audits are performed in along with other risk assessment techniques like What-if Analysis, Hazard and Operability (HAZOP) Analysis, Failure Modes and Effects Analysis (FMEA), Layers of Protection Analysis (LOPA) etc. Whenever possible, hazards should be eliminated at the early stage of the process development. Kletz et. al. have defined this concept as inherent safety. This includes the consideration of alternative manufacturing processes and reducing the use hazardous chemicals in the process.

Process Simulation

There is a tough competition amongst the world's premier LNG organizations for reducing the costs and increasing the profitability, while keeping the safety standards high. Although there has always been the use of computers in the process industry in the past, but due to exponential rise in information technology, sophisticated software like Aspen Engineering Suite and PRO/II for process design and simulation has become indispensable tool in the hands of process engineers (Mortazavi, et al., 2012) .

The main advantage of process simulation is that it has reduced the need for pilot plant runs. Both steady state and unsteady state operations can be simulated with accuracy, if the right combination of chemistry, thermodynamic package and other process variables are chosen to run the simulation. Another advantage is that various experimental routes may be simulated without putting any lives to danger, especially when dealing with hazardous chemicals (Dauber & Span, 2012).

SAFETY ENHANCEMENT IN LNG TERMINALS

In 1944, Cleveland, Ohio, USA, a tank failed in a peak-shaving plant causing the contents to spill in the street and storm sewer system. The resulting fire and explosion from the spill killed 128 people. Subsequent investigation found that the material of construction of the tank was alloy steel with low nickel content, which caused the tank to turn brittle when it was exposed to extreme low temperature of LNG. In 2005, Nigeria, a 28 inch underground LNG pipeline exploded causing fire in approximately 27 Sq. Km (CH.IV International, 2012).

Another accident at LNG liquefaction and export facility on January 19, 2004, Skikda, Algeria, resulted in 27 deaths and injured 72 people. There was a leak in the refrigerant system of the LNG facility and the resulting vapor cloud was drawn into a nearby high pressure steam boiler. The rapidly increasing boiler pressure could not be handled by the boiler's safety valve, causing the rupture of steam drum. Consequently, nearby gas caught fire and explosion occured. U.S. Federal Energy Regulatory Commission (FERC) report concluded that a lack of automatic equipment shutdown devices and hazard devices, along with local ignition sources was the cause of the accident (CH.IV International, 2012).

Enhancing the safety of the LNG terminal is a challenging task for the design engineers. Evaluation of risk in the early design stage is very important to avoid costly changes in the later stage of the project. Dow Fire and Explosion Hazard Index (DFEHI) and Mond Index (MI) are used to assess risk due to fire and explosion hazards. Calculation of these indices is a time consuming and complex methodology which requires a lot of detail, which is not available at the early design stage.

Inherent Safety Index (ISI), as proposed by Anna-Mari et al, requires less detail as compared to DFEHI and MI. The Inherent Safety Index does not take into account the flow rate of the material in the consideration. The Comprehensive Inherent Safety Index (CISI) takes into account the flow rates of the chemicals involved in the process in order to better assess the risk (Gangadharan, Singh, Cheng, & Lou, 2013). Since CISI provides more conservative results as compared to other indices in the early design stage, it can be used to assess risk for designing the LNG terminal.

A Safety Instrumented Function (SIF) in a process, such as LNG terminal, senses hazardous condition and automatically takes necessary actions to move the system to towards a safe state. Safety Instrumented System (SIS) includes safety interlocks, gas and fire systems, and emergency shutdown systems. There are many SIS in an industrial process and they implement various SIFs in order to make the process safe to operate. The entire life cycle requirements of an SIS are established by IEC 61511. The Safety Life Cycle (SLC) of an SIS includes:

- Design,
- Specification,
- Implementation,
- Operation,
- Maintenance,
- Modification.

The main steps for SLC of an SIS are:

1. Determine whether a SIS is required, if "yes", then,
2. Determine target Safety Integrity Levels (SIL) for each SIF that is implemented.

There is lack of an adequate guidance and methodology for conducting these two steps. Hector Javier et al in their study (Cruz-Campa & Cruz-Gomez, 2010) have suggested a methodology to determine the targeted SIS and SIL from HAZOP study. A high pressure sour gas amine treatment process was chosen as a working example. The methodology includes the following steps:

Step 1: Identify a hazardous event and asses its severity.
Step 2: Identify the initiating event and assess its frequency.
Step 3: Identify the applicable Independent Protection Layers (IPL) and evaluate their effectiveness.
Step 4: Calculate expected frequency for the hazardous event.
Step 5: Determine the need for additional Layers of Protection and the Required SIL, if a SIS is recommended.

It was found that many scenarios related to risk in process plants can be addressed by improving process design and instrumentation to mitigate the frequency and magnitude of process deviations so that there is less dependence on safety systems.

Hazard and Operability (HAZOP) studies, have become industry standard for safety assessment in chemical plants, however, in a survey conducted by Ian Duguid, it was found that about half the number of incidents in the survey could have been avoided if the HAZOP study was well-conducted (Duguid, 2001). The main limitation of a HAZOP study is that it excludes anything that may not be detected through operating procedures, P&IDs and backup data. It greatly relies on the skill and judgment of the team conducting the HAZOP study. For example, there is absolutely no way to deal with some human errors, straight mechanical failures or if the material of construction is not properly specified (Duguid, 2001). G Unni Krishnan in his study (Krishnan, 2005) suggests a fifteen point checklist to conduct a productive HAZOP study.

EFFICIENCY ENHANCEMENT IN LNG TERMINALS

The use of Aspen Engineering Suite to simulate both steady state and unsteady state processes involved in the operation LNG terminals (Contreras, 2005) has become common practice in the industry. The composition of natural gas depends upon its source and the kind of processes it has gone through for sweetening and purification. For the purpose of simulation, this composition can easily be obtained and entered in the simulator. Once the process variables are put in the simulation, they can be further optimized by experimenting with different values input by the user or generated by complex algorithms programmed by the design engineer.

Due to high energy demand of the refrigeration cycles used in the liquefaction of natural gas, energy efficiency is an important factor for economical operation of the plant. To liquefy one kilogram of natural gas, about 1188 kJ of energy is required (Finn, Johnson, & Tomlinson, 1999) and it varies based upon the site location and the kind of liquefaction cycle used in the process.

Improving the Refrigeration Cycle Efficiency

The efficiency of the refrigeration cycle can be improved by optimizing the mass flow rate, composition and pressure of the refrigerant used in that cycle. The Joule Thomson valve is used for LNG expansion in the propane precooled mixed refrigerant process (C3MR). The benefits of using turbine expander for achieving the desired refrigeration in place of the Joule Thompson valve have been suggested by (Kanoglu, 2001). Liquid turbines may also be used instead of LNG expansion valves and the effect of such replacement has been studied by Renaudin (Renaudin, 1995). Most of these studies focus on increasing the efficiency of the mixed refrigerant cycle and recovering energy from expansion of LNG. However, the use of expanders for enhancing propane cycle in the APCI process and LNG expansion process has not been considered in the past, (Mortazavi, et al., 2012) in the simulation with Aspen software have studied the effect of replacing the expansion valves with expanders on performance of the entire APCI LNG plant. They have shown that the LNG production increase, expansion work recovery, and compressor power reduction can be achieved up to 1.24 MW, 3.82 MW and 2.68 MW respectively if two-phase expanders were used in the processes for reducing the expansion losses. They have used Peng-Robinson-Boston-Mathias thermodynamic equation of state for the modeling of the properties of the chemical components.

For achieving higher accuracies for simulation of certain components including natural gas in Aspen Plus, Kunz and Wagner introduced GERG-2008 equation of state. Since this is a new reference equation, Dauber et al (Dauber & Span, 2012) tested it for validating the expected increase in accuracy for industrial application, and have compared it with other equations of state such as Peng-Robinson, Redlich-Kwong-Soave and Lee-Kesler-Plocker. Simplified Snohvit LNG sub cooling process was used to conduct the study. It was found that the simulation of liquefaction of natural gas, when modeled with GERG-2008 equation of state, can reduce the uncertainty level in the process. There was small deviation of results by GERG-2008 when compared with actual experimental data of the main cryogenic heat exchanger and the total power of the compressor of the liquefaction plant located in Norway.

Improving the Boil-Off Gas (BOG) Recondensation System

Heat ingress from the surroundings to LNG system cannot be fully avoided. Consequently, some of the LNG is vaporized and is called boil-off gas (BOG). BOG is generated in the storage tanks, pipelines and the cargo jetties during transportation, loading and unloading. Therefore, it becomes essential to recover the BOG to increase the efficiency. Condensing BOG has 30-60% higher energy utilization efficiency as compared to compressing BOG directly into the natural gas distribution pipeline (Nobuhiro, Mabuchi (Osaka Gas Co. Ltd., Jpn), 2006). The efficiency of the recondensation process primarily depends on the ratio of condensing LNG to boil-off gas and the operating pressure (Li, Chen, & Chein, 2012). A BOG recondensation system at an LNG receiving terminal is shown in the Figure 1.

Improving the BOG Compressor Design

Many different compressor designs have been used for boil-off gas applications. These designs are of following types:

- Reciprocating compressors,
- Centrifugal compressors (overhung design with variable inlet guide vane IGV),
- Centrifugal compressors (between the bearing machines with variable IGV),
- Dry rotary screw compressors.

Fixed speed motors are normally used as drives for these compressors, however, variable frequency drives may also be used in some applications. Variable IGVs are often used in the centrifugal compressor designs to provide turndown.

Head is relatively constant in a typical compressor for the BOG service. Therefore, flow control is normally done by IGV movement and not by throttling to improve the efficiency of the process. Discharge throttling wastes more energy than IGV application or recycle control.

Figure 1. LNG receiving terminal

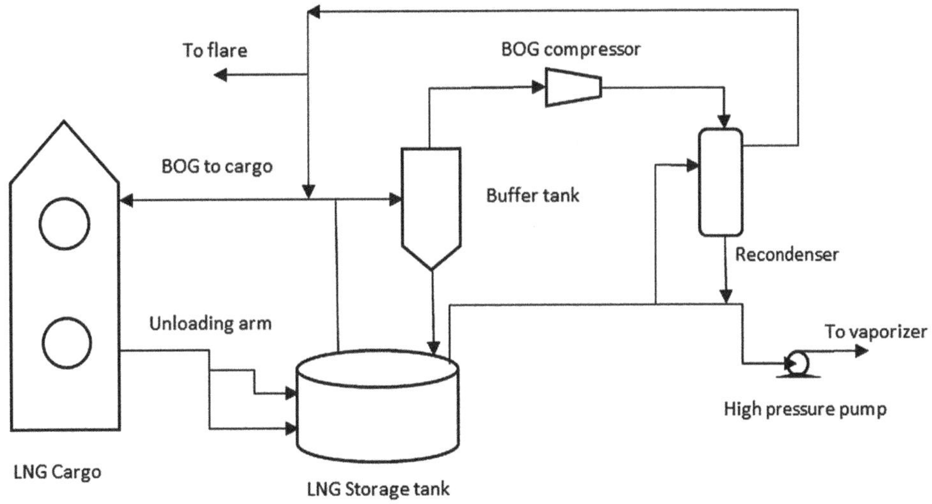

Selection of the right kind of BOG compressor is very important for the system hydraulics and thermodynamics because compressor control directly impacts the turndown, discharge temperature and head/flow characteristics. Consequently, the BOG compressor curves should be integrated within the total BOG system calculation (Chaker, Meher-Homji, Pillai, Bhattacharya, & Messersmith, 2011).

Improving the On-Board BOG Reliquefaction System

LNG cargo ships also generate BOG while at sea. Approximately 3% of the LNG cargo vaporizes due to unavoidable heat leakages from the surroundings. A BOG reliquefaction system developed by Cryostar for large on-board applications utilizes companders for effective BOG reliquefaction during shipping. A compander is a cryogenic expander coupled with a set of compression stages including intercoolers and aftercoolers. Cold power to condense BOG is produced by closed loop N2 reverse Brayton cycle in the Cryostar system (Chaker, Meher-Homji, Pillai, Bhattacharya, & Messersmith, 2011).

CONCLUSION

LNG terminals continue to evolve and grow more complex, as new technologies are developed. Safety is a primary concern for operating the terminal and the use of computer simulation software for testing these new technologies before they are actually implemented has become a standard. LNG terminals are always prone to accidents or have security threat from terrorists, and modern safety mechanisms that are in place usually are robust enough to deal with such scenarios. However, the reliability of safety mechanisms has to be periodically accessed and constantly be improved upon by various risk assessment methodologies. The documentation related to process data and diagrams need to be constantly updated in order to provide an accurate risk assessment.

The BOG management problem becomes more complex with the increase in the size of the LNG plants. Turndowns, system hydraulics and process transients are some of the major problems along with the challenges of compressor selection for the process. Different approaches should be considered for compressor selection including but not limited to turndown capability, flow capacity and pressure ratio.

Dynamic simulation provides better understanding of the unsteady state behavior. Once the dynamic model for the process is created, it can be used to evaluate large number of scenarios and studying existing BOG network systems.

REFERENCES

ABSG, C. I. (2004). *Consequence assessment methods for incidents involving releases from liquefied natural gas carriers.* Federal Energy Regulatory Commision.

Safety History. (2012, February 8). CH. IV International. Retrieved from]http://www.ch-iv.com/assets/documents/safety-history.pdf

Chaker, M., Meher-Homji, C. B., Pillai, P., Bhattacharya, D., & Messersmith, D. (2011). Application of Boil Off Gas Compressors in Liquefied Natural Gas Plants. *Journal of Engineering for Gas Turbines and Power.*

ConocoPhillips and EPA's Natural Gas STAR Program. (2006, May 25). Retrieved from Environment Protection Agency website: http://www.epa.gov/gasstar/documents/lng_opportunities.pdf

Contreras, J. F. (2005, May). Dynamic simulation: a case study. *Hydrocarbon Engineering*. Retrieved from Aspentech: www.aspentech.com

Cramer, H. R. (1991). *Risk assessment and risk management for the chemical process industry*. New York: Van Nostrand Reinhold.

Cruz-Campa, H. J., & Cruz-Gomez, M. J. (2010, March). Determine SIS and SIL using HAZOPS. *Process Safety Progress*, 22-31.

Dauber, F., & Span, R. (2012). Achieving higher accuracies for process simulations by implementing the new reference equation for natural gases. *Computers & Chemical Engineering*, *37*, 15–21. doi:10.1016/j.compchemeng.2011.09.009

Duguid, I. (2001, July). Take this safety database to heart.

Federal Energy Regulatory Commission. (2015). *Energy Primer: A Handbook of Energy Market Basics*. The Division of Energy Market Oversight.

Finn, A., Johnson, G., & Tomlinson, T. (1999). Developments in natural gas liquefaction. *Hydrocarbon Process*, 47-59.

Gangadharan, P., Singh, R., Cheng, F., & Lou, H. (2013). Novel methodology for inherent safety assessment in the process design stage. *Industrial & Engineering Chemistry Research*, *52*(17), 5921–5933. doi:10.1021/ie303163y

Kanoglu, M. (2001). Cryogenic turbine efficiencies. *Exergy Int. J*, 238-253.

Krishnan, G. (2005). What HAZOP studies cannot do. *Hydrocarbon Processing*, 93-95.

Li, Y., Chen, X., & Chein, M.-H. (2012). Flexible and cost-effective optimization of BOG (boil-off gas) recondensation process at LNG terminals. *Chemical Engineering Research & Design*, *90*(10), 1500–1505. doi:10.1016/j.cherd.2012.01.013

Mortazavi, A., Somers, C., Hwang, Y., Radermacher, R., Rodgers, P., & Al-Hashimi, S. (2012). Performance enhancement of propane pre-cooled mixed refrigerant LNG plant. *Applied Energy*, *93*, 125–131. doi:10.1016/j.apenergy.2011.05.009

Nibbelke, R., Kauffman, S., & Pek, B. (2002, July 8). Double mixed refrigerant LNG process provides viable alternative for tropocal conditions. *Oil & Gas Journal*, *100*(27), 64.

Nobuhiro, M. (2006). BOG reliquefaction with cold energy storage and utilization of LNG cold energy using cascade process. *Energy Resource*, 334-336.

Renaudin, G. (1995). Improvement of natural gas liquefaction processes by using liquid turbines. *Proceedings of the Eleventh International Conference on Liquefied Natural Gas*. Chicago, Institute of Gas Technology.

Tixier, J. D., Dusserre, G., Salvi, O., & Gaston, D. (2002). Review of 62 risk analysis methodologies of industrial plants. *Journal of Loss Prevention in the Process Industries*, *15*(4), 291–303. doi:10.1016/S0950-4230(02)00008-6

Independent Statistics and Analysis. (2014, May 7). *U.S Energy Information Administration*. Retrieved from http://www.eia.gov/forecasts/aeo/

ADDITIONAL READING

Cleaver, P., Johnson, M., & Ho, B. (2007). A summary of some experimental data on LNG safety. *Journal of Hazardous Materials*, *140*(3), 429–438. doi:10.1016/j.jhazmat.2006.10.047 PMID:17129671

Contreras, J. F. (2005, May). Dynamic simulation: a case study. Hydrocarbon Engineering. *Aspentech*. Retrieved from www.aspentech.com

OPTHAZOP -. An effective and optimum approach for HAZOP study. (n. d.). In F. A. Khan, Risk Assessment In Chemical Process Industries: Advanced Techniques (pp. 47-49).

Pitblado, R. M., & Woodward, J. L. (2011). Highlights of LNG risk technology. *Journal of Loss Prevention in the Process Industries*, *24*(6), 827–836. doi:10.1016/j.jlp.2011.06.009

Rasel, M.A.K. & Richmond, P.C. (2014). Improve safety and reliability with dynamic simulation.

Chapter 8
Advances in Catalytic Conversion of Syngas to Ethanol and Higher Alcohols

Jie Sun
Xiamen University, China

Jingdong Lin
Xiamen University, China

Shaolong Wan
Xiamen University, China

Yong Wang
Pacific Northwest National Laboratory, USA

ABSTRACT

Ethanol and higher alcohols (C2+OH) have attracted much attention owing to their wide applications. They can be produced from syngas using homogeneous and heterogeneous catalysts. Although homogeneous catalysts exhibit high productivity and selectivity of C2+OH, difficulties in separating and recycling homogeneous catalysts remain challenging. Among heterogeneous catalysts, Rh-based catalysts show promising higher selectivity of C2+OH. However, prohibitive cost of Rh metal hinders its large-scale application. Non-noble metal based heterogeneous catalysts include modified methanol synthesis catalysts, modified Fischer-Tropsch (F-T) synthesis catalysts, and Mo/MoS2-based catalysts. Compared with the modified F-T synthesis catalysts and Mo/MoS2-based catalysts, production of undesired byproducts on modified Cu-based catalysts can be well suppressed. Here, the influences of additives and supports on catalytic activity of modified Cu-based catalysts are discussed. Reaction mechanism and development of novel reactors are also included.

INTRODUCTION

Ever since 1970s, energy crisis has been one of the hottest topics (Forzatti, Tronconi, & Pasquon, 1991). Recent nuclear disaster in Japan in 2011 has pulled the trigger on the reconsideration of the security of nuclear energy, leading to perpetual termination of several nuclear power plants in Europe (Benichou & Mayr, 2014). Meanwhile, increasing demand of crude oil has triggered a sharp decline to the oil reserves worldwide, leading to accelerated search for alternative fuels. Recent advancement in the shale gas in the United States will accelerate the transformation in energy resource structure to decrease the severity of current oil dependence ("The shale revolution," 2009).

DOI: 10.4018/978-1-4666-9975-5.ch008

Integrated utilization of coal, natural gas and biomass has received increasing attention recent years. Abundant reserves of coal and natural gas can be converted into syngas. In South Africa and some other countries that are short of oil, the technology for gasification of coal has been well developed for decades (van Dyk, Keyser, & Coertzen, 2006). Syngas can be obtained from oil residues, some solid and/or liquid carbonaceous wastes and non-food biomass, as well (Göransson, Söderlind, He, & Zhang, 2011; He et al., 2009). Large reserves of biomass, especially abundant non-food based lignocelluloses (e.g. weed, bark and wheat-straw etc.), could be converted to syngas through gasification process and subsequent purification so that the obtained syngas could satisfy the specific requirements for ethanol and higher alcohol ($C_{2+}OH$) production (Göransson et al., 2011; Higman & Tam, 2013). The process of direct gasification includes low temperature gasification and high temperature gasification (Göransson et al., 2011). Low temperature gasification process can be performed under 900 °C within a fixed-bed reactor or a fluidized-bed reactor with flowing air or other oxidizing agent, while the high temperature gasification are conducted above 1300 °C in an fluidized gasifier which has already been adopted to industrial production in some coal-fired power stations.

As schematically shown in Figure 1, Syngas can be converted into various chemicals, such as methanol, ethanol, higher alcohols, alkanes (F-T process) etc. (Corma, 2014) Production of methanol has been industrialized for decades worldwide, basically using Cu/ZnO-based heterogeneous catalysts at 220-270 °C and 5-15 MPa (Behrens et al., 2012). Based on methanol, various value-added chemicals can be obtained, e.g. dimethyl ether (DME) and lower olefins etc. Different from methanol, $C_{2+}OH$ can be supplied to combustion engine as fuel directly or be added into gasoline to improve the octane value as fuel additives in place of methyl tert-butyl ether (MTBE) (Subramani & Gangwal, 2008). Compared with alkane products (mainly as fuel) obtained through conventional F-T synthesis process, $C_{2+}OH$ are much more versatile products. Ramayya et al. studied the dehydration process of several alcohols (e.g. ethanol, propanol etc.) catalyzed by Arrhenius acids (such as H_2SO_4 and HCl) in supercritical water (Ramayya, Brittain, DeAlmeida, Mok, & Antal Jr, 1987). It was reported that high selectivity of lower olefins ($\geq$ 97%) was obtained within 1-2 min when catalyzed by 0.05 M HCl. When the dehydration process of higher alcohols was catalyzed heterogeneously using Al_2O_3 and/or zeolites, it was suggested that the selectivity of lower olefins was 97-100% under mild reaction condition (Bi, Guo, Liu, & Wang, 2010; G. Chen, Li, Jiao, & Yuan, 2007). Zhang et al. reviewed recent advances in dehydration of ethanol to ethylene (Zhang & Yu, 2013). They suggested that this process was more economically feasible over activated Al_2O_3-based catalysts considering the stability of catalysts, the process of which had already been applied in the industrial plant. In addition, ethanol is environmentally friendly and relatively easier to store. Hence, it has also been selected as one promising candidate to store and transport hydrogen (Davidson et al., 2014). Steaming reforming of $C_{2+}OH$ can provide on-site hydrogen source for fuel cell and/or some small-sized and scattered chemical processes (Ferencz et al., 2014; Wei, Sun, Li, Datye, & Wang, 2012).

From a commercial viewpoint, both the yield/selectivity of higher alcohols, and the cost of catalysts should be considered in evaluating different catalysts. The required syngas conversion should be as high as 5-10% with alcohol productivity of about 0.2-0.5 g/(g_{cat}•h) in order to ensure economic viability (Knifton, Lin, Storm, & Wong, 1993). Since 1980s, plenty of patents focusing on higher alcohol synthesis from syngas have been filed by some famous chemical companies and institutes, such as Dow Chemical and Institut Francais du Petrole (IFP) etc. (Subramani & Gangwal, 2008; Xiaoding, Doesburg, & Scholten, 1987). Up to date, commercial success has not been realized in this field because of the low yield of higher alcohols from syngas and the decline in oil prices recent years. However, with increasing con-

Figure 1. Conversion of syngas to fuel, fuel additives, and some other value-added chemicals

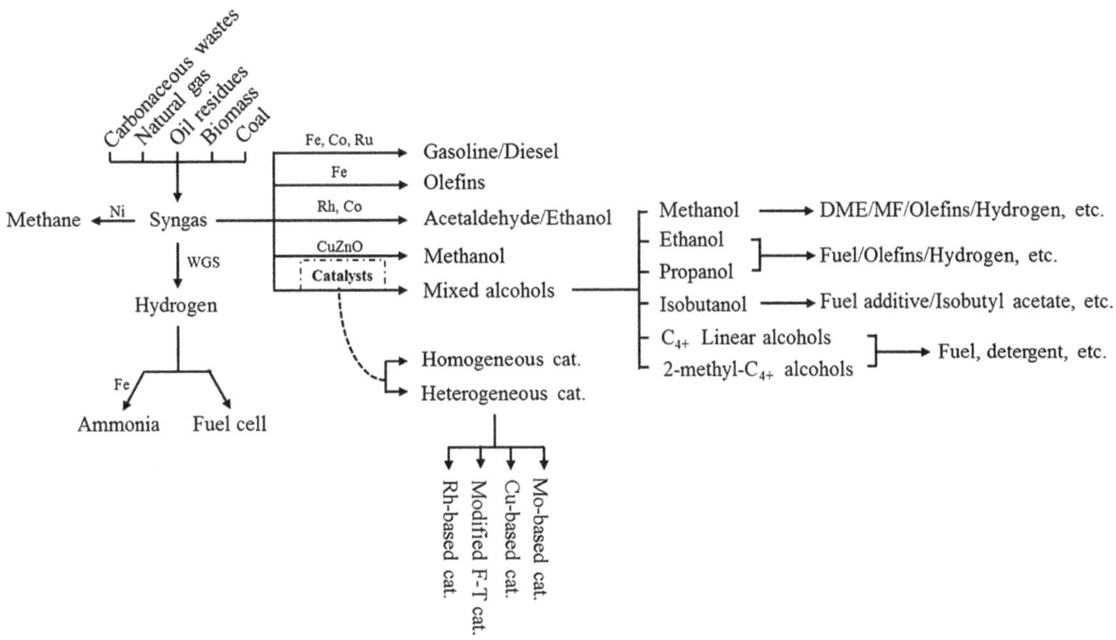

cern about the depletion of oil reserves and the price fluctuation in the future, synthesis of $C_{2+}OH$ has been pushed forward to the research frontier once again (Sun, Wan, Wang, Lin, & Wang, 2015; Walter, Schubert, Kleist, & Grunwaldt, 2015).

Recent studies indicate that the value-added C_{3+} alcohols are more favorable with small addition of methanol and/or ethanol to the reaction system over Cu-based catalysts which shows great potential in commercial application (Nielsen, Temel, & Beato, 2014; Walter et al., 2015). In addition to recent development in catalyst design, novel reactors (such as slurry phase reactor and microchannel reactor etc.) have also been studied in this field (Lebarbier et al., 2011; Zuo, Wang, Yu, Han, & Huang, 2014). Heat transfer is much faster using these novel reactors, so that the formation of hotpots in the catalyst bed can be well suppressed. Different new supports, novel strategies for catalyst design, and some mixed-mode catalysts have received increasing attention recent years which also shows promising improvement towards higher alcohol synthesis (Prieto et al., 2014; J.-J. Wang et al., 2013; Y. Xiang et al., 2013; S. F. Zaman & Smith, 2011). Herein, we provide a review on $C_{2+}OH$ synthesis direct from syngas, and shed light on recent development of different types of catalysts, as well as the development of novel reactor designs to show promising industrial application of this chemical process in the future.

DEVELOPMENT OF CATALYSTS

Homogeneous Catalysts

$C_{2+}OH$ can be synthesized using homogeneous catalysts direct from syngas and/or methanol homologation. Homogeneous CO hydrogenation can be realized at lower temperature (usually < 220 °C) which exhibits higher selectivity toward $C_{2+}OH$. Many homogeneous catalysts, such as Fe, Ru, Co, Ir and

Rh etc. metal complexes, alone or in combination, have been reported in the previous studies (Blank, Hembre, Ponasik, & Cole-Hamilton, 2014; Kosaric et al., 2000; Maitlis, 2003). Halide promoters, such as iodides, are usually introduced to enhance reaction activity and the production of ethanol (Blank, Hembre, Ponasik, & Cole-Hamilton, 2013; West, Miller, Labinger, & Bercaw, 2011). However, for most homogeneous catalyst, high reaction pressure was required, usually higher than 100 MPa (West et al., 2011). When $Ru_3(CO)_{12}$ was studied, it was found that the reaction pressure could be as low as 10 MPa. Usually, $[Bu_4P]Br$ was used as promoter in homogeneous Ru catalysts. With further assistance of $Co_2(CO)_8$ complex, turnover frequency of ethanol could be increased from 1.6×10^{-3} s^{-1} to 4.6×10^{-3} s^{-1} (Knifton, Grigsby, & Lin, 1984). It is worth noting that ethylene glycol (EG) synthesis was very favorable using Rh/P-i-Pr_3 and Rh/P-n-Bu_3 catalysts (Hara, Watanabe, Wada, & Onoda, 1989). The space time yield of EG was as high as 280 g/(L_{cat}•h) with a selectivity of 76% (Whyman, Wright, Iggo, & Heaton, 2002).

Ethanol synthesis from methanol homologation has been extensively studied using Co-, Ru-containing complexes or in combination since 1950s (Subramani & Gangwal, 2008). Research from TDA research Inc. showed that ethanol production rate was as high as 0.26 g/(g_{cat}•h) at 220 °C, 8.5 MPa, with very small amount of 1-propanol and dimethyl ether etc. as co-products. However, the details of catalysts components were not provided (Srinivas, Martin, Gebhard, & Mundschau, 2013). Recently, a combination of homogeneous ethanol synthesis catalysts with heterogeneous methanol synthesis catalysts was achieved by Krause et al. in Argonne National Laboratory (Krause, Rathke, & Chen, 2013). First, methanol was produced over Cu/ZnO/Al_2O_3 catalysts (mixed with Co-Mo oxides) from syngas at 220-270 °C, 5MPa. Then, methanol homologation occurred with $Mn_2(CO)_{10}$ catalysts at similar temperature range and higher pressure.

The major advantage of homogeneous catalysts is their much higher selectivity. However, big challenges also exist, e.g. much higher pressure required and the difficulties in handing the toxic transition metal carbonyl complexes (Krause et al., 2013). Therefore, more recently, very few papers in this field have been published and up to date no commercial success in homogeneous catalysis has been made (West et al., 2011).

Heterogeneous Catalysts

- **Catalysts and Primary Reactions:** Heterogeneous catalysts have attracted much more attention when compared with homogeneous counterparts. As exhibited in Figure 2, four types of heterogeneous catalysts have been developed for the synthesis of C_{2+}OH: a) Rh-based catalysts; b) modified methanol synthesis catalysts, typically Cu-based catalysts; c) modified Fischer-Tropsch (F-T) catalysts and d) Mo-based catalysts. Whether the four types of catalysts were investigated separately or a combination of different catalysts were evaluated, it has been demonstrated that numerous reactions exist in the synthesis of C_{2+}OH (Suib, 2013). Recent study in thermodynamics has shown that the conversion of syngas to C_{2+}OH is thermodynamically favorable, however, large amount of by-products were produced simultaneously (París et al., 2015). Hence, it was inferred that the formation of C_{2+}OH was kinetically controlled (Gupta, Smith, & Spivey, 2011; Spivey & Egbebi, 2007; Subramani & Gangwal, 2008). Briefly lower alkanes and CO_2 are two primary undesirable products. According to previous experimental results, it is worth noting that alkanes were the primary by-products on Rh-based catalysts, modified F-T catalysts, and Mo-based catalysts, in contrary to that on Cu-based catalysts, as shown in Table 1 (K. Fang et al., 2009; Gao, Mo, Chien, Torres, & Goodwin Jr, 2009; D. Jiang et al., 2007; Ma, Li, Lin, & Zhang, 2010; Ojeda,

Figure 2. Four different types of catalysts for the synthesis of $C_{2+}OH$ (Products marked in italic are coproduced with $C_{2+}OH$.)

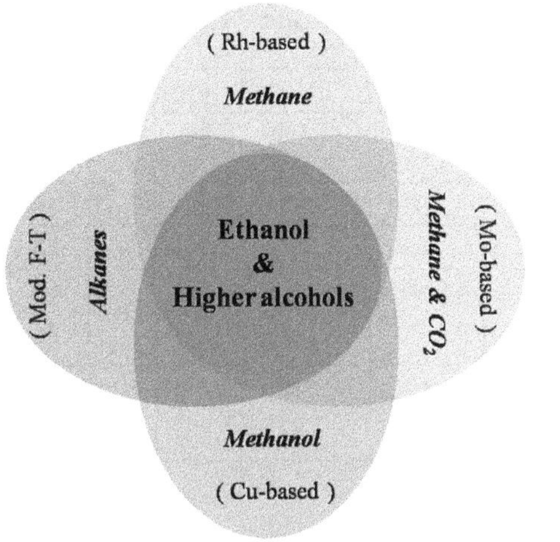

Granados, et al., 2004). A large amount of the undesired CO_2 was formed on Cu-based catalysts or Mo-based catalysts, which is generally ascribed to the water-gas-shift (WGS) reaction (K. Fang et al., 2009; Hilmen, Xu, Gines, & Iglesia, 1998; Kulawska & Skrzypek, 2001; Spivey & Egbebi, 2007). A general comparison of the four types of catalysts is also summarized in Table 1.

Rh-Based Catalysts

Previous woks have unequivocally indicated that promoted Rh-based catalysts exhibit better catalytic performances (e.g. higher selectivity towards C_{2+} oxygenates) (Gao, Mo, Chien, et al., 2009; Ojeda, Rojas, et al., 2004; Tanaka, Kaino, Okumura, Kizuka, & Tomishige, 2009). Prohibitive cost of Rh has always been a concern in any potential large-scale commercial applications. Earlier work showed that on Rh-

Table 1. Primary reactions involved in the catalytic process over different types of heterogeneous catalysts

Catalysts	Dominant Reactions	Advantages	Disadvantages
Rh-based catalysts	$2CO + 4H_2 = C_2H_5OH + H_2O$	Higher $C_{2+}OH$ selectivity	Higher cost; low yield
	$CO + 3H_2 = CH_4 + H_2O$		
Cu-based catalysts	$CO + 2H_2 = CH_3OH$		
	$nCO + 2nH_2 = C_nH_{2n+1}OH^{a} + (n-1)\ H_2O\ (6 \geq n \geq 2)$	Low cost; high yield	Large amount of methanol
	$CO + H_2O = CO_2 + H_2$		
Modified F-T catalysts	$nCO + (2n+1)\ H_2 = C_nH_{2n+2}^{b} + nH_2O\ (n \geq 2)$		
	$nCO + 2nH_2 = C_nH_{2n+1}OH^{c} + (n-1)\ H_2O\ (14 \geq n \geq 2)$	Low cost	Large amount of alkanes
Mo/MoS_2-based catalysts	$CO + 3H_2 = CH_4 + H_2O$	Low cost; high yield; sulfur-resistance	Large amount of alkanes and CO_2
	$nCO + 2nH_2 = C_nH_{2n+1}OH^{c} + (n-1)\ H_2O\ (6 \geq n \geq 1)$		
	$CO + H_2O = CO_2 + H_2$		

[a]$C_nH_{2n+1}OH$ = Ethanol, propanol, and 2-methyl branched alcohols. [b] C_nH_{2n+2} = Linear alkanes. [c] $C_nH_{2n+1}OH$ = Linear primary alcohols.

Table 2. Catalytic performances of some selected Rh-based catalysts with various promoters for $C_{2+}OH$ synthesis

Catalysts (wt%)	T	P	Feed Gas	GHSV	Conversion	Selectivity (%)						Ref.
	°C	MPa	H_2/CO	$L\cdot(L_{Cat.}\cdot h)$	CO (%)	CH_4	CO_2	HC_{2+}	MeOH	EtOH	C_{2+}Oxy.	
2%Mn/3%Rh/CNTs	290	2.0	2:1	-	4.0	48.9	0	2.9	6.5	35.2	-	(J. Liu et al., 2014)
5.56%Rh/1.69%Mn/ SiO_2	277	8.3	1.9	7500	20.8	31.7	-	16.2	0.5	-	51.6	(Mei et al., 2010)
	302			11,000	20.5	47.1	-	12.4	0.9	-	39.6	
2% Rh/1% Fe/SiO_2	270	2.0	1	8,000	2.7	35.5	-	20.9	13.5	21.8	8.4	(Haider, Gogate, & Davis, 2009)
2% Rh/5% Fe/TiO_2					6.2	35.3	-	7.0	4.8	37.2	15.6	
1.5%Rh/2.6%La/1.5%V/ SiO_2	270	1.4	2	9000	7.9	15.4	3.1	-	5.0	51.8	19.1	(Nachal D. Subramanian et al., 2010)

based catalysts without promoters, hydrocarbons, primarily methane is the main product and very limited C_{2+}oxygenates are produced (Katzer et al., 1981; Nachal D. Subramanian et al., 2010). This is probably due to the complete dissociation of adsorbed CO on Rh surface followed by subsequent hydrogenation to produce hydrocarbons, as opposed to the desired CO insertion reaction to produce the intermediates leading to C_{2+} oxygenate formation.

Recent efforts mainly focused on the design and synthesis of novel nanocatalysts using different types of promoters and advanced support materials which use less amount of Rh but exhibit higher activity. In the following context, we summarize the recent studies using different types of promoters and supports to enhance the performances of Rh-based catalysts for $C_{2+}OH$ synthesis, particularly focusing on the interactions between promoters and Rh. One of the main effects of promoter is to stabilize the Rh^+ species with the assistance of supports, leading to a more appropriate ratio of Rh^+/Rh^0 which was considered to be of vital importance for the synthesis of $C_{2+}OH$ (Abdelsayed, Shekhawat, Poston Jr, & Spivey, 2013; Han, Mao, Yu, Guo, & Lu, 2013). Some selected Rh-based catalysts in recent studies are summarized in Table 2. It can be seen that ethanol dominates among all the alcohol products. When promoted with La and V promoters, the selectivity of ethanol is higher than 50%.

- **Supports:** Different surface chemical properties of supports can influence the interaction between support and the active center, which plays a significant role on the catalysis. In addition, the dispersion of nanoparticles could be significantly affected by the physicochemical properties of supports. Consequently, as for the active Rh centers, both the activity of CO hydrogenation and the distribution of reaction products could be influenced by the supports.

Various types of supports, such as Al_2O_3, ZrO_2, SiO_2, TiO_2, MgO and carbon-based materials etc. (Burch, Loader, & Cruise, 1996; Egbebi, Schwartz, Overbury, & Spivey, 2010; Fan, Chen, Pan, & Bao, 2009; Gogate & Davis, 2009; Katzer et al., 1981; Tanaka et al., 2009), have been studied to improve the dispersion of Rh particles and to further verify the impact of support on the distribution of products. Recently some novel supports have drawn plenty of attention due to their unique properties, which not only can affect the dispersion of the active centers, but also introduce a beneficial interaction between supports and the active metal, facilitating the formation of $C_{2+}OH$.

Yu et al investigated the catalytic performances of Rh/Mn/Li supported on SiO_2, wherein the SiO_2 support was calcined under different temperatures to tune the amount of Si–OH on the support. It was found that the amount of Si–OH on the surface of SiO_2 support has a vital influence on the dispersion of Rh, CO adsorption and even the interaction between Rh and Mn, which ultimately affects the formation of C_{2+} oxygenates (Yu, Mao, Han, Guo, & Lu, 2013).

Han and his coworkers examined different types of supports, SiO_2, ZrO_2 and a combination of their mixtures for the Rh-based catalysts (Han et al., 2013). Interestingly, when Rn/Mn/Li was supported on the mixed silicon and zirconium oxides with Si/Zr = 1:3 (molar ratio), the yield of C_{2+} oxygenates was significantly enhanced which was attributed to the improved dispersion of Rh and higher concentration of surface Rh^+ stabilized by the mixed SiO_2/ZrO_2 oxides.

Lanthanum zirconate pyrochlores ($La_2Zr_2O_7$) were also used as supports for Rh-based catalysts by Abdelsayed and Spivey et al. (Abdelsayed et al., 2013). Two types of catalysts were synthesized: 1) Rh was supported on the surface of the pyrochlores with traditional wet impregnation approach, and 2) Rh was homogeneously doped into the $La_2Zr_2O_7$ lattice. It was found that surface Rh of the second type of catalysts was much easier to be reduced because of the instability of the resultant perovskite phase ($LaRhO_3$), leaving less partially reduced $Rh^{\delta+}$ species on the surface. Experimental results manifested that the first type of catalysts exhibited relatively higher selectivity towards ethanol, wherein the higher catalytic activity was attributed to their larger amount of Rh^+/Rh^0 active sites.

Carbon nanotubes (CNTs) supported Rh/Mn/Li/Fe catalysts were investigated by Fan et al. (Fan et al., 2009). Three other carbon-based materials were also examined as comparison, namely mesoporous CMK-3, carbon black (CB), and activated carbon (AC). The highest overall activity and yield of C_2 oxygenates were found on CNTs supported Rh/Mn/Li/Fe catalysts. Both the particle size and its distribution were considered to contribute significantly to the higher oxygenate formation. In addition, the nanochannels and graphitic structure of CNTs were also hypothesized to have promotional effect on the activity. When $Rh_1/Mn_1/Li_{0.075}/Fe_{0.05}$ (in a weight ratio) was mainly deposited inside CNTs, it was observed that the yield of C_2 oxygenates was as high as 84.4 $mol{\cdot}mol^{-1}{}_{Rh}{\cdot}h^{-1}$ (including 76% ethanol) (X. L. Pan et al., 2007) and the undesired methane was significantly suppressed to about 15% ◦ As a comparison, 41% selectivity to methane was observed when the same active constituents were supported on silica. The authors provided two primary reasons to illustrate the superb catalytic performance of Rh-based catalysts inside CNTs. Firstly, compared with the electro-abundant exterior surface, the interior surface of CNTs was electro-insufficient which facilitated the activation of CO with the assistance of Mn promoter. Secondly, there probably existed a confinement effect inside the CNTs that may bring about an increase in the concentration of feed gas and result in a peculiar interaction between the encapsulated particles and the internal surfaces of CNTs (W. Chen, Pan, Willinger, Su, & Bao, 2006; X. Pan & Bao, 2011).

- **Promoters:** It is widely accepted that the selectivity towards ethanol and some other C_{2+} oxygenates could be increased by introducing various promoters, such as alkali oxides and transition metal oxides etc. (Egbebi et al., 2010; Gogate & Davis, 2009; Haider et al., 2009; D. Jiang et al., 2007; Mo, Gao, Umnajkaseam, & Goodwin Jr, 2009; J. Wang, Zhang, & Wang, 2011; Yin et al., 2003; S. F. Zaman & Smith, 2011). It is worth mentioning that the CO conversion could also be influenced by certain promoters at least to a certain degree, as discussed in the work of Huang and Wang et al. (Y. Huang et al., 2012; Y. Wang, Luo, Liang, & Bao, 2000).

Promotional effect of alkali on Rh-based catalysts has been extensively studied. However, the specific effect of alkali promoter is still unclear. Li was considered to be a very desirable alkali promoter and plenty of attention has been devoted to the study of its influence on the electron density and the corresponding effect on the absorption and dissociation of CO on the surface of Rh (Egbebi et al., 2010; Luo et al., 1997). Production of methane could be suppressed to a relatively lower level because of the reduced hydrogenation capability of Rh caused by alkali additives (Y. Wang et al., 2000).

Vanadia promoters, as well as rare earth oxide promoters (e.g. Lanthana etc.), could also be used to enhance the selectivity of C_{2+} oxygenates, especially ethanol (Gao, Mo, Chien, et al., 2009; Nachal D. Subramanian et al., 2010). Subramanian et al. systematically studied the $Rh/La/V/SiO_2$ catalysts through optimization of the reaction conditions, and as high as 51.8% selectivity toethanol was reported at 1.4 MPa and 270 °C (Nachal D. Subramanian et al., 2010).

On 2 wt% Rh/Al_2O_3 promoted with 10 wt% Fe, the selectivity of ethanol was enhanced up to 50%. The authors concluded that close interaction is generated between the active centers and the promoters, as verified by characterizations using chemisorption and temperature-programed reduction (TPR) (Burch & Hayes, 1997).

Silica and titania supported Rh catalysts, with/without Fe promoter, were synthesized for the direct synthesis of $C_{2+}OH$ in the study of Haider et al. (Haider et al., 2009). It was reported that when 2 wt% Rh/SiO_2 was promoted with 1 wt% Fe, the selectivity of ethanol was increased from 11.1% to 21.8%. When 5 wt% Fe was added to 2 wt% Rh/TiO_2, the selectivity was further improved to 37.2%. CH_4 was found as the major undesired product.

In order to probe the interactions between Rh and the Mn promoter, CNTs supported Rh catalysts with different amount of Mn were prepared and investigated by Liu etc. It was concluded that no Rh/Mn bimetallic alloy was formed, and higher Mn loading resulted in higher degree of Rh/Mn interaction which facilitated ethanol production (J. Liu et al., 2014). However, opposite conclusion was drawn by Mei etc. that Rh/Mn alloy was generated on SiO_2 supported Rh/Mn catalysts through a combination of experimental study with density functional theory (DFT) calculations (Mei et al., 2010). The binary alloy was deemed of great importance, since it can not only stabilize the catalysts under the strong reducing reaction conditions but also lower the insertion reaction barrier between CO and CH_x species, resulting in a significant improvement of the stability of the supported catalysts and the selectivity of C_{2+} oxygenates.

The influence of Mn on Rh-based catalysts was recently studied by Liu et al. using $Rh/Mn/SiO_2$ catalysts synthesized with a strong electrostatic adsorption (SEA) method (J. Liu et al., 2013). It was found that the catalysts prepared by SEA technique exhibited higher extent of the desired interaction which was critical in the production of $C_{2+}OH$. However, no detailed description about the interaction was provided.

Rh-based catalysts promoted by dual additives are of great interest as well. Mn and Li promoted Rh/TiO_2 catalysts were studied by Schwartz et al. (Schwartz, Campos, Egbebi, Spivey, & Overbury, 2011). The selectivity of C_{2+} oxygenates was reported to follow the order of: $Rh/Mn/Li/TiO_2 > Rh/Li/TiO_2 > Rh/Mn/TiO_2 > Rh/TiO_2$. However, characterization results of extended X-ray absorption fine structure (EXAFS) showed that neither Rh/Li nor Rh/Mn alloy were generated.

Promoted Rh-based catalysts are typically synthesized by incipient wetness impregnation method, followed by reduction in the slurry state or through calcination combined with thermal reduction by hydrogen (Egbebi et al., 2010; Haider et al., 2009; Han et al., 2013; Mo et al., 2009). However, the particle size of Rh suffers high fluctuation and the achievement of monodispersed Rh particles is even more challenging upon the introduction of promoters (Y. Huang et al., 2012). To solve the problem, Huang et al. developed a novel nanocatalyst, i.e. Rh nanoparticles modified by manganese and stabilized by mesoporous SiO_2. This catalyst was found to be highly selective towards C_{2+} oxygenates (75.4%) with methanol formation being dramatically suppressed. Both the catalytic activities and the selectivity towards $C_{2+}OH$ were drastically affected by Mn promoters which was in consistent with the previous study (Y. Wang et al., 2000).

- **Reaction Mechanism:** As mentioned above, unpromoted Rh-based catalysts showed very low selectivity towards ethanol and other C_{2+} oxygenates, with methane being the main undesired by-product. With the addition of a small amount of the elaborately selected promoters, such as Fe, Mn, Li, V etc., both the activity and selectivity towards ethanol generation were significantly improved (Egbebi et al., 2010; Haider et al., 2009; Mo et al., 2009). Generally Rh-based modified catalysts were prepared via co-impregnation or stepwise impregnation techniques with the precursors of promoters. It was speculated that the complete dissociation of CO was suppressed, with an increase of the non-dissociative counterparts, thereby, the insertion reaction of CO to CH_x species became much more facile (Schwartz et al., 2011; Y. Wang et al., 2000). Although the formation of Rh/M alloy (M = promoters) still subjects to intensive controversy, it is well accepted that there exists a strong interaction between Rh and the promoters. The speculated active site was considered to involve both Rh center and promoters (Subramani & Gangwal, 2008; Y. Wang et al., 2000).

Earlier work by Treviño et al. indicated that surface acetate located on the interface of Rh and Mn appeared to be one of the proper intermediates (Treviño, Hyeon, & Sachtler, 1997). Wang et al. proposed a geometrical structure of active site, $(Rh^0_x Rh^+_y)$–O–M^{n+} (x >> y), where Rh could not be fully reduced and a few of Rh cations (Rh^+) remained which are stabilized by promoter ions (M^{n+}) (Y. Wang et al., 2000). They further concluded that the promoter ion could interact with the absorbed CO, and the on-top bonded CO with carbon atom linked with Rh could be tilted to a slightly or completely lying-down position, as depicted in Figure 3. In the presence of promoters, CO bond could be more readily to be activated and hydrogenated to form H_xCO which facilitates the subsequent insertion of CO and/or CH_x for the generation of C_{2+} oxygenates. Meanwhile, the absorbed CO can also be continuously hydrogenated to form methanol with the retention of O atoms.

Various reaction pathways along with diverse intermediates have been postulated and extensively discussed. Plenty of attention has been paid to the initial C_2 intermediates which were widely assumed as the precursors of the target products, ethanol and/or other C_{2+} oxygenates. Ketene ($H_2C=C=O$) and acetyl ($H_3C–C=O$) seem to be two possible intermediates (H. Wang, 1993; Takeuchi & Katzer, 1982). The former was considered to be generated from the insertion of CO to CH_2, while the latter through the combination of CO with CH_3 (H. Y. Wang, Liu, Fu, Wan, & Tsai, 1992; H. Y. Wang, Liu, Fu, Zhang, & Tsai, 1992; Y. Wang et al., 2000).

Figure 3. The dissociation of CO bonded over Mn promoted Rh-based catalysts (Adapted from (Y. Wang et al., 2000))

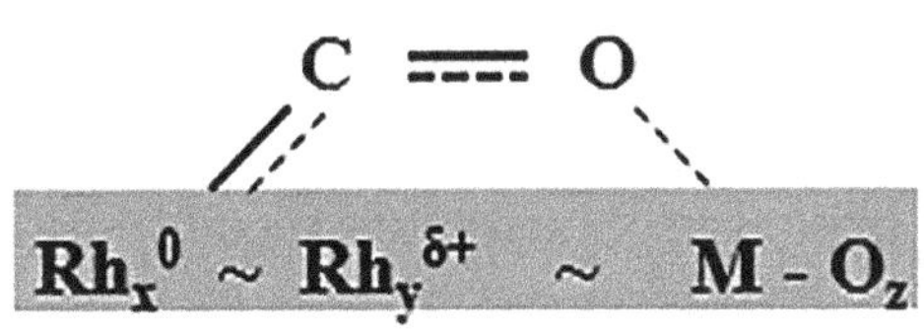

Figure 4. Simplified reaction routes for ethanol synthesis over promoted Rh-based catalysts (M = promoters)

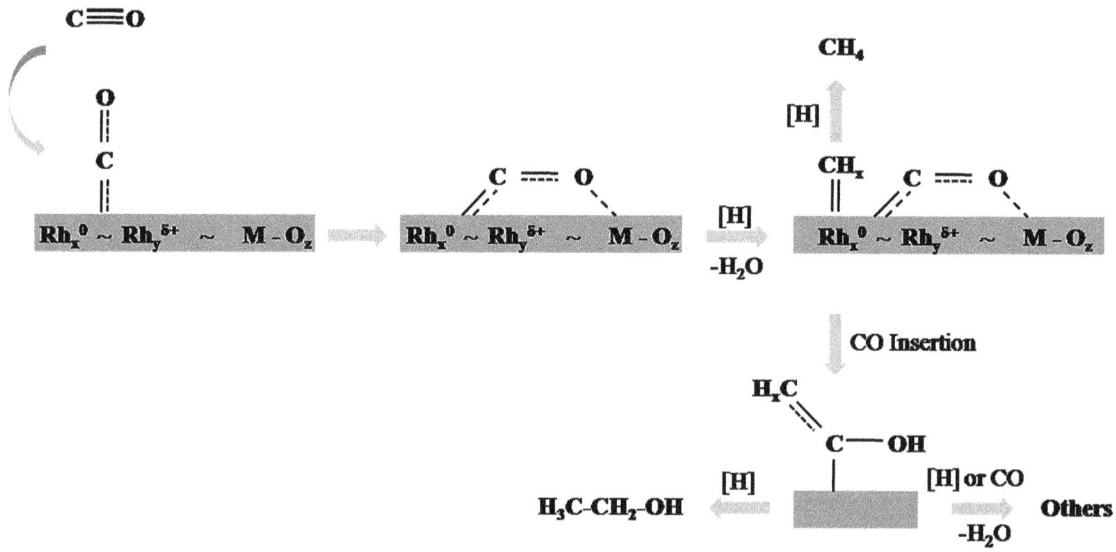

Generally the addition of promoters into Rh-based catalysts could facilitate the CO insertion reaction (as described in Figure 4), and accordingly suppress the production of the undesired methane product (Egbebi et al., 2010; Gao, Mo, & Goodwin Jr, 2009; Mo et al., 2009). The formation of methane is primarily attributed to continuous hydrogenation of CH_x, which competes against the insertion of CO into CH_x (Gronchi, Tempesti, & Mazzocchia, 1994).

Density functional theory (DFT) study was performed by Choi et al. to study the mechanism of ethanol formation from syngas on Rh (111) (Choi & Liu, 2009). It was found that ethanol forms though the insertion of CO into the C_1 species derived from the hydrogenation of formyl species, which was consistent with the experimental results. They also reported that the rate of ethanol production was affected by two factors, the formation barrier for methane and that for the initial C–C bond. The generation of formyl species derived from hydrogenation of CO was suggested to be the rate-determining step, which strongly depends upon the morphology and/or particle size of Rh clusters, types of promoters and supports.

Modified Methanol Synthesis Catalysts

Cu-based catalysts ($Cu/ZnO/Al_2O_3$) have already been utilized to produce methanol industrially from synthesis gas for decades (Behrens et al., 2012). During the preparation of Cu-based methanol synthesis catalysts, traces of alkali residuals from the precipitant (alkali carbonates or bicarbonates) remained in the finished catalyst (Gadewar, VICENTE, Stoimenov, & Julka, 2015). The presence of alkali was found to lead to the formation of a small amount of $C_{2+}OH$, which has since triggered much attention to the alkali-modified Cu-based catalysts for $C_{2+}OH$ synthesis (Hofstadt, Schneider, Bock, & Kochloefl, 1983; Smith & Anderson, 1983, 1984). Cu-based catalysts produce much less byproducts (such as methane and higher alkanes) compared with other three types of catalysts, and they have superior cost advantage compared to Rh (Jones, Agrawal, & Nguyen, 2013; Onda, Ogo, & Yanagisawa, 2011). Consequently, with the increasing interest in higher alcohol production, modified Cu-based methanol catalysts have been extensive studied (White, Wilson, Berggren, & Zubrin, 2014).

Early research primarily focused on the effect of alkali promoters on Cu-based catalysts. Non-alkali promoters, such as alkaline earth metals, rare earth metals, as well as transition metals, were later on studied to maximize the C_{2+} oxygenates yield (Walter et al., 2015). Both high-temperature methanol synthesis catalysts (based upon ZnO/Cr_2O_3) and low-temperature counterparts (Cu/ZnO-based catalysts) were studied (Calverley & Smith, 1991; Camposmartin, Guerreroruiz, & Fierro, 1995; Herman, 2000; Vedage G, Himelfarb P, Simmons G, & Klier, 1985). The former one modified with K_2O showed high selectivity to methanol and 2-methyl-1-propanol (isobutanol) when operated at 420-480 °C, 5.0-30.0 MPa or even higher pressure. Large amount of CO_2 was produced, along with small quantities of alkanes, aldehydes, esters and ketones (Temel, Nielsen, & Beato, 2013).

Since 1980s, modified low-temperature Cu/ZnO-based catalysts have received more attention and been comprehensively investigated (Tijm & Stevens, 2014). Therefore, in this section we mainly focus on the promoted binary Cu/ZnO and ternary Cu/ZnO/M (M = Al, Cr etc.) catalysts. The effect of promoters and the well accepted mechanisms proposed in the previous literatures will be discussed in detail.

- **Alkali Promoters:** In the catalytic conversion of syngas to $C_{2+}OH$ upon promoted Cu/ZnO-based catalysts, different operating parameters, such as temperature, space velocity, the ratio of H_2/CO, and pressure etc., were found to play an important role in the product distributions (Boz, Sahibzada, & Metcalfe, 1994; Calverley & Smith, 1991; Gadewar et al., 2015). Among these parameters, temperature seems to be the most important factor in affecting product distributions (Brown Bourzutschky, Homs, & Bell, 1990b; Calverley & Smith, 1991; Majocchi et al., 1998). For instance, the yield of $C_{2+}OH$ was very low or even below the detection limit when reaction was operated at lower temperature (< 250 °C). Nunan et al. probed the promotion of Cs on Cu/ZnO at 250 °C and reported that the influence of Cs doping mainly increased the formation of methanol, but with negligible promotional effect on the formation of ethanol (less than 1.1%) (John Nunan, Klier, Young, Himelfarb, & Herman, 1986).

Thereafter, numerous studies found that the most applicable temperature range for C_{2+} alcohol production was 270-310 °C while higher temperature (> 310 °C) caused formation of a large amount of undesired carbon dioxide and higher alkanes (Spivey & Egbebi, 2007). Within the narrow temperature range, different alkali metals were added into Cu/ZnO system to explore their promoting effect. These promoters were added mainly via impregnation techniques, either over the calcined catalyst or on the further reduced CuO/ZnO catalysts (J. G. Nunan et al., 1988). The obtained catalysts were recalcined to ensure sufficient interaction between alkali promoters and CuO/ZnO. Generally, the promotional effect of alkali additives follows the order of: Cs > Rb > K > Na > Li (Vedage G et al., 1985). It should be noted that and Cs have received much more attention since it was found to enhance the formation of $C_{2+}OH$ significantly and efficiently suppress the formation of the undesired alkanes (Walter et al., 2015).

Calverley et al. investigated the effect of K on the $Cu/ZnO/Cr_2O_3$ catalysts (Calverley & Smith, 1991). They claimed that 2-methyl branched aliphatic alcohols, such as 2-methyl-1-propanol and 2-methyl-1-butanol etc., were the dominant primary higher oxygenates. Boz et al. studied the kinetics of higher alcohol formation over $K/CuO/ZnO/Al_2O_3$catalysts, where addition of 0.5 wt% K_2O clearly improved the yield of isobutanol, which was consistent with the results of Anderson and his coworkers (Boz et al., 1994; Smith & Anderson, 1983, 1984). Binary Cu/ZnO catalysts modified with all the five different alkali hydroxides and Ba hydroxides were systematically investigated by Vedage et al. (Vedage G et al., 1985). Both methanol and higher alcohols were found to be markedly enhanced. It was assumed that CsOH moiety could accelerate the β_C addition reaction to generate branched higher alcohols. The improvement in methanol production was attributed to the more feasible formation of surface formate which was considered as a key intermediate for methanol formation.

Addition of structural promoters Cr or Al to Cu/ZnO catalysts increased the yield of hydrocarbons and dimethyl ether (DME), which was attributed to the stronger acidity sites introduced by Cr_2O_3 and/or Al_2O_3 (Kurtz, Wilmer, Genger, Hinrichsen, & Muhler, 2003; J. G. Nunan, Herman, & Klier, 1989). Consequently, it appeared that more alkali promoters are required to neutralize the acidity on the surface (J. G. Nunan et al., 1989). The overall concentration of hydrocarbons, esters, and DME in the products was further confirmed to vary linearly with the surface acidity of the evaluated catalysts with varied amounts of alkali. Hilmen and Xu et al. also studied the effect of Cs on the Cu/ZnO/Al_2O_3 and they found that excess Cs prohibited alcohol synthesis by blocking the active sites on the Cu-based catalysts (Hilmen et al., 1998).

- **Non-Alkali Promoters:** Transition metal promoters, such as Fe, Co, Ru etc., have also been introduced to the Cu-based catalysts due to their superb performances in the carbon chain growth as demonstrated in Fischer-Tropsch synthesis (Beiramar, Griboval-Constant, & Khodakov, 2014; Lin, Gan, Huang, Pan, & Noller, 1988; Run Xu et al., 2004). A series of Cu/ZnO/Al_2O_3 modified with Fe, Co, and Ni were prepared using co-precipitation from metal nitrate precursors to study the light alcohol synthesis from syngas by Lin et al.(Lin et al., 1988). They concluded that Fe was the most suitable additive. Further increase in Fe loading, however, reduces the yield of C_2-C_5 alcohols when Fe_2O_3/CuO weight ratio was above 66.7%. Similarly, Cu/Mn/ZrO_2 catalysts promoted with Fe were synthesized using wetness impregnation and co-precipitation methods respectively by Xu et al.(Run Xu et al., 2004). The interaction between Fe and Cu/Mn/ZrO_2 catalysts prepared using the former method was found to be weaker than that prepared using the latter approach. It was also found that the mixed alcohol formation does not follow Anderson-Schulz–Flory (ASF) distribution, whereas the hydrocarbon distributions fit the equation quite well. Consequently, they speculated that the active sites for alcohol production and hydrocarbon were different.

Some other non-alkali promoters, such as La, Ce, Th, Mn etc., were also studied individually or in combination with alkali oxides (Brown Bourzutschky, Homs, & Bell, 1990a; Hofstadt et al., 1983; J. G. Nunan et al., 1989; Slaa, van Ommen, & Ross, 1995). Mn or Ce promoted Cu/ZnO/Al_2O_3 catalysts with a Cu/Zn/A1 ratio of 9:9:2 were evaluated at 280 °C and H_2/CO = 1:1 by Slaa et al.,(Slaa et al., 1995). Only isobutanol in the higher alcohol products was improved noticeably by Mn. CO_2 and alkanes were increased accompanied with a lower yield of methanol and higher alcohols on Ce promoted catalyst. However, they also observed a higher yield of isobutanol when 2 mol% Ce was doped via co-precipitation process. The introduction of La_2O_3 into Cu/SiO_2 could enhance both the conversion of CO and the yield of alcohols (Brown Bourzutschky et al., 1990a). It was speculated that the presence of La_2O_3 facilitated the stabilization of Cu^+, which was considered essential for higher alcohol synthesis. After a detailed review of Cu-based catalysts, Gupta et al. summarized that both Cu^+ and Cu^0 were indispensable for ethanol formation (Gupta et al., 2011). It was inferred that CH_3O intermediates tended to be generated on Cu^+ sites, while Cu^0 favored CH_2 species formation. The coalition of CH_3O with CH_2 resulted in the generation of C_2 oxygenates, which can be subsequently converted to ethanol through moderate hydrogenation. This hypothesis aforementioned was consistent with the experimental results observed in the investigation of some non-alkali promoters like Th, Cr and Ga etc. (Hofstadt et al., 1983).

Recently, based on $Cu/ZnO/Al_2O_3$ catalysts, Beiramar et al. systematically studied the influence of some transition metal promoters (Fe, Co, and Ru) and some other non-alkali promoters (Zr, Mo, Mg, Mn, and Cr) on the catalytic performance for alcohol synthesis (Beiramar et al., 2014). A set of tests were conducted under the selected reaction conditionsat 280 °C and 320 °C, $H_2/CO = 2$ and 2 MPa. They concluded that C_{2+} alcohol productivity was slightly enhanced when Cr, Zr, Mg and Mn were utilized as promoters. However, Fe promoted catalyst dramatically increased CO conversion from 6.1% to 69.2% and the C_{2+} alcohol productivity from 6 to 30 mg/(g_{cat}·h), when the reaction temperature was increased from 280 °C to 320 °C. Such drastic changes in activity were attributed to the formation of iron carbides at 320 °C under CO/H_2 atmosphere.

- **Reaction Mechanism:** During the catalytic conversion of syngas to $C_{2+}OH$, the distribution of higher alcohols varied remarkably with the modification of various promoters upon Cu-based catalysts, which deeply complicated the reaction mechanism (Sun et al., 2015). To simplify the complex reaction pathways, researchers mainly focused on the generation of the major C_{2+} alcohols (e.g. ethanol, 1-propanol, and isobutanol), among which the formation of initial C-C bond and the corresponding C_1 intermediates were regarded as the key step. In the following context we summarize some well-accepted mechanisms over Cu-base catalysts with alkali promoters (Campos et al., 1996; Elliott & Pennella, 1988; M. Xu & Iglesia, 1999).

Ethanol was once considered to be derived from the condensation of methanol (equation 1) or methanol homologation reaction (equation 2) in some literatures where initial C–C bond formation was deemed as the rate-determining step (J. G. Nunan et al., 1988; M. Xu & Iglesia, 1999).

$$2CH_3OH = C_2H_5OH + H_2O \quad (1)$$

$$CH_3OH + CO + 2H_2 = C_2H_5OH + H_2O \quad (2)$$

Previously, both methyl ether and methyl formate were assumed as the intermediates. For example, Frolich and Graves et al. speculated that ethanol was formed though the rearrangement of methyl ether followed by successive hydrogenation (Frolich, 1931; Graves, 1931). However, this pathway was cast in doubt by the theoretical investigations. Murrell et al. indicated that the activation energy barrier of the rearrangement of methyl formate to generate C–C bond was too high (Murrell, 1968).

Later on, formaldehyde was proposed as a key intermediate over Cu/ZnO-based catalysts. In the work of Elliott et al., ^{13}C labeled methanol was added to the feed gas to probe the distribution of ^{13}C-containing ethanol in the products (Elliott & Pennella, 1988). After analyzing the labeled composition of the products at zero residence time, it was found that the concentration of labeled ethanol (doubly and/or singly) in the products was closely related to the concentration of ^{13}C labeled methanol. The amount of doubly labeled ethanol was much higher than that of the singly labeled/unlabeled ethanol at lower contact time. This was contradictory to the hypothesis by Mazanec et al. which suggested that the initial C-C was formed through the insertion of CO into the surface-bound H_xCO intermediates (Mazanec, 1986). Therefore, they inferred that there might exist one common intermediate shared by ethanol and methanol synchronously.

Figure 5. Mechanism for the synthesis of ethanol over Cs/Cu/ZnO catalysts (Adapted from (J. G. Nunan et al., 1988))

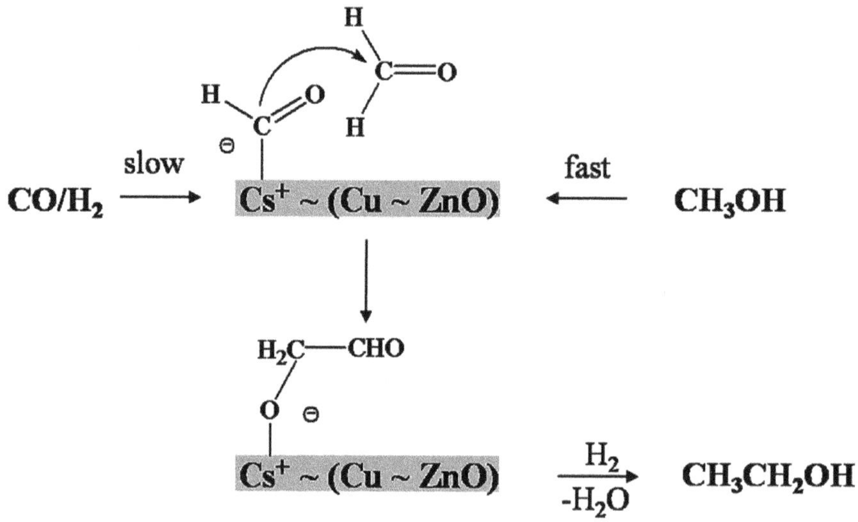

In another study, Nunan et al. found that ethanol was mainly composed by two labeled ^{13}C over $Cs/Cu/ZnO/Al_2O_3$ catalysts. Ruling out the aforementioned reaction pathway though homologation of methanol with the insertion of CO, they proposed that the adsorbed formyl was required for the initial formation of C–C bond (Klier et al., 1988; J. G. Nunan et al., 1988). The evidence over the existence of chemisorbed formyl was cited from the work by Deluzarche et al. through trapping reactions using CH_3I (Deluzarche, Hindermann, Kiennemann, & Kieffer, 1985). The proposed pathway is shown in Figure 5, where nucleophilic attack reactions occur between the surface-adsorbed formyl species with formaldehydes. In addition, it was also suggested that the coupling of the formyl with methanol cannot be excluded, albeit there exists steric effect which counted against the nucleophilic attacking reaction (J. G. Nunan et al., 1988).

Xu and Iglesia et al. stated that the mechanism of the initial C-C formation over different Cu-based catalysts is different. Over $K/Cu/Mg_5CeO_x$ using $^{13}CO/H_2/^{12}CH_3OH$, the predominant route to ethanol was attributed to the hydrogenolysis of methyl acetate containing a methyl group from $^{12}CH_3OH$ and a doubly labeled (from ^{13}CO) acetate group. Thus, the C–C bond-forming step involved reaction of species formed directly from ^{13}CO while methanol was initially much less involved (M. Xu & Iglesia, 1998). However, upon Cs promoted $Cu/ZnO/Al_2O_3$ catalysts, it favored to proceed through coupling of two methanol-derived intermediates, as shown in Figure 5 (M. Xu & Iglesia, 1999). The difference was attributed to that Cs^+ could better stabilize the absorbed formyl species than K^+. It was illustrated that both the basic anions (O^{2-}) and the acidic cations (K^+) in the K^+–O^{2-} cation-anion pairs were relatively insufficient compared with that in the Cs^+–O^{2-} constituents. Thus, the C–H bonds in the formaldehyde could be activated much more readily with the assistance of Cs^+–O^{2-} cation-anion pairs to generate the chemisorbed formyl species which were required for the formation of initial C–C bonds through subsequent nucleophilic attack reactions.

Reaction pathways were much more complicated for the formation of C_{3+} alcohols. Various routes may be possible for the chain growth process. In the work of Calverley et al., higher alcohols were believed to be generated at the interface between copper and alkali (Calverley & Smith, 1991). When binary Cu-based catalysts were modified with alkali, the yields of methanol, ethanol and methyl formate reached a maximum with the increase of alkali loading (John Nunan et al., 1986; Vedage G et al., 1985). It was considered that the introduction of alkali not only facilitated the activation of CO, but also provided basic sites that were required for aldol condensation reactions to produce C_{3+} oxygenates (Campos et al., 1996; Elliott & Pennella, 1988; Klier et al., 1988; Majocchi et al., 1998; JohnG Nunan, Bogdan, Herman, & Klier, 1989; M. Xu & Iglesia, 1999). In addition, a balanced Cu^{+}/Cu^{0} sites can be achieved with alkali addition, which was considered to be of vital importance in the $C_{2+}OH$ synthesis (Boz et al., 1994; Sheffer & King, 1989).

Reaction routes in Figure 6 were proposed in the work of Hilmen et al., which coincided with those suggested by some other scholars (Breman, Beenackers, Schuurman, & Oesterholt, 1995; Hilmen et al., 1998; Smith & Anderson, 1983, 1984). Linear terminal alcohols, C_n–OH (usually $3 \leq n \leq 6$), could be generated from the insertion of C_1 intermediate into the C–O bond of C_{n-1}–OH the by the end of carbon atom of C_1, and/or by aldol condensation where the C_{2+} alcohols attacked the β_C of ethanol. Likewise, the primary 2-methyl branched aliphatic alcohols could be produced either though the attacking of C_1 species on the β_C of C_{2+} oxygenates or some other reaction routes. Take 2-methyl-1-butanol for instance, it could be generated by the attacking of C_1 intermediates on the β_C of linear C_4 oxygenates or by the condensation of C_2 oxygenates with C_3 oxygenates.

As mentioned above, the rate-determining step is the formation of initial C-C bond where the key C_1 intermediate can be obtained easier from methanol than from CO (Gupta et al., 2011). In addition, $C_{3+}OH$ are mainly derived from aldol condensation reaction which is much easier compared with the initial C-C formation (Subramani & Gangwal, 2008). Based on these conclusions, recent studies mainly focus on addition of methanol and/or ethanol to increase the productivity of $C_{2+}OH$ (Gadewar et al., 2015). Table 3 shows that when methanol and/or ethanol was added to syngas, the production of $C_{2+}OH$ could be as high as 1000 g/(kg_{cat} • h) which exhibited very promising potential in industrial application (Temel et al., 2013).

Figure 6. Simplified reaction routes for $C_{2+}OH$ synthesis over Cu-based catalysts promoted with alkali oxides (Adapted from (Hilmen et al., 1998))

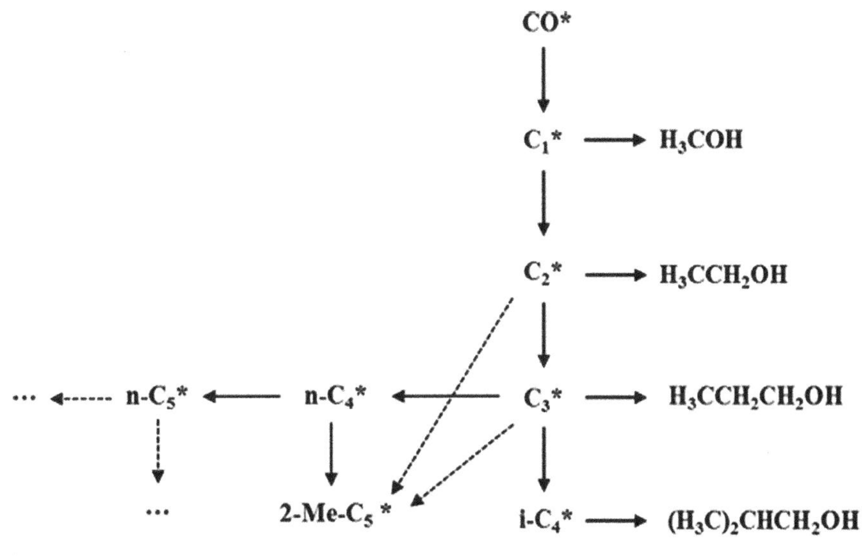

Table 3. Effect of inlet composition on the higher alcohols production: $H_2/CO = 0.5$, 10 MPa, GHSV = 20000 Nl/kg·cat/h, 10 g catalyst. (Temel et al., 2013)

	Source Alcohols		
	Methanol	Methanol Ethanol	Methanol 1-Propanol
Example	3.1	3.2	3.1
Methanol (inlet, g/h)	23.7	23.5	23.9
Ethanol (inlet, g/h)	-	112.4	-
1-Propanol (inlet, g/h)	-	-	149.6
CO % conversion	5	5	0.01
	Exit Composition (g/(kg_{cat} • h))		
Methanol	894	1693.4	1498.8
Ethanol	46.9	7794.9	0
1-Propanol	43.3	448.4	12357.8
2-Propanol	0	15.8	58.9
Isobutanol	44.2	47.1	376
Other butanols	12.3	406	226.9
Pentanol	23.1	101.3	162.4
Hexanols and higher	9.2	8.6	76.5
Total (Excluding reactant alcohols)	179	1027.2	900.7

Modified Fischer-Tropsch Catalysts

Fischer-Tropsch (F-T) synthesis process is used for hydrocarbon production from syngas. Usually Fe, Co, Ni, and Ru were used in the F-T catalysts as the primary components, with or without promoters (Enger, Fossan, Borg, Rytter, & Holmen, 2011; Jahangiri, Bennett, Mahjoubi, Wilson, & Gu, 2014; Keyvanloo et al., 2014; Koh, Koo, Yu, Lim, & Bae, 2014). When doped with some metals that cannot readily dissociate CO, such as Cu, Ir, Re, etc. (Courty, Durand, Freund, & Sugier, 1982; Nachal Devi Subramanian, Balaji, Kumar, & Spivey, 2009; Tien-Thao, Zahedi-Niaki, Alamdari, & Kaliaguine, 2007; K. Xiao, Bao, et al., 2013; K. Xiao, Qi, et al., 2013; Yang, Zhu, Hou, Zhou, & Su, 2011), F-T catalysts exhibited an increased selectivity towards higher oxygenates, especially the linear C_1-C_6 terminal alcohols complying with the Anderson-Schulz-Flory (ASF) distribution (Courty et al., 1982; Graham et al., 2014).

Since 1980s, numerous novel catalysts based on Co-Cu have been developed following the pioneering investigation by IFP (Keim, 1987). Various promoters, for instance, La, Re, Ir, Cu, Mn, Cr, etc., combined with alkali promoters have been studied to improve the yields of higher alcohols (Y. Z. Fang, Liu, & Zhang, 2011; Lebarbier et al., 2011; Yang et al., 2011). Generally, the reducibility of Co oxides was significantly affected by the promoters added. Co/M bulk and/or surface alloy (M = promoters) were detected as an indication of the strong interaction between Co and some of the promoters, which was considered to play a pivotal role in higher alcohol production. Most of the modified F-T catalysts for higher alcohol synthesis were prepared using co-impregnation or co-precipitation techniques, followed by direct reduction in the slurry using reductant solution, or through a combination of calcinations and subsequent reduction by pure or diluted hydrogen.

- **Cu-Modified F-T Catalysts:** Cu-modified F-T catalyst is usually denoted as Cu/M (M = Fe, Co etc.) which is considered as one of the most promising catalysts owing to its higher efficient productivity towards ethanol and higher oxygenates (K. Fang et al., 2009; Y. Xiang et al., 2013; K. Xiao, Qi, et al., 2013). Generally, the intriguing influence of Cu additives was to decrease the reducibility of Co particles and increase their dispersion (Baker, Burch, Hibble, & Loader, 1990). The mechanism of $C_{2+}OH$ synthesis over modified F-T catalysts is similar to that of Rh-based catalysts, briefly, the insertion of CO into the CH_x species to form the initial C–C bond. With the assistance of hydrogenation, triple bonds in CO could be dissociated readily over metals for F-T synthesis, such as Fe, Co, Ru etc., resulting in the formation of surface CH_x species. The introduction of Cu additives could facilitate the non-dissociative activation of CO, which was required for the insertion reaction mentioned above for $C_{2+}OH$ formation.

The active sites of copper-containing F-T catalysts remain controversial. Taking Cu/Co catalysts as an example, some researchers considered that Cu/Co alloy is the primary reactive site (Dalmon, Chaumette, & Mirodatos, 1992), while others suggested that the active sites consists of metallic Cu^0 and partially reduced cobalt particles (Baker et al., 1990; Sheffer, Jacobson, & King, 1989).

A series of studies were conducted on Cu/Fe (K. Xiao, Bao, et al., 2013), Cu/Co and Cu/Ni binary catalysts for the synthesis of $C_{2+}OH$ (Fraga & Jordao, 1998; Y. Y. Liu, Murata, Inaba, Takahara, & Okabe, 2013; Mouaddib, Perrichon, & Martin, 1994; Nachal Devi Subramanian et al., 2009). Xiao et al. concluded that the introduction of Cu had different effects on those F-T metals (K. Xiao, Qi, et al., 2013). The activity of CO hydrogenation followed the order of Cu/Ni < Cu/Co < Cu/Fe. High activity of the Cu/Fe catalysts was attributed to the presence of monometallic Fe which was separated from the Cu/Fe binary nanoparticles. Therefore, Cu/Fe catalysts exhibited similar selectivity to F-T synthesis because of the separated Fe nanoparticles. Slight separation of Co from Cu/Co was also observed, which led to the formation of Cu@Co core-shell structure co-existing with Cu/Co alloy. Compared to the other two binary catalysts, the interaction between Cu and Ni seemed to be much stronger, leading to almost no phase separation in the Cu/Ni catalysts.

Alumina-supported Cu/Co catalysts with different ratio of Cu/Co were studied by Wang et al. (J. Wang, Chernavskii, Khodakov, & Wang, 2012). Introduction of Cu significantly decreased the reducibility of Co and lead to the formation of bimetallic Cu/Co nanoparticles, which was speculated to be critical to the synthesis of higher alcohols. $CuCo_2O_4$ mixed spinels were detected in the calcined catalysts indicating the existence of strong interaction between Cu and Co in nanoscale. The strong affinity was also demonstrated by the characterization using magnetic measurements. To further clarify the synergistic effect between Cu and Co, a series of well-controlled Co@Cu core-shell and Co/Cu mixed nanoparticles were prepared (Nachal Devi Subramanian et al., 2009). It was concluded that mixed Co/Cu catalysts exhibited a higher selectivity towards ethanol and other oxygenates than the core-shell catalysts. Large amounts of hydrocarbons were detected on the core-shell catalysts, because of the enrichment of Co on the surface which favored the traditional F-T synthesis towards alkanes.

Other components, such as alkali oxides (de Aquino & Gomez Cobo, 2001; Tien-Thao, Hassan Zahedi-Niaki, Alamdari, & Kaliaguine, 2007), La (Lebarbier et al., 2011), Mn (Y. Xiang et al., 2013), Cr, and Mo etc. (Prieto et al., 2014), were used to further improve the yields of higher alcohols based on binary Cu/Co catalysts. Much attention has been shifted to deeply understand the microstructure of the well-developed catalysts with advanced characterizations in combination with theoretical calculations. On the ternary Co/Cu/Mn core-shell nanocatalysts, Xiang et al. found that the formation of linear

C_8~C_{14}OH rather than C_2~C_6OH was more favored (Y. Xiang et al., 2013). 3D tomographic reconstructions through atom probe microscopy visualized the detailed structure and atom distribution in the core-shell nanocatalysts. Detailed catalytic performance of Co/Cu/Mn with various amounts of constituents was displayed in Figure 7. The superb catalytic performance for long chain primary alcohols was ascribed to the unique structure of the catalyst, namely Co-rich core in close combination with Cu-dominated ternary Co/Cu/Mn shell.

Recently, an innovative method of preparing the desired Cu/Co alloy catalyst was reported by Prieto et al. (Prieto et al., 2014). In their research, the formation of Cu/Co alloy on the surface of bulk catalysts was attributed to the simultaneous exsolution of Co^{2+} and Cu^{2+} from the metal molybdate crystalline precursors during the reduction by H_2. Catalytic performance of the novel Cu/Co/Mo catalysts was tested after modification of K in comparison with the reference of Cu/Co/Cr catalysts in Figure 8. The optimal catalyst with a Cu/(Cu+Co) ratio of 0.3 gave a space-time-yield of C_{2+} alcohols as high as 27 mmol/(g_{cat}·h).

Figure 7. Detailed catalytic performance of Co/Cu/Mn catalysts. (Copyright 2013, American Chemical Society. Used with permission from (Y. Xiang et al., 2013))

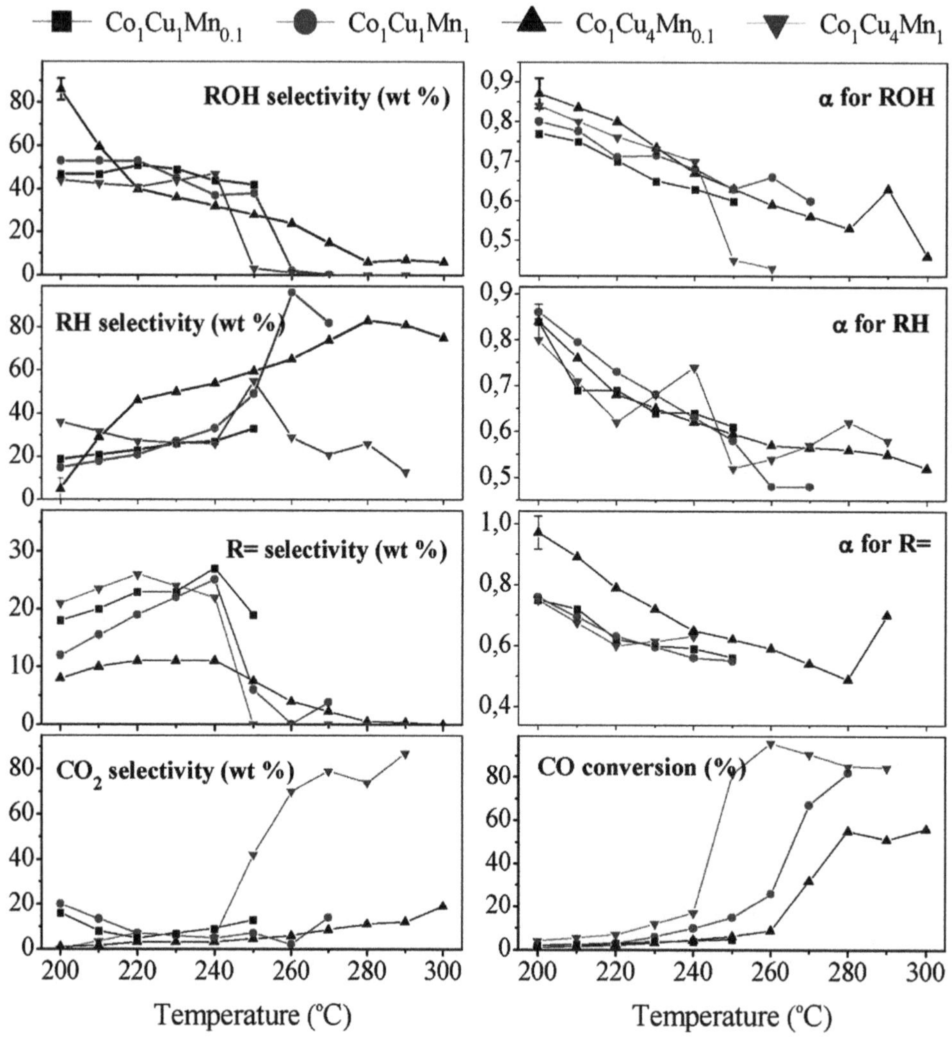

Figure 8. CO hydrogenation over K/Cu/Co/MoO$_x$ (open circles and solid bars) and K/Cu/Co/CrO$_x$ (full circle and dashed bars) catalysts (Experimental conditions: 4.0 MPa; 270 °C; H$_2$/CO = 1) (Copyright 2014, Wiley-VCH Verlag GmbH & Co. KGaA, Weinheim. Used with permission from (Prieto et al., 2014))

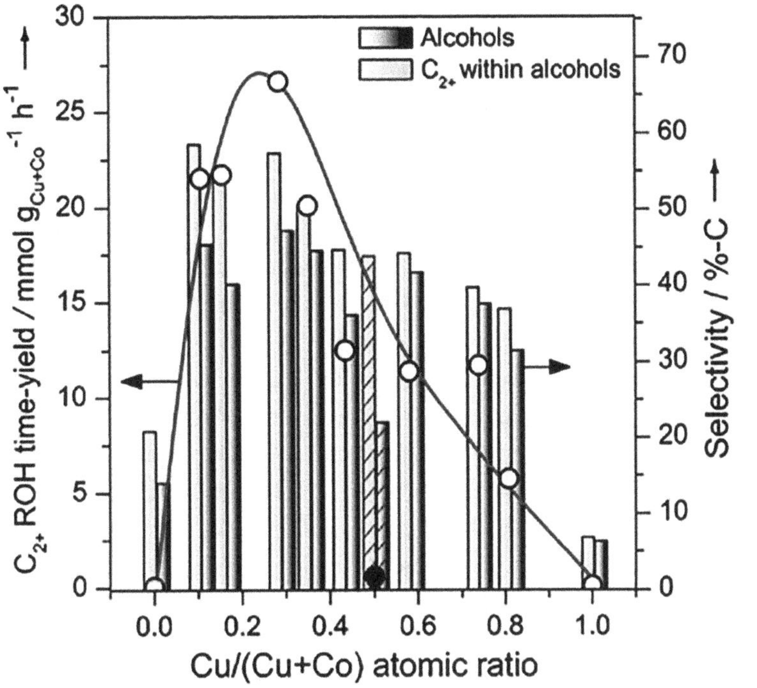

- **Influence of Supports:** Porous SiO_2 and Al_2O_3 are typically used as supports to provide a better dispersion of metals. The strong interaction between metal and carrier generally could prolong the durability for the modified F-T catalysts under severe reaction conditions. A strong metal-support interaction was ascribed to the formation of various types of spinels, such as silicates, aluminates (Bessell, 1993), titanates (Spencer, 1985), and molybdates etc. (Prieto et al., 2014).

Recently, supported Cu/Co catalysts over alumina, silica and carbon nanotubes (CNTs) were prepared and evaluated by Wang et al. (J. Wang, Chernavskii, Wang, & Khodakov, 2013). Both $CoAl_2O_4$ and $CuAl_2O_4$ were detected in the Al_2O_3 supported bimetallic catalysts, suggesting the penetration of Cu and Co atoms into the aluminate matrix (Dalmon et al., 1992). However, obvious separation of CuO and Co_2O_3 occurred when SiO_2 and CNTs were used as carriers, due to the comparatively weaker interaction between metal and supports. In the research of Lee et al., the influence of three different supports ZnO, MgO and Al_2O_3 on Cu/Co catalysts was examined (Lee, Reddy, Jung, Yang, & Moon, 2014). ZnO support was considered to be more preferred for higher alcohol production compared with Al_2O_3 and MgO. More byproducts including dimethyl ether and hydrocarbons were generated when Al_2O_3 was used owing to the acidic sites present.

Some other novel supports were also synthesized to better stabilize the desired Cu/Co particles. In the work of Fang et al., perovskite $LaFe_{0.7}Cu_{0.3}O_3$ supported Co_2O_3 was prepared. During the reduction process, Cu/Co alloy and Cu@Co core-shell structures were generated in situ following the simultaneous reduction of surface Co_3O_4 and CuO_x from the bulk of the supports (Y. Z. Fang et al., 2011). When

6%Co_3O_4/$LaFe_{0.7}Cu_{0.3}O_3$ was tested at 300 °C, 3 MPa, the CO conversion was as high as 56%. The selectivity of total alcohols was 43% including methanol and ethanol 55% and 45%, respectively. In addition, the catalysts kept stable during the test for 200h.

Mo/MoS_2-Based Catalysts

Since the pioneering work in 1980s by Murchison et al., Mo-based catalysts have received significant attention (Phillips, Ternan, & Editors, 1988). When MoO_x catalysts alone were utilized for the hydrogenation of CO, both the activity and the yield of mixed alcohols were low with hydrocarbons as the major products (Juan & Damiani, 1992; S. Zaman & Smith, 2012). The $C_{2+}OH$ and hydrocarbon products formed over Mo-based catalysts generally follow the standard Anderson-Schulz-Flory (ASF) distribution (Bao et al., 2009; K. Fang et al., 2009; Santos et al., 2013). To enhance the activity and improve the yield of the desired products, Mo-based catalysts were often modified with S, C, or P elements and promoted with alkali metals and/or transition metals (Fe, Co, Ni etc.) (S. Zaman & Smith, 2012). Recent development of MoO_2, Mo_2C, Mo_2N, MoP and MoS_2 catalysts will be discussed respectively in the following content.

Molybdenum oxide catalysts were typically supported on SiO_2, Al_2O_3, active carbon or carbon etc., and doubly promoted with 3d transition metals and alkali oxides (Juan & Damiani, 1992; Ma et al., 2010; Storm, 1994; F.-S. Xiao, Fukuoka, Ichikawa, Henderson, & Shriver, 1992). Ni/Mo/K Catalysts supported on multi-walled carbon nanotubes were studied by Ma and coworkers (Ma et al., 2010). The overall selectivity to C_1~C_3 alcohols and dimethyl ether was as high as 59% (on a CO_2-free basis), with the space-time yield of 205 mg/(g_{cat}·h). Mo^{4+} and partially reduced $Ni^{\delta+}$ species were considered to be the key constituents of the active sites, which was consistent with some previous literatures (Bao et al., 2009; Muramatsu, Tatsumi, & Tominaga, 1987; Tatsumi, Muramatsu, & Tominaga, 1986). They emphasized that MCNTs, instead of alkali, significantly caused an increase in the concentration of active sites (Muramatsu et al., 1987).

Mo_2C catalysts are interesting due to their unique capability of hydrogenation (N. Wang et al., 2010; M. Xiang, Li, Li, Zhong, & Sun, 2007). Both alkali compounds (usually K_2CO_3) and transition metals were introduced into Mo_2C catalysts to improve the yield of $C_{2+}OH$. Xiang et al. concluded that $C_{2+}OH$ were dramatically increased and the selectivity to methanol was significantly reduced when Mo_2C catalysts were modified with Fe, Co, and Ni promoters (M. Xiang et al., 2007). Co_3Mo_3C compounds were detected when Co was added into Mo_2C system, which was considered as the probable active centers for $C_{2+}OH$ synthesis. Similar conclusion was also made in some other literatures, where the penetration of Ni into the crystal matrix of Mo_2C occurred to facilitate $C_{2+}OH$ generation (N. Wang et al., 2010).

Only sporadic practice has been conducted over Mo_2N and MoP catalysts for $C_{2+}OH$ synthesis (Ranhotra, Bell, & Reimer, 1987; S. F. Zaman & Smith, 2010). Recently Zaman and coworkers developed a series of SiO_2 supported K-promoted MoP catalyst, wherethe spce-time yield of oxygenates was found as high as 147.2 mg/(g_{cat}·h) over 5 wt% K-promoted 15 wt% MoP catalysts (S. F. Zaman & Smith, 2010). The highest selectivity to C_{2+} oxygenates (72.6 C atom%) was obtained on the 5 wt% K/10 wt% MoP/SiO_2 catalysts at 275 °C, H_2/CO = 1, with the undesired CH_4 being well suppressed (9.7 C atom%) (S. F. Zaman & Smith, 2009).

Among the various Mo-based Catalysts studied, MoS_2–based catalysts have been extensively studied due to their preferable sulfur-resistance properties, higher yield towards $C_{2+}OH$ and effective suppression of carbon deposition (K. Fang et al., 2009; Santos et al., 2013; S. Zaman & Smith, 2012). MoS_2 alone, similar to the unmodified MoO_x-based catalysts, would only result in dominant undesired hydrocarbons in the products, especially methane (Z. Liu et al., 1997). It was conceived that the alkali dopants could neutralize the acidic sites on the surface of MoS_2 and stabilize the partially reduced Mo^{4+}, thus leading to the decrease of hydrocarbon production and the enhancement of $C_{2+}OH$ synthesis (Tatsumi et al., 1986; Woo et al., 1993; S. Zaman & Smith, 2012). Recently some researchers proposed that the modification of alkali may also alter the electronic properties of the catalysts, which facilitated the aldol condensation reaction (Santos et al., 2013). However, others claimed that there was no occurrence of aldol condensation reactions over Co/MoS_2 catalysts (Toyoda, Minami, & Qian, 2013).

In some studies, the promotional effect of various alkali oxides was reported to increase following the order of K < Rb < Cs (Koizumi, Murai, Ozaki, & Yamada, 2004). Nevertheless, some others added that the physicochemical properties of the catalysts also influenced the rank of promotional effect, for instance, the surface area of MoS_2 or the dosage of alkali [139]. Consequently, sometimes the promotional effect followed the completely opposite order Cs < Rb < K in the work by Liu and coworkers (Iranmahboob, Toghiani, & Hill, 2003; Z. Liu et al., 1997; Woo et al., 1993).The yield of $C_{2+}OH$ over MoS_2-based catalysts can be increased with transition metals, usually (Fe, Co, and Ni) (Iranmahboob et al., 2003; Surisetty, Dalai, & Kozinski, 2010b), and some noble metals as well (e.g. Rh and Ir etc.) (Li, Fu, & Jiang, 1999; Matsumoto, Kim, Kishida, Nagata, & Wakabayashi, 1994; Surisetty, Dalai, & Kozinski, 2010a). It was presumed that $C_{2+}OH$ synthesis was initiated with the insertion of non-dissociative CO into the CH_x species that were derived from the successive hydrogenation of dissociated CO, which was analogous to the mechanism over Rh-based catalysts aforementioned (Santos et al., 2013). The dissociation of CO mainly occurred on the transition metal (Fe, Co, Ni etc.) sites which led to the formation of CH_x species, while the non-dissociative CO on the MoS_2 or K/MoS_2 sites (K. Fang et al., 2009; M. Huang & Cho, 2009). Recently, K/Co/MoS_2/Al_2O_3 catalysts were prepared and evaluated in the study by Toyoda and coworkers (Toyoda et al., 2013). Detailed catalytic performance was provided in Table 4. The highest selectivity towards $C_{2+}OH$ (approximately 40%) was obtained over $K_{0.6}$/$Co_{0.5}$/MoS_2/Al_2O_3 catalysts with the support precalcined at 1000 °C. On the Co-promoted MoS_2 catalysts, the absorbed CO was speculated to be dissociated on the Co_xS_y sites while not be dissociated on the MoS_2 sites. As discussed above, both the dissociation and non-dissociation of CO were indispensable for the synthesis of $C_{2+}OH$. MCNTs supported K/MoS_2 catalysts with Ni were reported by Wang et al. (J.-J. Wang et al., 2013). They argued that the overall yield of oxygenates was approximate 113 mg/(g_{cat}·h) with the corresponding selectivity up to 64% (on a CO_2-free basis) on the catalyst a Ni/Mo molar ratio of being 0.5:1 at 320 °C, 8.0 MPa. Similarly, noble metals such as Rh were also added into the MoS_2-based catalysts (Surisetty et al., 2010a), where Mo (VI) became more facile to be reduced to Mo (IV) and the catalytic performance was significantly improved. The maximum yield of total alcohols reached 0.211 g/(g_{cat}·h) with 16% of ethanol over 1.5 wt% of Rh-containing catalysts at 320 °C, 8.27 MPa.

Table 4. Catalytic performance of K/Co/MoS_2/Al_2O_3 catalysts [a] (from (Toyoda et al., 2013))

T	CO Conversion	Selectivity(%)						α*	STY of $C_{2+}OH$
°C	(%)	CO_2	MeOH	EtOH	PrOH	BuOH	Alkanes		mol/(kg_{cat}·h)
260	4.4	33.2	8.9	31.6	4.1	1.1	21.1	0.19	1.8
300	6.3	32.2	6.7	29.3	4.0	0.6	27.1	0.21	2.4
320	7.1	34.5	5.4	21.3	3.8	1.1	33.9	0.28	2.0

* Chain growth probability; [a] Reaction conditions: GHSV = 5000h^{-1}, H_2/CO = 1

Stability Evaluation

During the catalytic conversion of syngas to $C_{2+}OH$, various chemical reactions occur synchronously, and almost all the reactions involved are highly exothermic, as indicated in equations 3-6 (Gupta et al., 2011; Spivey & Egbebi, 2007; Subramani & Gangwal, 2008; Xiaoding et al., 1987). Thus, stability of catalysts is a quite critical factor considering the industrial application in the future. As aforementioned, separation of the homogeneous catalysts (i.e. toxic metal carbonyl complexes) from products for reuse is quite challenging. As for heterogeneous catalysts, the stability tests of Rh-based catalysts are scarcely studied and few patents about Rh-based ones have been fielded, probably because of the prohibitive cost of this noble metal. For modified F-T and/or Mo-based catalysts, usually reaction processes are conducted at 220-260 °C. When Cu/Co nanoalloy catalysts were tested, it was found that phase separation occurred gradually, and then, production of alkanes increased (K. Xiao, Qi, et al., 2013). Nanoscale evolution of Cu/Co core-shell nanoparticles was reported by Carenco et al. (Carenco et al., 2013) Formation of Cu-rich nanoparticles was assumed to Co extraction via Co carbonyl species. Mo-based catalysts showed quite a desirable durability during the tests. It was reported that an innovated Ni/MoS_2 catalyst could keep high activity (0.40 $g/ml_{cat}•h$ of total alcohols and 0.16 $g/(ml_{cat}•h)$ of higher alcohols) after 2000 h (K. Fang et al., 2009).

Compare with high-temperature methanol synthesis catalysts (ZnO/Cr_2O_3-based), stability of alkali modified low-temperature catalysts (Cu/ZnO-based) have attracted more attention. When Cu/ZnO-based catalysts are tested, higher reaction temperature is needed (280-320 °C) (Spivey & Egbebi, 2007). Fortunately, stability of Cu/ZnO can be promoted with the help of alkali additives (G. A, P. B, G. W, & K, 1985; Sun et al., 2015). The activity of Cs_2O-promoted $Cu/ZnO/Al_2O_3$ catalysts suffered only 7% of activity after 80h (Sun et al., 2015). Recently, different novel reactors have been introduced to facilitate heat transfer to prevent the catalysts from sintering, as shown in the following context.

$$2CO + 4H_2 = C_2H_5OH + H_2O; \quad (3)$$

$$\Delta H_r° = -253.6 \text{ kJ/mol}$$

$$CO + 2H_2 = CH_3OH; \quad (4)$$

$$\Delta H_r° = -90.4 \text{ kJ/mol}$$

$$CO + 3H_2 = CH_4 + H_2O; \quad (5)$$

$$\Delta H_r° = -205.9 \text{ kJ/mol}$$

$$CO + H_2O = CO_2 + H_2; \quad (6)$$

$$\Delta H_r° = -41.1 \text{ kJ/mol}$$

Novel Reactors

Synthesis of $C_{2+}OH$ from syngas is highly exothermic, as aforementioned. Accordingly, efficient heat transfer in the reactor is desired to prevent the formation of hotspots within the catalyst bed which may increase the formation of undesired alkane products and cause deactivation due to sintering especially when using of low-melting-point metals as catalysts (K. Fang et al., 2009). Several novel reactors with high efficiency in heat and/or mass transfer have been developed, including slurry phase reactors (Herman, 2000; Verkerk, Jaeger, Finkeldei, & Keim, 1999; Zuo et al., 2014) and microchannel reactors etc. (Cao, Hu, Li, Wilcox, & Wang, 2009; Hu et al., 2005).

Slurry phase reactors have the advantages of heat transfer, accurate temperature control, and amenable to large-scale applications (Herman, 2000). Since first utilized in methanol synthesis in 1970s, slurry phase reactors have received much attention in F-T synthesis and $C_{2+}OH$ synthesis reactions etc. When slurry $Cu/ZnO/Al_2O_3$ catalysts or finely crushed solid $Cu/ZnO/Al_2O_3$ catalysts were evaluated in slurry bed reactor (Breman et al., 1995; Zuo et al., 2014), it appeared that the distribution of products was substantially impacted by the slurry solvent (usually liquid paraffin etc.). For instance, relatively higher ratio of (1-alcohol)/(2-methyl-l-alcohol) ($n \geq 4$) was obtained when using gas-slurry bed reactor (Breman et al., 1995).

Unlike the slurry phase reactor, liquid holdup is less an issue when microchannel reactor is used (Hu et al., 2005). Moreover, higher gas velocity, efficient heat removal and more precise temperature control could be achieved (Cao et al., 2009; Lerou, Tonkovich, Silva, Perry, & McDaniel, 2010). Microchannel reactor has been applied in F-T synthesis (Cao et al., 2009), the production of DME (Hu et al., 2005), or in mixed $C_{2+}OH$ synthesis etc. (Hu et al., 2007). The reactor was depicted in Figure 9, with typical channel dimensions of 5.08 cm × 0.94 cm × 0.15 cm. When the structured $Rh/Mn/SiO_2$ catalysts were evaluated inside the microchannel reactor at a relatively high throughput (GHSV = 20,000 h^{-1}), it was found that the reaction temperature was still under control with only a small increase (from 31.1% to 36.5%) in methane selectivity. This was attributed to the highly efficient heat removal and the unique properties of the integrated microchannel reactor system (Hu et al., 2007).

Figure 9. Microchannel reactor assembly (from (Cao et al., 2009))

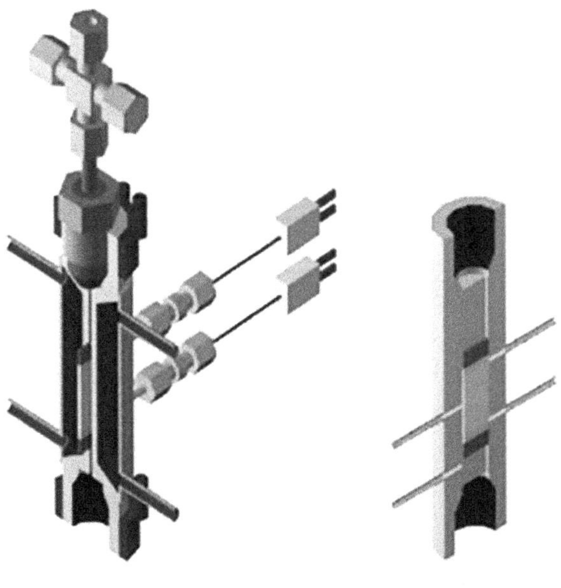

Besides the novel reactor design mentioned above, some other innovative approaches based on the traditional fixed-bed reactor have been developed as well, such as the introduction of supercritical fluids (T. Jiang, Niu, & Zhong, 2001; Reubroycharoen, Yoneyama, Vitidsant, & Tsubaki, 2003; Rui Xu, Zhang, & Roberts, 2013) and the design of double-bed reactors (Burcham, Herman, & Klier, 1998; Guo et al., 2013). Xu et al. investigated the influence of supercritical hexanes on the performance of K/Cu/ZnO/Al_2O_3 catalysts within a fixed-bed reactor. They concluded that both the reaction heat and the products (methanol, ethanol etc.) could be removed promptly during the exothermic process, which could mitigate the formation of hotspots and positively shift the reaction equilibrium to achieve a higher CO conversion (Rui Xu et al., 2013). Burcham and coworkers examined the 3 mol% Cs/Cu/ZnO/Cr_2O_3 catalysts over two tandem beds maintained at 325 °C and 340 °C, respectively (Burcham et al., 1998). As aforementioned, methanol can be further converted into $C_{2+}OH$ under higher reaction temperature. Hence, the yield of isobutanol was as high as 202 g/(kg_{cat}·h). Double-bed reactor with two types of different catalysts has also been explored, either at the same or at different temperature (Beretta, Sun, Herman, & Klier, 1996; Guo et al., 2013). It was interesting that when the syngas passed through K/Cu/Fe/Co catalysts first and then through Cu/Zn/Al catalysts, CO conversion was dramatically enhanced to 48.3% from ~25% under the same conditions (T = 280 °C, P = 5.5 MPa. This was ascribed to the thermal coupling effect and product conversion coupling effect within the two-stage reactor (Guo et al., 2013).

CONCLUSION AND PERSPECTIVES

Product distribution is strongly dependent on the absorption and dissociation of CO over various types of heterogeneous catalysts. Initial C-C formation requires bifunctional sites for the coupling of undissociated CO (with no complete break of C-O bond) with CH_x (x = 1, 2 or 3) species over Rh-based catalysts, modified F-T catalysts, and Mo-based catalysts. For alkali-promoted Cu-based catalysts, bonded formyl species are considered as key intermediates. Subsequent nucleophilic attack reactions between C_1 intermediates result in the formation of initial C-C bond of ethanol.

Compared with heterogeneous catalysts, homogeneous catalysts exhibit much higher selectivity towards $C_{2+}OH$. The major challenges on the way towards large-scale application for homogeneous catalysts are recycling and controlling the toxic complexes. Recent advances indicate that space time yield of $C_{2+}OH$ over heterogeneous catalysts can be as high as 200-500 g/(kg_{cat}·h) for all the four types of heterogeneous catalysts, with CO conversion as high as 10-40%. Rh-based catalysts exhibit promising higher selectivity towards $C_{2+}OH$. Nevertheless, prohibitive cost of Rh metal hinders its large-scale application. Therefore, non-noble metal based catalysts have attracted much more attention in commercial application. Recently, mixed alcohol production capacity of up to 5,000 ton/a was demonstrated by the Institute of Coal Chemistry and Shenhua Coal to Oil and Chemical co., LTD in China over novel Cu-based catalysts for the conversion of coal-derived syngas.

Further research is needed to increase the productivity of $C_{2+}OH$ to a higher level and effectively reduce the production of byproducts (such as alkanes and CO_2) for a higher selectivity of $C_{2+}OH$. New design of Rh-based catalysts, non-noble metal based catalysts, and novel reactors is also needed for further improvement of this process. Modified Cu-based catalysts exhibit great potential in large-scale application in the near future owing to their lower cost, relatively higher selectivity and productivity towards $C_{2+}OH$.

REFERENCES

Abdelsayed, V., Shekhawat, D., Poston, J. A. Jr, & Spivey, J. J. (2013). Synthesis, characterization, and catalytic activity of Rh-based lanthanum zirconate pyrochlores for higher alcohol synthesis. *Catalysis Today*, *207*(0), 65–73. doi:10.1016/j.cattod.2012.07.011

Baker, J. E., Burch, R., Hibble, S. J., & Loader, P. K. (1990). Properties of silica-supported Cu/Co bimetallic catalysts in the synthesis of higher alcohols. *Applied Catalysis*, *65*(2), 281–292. doi:10.1016/S0166-9834(00)81603-6

Bao, J., Sun, Z.-H., Fu, Y.-L., Bian, G.-Z., Zhang, Y., & Tsubaki, N. (2009). Mixed Alcohol Synthesis from Syngas on K-Co-Mo/C Catalyst Prepared by a Sol-Gel Method. *Topics in Catalysis*, *52*(6-7), 789–794. doi:10.1007/s11244-009-9200-x

Behrens, M., Studt, F., Kasatkin, I., Kühl, S., Hävecker, M., Abild-Pedersen, F., & Schlögl, R. (2012). The Active Site of Methanol Synthesis over Cu/ZnO/Al_2O_3 Industrial Catalysts. *Science*, *336*(6083), 893–897. doi:10.1126/science.1219831 PMID:22517324

Beiramar, J. M., Griboval-Constant, A., & Khodakov, A. Y. (2014). Effects of Metal Promotion on the Performance of CuZnAl Catalysts for Alcohol Synthesis. *ChemCatChem*, *6*(6), 1788–1793. doi:10.1002/cctc.201402037

Benichou, L., & Mayr, S. (2014). Rogeaulito: a world energy scenario modeling tool for transparent energy system thinking. *Frontiers in Energy Research, 1.*

Beretta, A., Sun, Q., Herman, R. G., & Klier, K. (1996). Production of Methanol and Isobutyl Alcohol Mixtures over Double-Bed Cesium-Promoted Cu/ZnO/Cr_2O_3 and ZnO/Cr_2O_3 Catalysts. *Industrial & Engineering Chemistry Research*, *35*(5), 1534–1542. doi:10.1021/ie9505219

Bessell, S. (1993). Support effects in cobalt-based fischer-tropsch catalysis. *Applied Catalysis A, General*, *96*(2), 253–268. doi:10.1016/0926-860X(90)80014-6

Bi, J., Guo, X., Liu, M., & Wang, X. (2010). High effective dehydration of bio-ethanol into ethylene over nanoscale HZSM-5 zeolite catalysts. *Catalysis Today*, *149*(1–2), 143–147. doi:10.1016/j.cattod.2009.04.016

Blank, J. H., Hembre, R., Ponasik, J., & Cole-Hamilton, D. J. (2013). A Tertiary Phosphonium Salt as a Promoter for the Hydrogenation of CO. *ChemCatChem*, *5*(5), 1075–1078. doi:10.1002/cctc.201200176

Blank, J. H., Hembre, R., Ponasik, J., & Cole-Hamilton, D. J. (2014). Alternative pathways in the ruthenium catalysed hydrogenation of CO to alcohols. *Catalysis Science & Technology*, *4*(1), 218–223. doi:10.1039/C3CY00423F

Boz, I., Sahibzada, M., & Metcalfe, I. S. (1994). Kinetics of the Higher Alcohol Synthesis over a K-promoted CuO/ZnO/Al_2O_3 Catalyst. *Industrial & Engineering Chemistry Research*, *33*(9), 2021–2028. doi:10.1021/ie00033a001

Breman, B. B., Beenackers, A. A. C. M., Schuurman, H. A., & Oesterholt, E. (1995). Kinetics of the gas-slurry methanol-higher alcohol synthesis from $CO/CO_2/H_2$ over a Cs-Cu/ZnO/Al_2O_3 catalyst, including simultaneous formation of methyl esters and hydrocarbons. *Catalysis Today*, *24*(1–2), 5–14. doi:10.1016/0920-5861(95)00097-Y

Brown Bourzutschky, J. A., Homs, N., & Bell, A. T. (1990a). Conversion of synthesis gas over $LaMn_{1-x}Cu_xO_{3+\lambda}$ perovskites and related copper catalysts. *Journal of Catalysis*, *124*(1), 52–72. doi:10.1016/0021-9517(90)90103-Q

Brown Bourzutschky, J. A., Homs, N., & Bell, A. T. (1990b). Hydrogenation of CO_2 and CO_2/CO mixtures over copper-containing catalysts. *Journal of Catalysis*, *124*(1), 73–85. doi:10.1016/0021-9517(90)90104-R

Burch, R., & Hayes, M. J. (1997). The Preparation and Characterisation of Fe-Promoted Al_2O_3-Supported Rh Catalysts for the Selective Production of Ethanol from Syngas. *Journal of Catalysis*, *165*(2), 249–261. doi:10.1006/jcat.1997.1482

Burch, R., Loader, P. K., & Cruise, N. A. (1996). An investigation of the deactivation of Rh/alumina catalysts under strong oxidising conditions. *Applied Catalysis A, General*, *147*(2), 375–394. doi:10.1016/S0926-860X(96)00212-8

Burcham, M. M., Herman, R. G., & Klier, K. (1998). Higher Alcohol Synthesis over Double Bed Cs–Cu/ZnO/Cr_2O_3 Catalysts: Optimizing the Yields of 2-Methyl-1-propanol (Isobutanol). *Industrial & Engineering Chemistry Research*, *37*(12), 4657–4668. doi:10.1021/ie9705620

Calverley, E. M., & Smith, K. T. (1991). The effects of carbon dioxide, methanol, and alkali promoter concentration on the higher alcohol synthesis over a Cu/ZnO/Cr_2O_3 catalyst. *Journal of Catalysis*, *130*(2), 616–626. doi:10.1016/0021-9517(91)90141-P

Campos, M. (1996). Promoter Effect of Cesium on C–C Bond Formation during Alcohol Synthesis from CO/H_2 over Cu/ZnO/Cr_2O_3Catalysts. *Journal of Catalysis*, *163*(2), 418–428. doi:10.1006/jcat.1996.0343

Camposmartin, J. M., Guerreroruiz, A., & Fierro, J. L. G. (1995). Structural and Surface Properties of CuO-ZnO-Cr_2O_3 Catalysts and Their Relationship with Selectivity to Higher Alcohol Synthesis. *Journal of Catalysis*, *156*(2), 208–218. doi:10.1006/jcat.1995.1248

Cao, C., Hu, J., Li, S., Wilcox, W., & Wang, Y. (2009). Intensified Fischer–Tropsch synthesis process with microchannel catalytic reactors. *Catalysis Today*, *140*(3–4), 149–156. doi:10.1016/j.cattod.2008.10.016

Carenco, S., Tuxen, A., Chintapalli, M., Pach, E., Escudero, C., Ewers, T. D., & Salmeron, M. (2013). Dealloying of Cobalt from CuCo Nanoparticles under Syngas Exposure. *The Journal of Physical Chemistry C*, *117*(12), 6259–6266. doi:10.1021/jp4000297

Chen, G., Li, S., Jiao, F., & Yuan, Q. (2007). Catalytic dehydration of bioethanol to ethylene over TiO_2/γ-Al_2O_3 catalysts in microchannel reactors. *Catalysis Today*, *125*(1–2), 111–119. doi:10.1016/j.cattod.2007.01.071

Chen, W., Pan, X. L., Willinger, M. G., Su, D. S., & Bao, X. H. (2006). Facile autoreduction of iron oxide/carbon nanotube encapsulates. *Journal of the American Chemical Society*, *128*(10), 3136–3137. doi:10.1021/ja0567211 PMID:16522081

Choi, Y., & Liu, P. (2009). Mechanism of Ethanol Synthesis from Syngas on Rh(111). *Journal of the American Chemical Society*, *131*(36), 13054–13061. doi:10.1021/ja903013x PMID:19702298

Corma, A. (2014). Introduction: Chemicals from Coal, Alkynes, and Biofuels. *Chemical Reviews*, *114*(3), 1545–1546. doi:10.1021/cr400719e PMID:24517085

Courty, P., Durand, D., Freund, E., & Sugier, A. (1982). C_1-C_6 alcohols from synthesis gas on copper-cobalt catalysts. *Journal of Molecular Catalysis*, *17*(2–3), 241–254. doi:10.1016/0304-5102(82)85035-9

Dalmon, J. A., Chaumette, P., & Mirodatos, C. (1992). Higher alcohols synthesis on cobalt based model catalysts. *Catalysis Today*, *15*(1), 101–127.

Davidson, S. D., Sun, J., Hong, Y., Karim, A. M., Datye, A. K., & Wang, Y. (2014). The effect of ZnO addition on Co/C catalyst for vapor and aqueous phase reforming of ethanol. *Catalysis Today*, *233*, 38–45. doi:10.1016/j.cattod.2013.12.044

De Aquino, A. D., & Gomez Cobo, A. J. (2001). Synthesis of higher alcohols with cobalt and copper based model catalysts: Effect of the alkaline metals. *Catalysis Today*, *65*(2–4), 209–216. doi:10.1016/S0920-5861(00)00575-7

Deluzarche, A., Hindermann, J. P., Kiennemann, A., & Kieffer, R. (1985). Application of chemical trapping to determination of surface species and to the study of their evolution under reaction conditions in heterogeneous catalysis. *Journal of Molecular Catalysis*, *31*(2), 225–250. doi:10.1016/0304-5102(85)85066-5

Egbebi, A., Schwartz, V., Overbury, S. H., & Spivey, J. J. (2010). Effect of Li Promoter on titania-supported Rh catalyst for ethanol formation from CO hydrogenation. *Catalysis Today*, *149*(1–2), 91–97. doi:10.1016/j.cattod.2009.07.104

Elliott, D. J., & Pennella, F. (1988). Mechanism of ethanol formation from synthesis gas over CuO/ZnO/Al_2O_3. *Journal of Catalysis*, *114*(1), 90–99. doi:10.1016/0021-9517(88)90011-5

Enger, B. C., Fossan, Å.-L., Borg, Ø., Rytter, E., & Holmen, A. (2011). Modified alumina as catalyst support for cobalt in the Fischer–Tropsch synthesis. *Journal of Catalysis*, *284*(1), 9–22. doi:10.1016/j.jcat.2011.08.008

Fan, Z., Chen, W., Pan, X., & Bao, X. (2009). Catalytic conversion of syngas into C_2 oxygenates over Rh-based catalysts—Effect of carbon supports. *Catalysis Today*, *147*(2), 86–93. doi:10.1016/j.cattod.2009.03.004

Fang, K., Li, D., Lin, M., Xiang, M., Wei, W., & Sun, Y. (2009). A short review of heterogeneous catalytic process for mixed alcohols synthesis via syngas. *Catalysis Today*, *147*(2), 133–138. doi:10.1016/j.cattod.2009.01.038

Fang, Y. Z., Liu, Y., & Zhang, L. H. (2011). $LaFeO_3$-supported nano Co-Cu catalysts for higher alcohol synthesis from syngas. *Applied Catalysis A, General*, *397*(1–2), 183–191. doi:10.1016/j.apcata.2011.02.032

Ferencz, Z., Erdőhelyi, A., Baán, K., Oszkó, A., Óvári, L., Kónya, Z., & Kiss, J. (2014). Effects of Support and Rh Additive on Co-Based Catalysts in the Ethanol Steam Reforming Reaction. *ACS Catalysis*, *4*(4), 1205–1218. doi:10.1021/cs500045z

Forzatti, P., Tronconi, E., & Pasquon, I. (1991). Higher Alcohol Synthesis. *Catalysis Reviews*, *33*(1-2), 109–168. doi:10.1080/01614949108020298

Fraga, M. A., & Jordao, E. (1998). Alcohol synthesis over bulk aluminium-nickel-copper based catalysts from syngas. *Reaction Kinetics and Catalysis Letters*, *64*(2), 331–336. doi:10.1007/BF02475353

Frolich, P. K. (1931). Pressure as a Tool in the Chemical Industry of the Future. *Industrial & Engineering Chemistry*, *23*(1), 111–115. doi:10.1021/ie50253a039

Vedage, G.A., Himelfarb, P.B., Simmons, G., & Klier, K. (1985). Alkali-Promoted Copper-Zinc Oxide Catalysts for Low Alcohol Synthesis. Solid State Chemistry in Catalysis (Vol. 279, pp. 295-312). American Chemical Society.

Gadewar, S. B., Vicente, B. C., Stoimenov, P. K., & Julka, V. (2015). *Production of higher alcohols*. Google Patents.

Gao, J., Mo, X., Chien, A. C.-Y., Torres, W., & Goodwin, J. G. Jr. (2009). CO hydrogenation on lanthana and vanadia doubly promoted Rh/SiO_2 catalysts. *Journal of Catalysis*, *262*(1), 119–126. doi:10.1016/j.jcat.2008.12.006

Gao, J., Mo, X., & Goodwin, J. G. Jr. (2009). La, V, and Fe promotion of Rh/SiO_2 for CO hydrogenation: Detailed analysis of kinetics and mechanism. *Journal of Catalysis*, *268*(1), 142–149. doi:10.1016/j.jcat.2009.09.012

Gogate, M. R., & Davis, R. J. (2009). X-ray Absorption Spectroscopy of an Fe-Promoted Rh/TiO_2 Catalyst for Synthesis of Ethanol from Synthesis Gas. *ChemCatChem*, *1*(2), 295–303. doi:10.1002/cctc.200900104

Göransson, K., Söderlind, U., He, J., & Zhang, W. (2011). Review of syngas production via biomass DFBGs. *Renewable & Sustainable Energy Reviews*, *15*(1), 482–492. doi:10.1016/j.rser.2010.09.032

Graham, U. M., Jacobs, G., Gnanamani, M. K., Lipka, S. M., Shafer, W. D., Swartz, C. R., & Davis, B. H. (2014). Fischer–Tropsch Synthesis: Higher Oxygenate Selectivity of Cobalt Catalysts Supported on Hydrothermal Carbons. *ACS Catalysis*, *4*(6), 1662–1672. doi:10.1021/cs400965t

Graves, G. D. (1931). Higher Alcohols Formed from Carbon Monoxide and Hydrogen1,2. *Industrial & Engineering Chemistry*, *23*(12), 1381–1385. doi:10.1021/ie50264a016

Gronchi, P., Tempesti, E., & Mazzocchia, C. (1994). Metal dispersion dependent selectivities for syngas conversion to ethanol on V_2O_3 supported rhodium. *Applied Catalysis A, General*, *120*(1), 115–126. doi:10.1016/0926-860X(94)80337-4

Guo, H., Li, S., Zhang, H., Peng, F., Xiong, L., Yang, J., & Chen, Y. (2013). Reaction Condition Optimization and Lumped Kinetics Study for Lower Alcohols Synthesis from Syngas Using a Two-Stage Bed Catalyst Combination System. *Industrial & Engineering Chemistry Research*, *53*(1), 123–131. doi:10.1021/ie402422p

Gupta, M., Smith, M. L., & Spivey, J. J. (2011). Heterogeneous Catalytic Conversion of Dry Syngas to Ethanol and Higher Alcohols on Cu-Based Catalysts. *ACS Catalysis*, *1*(6), 641-656.

Wang, H., Fu, J. L. J., Cai, J., Zhang, H., & Cai, Q. (1993)... *Journal of Natural Gas Chemistry*, *2*, 13–18.

Haider, M. A., Gogate, M. R., & Davis, R. J. (2009). Fe-promotion of supported Rh catalysts for direct conversion of syngas to ethanol. Journal of Catalysis, 261(1), 9–16. doi:10.1016/j.jcat.2008.10.013 doi:10.1016/j.jcat.2008.10.013

Han, L., Mao, D., Yu, J., Guo, Q., & Lu, G. (2013). C2-oxygenates synthesis through CO hydrogenation on SiO2-ZrO2 supported Rh-based catalyst: The effect of support. Applied Catalysis A, General, 454(0), 81–87. doi:10.1016/j.apcata.2013.01.008 doi:10.1016/j.apcata.2013.01.008

Hara, Y., Watanabe, E., Wada, K., & Onoda, T. (1989). Effective hydrogenation of carbon monoxide to ethylene glycol using rhodium-trialkylphosphine-nitrogen base catalyst: Critical role of nitrogen base promoter in controlling selectivity. Journal of Organometallic Chemistry, 359(1), 97–104. doi:10.1016/0022-328X(89)85255-6 doi:10.1016/0022-328X(89)85255-6

He, M., Xiao, B., Hu, Z., Liu, S., Guo, X., & Luo, S. (2009). Syngas production from catalytic gasification of waste polyethylene: Influence of temperature on gas yield and composition. International Journal of Hydrogen Energy, 34(3), 1342–1348. doi:10.1016/j.ijhydene.2008.12.023 doi:10.1016/j.ijhydene.2008.12.023

Herman, R. G. (2000). Advances in catalytic synthesis and utilization of higher alcohols. Catalysis Today, 55(3), 233–245. doi:10.1016/S0920-5861(99)00246-1 doi:10.1016/S0920-5861(99)00246-1

Higman, C., & Tam, S. (2013). Advances in Coal Gasification, Hydrogenation, and Gas Treating for the Production of Chemicals and Fuels. Chemical Reviews, 114(3), 1673–1708. doi:10.1021/cr400202m PubMed doi:10.1021/cr400202m PMID:24144161

Hilmen, A.-M., Xu, M., Gines, M. J. L., & Iglesia, E. (1998). Synthesis of higher alcohols on copper catalysts supported on alkali-promoted basic oxides. Applied Catalysis A, General, 169(2), 355–372. doi:10.1016/S0926-860X(98)00025-8 doi:10.1016/S0926-860X(98)00025-8

Hofstadt, C. E., Schneider, M., Bock, O., & Kochloefl, K. (1983). Effect of preparation methods and promoters on activity and selectivity of cu-zno-al2o3-k catalysts in aliphatic alcohols synthesis from CO and H2. In P. G. G. Poncelet & P. A. Jacobs (Eds.), Studies in Surface Science and Catalysis (Vol. Volume 16, pp. 709-721): Elsevier.

Hu, J., Wang, Y., Cao, C., Elliott, D. C., Stevens, D. J., & White, J. F. (2005). Conversion of Biomass Syngas to DME Using a Microchannel Reactor. Industrial & Engineering Chemistry Research, 44(6), 1722–1727. doi:10.1021/ie0492707 doi:10.1021/ie0492707

Hu, J., Wang, Y., Cao, C., Elliott, D. C., Stevens, D. J., & White, J. F. (2007). Conversion of biomass-derived syngas to alcohols and C2 oxygenates using supported Rh catalysts in a microchannel reactor. Catalysis Today, 120(1), 90–95. doi:10.1016/j.cattod.2006.07.006 doi:10.1016/j.cattod.2006.07.006

Huang, M., & Cho, K. (2009). Density Functional Theory Study of CO Hydrogenation on a MoS2 Surface. The Journal of Physical Chemistry C, 113(13), 5238–5243. doi:10.1021/jp807705y doi:10.1021/jp807705y

Huang, Y., Deng, W., Guo, E., Chung, P.-W., Chen, S., Trewyn, B. G., & Lin, V. S. Y. (2012). Mesoporous Silica Nanoparticle-Stabilized and Manganese-Modified Rhodium Nanoparticles as Catalysts for Highly Selective Synthesis of Ethanol and Acetaldehyde from Syngas. ChemCatChem, 4(5), 674–680. doi:10.1002/cctc.201100460 doi:10.1002/cctc.201100460

Iranmahboob, J., Toghiani, H., & Hill, D. O. (2003). Dispersion of alkali on the surface of CO-MoS2/clay catalyst: A comparison of K and Cs as a promoter for synthesis of alcohol. Applied Catalysis A, General, 247(2), 207–218. doi:10.1016/S0926-860X(03)00092-9 doi:10.1016/S0926-860X(03)00092-9

Jahangiri, H., Bennett, J., Mahjoubi, P., Wilson, K., & Gu, S. (2014). A review of advanced catalyst development for Fischer-Tropsch synthesis of hydrocarbons from biomass derived syn-gas. Catalysis Science & Technology, 4(8), 2210–2229. doi:10.1039/C4CY00327F doi:10.1039/C4CY00327F

Jiang, D., Ding, Y., Pan, Z., Li, X., Jiao, G., Li, J., & Luo, H. (2007). Roles of chlorine in the CO hydrogenation to C2-oxygenates over Rh-Mn-Li/SiO2 catalysts. Applied Catalysis A, General, 331(0), 70–77. doi:10.1016/j.apcata.2007.06.038 doi:10.1016/j.apcata.2007.06.038

Jiang, T., Niu, Y., & Zhong, B. (2001). Synthesis of higher alcohols from syngas over Zn–Cr–K catalyst in supercritical fluids. Fuel Processing Technology, 73(3), 175–183. doi:10.1016/S0378-3820(01)00178-3 doi:10.1016/S0378-3820(01)00178-3

Jones, C. W., Agrawal, P. K., & Nguyen, T. T. (2013). Catalyst compositions for converting syngas to produce higher alcohols. Google Patents.

Juan, A., & Damiani, D. E. (1992). Alumina-supported Ru-Mo Catalysts for the syngas reaction. Catalysis Today, 15(3–4), 469–480. doi:10.1016/0920-5861(92)85012-B doi:10.1016/0920-5861(92)85012-B

Katzer, J. R., Sleight, A. W., Gajardo, P., Michel, J. B., Gleason, E. F., & McMillan, S. (1981). The Role of The Support in CO Hydrogenation Selectivity of Supported Rhodium. Faraday Discussions, 72, 121–133. doi:10.1039/dc9817200121 doi:10.1039/dc9817200121

Keim, W. (1987). Synthesis Gas Industrial Chemicals via C1 Processes (Vol. 328, pp. 1–16). American Chemical Society. doi:10.1021/bk-1987-0328.ch001 doi:10.1021/bk-1987-0328.ch001

Keyvanloo, K., Mardkhe, M. K., Alam, T. M., Bartholomew, C. H., Woodfield, B. F., & Hecker, W. C. (2014). Supported Iron Fischer–Tropsch Catalyst: Superior Activity and Stability Using a Thermally Stable Silica-Doped Alumina Support. ACS Catalysis, 4(4), 1071–1077. doi:10.1021/cs401242d doi:10.1021/cs401242d

Klier, K., Herman, R. G., Nunan, J. G., Smith, K. J., Bogdan, C. E., Young, C. W., & Santiesteban, J. G. (1988). Mechanism of Methanol and Higher Oxygenate Synthesis. In Studies in Surface Science and Catalysis (Vol. 36, pp. 109-125). Elsevier. doi:10.1016/S0167-2991(09)60506-1 doi:10.1016/S0167-2991(09)60506-1

Knifton, J. F., Grigsby, R. A., & Lin, J. J. (1984). Syngas reactions. 6. Aliphatic alcohols and esters from synthesis gas. Organometallics, 3(1), 62–69. doi:10.1021/om00079a012 doi:10.1021/om00079a012

Knifton, J. F., Lin, J. J., Storm, D. A., & Wong, S. F. (1993). New synthesis gas chemistry. Catalysis Today, 18(4), 355-384.

Koh, T., Koo, H. M., Yu, T., Lim, B., & Bae, J. W. (2014). Roles of Ruthenium–Support Interactions of Size-Controlled Ruthenium Nanoparticles for the Product Distribution of Fischer–Tropsch Synthesis. *ACS Catalysis*, *4*(4), 1054–1060.

Koizumi, N., Murai, K., Ozaki, T., & Yamada, M. (2004). Development of sulfur tolerant catalysts for the synthesis of high quality transportation fuels. Catalysis Today, 89(4), 465–478. doi:10.1016/j.cattod.2004.02.002 doi:10.1016/j.cattod.2004.02.002

Kosaric, N., Duvnjak, Z., Farkas, A., Sahm, H., Bringer-Meyer, S., Goebel, O., & Mayer, D. (2000). Ethanol. In Ullmann's Encyclopedia of Industrial Chemistry. Wiley-VCH Verlag GmbH & Co. KGaA.

Krause, T. R., Rathke, J. W., & Chen, M. J. (2013). Process for producing ethanol from syngas: Google Patents.

Kulawska, M., & Skrzypek, J. (2001). Kinetics of the synthesis of higher aliphatic alcohols from syngas over a modified methanol synthesis catalyst. Chemical Engineering and Processing: Process Intensification, 40(1), 33-40.

Kurtz, M., Wilmer, H., Genger, T., Hinrichsen, O., & Muhler, M. (2003). Deactivation of Supported Copper Catalysts for Methanol Synthesis. *Catalysis Letters*, *86*(1-3), 77–80.

Lebarbier, V. M., Mei, D., Kim, D. H., Andersen, A., Male, J. L., Holladay, J. E., & Wang, Y. (2011). Effects of La2O3 on the mixed higher alcohols synthesis from syngas over co catalysts: a combined theoretical and experimental study. The Journal of Physical Chemistry C, 115(35), 17440–17451. doi:10.1021/jp204003q doi:10.1021/jp204003q

Lee, J. H., Reddy, K. H., Jung, J. S., Yang, E.-H., & Moon, D. J. (2014). Role of support on higher alcohol synthesis from syngas. Applied Catalysis A, General, 480(0), 128–133. doi:10.1016/j.apcata.2014.04.026 doi:10.1016/j.apcata.2014.04.026

Lerou, J. J., Tonkovich, A. L., Silva, L., Perry, S., & McDaniel, J. (2010). Microchannel reactor architecture enables greener processes. Chemical Engineering Science, 65(1), 380-385.

Li, Z.-r., Fu, Y.-l., & Jiang, M. (1999). Structures and performance of Rh–Mo–K/Al2O3 catalysts used for mixed alcohol synthesis from synthesis gas. *Applied Catalysis A, General*, *187*(2), 187–198.

Lin, W. M., Gan, S. F., Huang, C. R., Pan, D. R., & Noller, H. (1988). Synthesis of light alcohols from carbon monoxide and hydrogen over copper-iron catalysts. Reaction Kinetics and Catalysis Letters, 37(2), 487–492. doi:10.1007/BF02062104 doi:10.1007/BF02062104

Liu, J., Guo, Z., Childers, D., Schweitzer, N., Marshall, C. L., Klie, R. F., & Meyer, R. J. (2014). Correlating the degree of metal–promoter interaction to ethanol selectivity over MnRh/CNTs CO hydrogenation catalysts. Journal of Catalysis, 313(0), 149–158.

Liu, J., Tao, R., Guo, Z., Regalbuto, J. R., Marshall, C. L., Klie, R. F., & Meyer, R. J. (2013). Selective Adsorption of Manganese onto Rhodium for Optimized Mn/Rh/SiO2 Alcohol Synthesis Catalysts. ChemCatChem, 5(12), 3665–3672. doi:10.1002/cctc.201300479 doi:10.1002/cctc.201300479

Liu, Y. Y., Murata, K., Inaba, M., Takahara, I., & Okabe, K. (2013). Mixed alcohols synthesis from syngas over Cs- and Ni-modified Cu/CeO2 catalysts. Fuel, 104, 62–69. doi:10.1016/j.fuel.2010.08.014 doi:10.1016/j.fuel.2010.08.014

Liu, Z., Li, X., Close, M. R., Kugler, E. L., Petersen, J. L., & Dadyburjor, D. B. (1997). Screening of Alkali-Promoted Vapor-Phase-Synthesized Molybdenum Sulfide Catalysts for the Production of Alcohols from Synthesis Gas. Industrial & Engineering Chemistry Research, 36(8), 3085–3093. doi:10.1021/ie960648d doi:10.1021/ie960648d

Luo, H.-Y., Lin, P.-Z., Xie, S.-B., Zhou, H.-W., Xu, C.-H., Huang, S.-Y., & Xin, Q. (1997). The role of Mn and Li promoters in supported rhodium catalysts in the formation of acetic acid and acetaldehyde. Journal of Molecular Catalysis A Chemical, 122(2–3), 115–123. doi:10.1016/S1381-1169(97)00012-5 doi:10.1016/S1381-1169(97)00012-5

Ma, C.-H., Li, H.-Y., Lin, G.-D., & Zhang, H.-B. (2010). MWCNT-Supported Ni-Mo-K Catalyst for Higher Alcohol Synthesis from Syngas. Catalysis Letters, 137(3-4), 171–179. doi:10.1007/s10562-010-0343-y doi:10.1007/s10562-010-0343-y

Maitlis, P. M. (2003). Metal catalysed CO hydrogenation: hetero- or homo-, what is the difference? Journal of Molecular Catalysis A: Chemical, 204–205, 54-61.

Majocchi, L., Lietti, L., Beretta, A., Forzatti, P., Micheli, E., & Tagliabue, L. (1998). Synthesis of short chain alcohols over a Cs-promoted Cu/ZnO/Cr2O3 catalyst. *Applied Catalysis A, General*, *166*(2), 393–405.

Matsumoto, T., Kim, W. Y., Kishida, M., Nagata, H., & Wakabayashi, K. (1994). Catalytic Hydrogenation of Carbon Monoxide over Silica-Supported Ir-MO-Rh Catalyst. Catalysis Letters, 24(3-4), 391–394. doi:10.1007/BF00811812 doi:10.1007/BF00811812

Mazanec, T. J. (1986). On the mechanism of higher alcohol formation over metal oxide catalysts: I. A rationale for branching in the synthesis of higher alcohols from syngas. Journal of Catalysis, 98(1), 115-125.

Mei, D., Rousseau, R., Kathmann, S. M., Glezakou, V.-A., Engelhard, M. H., Jiang, W., & Stevens, D. J. et al. (2010). Ethanol synthesis from syngas over Rh-based/SiO2 catalysts: A combined experimental and theoretical modeling study. *Journal of Catalysis*, *271*(2), 325–342.

Mo, X., Gao, J., Umnajkaseam, N., & Goodwin, J. G., Jr. (2009). La, V, and Fe promotion of Rh/SiO2 for CO hydrogenation: Effect on adsorption and reaction. Journal of Catalysis, 267(2), 167–176. doi:10.1016/j.jcat.2009.08.007 doi:10.1016/j.jcat.2009.08.007

Mouaddib, N., Perrichon, V., & Martin, G. A. (1994). Characterization of copper-cobalt catalysts for alcohol synthesis from syngas. Applied Catalysis A, General, 118(1), 63–72. doi:10.1016/0926-860X(94)80089-8 doi:10.1016/0926-860X(94)80089-8

Muramatsu, A., Tatsumi, T., & Tominaga, H.-. (1987). Mixed Alcohol Synthesis from CO/H2 by Use of KCl-Promoted Mo/SiO2 Catalysts. Bulletin of the Chemical Society of Japan, 60(9), 3157–3161. doi:10.1246/bcsj.60.3157 doi:10.1246/bcsj.60.3157

Murrell, J. N. L., & Laidler, K. J. (1968). Symmetries of activated complexes. Faraday Transactions, 64, 371. doi:10.1039/tf9686400371 doi:10.1039/tf9686400371

Nielsen, P. E. H., Temel, B., & Beato, P. (2014). Process for the preparation of ethanol and higher alcohols. Google Patents.

Nunan, J., Bogdan, C., Herman, R., & Klier, K. (1989). Efficient carbon-carbon bond formation in ethanol homologation by CO/H2 with specific C1 oxygen retention over Cs/Cu/ZnO catalysts. Catalysis Letters, 2(1), 49–55. doi:10.1007/BF00765330 doi:10.1007/BF00765330

Nunan, J., Klier, K., Young, C.-W., Himelfarb, P. B., & Herman, R. G. (1986). Promotion of methanol synthesis over Cu/ZnO catalysts by doping with caesium. Journal of the Chemical Society, Chemical Communications, 1986(3), 193–195. doi:10.1039/c39860000193 doi:10.1039/c39860000193

Nunan, J. G., Bogdan, C. E., Klier, K., Smith, K. J., Young, C.-W., & Herman, R. G. (1988). Methanol and C2 oxygenate synthesis over cesium doped CuZnO and Cu/ZnO/Al2O3 catalysts: A study of selectivity and 13C incorporation patterns. Journal of Catalysis, 113(2), 410–433. doi:10.1016/0021-9517(88)90268-0 doi:10.1016/0021-9517(88)90268-0

Nunan, J. G., Herman, R. G., & Klier, K. (1989). Higher alcohol and oxygenate synthesis over Cs/Cu/ZnO/M2O3 (M = Al, Cr) catalysts. Journal of Catalysis, 116(1), 222–229. doi:10.1016/0021-9517(89)90087-0 doi:10.1016/0021-9517(89)90087-0

Ojeda, M., Granados, M. L., Rojas, S., Terreros, P., García-García, F. J., & Fierro, J. L. G. (2004). Manganese-promoted Rh/Al2O3 for C2-oxygenates synthesis from syngas: Effect of manganese loading. Applied Catalysis A, General, 261(1), 47–55. doi:10.1016/j.apcata.2003.10.033 doi:10.1016/j.apcata.2003.10.033

Ojeda, M., Rojas, S., Boutonnet, M., Pérez-Alonso, F. J., Javier García-García, F., & Fierro, J. L. G. (2004). Synthesis of Rh nano-particles by the microemulsion technology: Particle size effect on the CO + H2 reaction. Applied Catalysis A, General, 274(1–2), 33–41. doi:10.1016/j.apcata.2004.05.014 doi:10.1016/j.apcata.2004.05.014

Onda, A., Ogo, S., & Yanagisawa, K. (2011). Catalyst and Alcohol Synthesis Method. Google Patents.

Pan, X., & Bao, X. (2011). The Effects of Confinement inside Carbon Nanotubes on Catalysis. Accounts of Chemical Research, 44(8), 553–562. doi:10.1021/ar100160t PubMed doi:10.1021/ar100160t PMID:21707038

Pan, X. L., Fan, Z. L., Chen, W., Ding, Y. J., Luo, H. Y., & Bao, X. H. (2007). Enhanced ethanol production inside carbon-nanotube reactors containing catalytic particles. Nature Materials, 6(7), 507–511. doi:10.1038/nmat1916 PubMed doi:10.1038/nmat1916 PMID:17515914

París, R. S., Lopez, L., Barrientos, J., Pardo, F., Boutonneta, M., & Järås, S. G. (2015). Catalytic conversion of biomass-derived synthesis gas to fuels. Catalysis, 62-143.

Phillips, M. J., & Ternan, M., & Editors. (1988). Catalysis: Theory to Practice. Proceedings of the 9th International Congress on Catalysis, Calgary.

Prieto, G., Beijer, S., Smith, M. L., He, M., Au, Y., Wang, Z., & de Jongh, P. E. (2014). Design and Synthesis of Copper–Cobalt Catalysts for the Selective Conversion of Synthesis Gas to Ethanol and Higher Alcohols. Angewandte Chemie International Edition, 53(25), 6397–6401. doi:10.1002/anie.201402680 PubMed doi:10.1002/anie.201402680 PMID:24827541

Ramayya, S., Brittain, A., DeAlmeida, C., Mok, W., & Antal, M. J., Jr. (1987). Acid-catalysed dehydration of alcohols in supercritical water. Fuel, 66(10), 1364–1371. doi:10.1016/0016-2361(87)90183-9 doi:10.1016/0016-2361(87)90183-9

Ranhotra, G. S., Bell, A. T., & Reimer, J. A. (1987). Catalysis over molybdenum carbides and nitrides: II. Studies of CO hydrogenation and C2H6 hydrogenolysis. Journal of Catalysis, 108(1), 40–49. doi:10.1016/0021-9517(87)90153-9 doi:10.1016/0021-9517(87)90153-9

Reubroycharoen, P., Yoneyama, Y., Vitidsant, T., & Tsubaki, N. (2003). Catalytically active supercritical fluid to accelerate methanol synthesis. Fuel, 82(18), 2255–2257. doi:10.1016/S0016-2361(03)00208-4 doi:10.1016/S0016-2361(03)00208-4

Santos, V. P., van der Linden, B., Chojecki, A., Budroni, G., Corthals, S., Shibata, H., & Gascon, J. (2013). Mechanistic Insight into the Synthesis of Higher Alcohols from Syngas: The Role of K Promotion on MoS2 Catalysts. ACS Catalysis, 3(7), 1634–1637. doi:10.1021/cs4003518 doi:10.1021/cs4003518

Schwartz, V., Campos, A., Egbebi, A., Spivey, J. J., & Overbury, S. H. (2011). EXAFS and FT-IR Characterization of Mn and Li Promoted Titania-Supported Rh Catalysts for CO Hydrogenation. ACS Catalysis, 1(10), 1298–1306. doi:10.1021/cs200281g doi:10.1021/cs200281g

The shale revolution. (2009). Nature, 460(7255), 551-552.

Sheffer, G. R., Jacobson, R. A., & King, T. S. (1989). Chemical nature of alkali-promoted copper-cobalt-chromium oxide higher alcohol catalysts. Journal of Catalysis, 116(1), 95–107. doi:10.1016/0021-9517(89)90078-X doi:10.1016/0021-9517(89)90078-X

Sheffer, G. R., & King, T. S. (1989). Differences in the promotional effect of the group IA elements on unsupported copper catalysts for carbon monoxide hydrogenation. Journal of Catalysis, 116(2), 488–497. doi:10.1016/0021-9517(89)90115-2 doi:10.1016/0021-9517(89)90115-2

Slaa, J. C., van Ommen, J. G., & Ross, J. R. H. (1995). The synthesis of alcohols using Cu/ZnO/A12O3 + (Ce or Mn) catalysts. Topics in Catalysis, 2(1-4), 79–89. doi:10.1007/BF01491957 doi:10.1007/BF01491957

Smith, K. J., & Anderson, R. B. (1983). The higher alcohol synthesis over promoted Cu/ZnO catalysts. Canadian Journal of Chemical Engineering, 61(1), 40–45. doi:10.1002/cjce.5450610106 doi:10.1002/cjce.5450610106

Smith, K. J., & Anderson, R. B. (1984). A chain growth scheme for the higher alcohols synthesis. Journal of Catalysis, 85(2), 428–436. doi:10.1016/0021-9517(84)90232-X doi:10.1016/0021-9517(84)90232-X

Spencer, M. S. (1985). Models of strong metal-support interaction (SMSI) in Pt on TiO2 catalysts. Journal of Catalysis, 93(2), 216–223. doi:10.1016/0021-9517(85)90169-1 doi:10.1016/0021-9517(85)90169-1

Spivey, J. J., & Egbebi, A. (2007). Heterogeneous catalytic synthesis of ethanol from biomass-derived syngas. Chemical Society Reviews, 36(9), 1514–1528. doi:10.1039/b414039g PubMed doi:10.1039/b414039g PMID:17660882

Srinivas, G., Martin, J., Gebhard, S. C., & Mundschau, M. V. (2013). Homologation of methanol for production of ethanol using homogeneous catalysis. Energy & Fuels, 58, 2.

Storm, D. A. (1994). The Production of Higher Alcohols from Syngas using Potassium-Promoted Co/Mo/Al2O3 and Rh/Co/Mo/Al2O3. [-CATL.]. Abstracts Of Papers Of the American Chemical Society, 207, 47.

Subramani, V., & Gangwal, S. K. (2008). A Review of Recent Literature to Search for an Efficient Catalytic Process for the Conversion of Syngas to Ethanol. Energy & Fuels, 22(2), 814–839. doi:10.1021/ef700411x doi:10.1021/ef700411x

Subramanian, N. D., Balaji, G., Kumar, C. S. S. R., & Spivey, J. J. (2009). Development of cobalt–copper nanoparticles as catalysts for higher alcohol synthesis from syngas. Catalysis Today, 147(2), 100–106. doi:10.1016/j.cattod.2009.02.027 doi:10.1016/j.cattod.2009.02.027

Subramanian, N. D., Gao, J., Mo, X., Goodwin, J. G., Jr., Torres, W., & Spivey, J. J. (2010). La and/or V oxide promoted Rh/SiO2 catalysts: Effect of temperature, H2/CO ratio, space velocity, and pressure on ethanol selectivity from syngas. Journal of Catalysis, 272(2), 204–209. doi:10.1016/j.jcat.2010.03.019 doi:10.1016/j.jcat.2010.03.019

Suib, S. L. (2013). New and Future Developments in Catalysis: Catalysis for Remediation and Environmental Concerns. Newnes.

Sun, J., Wan, S., Wang, F., Lin, J., & Wang, Y. (2015). Selective Synthesis of Methanol and Higher Alcohols over Cs/Cu/ZnO/Al2O3 Catalysts. Industrial & Engineering Chemistry Research, 54(32), 7841–7851. doi:10.1021/acs.iecr.5b01927 doi:10.1021/acs.iecr.5b01927

Surisetty, V. R., Dalai, A. K., & Kozinski, J. (2010a). Effect of Rh promoter on MWCNT-supported alkali-modified MoS2 catalysts for higher alcohols synthesis from CO hydrogenation. Applied Catalysis A, General, 381(1–2), 282–288. doi:10.1016/j.apcata.2010.04.036 doi:10.1016/j.apcata.2010.04.036

Surisetty, V. R., Dalai, A. K., & Kozinski, J. (2010b). Synthesis of higher alcohols from synthesis gas over Co-promoted alkali-modified MoS2 catalysts supported on MWCNTs. Applied Catalysis A, General, 385(1–2), 153–162. doi:10.1016/j.apcata.2010.07.009 doi:10.1016/j.apcata.2010.07.009

Takeuchi, A., & Katzer, J. R. (1982). Ethanol formation mechanism from carbon monoxide + molecular hydrogen. Journal of Physical Chemistry, 86(13), 2438–2441. doi:10.1021/j100210a038 doi:10.1021/j100210a038

Tanaka, H., Kaino, R., Okumura, K., Kizuka, T., & Tomishige, K. (2009). Catalytic performance and characterization of Rh–CeO2/MgO catalysts for the catalytic partial oxidation of methane at short contact time. Journal of Catalysis, 268(1), 1–8. doi:10.1016/j.jcat.2009.08.008 doi:10.1016/j.jcat.2009.08.008

Tatsumi, T. (1986). Importance of sequence of impregnation in the activity development of alkali-promoted mo catalysts for alcohol synthesis from CO-H2. Journal of Catalysis, 101(2), 553–556. doi:10.1016/0021-9517(86)90287-3 doi:10.1016/0021-9517(86)90287-3

Temel, B., Nielsen, P. E. H., & Beato, P. (2013). Process for the selective preparation of 1-propanol, iso-butanol and other C3+ alcohols from synthesis gas and methanol. Google Patents.

Tien-Thao, N., Hassan Zahedi-Niaki, M., Alamdari, H., & Kaliaguine, S. (2007). Effect of alkali additives over nanocrystalline Co–Cu-based perovskites as catalysts for higher-alcohol synthesis. Journal of Catalysis, 245(2), 348–357. doi:10.1016/j.jcat.2006.10.026 doi:10.1016/j.jcat.2006.10.026

Tien-Thao, N., Zahedi-Niaki, M. H., Alamdari, H., & Kaliaguine, S. (2007). Conversion of syngas to higher alcohols over nanosized LaCo0.7Cu0.3O3 perovskite precursors. Applied Catalysis A, General, 326(2), 152–163. doi:10.1016/j.apcata.2007.04.009 doi:10.1016/j.apcata.2007.04.009

Tijm, P. J., & Stevens, R. R. (2014). Production of mixed alcohols from synthesis gas. Google Patents.

Toyoda, T., Minami, T., & Qian, E. W. (2013). Mixed Alcohol Synthesis over Sulfided Molybdenum-Based Catalysts. Energy & Fuels, 27(7), 3769–3777. doi:10.1021/ef400262a doi:10.1021/ef400262a

Treviño, H., Hyeon, T., & Sachtler, W. M. H. (1997). A Novel Concept for the Mechanism of Higher Oxygenate Formation from Synthesis Gas over MnO-Promoted Rhodium Catalysts. Journal of Catalysis, 170(2), 236–243. doi:10.1006/jcat.1997.1756 doi:10.1006/jcat.1997.1756

van Dyk, J. C., Keyser, M. J., & Coertzen, M. (2006). Syngas production from South African coal sources using Sasol–Lurgi gasifiers. International Journal of Coal Geology, 65(3–4), 243–253. doi:10.1016/j.coal.2005.05.007 doi:10.1016/j.coal.2005.05.007

Vedage, G. A., Himelfarb, P. B., Simmons, G. W., & Klier, K. (1985). Alkali-Promoted Copper-Zinc Oxide Catalysts for Low Alcohol Synthesis Solid State Chemistry in Catalysis (Vol. 279, pp. 295–312). American Chemical Society.

Verkerk, K. A. N., Jaeger, B., Finkeldei, C.-H., & Keim, W. (1999). Recent developments in isobutanol synthesis from synthesis gas. Applied Catalysis A, General, 186(1–2), 407–431. doi:10.1016/S0926-860X(99)00158-1 doi:10.1016/S0926-860X(99)00158-1

Walter, K. M., Schubert, M., Kleist, W., & Grunwaldt, J.-D. (2015). Effect of the Addition of Ethanol to Synthesis Gas on the Production of Higher Alcohols over Cs and Ru Modified Cu/ZnO Catalysts. Industrial & Engineering Chemistry Research, 54(5), 1452–1463. doi:10.1021/ie504066f doi:10.1021/ie504066f

Wang, H. Y., Liu, J. P., Fu, J. K., Wan, H. L., & Tsai, K. R. (1992). Study on the Mechanism of Ethanol Synthesis from Syngas by Insitu Chemical Trapping and Isotopic Exchange Reactions. Catalysis Letters, 12(1-3), 87–96. doi:10.1007/BF00767191 doi:10.1007/BF00767191

Wang, H. Y., Liu, J. P., Fu, J. K., Zhang, H. B., & Tsai, K. R. (1992). Study on the Method of Chemical Trapping for Formyl Intermediates. Research on Chemical Intermediates, 17(3), 233–242. doi:10.1163/156856792X00155 doi:10.1163/156856792X00155

Wang, J.-J., Xie, J.-R., Huang, Y.-H., Chen, B.-H., Lin, G.-D., & Zhang, H.-B. (2013). An efficient Ni–Mo–K sulfide catalyst doped with CNTs for conversion of syngas to ethanol and higher alcohols. Applied Catalysis A, General, 468(0), 44–51. doi:10.1016/j.apcata.2013.08.026 doi:10.1016/j.apcata.2013.08.026

Wang, J., Chernavskii, P. A., Khodakov, A. Y., & Wang, Y. (2012). Structure and catalytic performance of alumina-supported copper–cobalt catalysts for carbon monoxide hydrogenation. Journal of Catalysis, 286(0), 51–61. doi:10.1016/j.jcat.2011.10.012 doi:10.1016/j.jcat.2011.10.012

Wang, J., Chernavskii, P. A., Wang, Y., & Khodakov, A. Y. (2013). Influence of the support and promotion on the structure and catalytic performance of copper–cobalt catalysts for carbon monoxide hydrogenation. Fuel, 103(0), 1111–1122. doi:10.1016/j.fuel.2012.07.055 doi:10.1016/j.fuel.2012.07.055

Wang, J., Zhang, Q., & Wang, Y. (2011). Rh-catalyzed syngas conversion to ethanol: Studies on the promoting effect of FeOx. Catalysis Today, 171(1), 257–265. doi:10.1016/j.cattod.2011.03.023 doi:10.1016/j.cattod.2011.03.023

Wang, N., Fang, K., Lin, M., Jiang, D., Li, D., & Sun, Y. (2010). Synthesis of Higher Alcohols from Syngas over Fe/K/β-Mo2C Catalyst. Catalysis Letters, 136(1-2), 9–13. doi:10.1007/s10562-010-0288-1 doi:10.1007/s10562-010-0288-1

Wang, Y., Luo, H., Liang, D., & Bao, X. (2000). Different Mechanisms for the Formation of Acetaldehyde and Ethanol on the Rh–Based Catalysts. Journal of Catalysis, 196(1), 46–55. doi:10.1006/jcat.2000.3026 doi:10.1006/jcat.2000.3026

Wei, Z., Sun, J., Li, Y., Datye, A. K., & Wang, Y. (2012). Bimetallic catalysts for hydrogen generation. Chemical Society Reviews, 41(24), 7994–8008. doi:10.1039/c2cs35201j PubMed doi:10.1039/c2cs35201j PMID:23011345

West, N. M., Miller, A. J. M., Labinger, J. A., & Bercaw, J. E. (2011). Homogeneous syngas conversion. Coordination Chemistry Reviews, 255(7–8), 881–898. doi:10.1016/j.ccr.2010.08.019 doi:10.1016/j.ccr.2010.08.019

White, E. B., Wilson, C., Berggren, M., & Zubrin, R. M. (2014). Methods of higher alcohol synthesis: Google Patents.

Whyman, R., Wright, A. P., Iggo, J. A., & Heaton, B. T. (2002). Carbon monoxide activation in homogeneously catalysed reactions: The nature and roles of catalytic promoters. Journal of the Chemical Society, Dalton Transactions: Inorganic Chemistry, (5): 771–777. doi:10.1039/b107940a doi:10.1039/b107940a

Woo, H. C., Park, T. Y., Kim, Y. G., Nam, I.-S., Lee, J. S., & Chung, J. S. (1993). Alkali-Promoted MoS2 Catalysts for Alcohol Synthesis: the Effect of Alkali Promotiom and Preparation Condition on Activity and Selectivity. In F. S. L. Guczi & T. P (Eds.), Studies in Surface Science and Catalysis (Vol. Volume 75, pp. 2749-2752). Elsevier.

Xiang, M., Li, D., Li, W., Zhong, B., & Sun, Y. (2007). Synthesis of higher alcohols from syngas over K/Co/β-Mo2C catalysts. Catalysis Communications, 8(3), 503–507. doi:10.1016/j.catcom.2006.07.029 doi:10.1016/j.catcom.2006.07.029

Xiang, Y., Chitry, V., Liddicoat, P., Felfer, P., Cairney, J., Ringer, S., & Kruse, N. (2013). Long-Chain Terminal Alcohols through Catalytic CO Hydrogenation. Journal of the American Chemical Society, 135(19), 7114–7117. doi:10.1021/ja402512r PubMed doi:10.1021/ja402512r PMID:23634891

Xiao, F.-S., Fukuoka, A., Ichikawa, M., Henderson, W., & Shriver, D. F. (1992). SiO2-supported bimetallic ruthenium cluster-derived catalysts for selectivity and activity control in CO hydrogenation toward C1-C5 alcohols. Journal of Molecular Catalysis, 74(1–3), 379–390. doi:10.1016/0304-5102(92)80255-F doi:10.1016/0304-5102(92)80255-F

Xiao, K., Bao, Z., Qi, X., Wang, X., Zhong, L., Fang, K., & Sun, Y. (2013). Structural evolution of CuFe bimetallic nanoparticles for higher alcohol synthesis. Journal of Molecular Catalysis A Chemical, 378(0), 319–325. doi:10.1016/j.molcata.2013.07.006 doi:10.1016/j.molcata.2013.07.006

Xiao, K., Qi, X., Bao, Z., Wang, X., Zhong, L., Fang, K., & Sun, Y. (2013). CuFe, CuCo and CuNi nanoparticles as catalysts for higher alcohol synthesis from syngas: A comparative study. Catalysis Science & Technology, 3(6), 1591–1602. doi:10.1039/c3cy00063j doi:10.1039/c3cy00063j

Xiaoding, X., Doesburg, E. B. M., & Scholten, J. J. F. (1987). Synthesis of higher alcohols from syngas - recently patented catalysts and tentative ideas on the mechanism. Catalysis Today, 2(1), 125–170. doi:10.1016/0920-5861(87)80002-0 doi:10.1016/0920-5861(87)80002-0

Xu, M., & Iglesia, E. (1998). Initial carbon–carbon bond formation during synthesis gas conversion to higher alcohols on K–Cu–Mg5CeOx catalysts. Catalysis Letters, 51(1/2), 47–52. doi:10.1023/A:1019016513428 doi:10.1023/A:1019016513428

Xu, M., & Iglesia, E. (1999). Carbon–Carbon Bond Formation Pathways in CO Hydrogenation to Higher Alcohols. Journal of Catalysis, 188(1), 125–131. doi:10.1006/jcat.1999.2650 doi:10.1006/jcat.1999.2650

Xu, R., Yang, C., Wei, W., Li, W.-h., Sun, Y.-h., & Hu, T.-d. (2004). Fe-modified CuMnZrO2 catalysts for higher alcohols synthesis from syngas. Journal of Molecular Catalysis A Chemical, 221(1–2), 51–58. doi:10.1016/j.molcata.2004.07.003 doi:10.1016/j.molcata.2004.07.003

Xu, R., Zhang, S., & Roberts, C. B. (2013). Mixed Alcohol Synthesis over a K Promoted Cu/ZnO/Al2O3 Catalyst in Supercritical Hexanes. Industrial & Engineering Chemistry Research, 52(41), 14514–14524. doi:10.1021/ie3024017 doi:10.1021/ie3024017

Yang, X., Zhu, X., Hou, R., Zhou, L., & Su, Y. (2011). The promotion effects of Pd on Fe–Cu–Co based catalyst for higher alcohols synthesis. Fuel Processing Technology, 92(10), 1876–1880. doi:10.1016/j.fuproc.2011.05.003 doi:10.1016/j.fuproc.2011.05.003

Yin, H., Ding, Y., Luo, H., Zhu, H., He, D., Xiong, J., & Lin, L. (2003). Influence of iron promoter on catalytic properties of Rh-Mn-Li/SiO2 for CO hydrogenation. Applied Catalysis A, General, 243(1), 155–164. doi:10.1016/S0926-860X(02)00560-4 doi:10.1016/S0926-860X(02)00560-4

Yu, J., Mao, D., Han, L., Guo, Q., & Lu, G. (2013). Synthesis of C2 oxygenates from syngas over monodispersed SiO2 supported Rh-based catalysts: Effect of calcination temperature of SiO2. Fuel Processing Technology, 106(0), 344–349. doi:10.1016/j.fuproc.2012.08.020 doi:10.1016/j.fuproc.2012.08.020

Zaman, S., & Smith, K. J. (2012). A Review of Molybdenum Catalysts for Synthesis Gas Conversion to Alcohols: Catalysts, Mechanisms and Kinetics. Catalysis Reviews, 54(1), 41–132. doi:10.1080/01614940.2012.627224 doi:10.1080/01614940.2012.627224

Zaman, S. F., & Smith, K. J. (2009). Synthesis gas conversion over MoP catalysts. Catalysis Communications, 10(5), 468–471. doi:10.1016/j.catcom.2008.10.022 doi:10.1016/j.catcom.2008.10.022

Zaman, S. F., & Smith, K. J. (2010). A study of K-promoted MoP–SiO2 catalysts for synthesis gas conversion. Applied Catalysis A, General, 378(1), 59–68. doi:10.1016/j.apcata.2010.01.046 doi:10.1016/j.apcata.2010.01.046

Zaman, S. F., & Smith, K. J. (2011). Synthesis gas conversion over a Rh–K–MoP/SiO2 catalyst. Catalysis Today, 171(1), 266–274. doi:10.1016/j.cattod.2011.02.017 doi:10.1016/j.cattod.2011.02.017

Zhang, M., & Yu, Y. (2013). Dehydration of Ethanol to Ethylene. Industrial & Engineering Chemistry Research, 52(28), 9505–9514. doi:10.1021/ie401157c doi:10.1021/ie401157c

Zuo, Z.-J., Wang, L., Yu, L.-M., Han, P.-D., & Huang, W. (2014). Experimental and Theoretical Studies of Ethanol Synthesis from Syngas over CuZnAl Catalysts without Other Promoters. The Journal of Physical Chemistry C, 118(24), 12890–12898. doi:10.1021/jp502828c doi:10.1021/jp502828c

Chapter 9
Preparation of Deep Hydrodesulfurzation Catalysts for Diesel Fuel using Organic Matrix Decomposition Method

Hamid Audah AlMegren
King Abdulaziz City for Science and Technology, Saudi Arabia

Sergio Gonzalez-Cortes
Oxford University, UK

Yu Huang
Guangzhou Boxenergy Tech Ltd, China

Haoyi Chen
Guangzhou Boxenergy Tech Ltd, China

Yangdong Qian
Oxford University, UK

Mohammed Alkinany
King Abdulaziz City of Science and Technology, Saudi Arabia

Saud Aldrees
King Abdulaziz City of Science and Technology, Saudi Arabia

Tiancun Xiao
Oxford University, UK

ABSTRACT

In this work, a series of supported CoMo or NiMo HDS catalysts have been prepared based on the organic-metal matrix decomposition method and tested for diesel deep HDS with minimum hydrogen consumption under relatively low hydrogen partial pressure conditions. The aim is to develop a HDS catalyst which can reduce sulphur in diesel fuel from 5000ppm down to 50 ppm in a single pass with minimum hydrogen consumption under the conditions of 340oC, 35 bar, LHSV 1.2 h-1 with low H2/oil ratio. The catalysts preparation process was monitored and the resultant catalyst samples before and after the HDS performance test have been characterised, some interesting results have been obtained. The presence of citric acid as organic additive/dispersing agent/chelating agent in the impregnation solution improved HDS activity compared to the equivalent CoMo catalyst prepared without citric acid, The order of activity of the cobalt precursors is Co citrate > Co acetate > Co nitrate.

DOI: 10.4018/978-1-4666-9975-5.ch009

1. BACKGROUND

Diesel is extensively and increasingly used as a main fuel for transportation systems due to its higher fuel-efficiency and longer mileage (Abro 2014; Chen, Wang, 2010; Ismagilov., 2009; Wasserscheid, 2003). However, the extensive use of diesel has caused increasing emissions of particulate, NO**x** and SO**x**, which are harmful to human health and also resulting in more eco-problems such as acid rain and smog. Sulfur, a natural part of the crude oil from which diesel fuel is derived, is one of the key causes of particulates or soot in diesel. Legislative regulations in many countries call for the production and use of more environmentally friendly transportation fuels with the lower contents of sulfur, nitrogen and aromatics, and indeed, since 2000s, many countries around the world has ordered to reduce the sulfur content of diesel fuel to ultra low levels (10–15 ppm) with the intention of lowering diesel engine's harmful exhaust emissions and improving air quality. The environmental issues related to sulfur in diesel, current and future diesel fuel would be the continuous focus in various countries around the world.

It is known that change from ordinary to ultra deep desulfurization is very complicated and problematic process for refinery given a built up reactor system, where there have been very limited space for revamp. The catalysts, origin of the fuel, reactor configuration, hydrogen availability and pressure all have significant effect on the sulphur removal during the hydrotreatment process. Some sulfur compounds that contain alkyl side chains in the 4- and 6-positions in the dibenzothiophene molecule close to the sulfur atom (e.g. 4, 6-dimethyl dibenzothiophene; 4-methyl, 6-ethyl dibenzothiophene) are difficult to desulfurize under conventional desulfurization conditions.

In recent years, researches on the fuels upgrading, including hydrodesulfurization (HDS), hydrodenitrogenation (HDN) and dearomatization, have become increasingly important, especially in the development of ULSD catalyst development. Our statistics of HDS catalyst survey shows that there had very few publications on diesel deep HDS work before 1992, but sharply increased to 25 publications in 1999, and 45 publications in 2004. These data were collected from the Scifinder data base searched by the keywords of diesel, deep HDS and catalysts. (Ali, 2014; Maity, 2004; Fujikawa, 2009; Kimura, Kiriyama,2006; Gao, 2011; Ho, 2004; Wan., 2010) Most of these studies have focused on exploring effective ways for the desulfurization of the least reactive sterically hindered alkyl DBTs. The role of catalysts and other key factors, such as feed quality, inhibition effects, process variables, kinetic and thermodynamic effects, reactor internals and feed distribution effects, influencing the desulfurization of the least reactive sulfur species have received increasing attention in recent years.

However, overviewing the catalysts for diesel deep HDS, the commercial available catalysts are either bulk metal oxide based catalyst (Chen, 2014) such as nebula or supported catalyst with ultrahigh metal loading (Wang, 2009), the catalyst has strength problem or the preparation process is very complex. These catalyst based on conventional impregnation gives more inert phase, which make the catalyst less active. In this work, we have developed a new catalyst preparation method, e.g., the organic-metal matrix decomposition method, which can easily control the active component loading, and adjust the active component interaction with the support. Also there have been little emission during the catalyst preparation., Series of CoMo and NiMo catalyst supported over various Al_2O_3 supports have been prepared and developed for diesel HDS.

2. EXPERIMENTAL

2.1 Catalyst Preparation

The CoMo HDS catalysts were prepared using three methods: (i). Pore volume (incipient wetness) impregnation, (ii). Wet impregnation and (iii) Sequential incipient wetness impregnation. The detailed preparation methods are described below.

2.1.1 Pore Volume Impregnation Method

The Mo precursor, ($(NH_4)_6Mo_7O_{24}.4H_2O$) was dissolved in deionised water to make specific Mo salt solutions. The pH of the solution was adjusted by adding ammonia solution. The Mo solution was used to impregnate a support (i.e. Al_2O_3, carbon or modified Al_2O_3) using pore volume impregnation (ie by adding the solution drop wise and hand mixing the support and the solution until all of the solution was taken up). The impregnated support was then dried and calcined at the required temperature. A second solution consisting of the Co salt (for example, $Co(NO_3)_2$, $Co(AC)_2$, etc) was deposited on the MoO_3/support in the same way, then dried or calcined to obtain CoO-MoO_3/support type HDS catalysts.

2.1.2 Wet Impregnation Method

An impregnation solution containing both the Mo salt ($(NH_4)_6Mo_7O_{24}.4H_2O$), Co salt ($CoCO_3$) and citric acid (whose structure is given in below figure), was prepared. The support (Al_2O_3 or modified Al_2O_3) was then added to the solution in a beaker. The solid was fully wetted for a period of 5-24 h. Excess impregnation solution was used, typically between 50 and 100% relative to the pickup volume of the support. The impregnated materials were dried/calcined at the required temperature. In some cases this whole process including drying, was repeated to achieve the required loading of Mo and Co.

2.1.3 Multi-Step Wet Impregnation for Modified CoMo Catalyst

The impregnation solution was prepared using the required amount of ammonium heptamolybdate ($(NH_4)_6Mo_7O_{24}.4H_2O$), cobalt carbonate ($CoCO_3$) and citric acid in deionised water. The resultant pH was between 2.4 and 2.6. For W-, P-, Ni-containing CoMo catalysts, ammonium metatungstate, phosphoric acid and nickel nitrate were used as the metal precursors. These metal precursors were added to the CoMo citrate containing impregnation solution. Drops of ammonia water solution were added to adjust the pH of the solution. The support (i.e. extrudate of alumina and modified alumina) was then wetted at room temperature for between 4 and 24 h. An excess of impregnation solution volume between 50 and 100% relative to the pickup volume of the support was used. Subsequently, the sample was dried at 110°C for 6 h and re-impregnated under similar condition with this series of catalysts. Finally, the sample was thermally treated at the required temperature condition (RT/Temp/hold: 1 °C/min / 110-350°C / 6 h). The nominal CoMo composition was determined by the pickup volume and also analysis thermal-gravimetric (TGA), after calcination (up to 500°C). The experimental composition was determined by ICP analysis

2.2 Catalyst Test Conditions

A stainless steel tubular reactor was loaded with 2.5 ml of catalyst precursor (particle size between 0.5 and 1 mm) mixed with equal volume of silicon carbide (SiC) and sandwiched between layers of SiC. After the leak test, the catalyst was treated under N_2 flow (50 mL.min^{-1}) for 60 min at 150 °C and then was soaked with sulphiding agent (i.e. straight run gas oil spiked with dimethyl disulfide to a total sulphur content of 4 wt. %) and hydrogen flow for 30 min at 150 °C. The liquid was continuously fed into the reactor using a Series II high pressure pump operating in the mode of volumetric flow. Subsequently, the liquid-phase sulphidation conditions were set: liquid hour space velocity (LHSV) of 2.4 h^{-1}, H_2/oil volume ratio of 300 N L/L and 3.0 MPa of hydrogen. The temperature was then increased at 0.5°C. min^{-1} up to 240 °C and held for 120 min. The temperature was then further risen at 0.5 °C.min^{-1} to 340 °C and held for 6 h. Then the reactor temperature was decreased up to 220 °C under sulphiding agent and H_2 flow condition.

After finishing the sulphidation process, the sulphiding agent was switched to the feedstock and then the operation conditions were changed to 340 °C, 35 bar, 250 NL/L and 1.2 h^{-1}.The sulphur and nitrogen analysis in the hydrotreated products was carried out on ANTEK 9000VNS instrument. The analysis is based on a combustion process coupled with UV fluoresce detector for sulphur-containing organic compounds (ASTM 5453-08) and chemiluminescence detector for nitrogen-containing hydrocarbon moleculcs (ASTM 4629-02).

The boiling point distribution was carried out using simulated distillation by gas chromatography. A standard mixture of alkane compounds was analysed in order to calibrate the retention time of each compound against the boiling point temperature. Then, the sample was analysed under similar condition as that used in the standard run. The GC profile is then integrated as a single peak and divided into time slices of 5 °C increments to get boiling point distribution.

2.3 Catalyst Characterisation

The textural properties (i.e. surface area, pore volume and pore distribution) of the catalyst oxide precursor were carried out by nitrogen physi- sorption at the nitrogen liquid temperature (-196°C). The sample (50-100 mg) was initially degassed at room temperature under vacuum (1x10^{-3} bar) for 2 h. Then the adsorbed volume of nitrogen at different relative pressure (P/Po) was measured in order to obtain an isotherm curve. It was obtained specific surface area with the Brunauer, Emmett, Teller (BET) method whereas pore volume and pore distribution were found out with the Barret-Joyner-Halenda (BJH) method.

The crystalline phases of the oxidic precursors and the sulfurized catalysts were identified with X-ray diffraction (XRD) in a Philips PW1710 diffractometer equipped with an X-ray tube (Cu Kα radiation $\lambda = 1.5406$ Å; 40 kV, 30 mA). The specimen were prepared by grinding a small amount of each sample using an agate mortar and pestle and then loaded into a flat sample holder. The data were collected in θ/2θ reflection mode, from 3 to 70° in 2θ, using steps of 0.05 °, with time per step of 1.25 s.

Thermogravimetric analysis (TGA) and differential scanning calorimetry (DSC) were simultaneously carried out on a SDT Q600 instrument. TGA-DSC profiles were recorded from room temperature to 750°C, using a 20 mL/min dry air flow and 5° C/min heating rate. For each analysis ca. 25–35 mg of sample was loaded into a small alumina crucible, using alumina as reference.

Measurements of dynamic viscosity of the impregnation solution upon impregnation process were carried out using a Vibro Viscometer SV-A1 instrument. This method is based upon the frequency change when a sensor plate is immersed in a fluid. This equipment was calibrated using an standard viscosity fluid (deionized water) at the laboratory temperature (ca. 18 °C)_

3. RESULT AND DISCUSSION

3.1. Influence of the Co-Impregnation Preparation Methods on Citrate-Prepared CoMo Catalysts

Table 1 shows HDS and HDN conversion using model feed and real FCC diesel feed for three catalyst preparation methods (i.e. pore volume impregnation, multi-step wet impregnation and single step wet impregnation with long time of impregnation). It must be noted that the compositions quoted are nominal, i.e the maximum possible assuming 100% uptake of the metals. In the case of pore volume impregnation the maximum is achieved. In the case of the 'wet' impregnation, only a fraction of the metal may be taken up from the solution.

The single step wet impregnated catalyst showed HDS conversion higher than the catalyst without organic containing solution. CoMo HDS catalysts showed better performance for the FCC diesel fuel while the catalyst prepared according to the literature (Gonzalez-Cortes, 2014) always showed high HDS conversion for model fuel. The HDS performance of CoMo catalysts using single step wet impregnation for long time is similar to the literature catalysts (Gonzalez-Cortes., 2014), but the trimetallic catalyst gave higher HDN conversion with model and real FCC gasoline fuels than the CoMo catalysts. This suggests that Ni type HDS catalysts are better than Co type HDS catalysts for HDN conversion, as it is known that Ni has higher hydrogenation ability (Brum Pereira, 1995; Wang., 2002).

Table 1. Performance of HDS CoMo catalysts using Al_2O_3 as support, reaction conditions: 340 °C, 35 bar, 250 NL/L (H_2/Oil ratio) and 7.2 h^{-1}

Sample	Nominal Composition %	Density (g/ml)	Preparation Method	Calcined at (°C)	S Conversion (%)		Desulfur Rate Based on Metal Loading
					Model Feed	FCC Feed	
Y-01	$31MoO_3$-7.7CoO	1.004	WIP	110	89	89	23
Y-02	$31MoO_3$-7.7CoO	0.974	WIP	200	89	89	23
Y-03	$13.9MoO_3$-3.5CoO	0.933	Single-step WIP	110	90	90	57
Y-04	$13.9MoO_3$-3.5CoO	0.891	Single-step WIP	180	91	90	62
Ref 1 cat.	9Ni15Mo30W	1.190			91	84	22

3.1.1 WIP, Multi-Step Wet Impregnation Method

To evaluate multi-step wet impregnation (WIP) and single step wet impregnation (single step WIP) with long time of impregnation for HDS catalyst performance, four HDS catalysts were prepared using Al_2O_3 support from Saint-Gobain. 31%MoO_3-7.7%CoO-100 and 31%MoO_3-7.7%CoO -200 were prepared using multi-step wet impregnation methods and Y-3 and Y-4were obtained using single step wet impregnation with Co and Mo solution for 24 hours impregnation. Four HDS catalysts were tested with Ref 1 catalyst on the HDS spider rig. The results obtained at 7.2 h^{-1} was listed in Table 1.

The results show that a single step wet impregnation, with nominally lower metal loading is in fact slightly more active than the catalysts made with multi-step wet impregnation. When one considers metal loading the activity per gram of metal is potentially much higher. This can be explained by the higher dispersion of the active component over the single step impregnation catalyst, while the total active sites per volume of the catalyst are much fewer.

3.2. Influence of the Co-Impregnation Preparation Time on Citrate-Prepared CoMo Catalysts

In order to study and to optimize impregnation time against catalyst performance, the dependence of the impregnation solution viscosity with impregnation time for Co, Mo citrate solution onto alumina (G250) at different operation mode were carried out (Figure 1). As it is easily understood that the active components will diffuse into the support in the impregnation step, and the diffusion rate may be relevant to the solution viscosity. Therefore the solution viscosity has significant effect on the catalyst preparation (Bentaleb,2014; Denisov, 1996; Van Dillen, 2003). Three separate experiments were conducted. In the first experiment shown in Figure 1A the initial dynamic viscosity of the impregnation solution (40 ml) was measured over 16 hours. The initial viscosity was 15.5 mPa.s. After 5 minutes, 10.0 g of alumina extrudate was added to the solution. An increase in the impregnation solution temperature from 19 to 21°C occurred. This is probably due to adsorption of water on alumina surface. The addition of alumina also led to an increase in the impregnation solution viscosity of 150% relative to the initial solution viscosity over ca 20 minutes. Then, impregnation solution temperature started dropping down according to the surrounding temperature as the solution viscosity was slightly increased. Note that a transient drop of viscosity was recorded between 24 and 25 min, owing to a possible effect of vibrator perturbation because of extrudate contact. After an hour of impregnation time, the solution viscosity started decreasing, owing to the deposition of Co and Mo citrate species onto alumina extrudate body. A fairly constant value of between 23 and 26 mPa.s was reached after 4.5 h of impregnation time.

It is noted that the final solution viscosity was nearly 60% higher than that of initial impregnation solution, most likely due to the increase of impregnation solution volume relative to the vibrator mark after adding alumina extrudate. Also, the hydrophilic property of the Al_2O_3 support would enable water in the solution quickly to absorb onto the support, and leading to a concentrated solution, therefore to increase the solute viscosity.

To verify this effect the_experiment was repeated, controlling impregnation solution (20 ml) liquid level relative to sensor plate level after adding 5 g of extrudate (Figure 1B). After measuring the dynamic viscosity of the impregnation for 10 minutes, 5 grams of alumina extrudate was added to the solution. Once the extrudate was deposited in the bottom of the beaker and the liquid level was adjusted to the vibrator mark the viscosity measurement were re-started with an operation dead time of 60 seconds. Note

Figure 1. Dependence of the impregnation solution viscosity (CoMo citrate solution at ca. 3 pH) and temperature versus time of impregnation

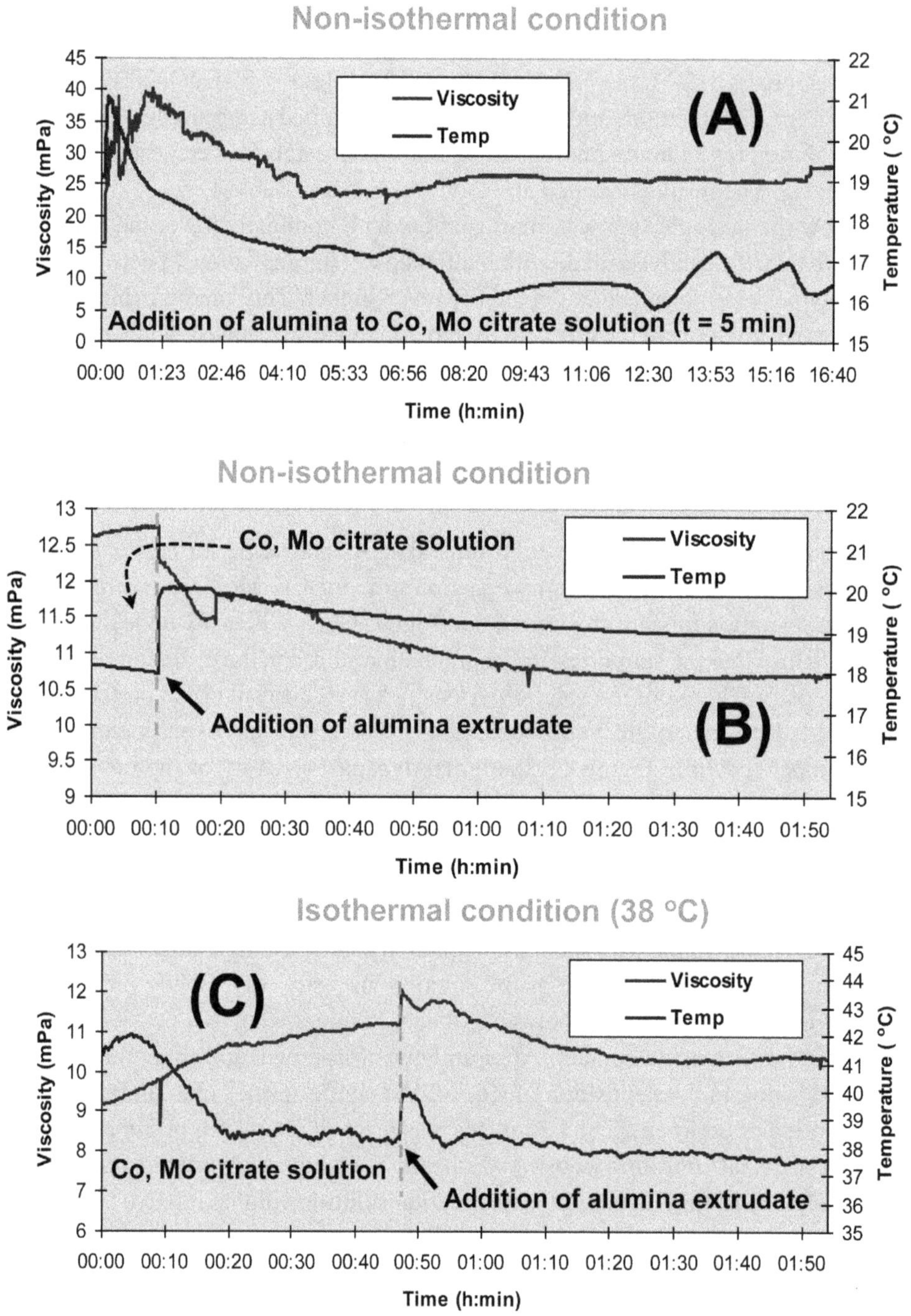

that the initial temperature of impregnation solution increased by nearly 2 °C. Afterwards, the temperature and viscosity dropped down upon impregnation time. This observation indicates that upon impregnation, the solution viscosity decreased as consequence of Co, Mo citrate deposition on alumina surface.

Figure 2. Dependence of residual S content of Real FCC diesel feed and bulk density of the catalyst precursor with impregnation time. Inset figure: Residual S contents versus bulk density. Catalytic reaction conducted at 340 °C, 35 bar, 250 NL/L (H_2/Oil ratio) and 7.2 h^{-1} with diesel feed

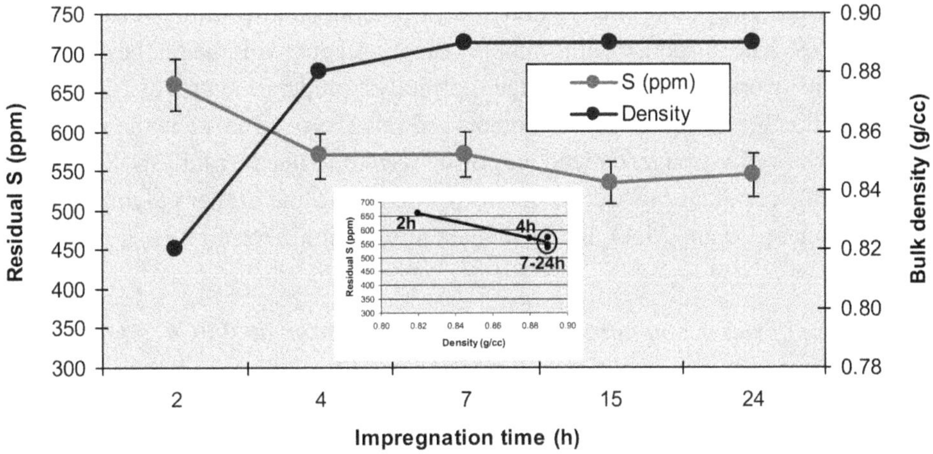

In order to determine the effect of temperature on impregnation solution viscosity an experiment was carried out under quasi-isothermal condition, (Figure 1C) using a similar procedure to the 2nd experiment. When the impregnation solution temperature was stable at 37-38°C, alumina extrudate was added. The impregnation solution temperature and viscosity increased rapidly and then decreased to get values fairly similar to the initial condition. Apparently, adsorption of Co, Mo and citrate is not facilitated at high impregnation temperature owing to the exothermicity of adsorption process..

The impregnation solution viscosity at low temperature (i.e. 17-18 °C) was ca. 12 mPa.s, which is significantly higher than dynamic viscosity of pure water (ca. 1 mPa.s). Upon short impregnation period (ca. 5 min), impregnation solution temperature increased nearly 10% relative to initial temperature as the dynamic viscosity dropped to 10.8 mPa.s after nearly 2 h of impregnation time. When impregnation solution temperature was increased up to 38 °C its dynamic viscosity decreased to ca. 10.5-11 mPa.s. After an hour of impregnation time the solution viscosity did not change significantly.

In order to find out the impact of the viscosity and the impregnation temperature changes on the catalyst performance, a large pore γ-alumina-supported CoMo catalysts were prepared by wet impregnation and tested simultaneously in the HDS spider (Figure 2). According to the sulphur content trend with impregnation time the CoMo catalyst showed an HDS activity apparently constant after 4 hours of impregnation time, which corresponds to the period when impregnation solution viscosity resulted fairly constant (Figure 1A).

These results suggest that after approximately 4 hours of impregnation a maximum loading of Co, Mo citrate species was adsorbed/deposited onto alumina extrudate. This is in agreement with the results of measurements of bulk density as a function of impregnation time. When one considers the relationship between residual sulphur content and bulk density (see inset figure), an apparent linear dependence is observed with a decrease of S content (ca. 14%) and increase of bulk density (7%) when impregnation time was changed from 2 to 4 hours. After that period both variables underwent smaller variation. Thus it is clear that impregnation time longer than 5-7 h appear not to have any improvement affect HDS activity as shown in (see Figure 1A). This indicates that a period of impregnation between 5 and 7 h would be sufficient to fully deposit Co and Mo citrate species onto alumina extrudate, especially with this large pore alumina where the diffusion of the active component salt encountering less resistance.

It has been shown that the activation temperatures and the composition of the active components loading exert significant effect on the active component loading and location over the catalyst. (Duan, 2011; Hernandez, 2009) The influence of the support shape and pore volume on the catalyst preparation has been studied in this work in order to optimize the catalyst preparation process. Figure 3 shows that the residue sulfur in the S-3 sample has low content, and relatively stable, so is the nitrogen content.

HDN conversion showed a stronger dependency on contact time and the catalysts. S-3 gives a relatively higher N content in the resultant products, which may be due to the higher calcination temperatures, given the same loading of Co and MoO_3 in the catalyst preparation. Overall it is shown that the residue

Figure 3. Dependence of sulfur and nitrogen content in the hydrotreated FCC diesel with the time on stream over S-3 catalyst, reaction conditions: conducted at 340 oC, 35 bar, 300 NL/L (H2/Oil ratio)

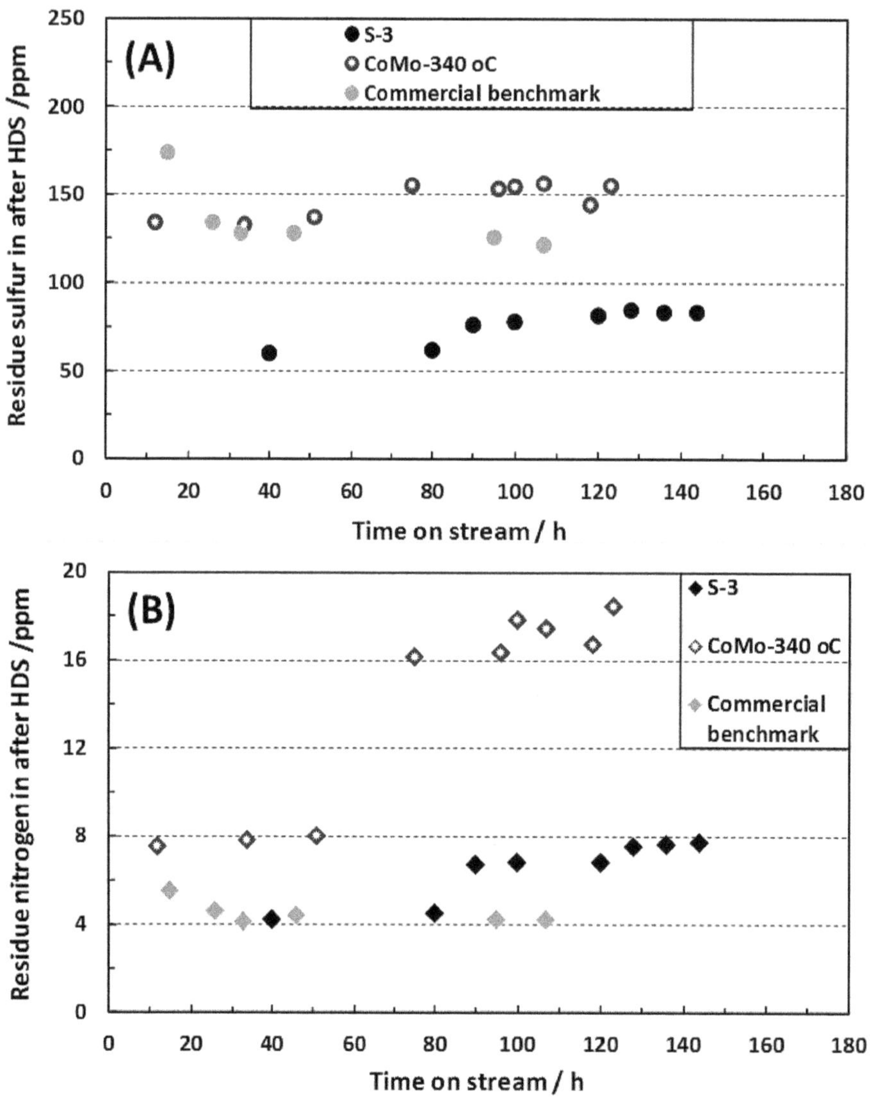

Figure 4. HDS and HDN decays for citrate-prepared CoMo catalysts, benchmark catalyst A and Ref 1 Cat NiMoW catalyst over model feed. Catalytic reaction conducted at 340 °C, 35 bar, 250 NL/L (H_2/Oil ratio) and variable LHSVs

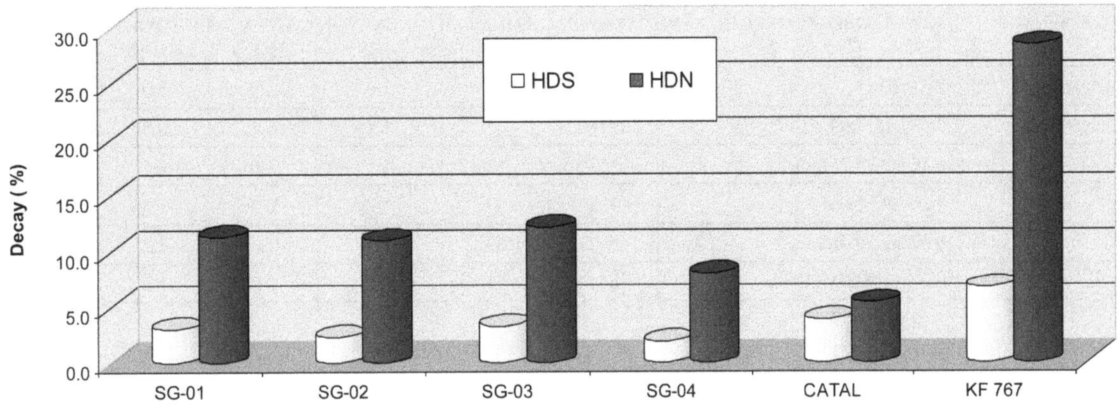

sulfur after the hydrotreatment is less than 50ppm, and tends to be stable in 100 hours of time on stream. Although the calcination temperature does have a significant effect on the catalyst performance, as low temperature may leads to weak interaction between Mo and the alumina support. (Duan, 2011; Rashidi, 2013)

In order to determine the catalyst for HDS stability we have compared the decay of HDS and HDN conversion after 10 day of spider operation time (Figure 4). This variable represents the drop of HDS and HDN conversions relative to the initial conversion after testing at different reaction conditions and then returning to the initial LHSV. Note that the benchmark catalyst A was determined at 19.2 h^{-1} while the other catalyst decays were determined at 7.2 h^{-1}.

The decay rates for HDS of the in house-prepared CoMo catalyst are lower than the Ref 1 catalyst and benchmark catalyst A. HDN decay was higher than HDS decay for CoMo catalysts. Based on the HDS test results with model feed, one can conclude that citrate-prepared CoMo catalysts are superior for HDS than commercial catalyst A and are comparable to the highly loaded Ref 1 NiMoW catalyst. However the ref 1 catalyst showed higher activity for HDN, maybe due to its higher nickel loading.

However, it is shown that the support shape itself has little effect on the catalyst HDS performance in both the model and real FCC diesel feeds.

High loading of CoMo, showed superior HDS and HDN conversions to the catalysts with low loading CoMo catalyst for benchmark catalyst A. The catalyst activity drops with the time on stream, but the citrate-prepared CoMo catalyst is clearly lower than the ref 1 catalyst and benchmark industrial catalyst A. HDN decay, on the other hand, was clearly higher than HDS decay for CoMo catalysts as ref 1 catalyst -prepared formulation showed fairly similar HDS and HDN decays.

The relative volume and weight activity were calculated assuming that both HDS and HDN reactions are pseudo-first order kinetics (n = 1), using equation (1) and commercial catalyst A as reference catalyst. C_F and C_P are S or N concentration on the feed and hydrotreated product, respectively, as k_i is the reaction rate constant.

$$\ln\left(\frac{C_F}{C_P}\right) = k_i\left(\frac{1}{LHSV}\right) \tag{1}$$

Table 2. Relative volume activity (RVA) and relative weight activity (RWA) for citrate-prepared CoMo catalysts, benchmark catalyst A and ref 1 NiMoW catalyst over real FCC diesel feed. Catalytic reaction conducted at 340 °C, 35 bar, 250 NL/L (H_2/oil ratio) and 7.2 h⁻

Catalyst	Composition (wt%)	S (ppm)	N (ppm)	RVA (HDS)	RVA (HDN)	RWA (HDS)	RWA (HDN)
Benchmark 1	Type II	1003	53	1.0	1.0	1.0	1.0
Benchmark II	$9NiO-15MoO_3-20WO_3$	939	47	1.1	1.3	0.8	1.0
S-1	$6.4CoO-25.7MoO_3$	553	53	1.3	1.0	1.3	1.0
S-2	$4.2CoO-18.0MoO_3$	725	57	1.2	0.8	1.3	0.8
S-3	$6.6CoO-26.4MoO_3$	562	53	1.3	1.0	1.2	0.9
S-4	$6.0CoO-22.0MoO_3$	664	55	1.2	0.9	1.3	0.8

According to tabulated results given in Table 2, it is clear that citrate-prepared CoMo catalysts showed significantly higher HDS volume and weight activity than the benchmark catalyst and also Ref 1 NiMoW. RWA for the ref 1 catalyst was markedly lower than CoMo catalysts, owing to its higher bulk density. In line with finding obtained on model feed, residual S content for S-01 and S-03 is similar, suggesting that same composition alumina support (trilobe versus extrudate) did not affect catalyst performance. Low CoMo loading catalysts (S-02 and S-04), did show lower performance than high CoMo loading catalyst. RVA and RWA for HDN reaction were comparable for high loading CoMo catalysts. However the NiMoW catalyst showed a significantly higher RVA than CoMo catalysts.

The conversions of 4,6 DM-DBT and fluorene can give valuable information about the main catalytic pathways. Fluorene is mainly converted through hydrogenation route (HYD) while DM_DBT can be also desulphurised through direct desulphurisation pathway (DDS).

The ref 1 NiMoW catalyst gave relatively high conversion of both model molecules (Table 3). High loading CoMo catalysts (S-01 and S-03) also showed high conversion of 4,6 DM-DBT relative to low loaded CoMo (S-02), however fluorene conversion was relatively low (22-25%) compared with the ref 1 catalyst (36%). This suggests that direct desulphurization pathway was the main catalytic route operating on the CoMo catalyst, particularly for high loading CoMo catalysts (type II Co-Mo-S phase), whereas the hydrogenation pathway was operating on the NiMoW catalyst in line with previous result (Gonzalez-Cortes et al., 2014).

Table 3. Conversion of 4,6 dimethyl-dibenzothiophene (4,6 DM-DBT) and fluorene determined by GC-FID-MS analysis for citrate-prepared CoMo catalysts, benchmark catalyst and ref 1 Catal-prepared NiMoW catalyst over model feed. Catalytic reaction conducted at 340 °C, 35 bar, 250 NL/L (H_2/Oil ratio) and 7.2 h^{-1}

Cata Code	Nominated Composition (wt%)	4,6 DM DBT Conversion (%)	Fluorene Conversion (%)
Benchmark Cat A	Type II catalyst, CoMo	71	23
Ref 1 Cata	$9NiO-15MoO_3-20WO_3$	59	36
S-01	$6.4CoO-25.7MoO_3$	26	25
S-02	$4.2CoO-18MoO_3$	61	22
S-03	$6.6CoO-26.4MoO_3$	35	25
S-04	$6.0CoO-22MoO_3$		23

Table 4. HDS activity comparison for CoMo catalysts prepared from Co nitrate and Co citrate. Catalytic reaction conducted at 340 °C, 35 bar, 250 NL/L (H_2/Oil ratio), 9.6 h^{-1} for model feed and 7.2 h^{-1} for Real FCC diesel feed

Sample	Co Precursor	Model Feed @ 9.6 h^{-1}		Real FCC Diesel Feed @ 7.2 h^{-1}	
		HDS Conversion (%)	Relative HDS Activity (% of Ref 1 Cat)*	HDS Conversion (%)	Relative HDS Activity (% of Ref 1 Cat)*
Industrial Benchmark	Unknown	79	--	83.0	--
Ref 1 Cat	-	90	--	84.0	--
D-1	Co nitrate	82	**134**	81.0	**160**
D-2	Co citrate	88	**155**	89.0	**194**

* Weight based and normalized to active metal oxide loading.

3.3. Influence of the Co Precursor on the HDS Performance Of Alumina-Supported CoMo Catalysts

The effect of changing citric acid was examined by preparing two catalysts by sequential incipient wetness impregnation (SIWI) of cobalt-containing solution on a MoOx/γ-alumina precursor. The same molybdenum oxide supported over alumina (trilobe γ-alumina) was taken as a precursor for the cobalt impregnation in both samples.

Sample D-1: Step 1: AHM → 250 °C; Step 2: $Co(NO_3)_2$ → 100 °C.

Sample D-2: Step 1: AHM → 250 °C; Step 2: $Co(CO_3)_2$ + citric acid → 100 °C.

These samples were tested for HDS of the model feed and the results are shown in Table 4.

The CoMo catalyst prepared by sequential impregnation with citric acid is markedly more active than the equivalent catalyst prepared with Co nitrate as precursor with both feeds. In order to get insight into the advantage effect of citric acid on HDS and HDN catalyst performance a CoMo citrate/γ-alumina catalyst (S-03) precursor prepared by wet co-impregnation and a Co nitrate-AHM/γ-alumina prepared by sequential incipient wetness impregnation were characterized by powder X-ray diffraction (Figure 5).

Figure 5. XRD patterns of CoMo/γ-alumina catalyst precursors treated at 110 °C and 340 °C

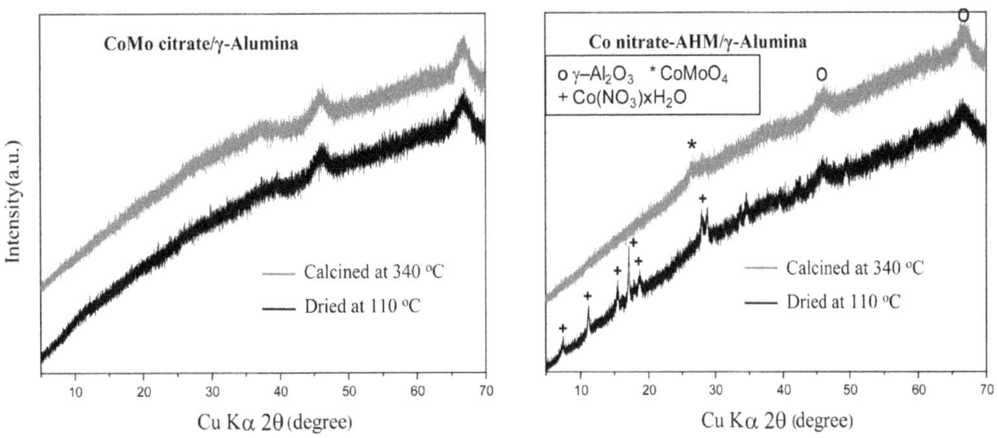

The XRD pattern for citrate-prepared CoMo catalyst showed only γ-alumina peaks and no XRD peak associated with Co or Mo species, even after calcination. This suggests that citric acid as organic additive (chelating agent, dispersing agent) facilitates the formation of amorphous Co and Mo species on alumina surface as was determined for Mo citrate on alumina (Klimov., 2010; Nikulshin, 2012). Equivalent CoMo catalyst prepared without citric acid showed strong XRD peaks associated with cobalt nitrate. After calcination at 340°C, cobalt molybdate was formed which is considered as one or less active phase precursor. This mixed-metal oxide suppresses the formation of Co-Mo-S active phase and hence reducing the catalyst activity.

The cobalt precursor effect was also determined on CoMo catalysts prepared by sequential incipient wetness impregnation without adding organic additive. The effect of cobalt precursor (acetate versus nitrate) on 9% CoO- 35% MoO_3/5%TiO_2-Al_2O_3 catalyst prepared under condition listed in Table 5. The acetate-containing catalyst is clearly superior to the nitrate catalyst. It is suggested that cobalt nitrate facilities the formation of Co_xS_y rather than Co-Mo-S upon sulphidation process (Fujikawa, 2006, 2007, 2009; Fujikawa., 2005; Nava, 2007) . It was also found that the acetate catalyst treated at low drying temperature (150 °C) was more active than the equivalent catalyst treated at 250 °C, indicating that $Co(CH_3COO)_2$ may facilitate the formation of Co_xS_y after treating the catalyst precursor at high calcination temperature. Cosmo Oil Co. prepared CoMo HDS catalyst using $CoCO_3$ and carboxylic acid, which depressed the sulphidation process of Co at low temperature, thereby increasing formation of the Co-Mo-S phase (Fujikawa, 2006, 2007, 2009; Fujikawa, 2005; Nava., 2007). Carboxylic acid may have a similar effect as CH_3COO- from $Co(CH_3COO)_2$.

Comparing data on Table 4 and 5, one can conclude that Co citrate appears to be a better precursor that Co acetate. Note, however, that these formulations were prepared by different preparation methods, which may also affect catalyst performance. Further work is needed to focus on preparing and testing equivalent formulations prepared under similar operation condition.

In summary, the cobalt precursor affects markedly the HDS and HDN performance of CoMo catalysts. Apparently the presence of citric acid facilitates the formation of amorphous Co and Mo species on alumina surface and hinders the formation of mixed-metal oxides. The order of performance of the catalyst made with different precursors is Co citrate > Co acetate > Co nitrate.

Table 5. HDS performance of CoMo catalysts prepared by using Co acetate and Co nitrate as Co source. Catalytic reaction conducted at 340 °C, 35 bar, 250 NL/L (H_2/Oil ratio) and 7.2 h^{-1}

Sample	Composition	Density (g/ml)	Co Precursor	S Conversion %			N Conversion %	
				Dry Condition	Model	Real FCC Feed	Model	Real FCC Model
					7.2	7.2	7.2	7.2
Y-09	$35MoO_3$-7.5CoO	1.057	$Co(NO_3)_2$	air	81	82	63	31
Y-10	$35MoO_3$-7.5CoO	1.023	$Co(NO_3)_2$	He	82	82	63	31
Y-12	$35MoO_3$-3.9CoO	0.998	$Co(CH_3COO)_2$	150	91	87	83	34
Y-13	$35MoO_3$-3.9CoO	1.004	$Co(CH_3COO)_2$	250	87	82	70	31

3.4. Effect of Drying/Calcination Temperature on CoMo Catalyst Performance

A series of catalysts supported on trilobe alumina were treated at different drying/calcination temperatures. Figure 6 shows that the residual S and N contents decreased with increasing contact time of model feed. Higher drying/calcination temperature did not seem to affect the residual S & N contents. At the highest contact time sulphur and nitrogen contents on the catalyst dried at 110 °C were slightly lower than with the equivalent catalyst treated at higher temperatures. Note that S-05 sample prepared by sequential impregnation (i.e. calcined at 275 °C and then at 110 °C after second impregnation) did show slightly higher residual S & N contents, most likely due to low loading of CoMo on alumina.

The influence of the drying/calcination temperature on HDS and HDN activities for citrate-prepared CoMo catalysts on industrial FCC diesel feed is given in Figure 7. Residual N content was not affected by thermal treatment on CoMo catalyst precursor (Figure 7a). Residual sulphur content, on the other hand, was slightly dependent upon thermal treatment. When one considers the catalyst density to find out HDS and HDN rate constants, assuming a pseudo-first kinetic order (see eq. 1), it was determined that both rate constants are fairly independent of the drying/calcination temperature. In conclusion, dry-

Figure 6. Dependence of residual sulphur (A) and nitrogen (B) contents on model feed with liquid contact time for citrate-prepared CoMo catalysts. Catalytic reaction conducted at 340 oC, 35 bar, 250 NL/L (H2/Oil ratio) and variable LHSVs

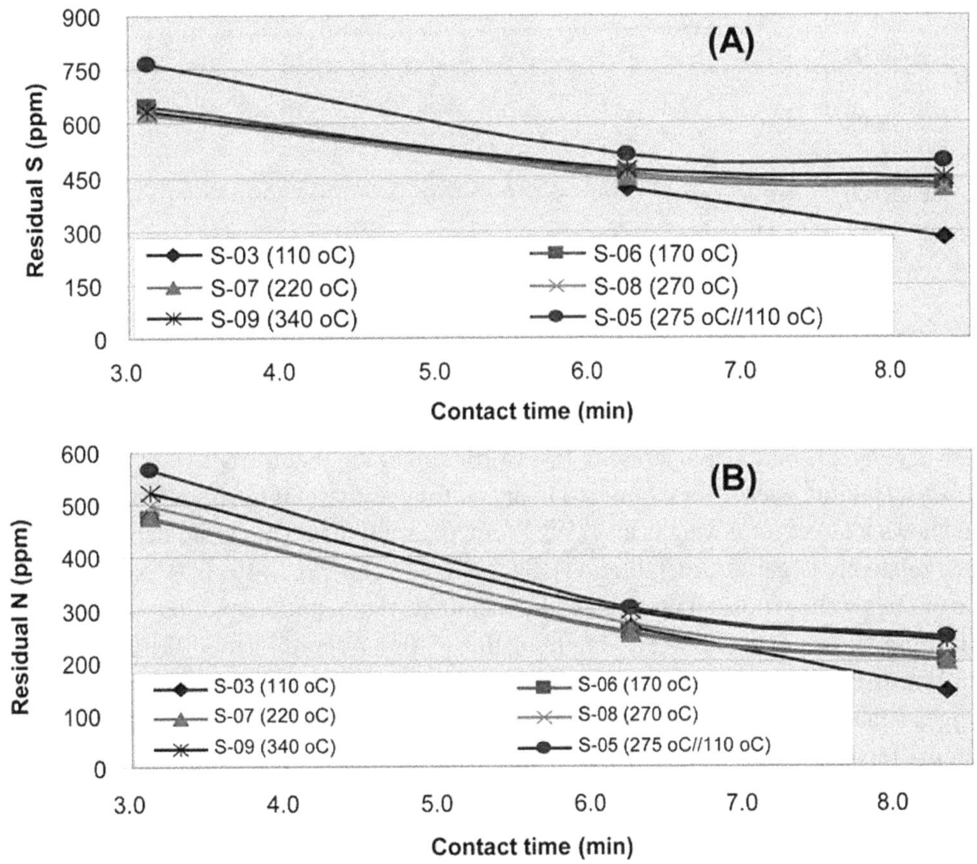

Figure 7. Dependence of residual S & N contents (A), HDS and HDN rate constants (B) on real FCC diesel feed with drying/calcination temperature for citrate-prepared CoMo catalysts. Catalytic reaction conducted at 340 °C, 35 bar, 250 NL/L (H_2/Oil ratio) and 7.2 h^{-1}

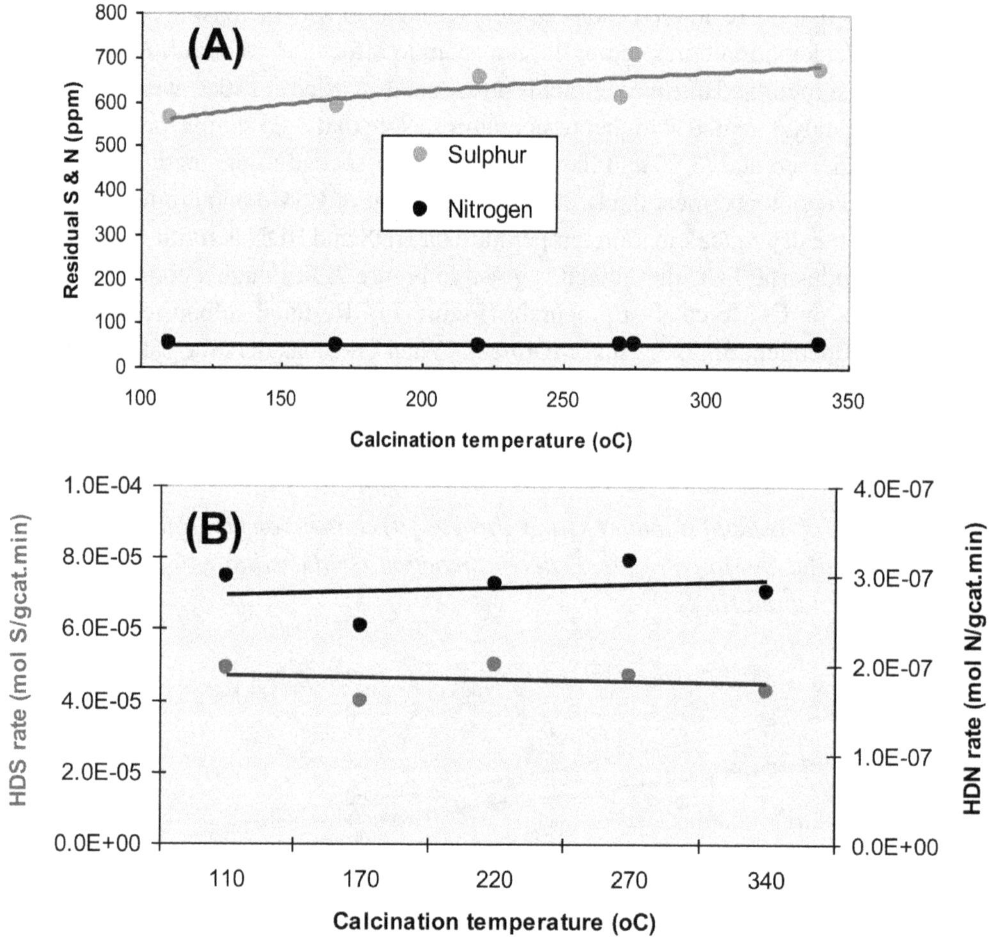

ing/calcination temperature did not significantly affect catalyst density-based HDS and HDN activities catalysts prepared by sequential wet impregnation. However, residual S concentration appears to increase slightly with drying/calcination temperature. It is worth remarking that further work needs to be done to find out catalyst stability over several hundred hours on time as thermal treatment may impact stability.

Table 6 shows a larger total weight loss (WL) for Co citrate than Mo citrate deposited on alumina, owing to the relatively larger CA/metal ratio (i.e. CA/Co = 1 and CA/Mo = 0.5). Note that citric acid; Mo and Co citrates were deposited on alumina by incipient wetness impregnation as CoMo citrate was deposited on alumina by sequential wet co-impregnation (S-03). According to S-03 nominal composition a total weight loss of ca. 34% is expected which is in on line with the obtained value (i.e. 32%). This finding suggests that high calcination temperature may reduce substantially residual carbon content on CoMo/alumina catalyst, which is reflected on the diminution of the density of the catalyst oxide precursor High calcination temperature seems not to significantly affect catalyst performance but this could improve catalytic stability. According to TGA-DSC profiles, Mo citrate attenuates the violent exothermic

Table 6. Partial and total weight losses of citrate-containing Co and/or Mo samples

Sample	Major WL Peak (°C)	T. Range (°C)	WL (wt. %)	Total WL (wt. %)
Mo citrate-Al_2O_3	228	160-360	14.6	21.0
Co citrate-Al_2O_3	280-300	180-320	19.8	29.0
Citric acid-Al_2O_3	185	130-275	17.7	32.0
S-03	230	165-370	19.8	32.0

reaction of Co citrate decomposition and apparently catalyzes the decomposition of Co citrate. Thermal decomposition of Co, Mo citrate starts at ca. 180 °C and would finish at calcination temperature above 350 °C, which was reflected on the diminution of the density of the catalyst oxide precursor.

The drying temperature effect on HDS activity for CoMo catalyst supported on modified and unmodified γ-alumina (Table 7) was investigated. The bulk density decreased with temperature due to water loss. We can rationalise this finding assuming that water is removed from alumina pore and citric acid partial decomposition could also occur with increasing temperature. The HDS catalyst performance at temperatures below 200°C did not change markedly. According to the literature for CoMo catalysts (Al-Dalama 2006; Bezergianni,, 2014; Fujikawa, 2009; Nava., 2007; Nikulshin, 2012), after sulphidation, Co-Mo-S phase or CoS+MoS_2 phase will be formed. If the HDS catalyst comprises mainly Co-Mo-S phase, it can produce ultra-low sulphur diesel (ULSD) with high performance and high stability. CoS+MoS_2 phases can be avoided when citric acid is used as organic additive since this can favour simultaneous sulphidation of Co and Mo species and hence facilitating Co-Mo-S phase. Note that HDS conversion did not vary with drying temperature, owing to the relatively narrow range of temperature (i.e- 110-180°C).

Another variable that can be also analysed is the effect of silica on pre-modified alumina for CoMo catalyst performance, Table 7. The support for Y-30 catalyst is similar than that used for Y-27 (i.e. Al_2O_3 PV075). However, Al_2O_3 support was initially modified using $Si(OC_2H_5)_4$ solution to obtain 2.5 wt. %

Table 7. HDS and HDN conversions of citrate-prepared CoMo catalysts supported on alumina and modified alumina. Catalytic reactions conducted at 340 °C, 35 bar, 250 NL/L (H_2/Oil ratio) and 7.2 h^{-1} with model and real FCC diesel feeds

			Density	Dry	S Conversion (%) Model Feed Real Feed		Rate Constant Based on Metal Loading
Sample	Composition	Support	g/ml	°C	LHSV 7.2	LHSV 7.2	LHSV 7.2h^{-1}
Y-27	17.1MoO_3-4.2CoO	Al_2O_3 beads	1.091	110	91	92	42
Y-28	12.4MoO_3-3.1CoO	Al_2O_3-B_2O_3	1.163	110	89	89	48
Y-29	13.9MoO_3-3.5CoO	Al_2O_3-extrudate	0.933	110	90	90	57
Y-30	16.3MoO_3-4.1CoO	SiO_2-Al_2O_3	1.223	110	92	94	34
Y-31	17.1MoO_3-4.2CoO	Al_2O_3	1.018	180	93	90	48
Y-32	12.4MoO_3-3.1CoO	Al_2O_3-extrudate	1.086	180	93	91	62
Y-33	13.9MoO_3-3.5CoO	SiO_2-Al_2O_3	0.891	180	91	90	62
Y-34	16.3MoO_3-4.1CoO	Al_2O_3	1.034	180	92	91	49

SiO_2-Al_2O_3, then this modified support was impregnated with CoMo citrate solution through a single step by wet impregnation with long time impregnation (i.e. 24 h). Comparing Y-30 and Y-27 catalyst performance, we found that Y-30 performance is superior than Y-27 for Real FCC diesel feed, suggesting that the modification of alumina support with silica has indeed a benefit effect on HDS activity in line with previous report (Duan, Wan, et al., 2011; Robinson, van Veen, de Beer, & van Santen, 1999). Taking into account this encourage result it is planned to modify alumina with different metal oxides such as ZrO_2, TiO_2, etc.

In order to study the effect of calcination temperature on CoMo catalyst prepared by sequential incipient wetness impregnation, the catalyst D-4 was calcined to 250 °C. Table 8 gives the catalytic results over the catalysts calcined at different temperature while sulfiding at the same temperature. A clear improvement in HDS activity was seen when the catalyst was treated at the higher temperature. Further testing and characterisation data is required to compare stability, as the two samples were not subjected to the same testing regime, and to determine the possible cause of this activity improvement. In summary, CoMo supported on silica-modified alumina prepared by wet co-impregnation and CoMo catalyst prepared by sequential incipient wetness impregnation and calcined at 250 °C showed improved HDS conversion. These two strategies could be also useful for improving catalyst stability.

3.5. Influence of the Impregnation Solution pH on CoMo Catalyst

In order to determine the effect of the impregnation solution pH on CoMo catalyst performance prepared by sequential incipient wetness impregnation (SIWI) and wet co-impregnation a different catalyst series were prepared and tested in the HDS spider.

The use of ammonium citrate in place of citric acid gave an impregnation solution of pH 9 for Co impregnation solution. The catalyst (D-8) was tested and the performance compared with that of D-3 (Table 9). An improvement in HDS activity was obtained following the change of citrate source. The use of the ammonium citrate rather than citric acid is believed to change the solution chemistry that occurs within the structure of the extrudate body upon impregnation. AHM will readily dissolve in a solution of cobalt citrate, but once heated to 250 °C (as in the preparation of the catalysts above) a partial decomposition occurs with loss of ammonium (Klimov et al., 2010; Yoosuk, Song, Kim, Ngamcharussrivichai, & Prasassarakich, 2010). The dissolution of this material is aided by the presence of ammonium cation (NH_4^+), and this may be provided by the ammonium citrate. Therefore an impregnation of a basic cobalt citrate solution is more likely to cause re-dissolution of the molybdenum species already supported on the alumina. This appears to have a slightly positive effect on the HDS activity.

Table 8. Effect of calcination temperature on HDS activity of a sequentially impregnated CoMo catalyst. Catalyst tests conducted at 340 °C, 35 bar, 250 NL/L (H_2/Oil ratio) at 9.6 and 7.2 h^{-1} with model and Real FCC diesel feeds

Sample	Calcination Temp of Catalyst (°C)	Model Feed @ 9.6 h^{-1}		Real FCC Diesel Feed @ 7.2 h^{-1}	
		HDS Conv. (%)	Relative HDS Activity (% of Ref 1 Cat)*	HDS Conv. (%)	Relative HDS Activity (% of Ref 1 Cat)*
D-4	100	87	164	89	208
D-9	250	90	194	92	246

Table 9. Influence of the pH of Co citrate solution on CoMo catalyst prepared by SIWI. Catalyst tests conducted at 340 °C, 35 bar, 250 NL/L (H_2/Oil ratio) at 9.6 and 7.2 h^{-1} with model and industrial FCC diesel feeds

Sample	pH of Cobalt Citrate Solution	Model Feed @ 9.6 h^{-1}		Real Disel Feed @ 7.2 h^{-1}	
		HDS Conv. (%)	Relative HDS Activity (% of Ref 1 Cat)*	HDS Conv. (%)	Relative HDS Activity (% of Ref 1 Cat)
D-3	3	87	156	89	200
D-8	9	89	173	90	220

A series of CoMo catalysts were also prepared by incipient wetness co-impregnation (IWCI) to investigate the influence of pH and cobalt to citric acid ratio (Table 10). It has been reported that a complex containing two citrate moieties per cobalt atom is formed at higher pH (Bentaleb et al., 2014; Fan et al., 2011; Fujikawa et al., 2006; Klimov et al., 2010). Note that previous preparations had all used cobalt: citric acid = 1:1 with a slight excess of citric acid. In each case it was necessary to employ several incipient wetness impregnation steps to achieve a MoO_3 loading of approximately 20%. It is worth remarking that the results listed on Table 9 were significantly inferior to that displayed on Table 10, indicating that co-impregnation preparation method gave clearly advantages compared with sequential impregnation method. The changes made to the pH of the impregnation solution and the citric acid to cobalt ratio into this range has no an apparent effect on the HDS activity of the catalysts as tested.

Table 11 shows the effect impregnation solution acidity on the HDS and HDN performance of citrate-prepared CoMo catalyst by wet co-impregnation. A CA(1)/Co molar ratio of one and CA(2)/Mo molar ratio of 0.5 to get CA/(Co+Mo) molar ratio of 0.67. Impregnation solution pH above natural pH (2-3) for preparing S-14 and S-15 does not have a strong effect on the HDS performances compared to S-01 catalyst. However, HDN performance seems to be significantly affected by impregnation solution pH. Note that a catalyst prepared in one impregnation step (S-13) with high CoO loading relative to MoO_3 did present 50% superior weight HDS activity and 30% weight HDN activity relative to commercial benchmark Cat A, which suggests that Co loading can be an additional variable to be optimized for preparing CoMo catalyst by wet co-impregnation, since repartition of Co and Mo citrate onto alumina may be affected by competitive adsorption of Mo and citrate anions and steric effect of citrate on Co adsorption.

Table 10. Effect of pH and CA/Co molar ratio on CoMo catalysts prepared by incipient wetness co-impregnation. Catalyst tests conducted at 340 °C, 35 bar, 250 NL/L (H_2/Oil ratio) at 9.6 and 7.2 h^{-1} with model and Real FCC diesel feeds

Sample	pH	Citric Acid: Cobalt Molar Ratio	Model Feed @ 9.6 h^{-1}		Real Diesel Feed @ 7.2 h^{-1}	
			HDS Conv. (%)	Relative HDS Activity (% of Ref 1 Cat)	HDS Conv. (%)	Relative HDS Activity (% of Ref 1 Cat)
D-10	7	1:1	88	203	89	268
D-11	5	1:1	88	201	88	258
D-12	8	2:1	90	192	91	265
D-13	2	2:1	89	198	90	264

Table 11. Effect of impregnation solution pH on citrate-prepared $CoMo/Al_2O_3$ catalysts using real FCC diesel Catalytic reaction conducted at 340 °C, 35 bar, 250 NL/L (H_2/Oil ratio) and 7.2 h^{-1}

Cat	Information	RVA (HDS)	RVA (HDN)	RWA(HDS)	RWA (HDN)
Bench A	Type II	1.0	1.0	1.0	1.0
S-12	pH 2.4 6.4CoO-25.7MoO_3	1.3	1.0	1.3	1.0
S-13	pH 2.4 5.5CoO-20.4MoO_3	1.4	1.2	1.5	1.3
S-14	pH 5.1 5.7CoO-22.7MoO_3	1.4	1.1	1.4	1.2
S-15	pH 6.8 6.4CoO-22.1MoO_3	1.4	1.1	1.4	1.2

The detailed information of the sample S- and D series were given in Table 3 and Tables 7-10.

Figure 8 shows that within the range of impregnation solution pH the HDS conversion (volume-based activity) for FCC diesel feed did not change significantly for series of catalyst prepared by similar preparation method. However, when one compares S-14 with D-11 (pH ~ 5) and S-15 with D-10 (pH ~ 7), there is a significant difference, most likely due to the preparation method and/or CA/Co ratio. Note that S- samples and D-3 and D-8 were prepared by wet co-impregnation by sequential impregnation whereas D-10 to13 were made by IW co-impregnation.

Further works need to be done to study the effect of CA/Co molar ratio. HDS activity of the catalysts prepared by co-impregnation was superior to the equivalent catalyst prepared by sequential impregnation as discussed previously.

Figure 8. Dependence of the HDS conversion of CoMo catalysts prepared by different impregnation methods versus impregnation solution pH. Catalytic reaction conducted at 340°C, 35 bar, 250 NL/L (H_2/Oil ratio) and 7.2 h^{-1}

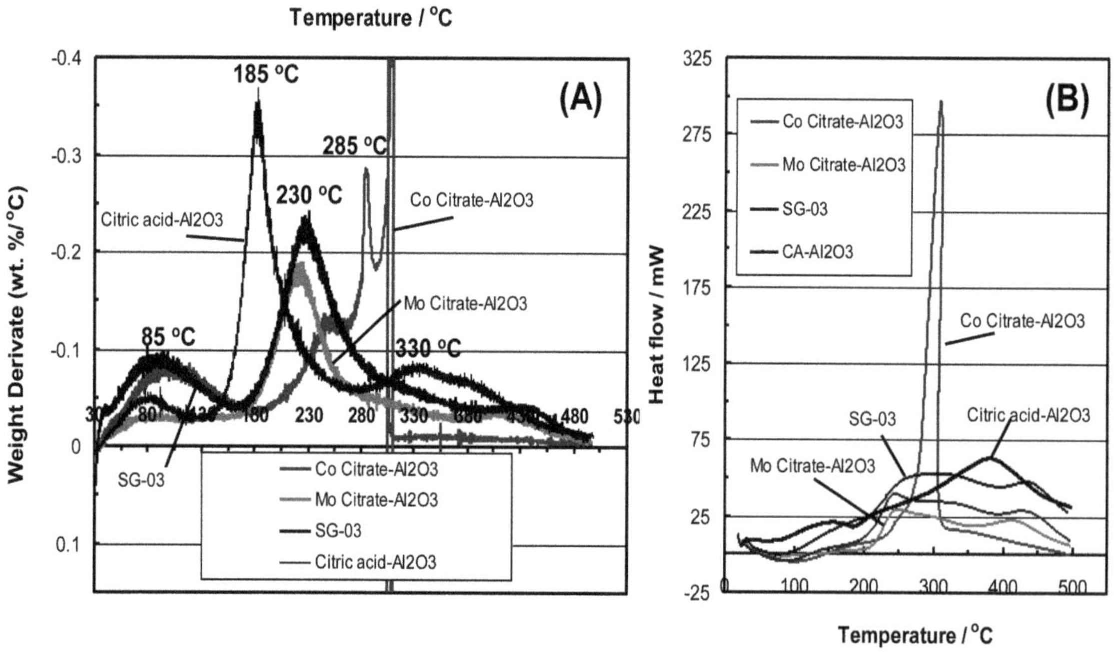

Figure 9. TGA profiles for γ-alumina-supported CoMo catalysts (maximum nominal composition 6% CoO-26% MoO_3) prepared by wet co-impregnation and sequential incipient wetness impregnation

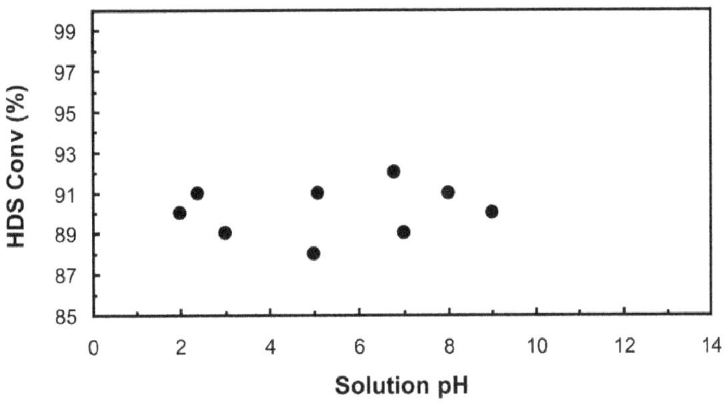

In order to get insight into the effect of preparation method (i.e. wet co-impregnation versus sequential incipient wetness impregnation) on the Co and Mo species deposited on alumina a series of CoMo catalysts were characterized by thermogravimetric analysis. Figure 9 compares the derivate of the TGA profile for a sample prepared by wet co-impregnation to sequential impregnation-prepared CoMo catalyst. There is a common major weight loss peak at 230 °C. However, SIWI-prepared CoMo catalyst showed a new strong weight loss peak at 335 °C, which was not displayed in the equivalent CoMo catalyst prepared by wet co-impregnation.

This finding reveals that preparation method can indeed affect CoMo distribution onto alumina extrudate. Apparently wet co-impregnation method facilitates a more uniform CoMo citrate species than SIWI method, which was reflected on a higher HDS catalyst performance (see inset table). In order to determine the cause of this new peak at 335 °C a series of CoMo samples with different Mo/Co molar ratio was prepared by SIWI and characterised by TGA (Figure 10). All the profiles showed two major peaks, whose positions and intensities are dependent upon Mo/Co ratio (inset figures).

The maximum temperature of the peak positions (i.e. LT and HT) were slightly shifted to lower temperature with increasing Mo/Co ratio (Figure 10a) because of weaker metal oxide precursor-alumina interaction. On the other hand, the LT/HT peak intensity ratio showed a fairly linear dependence with Mo/Co ratio (Figure 10b), suggesting that the low temperature peak is mainly associated with Mo species decomposition as the high temperature peak is related with Co citrate. However, when one considers the straight line equation [LT/HT = 0.4827*(Mo/Co) + 0.4678], it is possible infers that Co species can also be decomposed at LT peak, as indeed occurred for WCI-prepared CoMo catalysts. Indeed, alumina-deposited cobalt citrate showed a weak peak at 240 °C, as was previously discussed. So, upon the CoMo catalyst preparation by sequential incipient wetness impregnation one can envisage that initially Mo is deposited onto the core extrudate body during the first Impregnation step. After a partial decomposition of AHM at 250 °C, the solid is then impregnating with cobalt citrate solution, which can partially re-dissolve Mo species and re-deposit as Mo (Co) citrate. It is likely that Co citrate is deposited on the edge of the extrudate body. Note that WCI catalyst are currently prepared in only one impregnation step whereas SIWI sample required two impregnation steps and also a calcination step so the former method could be also attractive based on catalyst manufacture cost.

Figure 10. TGA profiles of γ-alumina supported CoMo catalysts with different Mo/Co ratio prepared by sequential incipient wetness impregnation (SIWI). Inset figures: (A) Dependence of the weight loss peak positions and (B) peak intensity ratio with Mo/Co ratio

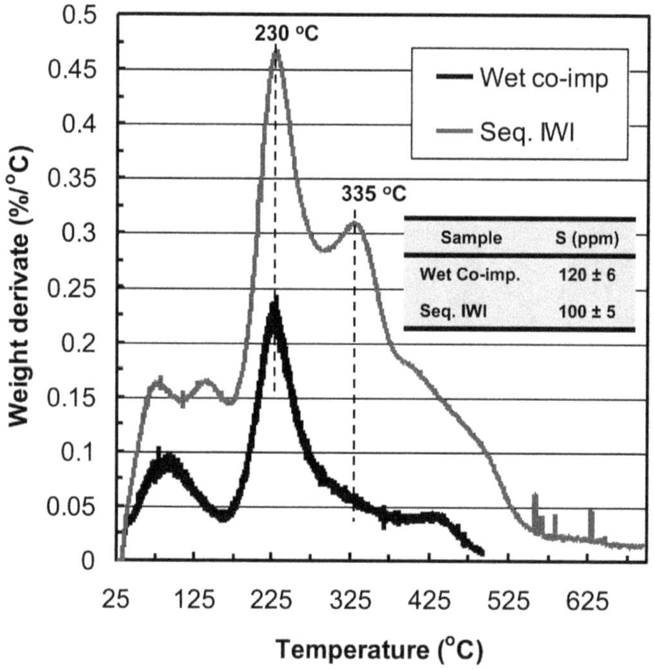

3.6. Effect of Organic Additives on CoMo Catalyst Performance

In previous work, an organic compound (urea) improved the stability of AHM and AMT (ammonium metatungstate) solution, an effect that was linked to improved HDS performance in a HDS catalyst (Al-Megren, 2009; Gonzalez-Cortes, 2004). A similar test was carried out for the impregnation of AHM on alumina and then Co citrate was deposited on the molybdenum bearing alumina to get CoMo/γ-alumina catalyst and the results are listed in Table 12.

A test with and without glycine was then carried out for a higher concentration (D-6 and D-7) AHM impregnation solution, assuming that stabilising effect may only be important with higher concentration, where an unstabilised solution may give precipitate on impregnation. However, as it is shown in Table 9 there was no significant effect at higher concentration either.

In order to find out the effect of an organic compound on Co nitrate impregnation solution rather than Mo-containing solution, a series of high loading CoMo/γ-alumina catalyst was prepared. Polyol organic molecules were selected for the reported ability of glycol to improve diesel HDS catalysts (Escobar et al., 2009). Note that the organic additive (OA) was added upon Co impregnation, after loading AHM on alumina. According to Table 13, the addition of an organic additive such as ethylene glycol (EG), polyethylene glycol (PEG) and glycerol (Gly) improved significantly not only the HDS but also HDN performance of γ-alumina-supported CoMo catalyst under model and Real FCC diesel feeds relative to no OA-prepared equivalent catalyst.

Table 12. Influence of organic compounds in impregnation of AHM on alumina for preparing CoMo/γ-alumina catalyst. Catalysts tested at 340 °C, 35 bar, 250 NL/L (H_2/Oil ratio), 9.6 h^{-1} (model feed) and 7.2 h^{-1} (Real FCC diesel feed)

Sample	Organic Added in AHM Impregnation	Model Feed @ 9.6 h^{-1}		FCC Diesel Feed @ 7.2 h^{-1}	
		HDS Conv. (%)	Relative HDS Activity (% of Ref Cat 1)*	HDS Conv. (%)	Relative HDS Activity (% of Ref Cat1)
D-2	Urea	87	168	89	202
D-3	None	87	156	89	200
D-4	Glycine	87	164	89	208
D-5	L-cysteine	84	143	85	195
D-6	None	88	134	88	165
D-7	Glycine	88	127	89	160

*Measured on a weight basis, normalized for nominal Mo content.

Comparing the data of Table 12 and Table 13 and considering nominal composition, low loading CoMo catalysts (ca. 6 wt. % CoO-24 wt. % MoO_3) showed slightly higher HDS conversion from real FCC diesel feed than high loading CoMo catalyst (ca. 8-9 wt. % CoO-35 wt. % MoO_3). The addition of organic additive upon Mo impregnation did not improve catalyst performance in contrast with OA-containing cobalt impregnation. However, that both series of catalysts were prepared with different organic additives, Co precursors and CoMo loadings. Nevertheless, we can consider that the addition of citric acid into cobalt impregnation solution rather than OA added upon Mo impregnation could be the main factor to improve CoMo catalyst performance. Since Mo-containing alumina sample was calcined to 250 °C, which can decompose partially the OA, a subsequent addition of Co citrate may re-disperse molybdate species and/or disperse Co species regardless OA used upon Mo impregnation. On the other hand, the addition of glycol-type organic molecules upon Co impregnation may facilitate the dispersion of oxidic clusters with optimal Co/Mo ratio (Chen, 2010; Nikulshin, 2012). In summary, organic additives such as urea and glycine used upon Mo impregnation did not alter HDS conversion as L-cysteine decreased HDS activity. The presence of organic additives such as citric acid, ethylene glycol, polyethylene glycol and glycerol in cobalt impregnation solution did improve significantly not only HDS but

Table 13. Dependence of HDS and HDN conversions on model and industrial FCC diesel feeds treated at 340 °C, 35 bar, 250 NL/L (H_2/Oil ratio) and 7.2 h^{-1}

Code	Organic Additive	Model Feed @ 7.2 h^{-1}		Real FCC Diesel Feed @ 7.2 h^{-1}	
		HDS Conv (%)	HDN Conv (%)	HDS Conv (%)	HDN Conv (%)
S-16	None	74	49	69	29
S-17	EG	86	76	82	36
S-18	PEG	85	73	82	34
S-19	Gly	85	72	82	34
Ref 1 Cat	Urea	94	100	82	52

also HDN activities. This is likely due to the formation of well dispersed Co species on MoOx-alumina surface, another reason may be that the citric acid can change the surface isothermal electric property, thus changing the metal and support interaction. .

3.7. Effect of Secondary Promoters such as Ni, P and W on γ-Alumina-Supported CoMo Catalyst

The relative activities of Co-Mo and Ni-Mo catalysts for the most important hydro-processing reactions are well known (Chen, 2010; Gao, 2011; Infantes-Molina, 2011; Navarrete., 2011):

DDS (direct desulphurization): Co-Mo > Ni-Mo

Hydrodenitrogenation: Ni-Mo > Co-Mo

HGA (hydrogenation): Ni-Mo > Co-Mo

In the case of the real FCC diesel feed under relatively low H2 partial pressure conditions, there is a much greater requirement for DDS than HDN or HGA, as the nitrogen content of the feed is low and small cetane number upgrading is required. Therefore Co-Mo is the obvious bimetallic system choice which would have less hydrogen consumption for olefin saturation. However, the FCC diesel feed is known to contain a fraction of more refractory sulphur compounds including 4,6-DMDBT and various analogues, which cannot be processed by the DDS pathway, and must be first hydrogenated to reduce steric hindrance. By substituting some CoO in the precursor with NiO it was believed that a greater HGA functionality would be included in the catalyst and this would allow the more refractory compounds to be desulphurised.

To this end a series of catalysts were compared with varying ratios of the two metals (Table 14). The catalysts were prepared as for D-3, but in the second impregnation cobalt carbonate was substituted with basic nickel carbonate as required.

From the model feed data the benefit of increasing NiO content on HDN conversion can be clearly seen. Note that since the nitrogen content of the FCC diesel feed is less than 100 ppm, the HDN conversion figures are susceptible to error, which may explain the absent of correlation with Ni (and/or Co) oxide loadings. Unfortunately it seems that the effect on the HDS conversion is negative. Indeed, when one considers the effect of increasing Ni/Co molar ratio on HDS conversion a directly proportional de-

Table 14. Influence of the Ni/Co molar ratio on Co(Ni)Mo catalyst performances. Catalysts tested at 340 °C, 35 bar, 250 NL/L (H_2/Oil ratio), 9.6 h^{-1} (model feed) and 7.2 h^{-1} (Real FCC diesel feed)

Sample	NiO (wt. %)	CoO (wt. %)	Ni/Co Molar Ratio	Model Feed @ 9.6 h^{-1}		FCC Diesel Feed @ 7.2 h^{-1}	
				HDS Conv. (%)	HDN Conv. (%)	HDS Conv. (%)	HDN Conv. (%)
D-3	0.0	6.0	0	87	70	89	36
D-14	0.9	4.5	0.2	88	72	88	33
D-15	3.1	2.9	1.1	88	77	85	35

Table 15. Effect of low loading of WO_3, NiO and P_2O_5 on HDS performance of citrate-prepared CoMo catalysts using FCC diesel feed and model feeds. Catalysts tested at 340 °C, 35 bar, 250 NL/L (H_2/Oil ratio), 9.6 h^{-1} (model feed) and 7.2 h^{-1} (Real FCC diesel feed)

Cat	Composition	Model Feed LHSV 9.6h^{-1}		FCC Diesel Feed LHSV 7.2h^{-1}	
		HDS Activity %	HDN Activity %	HDS Activity %	HDN Activity %
Benchmark	Type II	79	64	83	34
S-01	6.4CoO-25.7MoO_3	90	72	91	34
S-03	6.6CoO-26.4MoO_3	90	74	91	34
S-10	6.0CoO-26.0MoO_3-0.7P_2O_5	89	73	92	38
S-11	6.2CoO-25.0MoO_3-1.0WO_3	89	73	90	36
S-12	8.5CoO-0.7NiO-37.0MoO_3	89	76	90	40

pendence was observed for FCC diesel feed, suggesting an apparent hinderence effect on HDS catalysts performance. A possible reason for this effect is the segregation of $(Co,Ni)_9S_2$ pentladite rather than the formation of Co(Ni)MoS active phase as increasing Ni loading, hence decreasing Co and Ni promoter effect on MoS_2.

Table 15 shows that there was no catalytic difference due to the supports; G250 extrudate and the BASF trilobe. Promoter such as WO_3, P_2O_5 and NiO did not affect HDS conversion. HDN activity appears to be improved for FCC diesel feed in the case of P_2O_5 and NiO addition. Further work should be done to evaluate higher loadings of P_2O_5, since it has been shown that P_2O_5 can improve the hydrogenation function of HDS catalysts (Ali, 2014; Ali, 2012).

Figure 11. Dependence of residual S & N contents upon time on stream for citrate-prepared CoMo catalyst (S-20) (A). Dependence of residual S & N contents versus reaction temperature (B). Catalytic test conducted at 35 bar, 250 NL/L (H_2/Oil ratio), 1.2 h^{-1} and variable temperature under real FCC diesel feed

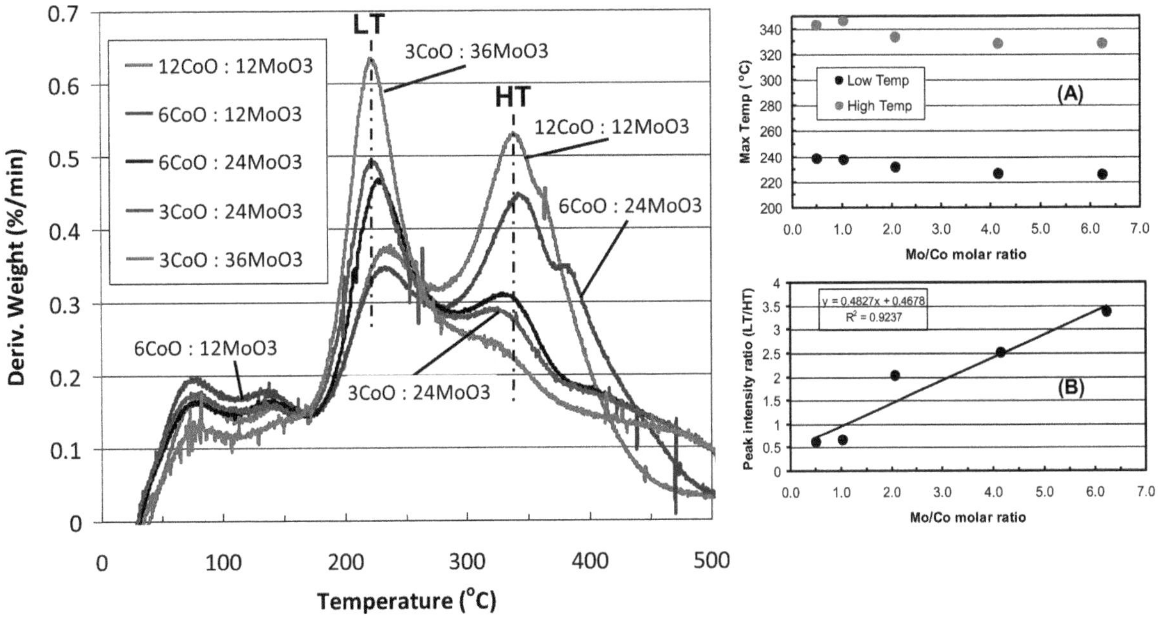

Figure 12. Boiling point distribution of the FCC diesel feed and liquid product obtained at different reaction temperature. Liquid product obtained at 35 bar, 250 NL/L (H_2/Oil ratio), 1.2 h^{-1} and variable temperature under FCC diesel feed. Inset figure: Cetane index versus hydrotreated temperature

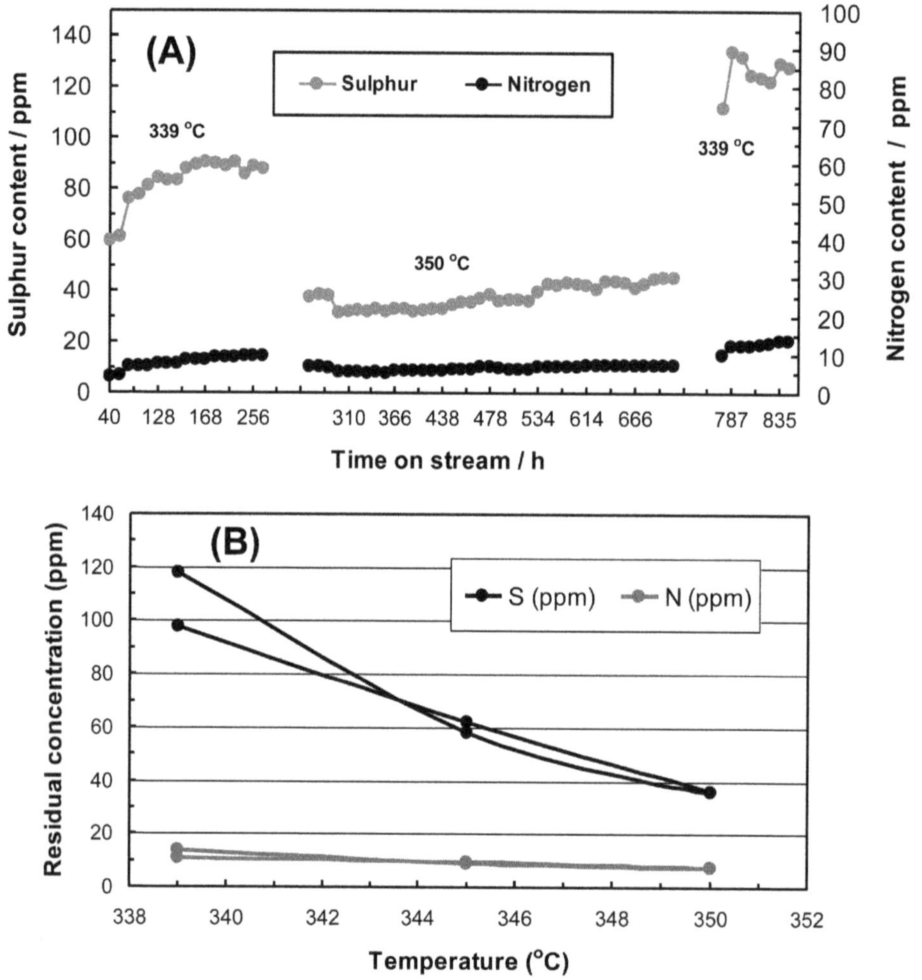

3.8. Stability Tests of Citrate-Prepared CoMo Catalysts on Small HDS Unit

In order to test CoMo catalyst stability and performance under FCC diesel lean hydrogen partial pressure conditions, a few potential good catalysts previously identified during HDS spider screening have been tested on the HDS small unit. For instance, Figure 11 shows results obtained with S-20 at different reaction temperatures. At 339 °C, the residual S content increased to 67% relative to initial level.

An Arrehnius equation activation energy of 120 ± 5 kJ/mol was calculated. The poor catalyst stability is partially attributed to the 'soft' thermal treatment (110°C) after impregnation which was done to obtain well-dispersed CoMo citrate species on alumina. Note that a high calcination temperature can also substantially increase residual S content, as was previously discussed (see Figure 7).

The liquid hydrotreated product was analysed by simulation distillation to determine the boiling point distribution (Figure 12). The feed and liquid products show a similar boiling point distribution, indicating that cracking was very insignificant. Cetane index determined from boiling point (BP) distribution data, according to ASTM standard D4737, was slightly at higher reaction temperature (see inset figure).

Figure 13. Residue S & N for industrial and different citrate-prepared CoMo catalysts (A) and for S-01 and S-03 samples at different H2/oil ratio (B) The catalysts were tested in an industrial pilot plant HDS rig at 340 oC, 35 bar, 250 NL/L (H2/Oil ratio) and 1.2 h-1 under FCC diesel feed as S-01 was tested at 400 NL/L (H2/Oil ratio).

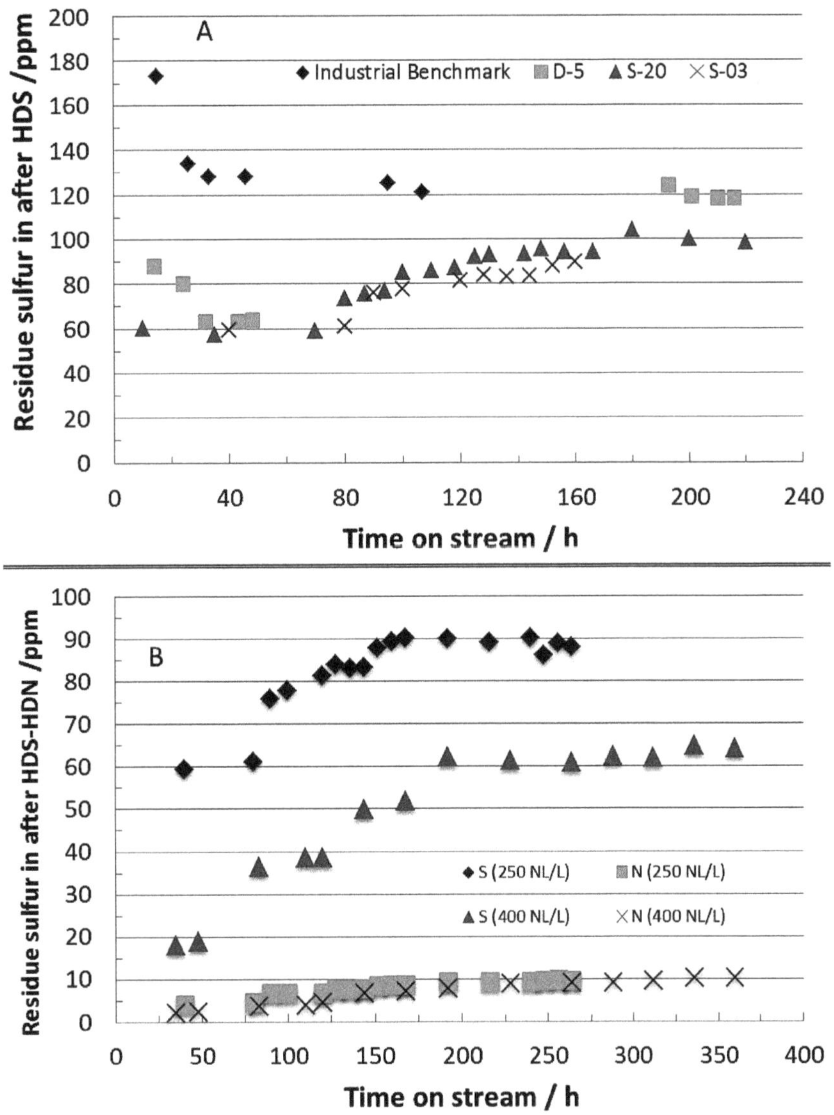

Figure 13 shows results of life tests in the small HDS rig on various selected catalysts. The industrial bench mark catalyst and D-5 showed some initial activation with time whereas the other catalysts showed initial deactivation. Note that Industrial bench mark catalysis pre-sulphided whereas the catalysts made in-house were sulphided in-situ, the benchmark catalyst became stabilised after 24 hours at 120 ppm residual S.

The in-house prepared catalysts using citric acid as the organic additives gave low initial residual S content, e.g., in the case of S-20 this was ca. 60 ppm below 75 hours on stream rising to about 90-100 ppm after 150 h of operation. Note that D-5 sample after 48 h of operation under FCC diesel condition showed ca. 60 ppm of residual sulphur. Therefore this sample was tested at 0.9 h^{-1} for ca. 120 h and subsequently the LHSV was increased up to 1.2 h^{-1} to get ca. 120 ppm of residual sulphur. This trend indicates that the catalyst prepared using this method also underwent deactivation upon time on stream.

Figure 13b compares results from catalyst S-01 made with bead Al_2O_3 support and S-03 made with extrudate trilobe support. These catalysts did not show marked differences in performance in the HDS rig. (see Table 2). Increasing the H_2/oil ratio from 250 to 400 reduced S levels by nearly 30% of residual sulphur content for S-01 relative to S-03 under apparent steady-state condition. Apparently, HDS catalyst activity can be increased at high H_2/oil ratio likely due to the improvement of hydrogen transfer reactions (Mani, 2011; Nava, 2004; Ramirez, 2004). At initial operation time, residual S concentration for citrate-prepared CoMo catalysts supported on big pore support was fairly close to the industrial operation target (ca. 60 ppm rather than 50 ppm –FCC diesel target-). In line with HDS spider results, citrate-prepared CoMo catalysts did give a significantly lower residual S content (90-100 ppm) compared with benchmark catalysts (ca. 120 ppm) under steady state condition. A further increase of H_2/oil ratio from 250 to 400 NL/L reduces significantly residual S content (nearly 30%) and tends to improve catalyst stability.

Figure 14 illustrated that the catalyst (17.3MoO_3-4.1CoO/Al_2O_3, dried at 110°C, calcined at using single wet co-impregnation for 24 h did show similar residual sulphur concentration (150-160 ppm) under steady state condition regardless of calcination/sulphidation temperature. This displayed lower HDS activity than benchmark catalyst (ca. 120 ppm). Nevertheless, at short time on stream (TOS) dried catalyst precursor showed low residual S level compared to industrial benchmark catalyst A. This indicates that citrate-prepared CoMo catalyst shows indeed higher intrinsic HDS activity than benchmark catalyst however this formulation was not stable.

Figure 14. Test results from small HDS rig at 340 °C, 35 bar, 250 NL/L (H_2/Oil ratio) and 1.2 h^{-1} under real FCC diesel feed. Effects of thermal treatment on citrate-prepared CoMo catalyst

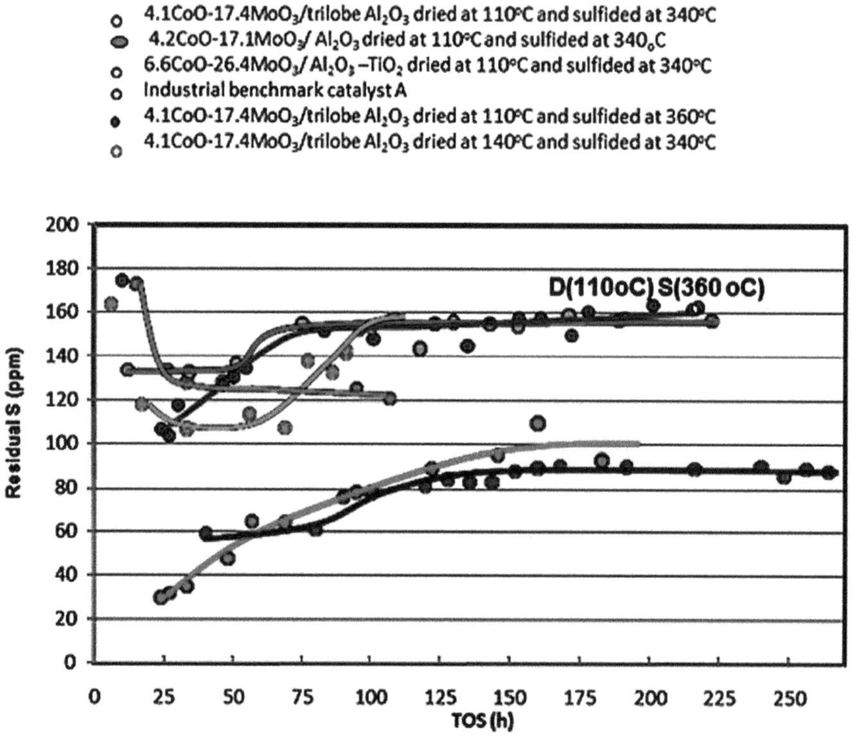

When one considers the trilobe alumina supported CoMo catalyst (i.e. S-03) prepared by sequential wet co-impregnation (5 h) and the sample prepared by single wet co-impregnation for 24 h and supported on PV075 alumina it is possible to infer that high nominal loading of CoMo on trilobe alumina and textural properties of PV075 (pore diameter ~ 90 Å) for HDS catalyst may play a very important role for obtaining lower residual S concentration (90-100 ppm) than bead-supported CoMo catalysts. Indeed, The Catalysts20 (Figure 13a) supported on bead Al_2O_3 using sequential wet co-impregnation did show lower residual S level (ca. 100 ppm) than equivalent CoMo catalyst also supported on bead Al_2O_3 prepared by single co-impregnation (ca.150-160 ppm), which clearly indicates that high CoMo loading improved markedly HDS activity. In conclusion drying, calcination and/or sulphidation temperature did not affect HDS activity of CoMo catalyst prepared by single wet impregnation under ultra low sulphur-production condition. Also, CoMo supported on Al_2O_3 (PV075, single wet co-impregnation) and CoMo supported on bead Al_2O_3 (seq. wet co-impregnation) showed lower residual S concentration than the benchmark catalyst.

3.9. Chemistry of Co and Mo Citrate in Solution and Upon Adsorption-Deposition on Alumina Extrudate

It is currently well-established that the stoichiometric ratios of Mo to citrate (2:1, 1:1, 1:2) is strongly dependent upon solution pH and most likely Mo concentration (Klimov., 2010; Nikulshin, 2012). A similar effect has been also determined for Co citrate complex, which was observed at pH 5 and 9 as at pH 1 only $[Co(H_2O)_6]^{2+}$ and citric acid were detected. Unfortunately, a similar study for Co and Mo citrate-containing solution is not available in the open literature. Nevertheless, a recent report revealed that a bimetallic CoMo citrate complex could be present in aqua solution when pH is between 1.5 and 3.5, according to ^{17}O and ^{13}C NMR results. A high HDS activity for straight run gas oil of citrate-prepared CoMo catalyst was attributed to the particular composition (Co/Mo = 0.5) and stability of the CoMo citrate complex (Klimov, 2010; Nikulshin, 2012). Researchers from Cosmo Oil Co., claimed that the improved hydrotreating performance of citrate-prepared CoMoP formulation is due to the chelating effect of citrate on cobalt that preventing its sulphidation at low temperatures, facilitating the interaction of Co ions with MoS_2-like structure and hence the synergy between Mo and Co (Fujikawa, 2009; Fujikawa, 1998; Fujikawa, 2006; Yumoto, 1997). Hence in the following work, two different scenarios are considered in more detail.

3.9.1. Wet Impregnation of Co, Mo Citrate Solution (pH = 2-3) onto Alumina Extrudate Body

Protonation and deproptonation of the surface hydroxyl groups are the common reactions when alumina is wetted into an aqueous solution as is illustrated in Figure 15.

The point of zero charge (PZC) of alumina is between 8 and 9 pH unit. The presence of water molecule on alumina surface increases the solution pH because of pH buffer effect (eq. 1).

$$Al_SOH + H^+ \rightarrow Al_SOH_2^+ \quad (1)$$

Figure 15. Dependence of the surface potential with solution pH for alumina support

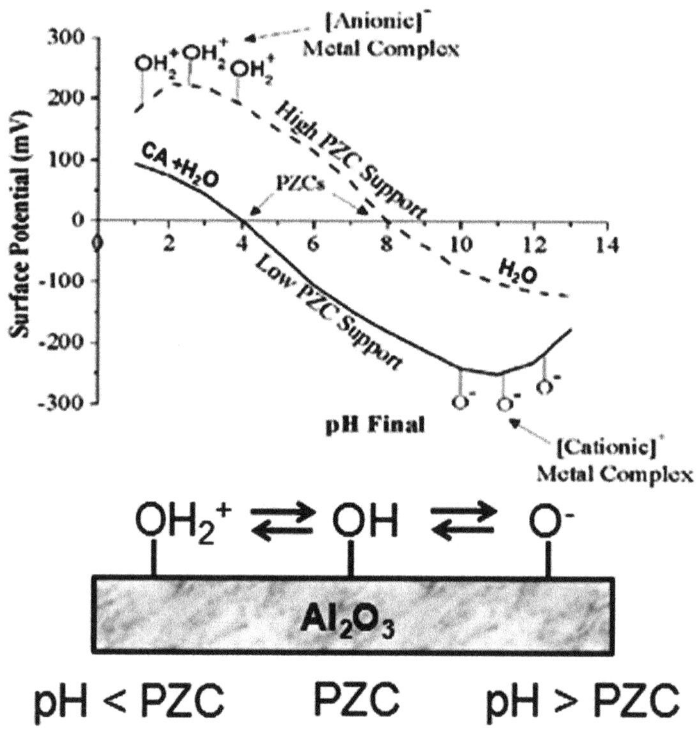

When alumina is wetted into a solution of citric acid (i.e. $CitrH_4$) at low pH (below 5), citrate anion can react with alumina surface, causing a highly negative charged surface and a shift of the PZC to lower pH values (Fig. 16) (Klimov, 2010). This interaction can take place on Lewis acid sites of alumina to release a molecule of water (eq. 2).

$$CitrH^{3-} + Al_SOH_2^+ \rightarrow [Al_S - CitrH]^{2-} + H_2O \quad (2)$$

On the other hand, when one considers a citric acid-containing Co (II) solution at low pH (below 5) it is expected that $[Co(H_2O)_6]^{2+}$ and citrate are the major ionic species present in aqua solution . Alumina can interacts with $[Co(H_2O)_6]^{2+}$ complex through water ligands exchange and support OH group to lead adsorption of Co (ii) aqua-complex on alumina surface (eq. 3).

$$\left[Co(H_2O)_6\right]^{2+} + Al_SOH \leftrightarrow \left[Al_SO-(H_2O)_5\right]^+ + H_3O^+ \quad (3)$$

It should be noted that according to eqs. 2 and 3, $[Co(H_2O)_6]^{2+}$ complex may have a weaker interaction that citrate with alumina surface.

Whether citric acid-containing Mo (VI) solution is considered, a variety of Mo citrate complex can be formed as consequence of variation in pH and Mo concentration. Note that Mo concentration in this work is between 1.7 and 2.0 mol/L and solution pH is between 2 and 3, so one would expect to get a similar complex in the impregnation solution. This complex can react with alumina surface to lead the adsorption of $[Mo_4O_{11}(CitrH)_2]^{4-}$ complex on alumina surface (eq. 4).

$$\left[Mo_4O_{11}(CitrH)2\right]^{4-} + Al_SOH_2^+ \rightarrow \left[Al_S\text{-}Mo_4O_{11}(CitrH)_2\right]^{3-} + H_2O \tag{4}$$

Upon the preparation of alumina-supported CoMo catalysts by wet impregnation a further complicated tri-component system (i.e. Co, Mo, citrate) is used. In this system a possible interaction of Mo citrate with Co (ii) aqua-complex to produce $Co_2[Mo_4O_{11}(CitrH)_2].H_2O$ should be considered, according to previous report. However, when one takes into account that possible CoMo precursor one can argues for the presence of Mo citrate and $[Co(H_2O)_6]^{2+}$ complexes in aqua solution, according to eq. 5.

$$\left[Co(H_2O)_6\right]_2\left[Mo_4O_{11}(CitrH)_2\right] \rightarrow \left[Mo_4O_{11}(CitrH)_2\right]^{4-} + 2\left[Co(H_2O)_6\right]^{2+} \tag{5}$$

These complexes may follow an adsorption process described by eqs. 3 and 4 if there is not free citrate in solution that may compete for acidic OH groups with Mo citrate complex. Indeed, in our catalyst preparation a CA/(Co+Mo) molar ratio of ca. 0.67 was used as Mo citrate complex has a molar ratio of 0.33, also under this pH (2-3) citrate appears to be free rather than forming a complex with Co (II). It is reasonably to conclude that citric acid could compete with $[Mo_4O_{11}(CitrH)_2]^{4-}$ complex for basic OH groups. The adsorption of these large anions on alumina surface can hinder the adsorption of $[Co(H_2O)_6]^{2+}$ complex because of steric effect (Klimov et al., 2010)

It is should be noted that MoO_3/CoO mass ratio is larger than nominal ratio. This Co and Mo analysis clearly indicates that upon impregnation process Co (II)-alumina interaction is weaker than Mo (II)-alumina interaction, owing to the favourable electrostatic interaction of Mo(VI) species with alumina surface. The type of alumina also appear to affect Mo(VI) adsorption. According to these results, catalyst composition needs to be changed by increasing Co/Mo ratio in the impregnation solution to get an appropriate Co/Mo ratio deposited on alumina surface and/or CA concentration can be reduced to improve Co adsorption. Note that after 100 h of time on steam, the catalyst prepared using single step impregnation with 4.1CoO-17.6MoO_3/Al_2O_3 (Y-43) showed superior HDS-HDN activities to the one with Y-27 and also industrial benchmark catalyst despite its lower Co and Mo loadings. This preparation method produces a CoMo formulation with superior performance that benchmark catalyst, despite having significantly lower Co and Mo loadings. Citrate-prepared CoMo catalysts showed fairly similar MoO_3/CoO mass ratio and slightly different CoMo loading. It is expected that such small variations may not cause such differences in residual S and N contents (see Table 16). Support textural properties may play an important role in this behaviour. Future work will be focused in this issue.

The $[Co(H_2O)_6]^{2+}$ complex seems to interact preferentially with surface neutral OH groups in the formation of inner-sphere Co(II) surface complexes (eq. 3) as $[Mo_4O_{11}(CitrH)_2]^{4-}$ complex prefers surface basic OH groups through electrostatic interaction (eq. 4). So, the adsorption of these complexes can be apparently governed by the surface density of these OH groups on alumina surface and solution pH. Considering a typical value of ionizable Al-OH group of 0.5 μmol/m^2 and an alumina surface area of

Table 16. Compositions of Co and Mo determined by ICP-MS for alumina-supported CoMo catalysts prepared by wet co-impregnation. Tests conducted at 340 °C, 35 bar, 250 NL/L (H_2/Oil ratio) and 1.2 h^{-1}

Sample	Nominal Composition			Measured Composition by ICP								
				Dried Ample		Calcined Sample						
	MoO_3	CoO	MoO_3/CoO Ratio	MoO_3	CoO	MoO_3	CoO	P_2O_5	MoO_3/ CoO Ratio	S (ppm)	N (ppm)
Y-24	17.4	4.1	4.2	16.2	3.3	23.1	4.7		4.9	150	14
Y-27	17.1	4.2	4.1	14.7	3.1	21.0	4.4		4.7		
Industrial bench mark						23.1	4.1	6.8	5.8	123	4

250 m^2/g, a total weight loading of ca. 6% MoO_3 on alumina would be expected (Al-Dalama, 2005; Hu, 1995; Lapidus, 2010; Luthra, 1987; Santhanam, 1994; Sarrazin, 1989; Wang, 1980; Weigold, 1983). Note that it is assumed Mo is exclusively deposited on alumina by adsorption (i.e. no precipitation). This Mo loading is significantly lower than that listed in Table 16, most likely due to the precipitation of Co and Mo citrate upon impregnation process. Indeed, the use of high concentrated CoMo citrate solution (supersaturate solution) can be apparently destabilized by alumina extrudate inducing metal oxide precipitation (alumina-assisted CoMo citrate precipitation).

This phenomenon (i.e. alumina-assisted CoMo citrate precipitation) appears to be quite severe, particularly after multiple impregnations on alumina body and also at long period of impregnation (above 4-5 h) with high concentrated CoMo citrate solution as is listed in Table 17. It is worth remarking that low Co, Mo and citric acid (CA) concentration impregnation was carried out for 24 h as high Co, Mo and CA concentration test was done for 5 h because of precipitation issue for longer period. After soaking 3g of each sample in 200 ml H_2O over 18 h, the soaked samples were then dried and the densities compared. It is clearly illustrated that nearly 80% of the deposited Co, Mo and CA on alumina body is removed after soaking process regardless of the Co, Mo and CA concentrations. Two important findings can be extracted from this experiment:

- Co and Mo citrate are highly soluble in deionised water.
- Nearly 80% of the material deposited on alumina is due to precipitation process rather than adsorption process.

Table 17. Densities and nominal compositions of Co, Mo and citrate post impregnation and post soaking processes

Mo (Mol/L)	Co (Mol/L)	CA (Mol/L)	Post-Impregnation		Post-Soaking	
			Density (g/ml)	Total Co, Mo and Citrate Loading (wt%)	Density (g/cc)	Total Co, Mo and Citrate Loading (wt%
1.62	0.77	1.6	0.82	37	0.57	9
2.55	1.22	2.52	0.92	44	0.58	10

3.9.2. Catalyst Preparation using Incipient Wetness Impregnation (IWI) of OA-Containing AHM Solution onto Alumina Extrudate Body Followed by IWI of Co Citrate Solution

Ammonium heptamolybdate (AHM) solution in water normally shows a pH between 5 and 6 and heptamolybdate anion ($Mo_7O_{24}^{6-}$) as the predominant specie. At higher pH, the formation of molybdate anion is facilitated, according to eq. 6.

$$Mo_7O_{24}^{6-} + H_2O \leftrightarrow 7MoO_4^{2-} + 8H^+ \quad (6)$$

Upon impregnation process, the reaction of protonation of the surface hydroxyl groups takes place (eq. 1), shifting the equilibrium of eq. 6 toward the formation of molybdate anion. There have been widely studied on the deposition of heptamolybdate and molybdate species onto alumina extrudate, as has been reported elsewhere (Al-Dalama, 2005; Hu, 1995; Luthra, 1987; Santhanam, 1994; Sarrazinl., 1989; Wang, 1980; Weigold, 1983). Herein, it was used urea, glycine and L-cysteine as organic additive upon AHM impregnation onto alumina extrudate. Urea and glycine (isoelectric point is ca. 6) may not affect very much the equilibrium reaction at room temperature (eq. 6) however these molecules could alter the interaction of molybdenum species with alumina and therefore their distribution/dispersion onto alumina extrudate, which was reflected on an improvement of the catalyst performance (see Table 12). On the other hand, L-cysteine can alter drastically the eq. 6, owing to its interaction with molybdate species to produce $MoOS_x$ species. This effect appears to be negative on catalyst performance (Table 12).

After deposition of molybdenum (VI) on alumina, the sample was calcined at 250 °C to partially decompose ammonium heptamolybdate. This would affect drastically the charge on the alumina surface on aqua solution since pH dropped 3 units after depositing Mo(VI) oxo species on alumina. Note that pure Al_2O_3 has pH at 8 as 15 wt. % MoO_x/Al_2O_3 displayed 5 pH. Upon addition of cobalt citrate solution to Mo-modified alumina two scenarios can be considered:

1. When a cobalt citrate impregnation solution at 3 pH is used; the positive charges on Mo-modified alumina surface (eq. 1) can facilitate the reaction of free citrate with Lewis acidic sites (eq. 2), weakling the interaction of $[Co(H_2O)_6]^{2+}$ complex with alumina. This effect could facilitate an eggshell distribution of Co (ii) into alumina extrudate, according to previous results (van de Water, 2005). It is worth mentioning that $[Co(H_2O)_6]^{2+}$ complex through water ligands exchange and support OH group could lead adsorption of Co (ii) aqua-complex on alumina surface (eq. 3). Another process that should be also considered it is the possible dissolution of Mo (VI) oxo species to favor $[H_xMo_yO_x]^{n-}$ during Co citrate impregnation, which would alter Mo (VI) dispersion onto extrudate body.
2. When cobalt citrate impregnation solution (pH = 9) is carried out by using ammonium citrate a marked increase in the solution viscosity was observed, owing to the formation of $[Co(Citr)(H_2O)_3]^{2-}$complex and/or $[Co(Citr)(H_2O)_{3-x}(NH_3)_x]^{2}$. $[Co(Citr)(H_2O)_3]^{2-}$ complex can react with OH groups through water ligands exchange and support OH group to lead adsorption of Co (ii) complex on alumina surface (eq. 7) (van de Water et al., 2005)

$$\left[Co(Citr)(H_2O)_3\right]^{2-} + Al_SOH \rightarrow \left[Al_SO\text{-}Co(Citr)(H_2O)_2\right]^{3-} + H_3O^+ \quad (7)$$

Apparently upon the Co impregnation the high pH can facilitate dissolution of Mo (VI) oxo species deposited on alumina, which could alter the Mo distribution into extrudate body.

4. CONCLUSION

- The preparation parameters has significant effect on the HDS catalyst performance, including the soaking time, organic addition, where citric acid is found to be the most favourable agent. The viscosity of citrate-containing CoMo Impregnation solution at low temperature (i.e. 17-18 °C) was ca. 12 mPa.s, which is significantly higher than dynamic viscosity of pure water (ca. 1 mPa.s). During impregnation of alumina the solution temperature dropped and the dynamic viscosity dropped to 10.8 mPa.s after nearly 2 h of impregnation time. With an impregnation solution temperature of 38° C the dynamic viscosity was 10.5-11 mPa.s. After an hour of impregnation time the solution viscosity did not change significantly.
- The wet co-impregnation (WCI) method apparently facilitates a more uniform Co, Mo citrate distribution onto alumina than sequential incipient wetness impregnation (SIWI) method. This is reflected in a higher HDS activity of WCI-prepared samples compared to SIWI-prepared catalyst. The co-impregnation method has been used in the pilot plant catalyst preparation and showed reproducible results.
- The presence of citric acid as organic additive/dispersing agent/chelating agent in the impregnation solution improved HDS activity compared to equivalent CoMo catalyst prepared without citric acid. Apparently the presence of citric acid facilitates the formation of amorphous Co and Mo species on alumina surface and hinders the formation of mixed-metal oxides. According to cobalt precursor the following decreasing HDS and HDN activities order was obtained: Co citrate > Co acetate > Co nitrat. The Citrate-prepared CoMo catalysts showed 20-30% superior activity on a weight basis for HDS than the industrial benchmark catalyst.
- In the impregnation stage, adding organic additives such as urea and glycine to Mo impregnation did not alter HDS conversion. One additive, L-cysteine decreased HDS activity. The presence of organic additives such as citric acid, ethylene glycol, polyethylene glycol and glycerol in the cobalt impregnation solution did improve significantly not only HDS but also HDN activities. Citrate-prepared CoMo catalyst deactivated with time on stream (ca. 70% relative to initial residual S concentration). Increasing reaction temperature also seems to decrease catalyst stability. The cetane index (CI) of the liquid hydrotreated product did increase slightly, relative to FCC diesel feed, as rising reaction temperature. This is an additional advantage of CoMo catalyst for improving diesel property (i.e. cetane number).
- Partial substitution of cobalt oxide for nickel oxide in alumina-supported CoMo catalyst decreased HDS conversion with Real FCC diesel feed. According to model feed data, HDS conversion did not vary as HDN activity increased with rising NiO loading. On the other hand, the addition of low loading of WO_3, P_2O_5 and NiO (i.e. 0.7-1.0 wt. %) to CoMo catalyst did not affect HDS performance. HDN conversion was clearly improved particularly when FCC diesel feed was used.

REFERENCES

Abro, R., Abdeltawab, A. A., Al-Deyab, S. S., Yu, G., Qazi, A. B., Gao, S., & Chen, X. (2014). A review of extractive desulfurization of fuel oils using ionic liquids. *RSC Adv.*, 4(67), 35302-35317. doi:10.1039/C4RA03478C

Al-Dalama, K., Aravind, B., & Stanislaus, A. (2005). Influence of complexing agents on the adsorption of molybdate and nickel ions on alumina. *Appl. Catal.*, A, 296(1), 49-53. doi:10.1016/j.apcata.2005.07.041

Al-Dalama, K., & Stanislaus, A. (2006). A Comparative Study of the Influence of Chelating Agents on the Hydrodesulfurization (HDS) Activity of Alumina and Silica-Alumina-Supported CoMo Catalysts. *Energy Fuels*, 20(5), 1777-1783. doi:10.1021/ef060125a

Al-Megren, H. A. (2009). Hydrodesulfurization of thiophene over bimetallic Ni-Mo sulfide catalysts prepared by different methods. *Arabian J. Sci. Eng. Sect. A*, *34*(1A), 55–66.

Ali, S. A. (2014). *Development of improved catalysts for deep HDS of diesel fuels.* Appl. Petrochem. Res. Doi: doi:10.1007/s13203-014-0082-x

Ali, S. A., Ahmed, S., Ahmed, K. W., & Al-Saleh, M. A. (2012). Simultaneous hydrodesulfurization of dibenzothiophene and substituted dibenzothiophene over phosphorus modified CoMo/Al2O3 catalysts. *Fuel Process. Technol.*, 98, 39-44. doi: 10.1016/j.fuproc.2012.01.027

Banerjee, S., Mani, K., Leonard, L., & Kokayeff, P. (2011). Distillate hydrotreating to ultra-low-sulfur diesel - the impact of aromatics. *Indian Chem. Eng.*, 53(3), 152-169. doi:10.1080/00194506.2011.696372

Bej, S. K., Maity, S. K., & Turaga, U. T. (2004). Search for an Efficient 4,6-DMDBT Hydrodesulfurization Catalyst: A Review of Recent Studies. *Energy Fuels*, 18(5), 1227-1237. doi:10.1021/ef030179+

Bentaleb, F., Che, M., Dubreuil, A.-C., Thomazeau, C., & Marceau, E. (2014). Influence of organic additives on the properties of impregnation solutions and on nickel oxide particle size for Ni/Al2O3 catalysts. *Catal. Today*, 235, 250-255. doi:10.1016/j.cattod.2014.02.020

Bezergianni, S., Dimitriadis, A., & Meletidis, G. (2014). Effectiveness of CoMo and NiMo catalysts on co-hydroprocessing of heavy atmospheric gas oil-waste cooking oil mixtures. *Fuel*, 125, 129-136. doi:10.1016/j.fuel.2014.02.010

Chen, T.-M., Wang, C.-M., Wang, I., & Tsai, T.-C. (2010). Promoter effect of vanadia on Co/Mo/Al2O3 catalyst for deep hydrodesulfurization via the hydrogenation reaction pathway. *J. Catal.*, 272(1), 28-36. doi:10.1016/j.jcat.2010.03.015

Chen, Y., Wang, L., Zhang, Y., Liu, T., Liu, X., Jiang, Z., & Li, C. (2014). A new multi-metallic bulk catalyst with high hydrodesulfurization activity of 4,6-DMDBT prepared using layered hydroxide salts as structural templates. *Appl. Catal.*, A, 474, 69-77. doi: 10.1016/j.apcata.2013.09.002

Denisov, A. A., Makarenko, V. A., & Shamrai, A. A. (1996). Impregnation as a method for control of the distribution of active components along the depth of a solid ceramic catalyst support, 3. Solutions and catalysts. Ekotekhnol. Resursosberezhenie(5-6), 55-61.

Duan, A., Fan, H., Zhao, Z., Jiang, G., & Liu, J. (2011). Hydrodesulfurization performances of NiMo supported on micro/mesoporous beta/MCM-48 composite catalyst. *Adv. Mater. Res.* (Durnten-Zurich, Switz.), 233-235(Pt. 1, Fundamental of Chemical Engineering), 827-832. doi:10.4028/www.scientific.net/AMR.233-235.827

Duan, A., Wan, G., Zhang, Y., Zhao, Z., Jiang, G., & Liu, J. (2011). Optimal synthesis of micro/mesoporous beta zeolite from kaolin clay and catalytic performance for hydrodesulfurization of diesel. *Catal. Today*, 175(1), 485-493. doi:10.1016/j.cattod.2011.03.044

Escobar, J., Barrera, M. C., Toledo, J. A., Cortes-Jacome, M. A., Angeles-Chavez, C., Nunez, S., . . . Pacheco, J. G. (2009). Effect of ethyleneglycol addition on the properties of P-doped NiMo/Al2O3 HDS catalysts: Part I. Materials preparation and characterization. *Appl. Catal.*, B, 88(3-4), 564-575. doi:10.1016/j.apcatb.2008.10.005

Fan, Y., Xiao, H., Shi, G., Liu, H., Qian, Y., Wang, T., . . . Bao, X. (2011). Citric acid-assisted hydrothermal method for preparing NiW/USY-Al2O3 ultradeep hydrodesulfurization catalysts. *J. Catal.*, 279(1), 27-35. doi:10.1016/j.jcat.2010.12.014

Fujikawa, T. (2006). Highly active CoMo hydrodesulfurization catalyst for the production of clean diesel fuels. *Catal. Surv. Asia*, 10(2), 89-97. doi:10.1007/s10563-006-9006-9

Fujikawa, T. (2007). Development of new CoMo HDS catalyst for ultra-low sulfur diesel fuel production. *J. Jpn. Pet. Inst.*, 50(5), 249-261. doi:10.1627/jpi.50.249

Fujikawa, T. (2009). Highly Active HDS Catalyst for Producing Ultra-low Sulfur Diesel Fuels. *Top. Catal.*, 52(6-7), 872-879. doi:10.1007/s11244-009-9228-y

Fujikawa, T., Chiyoda, O., Tsukagoshi, M., Idei, K., & Takehara, S. (1998). Development of a high activity HDS catalyst for diesel fuel: from basic research to commercial experience. *Catal. Today*, 45(1-4), 307-312. doi:10.1016/S0920-5861(98)00235-1

Fujikawa, T., Kato, M., Ebihara, T., Hagiwara, K., Kubota, T., & Okamoto, Y. (2005). Development of highly active Co-Mo catalysts with phosphorus and citric acid for ultra-deep desulfurization of diesel fractions (part 2) characterization of active sites. *J. Jpn. Pet. Inst.*, 48(2), 114-120. doi:10.1627/jpi.48.114

Fujikawa, T., Kimura, H., Kiriyama, K., & Hagiwara, K. (2006). Development of ultra-deep HDS catalyst for production of clean diesel fuels. *Catal. Today*, 111(3-4), 188-193. doi:10.1016/j.cattod.2005.10.024

Gao, Q., Ofosu, T. N. K., Ma, S.-G., Komvokis, V. G., Williams, C. T., & Segawa, K. (2011). Catalyst development for ultra-deep hydrodesulfurization (HDS) of dibenzothiophenes. I: Effects of Ni promotion in molybdenum-based catalysts. *Catal. Today*, 164(1), 538-543. doi:10.1016/j.cattod.2010.10.016

Gonzalez-Cortes, S. L., Rugmini, S., Xiao, T., Green, M. L. H., Rodulfo-Baechler, S. M., & Imbert, F. E. (2014). Deep hydrotreating of different feedstocks over a highly active Al2O3-supported NiMoW sulfide catalyst. *Appl. Catal.*, A, 475, 270-281. doi:10.1016/j.apcata.2014.01.045

Gonzalez-Cortes, S. L., Xiao, T.-C., Costa, P. M. F. J., Fontal, B., & Green, M. L. H. (2004). Urea-organic matrix method: an alternative approach to prepare Co-MoS2/γ-Al2O3 HDS catalyst. Appl. *Catal., A*, 270(1-2), 209-222. doi:10.1016/j.apcata.2004.05.006

Ho, T. C. (2004). Deep HDS of diesel fuel: chemistry and catalysis. *Catal. Today*, 98(1-2), 3-18. doi:10.1016/j.cattod.2004.07.048

Hu, H., Wachs, I. E., & Bare, S. R. (1995). Surface Structures of Supported Molybdenum Oxide Catalysts: Characterization by Raman and Mo L3-Edge XANES. *J. Phys. Chem.*, 99(27), 10897-10910. doi:10.1021/j100027a034

Infantes-Molina, A., Romero-Perez, A., Sanchez-Gonzalez, V., Pawelec, B., Fierro, J. L. G., Jimenez-Lopez, A., & Rodriguez-Castellon, E. (2011). Role of Cs on Hydrodesulfurization Activity of RuS2 Catalysts Supported on a Mesoporous SBA-15 Type Material. *ACS Catal.*, 1(3), 175-186. doi:10.1021/cs100053e

Ismagilov, Z. R., Yashnik, S. A., Startsev, A. N., Boronin, A. I., Stadnichenko, A. I., Kriventsov, V. V., . . . Moulijn, J. A. (2009). Deep desulphurization of diesel fuels on bifunctional monolithic nanostructured Pt-zeolite catalysts. *Catal. Today*, 144(3-4), 235-250. doi:10.1016/j.cattod.2009.01.030

Jess, A., Wasserscheid, P., & Esser, J. (2003). Hydrogen-free deep desulfurization of diesel oil by extraction with ionic liquids. *Proceedings of the DGMK-Conference on Innovation in the Manufacture and Use of Hydrogen* (pp. 313-320).

Klimov, O. V., Pashigreva, A. V., Fedotov, M. A., Kochubey, D. I., Chesalov, Y. A., Bukhtiyarova, G. A., & Noskov, A. S. (2010). Co-Mo catalysts for ultra-deep HDS of diesel fuels prepared via synthesis of bimetallic surface compounds. *J. Mol. Catal. A: Chem.*, 322(1-2), 80-89. doi:10.1016/j.molcata.2010.02.020

Lapidus, A. L., Mikhailov, M. N., & Kustov, L. M. (2010). Interaction of a Co6 cluster with oxides of different nature: a quantum chemical study. *Russ. Chem. Bull.*, 58(2), 273-279. doi:10.1007/s11172-010-0001-y

Luthra, N. P., & Cheng, W. C. (1987). Molybdenum-95 NMR study of the adsorption of molybdates on alumina. *J. Catal.*, 107(1), 154-160. doi:10.1016/0021-9517(87)90280-6

Macias Hernandez, M. J., Morales, R. D., & Ramirez-Lopez, A. (2009). Simulation of the effectiveness factor for a tri-lobular catalyst on the hydrodesulfurization of diesel. *Int. J. Chem. React. Eng.*, 7. doi:10.2202/1542-6580.1806

Mirodatos, C., Brum Pereira, E., Gomez Cobo, A., Dalmon, J. A., & Martin, G. A. (1995). CO hydrogenation over Ni- and Co-based catalysts: influence of alkali addition on morphological and catalytic properties. *Top. Catal.*, 2(1-4, Fischer-Tropsch and Methanol Synthesis), 183-192. doi:10.1007/BF01491966

Nava, J. A. O., & Krishna, R. (2004). In-situ stripping of H2S in gasoil hydrodesulphurization reactor design considerations. *Chem. Eng. Res. Des.*, 82(A2), 208-214. doi:10.1205/026387604772992783

Nava, R., Ortega, R. A., Alonso, G., Ornelas, C., Pawelec, B., & Fierro, J. L. G. (2007). CoMo/Ti-SBA-15 catalysts for dibenzothiophene desulfurization. *Catal. Today*, 127(1-4), 70-84. doi:10.1016/j.cattod.2007.02.034

Navarrete, C., Garcia, R., Sepulveda, C., Gil-Llambias, F. J., Fierro, J. L. G., & Escalona, N. (2011). Deep Hydrodesulphurization via Hydrogen Spillover. *Catal. Lett.*, 141(12), 1796-1802. doi:10.1007/s10562-011-0717-9

Nikulshin, P. A., Mozhaev, A. V., Pimerzin, A. A., Konovalov, V. V., & Pimerzin, A. A. (2012). CoMo/Al2O3 catalysts prepared on the basis of Co2Mo10-heteropolyacid and cobalt citrate: Effect of Co/Mo ratio. *Fuel*, 100, 24-33. doi:10.1016/j.fuel.2011.11.028

Ramirez, J., Macias, G., Cedeno, L., Gutierrez-Alejandre, A., Cuevas, R., & Castillo, P. (2004). The role of titania in supported Mo, CoMo, NiMo, and NiW hydrodesulfurization catalysts: analysis of past and new evidences. *Catal. Today*, 98(1-2), 19-30. doi:10.1016/j.cattod.2004.07.050

Rashidi, F., Sasaki, T., Rashidi, A. M., Nemati Kharat, A., & Jozani, K. J. (2013). Ultradeep hydrodesulfurization of diesel fuels using highly efficient nanoalumina-supported catalysts: Impact of support, phosphorus, and/or boron on the structure and catalytic activity. *J. Catal.*, 299, 321-335. doi:10.1016/j.jcat.2012.11.012

Robinson, W. R. A. M., van Veen, J. A. R., de Beer, V. H. J., & van Santen, R. A. (1999). Development of deep hydrodesulfurization catalysts I. CoMo and NiMo catalysts tested with (substituted) dibenzothiophene. *Fuel Process. Technol.*, 61(1-2), 89-101. doi:10.1016/S0378-3820(99)00033-8

Santhanam, N., Conforti, T. A., Spieker, W., & Regalbuto, J. R. (1994). Nature of metal catalyst precursors adsorbed onto oxide supports. *Catal. Today*, 21(1), 141-156. doi:10.1016/0920-5861(94)80040-5

Sarrazin, P., Mouchel, B., & Kasztelan, S. (1989). Molybdenum-95 NMR study of the interaction of heptamolybdate solutions with alumina and silica. *J. Phys. Chem.*, 93(2), 904-908. doi:10.1021/j100339a069

van de Water, L. G. A., Bergwerff, J. A., Leliveld, B. G., Weckhuysen, B. M., & de Jong, K. P. (2005). Insights into the Preparation of Supported Catalysts: A Spatially Resolved Raman and UV-Vis Spectroscopic Study into the Drying Process of CoMo/γ-Al2O3 Catalyst Bodies. *J. Phys. Chem. B*, 109(30), 14513-14522. doi:10.1021/jp051037e

Van Dillen, A. J., Terorde, R. J. A. M., Lensveld, D. J., Geus, J. W., & De Jong, K. P. (2003). Synthesis of supported catalysts by impregnation and drying using aqueous chelated metal complexes. *J. Catal.*, 216(1-2), 257-264. doi:10.1016/S0021-9517(02)00130-6

Wan, G., Duan, A., Zhang, Y., Zhao, Z., Jiang, G., Zhang, D., . . . Chung, K. (2010). NiW/AMBT catalysts for the production of ultra-low sulfur diesel. *Catal. Today*, 158(3-4), 521-529. doi:10.1016/j.cattod.2010.08.021

Wang, A., Wang, Y., Kabe, T., Chen, Y., Ishihara, A., Qian, W., & Yao, P. (2002). Hydrodesulfurization of Dibenzothiophene over Siliceous MCM-41-Supported Catalysts. *J. Catal.*, 210(2), 319-327. doi:10.1006/jcat.2002.3674

Wang, L., & Hall, W. K. (1980). On the genesis of molybdena-alumina catalyst. *J. Catal.*, 66(1), 251-255. doi: 10.1016/0021-9517(80)90029-9

Wang, L., Zhang, Y., Zhang, Y., Jiang, Z., & Li, C. (2009). Ultra-Deep Hydrodesulfurization of Diesel Fuels on Trimetallic NiMoW Sulfide Catalysts. *Chem. - Eur. J.*, 15(46), 12571-12575. doi:10.1002/chem.200901997

Weigold, H. (1983). The nature of molybdenum oxide species mounted on alumina: an oligomer model. *J. Catal.*, 83(1), 85-98. doi:10.1016/0021-9517(83)90032-5

Yoosuk, B., Song, C., Kim, J. H., Ngamcharussrivichai, C., & Prasassarakich, P. (2010). Effects of preparation conditions in hydrothermal synthesis of highly active unsupported NiMo sulfide catalysts for simultaneous hydrodesulfurization of dibenzothiophene and 4,6-dimethyldibenzothiophene. *Catal. Today*, 149(1-2), 52-61. doi:10.1016/j.cattod.2009.05.001

Yumoto, M., Usui, K., Watanabe, K., Idei, K., & Yamazaki, H. (1997). Development of a Cosmo deep HDS catalyst for diesel fuel. *Catal. Today*, 35(1-2), 45-50. doi:10.1016/S0920-5861(96)00133-2

Chapter 10
Desulphurization of Fuel Oils Using Ionic Liquids

Abdul Waheed Bhutto
Dawood University of Engineering and Technology, Pakistan

Tauqeer Abbas
COMSATS Institute of Information Technology, Pakistan

Rashid Abro
Beijing University of Chemical Technology, China

Guangren Yu
Beijing University of Chemical Technology, China

Xiaochun Chen
Beijing University of Chemical Technology, China

ABSTRACT

Hydrodesulphurization (HDS) is a standard process for removing sulphur compounds in fuel oils in industry. HDS is effective to remove simple aliphatic sulphur compounds while less effective to remove thiophenes, dibenzothiophenes, and their derivatives because of sterically hindered adsorption on catalyst surface. Application of ionic liquids (ILs, a new class of compounds) substituting for traditional volatile organic solvents in extractive desulphurization (EDS) or oxidative desulphurization (ODS), have been being studied intensively in the latest decades, and many very promising results have been obtained, showing a good prospect as complement method to HDS. In this chapter, these fresh research results of EDS and ODS using ILs are summarized along with comprehensive discussions on diversified desulphurization factors along with some potential problems. It can be inferred that ILs are a class of potential ideal solvents to realize clean fuel oil in future although some problems come too.

INTRODUCTION[1]

Deep desulphurization of fuels has become a very important research subject worldwide, due to environmental concern and the upcoming stricter sulphur legislative regulation. Oil refineries are finding ways to get around a bottleneck caused by the presence of heterocyclic sulphur compounds in fuels, which emits oxides of sulphur (SOx) as environmental pollutants during combustion (Mei et al, 2003). The task to minimize sulphur content in fuel has hastened in recent years due to growing environmental awareness from the public and more importantly, demanding statutory limit of these compounds in fuels. Europe,

DOI: 10.4018/978-1-4666-9975-5.ch010

Figure 1. Previous, current, and proposed sulphur limitations in diesel (a) and gasoline (b) in different countries

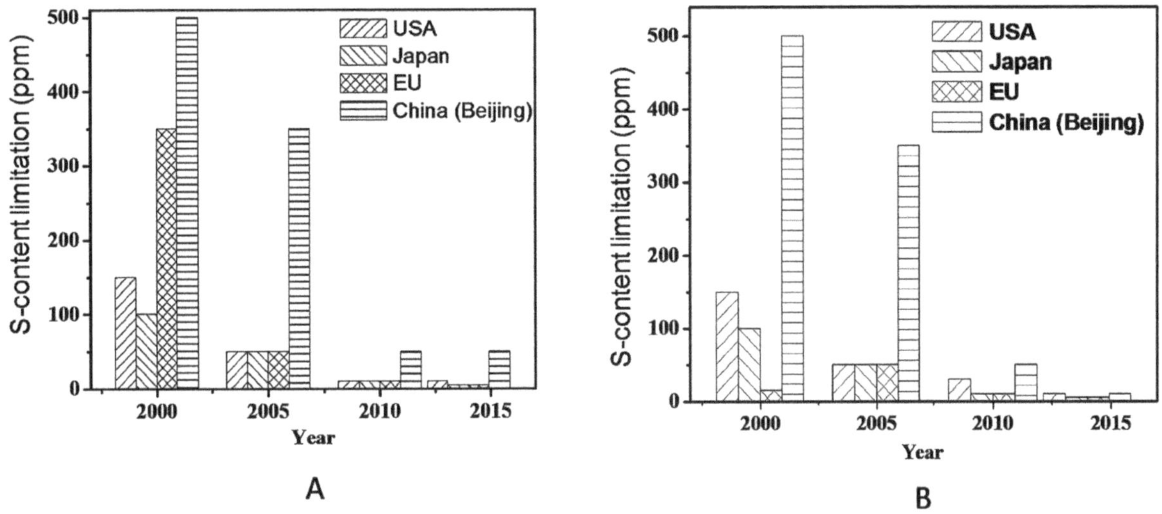

the US and other mature economies have been forced to lower the content in fuel through legislation (Figure 1a and 1a for previous, current and proposed sulphur levels in diesel and gasoline in developed countries). In 1998, the European directive on transportation fuels limited the sulphur content to 150 ppm for gasoline and 350 ppm for diesel and just five years later, that number was reduced to 50 ppm. Now, it stands at <10 ppm. The United States and Japan have put tax on transportation fuel exceeding 10 ppm Sulphur content (Essar et al, 2004).

Fuel oils contain variety of sulphur compounds. These sulphur compounds are both aliphatic (thiols, sulfides, disulfides etc) and aromatic (TS, BT, DBT and their derivatives).The structures of those sulphur compounds are shown in the Figure 2.

The hydrodesulphurization (HDS) method is widely used in industry for the desulphurization of diesel fuels, where sulphur compounds are catalytically converted into H_2S and corresponding hydrocarbons, which are subsequently separated from oils and catalytically oxidized into elemental sulphur in the Claus

Figure 2. Sulphur compounds in fuel oils

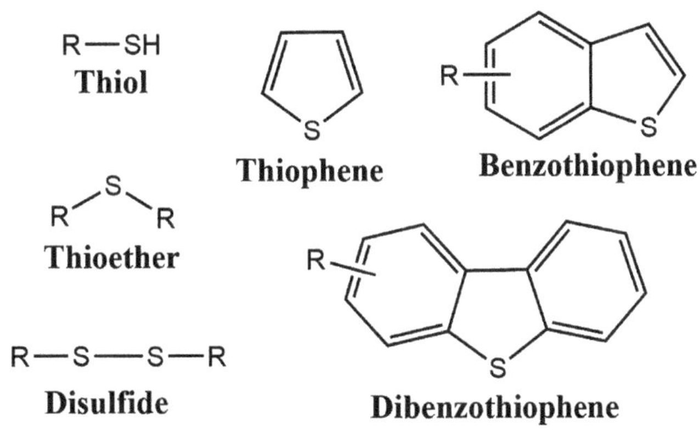

process. HDS achieved by catalytic processes operates at elevated temperatures (300 °C to 340 °C) and pressures (20 to 100 atm of H_2) with Co-Mo/Al_2O_3 or the Ni-Mo/Al_2O_3 catalyst (Kabe et al, 1993 & Pedarnera et al, 2003). The HDS process is highly efficient in removing thiols, sulfides, and disulfides but is less effective for aromatic thiophenes and thiophene derivatives. Thus, the sulphur compounds that remain in the transportation fuels are thiophene (TS), benzothiophene (BT), dibenzothiophene (DBT), and their alkylated derivatives. The reactor size of HDS unit needs to be increased by factors of 5 to 15. The severe operation conditions, expensive equipment, high hydrogen consumption are the other disadvantages hamper the application of HDS in deep desulphurization (F.-L. Yu, Wang, Liu, Xie, & Yu, 2014). Unfortunately, more severe conditions result not only in a higher level of desulphurization but also in undesired side reactions. When fluid catalytic cracking (FCC) gasoline is desulphurization at higher pressure, many olefins are saturated and the octane number decreases. Higher temperature processing leads to increased coke formation and subsequent catalyst deactivation. It is also important to note that in practice the severity of the operating conditions is limited by the HDS unit design (Babich & Moulijn, 2003 and Zhang et al, 2007).). Faced with the severely high costs of compliance, a surprising number of petroleum refiners are seriously considering reducing or eliminating production of transportation fuels (Yang, Hernández-Maldonado, & Yang, 2003). To address these drawbacks, the alternative methods such as extraction, oxidation, adsorption, and bio desulphurization are under considerations. Status and details about these technologies are mentioned in section "technologies for desulphurization of fuel oils".

Since refractory aromatic thiophenes, such as DBT and its derivatives are easily removed by the Oxidative Desulphurization (ODS) and Extractive Desulphurization (EDS) (Campos-Martin, Capel-Sanchez, Perez-Presas, & Fierro, 2010; Zhu et al., 2013), accordingly, These technologies are regarded as a promising strategy to achieve an ultralow sulphur level. In EDS, sulphur compounds are separated from fuel oils using solvents. ODS typically includes two steps. In first step, sulphides are oxidized into sulfoxides or sulfones by oxidant. In second step, these sulfoxides or sulfones are separated from fuel oils based on their increased polarity (Abro et al., 2014). Thus, desulphurization process includes two stages: (i) oxidation in a first step ;(ii) separation operations in a second step. Preferably, oxidized compounds can be extracted from the light oil by contacting oxidized light oil with a non-miscible polar solvent. The use of molecular organic solvents such as Methanol, Diethylene glycol (DEG), Triethylene glycol (TEG), Furfural, Furfual alcohol, Dimethyl sulfoxide (DMSO), Dimethyl formamide (DMF) and acetonitrile in EDS and ODS have been patented; some problems, however, exist in employing these organic solvents such as loss, contamination, regeneration, recycling and safety issues stemming from the volatility nature of organic solvents.

In the past few decades, Ionic liquids (ILs) in EDS and ODS has drawn increasing attention. Here the oxidized products are removed using ILs. Recent studies have used a series of acidic ILs as both extractant and catalyst (instead of organic acids) to extract and oxidize sulphur compounds in fuels.

Recently, ILs were widely applied to catalyze desulphurization reactions and separation sulphur compound from oil phase have shown positive results (C. Zhang, Pan, Wang, & Liu, 2012). IL-based technology has various advantages, such as wide liquid range, negligible vapor pressure, and high stability. ILs also show good extraction ability for aromatic sulphur-containing compounds, and are immiscible with aliphatic liquids such as fuel oil (Yansheng et al., 2011). In some cases, in order to increase the efficiency of sulphur removal, the extraction using ILs is combined with oxidation desulphurization process. ILs can be recycled number of times without a significant decrease in the sulphur removal. Homogeneous catalytic systems need large amount of ILs that are costly and may affect the economic viability of a potential process. The recovery of the ILs is another problem. To solve these problems, many groups have committed to heterogeneous catalysis system and turned to studying supported ILs. ILs, as green solvents have been studied widely recently due to their appealing properties such as large liquid range, high ionic conductivity, negligible vapor pressure, high thermal stability, large electrochemical window, and ability to solvate compounds of extensively varying polarity. Utilizing ionic liquids is one of the goals of green chemistry because they create a cleaner and more sustainable chemistry and are receiving increasing interest as environmental friendly solvents for many synthetic and catalytic processes (Docherty et al, 2005).

ALTERNATIVE TECHNOLOGIES FOR DESULPHURIZATION OF FUEL OILS

Adsorptive Desulphurization (ADS)

In ADS, sulphur compounds in fuel oils are removed through being adsorbed on one adsorbent that is typically positioned on non-reactive and porous substrate. ADS can be direct or reactive, depending on interaction nature between sulphur compound and adsorbent. The choice of adsorbent is one main factor influencing adsorptive efficiency. Different types of adsorbents were investigated, e.g., activated carbon (Ania et al, 2007), alumina (Larrubia et al, 2002), zeolites (Ngamcharussrivichai et al, 2008), zirconia (Baeza et al, 2008) and silica gel among which activated carbon is the mostly studied because of its effective surface area with good porosity. Lee and his co-workers investigated 10 type of activated carbon produced from different natural raw materials and observed that coconut-shell based activated carbon was more efficient than coal based and wood based activated carbon (Lee et al, 2002). S-Zorbwas developed by Conoco Phillips Petroleum Company, in which sulphur compounds were removed by reactive adsorption at high temperature and low hydrogen pressure. Table 1 shows different adsorbents used in ADS at optimized conditions.

One challenge in ADS is to find good adsorbents that selectively and efficiently adsorb sulphur compounds while not adsorb hydrocarbon and olefin; aromatic sulphur compounds, however, are similar to olefenic in some properties such as presence of more than two double bonds which atoms of metals can easily interact. Another challenge is the regeneration of adsorbents completely after adsorption. In addition, the adsorption efficiency of many adsorbents were found to be low, e.g., 1~4 mg sulphur compounds were adsorbed per g clay material which means huge amounts of adsorbent required.

Table 1. Different adsorbents used in ADS at optimized conditions (Abro et al., 2014)

Adsorbents	Model Oil	S-Comp.	Initial S: (ppm)	Temp. (K)	Pressure (atm)	S-Removal
Activated carbon	Gas oil	DBT	178	198	1	95%
	Gas oil	TS	300	243	1.5	88%
Alumina	Gas oil	DBT	700	393	n/a	30%
Zeolites from coal	Hexane	TS, BT	500	303	1	63%
$NiMoP/Al_2O_3$	Hexane	DBT	450	600	40	56%
Gallium +Y -zeolite	Nonane	DBT	500	333	n/a	97%
Cu-zirconia	Octane	TS	2000	180	n/a	99%
Ruthenium complexes	Hexane	DBT	40	298	1	55%

Table 2. BioDesulphurization by different microorganisms (Abro et al., 2014)

Microorganism	Model Oil	Initial S: (ppm)	Temp (K)	S-Removal
Gordoniaalkanivorans RIPI90A	Hexadecane	320	303	90%
Mycobacterium sp. ZD-19	Hexadecane	92	303	100%
Mycobacterium goodii X7B	Tertradecane	200	313	99%
Rhodococcuserythropolis IGTS8	Hexadecane	100	303	80%
Gordoniaalkanivorans strain 1B	Heptane	100	308	63%
Bacillus subtilis WU-S2B	Tridecane	100	323	50%
Pseudomonas stutzeri UP-1	Hexadecane	500	304	74%
Sphingomonassubarctica T7b	Gas oil	280	300	94%
Bacterium, strain RIPI-22	Hexadecane	100	303	77%
Pseudomonas delafieldii R-8	Diesel oil	591	303	47%

Bio Desulphurization (BDS)

BDS is considered as a green process, in which microorganisms are used to remove sulphur atom from organic compounds. The microorganisms have the natural ability to consume aromatic sulphur compounds as their energy source and metabolize organo-sulphur compounds by reductive Carbon–Sulphur bond cleavage, oxidative Carbon–Sulphur bond cleavage, or oxidative Carbon–Carbo bond cleavage. Generally, BDS takes place by two pathways: the Kodama pathway, in which the first attack is intended against one of the carbon atoms; the 4S pathway, in which early catalysis is directed against the axis of sulphur. Table 2 summarizes the results of BDS at optimized conditions.

Slow biodegradation and suitability of biocatalyst are the challenges in BDS. The logistics of sanitary storage, handling, shipment, storage, and use of living microbes within the refinery atmosphere are also problematic, because they are very sensitive in nature (Martin et al, 2010).

Oxidative Desulphurization (ODS)

ODS is able to remove sulphur compounds from fuel oils at mild conditions. ODS typically includes two steps. In first step, sulfides are oxidized into sulfoxides or sulfones by oxidant. In second step, these sulfoxides or sulfones are separated from fuel oils by solvent extraction based on their increased polarity. The used solvent is recovered through distillation. Nitrogen dioxide (NO_2) was used first time as oxidant in ODS followed by extraction with methanol. However, NO_2 led to the formation of huge quantity of residue and waste (Baxendale et al, 1946). Many other organic and inorganic acids such as formic, acetic and formic acids were used as catalysts in ODS technology, while hydrogen peroxide (H_2O_2) was used commonly in ODS as an oxidant. ILs as alternative to traditional organic solvents were also used in ODS.

Selectivity of oxidant, and extractive solvent are two challenges in ODS. Some oxidants resulted in unwanted side reactions that decrease the mass and quality of fuel oils. Selection of incorrect solvent may cause eliminating hydrocarbon compounds such as aromatic and olefin from the liquid (Ali et al, 2006).

Extractive Desulphurization (EDS)

As mentioned in introduction, EDS is a desirable method where desulphurization is performed at mild and simple conditions (low temperature, low pressure) without catalyst and hydrogen required. More important, EDS selectively extract sulphur compounds from fuel oils while does not destruct the other compounds in fuel oils, and the extracted compounds can be re-used as raw materials. EDS is based on different partition of sulphur compounds in oil phase and extractant phase. The choice of extractive solvent is one main factor in EDS. As mentioned above, there are a few problems when volatile solvents such as polyalkylene glycol, imidazolidinone, pyrimidinone and dimethylsulfoxide are used. The ideality of solvent depends on its efficiency, recyclability, and reusability, therefore the important assignment for researchers is to discover a technique of substituting volatile organic compounds with green solvents. ILs technology is attracting widespread attention in separation processes. Various types of ILs have been employed in recent studies involving as a solvent in EDS.

Desulphurization Using ILs

Ionic Liquids (ILs)

In the last few decades, ILs are classified as a vast group of chemical compounds, constituted entirely of ions (organic cation and organic or inorganic anion), which are liquid at, or close to, room temperature. Their ionic nature imparts unique properties to these liquids. Generally, they have high conductivity, high thermal stability, low flammability and very low (if any) volatility. This latter property reduces explosion risks and toxic effects for processes, two risky features that are usually linked with the use of volatile organic solvents (VOCs). The often-stressed "green" character of ILs is indeed largely related to their negligible vapour pressure and inability to produce air pollution (Cevasco et al, 2014). The efficiency of ILs for any process strongly depends on cation–anion structure: careful design or selection of the IL anion and cation should always be able to guarantee both high process performance and sustainability (Petkovic et al, 2011). It is noteworthy that the main advantage of ILs relies on the possibility to modify all their properties (*e.g.* viscosity, conductivity, hydrophilicity, hydrophobicity, polarity, hydrogen bond ability) through the selection of appropriate anion and cation combinations or through the introduction of specific functional groups on cation or anion (Chiappe et al, 2005 and Welton et al, 1999). Well-known cations are imidazolium, pyridiniuim, isoquinolonium, ammonium, phosphonium and sulfonium as shown in Figure 3.

ODS Using ILS

Recently, some researchers used functional Brønsted acidic ILs (Figure 4a) as catalysts instead of traditional organic acids (at the same time, the ILs are also extractants), and 30 wt % H_2O_2 as oxidant, to oxidatively remove the thiophenic sulphur compounds from diesel fuel, as shown in Figure 4b.

It was interestingly found that the Brønsted acidic ILs presented good catalytic activity and the thiophenic sulphur compounds can be almost completely removed from diesel fuel. As a typical example, for the [CH_2COOHMIM][NTf_2]/ H_2O_2 system where [CH_2COOHMIM] is 1-methyl-3-ethylcarboxylic acid imidazolium, the sulphur content was reduced from 1000 to <10 ppm with 99% Sulphur removal (IL, 1 mL; model oil composed of DBT and n-octane, 1 g; 30 wt % H_2O_2, 0.25 mL; >50 °C; 18 h) (Liss-

Figure 3. Different structures for IL cations

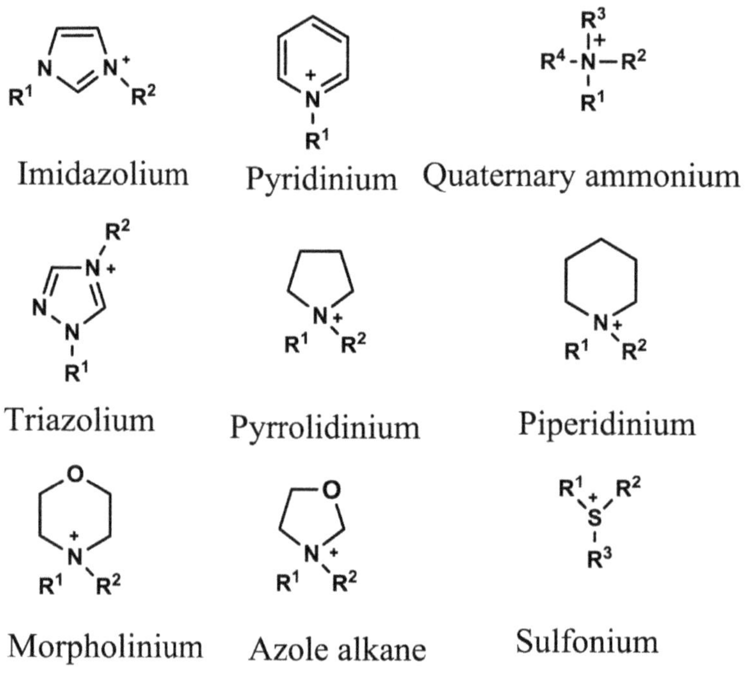

ner, de Souza, Ferrera, & Dupont, 2009). In such a method, the additional catalysts are not needed and the aforementioned problems from catalysts do not exist (G. Yu et al., 2011). For [BMIM][HSO_4]. The sulfur removal of DBT containing model diesel could reach 99.6% at room temperature after 90 min (H. Gao, 2010). The removal efficiency of different ILs is given in Figure 5. It is interestingly found that not only Brønsted acidic ILs but also Lewis acidic ILs are capable of effectively removing DBT in model diesel fuel. For the Brønsted acidic ILs, the sulphur removal ability follows the order [BMIM][HSO_4] > [SO_3H–BMIM][HSO_4] > [HMIM][HSO_4] > [CH_2COOHMIM][HSO_4] (Yu et al., 2011). A sulphur removal efficiency of 100% is obtained for[BMIM][HSO_4] after 7 h (G. Yu et al., 2011). Same results were obtained when a series of Lewis acidic ILs, dialkylpyridinium tetrachloroferrates [$C_4{}^3$MPy] ($FeCl_4$, [$C_6{}^3$MPy] $FeCl_4$, and [$C_8{}^3$MPy] $FeCl_4$) served as both extractant and catalyst, with 30 wt% H_2O_2

Figure 4. (A&B)

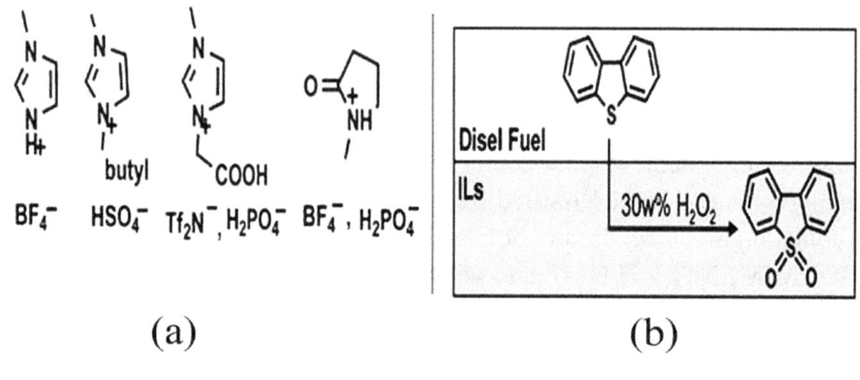

solution as oxidant in 10 min at 298 K. The ILs could be recycled twice with negligible activity loss. The kinetics of oxidative desulfurization of DBT by H_2O_2 and $[C_4{}^3MPy]$ $FeCl_4$ is first-order with an apparent rate constant of 0.9951 min^{-1} at 1/3 IL/oil mass ratio, and affected by the IL/oil mass ratios. $[C_8{}^3MPy]$ $FeCl_4$ was effective for the ODS of actual gasoline, with an S-removal of 44% at 1/3 IL/gasoline mass ratio (Yi Nie et al, 2013).

Recently, acidic ILs have led to very good results in ODS of fuels without the need of addition of an external catalyst to the medium, since the intrinsic acidic character of the ILs allows them to act as both catalysts and solvents. The acetic acid-based ILs are used as both phase-transfer catalysts and extractants, which showed good oxidative ability. It also eliminates the requirement of additional catalyst. One of its catalytic roles is to decompose hydrogen peroxide to form hydroxyl radicals. Cooper and co-workers (M. S. Cooper, Heaney, Newbold, & Sanderson, 1990) observed that a coordination compound could be generated between hydrogen peroxide and an amide, such as urea; then the hydrogen peroxide on the coordination compound decomposes to produce hydroxyl radicals that are strong oxidizing agents. Likewise, in the present case, DBT in the IL phase was oxidized tosulfone by hydroxyl radicals. sulfone does not exist in the oil phase because of the high polarity of IL. DBT sulfone was detected in the IL phase. Within 6 h of oxidation the DBT content had decreased from 758 to 7.8 ppm, indicating 99% extraction of DBT from the model light oil while mere extraction of DBT using IL $[BMIM][PF_6]$ – has shown only 47% extraction. These experimental findings demonstrate that a combination of chemical oxidation and solvent extraction in a water-immiscible IL has the potential to completely remove sulphur compounds from light oil in one-step. Kulkarni and co-workers (Kulkarni & Afonso, 2010) also applied this oxidation/extraction system to actual light oil containing a sulphur content of 8040 ppm. After 10 h, in the $[BMIM][PF_6]$ – extraction/oxidation system, the decrease of sulphur content was observed from 8040 to 1300 ppm.

The catalytic ODS of thiophene in n-heptane could reach 87.5% under the optimal reaction conditions. The used ILs were regenerated by re-extraction in dichloromethane, then dichloromethane was removed under vacuum. $[Otbim][CH_3COO]$ – containing eight-carbon side chain has of best catalytic activity (Liang, Zhang, Li, & Zhang, 2013).

Figure 5.

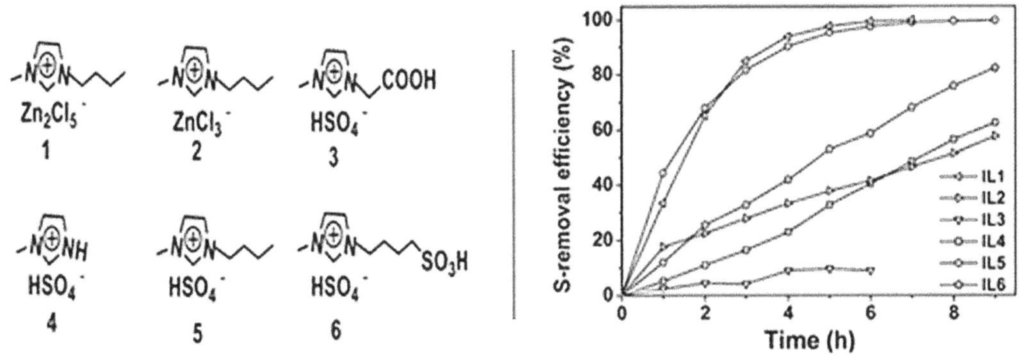

(a) (b)

Nonetheless, some of these ILs, for instance those with chloroaluminate anions, are extremely sensitive to moisture and must be handled either in vacuum or under inert atmosphere (Hallett & Welton, 2011); which is quite an unrealistic operating condition in a scaled-up industrial process for desulphurization of fuels. Other acidic ILs, such as those with the hydrogen sulfate and hydrogen phosphate anions, lose their protons in aqueous solutions to form acidic by-products. Moreover, corrosive and harmful acids are used for their synthesis(Rodríguez-Cabo, Rodríguez, Rodil, Arce, & Soto, 2014).

The relatively lower electron density of thiophene made ILs extraction after OSD process ineffective when attempting to desulphurize gasoline. However, [EMIM][NTf_2] would be the optimum ionic liquid for ODS of diesel (Rodríguez-Cabo et al., 2014). However, sulphur extraction is not the only parameter to take into account. Co-extraction of (un) desired compounds, integration of the process in the refinery, and life cycle analyses, among others, are further aspects to take into account in the decision-making prior to implementation of the proposed alternatives.

ILs as Catalytic Solvent

ILs containing metal halide anions has shown positive results in EDS. However, processes that focused on homogeneous catalytic systems need large amount of ILs that are costly and may affect the economic viability of a potential process. What's more, the recovery of the ILs is another problem. To solve these problems, many groups have committed to heterogeneous catalysis system and turned to studying supported ionic liquid. Immobilizing a certain amount of ionic liquid on solid carriers could obtain ionic liquid supported catalysts. The obtained catalysts combined the advantages of ILs with the easy separation and recovery performances of the supported materials. Xun and co-worker (Xun et al., 2014) in their study of the synthesis of metal-based ionic liquid supported catalyst and its application in catalytic ODS of fuels prepared the heterogeneous catalyst by embedding [BMIM][$FeCl_4$] IL in silica gel. Scheme given in Figure 6 was followed to synthesize the [BMIM][$FeCl_4$]/silica gel catalysts. The amount of [BMIM][$FeCl_4$] decreased sharply to reach deep desulphurization and solve the problems of recycle. The catalytic oxidation reaction of different sulphur-containing compounds decreased in the order of DBT > BT > 4,6-DMDBT(Xun et al., 2014).

Figure 6. The synthesis process of [Bmim]$FeCl_4$/silica gel catalyst(Xun et al., 2014)

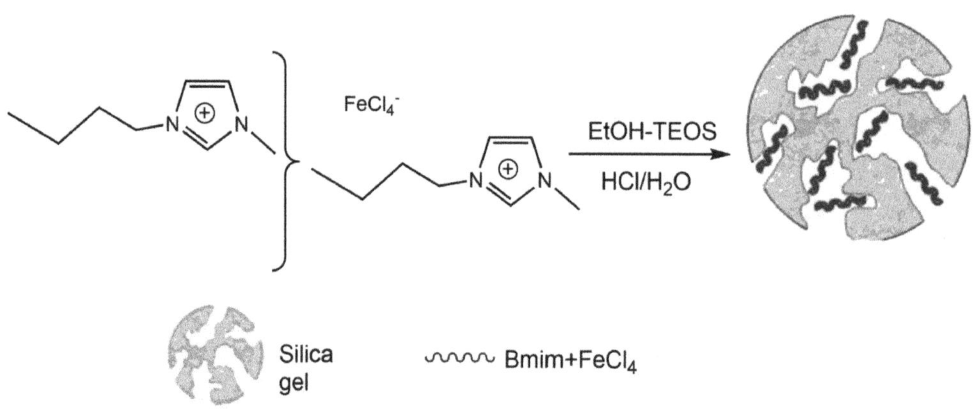

As a supported catalyst, $[BMIM]_3PMo_{12}O_{40}/SiO_2$ could be easily separated by centrifugation and reused without deactivation(J. Zhang, Wang, Li, & Ma, 2011). Different desulphurization systems were studied and ECODS system containing 30% $[BMIM][FeCl_4]$/silica gel, H_2O_2 and $[OMIM][BF_4]$ showed the highest catalytic activity. The sulphur removal of DBT could reach 97.3% under the optimum conditions.

Effect of Reaction Temperature on S-Removal

The reaction temperature in ODS plays vital role. When the reaction temperature for $[C_6^3mpy]$ Cl/$FeCl_3$ increases S-removal increases. However, when the temperature increases above 308 K, the S-removal decreases sharply. DBT is easier to remove than BT at 298 K by $[C_6^3mpy]Cl/FeCl_3$ (Y. Dong et al, 2013). For $[Bmim]Cl/3ZnCl_2$, the highest S-removal efficiency of 99.9% is obtained at 45 °C after 7 h, but only 65.7% at 30 °C and 89% at 60 °C; what's more, the S-removal is 99.9% after 3 h at 45 °C, 47.4% at 30 °C and 62.1% at 60 °C (X, Chen et al, 2012). It was found out the optimal temperature at 60 °C, for ODS of DBT in model oils with Brönsted acidic IL $[HNMP][H_2PO_4]$ and $[BMIM][HSO_4]$ as catalytic solvent and H_2O_2, as oxidant (Zhao et al, 2009 and Zhang et al, 2010). When temperature was raised to 70 °C or 80 °C, the desulfurization decreased to 99% and 96%, respectively (Zhang et al, 2010). These studies suggest a high temperature is not beneficial for the removal of DBT. This might be because the decomposition of H_2O_2 was accelerated at high temperatures (L. Lu et al, 2006). Therefore, ODS using ILs could be easily carried out at moderate (40 °C – 60 °C) temperature. The optimal temperatures depends upon nature of ILs (G. Yu et al, 2012).

Mass Ratio of IL to Model Fuel

The IL/oil mass ratio is one important factor for the selectivity of ILs for ODS desulfurization. As the cost of ILs is high, it is preferred that a minimum quantity is utilized in fuel desulfurization, but it is observed from the literature that the S-removal efficiency decreases with a decreasing IL/oil mass ratio. For instance, the mass ratios of $[BMIM][Cl/3ZnCl_2]$ to oil were tested at 60 minutes and the order of S extraction in ODS efficiency follows the order 1: 5 < 1: 4 < 1: 3 < 1: 2 < 1: 1 < 2: 1 (w/w) IL–oil. Despite the increasing efficiency with an increasing IL/oil mass ratio, it is found that the increase does not have the same rate for all ILs. This typically depends on how the individual chemical natures of ILs are related to their extraction abilities. According to the study by Mochizuki and Sugawara the extraction yield of DBT increased linearly with an increase in the length of alkyl chains and the mass ratio of the IL to the model fuel (Mochizuki and Sugawara et al). Apart from the positive effects of increasing the IL/oil mass ratio, the IL–oil mass ratio has to be carefully selected based on a compromise between S removal and oil recovery. Moreover, the regeneration of ILs may also be considered for the process as this highly affects the overall process cost. The ILs/oil volume ratio is among the most important technical bottlenecks in reducing operational costs associated with handling, separation and disposal of water, ideally the volume ratios of solvent/oil should be minimized.

Concentration of H_2O_2 as Oxidant

The concentration of H_2O_2 has a significant influence on desulfurization efficiency. Mostly, the S-removal efficiency increases with oxidant to S- compound (O/S) molar ratio. The effect of O/S molar ratio on desulfurization for $[BMIM][Cl/3ZnCl_2]$ was determined by Chen et al in 2012, the S-removal efficiency increases from 18.6% when the molar ratio of O/S is 2, to the maximum value of 99.9% when the molar ratio of O/S is 8; but exceeding the O/S molar ratio of 8, the S-removal efficiency decreases with increasing O/S molar ratio, e.g., the value is reduced to 49.3% when the molar ratio of O/S is 9. According to the stoichiometry of the ODS reaction, 2 mol of H_2O_2 is consumed for every 1 mol of DBT to give the corresponding sulfone, i.e., DBT-sulfone ($DBTO_2$), thus the molar ratio of O/S is 2; however, the highest S-removal of 99.9% is obtained at the O/S molar ratio of 8, instead of 2,i.e., the stoichiometric H_2O_2 is not enough to oxidize all the DBT extracted into the IL phase; this is because of the fact that the decomposition of some H_2O_2 occurs in ODS process and excess H_2O_2 is required to ensure the DBT oxidization, which are also indicated in several papers in literature. At the same time, it is neither desirable when the amount of H_2O_2 solution is too much, because excessive H_2O_2 solution will dilute the IL and influence the extraction of DBT from diesel fuel (X, Chen, 2015).

Influence of Molar Ratio of ILs on S-Removal

Quantity of ILs has an important effect on the removal of DBT. The S-removal efficiency was increased from 49.1% to 99.9% when the mass ratio of IL/oil was increased from 1:5 to 1:2. However, the S-removal efficiency came down to 87.8% when mass ratio of IL/oil was further increased to 1:1(X, Chen et al 2012). The reason being that when O/S molar ratio was reduced with the increase in IL/oil mass ratio, the IL get capacity to extract more DBT(D. Liu et al, 2008), but the amount of H_2O_2 was kept constant so the O/S molar ratio is reduced, less S-compounds were oxidized. Similar results were obtained by both Zhang et.al, 2010 and Chen, X., et al, 2015.

Mass Transfer between Phases

The ODS is essentially a liquid–liquid heterogeneous system. These systems are well known to be limited by mass transfer, and emulsification of the reaction mixture, which governs the interfacial area, plays a crucial role in determination of overall kinetics of ODS. The increase in the mass transfer clearly improves the overall sulphur removal rate. Some attempts have been made employing microemulsions (Li C 2004; Lu HY, 2006) and microstructured reactors (Huang, 2008). Microemulsions need intense development to be able to be applied in a refinery process. A very interesting concept is the use of microstructured reactors. Another interesting alternative is the use of ultrasound-assisted oxidative desulfurization (UAOD). UAOD can enhance the mass transfer in heterogeneous systems and can improve kinetics. The implosion of microbubbles, i.e. cavities, results in desirable chemical effects by generation of free radicals with a great propensity for the reaction (Margeta In Press).

Regeneration of Used ILs after ODS

After each ODS process finished, the IL is recovered and reused without significant loss of activity. Regeneration of ILs is an important feature for selection of ILs. It can compensate for the high cost of ILs. The methods of regeneration of S-loaded ILs may vary on the basis of the nature of the ILs and S-compounds. The regeneration of used hydrophilic ILs is carried out by dilution with water followed by simple distillation. According to the study by Zhang et al in 2010 that [BMIM][HSO_4] can be recycled five times with a slight loss in activity, e.g., DBT removal dropped from 100% to 95% under the same experimental conditions. The study by Zhao et al, 2007 suggest that IL [Hnmp]BF_4 can be recycled 7 times without a significant decrease in activity. $[CH_2COOHPy][HSO_4]$ can be recycled 9 times without a significant decrease in the sulfur removal (C. Zhang, et al, 2012). During the recycling some IL was wasted when IL was separated and regenerated which result in little drop in DBT removal from the oil phase with the increase of the cycle (Wang et al, 2009). The decrease of S-removal to the increase of oxidation product $DBTO_2$ that accumulates in the IL phase (Zhang et al, 2009).

Catalytic ODS Using IL Emulsion System

Ge and coworkers has developed a catalytic oxidation ionic liquid emulsion desulphurization system, which composed of water-immiscible $[BMIM][PF_6]$, 30 wt % H_2O_2, and an amphiphilic catalyst ($[C_{18}H_{37}N(CH3)_3]_7[PW_{11}O_{39}]$). Here $[C_{18}H_{37}N(CH_3)_3]_7[PW_{11}O_{39}]$ behave both an emulsifying agent and a catalyst instead of only a surfactant. This kind of catalyst not only maintains the emulsion droplets stable but also provides higher interfacial surface area where the oxidation of organic sulphur molecules takes place. During the reaction, the ionic liquid emulsion functions as highly dispersed micro reactors. The sulphur-containing compounds in the model oil were first extracted into the ionic liquid emulsion phase and then oxidized to their corresponding sulfones by $[C_{18}H_{37}N(CH3)_3]_7[PW_{11}O_{39}]$ in the H_2O_2/$[BMIM][PF_6]$ interface. The sulfones accumulated in the $[BMIM][PF_6]$ phase. After reaction, the desulphurization system quickly divided into two layers; the deep desulphurization can be achieved in this way. After the reaction, the reaction system was still a biphasic system, so the oil could be separated by simple decantation from the biphasic system of the ionic liquid easily. The used ILs, whose structure does not change and the recovered IL can be reused for further emulsion catalytic oxidation cycles. For example $[Bmim]PF_6$ emulsion desulphurization system could be recycled five times with an unnoticeable decrease in catalytic activity.

The milky-white ionic liquid emulsions are formed when H_2O_2 and $[C_{18}H_{37}N(CH_3)_3]_7[PW_{11}O_{39}]$ were added into $[BMIM][PF_6]$. The catalyst is distributed in the interface of H_2O_2 and water-immiscible ILs. In this emulsion reaction system, the catalyst molecule acts as an emulsifying agent, could be uniformly distributed in the interface of H_2O_2–ionic liquid, and forms a film around the dispersed ionic liquid droplets (Figure 7).

Consequently, the lipophilic quaternary ammonium cations of the amphiphilic catalyst would lie on the oil side and the hydrophilic hetero poly anions would lie on the H_2O_2 side. Since hetero poly anions ($[PW_{11}O_{39}]_7$ depolymerized into several smaller active species including $[(nC_8H_{17})_3NCH_3]_3\{PO_4[MoO(O_2)_2]_4\}$, $[(n\text{-}C_{12}H_{25})_3NCH_3]_3\{PO_4[MoO(O_2)_2]_4\}$, $[(n\text{-}C_8H_{17})_3NCH_3]_3\{PO_4[WO(O_2)_2]_4\}$ and $[(n\text{-}C_{12}H_{25})_3NCH_3]_3\{PO_4[WO(O_2)_2]_4\}$ in the presence of H_2O_2, because the ionic liquid emulsion was immiscible with n-octane and formed a biphasic system. The sulphur-containing compounds in the model oil were first extracted from oil phase into IL phase and oxidized to their corresponding sulfones by the active species simultaneously; the sulfones accumulated in IL phase.

After reaction, the desulphurization system quickly divided into two layers. In this way, the catalyst in the emulsion droplets can be readily separated from the oil and recycled, and this ionic liquid emulsion system could be recycled five times without an obvious decrease in activity.

Table 3 shows four kinds of different desulphurization systems, such as the following: extraction, extraction combined with chemical oxidation, catalytic oxidation without ionic liquid extraction, and extraction coupled with catalytic oxidation in the conditions of hydrophobic ionic liquid. In an extraction system, when using ionic liquid $[BMIM][PF_6]$ solely as the extractant for removing DBT-containing in model oil, the sulphur removal only reached 14.8%, with addition of H_2O_2 in $[BMIM][PF_6]$ the sulphur

Figure 7. Catalytic oxidation of DBT in IL emulsion system (Ge, Zhou, Yang, & Xue, 2011)

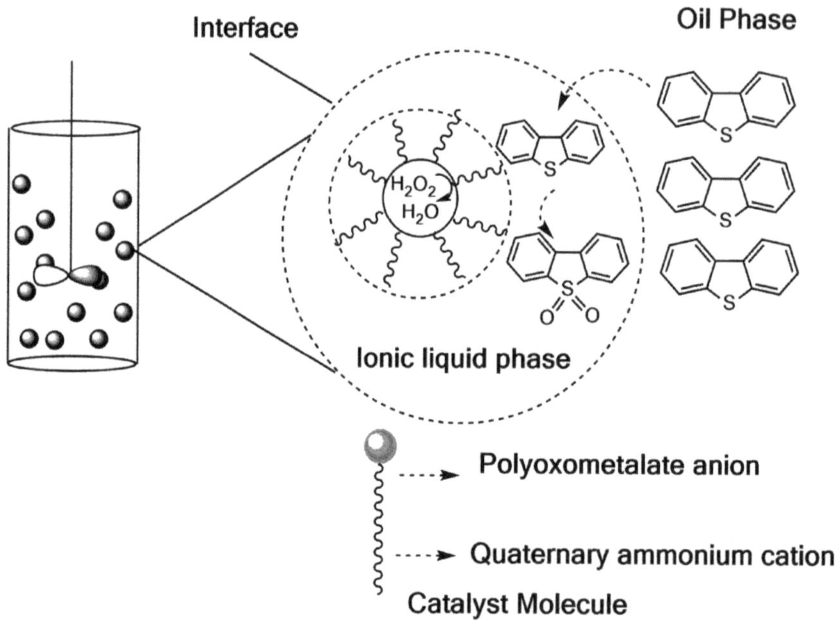

Table 3. Sulphur Nernst partition coefficient (KN) for extraction with ILs (Wasserscheid & Jess, 2004)

IL	KN in mg(S) kg(IL)-1/mg(S) kg(oil)-1
[BMIM]Cl/$AlCl_3$	4.0 a
[BMIM][BF_4]	0.7 a
[BMIM][PF_6]	0.9 b
[BMIM][$OcSO_4$]	1.9 a
[EMIM][$EtSO_4$]	0.8 a
[MMIM][Me_2PO_4]	0.7 a

a Model oil: 500 ppm Sulphur as DBT in n-dodecane; mass ratio 1: 1, mixing time 15 min, room temperature. b 60 °C.

removal increased to 30.7%, on the other hand, when catalyst, H_2O_2, and hydrophobic ionic liquid were employed together, the removal of DBT increased sharply (99.1%). However, in this experiment, the catalyst with long alkyl can selectively oxidize DBT into their corresponding sulfones without hydrophobic ionic liquid, then the sulfones could be removed by VOCs as an extractant in the following process. Based on the results gained herein, the catalytic oxidation desulphurization system, which composed of [Bmim]PF_6, 30 wt % H_2O_2, and $[C_{18}H_{37}N(CH_3)_3]_7[PW_{11}O_{39}]$, was superior to the simple extraction with IL or the oxidation/extraction without catalyst.

For a given design to be commercialized, it needs to meet quite a stringent set of criteria, some of the more important of which include(Ito & van Veen, 2006):

- **Capital Cost:** E.g., the less unit operations (less pieces of equipment) the better, no expensive materials; preferably a single product stream for diesel.

- **Operation Cost:** E.g., minimize the consumption of hydrogen or expensive chemicals, minimize the generation of waste streams (esp. dilute ones).
- Product volume should be high, e.g., >99%.
- Process cycle life should be reasonably long, e.g., four plus years in general, and five plus years for FCC gasoline post treatment units in order to match the FCC cycle length.
- Technical complexity should be avoided as much as possible: e.g., no difficult to operate units (minimizing down time), no difficult catalyst(s) (tolerant of upsets).
- Feed flexibility.
- Overall value to refiner.

EDS USING ILs

ILs are ideal candidates as extracting agent for EDS with low melting salts (<100 °C) and measurable vapor pressure. The basic concept of such an extraction with ILs is outlined in Figure 8.

The use of ILs for the selective extraction of sulphur compounds from fuels was reported for the first time in 2001 by Bosmann and co-workers (Bosmann et al., 2001). Since then, the desulphurization ability of many ILs has been tested by many researchers; Table 7 shows the list of ILs used EDS of fuel oils.

In the extractive process, the ILs [EMIM][BF_4], [BMIM][PF_6] and [BMIM][BF_4] showed remarkable selectivity for the absorption of aromatic sulphur containing molecules from the model fuel (Kulkarni & Afonso, 2010). The preferential absorption of thiophene and DBT over other aromatic hydrocarbons suggests that compounds with higher aromatic π-electron density in C_5 rings are favourably absorbed. The cation and anion structure and size in the ILs are important parameters affecting the absorption capacity for aromatic compounds. These ILs are moisture insensitive at low temperature, thermally stable under the distillation conditions, and readily regenerated for reuse after the distillation. The absorbed aromatic sulphur containing compounds were quantitatively recovered during the regeneration. However, it requires several extraction stages to completely remove the sulphur compounds(Kulkarni & Afonso, 2010).

Figure 8. Graphical basic concept EDS of fuel oils using ILs

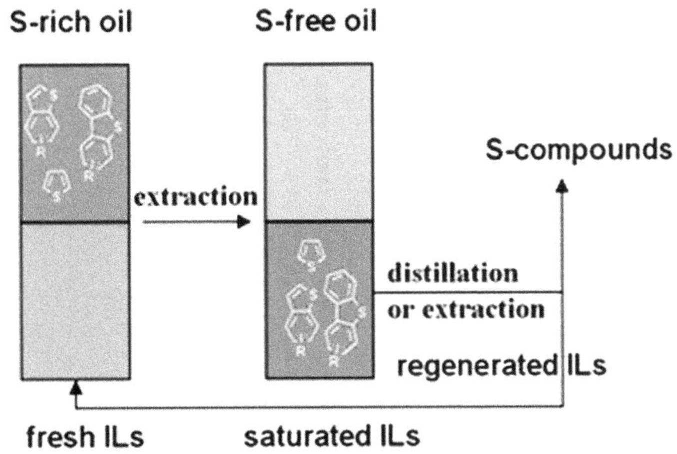

Thermodynamic Extractive Equilibrium

Thermodynamics extractive equilibrium is a very important aspect of any process because short time favors high product yield or small size of contact equipment in industry. Thermodynamics extractive equilibrium of ILs-EDS is directly related to ILs viscosity for both model and real fuel oils. High viscous

Figure 9. Desulphurization efficiency vs extraction time for the Desulphurization of a model oil using high viscosity ILs

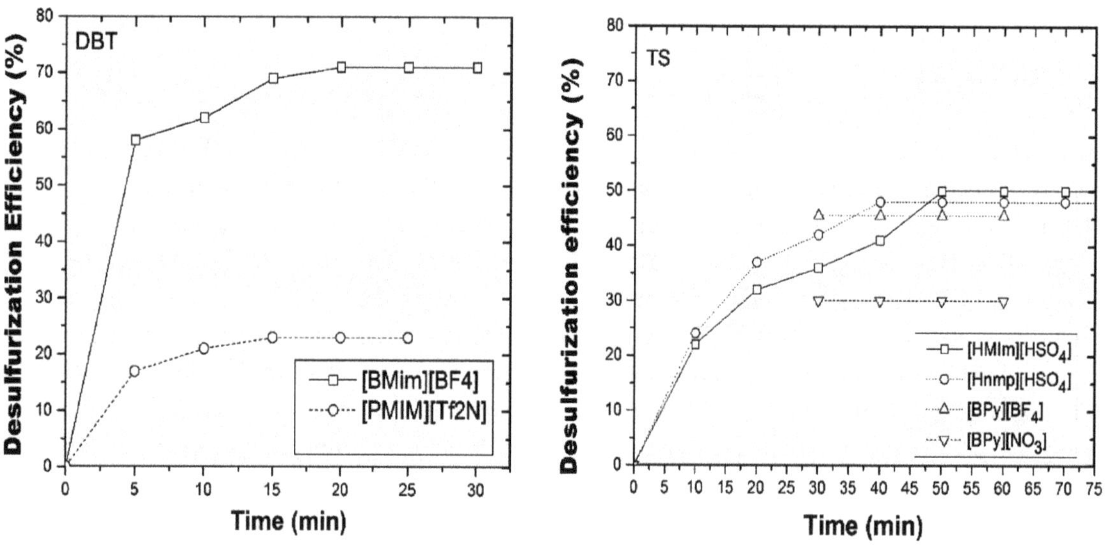

Figure 10. Desulphurization efficiency vs extraction time for the Desulphurization of a model oil using low viscosity ILs

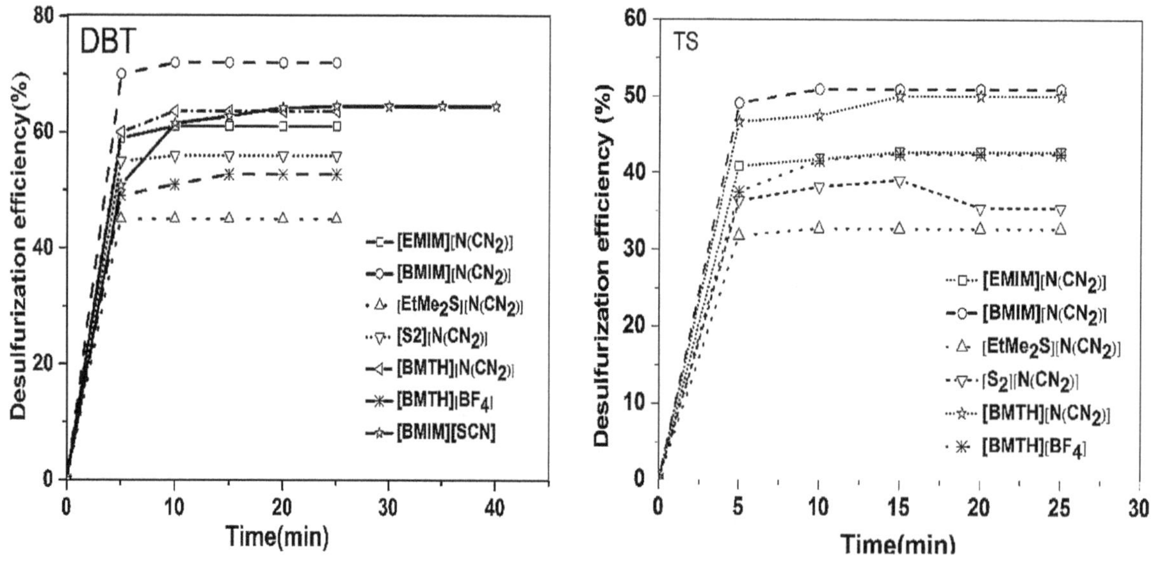

Table 4. ILs used in EDS of fuel oils(Abro et al., 2014)

Name of IL	Formula of ILs
N-methyl-n-methylimidazolium dimethyl phosphate	[MMIM][DMP]
N-ethyl-n-methylimidazolium diethyl phosphate	[EMIM][DEP]
N-butyl-n-methylimidazoliumdibutyl phosphate	[BMIM][DBP]
1-ethyl-2,3-trimethyleneimidazolium chloride	[EtMIM]Cl
N-ethyl-N-methyl-imidazoliumdimethylphosphate	[EMIM][DMP]
N-ethyl-N-ethyl-imidazoliumdiethylphosphate	[EEIM][DEP]
N-butyl-N-ethyl-imidazoliumdibutylphosphate	[BEIM][DBP]
1-ethyl-3-methylimidazolium tetrafluoroborate	[EMIM][BF_4]
1-butyl-3-methylimidazolium hexafluorophosphate	[BMIM][PF_6]
1-butyl-3-methylimidazolium tetrafluoroborate	[BMIM][BF_4]
1-methyl-3- octylimidazoliumtetrafluoroborate	[OMIM] [BF_4]
1-methyl-3-octylimidazolium bis [trifluoromethylsulfonyl]imide	[OMIM][NTf_2]
l-butyl-3-ethylimidazolium ethyl sulfate	[BEIMEt[[SO_4]
l-ethyl-3-ethylimidazolium ethyl sulfate	[EIMEt][SO_4]
l-ethyl-3-methylimidazolium ethyl sulfate	[EMIMET][SO_4]
l-ethyl-3-methylimidazolium methyl sulfate	[EMIMMe][SO_4]
l-butyl-3-methylimidazolium methyl sulfate	[BMIMMe][SO_4]
1,3-dimethylimidazolium methyl sulfate	[MMIMMe][SO_4]
1-nbutyl- 3 methylimidazoliumoctylsulfate	[BMIM] [$OcSO_4$]
1-ethyl-3-methyl-imidazolium ethylsulfate	[EMIM][$EtSO_4$]
1-butyl-2,3-dimethyl imidazolium chloride	[BDMI][Cl/$FeCl_3$]
1-hexyl-2,3-dimetheyl imidazolium chloride	[HDMICl][FeCl]
1-butyl-3-methylimdazolium dicyanamide	[BMI][$N(CN)_2$]
1-ethyl-3-methylimdazolium dicyanamide	[EMI][$N(CN)_2$]
Ethylated tetrahydrothiopheniumdicyanamide	[S_2][$N(CN)_2$]
Ethyldimethylsulfoniumdicyanamide	[$EtMe_2S$][$N(CN)_2$]
1-ethyl-3-methylimidazolium ethyl sulfate	[EMIM] [$EtSO_4$]
1-pentyl-3-methylimidazolium bis(trifluoromethylsulfonyl)imide	[PMIM] [Tf_2N]
1-benzyl-3-methylimidazolium bis(trifluoromethylsulfonyl)imide	[BzMIM] [Tf_2N]
1-heptyl-2,3dimethylimidazoliumbis (trifluoromethylsulfonyl) imide	[HpMMIM] [Tf_2N]
1-decyl-2,3-dimethylimidazoliumbis (trifluoromethylsulfonyl) imide	[OMMIM] [Tf_2N]
1-hexyl-3,5-dimethylpyridinium bis(trifluoromethylsulfonyl)imide	[HMMPy] [Tf_2N]
N-butylpyridiniumtetrafluoroborate	[BPy][BF_4]
N-hexylpyridiniumtetrafluoroborate	[HPy][BF_4]
N-octylpyridiniumtetrafluoroborate	[OPy][BF_4]
N-butyl-pyridinium nitrate	[BPy][NO_3]
N-ethyl-pyridinium nitrate	[EPy][NO_3]
N-butyl-pyridiniumtetrafluoroborate	[BPy][BF_4]
N-ethyl-pyridiniumtetrafluoroborate	[EPy][BF_4]

continued on following page

Table 4. Continued

Name of IL	Formula of ILs
N-ethyl-pyridinium acetate	[EPy][Ac]
N-butyl-pyridinium acetate	[BPy][Ac]
3-methyl-N-butyl-pyridinum dicyanamide	[MBPy][$N(CN)_2$]
4-methyl-N-butyl-pyridinum dicyanamide	[MBPy][$N(CN)_2$]
4-methyl-N-butyl-pyridinum thiocyanate	[MBPy][SCN]
N-butyl-N-methylimidazolium tricyanomethanide	[BMIM][$C(CN)_3$]
N-butyl-N-methylimidazolium dicyanamide	[BMIM][$N(CN)_2$]
1-butyl- 3- methylimidazoliumthiocyanate	[BMIM][SCN]
1-butyl- 3- methylimidazolium chloride	[BMIM]Cl/FeCl
1-octyl-3-methylimidazolium chloride	[OMIM]Cl/FeCl
3-butyl-4-methylthiazolium dicyanamide	[BMTH][DCA]
3-butyl-4-methylthiazolium thiocyanate	[BMTH][SCN]
3-butyl-4-methylthiazolium hexafluorophosphate	[BMTH][PF_6]
3-butyl-4-methylthiazolium tetrafluoroborate	[BMTH][BF_4]
1-butyl-1-methylpyrrolidinium tri⟩uoromethanesulfonate	[BMPYR][CF_3SO]
1-butyl-1-methylpyrrolidinium tricyanomethanide	[BMPYR][TCM]
1-hexyl-3-methylimidazolium tetracyanoborate,	[HMIM][TCB]
1-methyl-3-methylpyrrolidone dimethyl phosphate	[MMP][DMP]
1-methyl-3-ethylpyrrolidone diethyl phosphate	[MEP][DEP]
1-hexyl-3-methylpyrrolidone dimethyl phosphate	[HMP][DMP]
1-hexyl-3-ethylpyrrolidone diethyl phosphate	[HEP][DEP]
1-butyl-3-methylimidazolium tetra⟩uoroborate	[BMIM][BF_4]
1-octyl-3-methylimidazolium tetra⟩uoroborate	[OMIM][BF_4]
1-butyl-3-methylimidazolium hexa⟩uorophosphate,	[BMIM][PF_6]
1-octyl-3-methylimidazolium hexa⟩uorophosphate	[OMIM][PF_6]
N-methylpiperazinium lactate	[C1 pi][Lac]
N-ethylpiperazinium lactate	[C2 pi][Lac]
N,N'-dimethylpiperaziniumdilactate	[C1C1pi][lac]
2-2-dimethylamino ethoxy propionate	[DMEE][CO_2Et_2]
3-dimethylamino propanenitrile propionate	[DMAPN][CO_2Et]
1-butyl-3-methylimidazolium copper chloride	[BMIM][$CuCl_2$]
1-butyl-3-methylimidazoliumtetrachloroaluminate	[BMIM]$AlCl_4$]
1-butyl-3-methylimidazolium octylsulfate	[BMIM][$OcSO_4$]
1-ethyl-3-methylimidazolium thiocyanate	[EMIM][SCN]
1-Alkyl-3-methyl Imidazolium methyl phosphate	[MMIM][MP]
1-Butyl-3-methylimidazolium trifluoromethansulfonate	[BMIM][OTf]
N-methylimidazolehydrogen sulfate	[HMIM][HSO_4]
N-methylpyrrolidonehydrogen sulfate	[HNMP][HSO_4]

Table 5. Sulphur extraction efficiency of different ILs at different mass ratio of IL-oil for real fuel oils (Abro et al., 2014)

ILs	Oil	Initial S: (ppm)	Time (min)	Temp. (K)	S-Extraction Efficiency for IL-Oil Mass Ratio				
					2:1	1:1	1:2	1:3	1:5
[BMIM][N(CN)$_2$]	Gasoline	300	20	298	22	21.5	11	10.3	9.4
	Diesel	380	20	298	45	30	27	19	19
[OMIM] [BF$_4$]	Diesel	711	20	298	n/a	29.9	18.1	n/a	10.1
		1572	20	298	n/a	39.8	24.7	n/a	16.1
[OPy] [BF$_4$]	Diesel	711	20	298	n/a	16.2	18.1	n/a	10.1
		1572	20	298	n/a	39.1	25.7	n/a	15.8
[BMIM][BF$_4$]	Diesel	1572	20	298	n/a	n/a	12.6	n/a	8.2

Table 6. Sulphur extraction efficiency of different ILs at different mass ratio of IL-oil for model fuel oils (Abro et al, 2014)

ILs	Oil	S-Comp.	Time (min)	Temp. (K)	S-Extraction Efficiency for IL-Oil Mass Ratio				
					2:1	1:1	1:2	1:3	1:5
[EMIM][MeSO$_4$]	*n*-hexane	DBT	20	298.1	n/a	70	40	n/a	15
[MMIM][MeSO$_4$]	*n*-hexane	DBT	20	298.1	n/a	40	23	n/a	10
[BMI][N(CN)$_2$]	*n*-hexane+ toluene	TS	20	298.1	62.6	47.5	29	22	17
	n-hexane	DBT	20	298.1	78.9	69.5	50	44	30
[BMTH] [N(CN)$_2$]	*n*-octane	DBT	20	298.1	66.8	49.6	30	20	15
	n-hexane	TS	20	298.1	60.4	39.8	30	20	18
[BPy][BF$_4$]	*n*-heptane + xylol	DBT	30	327	n/a	45.5	28.6	16.9	n/a
[BPy][NO$_3$]	*n*-heptane + xylol	DBT	30	327	n/a	30.1	19	9.4	n/a
[BPy][Ac]	*n*-heptane + xylol	DBT	30	327	n/a	32.1	20.1	10.	n/a
[EPy][BF$_4$]	*n*-heptane + xylol	DBT	30	327	n/a	21.8	12.6	8.9	n/a
[EPy][NO$_3$]	*n*-heptane + xylol	DBT	30	327	n/a	27.1	17.3	13.8	n/a
[BPy][BF$_4$]	*n*-dodecane	TS	15	298.1	n/a	37.5	n/a	9.37	5
		DBT	15	298.1	n/a	40.6	n/a	18.7	12.5
[HPy][BF$_4$]	*n*-dodacane	TS	15	298.1	n/a	41.9	n/a	17.5	6.25
		DBT	15	298.1	n/a	59.4	n/a	31.2	25
[OPy][BF$_4$]	*n*-dodacane	TS	15	298.1	n/a	43.7	n/a	18.7	6.25
		DBT	15	298.1	n/a	62.5	n/a	37.5	31.5
[OMIM][BF$_4$]	*n*-hexane	DBT	20	298.1	n/a	39.8	24.7	n/a	16.1
[PMIM][Tf$_2$N]	*n*-hexane	DBT	20	298.1	n/a	52	37	n/a	20
[C1pi][Lac]	*n*-heptane	TS	30	298.1	46	16	16	11	n/a
	n-heptane	DBT	30	298.1	58.2	43.3	32	24	n/a
[NMP][DMP]	*n*-octane	DBT	20	298.1	n/a	87.5	75.1	66.2	51.6
		BT	20	298.1	n/a	73.2	61.3	51.4	34.4
		TS	20	298.1	n/a	72.1	50.2	42.5	25.3

ILs such as [BMIM][OcSO$_4$] [BMIM][CH$_3$CO$_4$], [HMIM][HSO$_4$] and [HNMP][SO$_4$] took 30-40 minutes to achieve the equilibrium while low viscous ILs such as [BMIM][SCN], [BMTH][N(CN)$_2$], [BMIM] [[N(CN)$_2$] and [BMTH][SCN] took less than 10 minutes to reach at equilibrium (Abro et al., 2014).

A possible reason behind these variations can be that the low viscous ILs had better dispersion in fuel oils and faster mass transfer of Sulphur compounds from the oil phase to the IL phase. Figures 9 and 10 shows DBT removal efficiency verses thermodynamic extractive equilibrium and TS removal efficiency verses thermodynamics extractive equilibrium by different ILs at mild conditions.

IL-Oil Mass Ratio

Sulphur extraction efficiency of different ILs at different mass ratio of IL-oil for real and model fuel oils are given in Tables 5 and 6 (Abro et al., 2014). IL-oil mass ratio is one of the most important factors for selectivity of ILs for desulphurization. As ILs' cost is high, it is preferred to utilize minimum quantity in fuel desulphurization, but it is observed from the literatures that the S-extraction efficiency decreases with decreasing IL-oil mass ratio. For instance mass ratios of [BMI][N(CN)$_2$] to oil were tested under atmospheric conditions for 20 minutes and the order of sulphur extraction efficiency follows the order 1:5<1:4<1:3<1:2<1:1<2:1 (w/w) IL-oil (Asumana et al, 2010). Meanwhile, the sulphur partition coefficients did not prove very sensitive to the mass ratio. Similar results were obtained when a series of [BF$_4$] and [MeSO$_4$]-based ILs were used to investigate the influence of mass ratio. Despite increasing behaviour of extraction efficiency with increasing IL-oil mass ratio, it is found that the extent of increase is not with same rate. This typically depends on the individual chemical nature of ILs related with their extraction abilities.

Sulphur partition is not sensitive for physical extraction. However, it might be sensitive for chemical extraction. Mostly the less viscous ILs such as [BMI][N(CN$_2$)], [EMI][N(CN$_2$)], [BMTH][N(CN$_2$)], [BMPYR][TCM], [HMIM][TCB], [C1pi][Lac], [C2pi][Lac] and [C1C1pi][Lac] showed less sensitivity as compared to high-viscous ILs such as [OMIM][BF$_4$], [EMI][BF$_4$], [EMIM][MeSO$_4$], [HEP][DEP] and [NMP][DMP] as per in Table 6.

Apart from the positive effect of increasing ILs-oil mass ratio, IL-oil mass ratio has to be selected carefully based on a compromise between Sulphur removal and oil recovery. Moreover the regeneration of ILs may also be considered for the process as this highly affects the overall process cost (Li et al,2013).

Initial Sulphur Content

For an ideal physically determined ILs-EDS technique, the initial sulphur content in the oil to be treated has little or no influence on the extraction efficiency. Notwithstanding, different IL systems tend to exhibit varied extraction mechanisms. For example, [BMI][N(CN)$_2$], [BMTH][N(CN)$_2$] and a [BF$_4$]-based IL showed decreasing sulphur extraction efficiency with increasing initial sulphur content in oil, although this influence was not pronounced. However, when Huang, C. at al. used [BMIM][Cu$_2$Cl$_3$] for the desulphurization of gasoline, increasing Sulphur extraction efficiency with increasing initial sulphur content was reported, and this influence was obvious, e.g. between 950 ppm/w sulphur content and 196 ppm/w sulphur content, 16.2% and 37.4% S-extraction efficiency was realized (Huang et al,2004). Table 7 demonstrates some examples of S-extraction efficiency of different ILs for different initial Sulphur contents.

Obviously, this insensitivity of sulphur extraction efficiency to initial sulphur content in fuel oils is valuable to industrially deal with oils with wide range of sulphur content.

Table 7. S-extraction efficiency of different ILs for different initial S-contents(Abro et al., 2014)

IL	Oil	S-Comp.	Temp. (K)	IL-Oil Mass Ratio	Time (min)	Initial S: (ppm)	S-Extraction Efficiency (%)
[BMI][N(CN)$_2$]	*n*-Hexane + Toluene	TS	298	1:1	20	200	52.1
						500	51.4
						800	49.8
						1500	48.1
						2000	45.1
	n-Hexane	DBT	298	1:1	20	200	70.9
						500	70.3
						800	69.2
						1500	67.1
						2000	67.9
[BMIM][Cu$_2$Cl$_3$]	Gasoline	DBT/TS	RT	1:1	20	196	37.4
						410	29.1
						680	21.6
						950	16.2
[BMTH][N(CN)$_2$]	*n*-Octane	DBT	298	1:1	20	203	64.3
						557	63.9
						944	63.1
						1440	62.8
						2927	61.6
	n-Hexane	TS	298	1:1	20	275	45.2
						549	44.1
						849	43.2
						1546	41.1
						2645	39.4

Temperature

Studies have shown that there is infinitesimal influence of temperature on S-extraction efficiency and the corresponding S-partition coefficients. Effect of temperature on S-removing efficiency for ILs is tabulated in Table 8 and Table 9. The literatures show that low viscous ILs, such as [BMI][N(CN)$_2$], [BMTH][N(CN)$_2$] and [BMIM][SCN], have a slight decreasing trend for sulpur extraction efficiency and the corresponding S-partition coefficient for increase in temperature. On the other hand the trend becomes reversed for high viscous ILs, such as [C$_2$Py][BF$_4$] and [BMIM][PF$_6$], where increase in temperature confirms slight increase in efficiency and partition coefficient. However, in some studies temperature has showed great effect on desulphurization, such as [HNMP][HSO$_4$] where increase in temperature from 400 °C to 700 °C corroborates great increase in S-extraction efficiency from 48% to 63% (Wang et al,2011).

The reason for these variations for high and low viscous ILs can be understood from the variation in viscosity of ILs on increasing temperature. With the increase of temperature for high viscous ILs the viscosity of ILs is decreased, due to which the ILs and the sulphur content in the oil have more possibil-

Table 8. S-extraction efficiency of different ILs at different temperatures for model fuel oils (Abro et al, 2014)

IL	Oil	S-Comp.	Time (min)	S-Extraction Efficiency for Temperatures (%)				
				20°C	25°C	35°C	45°C	55°C
[BMI][N(CN)$_2$]	*n*-Hexane + Toluene	TS	20	46.7	46.2	45.3	43.2	40.6
	n-Hexane	DBT	20	66.8	65.4	63.5	61.7	59.9
[BMTH][N(CN)$_2$]	*n*-Octane	DBT	20	55.3	55.1	53.3	52.8	51.9
	n-Hexane	TS	20	38.5	38.4	37.8	36.1	35.4
[BMI][SCN]	Dodacane	DBT	20	n/a	n/a	64	62	60
[EMI][N(CN)$_2$]	Gasoline	DBT	20	41	40.7	38.9	37.4	36.2
[BPy][BF$_4$]	*n*-Heptane + xylol	DBT	30	n/a	45.5	45.9	46.5	47
[C$_1$pi][Lac]	*n*-Heptane	DBT	30	21	20	19	18	18
	n-Heptane	TS	30	39	40	43	40	39
[NMP][DMP]	*n*-Octane	DBT	20	n/a	87.5	86.2	86.1	85.8
		BT	20	n/a	73.2	72.1	71.1	70.2
		TS	20	n/a	72.1	71.7	71.1	70.8
[HNMP][HSO$_4$]	N-octane	TS	20	n/a	n/a	47.1	53.2	60.1

Table 9. S-extraction efficiency of different ILs at different temperatures for real fuel oils (Abro et al, 2014)

ILs	Oil	Initial S: (ppm)	IL-Oil Mass Ratio	S-Extraction Efficiency for Temperatures (%)				
				20°C	25°C	35°C	45°C	50°C
[HMMPY][NTf$_2$]	Diesel	10000	1:1	n/a	24	n/a	n/a	62.2
[EMIM] [NTf$_2$]	Diesel	10000	1:1	n/a	11	n/a	n/a	50
[EMI][OAc]	Diesel	10000	1:1	n/a	10.6	n/a	n/a	13
[EMI][Et$_2$PO$_4$]	Diesel	10000	1:1	n/a	13.5	n/a	n/a	17
[BMIM][N(CN)$_2$]	Gasoline	300	1:1	18	12.1	12	11.1	11
	Diesel	400	1:1	31.8	28.1	27.5	27	27

ity to interact with each other, which leads the high sulphur removal efficiency. However, the impact of temperature on the desulphurization is limited. The conclusive remark in this regard is that low viscous ILs has more advantages over high viscous ILs in terms of energy consumption and S-extraction efficiency at room temperature (abro et al, 2014, Asumana et al, 2010 & 2013).

Regeneration

Regeneration of ILs after extraction is a significant feature for selectivity of ILs. Regeneration and recyclability can compensate for the high cost of ILs. The methods of regeneration of sulphur loaded ILs may vary based on nature of ILs and sulphur compounds. Mostly regeneration of hydrophobic ILs are done by distillation followed by adsorption, and regeneration of hydrophilic ILs by dilution process with water followed by distillation. Original structure of ILs mostly found unchanged after regeneration, which is best sign for ILs. During regeneration, sulphur compounds in the vicinity of solvent molecules are repelled; after regeneration, ILs are reused without a noticeable decline in efficiency for model oils as demonstrated in the literatures (Anugwom et al, 2011 & Endres et al, 2004). So many ILs such as [BMI][N(CN)$_2$], [BMTH][N(CN)$_2$], [EMI][N(CN)$_2$], [MBPy][N(CN)$_2$], [NMP][DMP], [C1pi][Lac], [MMIM][DMP], [EMIM][DEP] and [BMIM][DBP], showed remarkable results after their regeneration but it was also found in the literatures that the low viscous ILs (such as [BMI][N(CN)$_2$], [BMTH][N(CN)$_2$], [EMI][N(CN)$_2$] and [MBPy][N(CN)$_2$]) can easily be regenerated as compared to high viscous ILs such as [NMP][DMP], [HNMP][SO$_4$], [BMIM][DMP] and [BEIM][DEP]. Table 10 shows sulphur removal efficiency verses regeneration cycle by different ILs at mild conditions.

Table 10. S-extraction efficiency vs regeneration cycle by different mild conditions for real fuel oils (Abro et al, 2014)

ILs	Oil	S-Comp.	Time (min)	Temp. (K)	IL-Oil Mass Ratio	S-Extraction Efficiency for Regeneration Cycles			
						1	2	3	4
[BMI][N(CN)$_2$]	*n*-Hexane + Toluene	TS	20	298	1:1	47.5	47.2	46.4	45.2
	n-Hexane	DBT	20	298	1:1	69.5	68	67	65.8
[BMTH] [N(CN)$_2$]	*n*-Octane	DBT	20	298	1:1	69.5	69.4	69.3	69
	n-Hexane	TS	20	298	1:1	49.6	49.5	49.5	49.5
[EMIM][N(CN)$_2$]	*n*-Hexane	TS	20	298	1:1	43	42.9	42.8	42.7
Et_3NHCl-$AlCl_3$	Benzene	DBT	20	293	1:1	40.2	39.1	n/a	n/a
	Toluene	TS	20	293	1:1	25.	25.0	n/a	n/a

Table 11. S-extraction efficiency of different ILs at multiple extraction steps for model fuel oils (Abro et al, 2014)

ILs	Oil	S-Comp.	Temp. (K)	IL-Oil Mass Ratio	Time (min)	S-Extraction Efficiency for Multiple Steps (%)			
						1	2	3	4
[DMIM][Tf$_2$N]	Diesel	DBT	RT	1:1	20	36.3	56.3	69.7	77.2
	Gasoline	TS	RT	1:1	20	17.1	29.8	40.9	50.5
[HpMMIM][Tf$_2$N]	Diesel	DBT	RT	1:1	20	31.9	53.7	59.1	70.4
	Gasoline	TS	RT	1:1	20	12.7	26.1	35.2	40.5
[HMMPy][Tf$_2$N]	Diesel	DBT	RT	1:1	20	35.6	59.1	71.3	92.6
	Gasoline	TS	RT	1:1	20	12.0	24.4	40.1	50.5
[BMTH] [N(CN)$_2$]	Diesel	DBT	293	1:1	20	50.3	81.8	94.5	99.1
	Gasoline	TS	293	1:1	20	58.1	61.8	79.1	90
[BMI][N(CN)$_2$]	Diesel	DBT	298	1:1	20	66.6	90	98.3	n/a
	Gasoline	TS	298	1:1	20	49.9	83.3	89.9	95.8
[EMIM][N(CN)$_2$]	Gasoline	TS	298	1:1	20	41.1	66	81.2	90.1
[NMP][DMP]	N-octane	TS	298	1:1	30	71	89.6	96	98

Table 12. S-extraction efficiency of different ILs for real fuel oils at multiple extraction steps (Abro et al, 2014)

ILs	Oil	Initial S: (ppm)	Temp. (K)	IL-Oil Mass Ratio	Time (min)	S-Extraction Efficiency for Multiple Steps (%)			
						1	2	3	4
[HMMPY][NTf_2]	Diesel	10000	298	1:1	n/a	24	32	41	n/a
	Diesel	10000	323	1:1	n/a	37.8	90	98	n/a
	Gasoline	240	298	1:1	n/a	20	25	26.7	n/a
[EMIM][OAc]	Diesel	10000	298	1:1	n/a	11	12.8	13	n/a
	Gasoline	240	298	1:1	n/a	38.7	38.9	38.2	n/a
[BMIM][$N(CN)_2$]	Diesel	400	298	1:1	20	32	45	52.5	60
	Gasoline	300	298	1:1	20	17.1	23.3	36.7	58.6

Multiple Extractions

In a single-step extraction, it is very difficult to reduce the sulphur content in oil to the statutory limit (say, 10 ppm), a fact that has necessitated multiple extractions (Wang et al, 2011).This may be due to sulphur compounds partition between fuel oils and ILs which makes a complete removal of sulphur compounds in a single extraction rather impossible. [EMIM][$N(CN)_2$] reduced thiophenic sulphur in model gasoline from ~500 ppm to ~300 ppm (41.7% efficiency) in a single extraction at 1:1(w/w) IL-oil, 25 °C for 20 minutes, and to ~15 ppm (about 97% efficiency) after five extraction cycles. In the same token, [BMI][$N(CN)_2$] realized an almost 100% sulphur removal after five cycles as opposed to a mere 49.2% in a single extraction (Asumana et al, 2010). [EMIM][SCN] removed DBT completely from model diesel in third step and removed 96% TS after three stages. These studies suggest that the multiple-extractive process is a remarkable way to reduce sulphur content to near or below the required limit as shown in Tables 11 and 12.

As it is already discussed that it is not possible to reduce sulphur content to the legislative limit in a single extraction step, but when we consider performances of different ILs for only single step extraction, low viscous ILs such as [BTH][$N(CN)_2$], [EMI][$N(CN)_2$] and [BMI][$N(CN)_2$] shows highest results for both sulphur content compounds i.e. DBT and TS at mild conditions (Asumana et al, 2010, Asumana et al, 2013 & Chen et al, 2012). However, some high viscous ILs also showed good results for sulphur content in single step at high temperature such as [NMP][DMP] and [HNMP][HSO_4].

On the basis of above discussion about different parameter for desulphurization of model fuel oils and real fuel oils, it is observed that the low viscous ILs gave considerable results. This is widely welcomed industrially as far as the generation of high production yield is concerned. On the influence of IL-oil mass ratio, efficiency decreases with decreasing IL-oil mass ratio. The influence of temperature on ILs-EDS of real fuel oils was also observed to be insignificant mainly due to the low viscosity, such as in the case of [EMI][$N(CN)_2$] and [BMI][$N(CN)_2$] (Asumana et al, 2010, Asumana et al, 2013 & Chen et al, 2012). Again, this favors industrial applications since less energy and thus low costs are required. EDS of real fuel oils is characterized by low S-selectivity (less than 65%), a situation that can be attributed to the presence of many and varied polyaromatic compounds and heterocycles which tend to hinder extraction

Table 13. ILs-EDS of real fuel oils at optimized conditions (Abro et al, 2014)

ILs	Oil	Initial S: (ppm)	IL-Oil Mass Ratio	Temp. (K)	Time (min)	S-Extractiion Efficiency (%)
[BMIM][N(CN)$_2$]	Gasoline	300	1:1	298	20	18.63
[BMIM][N(CN)$_2$]	Diesel	380	1:1	298	20	32.82
[HPy][BF$_4$]	Diesel	83.2	1:1	RT	15	16.7
[OPy][BF$_4$]	Diesel	83.2	1:1	RT	15	24.3
[BMIM][BF$_4$]	Diesel	1572	1:5	RT	20	8.23
[C$_4$Py][BF$_4$]	Diesel	1572	1:5	RT	20	7
[C$_8$Py][BF$_4$]	Diesel	1710	1:5	RT	20	8.16
[BMIM][BF$_4$]	Diesel	385	1:1	303	30	55.8
			1:3	303	30	40.1
			1:5	303	30	27.7
[BMIM][BF$_4$]	Gasoline	180	1:1	303	30	70
			1:3	303	30	59.2
			1:5	303	30	44.8
[EMIM][OAc]	Gasoline	240	1:1	298	20	38.7
	Diesel	1000	1:1	298	20	12
[HMpy][NTf$_2$]	Gasoline	240	1:1	298	20	30
	Diesel	1000	1:1	298	20	37.8

and complicate analysis. This reality showed the way for us to think that which IL could have healthier sulphur removal efficiency, since their ability to extract mercaptans with respect to the other ILs.

As opposed to model fuel oils, ILs-EDS technique for real fuel oils is complicated due to the presence of wide-ranging amount of heterocyclic and polymatic compounds, which are derivates of thiophene and mercaptans, and only a limited number of literatures reported the results for real fuel oils (Abro et al 2014). Table 13 summarizes the results of ILs-EDS for real fuel oils at optimized conditions.

CHALLENGES AND LIMITATION OF ILS IN DESULPHURIZATION

The compulsory low sulphur limit in fuel can only be met by HDS through extreme operation conditions in terms of pressure and residence time. Therewith, the process becomes uneconomic. Aiming at the replacement or supplement of the present inefficient and expensive HDS, to remove those heterocyclic S-compounds to produce clean ultra-low S or S-free fuel oils, ODS using ILs have been intensively studied recently. The ODS process is usually carried out under mild conditions that present competitiveness over the conventional HDS process.

Although the research results presented and discussed above indicate a good perspective for such a new method. There are some issues to be considered such as treatment of waste sulphones as oxidation product of S-compounds, cost of ILs and selectivity of ODS. In addition, most studies performed currently are focused on the investigation of model fuel oils, and researchers are encouraged to investigate the desulfurization of real feedstock such as gasoline and diesel fuels. Compared with many studies on desulfurization experiments, the desulfurization mechanism are rather scarce, and some computer simulations coupled with spectrum characterization studies will favour uncovering the mechanism. In near future, to promote the final pilot test or industrial application of such a new technology, some studies are desired, e.g., development of cheaper and more active ILs; set-up of kilogram/ton-scale of desulfurization process; optimization of operation conditions; investigations of loss, lifetime/stability, regeneration and reusability of ILs; evaluations on oil yield and oil quality.

Remarks

As the demand for cleaner oils continues to raise, more stringent legislations to lower sulphur content in fuel oils is expected. Thus, these aforementioned alternative techniques will continue to be fully studied because of the inability of the traditional HDS technology to meet growing demands. The study suggests deep ODS using Lewis acidic ILs can guide the deep Desulphurization of fuel oil. In ODS an optimization of the oxidation and extraction conditions was done, taking into account catalyst type, ratio of catalyst and oxidant, temperature, mixing and settling times, etc. ODS with homogeneous catalytic systems need large amount of ILs that are costly and may affect the economic viability of a potential process. The recovery of the ILs is another problem. To solve these problems, many groups have committed to heterogeneous catalysis system and turned to studying supported ILs. Immobilizing a certain amount of ILs on solid carriers combined the advantages of ILs with the easy separation and recovery performances of the supported materials.

ILs-EDS is potential preferable method to remove thiophenic sulphur compounds than traditional HDS technology. But sulphur extraction efficiency is not very satisfactory for high viscous ILs. They usually requires about ~5 extraction steps with longer thermal equilibrium time (eg., 30 minutes) at higher temperatures to lower the S-content to <10ppm. Low viscous ILs are proven to be desired solvents in heterogeneous liquid-liquid systems because they facilitate faster mass, require less equilibrium time and lower temperature, and offer the necessary and limited power requirements for mixing and regeneration. Since ILs are very expensive, a number of factors like thermal/chemical stability, ability to be regenerated and recycled, mutual solubility with oils, etc need to be considered during the selection. ILs are regenerable with negligible activity loss after a limited cycles. In real fuel oils check, similar conclusions are obtained; while the sulphur removal efficiency is lowered, which is ascribed to a complex composition of real fuel oil. Because extraction of sulphur compounds from oils by ILs is not a pure physical interaction process, ILs can be tailored to give better desulphurization performance for real fuel oils in future.

For the real fuel oils, low viscous ILs gave considerable results. In addition, the influence of temperature on ILs-EDS of real fuel oils is also insignificant mainly due to the low viscosity. This favors industrial applications since less energy and thus low costs are required. However, the EDS of real fuel oils are characterized by low S-selectivity (less than 65%), a situation that can be attributed to the presence of many and varied polyaromatic compounds and heterocycles which tend to hinder extraction and complicate analysis. More studies such as expanding IL species, more check of real fuel oils and desulphurization mechanism are waiting for us to do for promoting their final industrial fate. For commercialization of alternate desulphurization technologies discussed here, it needs to meet quite a stringent set of criteria, some of the more important of which include,

1. Less unit operations,
2. Minimum consumption of hydrogen or expensive chemicals,
3. Minimum generation of waste streams,
4. High product volume (>99%) of required purity,
5. Long process cycle life,
6. Feed flexibility, and
7. Overall value to refiner.

REFERENCES

Abro, R., Gao, S., Chen, X., Yu, G., Abdeltawabb, A. A., & Al-Deyabb, S. S. (2015). Oxidative Desulfurization of Gasoline by Ionic Liquids Coupled with Extraction by Organic Solvents. *Journal of the Brazilian Chemical Society*. doi:10.5935/0103-5053.20150355

Abro, R., Abdeltawab, A. A., Al-Deyab, S. S., Yu, G., Qazi, A. B., Gao, S., & Chen, X. (2014). A review for extractive desulphurization of fuel oils using ILs. *RSC Advances*, *4*(67), 35302–35317. doi:10.1039/C4RA03478C

Ali, M. F., Al-Malki, A., El-Ali, B., Martinie, G., & Siddiqui, M. N. (2006). Deep desulphurization of gasoline and diesel fuels using non-hydrogen consuming techniques. *Fuel*, *85*(10-11), 1354–1363. doi:10.1016/j.fuel.2005.12.006

Amarasekara, S. A., & Owereh, O. S. (2010). Synthesis of a sulfonic acid functionalized acidic ionic liquid modified silica catalyst and applications in the hydrolysis of cellulose. *Catalysis Communications*, *11*(13), 1072–1075. doi:10.1016/j.catcom.2010.05.012

Ania, C. O., Parra, J. B., Arenillas, A., Rubiera, F., Bandosz, T. J., & Pis, J. J. (2007). On the mechanism of reactive adsorption of dibenzothiophene on organic waste derived carbons. *Applied Surface Science*, *253*(13), 5899–5903. doi:10.1016/j.apsusc.2006.12.065

Anugwom, I., Salmi, T., & Mikkola, J.-P. (2011). Ionic liquid assisted extraction of nitrogen and sulphur-containing air pollutants from model oil and regeneration of the spent ionic liquid. *Journal of Environmental Protection*, *2*(06), 796–802. doi:10.4236/jep.2011.26091

Asumana, C., Haque, M. R., Yu, L., Wu, X., Chen, X., & Yu, G. (2013). Desulphurization of real fuel oils by extraction with ILs. *Separation Science and Technology*, *48*(17), 2582–2588. doi:10.1080/01496395.2013.804559

Asumana, C., Yu, G., Li, X., Zhao, J., Liu, G., & Chen, X. (2010). Extractive desulfurization of fuel oils with low-viscosity dicyanamide-based ionic liquids. *Green Chemistry*, *12*(11), 2030–2037. doi:10.1039/c0gc00118j

Babich, I. V., & Moulijn, J. A. (2003). Science and technology of novel processes for deep desulphurization of oil refinery streams: A review. *Fuel*, *82*(6), 607–631. doi:10.1016/S0016-2361(02)00324-1

Baeza, P., Aguila, G., Gracia, F., & Araya, P. (2008). Desulfurization by adsorption with copper supported on zirconia. *Catalysis communications*, 9(5), 751-755.

Baxendale, J. H., Evans, M. G., & Kilham, J. K. (1946). The kinetics of polymerisation reactions in aqueous solution. *Transactions of the Faraday Society*, *42*, 668–675. doi:10.1039/tf9464200668

Bosmann, A., Datsevich, L., Jess, A., Lauter, A., Schmitz, C., & Wasserscheid, P. (2001). Deep desulphurization of diesel fuel by extraction with ILs. Chemica*l Communications* (23), 2494-2495. doi:10.1039/B108411A

Bourlinos, B. A., Herrera, R., Chalkias, N., Jiang, D. D., Zhang, Q., Archer, L. A., & Giannelis, E. P. (2005). Surface-Functionalized Nanoparticles with Liquid-Like Behavior. *Advanced Materials*, *17*(2), 234–237. doi:10.1002/adma.200401060

Campos-Martin, J. M., Capel-Sanchez, M. C., Perez-Presas, P., & Fierro, J. L. G. (2010). Oxidative processes of desulphurization of liquid fuels. *Journal of Chemical Technology and Biotechnology (Oxford, Oxfordshire)*, *85*(7), 879–890. doi:10.1002/jctb.2371

Cavani, F., Ferroni, L., Frattini, A., Lucarelli, C., Mazzini, A., Raabova, K., & Babini, P. et al. (2011). Evidence for the presence of alternative mechanisms in the oxidation of cyclohexanone to adipic acid with oxygen, catalysed by Keggin polyoxometalates. *Applied Catalysis A, General*, *391*(1-2), 118–124. doi:10.1016/j.apcata.2010.04.032

Cevasco, G., & Chiappe, C. (2014). Are ionic liquids a proper solution to current environmental challenges? *Green Chemistry*, *16*(5), 2375–2385.

Chen, X., Guan, Y., Abdeltawab, A. A., Al-Deyab, S. S., Yuan, X., Wang, C., & Yu, G. (2015). Using functional acidic ionic liquids as both extractant and catalyst in oxidative desulfurization of diesel fuel: An investigation of real feedstock. *Fuel*, *146*, 6–12. doi:10.1016/j.fuel.2014.12.091

Chen, X., Song, D., Asumana, C., & Yu, G. (2012). Deep oxidative Desulphurization of diesel fuels by Lewis acidic ILs based on 1-*n*-butyl-3-methylimidazolium metal chloride. *Journal of Molecular Catalysis A Chemical*, *359*, 8–13. doi:10.1016/j.molcata.2012.03.014

Cooper, B., & Knudsen, K. (2006). Ultra deep desulfurization of diesel: How an understanding of the underlying kinetics can reduce investment costs. *Practical Advances in Petroleum Processing*: 297-316.

Cooper, M.S., & Heaney, , HNewbold, A.J.,, & Sanderson, , W.R. (1991). Oxidation Reactions Using Urea-Hydrogen Peroxide. A Safe Alternative to Anhydrous Hydrogen Peroxide. *ChemInform*, *22*(5).

Cortes-Jácome, M. A., Morales, M., Angeles Chavez, C., Ramírez-Verduzco, L. F., López-Salinas, E., & Toledo-Antonio, J. A. (2007). WO x/TiO2 catalysts via titania nanotubes for the oxidation of dibenzothiophene. *Chemistry of Materials*, *19*(26), 6605–6614. doi:10.1021/cm702010k

Docherty, K. M., & Kulpa, C. F. Jr. (2005). Toxicity and antimicrobial activity of imidazolium and pyridinium ionic liquids. *Green Chemistry*, *7*(4), 185–189. doi:10.1039/b419172b

Dong, Y., Nie, Y., & Zhou, Q. (2013). Highly efficient oxidative desulfurization of fuels by Lewis acidic ionic liquids based on iron chloride. *Chemical Engineering & Technology*, *36*(3), 435–442. doi:10.1002/ceat.201200570

Endres, F. (2004). Ionic Liquids: Promising solvents for electrochemistry. *Impact Factor*, 255-284. DOI:10.1524/zpch.218.2.255.25920

Eßer, J., Wasserscheid, P., & Jess, A. (2004). Deep desulfurization of oil refinery streams by extraction with ionic liquids. *Green Chemistry*, *6*(7), 316–322. doi:10.1039/b407028c

Ge, J., Zhou, Y., Yang, Y., & Xue, M. (2011). Catalytic oxidative desulphurization of gasoline using ionic liquid emulsion system. *Industrial & Engineering Chemistry Research*, *50*(24), 13686–13692. doi:10.1021/ie201325e

Giernoth, R. (2010). Task-Specific Ionic Liquids. *Angewandte Chemie International Edition*, *49*(16), 2834–2839. doi:10.1002/anie.200905981 PMID:20229544

Gui, J., Liu, D., Sun, Z., Liu, D., Min, D., Song, B., & Peng, X. (2010). Deep oxidative desulfurization with task-specific ionic liquids: An experimental and computational study. *Journal of Molecular Catalysis A Chemical*, *331*(1), 64–70. doi:10.1016/j.molcata.2010.08.003

Hallett, J. P., & Welton, T. (2011). Room-temperature ILs: Solvents for synthesis and catalysis. 2. *Chemical Reviews*, *111*(5), 3508–3576. doi:10.1021/cr1003248 PMID:21469639

http://www.iea-amf.org/content/fuel_information/diesel_gasoline, http://ec.europa.eu/, (Essar et al, 2004).

Huang, C.-P., Liu, Z.-C., Xu, C.-M., Chen, B.-H., & Liu, Y.-F. (2004). Effects of additives on the properties of chloroaluminate ionic liquids catalyst for alkylation of isobutane and butene. *Applied Catalysis A, General*, *277*(1), 41–43. doi:10.1016/j.apcata.2004.08.019

Huang, D., Lu, Y. C., Wang, Y. J., Yang, L., & Luo, G. S. (2008). Intensification of catalytic oxidation with a T-junction microchannel reactor for deep desulfurization. *Industrial & Engineering Chemistry Research*, *47*(11), 3870–3875. doi:10.1021/ie701781r

Petkovic, M., Seddon, K. R., Rebelo, L. P. N., & Silva Pereira, C. (2011). Ionic liquids: A pathway to environmental acceptability. *Chemical Society Reviews*, *40*(3), 1383–1403. doi:10.1039/C004968A PMID:21116514

Ito, E., & van Veen, J. A. R. (2006). On novel processes for removing sulphur from refinery streams. *Catalysis Today*, *116*(4), 446–460. doi:10.1016/j.cattod.2006.06.040

Kabe, T., Qian, W. H., Ogawa, S., & Ishihara, A. (1993). Mechanism of Hydro Desulphurization of Dibenzothiophene on Co-Mo/Al2O 3 and Co/Al 2O3 Catalyst by the Use of Radioisotope sup S Tracer. *Journal of Catalysis*, *143*(1), 239–248. doi:10.1006/jcat.1993.1269

Komintarachat, C., & Trakarnpruk, W. (2006). Oxidative desulfurization using polyoxometalates. *Industrial & Engineering Chemistry Research*, *45*(6), 1853–1856. doi:10.1021/ie051199x

Kulkarni, Prashant S., & Afonso, Carlos A. M. (2010). Deep desulphurization of diesel fuel using ILs: Current status and future challenges. *Green Chemistry,* 12(7), 1139-1149. doi:10.1039/C002113J

Larrubia, M. A., Gutièrrez-Alejandre, A., Ramìrez, J., & Busca, G. (2002). A FT-IR study of the adsorption of indole, carbazole, benzothiophene, dibenzothiophene and 4, 6-dibenzothiophene over solid adsorbents and catalysts. *Applied Catalysis A, General*, *224*(1-2), 167–178. doi:10.1016/S0926-860X(01)00769-4

Lee, S.H.D., Kumar, R., & Krumpelt, M. (2002). Sulfur removal from diesel fuel-contaminated methanol. *Separation and Purification Technology*, *26*(2), 247–258.

Li, C., Jiang, Z., Gao, J., Yang, Y., Wang, S., Tian, F., & Han, C. et al. (2004). Ultra-Deep Desulfurization of Diesel: Oxidation with a Recoverable Catalyst Assembled in Emulsion. *Chemistry (Weinheim an der Bergstrasse, Germany)*, *10*(9), 2277–2280. PMID:15112217

Li, C., Li, D., Zou, S., Li, Z., Yin, J., Wang, A., & Zhao, Q. et al. (2013). Extraction desulfurization process of fuels with ammonium-based deep eutectic solvents. *Green Chemistry*, *15*(10), 2793–2799. doi:10.1039/c3gc41067f

Liang, W., Zhang, S., Li, H., & Zhang, G. (2013). Oxidative desulphurization of simulated gasoline catalyzed by acetic acid-based ILs at room temperature. *Fuel Processing Technology*, *109*(0), 27–31. doi:10.1016/j.fuproc.2012.09.034

Lissner, E., de Souza, W. F., Ferrera, B., & Dupont, J. (2009). Oxidative desulphurization of fuels with task-specific ILs. *ChemSusChem*, *2*(10), 962–964. doi:10.1002/cssc.200900171 PMID:19795435

Liu, D., Gui, J., Song, L., Zhang, X., & Sun, Z. (2008). Deep desulfurization of diesel fuel by extraction with task-specific ionic liquids. *Petroleum Science and Technology*, *26*(9), 973–982. doi:10.1080/10916460600695496

Lü, H., Gao, J., Jiang, Z., Jing, F., Yang, Y., Wang, G., & Li, C. (2006). Ultra-deep desulfurization of diesel by selective oxidation with $[C_{18}H_{37}N(CH_3)_3]_4[H_2NaPW_{10}O_{36}]$ catalyst assembled in emulsion droplets. *Journal of Catalysis*, *239*(2), 369–375. doi:10.1016/j.jcat.2006.01.025

Lu, L., Cheng, S., Gao, J., Gao, G., & He, M.-y, . (2007). Deep oxidative desulfurization of fuels catalyzed by ionic liquid in the presence of H2O2. *Energy & Fuels*, *21*(1), 383–384. doi:10.1021/ef060345o

Lu, W., Fadeev, A. G., Qi, B. H., Smela, E., Mattes, B. R., & Ding, J. et al.. (2002). Use of ILs for pi-conjugated polymer electrochemical devices. *Science*, *297*(5583), 983–987. doi:10.1126/science.1072651 PMID:12098704

Mei, H., Mei, B. W., & Yen, T. F. (2003). A new method for obtaining ultra-low sulphur diesel fuel via ultrasound assisted oxidative Desulphurization. *Fuel*, *82*(4), 405–414. doi:10.1016/S0016-2361(02)00318-6

Mochizuki, Y., & Sugawara, K. (2008). Removal of organic sulfur from hydrocarbon resources using ionic liquids. *Energy & Fuels*, *22*(5), 3303–3307. doi:10.1021/ef800400k

Ngamcharussrivichai, C., Chatratananon, C., Nuntang, S., & Prasassarakich, P. (2008). Adsorptive removal of thiophene and benzothiophene over zeolites from Mae Moh coal fly ash. *Fuel*, *87*(10-11), 2347–2351. doi:10.1016/j.fuel.2007.10.003

Otsuki, S., Nonaka, T., Takashima, N., Qian, W., Ishihara, A., Imai, T., & Kabe, T. (2000). Oxidative desulfurization of light gas oil and vacuum gas oil by oxidation and solvent extraction. *Energy & Fuels*, *14*(6), 1232–1239. doi:10.1021/ef000096i

Pedernera, E., Reimert, R., Nguyen, N. L., & van Buren, V. (2003). Deep Desulphurization of middle distillates: Process adaptation to oil fractions' compositions. *Catalysis Today*, *79*, 371–381. doi:10.1016/S0920-5861(03)00066-X

Rodríguez-Cabo, B., Rodríguez, H., Rodil, E., Arce, A., & Soto, A. (2014). Extractive and oxidative-extractive desulfurization of fuels with ionic liquids. *Fuel*, *117*, 882–889. doi:10.1016/j.fuel.2013.10.012

Wang, J., Zhao, D., & Li, K. (2009). Oxidative desulfurization of dibenzothiophene catalyzed by Brönsted acid ionic liquid. *Energy & Fuels*, *23*(8), 3831–3834. doi:10.1021/ef900251a

Wang, X., Han, M., Wan, H., Yang, C., & Guan, G. (2011). Study on extraction of thiophene from model gasoline with brønsted acidic ionic liquids. *Frontiers of Chemical Science and Engineering*, *5*(1), 107–112. doi:10.1007/s11705-010-0539-0

Wasserscheid, P., & Jess, A. (2004). Deep Desulphurization of oil refinery streams by extraction with ILs. *Green Chemistry*, *6*(7), 316–322. doi:10.1039/B407028C

Wasserscheid, P., van Hal, R., & Bösmann, A. (2002). 1-n-Butyl-3-methylimidazolium ([bmim]) octylsulfate an even 'greener' ionic liquid. *Green Chemistry*, *4*(4), 400–404. doi:10.1039/b205425f

Weingärtner, H. (2008). Understanding ionic liquids at the molecular level: Facts, problems, and controversies. *Angewandte Chemie International Edition*, *47*(4), 654–670. doi:10.1002/anie.200604951 PMID:17994652

Welton, T. (1999). Room-temperature ILs. Solvents for synthesis and catalysis. *Chemical Reviews*, *99*(8), 2071–2084. doi:10.1021/cr980032t PMID:11849019

Xun, S., Zhu, W., Zheng, D., Zhang, L., Liu, H., Yin, S., & Li, H. et al. (2014). Synthesis of metal-based ionic liquid supported catalyst and its application in catalytic oxidative desulphurization of fuels. *Fuel*, *136*(0), 358–365. doi:10.1016/j.fuel.2014.07.029

Yang, R. T., Hernández-Maldonado, A. J., & Yang, F. H. (2003). Desulfurization of transportation fuels with zeolites under ambient conditions. *Science*, *301*(5629), 79–81. doi:10.1126/science.1085088 PMID:12843388

Yansheng, C., Changping, L., Qingzhu, J., Qingshan, L., Peifang, Y., Xiumei, L., & Welz-Biermann, U. (2011). Desulfurization by oxidation combined with extraction using acidic room-temperature ionic liquids. *Green Chemistry*, *13*(5), 1224–1229. doi:10.1039/c0gc00745e

Yu, F.-L., Wang, Y.-Y., Liu, C.-Y., Xie, C.-X., & Yu, S.-T. (2014). Oxidative desulfurization of fuels catalyzed by ammonium oxidative-thermoregulated bifunctional ionic liquids. *Chemical Engineering Journal*, *255*, 372–376. doi:10.1016/j.cej.2014.06.087

Yu, G., Li, X., Liu, X., Asumana, C., & Chen, X. (2011). Deep desulfurization of fuel oils using low-viscosity 1-ethyl-3-methylimidazolium dicyanamide ionic liquid. *Industrial & Engineering Chemistry Research*, *50*(4), 2236–2244. doi:10.1021/ie1013103

Yu, G., Zhao, D., Wen, L., Yang, S., & Chen, X. (2012). Viscosity of ionic liquids: Database, observation, and quantitative structure-property relationship analysis. *AIChE Journal. American Institute of Chemical Engineers*, *58*(9), 2885–2899. doi:10.1002/aic.12786

Yu, G., Zhao, J., Song, D., Asumana, C., Zhang, X., & Chen, X. (2011). Deep oxidative desulfurization of diesel fuels by acidic ionic liquids. *Industrial & Engineering Chemistry Research*, *50*(20), 11690–11697. doi:10.1021/ie200735p

Zhang, C., Pan, X., Wang, F., & Liu, X. (2012). Extraction–oxidation desulfurization by pyridinium-based task-specific ionic liquids. *Fuel*, *102*, 580–584. doi:10.1016/j.fuel.2012.07.040

Zhang, J., Wang, A., Li, X., & Ma, X. (2011). Oxidative desulfurization of dibenzothiophene and diesel over [Bmim] 3 PMo 12 O 40. *Journal of Catalysis*, *279*(2), 269–275. doi:10.1016/j.jcat.2011.01.016

Zhang, J., Zhu, W., Li, H., Jiang, W., Jiang, Y., Huang, W., & Yan, Y. (2009). Deep oxidative desulfurization of fuels by Fenton-like reagent in ionic liquids. *Green Chemistry*, *11*(11), 1801–1807. doi:10.1039/b914130h

Zhang, W., Xu, K., Zhang, Q., Liu, D., Wu, S., Verpoort, F., & Song, X.-M. (2010). Oxidative desulfurization of dibenzothiophene catalyzed by ionic liquid [BMIm] HSO4. *Industrial & Engineering Chemistry Research*, *49*(22), 11760–11763. doi:10.1021/ie100957k

Zhao, Di-shun, Zhi-min Sun, Fa-tang Li, & Hai-dan Shan. (2009). Optimization of oxidative desulfurization of dibenzothiophene using acidic ionic liquid as catalytic solvent. *Journal of fuel chemistry and technology,* 37(2), 194-198. doi:10.1016/S18725813(09)60015-3

Zhu, W., Wu, P., Chao, Y., Li, H., Zou, F., Xun, S., & Zhao, Z. et al. (2013). A Novel Reaction-Controlled Foam-Type Polyoxometalate Catalyst for Deep Oxidative Desulfurization of Fuels. *Industrial & Engineering Chemistry Research*, *52*(49), 17399–17406. doi:10.1021/ie402513x

(2011). Zhu, WenShuai, Huaming Li, QingQing Gu, Peiwen Wu, Guopeng Zhu, Yongsheng Yan, & Guangying Chen. Kinetics and mechanism for oxidative desulfurization of fuels catalyzed by peroxomolybdenum amino acid complexes in water-immiscible ionic liquids. *Journal of Molecular Catalysis A Chemical*, *336*(1), 16–22.

ENDNOTE

1 Abdul Waheed Bhutto and Rashid Abro have equal contributions to this chapter.

Chapter 11
Catalysis with Room Temperature Ionic Liquids Mediated Metal Nanoparticles

Youquan Deng
Lanzhou Institute of Chemical Physics, Chinese Academy of Sciences, China

Shimin Liu
Lanzhou Institute of Chemical Physics, Chinese Academy of Sciences, China

ABSTRACT

There has been still a growing interest in ionic liquids (ILs) for catalysis, in particular considering their tremendous success made over the past decades. Indeed, ILs for catalysis has been a main subject in chemistry and technology, and a rough estimate is that more than 6000 publications involving IL catalysis have been reported in the past fifteen years. Since there have been a large number of excellent reviews and books concerning the catalysis in ILs, in this chapter the authors mainly focus on the IL immobilized nano- or super-fine metal particles for catalysis, which could bridge or fill the gap between homogeneous and heterogeneous catalysis. Detailed IL-immobilized catalyst preparation, characterization and their application in hydrogenation, C-C coupling, oxidation, etc. will be discussed.

INTRODUCTION

Over the past 30 years, perhaps the most exciting scientific and technological progress in chemical science and chemical engineering is the discovery and recognition of ionic liquids (ILs). As a matter of fact, ILs are not new and the first room temperature IL $[EtNH_3][NO_3]$ that has a melting point of 12 °C was described as early as 1914, and was primarily prepared for electric conductivity measurements (Sugden & Wilkins, 1929). Generally, RTILs are defined as those molten salts composing of bulky organic cations (such as imidazolium, pyridinium, quaternary ammonium, phosphonium, *etc.*,) and inorganic or organic anions, which melt around 100°C or below as an arbitrary temperature limit. A series of typical ILs are shown in Figure 1.

DOI: 10.4018/978-1-4666-9975-5.ch011

Figure 1. Typical structure of commercially available ILs

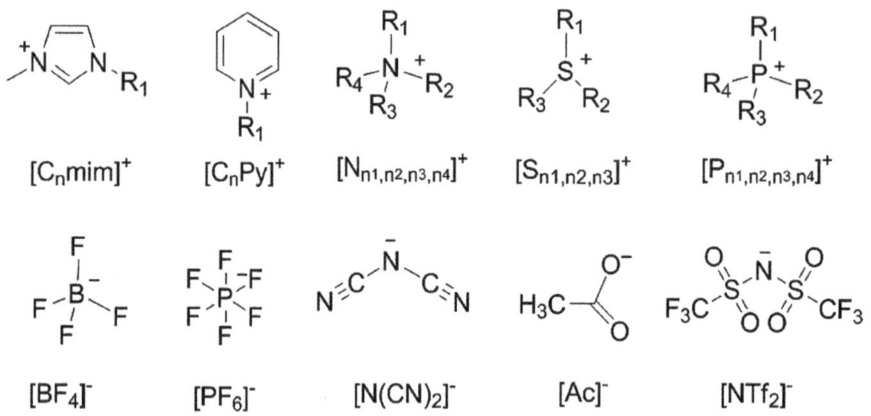

Catalytic reactions in ILs have been examined for about 30 years, and the first example of RTILs for catalysis is using dialkylimidazolium chloroaluminate ILs as catalysts for Friedel-Crafts reactions in 1986 (Boon, Levisky, Pflug, & Wilkes, 1986). Although the chloroaluminate ILs are relatively cheaper and exhibit good catalytic performance, they are air and water sensitive, which limited their further application in catalysis as well as in many other fields. In the 1990s, a significant milestone was passed as the imidazolium salts incorporated the tetrafluoroborate or hexafluorophosphate anions to form air- and water-stable ILs (Wilkes & Zaworotko, 1992). Since then, ILs began to bring a green revolution in the fields of catalytic science and all kinds of 'green' catalytic reactions or processes in or by ILs were gradually established.

No doubt that ILs have achieved great success in catalysis by acting as reaction media, catalysts, or both. Most often the reasons for this achievement include: 1) ILs could control both the direction and selectivity of chemical reactions due to their special "ionic environment", including stronger electrostatic field, peculiar micro-environment, existence of multidimensional weak interactions; 2) ILs could provide a two-phase systems for reaction, or the catalyst could be entrapped or ''immobilized" with ILs allowing catalytic system recycle; 3) ILs could be designed by varying the structures of both the anion and cation to modulate the physical and chemical properties, which made them apply to various catalytic systems; 4) ILs could endow a catalytic reaction with green character compared with organic solvents because of their lower vapor pressure and less combustibility (Hallett & Welton, 2011; Parvulescu & Hardacre, 2007; Welton, 1999; Q. Zhang, Zhang, & Deng, 2011). In a word, using ILs as catalysts or reaction media or both, unexpected catalytic performance, new reactions and new catalyst materials could be achieved, and even the current situation of catalytic science and technology may be changed greatly.

Up to now, ILs have pervaded in nearly every branch of catalytic science and technology, and in the past ten years the numbers of IL publications for catalysis have been significantly increasing. It is to say that ILs have been successfully involved in conventional homogeneous and heterogeneous catalysis. Furthermore, the incorporation of ILs into nano- or super-fine particle catalysis could also provide a possibility or platform to explore those fascinating catalysis, which could bridge or fill the gap between homogeneous and heterogeneous catalysis.

In this chapter, we will focus on IL immobilized nano- or super-fine particles, and the IL-immobilized nanocatalyst preparation, characterization and their application in various reactions will be discussed in detail. Moreover, the catalytic mechanism of so-called metal nanoparticles (MNPs) as reservoir - leaching - catalytically active metal complexes may vividly display the possible relations between homogeneous and heterogeneous catalysis. Finally, the future development of such IL-incorporated MNPs in the system hybridizing heterogeneous and homogeneous catalysis will be shortly prospected.

CATALYSIS WITH IL-INCORPORATED MNPS

Nanocatalysis has received an increasing attention in scientific research and industrial applications during the last fifteen years, which involves the use of nano-structured materials as catalysts, such as nano particles, nano films and nano porous solids, *etc*. Compared to the bulk counterparts, nano-structured catalytic materials with small diameters and a narrow size distribution have unique properties due to their inherent large surface-to-volume ratio and quantum size effects, which could bring about unexpected catalytic performance in many reactions. Solid supported MNPs catalysts have been widely studied in various processes, and the roles of solid support are the supporter of catalytically active site and the stabilizer of the MNPs. Unfortunately, only part of the MNPs with suitable interaction with the solid supports can form the active sites in this system owing to the rigid properties of MNPs and solid support. In theory, it can construct active site accessible in three-dimensional to reactant molecules if the MNPs are supported by a liquid as support, which might enhance the utilization efficiency of the MNPs greatly (Zahmakiran & Ozkar, 2011). However, the absence of support commonly makes the MNPs inactive and unstable. In this context, ILs may offer a great chance to build active and stable MNPs catalysts with a liquid support because of their nonvolatility, highly thermal stability, tunable structure and ionic environment. Indeed, ILs could act as the "solvents", stabilizers, ligands, and supports for MNPs (Luska, Migowski, & Leitner, 2015; K. L. Luska & A. Moores, 2012; Scholten, Leal, & Dupont, 2012). Moreover, the interaction between ILs and MNPs can be tuned by controlling the structure of ILs, which will generate ILs modified and stabilized MNPs with different electron state and also different catalytic activity, Figure 2. Therefore, a 'solid-state' liquid supported MNPs catalyst with active sites accessible in three-dimensional to reactant molecules can be achieved if combining ILs and MNPs together.

Except the potential improvement in catalytic efficiency, more importantly, this novel concept of IL-supported or mediated MNPs catalysts presents an ideal system to bridge homogeneous and heterogeneous catalytic systems. The MNPs dispersed inside ILs are essentially heterogeneous catalysts.

Figure 2. Schematic illustration of IL-supported or mediated MNPs catalysts

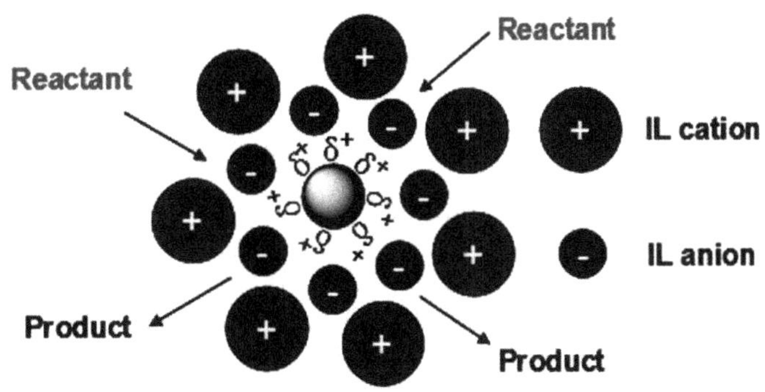

However, far different from the traditional solid supported MNPs catalysts, the MNPs can move freely on or in the support and can be accessible to the reactants in three-dimensions. These are exactly the fundamental characteristics of homogeneous catalysts. Therefore, the ILs incorporated MNPs catalysts are macroscopically homogeneous but microscopically heterogeneous systems containing the major features of these two branches of catalysis. The study of ILs supported MNPs catalysts may find a way to understand the relationship between heterogeneous and homogeneous catalysis.

Preparation of MNPs in ILs

In general, soluble and stable MNPs can be prepared in the presence of ILs, and the main advantage of this strategy is due to the ILs multiple roles in synthesis processes. IL could act as the solvent, NPs stabilizer, ligand, and immobilizing agent, *etc.* (Luska, Migowski, & Leitner, 2015; K. L. Luska & A. Moores, 2012; Scholten, Leal, & Dupont, 2012). In this way, MNPs can be prepared from metal precursors by different physical-chemical methods, such as chemical reduction, decomposition of metal salts or metal complexes, sputter deposition technique, *etc.* Therein, the chemical reduction of metal salts is the most widely used method to prepare MNPs in ILs. The size and shape of the MNPs depends on several factors, including the reaction conditions, the composition of metal precursor, reducing agents, and the nature of ILs, *etc.* (J. Dupont & Scholten, 2010). The often used reducing agents for the synthesis of MNPs cover IL itself, hydrogen, hydrides, *etc.* Moreover, the preparation methods for MNPs in ILs *via* decomposition of precursors and sputter deposition technique will also be introduced in this section.

Chemical Reduction in ILs

ILs Themselves as the Reducing Agent

In some cases, ILs themselves can behave as the reducing agent, especially for the imidazolium-based ILs. It was found that, in Heck reaction, the clusters of zero-valent Pd NPs can be formed and stabilized in ILs through the formation of Pd-biscarbene complexes under ultrasonic irradiation (Deshmukh, Rajagopal, & Srinivasan, 2001). In this way, the [BMIm][BF_4] itself was the reducing agent, and the reductive character of the imidazolium based ILs should be related to the C=C olefinic bonds of the imidazolium cation. TEM analyses showed that the presence of monodispersed grains nearly spherical in shape with average size of ~20 nm, which composed of dispersed particles of approximately 1 nm size. Moreover, the clusters of Pd NPs in ILs were found to be stable even after storage for a week.

Large-size single-crystal gold nanosheets could be successfully prepared in [BMIm][BF_4] by microwave heating of $HAuCl_4$ without any additional template agent (Li et al., 2005). SEM images showed that the products were mainly gold sheet with irregular shapes, such as the regular triangle or hexagon shapes. The thickness of gold nanosheets was estimated about 50 nm, Figure 3. If the IL [BMim][PF_6] was used as the medium, large-scale gold nanosheets with large size were also obtained. In this process, ILs not only worked as reducing agents, but also acted as a template agent for the growth of Au.

Hydrogen and Hydrides as the Reducing Agents

H_2 is a commonly used reducing agent for the preparation of MNPs in ILs. Typically, Rh and Ir NPs could be easily prepared from the hydrogen reduction of $[IrCl(COD)]_2$ (COD = 1,5-cyclooctadiene) or $RhCl_3$ in [BMIm][PF_6]. It is noteworthy that the MNPs should be synthesized in dry IL because the presence

Figure 3. SEM images of the gold nanosheets prepared via microwave heating of $HAuCl_4$ in [BMIm] [BF_4], a) the scale bar is 10 μm; b) the scale bar is 50 nm (Reprinted with permission from ref. (Li et al., 2005). Copyright 2005 American Chemical Society)

(a) (b)

of the water would result in the partial decomposition of IL with the formation of phosphates, HF and transition-metal fluorides. TEM micrographs showed that the separated Rh and Ir NPs possessed mean diameters of 2.3 ± 0.6 nm and 2.1 ± 0.3 nm, respectively (Fonseca, Umpierre, Fichtner, Teixeira, & Dupont, 2003), Figure 4. Moreover, the isolated Ir and Rh NPs could be used as solids ("solventless") or re-dispersed in [BMIm][PF_6] ("biphasic") or in acetone ("homogeneous") for the hydrogenation of benzene under mild reaction conditions.

Using the functionalized ILs (FILs) as the media for preparation of MNPs could provide a metal-binding moiety to affect some properties of the resulting MNPs, including the size, shape and degree of aggregation, etc. More importantly, the introduction of functional groups could change the surrounding electronic characteristics of the MNPs, which might remarkably influence the catalytic performance of the ILs/MNPs system. For example, the Pd NPs could be in situ prepared by reduction of $Pd(OAc)_2$ with atmospheric pressure H_2 and stabilized by 2,2'-dipyridylamine-functionalized imidazolium IL (Hu et al., 2008), Figure 5a. TEM characterization displayed morphology of nearly spherical Pd NPs with an average diameter of 5-6 nm, Figure 5b. It was found that Pd NPs were stabilized effectively by the bidentate nitrogen ligands attached covalently to the IL cations. These FIL-supported Pd NPs were shown to be highly efficient multiphase catalysts for the catalytic conversion of cyclohexene into cyclohexane with up to 100% yield. The excellent catalytic activity of the present system could be attributed to the good stabilization and high dispersion of Pd NPs.

Hydrides (mainly $NaBH_4$) are also often used as reducing agents for the generation of MNPs. In general, $NaBH_4$ is not likely to be used in ILs because it would deprotonate the imidazolium cation to generate carbenes binding with metal surface, and produce various by-products that is difficult to remove from the ILs (J. Dupont & Scholten, 2010). But the MNPs could be conveniently synthesized with ILs using $NaBH_4$ as the reducing agent in the presence of some stabilizers or FILs. For example, irregularly spherical-shaped Au NPs with diameters of 3.7 ± 0.8 nm could be generated from the $NaBH_4$ reduction of $HAuCl_4$ in [BMIm][PF_6] involving low levels of 1-methylimidazole (0.1 mM) (Dash & Scott, 2009). In the absence of 1-methylimidazole, bigger Au NPs with diameters of 4.4 ± 1.2 nm could be obtained, Figure 6, and they would aggregate rapidly in several days. In this system, 1-methylimidazole ligand could help slow the NPs growth in solution and behave as a stabilizer to improve the stability of the final Au NPs.

Figure 4. a) TEM micrographs of the Rh NPs (left) and Ir NPs (right); b) histogram illustrating the M NPs size distribution, Rh NPs (left) and Ir NPs (right) (Reprinted with permission from ref. (Fonseca et al., 2003). Copyright 2003 WILEY-VCH Verlag GmbH & Co. KGaA)

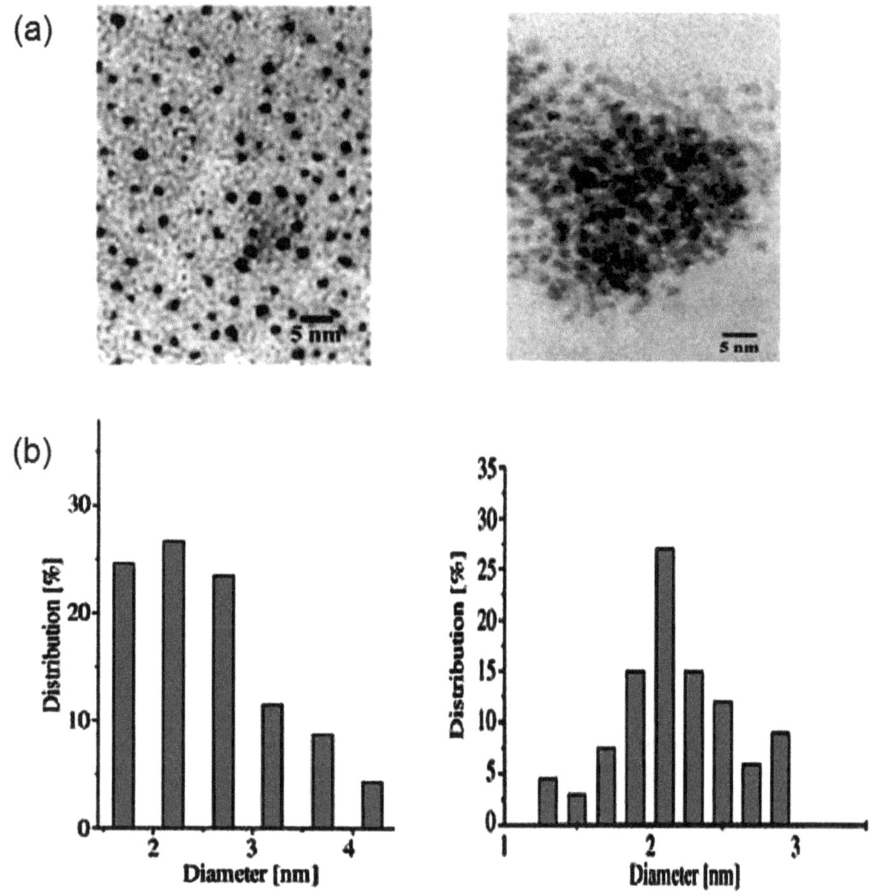

Figure 5. a) 2,2'-dipyridylamine-functionalized imidazolium IL stabilized Pd NPs; b) TEM micrographs of the resulting Pd NPs (Reprinted with permission from ref. (Hu et al., 2008). Copyright 2008 WILEY-VCH Verlag GmbH & Co. KGaA)

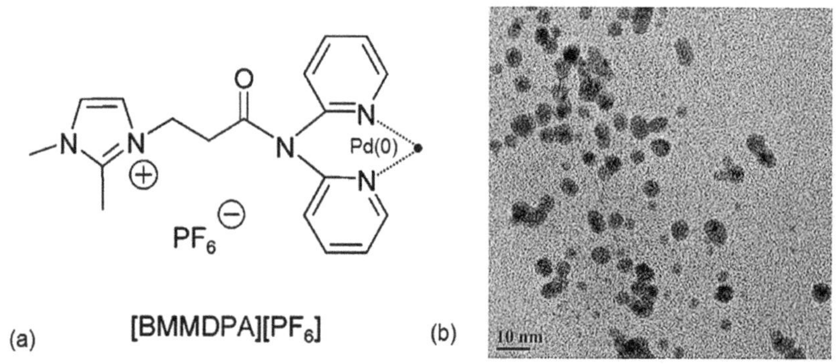

Figure 6. TEM images of Au NPs synthesized in [BMIM][PF$_6$] with (A) no additives added and (B) 1.0 mM 1-methylimidazole added (Reprinted with permission from ref. (Dash & Scott, 2009). Copyright 2009 Royal Society of Chemistry)

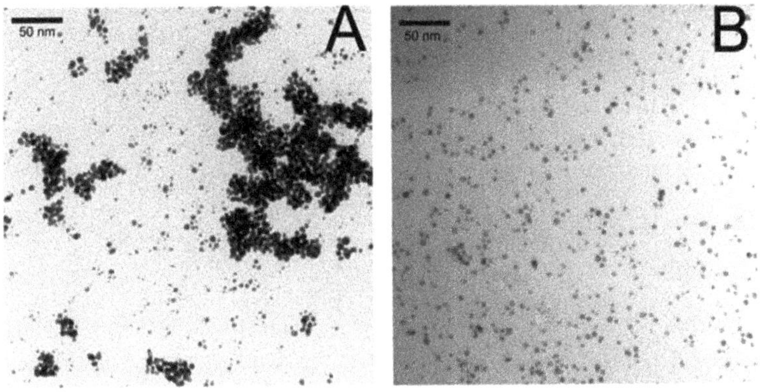

Using FILs as stabilizing agents is another strategy for synthesis of MNPs *via* the reduction method with $NaBH_4$. It was found that FILs containing thiol (Itoh, Naka, & Chujo, 2004), amino and carboxyl groups (H. Zhang & Cui, 2009) could be used for the generation of MNPs. The Au and Pt NPs could be readily obtained in the presence of thiol-FILs in water by treating $HAuCl_4$ and $Na_2Pt(OH)_6$ with $NaBH_4$ (Kim, Demberelnyamba, & Lee, 2004). The formed MNPs were crystalline structures with face-centered cubic packing arrangements and had small sizes (2-3.5 nm) and uniform distributions. Interestingly, the MNPs size and uniformity could be controlled, which depended on the number of thiol groups and their position in ILs. In this case, thiol-FILs surrounding the NP surface could act as the most effective stabilizers of Au and Pt NPs, thus becoming highly dispersible in aqueous media.

Hydrazine monohydrate is a relatively strong reducing agent, which could be employed in the preparation of Au NPs from $HAuCl_4$ precursor in an ether-functionalized IL (Schrekker et al., 2007). Although the relatively strong reducing agent was used, only innocuous byproducts were formed during the preparation process. TEM and XRD analyses showed that the mean diameter of Au NPs was 7.5 nm. It is claimed that there were two stabilization modes for Au NPs to prevent further aggregation, Figure 7, e.g. i) the steric stabilization component due to the arrangement of the tail attached to the imidazolium ring; ii) the cation stabilization component *via* the parallel coordinated imidazolium ring. Besides, the non-noble Cu NPs could be also prepared from $CuSO_4$ in the solvents of ethylene glycol (EG) and [BMIm][BF$_4$] using hydrazine hydrate as the reducing agent (Dewan, Kumar, Saxena, De, & Mozumdar, 2012). The mean diameter of obtained Cu NPs is 40 ± 3 nm with polydispersity index of 0.203. It is worth to note that the formed Cu NPs showed excellent long-term stability over a period of one year.

Other Reducing Agents

Other reducing agents such as CO, trisodium citrate, *etc.*, have also been used to prepare MNPs in ILs. Au NPs can be easily prepared through carbonylation and reduction of hydrochloroauric acid with CO and H_2O in [BMIM][BF$_4$] (Guo, Shi, Gu, Yang, & Deng, 2005). In this case, the reduction rates and Au NPs size could be controlled by regulating the water proportion. With increasing the amount of water added, the Au NPs with average diameters of 1.7, 6.4, and 12.8 nm could be obtained, Figure 8. CO in reaction could serve both as a ligand to form some Au complex and as a reducing agent for facilitat-

Figure 7. Imidazolium cation coordination and stabilization mode (Reprinted with permission from ref. (Schrekker et al., 2007). Copyright 2007 Elsevier)

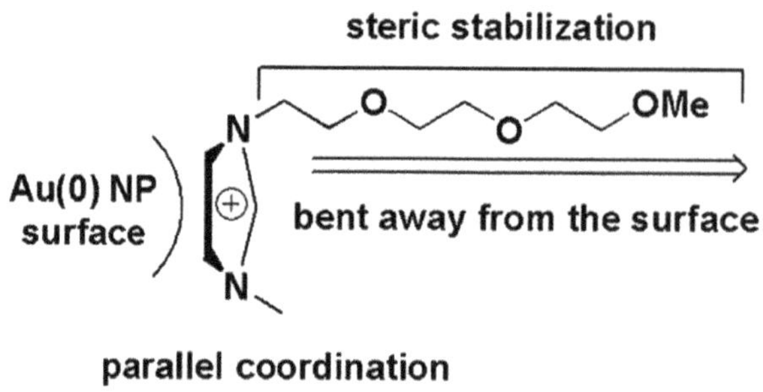

Figure 8. TEM images of Au NPs synthesized in ILs (added with different amount of water, then transferring Au NPs from IL to benzene containing n-tetradecylamine as stabilizer) (A) 0.2mL; (B) 2mL; (C) 10mL (Reprinted from ref. (Guo et al., 2005) with permission from the author)

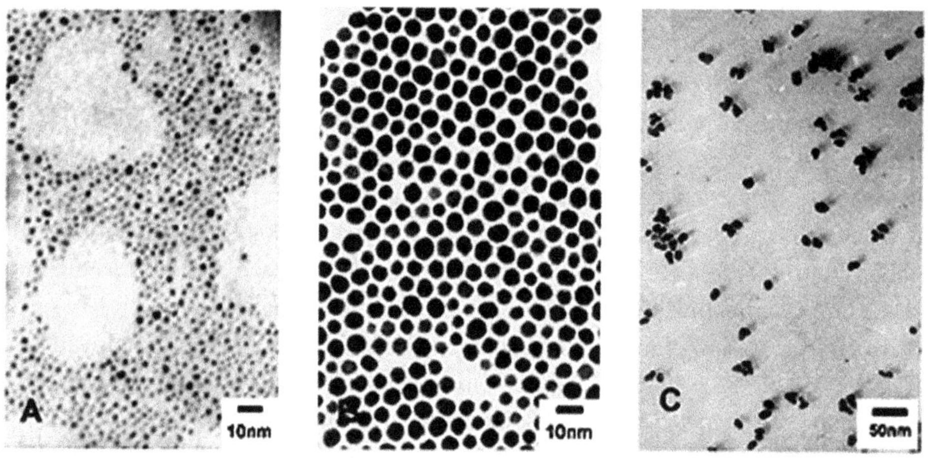

ing the formation of the Au NPs. After the formation of Au NPs, alkylamine as stabilizer was added to prevent the aggregation of Au NPs because the electrostatic stabilization of [BMIm][BF_4] was too weak to stabilize Au NPs for a long time. At the same time, the Au NPs could be easily transferred into toluene by extraction.

The size and shape of the prepared MNPs could be controlled by the nature of the reducing agents. For example, in the two-step synthesis method for preparation of rod-like Au NPs in [EMIM][$EtSO_4$] (Ryu, Sanchez, Keul, Raj, & Bockstaller, 2008), the size and shape of the Au nano-structures could be modulated by the amount of reducing agent in each step. In this strategy, trisodium citrate and sodium borohydride as the relatively strong reducing agents were first used to generate the seed crystals with different size, Figure 9. And then, ascorbic acid stabilized Au^+ species in [EMIm][$EtSO_4$] was added. Finally, the nanocrystals were injected into a "secondary-growth solution" comprising an IL solution of Au^+, Ag^+ and weak reducing agent ascorbic acid to form the target rod-like Au NPs.

Figure 9. Two-step synthesis of gold nano-rods using ILs as the shape-regulating medium (Reprinted with permission from ref. (Ryu et al., 2008). Copyright 2008 WILEY-VCH Verlag GmbH & Co. KGaA)

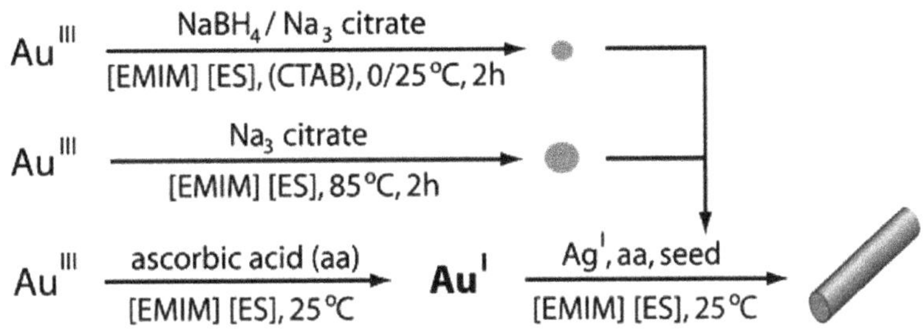

Direct Decomposition from Precursors in ILs

The direct decomposition of metal complex precursors in ILs is a simple and effective strategy to prepare MNPs. In this way, the role of ILs is stabilizing the MNPs through the ionic charge, high polarity interactions, high dielectric constant and supramolecular network. According to the DLVO (Derjaugin–Landau–Verwey–Overbeek) theory, ILs can provide an electrostatic protection in the form of a so called ''protective shell'' for MNPs avoiding extra stabilizing molecules (Krämer, Redel, Thomann, & Janiak, 2008). Another advantage of this method is the clean process for MNPs synthesis, because the volatile small fragments released during the decomposition can be easily removed under reduced pressure.

In imidazolium based ILs, Ni NPs composing of Ni metal core with a cap of NiO could be simply synthesized through the thermal decomposition of [bis(1,5-cyclooctadiene) nickel(0)] organometallic precursor (Migowski et al., 2007). It was found that the diameter, size-distribution and the metal/metal oxide ratio of Ni NPs depended on the structural arrangement of ILs. There were slight decreases in both NPs diameter and the size-distribution with the increase of the numbers of carbons on cations (from 5.9 ± 1.4 nm in [BMIm][NTf_2] to 4.9 ± 0.9 nm in [C_{14}MIm][NTf_2]), and then an increase in diameter and size-distribution for the material in [C_{16}MIm][NTf_2] (5.5 ± 1.1 nm). Moreover, the obtained Ni NPs dispersed in ILs could be used as catalysts for the hydrogenation of olefins under relatively mild reaction conditions.

By virtue of the unique ionic environment in ILs, a series of transition-metal NPs could be prepared in ILs *via* thermal, photolytic decomposition or microwave (MW) irradiation from metal-carbonyl precursors (Redel, Thomann, & Janiak, 2008; C. Vollmer et al., 2010), Figure 10. In this strategy, only gaseous by-product CO produced, which could be rapidly expelled from the reaction medium owing to the low solubility in ILs. Stable Cr, Mo and W NPs with a very small and uniform size of 1-1.5 nm were obtained reproducibly by the thermal or photolytic decomposition in [BMIm][BF_4], and the particle size increased with the molecular volume of the IL anion. As compared with the MNPs prepared through thermal decomposition, the method of photolytic decomposition may produce larger particles due to the faster decomposition rate in ILs. Moreover, the microwave (MW) irradiation technology was also used for the preparation of MNPs in ILs. As compared with UV-photolytic (1000W, 15min) and thermal decomposition (180-250°C, 6-12h), the MW-aided prepared MNPs displayed smaller size (0.8-5 nm) (C. Vollmer et al., 2010), And, the obtained Ru, Rh and Ir NPs/IL dispersions were shown to be highly efficient catalysts for the biphasic liquid-liquid hydrogenation of cyclohexene, and almost quantitative conversion could be achieved with a good recyclability upon 3 runs without metal leaching.

Figure 10. Microwave, photolytic, and thermal decomposition of metal carbonyls to MNPs (Reprinted with permission from ref. (C. Vollmer et al., 2010). Copyright 2010 WILEY-VCH Verlag GmbH & Co. KGaA)

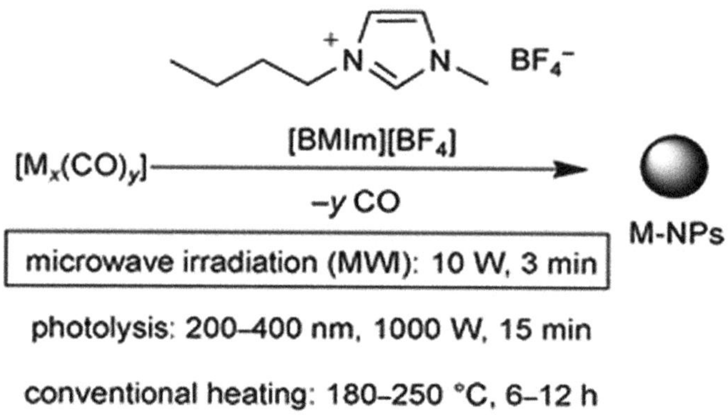

Sputter Deposition Method

The sputter deposition technique has also shown its high efficiency to prepare MNPs in ILs due to its operation simplicity and the reproducibility with relatively uniform sizes, which smartly combines the advantages of sputter deposition to generate atoms or small clusters and the extremely low vapor pressure of ILs. The first study on the sputter deposition technique using ILs as capture media for MNPs preparation was reported by Torimoto et al (Tsukasa Torimoto et al., 2006). Highly dispersive Au NPs with size of 5.5 ± 0.86 nm were prepared in [EMIm][BF_4] without any additional stabilizing agents, Figure 11. The size of MNPs depends on the IL type, *e.g.* the size of resulting Au NPs in [N_{1113}][NTf_2] (1.9 ± 0.86 nm) is much smaller than that in [EMIm][BF_4]. The influence factors regarding the properties of Au NPs were also studied. To be clear, the kinetics of aggregation and sedimentation scale could be affected by the viscosity of the ILs, and small amounts of water have a large detrimental influence on the stability

Figure 11. The schematic diagram of the sputtering apparatus (Reprinted with permission from ref. (Hatakeyama et al., 2011). Copyright 2011 Royal Society of Chemistry)

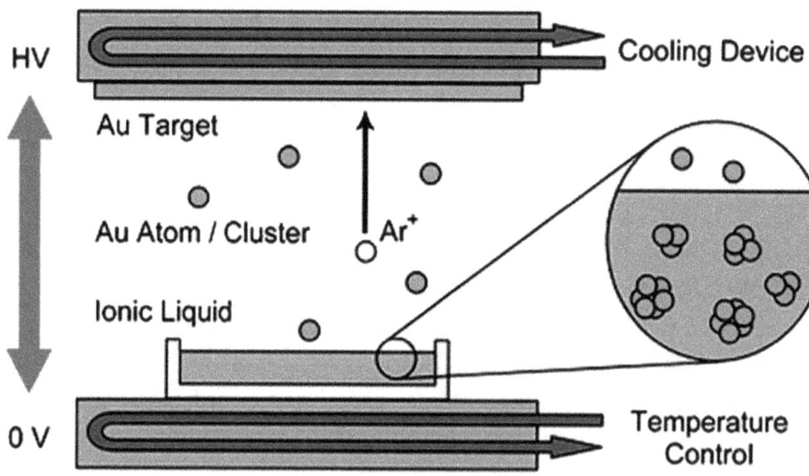

of the colloidal suspensions of the Au NPs in ILs (Vanecht et al., 2012). The temperature of the target and applied voltages, on the other hand, also has obvious influence on the size of Au NPs generated in ILs. For the generation of smaller NPs, lower temperature of the target, higher applied voltage and lower temperature of the capture medium may be desirable (Hatakeyama, Onishi, & Nishikawa, 2011).

It has been known that alloy NPs exhibit unique physicochemical properties, thus they are potential catalysts for a variety of reactions (B. Zhang, Zhan, Zhao, & Li, 2015). Fortunately, the IL chemistry coupled with sputter deposition technique has also provided a unique platform or strategy towards the design of binary alloy NPs, such as AuAg (Okazaki et al., 2008), AuPd (Hirano et al., 2013), AuPt (Suzuki et al., 2012), AuCu (König, Richter, Siegel, Mudring, & Ludwig, 2014), *etc.* More importantly, the chemical composition of the resulting NPs could be fine-tuned by changing the individual surface areas of the metal plates in binary metal targets used for sputter (T. Torimoto et al., 2015).

Homogeneously alloyed bimetallic particles of AuAg could be readily synthesized by simultaneous sputter deposition of Au and Ag in [BMIm][PF_6] without any additional stabilizing agents (Okazaki et al., 2008). It is found that the size distributions of the NPs changed with the alloy chemical composition *e.g.*, the NPs average diameters of pure Au (f_{Au} =1.0), AuAg (f_{Au} =0.5) and pure Ag (f_{Au} =0.5) was 2.6 ± 0.3 nm, 3.1 ± 0.6 nm and 6.0 ± 1.5nm, respectively, Figure 12. Noteworthy, the chemical composition and optical properties of the alloy NPs were easily controlled just by varying the area ratio of the individual pure metal foils in the sputtering targets. Furthermore, the AuPd alloy with an average size of *ca.* 2 nm could be also prepared through the same process in [BMIm][NTf_2] (Hirano et al., 2013). When the NPs were immobilized on highly oriented pyrolytic graphite (HOPG) surface by heat treatment to form a electrode, a slight increase in particles size (*ca.* 2-7 nm) was observed. Interestingly, the electrocatalytic activity for ethanol oxidation of the prepared electrodes could also be controlled by varying the chemi-

Figure 12. TEM images (a) and size distributions (b) of AuAg alloy NPs obtained by the sputter deposition in [BMIm][PF_6] with f_{Au} =0, 0.50 and 1.0. The bars in the pictures represent a length of 5 nm (Reprinted with permission from ref. (Okazaki et al., 2008). Copyright 2007 Royal Society of Chemistry)

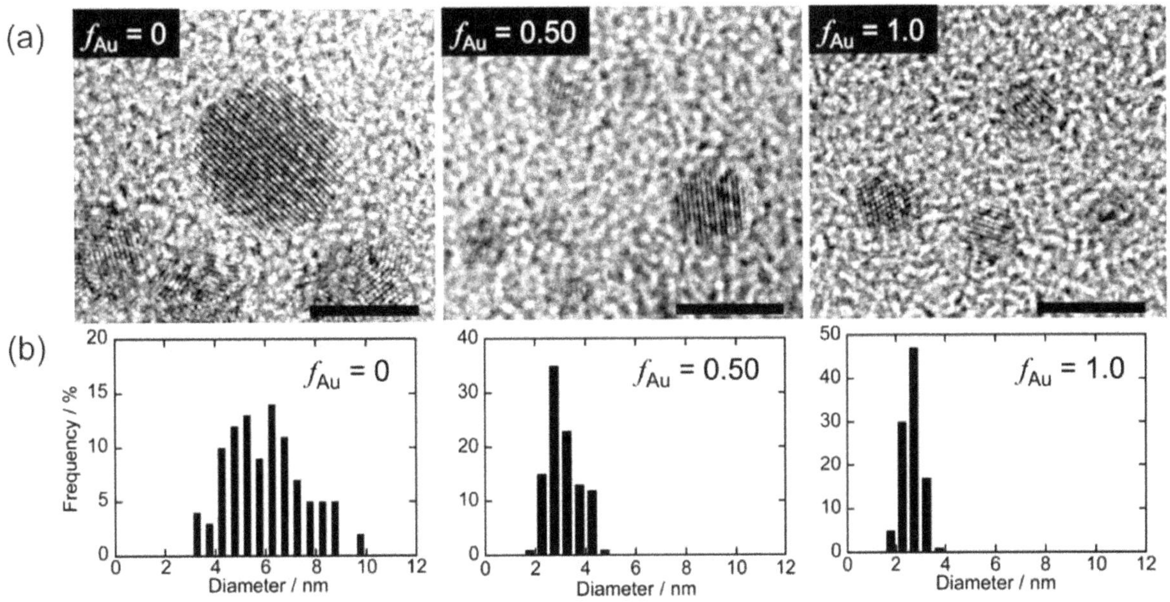

cal composition, and particles with an Au fraction of 0.61 showed the optimum activity. It is concluded that the physico-chemical properties of alloyed NPs such as size distributions, chemical compositions, optical properties, even catalytic activities could be easily modulated in the sputter deposition technique combining with ILs.

To compare the various preparation methods for MNPs conveniently, the advantages and disadvantages of the techniques/methods are summarized in Table 1.

Redispersion of MNPs in ILs

Although MNPs catalysts are highly active initially, they often tend to agglomeration and sintering after a finite number of cycles, which lead to subsequent loss of surface area and catalytic activity. Therefore, finding a feasible path to regenerate the deactivated NPs is more desirable. Scott *et al.* presented a facile, one-pot method to redisperse the sintered NPs in tetraalkylphosphonium chloride ($[P_{6,6,6,14}]$ Cl) *via* halide-promoted oxidation (A. Banerjee, Theron, & Scott, 2013). This protocol has been proved to be workable for a wide variety of transition MNPs, including Au, Pd, Pt, Ru, Rh, Ni, Co, Fe, Ag, Cu. For instance, in the Au NPs redispersion process, large sintered NPs in the $[P_{6,6,6,14}]$ Cl (generated by heating at 150 °C under nitrogen for 24 hours) were oxidized, followed by re-reduction with $LiBH_4$ to produce small Au NPs, which was comparable in size to the ones initially formed. TEM images showed that the initial sintered Au NPs were 8.9 ± 2.7 nm in size, while the particle size decreased to 5.5 ± 1.4 nm after redispersion. Moreover, the activity of redispersed Ni NPs (44% yield) was nearly equal to the initial one before sintering (46% yield) in the test of cyclohexene hydrogenation.

Table 1. The pros and cons of the techniques/methods for the syntheses of MNPs

Preparation Methods	Pros	Cons
Chemical reduction	• This method is generally simple to implement. • A wide range of MNPs could be prepared in this way. • The size of prepared MNPs can be controlled by altering the preparative condition.	• In some cases, the additional stabilizing agents or functional ILs have to be involved for preventing the agglomeration. • Metal precursors or reducing agents may generate by-products that act as extra-stabilisers or poisons.
Direct decomposition from precursors	• In general, no extra stabilizers or capping molecules are needed to achieve stable particle in ILs. • The released fragments during decomposition can be easily removed, which could give bare metal atoms with a contaminant-free surface. • This synthetic process could be finished rapidly by virtue of microwave.	• In the thermal decomposition process, relatively high reaction temperatures should be used to decompose the metal carbonyl compounds. • The ILs may decompose at high temperature, which would poison the MNPs. • The type of MNPs might be limited by the organometallic precursors.
Sputter deposition technique	• This is an extremely clean method to synthesize MNPs. • No additional stabilisers and reducing agents are necessary using this method. • It enables to prepare binary and multinary alloy particles with a desired chemical composition.	• This technique is often limited by the apparatus. • Even very small amounts of water in ILs have a large influence on the stability of the MNPs, intense and fast aggregation would take place in presence of water.

Characterization of IL-Incorporated MNPs

It is feasible to *in situ* characterize ILs using vacuum techniques due to the negligible vapor pressure of ILs. Indeed, several modern techniques have been employed to explore the corresponding information of MNPs either dispersed in the ILs or after isolation. For example, SEM images could give the information of morphology for the MNPs with a relatively large size; TEM analysis could be used to examine the shape and size of the MNPs; XPS could provide the surface information to reveal the interactions between ILs and MNPs; XRD could check the crystalline nature of the MNPs and estimate the mean diameter. The integration of various characterization methods is essential to deeply understand the microstructures- property relationship as well as assembly mechanism of the ILs/MNPs catalytic system.

SEM and TEM

A hierarchical, three-fold symmetrical, single-crystalline gold dendrites were employed as the model, which were synthesized by the reaction between a zinc plate and a solution of $HAuCl_4$ in [BMIM][PF_6] (Qin et al., 2008). Noteworthy, the SEM images could reveal the growth process of gold products on the zinc substrate, i.e., the gold nanocrystals with particle size of 50-200 nm formed initially on the substrate through a direct surface reaction, Figure 13a; and then the subsequent crystal growth preferentially took place on the preformed gold crystals through a primary cell reaction, Figure 13b; the hyperbranched dendrites nanostructure finally grew up with the length of >10 μm, Figure 13c, d. Moreover, each branch had a three-fold symmetrical structure with three groups of parallel rod-like leaves (~50 nm in diameter) grown on the branch symmetrically, Figure 13c.

Figure 13. SEM images of gold products for different reaction time in [BMIM][PF_6]; reaction time: (a) 10 min, (b) 1 h, (c and d) 4.5 h (Reprinted with permission from ref. (Qin et al., 2008). Copyright 2008 American Chemical Society)

Figure 14. a) TEM images of the branch of a gold dendrite with rod-like leaves; b) HRTEM images of the leaf tip. Insets show the ED patterns corresponding to the whole area of the related image. (Reprinted with permission from ref. (Qin et al., 2008). Copyright 2008 American Chemical Society)

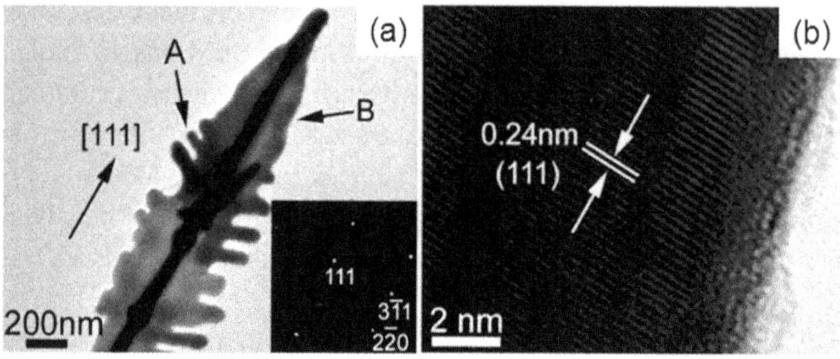

TEM images indicated that the dendrite consists of three groups of parallel branches stands on the copper grid with two groups of parallel branches touching the grid and one group of parallel branches extending upward, which is consistent with the presence of a dark line along the trunk axis (Qin et al., 2008), Figure 14a. The selected area electron diffraction (ED) pattern of the trunk tip suggested that Au dendrite is a single crystal with the trunk grown along the [111] direction. The single crystalline nature of the dendrite was further demonstrated by HRTEM investigation. The leaf tip and the branch edge exhibited lattice fringes with the same orientation and the same spacing (i.e., 0.24 nm,), which was in good agreement with the d values of the (111) lattice spacing of cubic gold crystals (0.235 nm), Figure 14b.

XPS and XRD

XPS analysis could revealed the surface information of the isolated NPs (Ir, Rh, Pt, Ru and Pd) prepared in ILs containing $[PF_6]^-$ and $[BF_4]^-$ anions, in which M-O and M-F contributions could be detected (Migowski & Dupont, 2007; Silveira et al., 2004; Umpierre, Machado, Fecher, Morais, & Dupont, 2005). It is found that the contribution of the M-F component is more distinct in the case of PF_6 anion than for BF_4 anion. According to the F 1s, P and B contributions in the XPS spectra, it is clearly indicated that even relatively low coordinating anions such as PF_6 and BF_4 could coordinate with the metal surface. Moreover, XPS experiments suggested that the MNPs surface is more susceptible to oxidation in air than the bulk metal and that this oxide layer may also be a significant source of stability of the MNPs (Fonseca et al., 2006). In the case of Pt NPs prepared in $[BMIm][PF_6]$, there was also a very small contribution of phosphorus (Scheeren et al., 2006). The F and P signals suggested that the isolated MNPs contained ILs residues. The Pt_{4f} spectra revealed the presence of three chemical states of Pt, i.e. Pt^0, Pt^{2+} and Pt^{4+}, on the NPs surface, which are related to Pt-F, Pt-O, Pt-O-F and other bonds, Figure 15. These results are a strong indication of the effective interaction of the IL with the metal surface, which may be responsible for the stabilization of the MNPs.

The XRD of the MNPs isolated from the ILs could be used to test the crystalline nature of the material and to estimate the mean diameter by means of the Debye-Scherrer equation by using Rietveld's refinements. Generally, the mean diameters of these isolated materials obtained by XRD are quite close

Figure 15. XPS of Pt NPs prepared in [BMIm][PF_6] showing the Pt 4f region with the fitting results. The insets of the figure showed the F 1s signal. (Reprinted with permission from ref. (Scheeren et al., 2006). Copyright 2006 American Chemical Society)

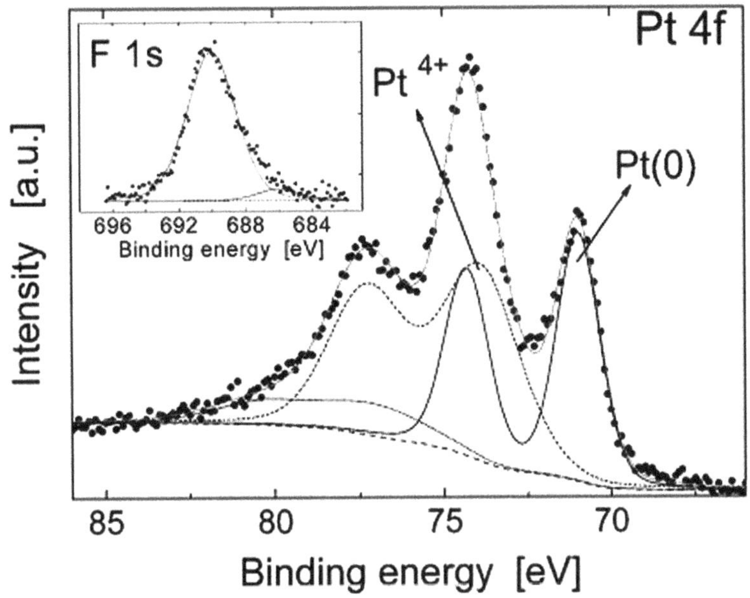

to that determined by TEM. For example, Ir NPs could be prepared *via* hydrogen reduction of $[Ir(cod)Cl]_2$ in three ILs, including [BMIm][BF_4], [BMIm][PF_6] and [BMIm][CF_3SO_3] (Fonseca et al., 2006). The XRD patterns of isolated Ir NPs showed the presence of broad reflection lines indicating the diffraction by very small crystals. And four diffraction peaks could be assigned to the Ir (111), (200), (220), and (311) reflection planes, Figure 16. It is found that the most representative reflections of Ir(0) were indexed as a face-centered cube (fcc) with unit cell parametera = 3.8784 Å. After calculating the mean diameter of Ir NPs using the Scherrer equation, the results were 2.6, 2.1, and 2.5nm for [BMIm][BF_4], [BMIm][PF_6] and [BMIm][CF_3SO_3], respectively, which were in good agreement with the values obtained from TEM technique, Figure 16.

MNPs/ILs in Catalysis

Hydrogenation Reactions

Catalytic hydrogenations as a kind of important reactions have wide application in industrial process for the synthesis of fine chemicals and also attract considerable interest in academic research (Scholten et al., 2012; Welton, 1999). In most cases, the hydrogenations as model reactions were employed to test the catalytic performance of IL-supported or mediated MNPs, which maybe due to the easy operation, availability of reaction as well as the usual insolubility of hydrogenation products (Parvulescu & Hardacre, 2007).

Figure 16. XRD patterns of the iridium in three ILs (left column) and the histograms illustrating the particle size distribution from TEM (right column); ILs: (a) [BMIm][BF$_4$], (b) [BMIm][PF$_6$] (c) [BMIm][CF$_3$SO$_3$] (Reprinted with permission from ref. (Fonseca et al., 2006). Copyright 2006 Elsevier)

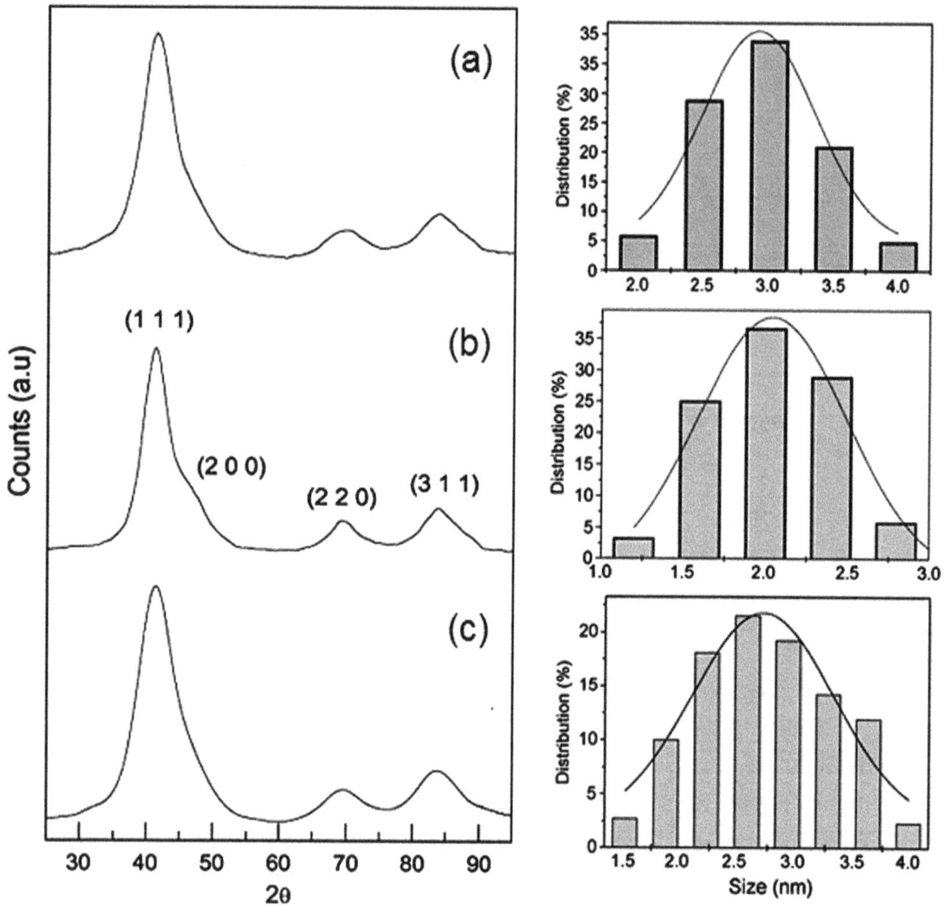

ILs-Incorporated Pd NPs

The hydrogenation of olefins is widely investigated to evaluate the catalytic activity of Pd NPs in ILs (Parvulescu & Hardacre, 2007). For example, the phenanthroline (Phen) ligand-protected Pd NPs generated *in situ* in [BMIm][PF$_6$] were very active for the hydrogenation of olefins under ambient reaction conditions (Huang et al., 2003). The cyclohexene and 1-hexene can be quantitatively converted to the corresponding products, and no deactivation was observed after 10 runs and the total TON (turnover numbers) exceeded 5000. The reason of the excellent catalytic performance was due to the inhibition of Phen ligand to the Pd NPs aggregation in IL. Moreover, the prepared catalysts were also very selective for the hydrogenation of 1,3-cyclohexadiene to cyclohexene. Only cyclohexene can be detected without byproduct in this process. The high selectivity resulted from the stronger adsorption of the alkadiene than alkene on the Pd NPs with smaller size (diameter in the range of 2-5 nm).

The PEG-functionalized IL ([PEG-2000-C_{12}MIM]$_{1.5}$[TPPT]) could be used as a moiety of thermoregulated Pd NPs catalytic system for hydrogenation (Zhu et al., 2013), Figure 17. In this system, both cation and anion of IL played crucial roles for the catalytic performance, i.e. the cations of the IL provide the thermoregulated property while the anions play a pivotal role in stabilizing and modifying NPs in ethyl acetate solvent. The Pd NPs size of catalyst was 2.3 ± 0.2 nm, and no apparent aggregation or agglomeration was observed after alkenes hydrogenation due to the strong coordination stability of the IL with the palladium center. Additionally, this catalytic system is also catalytically active for the hydrogenation of a wide range of the substrates bearing different functional groups, including alkenes, α, β-unsaturated aldehydes, with excellent activity (convention 92%~100%) and recyclability (10 runs) under mild conditions (45 °C, H_2 at atmospheric pressure).

The concept of two-phase catalysis could be used to promote the biodiesel hydrogenation using IL-mediated Pd NPs as catalysts, which was proven to be an efficient, clean, selective and eco-friendly strategy (Carvalho et al., 2011). In this catalytic system, one phase was composed by $Pd(OAc)_2$ and [BMIm][BF_4], and another phase was biodiesel. The *in situ* formation and stabilization of Pd NPs in ILs occurred during the reaction. The introduction of organic layer/IL biphasic system promoted the partial hydrogenation of biodiesel and the isomerization reaction was retarded significantly. And the

Figure 17. The FIL ([PEG-2000-C_{12}MIM]$_{1.5}$[TPPT])-stabilized/modified Pd NPs (Reprinted with permission from ref. (Zhu et al., 2013). Copyright 2013 WILEY-VCH Verlag GmbH & Co. KGaA)

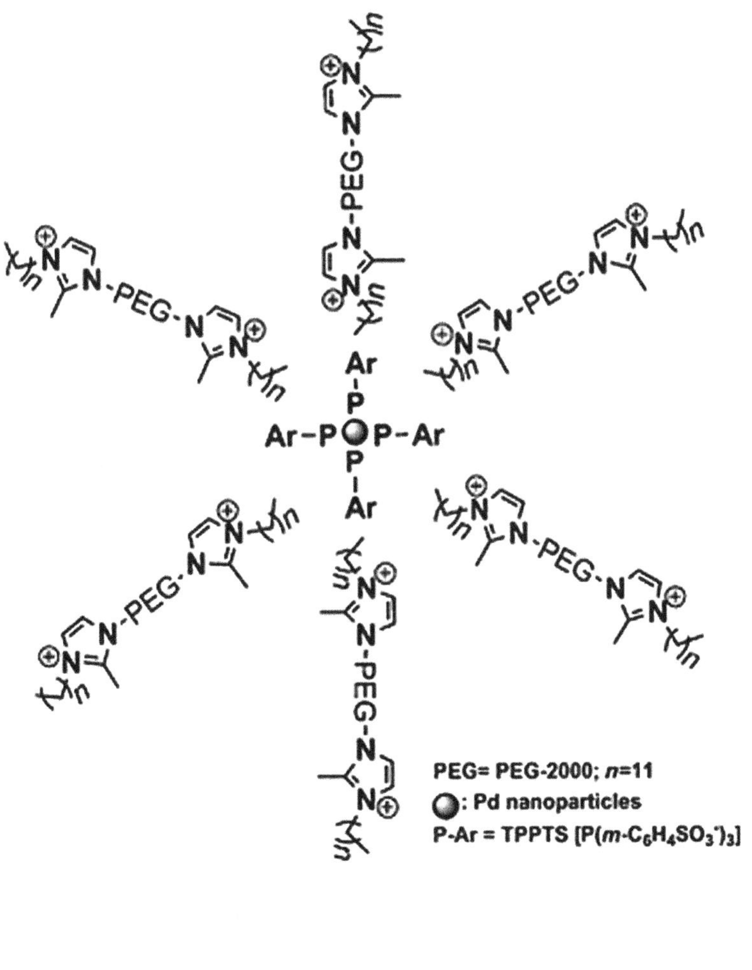

Figure 18. Synthesis of the IL supports containing hybrid organosilicas (Reprinted with permission from ref. (L. Luza et al., 2014). Copyright 2014 Royal Society of Chemistry)

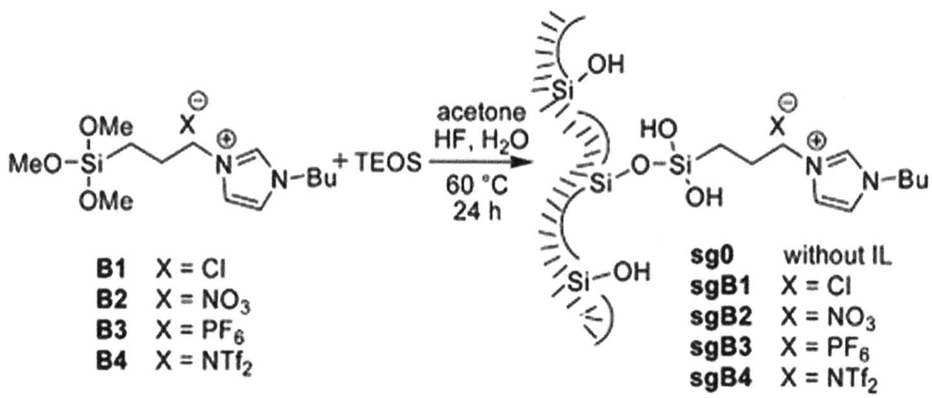

ionic phase containing the catalyst could be reused up to three times without significant loss in its catalytic performance. It is claimed that the enhanced hydrogenation selectivity were mainly due to the IL multiple effects. The IL functions as (i) a media for an additional stabilization of charged intermediates, (ii) a media to avoid isomerization due to a steric effect, and (iii) a media to allow a higher solubility of dienes (or poly-enes) instead of mono-enes, which resulted in a high and important selection, mostly based on solubility parameters.

The supported IL phase (SILP) is emerging efficient strategy for the immobilization of homotopic transition metal catalysts, and can be extended to IL soluble transition MNPs (Leandro Luza et al., 2013). For example, several kinds of IL containing hybrid organosilicas supported Pd NPs were prepared (sgB1-sgB4) via simple sputtering-deposition (L. Luza et al., 2014), Figure 18. The effect of IL anions, including Cl^-, NO_3^-, PF_6^-, NTf_2^-, on the Pd NPs properties and catalytic activity was studied through a model reaction of 1,3-cyclohexadiene hydrogenation. Interestingly, the size of MNPs is not directly controlled by the nature of the IL, and higher Pd concentration at the NPs surface was observed in the supports with smaller pore diameter (containing hydrophobic ILs) than those supports with the largest pore diameter (containing hydrophilic ILs). As for the activity of various MNPs, the TOF (turnover frequency) order was as follows: Pd/sgB3 > Pd/sgB4 > Pd/sg0 > Pd/sgB2 > Pd/sgB1, indicating that IL hydrophobicity may play a critical role in the hydrogenation of dienes by controlling diene access to the active sites of MNPs surface with a lesser effect on the selectivity.

ILs have also been widely used as excellent precursors to produce mesoporous N-doped carbon materials. By virtue of carbonization at high temperature, the ILs converted into novel carbon materials, in which the N atom was introduced into the carbon architecture to enhance the chemical, electrical, and functional properties of target materials. For example, a novel kind of N-doped carbon supported Pd NPs catalyst ($Pd@CN_{0.132}$) was prepared through the carbonization of 3-methyl-1-butylpyridine dicyanamide ([3-MBP][DCA]) at high temperature (X. Xu et al., 2012). In this case, the N-doped carbon with both narrowly distributed mesoporosity and local N-C-N structural motifs was obtained from the precursor mixture of 3-MBP-dca and 12 nm SiO_2 NPs after condensation at 900 °C for 1 h, and removal of SiO_2 template; then the Pd NPs were loaded on the N-doped carbon by an ultrasound-assisted method. HR-TEM images showed that the resulting Pd NPs with the diameter size of 4.1nm were well-dispersed on the surface of $CN_{0.132}$. $Pd@CN_{0.132}$ catalyst exhibited relatively high activity for the hydrodeoxygenation

Figure 19. The process of synthesis for Pd@$CN_{0.132}$ and as catalyst for hydrogenation of vanillin (Reprinted with permission from ref. (X. Xu et al., 2012). Copyright 2012 American Chemical Society)

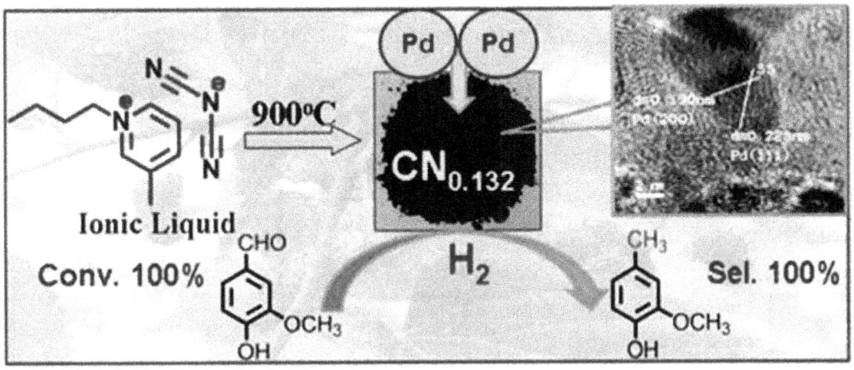

of vanillin, Figure 19. Nearly 100% conversion of vanillin with 100% selectivity of 2-methoxy-4-methylphenol were achieved in aqueous media at mild conditions (H_2 pressure 1.0 MPa, 150°C, 6 h), and no activity loss was observed after six recycles, suggesting its promising potential for the biofuel upgrade process. The high catalytic performance may be attributed to the special structure of the N-doped carbon-metal heterojunction, which led to a very stable and uniform dispersion of Pd NPs, additional electronic activation of the metal NPs, and good dispersion of the catalyst in water.

The integration of ILs technology and supercritical CO_2 ($ScCO_2$) could facilitate a green pathway for solventless Pd NPs catalyzed acetophenone hydrogenations (Jutz, Andanson, & Baiker, 2009). The major advantages for this strategy include the simple catalyst synthesis and efficient product separation. Pd NPs were synthesized by the hydrogen reduction of the palladium(II) acetylacetonate in [BMIm][PF_6] followed by the removal of the metal ligands by $ScCO_2$. The prepared Pd NPs with a narrow size distribution of 3.5 ± 0.6 nm exhibited excellent catalytic performance in the hydrogenation of acetophenone to 1-phenylethanol, and >90% conversion and selectivity could be achieved. In the catalytic cycle the hydrogenation reaction is followed by $ScCO_2$ extraction of the product, after which new substrate can be added to restart the cycle, and no noticeable deactivation of the catalyst was observed after six cycles, Figure 20.

ILs-Incorporated Pt

Stable and isolable Pt NPs with 2-3 nm diameters could be prepared by reducing $Pt_2(dba)_3$ in [BMIM][PF_6] with H_2 and used for alkenes and arenes hydrogenation to alkanes (Scheeren, Machado, Dupont, Fichtner, & Texeira, 2003). When cyclohexene and 1,3-cyclohexadiene were employed as the substrates, the conversions could reach 100% with TOF numbers of 625^{-1} and 156 h^{-1}, respectively. The results showed that the catalytic activity of such Pt NPs was higher than that obtained from the classical PtO_2 catalyst under the same reaction conditions. Interestingly, in this catalytic system, no unsaturated intermediates (cyclohexadienes or cyclohexenes) could be detected in the hydrogenation of benzene or toluene even at earlier stages of the reaction. Moreover, the IL/Pd NPs were recyclable systems for the solventless or biphasic hydrogenation of alkenes and arenes.

Figure 20. Schematic view of the complete "green" pathway. The catalyst preparation step consists of the formation of the NPs in the IL, followed by $ScCO_2$ extraction of the precursor ligand. In the catalytic cycle the hydrogenation reaction is followed by $ScCO_2$ extraction of the product, after which new substrate can be added to restart the cycle (Reprinted with permission from ref. (Jutz et al., 2009). Copyright 2009 Elsevier)

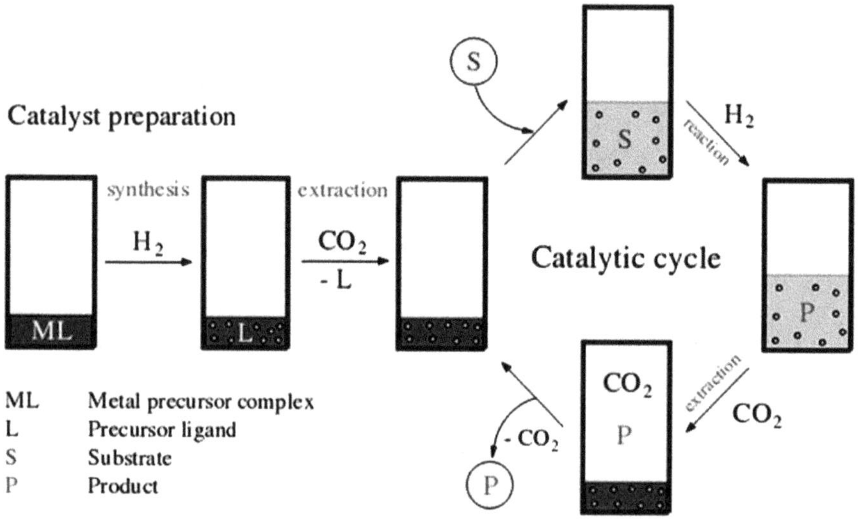

The ILs/Pt NPs catalysts were also active for enantioselective hydrogenations of methyl benzoylformate (MBF) to (R)-Methyl Mandelate (Beier, Andanson, Mallat, Krumeich, & Baiker, 2012). The Pt NPs with a mean size of 2-3 nm and narrow size distribution were obtained by stabilization with water-insoluble ILs or both an IL and cinchonidine (CD) as chiral costabilizer. Noteworthy, the particle size of NPs could be tuned by the costabilizer, *i.e.*, smaller NPs with narrow distribution could be obtained in ILs if use CD as the costabilizer. After optimization, 78% of methyl benzoylformate was hydrogenated with 78% *ee* and 3100 h^{-1} TOF number. The catalyst provided higher rate and *ee* in the hydrogenation of MBF than the widely used commercially available Pt/Al_2O_3 as reference catalyst (TOF = 1600 h^{-1}) without any preactivation or addition of an organic solvent. Though the structures of ILs showed small effect on the Pt NPs size, the catalytic activity was obviously affected and followed the order [BMIm][PF_6]< [BMIm][BF_4]<[BMPy][BF_4]<[BMIm][FEP] (FEP=tris(pentafluoroethyl)trifluorophosphate).

ILs-Incorporated Ru NPs

Ru NPs in ILs are quite effective catalysts for the hydrogenation (Abhinandan Banerjee & Scott, 2015). The relationship between size and catalytic performance of Ru NPs in [BMIm][NTf_2] has been studied using the hydrogenation of 1,3-cyclohexadiene and cyclohexene as probe reactions (Campbell et al., 2010). The tailor-made, size-controlled and zero-valent Ru NPs (1-3 nm) were generated from the decomposition of [Ru(COD)(COT)] under H_2 in [BMIm][NTf_2] by varying the experimental conditions. In this case, about 97% selectivity of cyclohexene with 66% conversion could be reached using 1.1 nm Ru NPs. Meanwhile, 80% selectivity of cyclohexene with 83% conversion was obtained using 2.9 nm Ru NPs. This enhancement in activity promoted by large Ru NPs in hydrogenation is related to the appropriate number of neighboring surface sites that facilitate π-bond coordination and activation of the conjugated diene. On the other hand, selectivity was explained on the basis of the type of active sites, which can be quite different, depending on the particle size. According to the simplified SYBYL model of 1,3-cyclohexadiene molecules coordinating to Ru NPs with 1.3 and 2.8 nm particle sizes, Figure 21, small Ru NPs have sites located mainly at the vertex and edge positions, which provide C=C coordination with further selective formation of alkene, whereas large particles tend to exhibit active sites at facial regions, which facilitate planar double C=C interactions and full diene hydrogenation. Furthermore, these Ru NPs showed a high level of recyclability with neither loss of activity nor significant agglomeration.

The influence of ionicity of ILs on the Ru NPs catalytic performance was also explored using biphasic hydrogenation of cyclohexene as model reaction (Kylie L. Luska & Audrey Moores, 2012). The Ru NPs were prepared in a series of imidazolium and phosphonium ILs *via* the reduction of Ru(2-methylallyl)$_2$(COD) with H_2. Under the same conditions, the conversions of cyclohexene could be range from 12% to 100% in various ILs media. According to the experimental results, the stability and catalytic activity of Ru NPs were highly dependent on the nature of ILs, which can be rationalized through a Walden plot analysis of the IL iconicity. That is, IL with a lower degree of ionicity would provide more stable NPs due to an increased association between IL partners and the formation of ionic aggregates. It was thought that the interaction between the IL's cation and anion may dominate within the ligand sphere of IL-stabilized Ru NPs. On the other hand, the ionicity of ILs also influences the mass-transfer in IL phase. For example, the hydrogenation of cyclohexene was favored in a less ionic system that would allow the substrate to move more freely.

ILs-Incorporated Rh NPs

The system of ILs/Rh NPs with bipyridine-based ligands showed good activity and selectivity in the hydrogenation of substituted aromatic compounds (Leger, Denicourt-Nowicki, Olivier-Bourbigou, & Roucoux, 2008). The catalytic performance of Rh NPs stabilized by various bipyridines ligands in non-aqueous ILs for the hydrogenation of styrene were investigated. The Rh NPs catalyst in the size range

Figure 21. SYBYL representations of CYD coordinated to the face of highly crystalline RuNPs of mean diameter 1.3nm (left) and 2.8nm (right) (Reprinted with permission from ref. (Campbell et al., 2010). Copyright 2010 Elsevier)

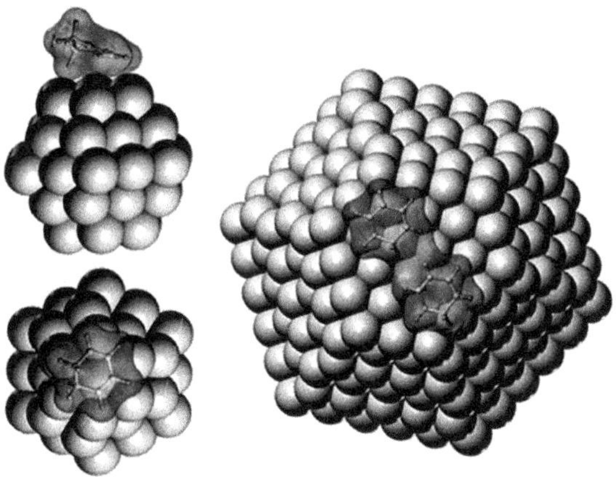

Figure 22. Representation of bipyridine coordination modes in MNPs (Reprinted with permission from ref. (Leger et al., 2008). Copyright 2008 American Chemical Society)

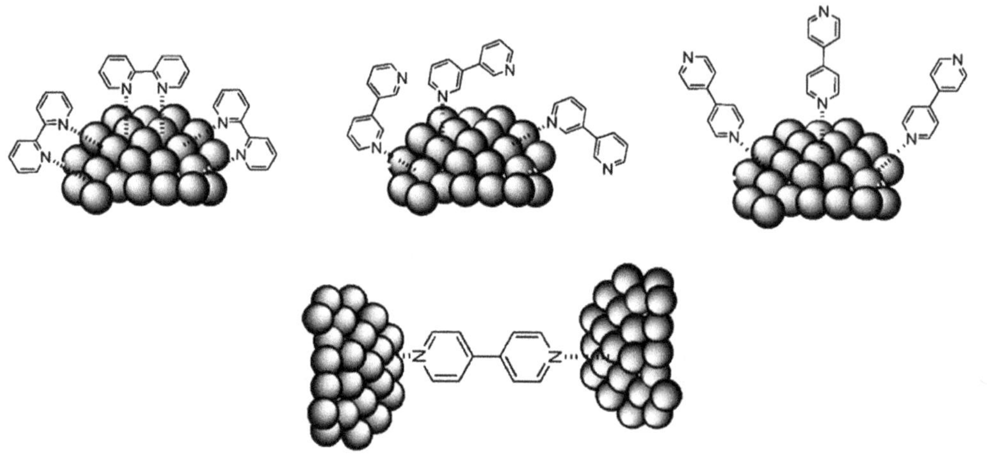

of 2.0-2.5 nm can be easily obtained in different ILs by chemical reduction of $RhCl_3$ using $NaBH_4$ with bipyridine-based ligands. In the styrene hydrogenation, 40% ethylbenzene and 60% ethylcyclohexane could be obtained using the Rh NPs with 2,2'-BPy as catalyst in [BMIm][PF_6]. However, only ethylcyclohaxane was generated using 3,3'-BPy and 4,4'-BPy stabilized Rh NPs as catalysts under the same conditions. The variation in catalytic performance might be derived from different coordination modes of these bipyridines to Rh NPs, Figure 22. In the case of 2,2'-BPy, it coordinates to NPs with both N-donor atoms yielding a chelate structure, which hindered the MNP surface, limiting access of the substrate and resulting in lower activity and selectivity. While in the case of 3,3'-BPy, and 4,4'-BPy, monodentate coordinated mode is dominated on the metal surface, which decreases steric hindrance and allows access of the substrate, thus increasing activity and selectivity.

Two bipyridine-functionalized ILs ([BIHB][NTf_2]$_2$ and [BIMB][NTf_2]$_2$) were synthesized, and used as Rh NPs stabilizer to catalyze the alkylbenzene hydrogenation in [BMIm][NTf_2] (R. R. Dykeman, Yan, Scopelliti, & Dyson, 2011), Figure 23. In the toluene hydrogenation to methylcyclohexane, the conversions were 100%, 68% and 18% using [BIHB][NTf_2], 2,2'-BPy and [BIMB][NTf_2] as ligands,

Figure 23. Syntheses of bipyridine functionalized ILs [BIHB][NTf_2] and [BIMB][NTf_2] (Reprinted with permission from ref. (R. R. Dykeman et al., 2011). Copyright 2011 American Chemical Society)

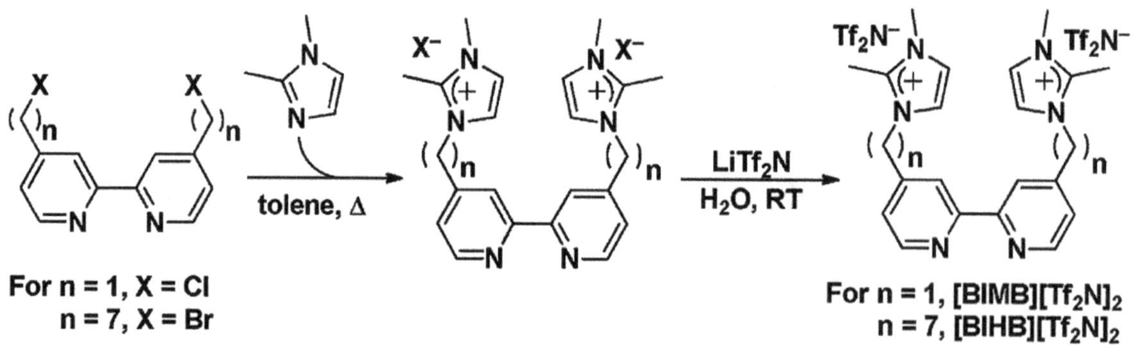

respectively. The results showed the catalytic activity was strongly influenced by the stabilizer employed and in the order of $[BIHB]^{2+}$ > bipy > $[BIMB]^{2+}$. The lower activity of the $[BIMB][NTf_2]$ supported NPs may stem from the weaker interaction of the ligand with the metal surface due to the electron withdrawing effect of the imidazolium cation and the close proximity of the positive charges to the NP surface.

A series of Rh NPs with different 4,4'-modified-bipy ligand stabilizers (with COOMe, Cl, Me, tert-butyl, MeO, or NH_2 substituents at the 4,4'-position) were prepared and used for hydrogenation reactions (Ryan R. Dykeman et al., 2012). However, it is found that the efficient ligand dissociation with electron-withdrawing substituent was also closely correlated to catalyst instability. In order to overcome this problem, an electron-poor "dimerized-bipy" stabilizer was further developed, and the size-distribution for the resulting MNPs was centered at about 1.5-2.0 nm, Figure 24a. The $[dibipy]^{2+}$-stabilized NPs showed higher catalytic activity than those bipy-stabilized Rh NPs in the reduction of toluene into methylcyclohexane, and no obvious decrease in activity was observed after five cycles. In catalytic applications with bipyridine-based stabilizer, each bipy unit may easily dissociate; however, overall, the dimeric structure is less prone to complete dissociation from the NP surface, Figure 24b. Thus, the synergistic effects may enable both higher activities and stabilities to be achieved in the catalytic reactions.

Generally, the strong covalent interaction between the ligands and the NP surface would decrease the number of available surface atoms for catalysis. The noncovalent stabilizer may be helpful for the preparation of highly catalytically active MNPs. The polymer ILs (PILs) have been proved to be a promising candidate as the noncovalent stabilizer. A series of Rh NPs were prepared using PIL poly-(3-((2,4-divinylcyclopentyl)methyl)-1,2-dimethyl-1H- imidazolium methanesulfonate, PIL1) as stabilizers, and the resulting catalysts were used in the hydrogenation of limonene (Biondi, Laporte, & Dyson, 2012). The stabilizers could be readily modified by anion exchange (2, NTf_2^-; 3, PF_6^-, Figure 25) to afford Rh NPs that could be dispersed in a wide range of organic solvents and ILs. The PIL stabilizer interacts with the NP surface mainly through noncovalent interactions, and thus their catalytic activity is not suppressed. In this system, excellent activities could be obtained in the selective hydrogenation of limonene i.e., 76% yield of carvomenthene with 100% selectivity could be achieved at 40 bar of hydrogen, 80 °C for 30 min.

Figure 24. (a) The structure of dimerized NP-bipy-stabilizer, $[dibipy][Tf_2N]_2$; (b) representation of the proposed role of bipy (top) and the dimerized bipy ligand (bottom) during the NP-catalyzed hydrogenation of benzene, which accounts for the higher stability of the dimeric stabilizer (Reprinted with permission from ref. (Ryan R. Dykeman et al., 2012). Copyright 2012 WILEY-VCH Verlag GmbH & Co. KGaA)

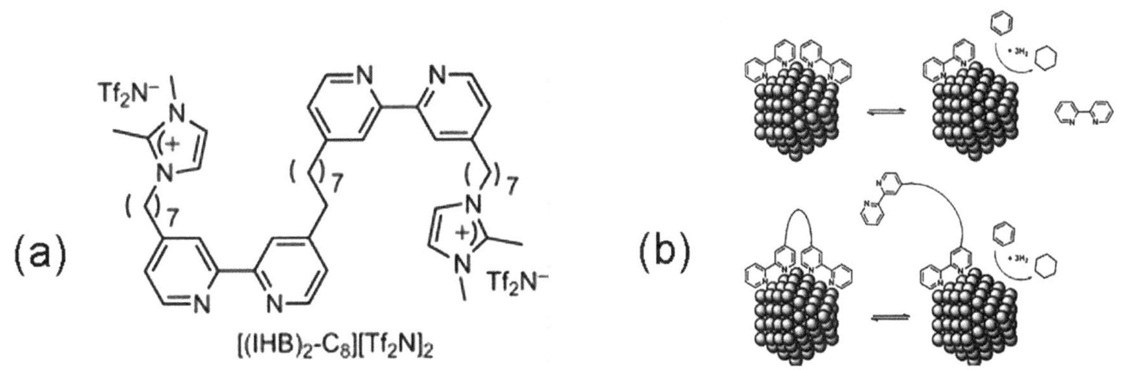

Figure 25. The used PIL1and the synthesis of PIL2 and PIL3 by anion exchange (Reprinted with permission from ref. (Biondi et al., 2012). Copyright 2012 WILEY-VCH Verlag GmbH & Co. KGaA)

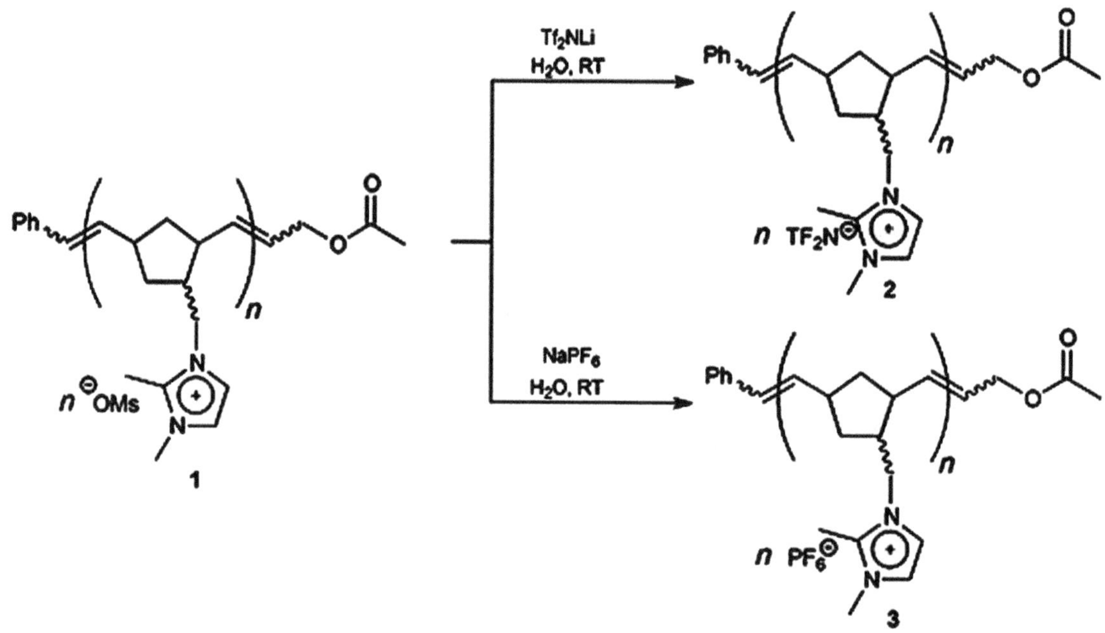

ILs-Incorporated Ir NPs

Ir NPs prepared by simple reduction of $[Ir(COD)Cl]_2$ in $[BMIm]PF_6$ with 2.0 ± 0.4 nm diameter could be active catalyst in the hydrogenation of olefins under mild reaction conditions in $[BMIm][PF_6]$ (Jairton Dupont, Fonseca, Umpierre, Fichtner, & Teixeira, 2002). Noticeably, the hydrogenation of 1-decene could be conducted at room temperature with a slightly positive pressure of hydrogen, 92% conversion was achieved at 30°C under 4 atm for 3h, and the Ir NPs in IL could maintain their efficiency for up to at least seven recycles. Moreover, the Ir NPs are also outstanding catalysts for the hydrogenation of arenes containing functional groups and they behave as heterogeneous catalysts in these reactions (Fonseca et al., 2003) .In the hydrogenation of cyclohexenone using Ir NPs as the catalysts (2.3 ± 0.4 nm in diameter), 90% cyclohexenone conversion could be obtained with 83% selectivity to cyclohexanol under improved reaction conditions (Jairton Dupont, Fonseca, & Scholten, 2004). However, in some cases, the addition of small amounts of imidazolium-based ILs would poison Ir NPs for acetone hydrogenation (Ott, Campbell, Seddon, & Finke, 2007). Isotope labeling studies suggested that the deactivation was caused by the coordination effect of NHC from the 2-position of imidazolium ring. Control experiments ruled out the negative effect of $[PF_6]^-$ because the use of $[NBu_4][PF_6]$ didn't cause the deactivation.

C-C Coupling Reactions

Suzuki Reaction

Suzuki reaction concerns the coupling of aryl- or vinyl-boronic acid with aryl- or vinyl-halide. The Suzuki coupling reaction of phenylboronic acid and aryl halides can be realized efficiently catalyzed by Pd NPs in tetraalkylammonium bromide (Calò, Nacci, Monopoli, & Montingelli, 2005), Figure 26.

Figure 26. Suzuki coupling reaction of phenylboronic acid and aryl halides

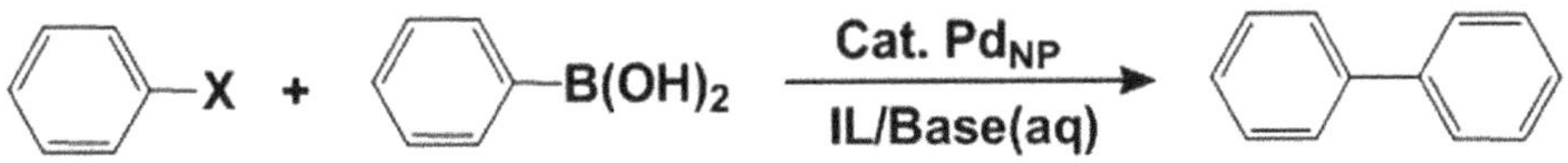

The Pd NPs were prepared by adding $Pd(OAc)_2$ into $[Bu_4N][OAc]$ and heated at 90 °C without addition of other reducing agent. The results showed that using $[Bu_4N][OH]$ as base increased the catalytic efficiency significantly, which may be due to the higher concentration of $[Bu_4N]^+$ into water, contributing through partitioning equilibrium, to keeping constant the concentration of the cations in the IL phase. Normally, using chlorobenzene as starting material, 83% yield of diphenyl benzene could be obtained in $[Heptyl_4N]Br$, which is higher than that in $[Bu_4N]Br$ (45% yield), probably due to the stronger stabilization of the Pd NPs provided by $[Heptyl_4N]Br$. Moreover, the catalytic system could be used at least 5 times with little loss in activity.

A series of nitrile-functionalized pyridium ILs could be used as more-effective immobilization solvent for Pd-catalyzed Suzuki reaction (Zhao, Fei, Geldbach, Scopelliti, & Dyson, 2004). Pd NPs with ~5 nm particle size were *in situ* formed in the reaction. Although all of the Pd complexes showed good catalytic activity with >80% yield in the Suzuki reaction of iodobenzene and phenylboronic acid, the incorporation of nitrile group could not visibly improve the catalytic activity. However, the use of nitrile functionalized ILs enabled the catalyst system to be remarkably reusable. For example, no deactivation was observed in $PdCl_2/[C_3CNPy][NTf_2]$ at the 9th runs while only <10% yield was obtained in $PdCl_2/[BPy][NTf_2]$ at the 5th run. Meanwhile, the inductive coupled plasma (ICP) spectroscopy suggested that the presence of the coordinating nitrile moiety in the IL led to a significant decrease in palladium leaching relative to simple N-alkylpyridinium ILs. Further work revealed that Pd NPs act as reservoirs for molecular Pd(II) species, assumed as the true active catalyst, and the nitrile-functionality of the ILs supports the stabilization of molecular intermediates via transient coordination of the nitrile group, as well as the protection of the metal NPs, also involving carbenes (Fei et al., 2007), Figure 27.

The IL-modified magnetic-supported Pd (Mag-IL-Pd) NPs were also prepared and used as the catalyst for Suzuki coupling reaction (Babak Karimi, Mansouri, & Vali, 2014). In order to achieve the magnetically isolable character for catalysts, magnetite/silica core/shell NPs were first prepared via a micro-emulsion method, and then triethylene glycol (TEG) functionalized IL was employed to modify the resultant magnetic NPs and subsequently simple ion exchange can give the target magnetic IL-modified MNPs, Figure 28. The obtained magnetic IL-Pd catalyst exhibited excellent catalytic performance in Suzuki reaction for a wide range of aryl halides with boronic acids in water. As it was expected, such catalysts could be easily separated from the aqueous phase by applying an external magnetic field. Quantitative yield could be achieved in 10 consecutive runs, and TON could be up to 39760 for the coupling reaction of 4-bromoanisole and phenylboronic acid. In this system, the immobilization of TEG-functionalized ILs can not only provide a good distribution of catalytically active species on the support, but also stabilize the Pd NPs to avoid aggregation.

Figure 27. Proposed formation of nitrile-IL stabilized Pd-NPs which act as reservoir for active molecular palladium species (Reprinted with permission from ref. (Fei et al., 2007). Copyright 2007 American Chemical Society)

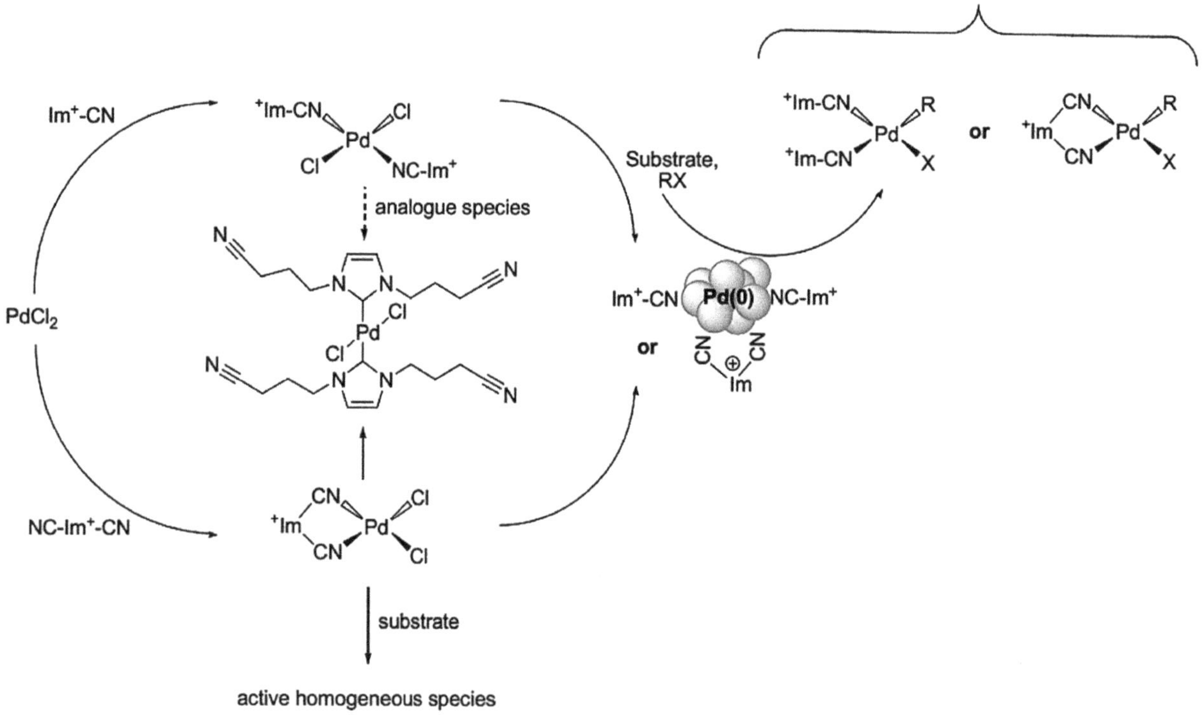

Mizoroki-Heck Reaction

In general, the Mizoroki-Heck reaction consists of the reaction of an unsaturated halide with alkenes in the presence of a base catalyzed by Pd catalyst. In 2001, Srinivasan et al. first reported the Heck reactions processed in the ILs/MNPs system using sodium acetate as base under ultrasonic irradiation (Deshmukh et al., 2001), Figure 29. It is observed that the use of [BBIm][Br] and [BBIm][BF_4] can significantly enhance the rate of the reactions in several substituted iodobenzenes and alkenes/alkynes. For example, 87% isolated yield of the ethyl cinnamate could be obtained in the Heck reaction of iodobenzene and ethyl acrylate in 1.5 h at 30 °C. Moreover, NMR/MS and *in situ* TEM analyses showed that the $Pd(OAc)_2$ precursor initially produced a Pd bis-carbene complex and then the complex was reduced to Pd NPs. In fact, the insights into the mechanism and role of Pd NPs dispersed in [BMIm][PF_6] during the Heck reaction have been investigated in detail (Cassol, Umpierre, Machado, Wolke, & Dupont, 2005). The TEM and ICP-AS analyses indicated that Pd NPs dispersed in IL act as a reservoir of catalytically active Pd species. It is highly probable that the reaction proceeds through the oxidative addition of PhI on the nanoparticles surface, and the oxidized Pd species thus formed are detached from the surface and are entering in the main catalytic cycle. The Pd(0) formed in the main catalytic cycle, after β-hydride and reductive elimination steps, can either continue in the catalytic, Figure 30.

Figure 28. Synthesis of magnetite/silica magnetic NPs and Mag-IL-Pd catalyst (Reprinted with permission from ref. (Babak Karimi et al., 2014). Copyright 2014 Royal Society of Chemistry)

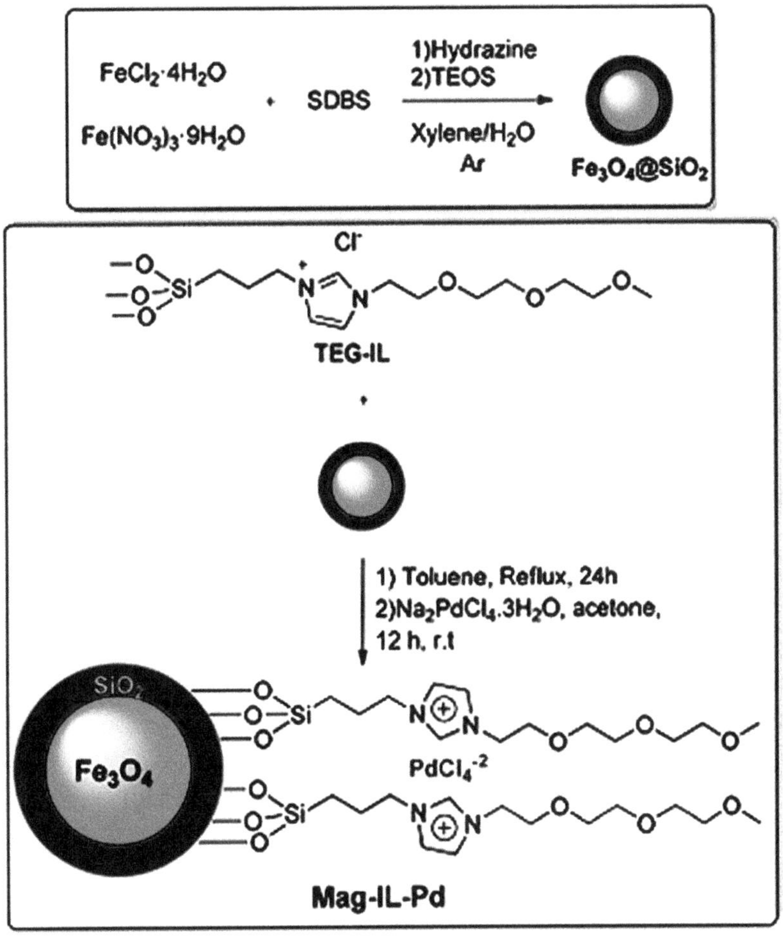

Figure 29. Heck reaction of substituted iodobenzenes and alkenes/alkynes in ILs at 30 °C under ultrasonic conditions

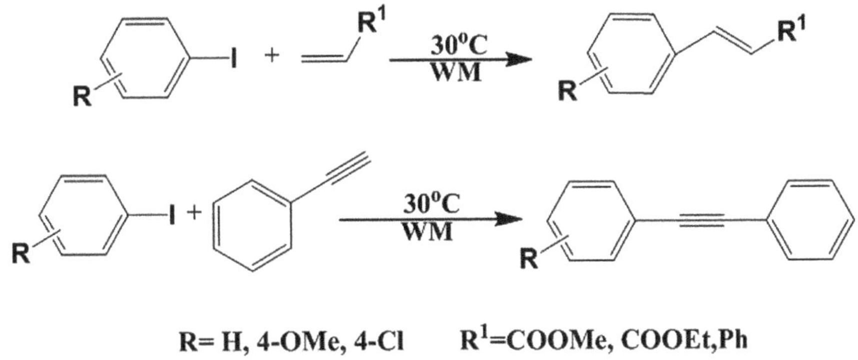

Figure 30. Possible pathways involved in the Heck reaction promoted by Pd NPs dispersed in imidazolium ILs (Reprinted with permission from ref. (Cassol et al., 2005). Copyright 2005 American Chemical Society)

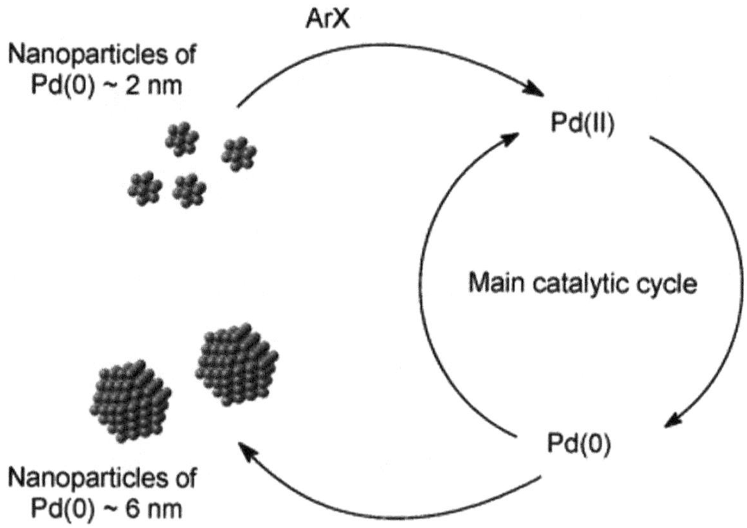

In $[Bu_4N]Br$ instead of imidazolium based ILs, it is found that Pd NPs were also able to catalyze the Heck reaction of aryl bromides with butyl methacrylate and α-methylstyrene using $[Bu_4N][OAc]$ as base (Calò, Nacci, Monopoli, Detomaso, & Iliade, 2003). In most cases, >99% conversion could be achieved at 120 °C, in which the formation of terminal olefins as a main product was observed. In contrast, reaction of *p*-bromoacetophenone led to the internal olefin. Whereas the solvent stabilizes the MNPs, the base is responsible for a fast neutralization of the PdH impeding the hydride readdition to reaction products and avoiding the olefin interconversion. The terminal olefins are efficiently converted into the more stable internal E isomers by using tetrabutylammonium pivalate as co-catalyst.

Except for the catalytic system of MNPs directly dispersing in ILs, the system of supported MNPs is an alternative strategy for Mizoroki-Heck coupling in ILs. For example, Pd NPs/ILs catalysts supported on chitosan were developed and behaved as efficient and recyclable heterogeneous catalyst in IL $[Bu_4N]Br$ for Heck reaction of aryl bromides and activated aryl chlorides (Calò et al., 2004). The efficiency of this catalyst was due to the stabilization of Pd colloids by the solvent and to a very fast PdH neutralization by the base. It is worthy that no reaction occurred in the case of imidazolium based ILs. The role of Pd NPs in heterogeneous system for Heck reaction was also explored using the Al_2O_3 supported Pd NPs catalyst in $[Bu_4N]Br$ as the model (Pryjomska-Ray, Gniewek, Trzeciak, Ziółkowski, & Tylus, 2006). In this case, the reversible transfer of Pd between solution and support was revealed. During the reaction, palladium was partially leached from the support, thereby forming $[Bu_4N][PdBr_4]$ complex that catalyzed Heck reaction or underwent re-adsorption on Al_2O_3. These soluble palladium complexes were partially reduced to soluble Pd NPs, which can also be anchored on Al_2O_3 giving active catalyst $Pd(0)/Al_2O_3$. Experimental evidence showed that heterogenized catalyst, $Pd(0)/Al_2O_3$, after oxidative addition of aryl halides, served as a source of soluble palladium species and colloidal NPs that were active as homogeneous catalysts. Possibly, both forms of palladium species were catalytically active as homogeneous catalysts or after readsorption on Al_2O_3 as heterogenized ones.

Stille Reaction

The Stille reaction involves the coupling of an organotin compound with an sp^2-hybridized organic halide catalyzed by Pd catalysts. Pd MNPs dispersed in tetraalkylammonium IL was proven to be an effective reaction medium for the Stille reaction between aryl halides and tributylphenyl stannane in [Heptyl$_4$N]Br (Calò et al., 2005), Figure 31a. In this way, >90% yields of biaryls could be achieved and five runs recycle with slight decrease in activity could be observed. Nitrile-functionalized ILs stabilized Pd NPs could be also used to catalyze Stille coupling reactions of tributylphenyl stannane and iodobenzene (Zhao et al., 2004). Similarly to Suzuki reaction in this system, the recycling experiments showed that the functionalized IL was also a far superior reaction medium when compared with a non-functionalized analog, which possible due to enhanced stabilization of the formed MNPs in the nitrile-functionalized ILs to prevent agglomeration. Moreover, symmetrical and asymmetrical nitrile-functionalized ILs incorporating nitrogen-containing anions such as $[NTf_2]^-$ and $[N(CN)_2]^-$ were employed in the Pd NPs-catalyzed Stille reactions (Chiappe, Pieraccini, Zhao, Fei, & Dyson, 2006). In the Stille coupling of bromobenzene and tributylvinylstannane, the incorporation of nitrile group in $[N(CN)_2]^-$ or $[NTf_2]^-$ based ILs had some effect on the kinetics of the reaction, and the increased efficiency of the nitrile-functionalized ILs might be derived from the interaction of the nitrile pendant group with the Pd metal centre.

Sonogashira Reaction

The Sonogashira reaction is based on the coupling of aryl halides and substituted alkynes promoted by transition metal catalysts. Pd nanowires *in situ* prepared in ILs have been showed to be active catalyst for Sonogashira coupling reaction (Gao et al., 2005). In the seed-growth method to prepare Pd nanowires, Au colloids as seeds with average diameter of 2.2 nm were first prepared *via* $HAuCl_4$ reduction by $NaBH_4$ using thiol-FIL as stabilizer; then H_2PdCl_4 was reduced with $NaBH_4$ in the solution of Au seeds forming the uniform Pd nanowires (diameter: 2-4 nm) under vigorous stirring at room temperature. The coupling of phenyl iodide and phenyl acetylene was successfully performed in the presence of CuI, PPh_3 and Pd nanowires within 7-12 hours at 75 °C, Figure 31b. In this case, total substrate conversions could be reached and the Pd nanowires catalyst could be reused several times. Interestingly, only 82% conversion of phenylacetylene was obtained using the bimetallic core/shell-NPs ($Au_{Core}Pd_{Shell}$) under similar conditions.

Figure 31. (a) Stille cross-coupling catalyzed by Pd NPs in IL; (b) Sonogashira reaction of iodobenzene and phenylacetylene catalyzed by Pd nanowires

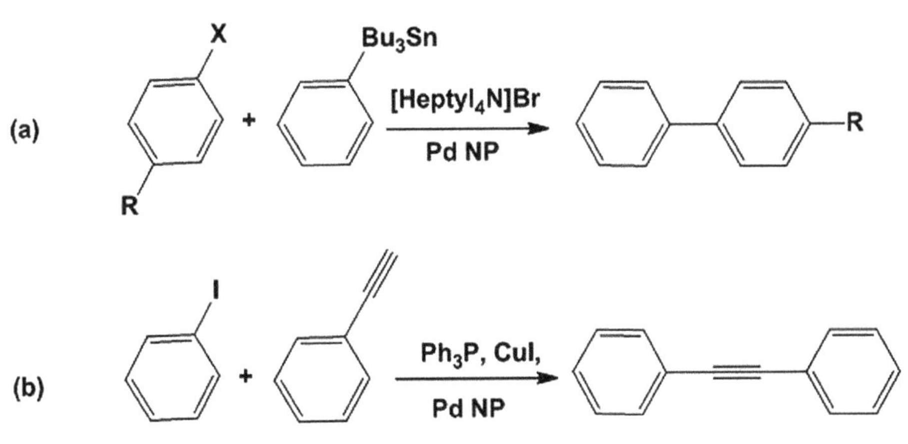

Ullmann Reaction

Ullmann-type reductive homocoupling of aryl, vinyl and heteroaryl halides could be promoted by Pd NPs using aldehyde as reductant in tetraalkylammonium ILs under very mild reaction conditions (T = 40-90 °C) (Calò, Nacci, Monopoli, & Cotugno, 2009). In this process, substituted aryl bromides and iodides could couple to biaryls with good conversions (70-90%) in absence of other additives. This procedure was very simple because the catalytically active species could immediately form *via* mixing the substrates and $Pd(AcO)_2$ together in the molten ILs. With propanal as the reducing agent, the work-up proved to be particularly easy, due to the volatility of the reaction by-product (acrolein). It was found that IL was crucial for this process because it behaved simultaneously as a base, reaction medium, and ligand for stabilization of the Pd-NPs, which behave as a reservoir for catalyst species. Interestingly, the choice of ILs could control the reaction selectivity, i.e., the Ullmann homocoupling occurred in the $[Bu_4N][AcO]$ medium while Heck reaction would take place in the presence of $[Bu_4N]Br$, Figure 32.

The combination of IL and $ScCO_2$ emerged as an efficient alternative to promote the Ullmann reaction in the presence of carbon-supported Pd NPs and aluminum as the reductant (Cheng, Tang, & Xu, 2010), Figure 33. In order to avoid the agglomeration of Pd NPs, the oxidized graphite was chosen as the support to immobilize the Pd NPs. The Pd NPs supported on the carbon platelets had a uniform and narrow particle size distribution with the size range of 2-10 nm. In the Pd NPs-catalyzed Ullmann reaction of aryl halides in $[BMImHSO_3C_4][HSO_4]$ IL/$ScCO_2$ system, up to 99% yields of biphenyl derivatives were achieved. After reaction, the Pd NPs catalyst and ILs could be easily separated and recycled for at least five runs without deactivation. In this strategy, the introduction of Brønsted-acidic imidazolium IL could replace the traditional active hydrogen donor readily with much enhanced product separation efficiency. Moreover, the utilization of $ScCO_2$ could also significantly promote the selectivity of the Pd NPs-catalyzed aluminium-induced reductive Ullmann reaction of aryl chloride. Therefore, the employment of IL/$ScCO_2$ system led to immobilization, activation, and stabilization of the Pd NPs catalyst that would be impossible in classical organic media separately.

Figure 32. Dependence of the reaction pathway on the nature of the IL anion (Reprinted with permission from ref. (Calò et al., 2009). Copyright 2009 WILEY-VCH Verlag GmbH & Co. KGaA)

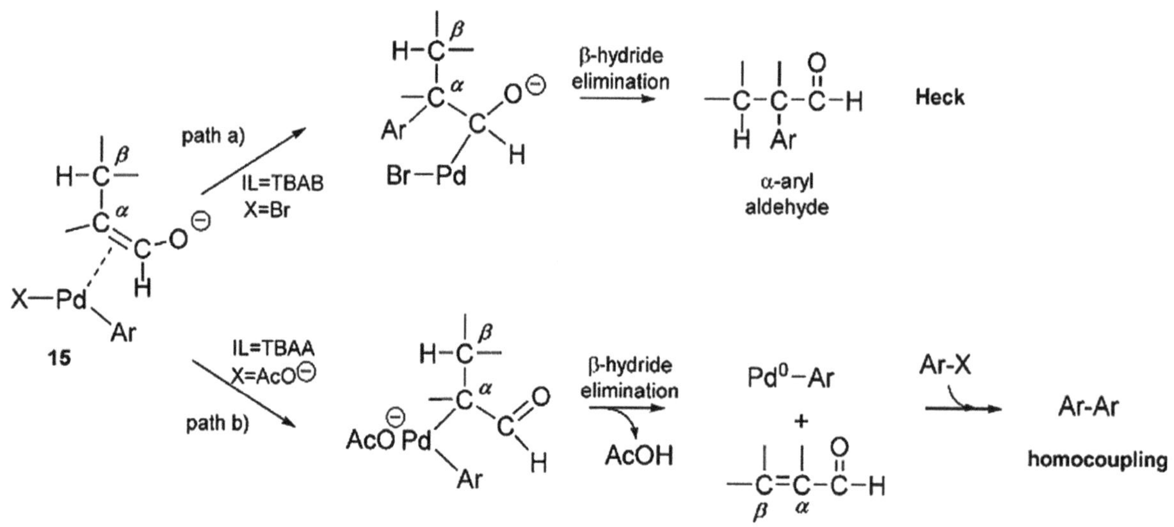

Figure 33. Illustration of Pd NPs/carbon-catalyzed Ullmann reaction in an IL/ScCO$_2$ system (Reprinted with permission from ref. (Cheng et al., 2010). Copyright 2010 WILEY-VCH Verlag GmbH & Co. KGaA)

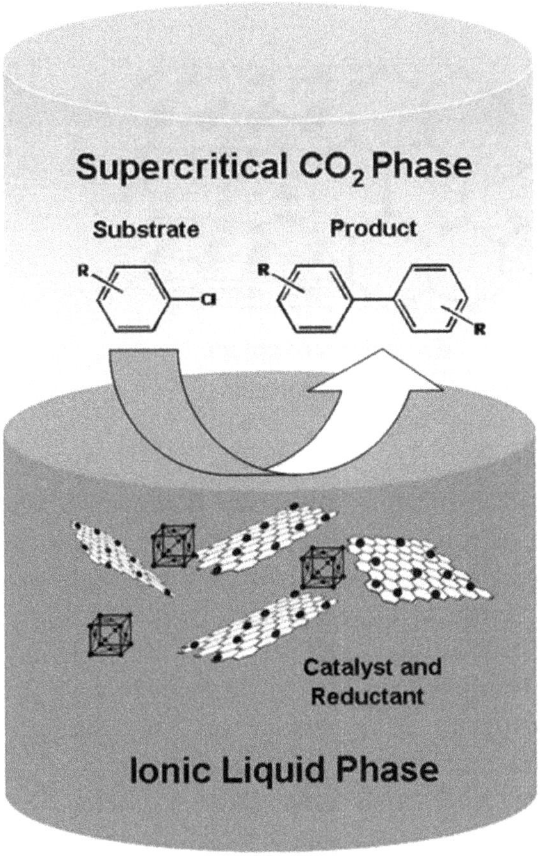

Oxidation

IL-supported or mediated MNPs (e.g., Pd, Au) catalysts have been also developed and used for the aerobic oxidation of alcohols, in which the ILs mainly enables the catalytic reactions more efficient and the recyclability or reusability much easier.

ILs-Incorporated Pd NPs

The use of palladium catalysts for the oxidation of alcohols to aldehydes and ketones with ILs as green media has been well documented, in which ILs are mainly used as solvents or catalysts, or both. SILP catalysis is an emerging concept in the catalytic science, which can significantly reduce the amount of high cost ILs in a typical process and meanwhile offer some advantages with regard to product isolation, catalyst recovery and recycling, *etc*.

The periodic mesoporous organosilica with alkyl imidazolium IL framework (PMO-IL) were synthesized and used for the immobilization and stabilization of Pd NPs (Pd@PMO-IL) for the catalytic aerobic oxidation of alcohols (B. Karimi, Elhamifar, Clark, & Hunt, 2011). Nitrogen sorption analysis

Figure 34. A possible model for the aerobic oxidation of alcohols using Pd@PMO-IL (Reprinted with permission from ref. (B. Karimi et al., 2011). Copyright 2011 Royal Society of Chemistry)

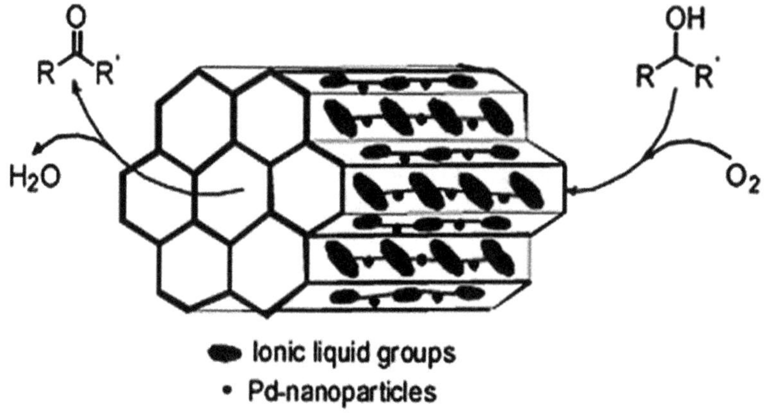

showed that BET surface areas, mesoporous volume and BJH pore diameter of the Pd@PMO-IL were 458 $m^2\ g^{-1}$, 0.88 $cm^3\ g^{-1}$ and 5.3 nm, respectively. In the catalytic reactions, several types of primary and secondary aromatic or aliphatic alcohols could be efficiently and rapidly oxidized to their corresponding carbonyl compounds in the presence of a low loading of Pd@PMO-IL catalyst. Moreover, the catalyst could be easily recovered and reused to give quantitative yield of carbonyl adducts even after 10 recycling runs. The high activity of this nanocatalyst may be attributed to the isolated IL nanoreactor with the mesochannels which could control the reaction mechanism by preventing the formation of agglomerated Pd NPs as well as stabilization of active catalytic Pd species, Figure 34.

Similarly, a nano-fibrillated mesoporous carbon (IFMC) material was prepared via carbonization of task-specific IL 1-methyl-3-phenethyl-1H-imidazolium hydrogen sulfate in the presence of SBA-15, which has shown its efficiency as the support for Pd NPs (Pd@IFMC) in the catalytic aerobic oxidation of alcohols (B. Karimi, Behzadnia, Bostina, & Vali, 2012). The resulting Pd NPs with the size range of 2~4 nm showed high catalytic activity to aerobic oxidation of containing variety of alcohols with either molecular oxygen or air under mild conditions. After six reaction runs, no noticeable deactivation was observed. Similar to above mentioned Pd@$CN_{0.132}$ catalyst, the excellent catalytic performance of Pd@IFMC may be attributed to the presence of nitrogen heteroatoms derived from the IL calcination in the carbon framework, thereby making the distribution of Pd NPs relatively uniform throughout the mesoporous structure and thus inhibiting the agglomeration of Pd NPs during the reaction.

ILs-Incorporated Au NPs

Gold catalyzed oxidation of alcohols has been considered as an attractive green process especially considering the use of clean reagents such as O_2 and the relatively lower price of gold than other noble metals such as Pt, and Pd. A series of Au NPs with 8 nm average diameter were prepared via thermal reduction from $KAuCl_4$ in the mixture of n-butylimidazole and [BMIm][BF_4], which showed high catalytic activities in the oxidation of 1-phenylethanol in a base free system (Hosseini-Monfared, Meyer, & Janiak, 2013), Figure 35. In combination with radical initiator Nhydroxyphthalimide (NHPI), the Au NPs/IL gave a good conversion (77%) and selectivity (58%) for the oxidation of 1-phenylethanol to acetophenone through formation of an α-hydroxy carbon radical. The maximum TOF could be achieved to 5220 based on the total number of moles of gold; if only considering the Au-NP surface atoms, the TOF could be up to 31300.

The encapsulated Au NPs inside porous polymeric liposomes based on polymerized ILs could also catalyze the selective aerobic oxidation of alcohol (Buaki, Aprile, Dhakshinamoorthy, Alvaro, & Garcia, 2009). Starting from functionalized imidazolium ILs bearing two C11 alkenyl chains with terminal C=C double bonds, the liposomes mixed with $HAuCl_4$ and then underwent photochemical self-assembly to form vesicle-like structures, Figure 36a. In this strategy, the positive charge on the imidazolium head can attract $AuCl_4^-$ as precursor for Au NPs through ion pairing inside the polymeric structure, Figure 36b. TEM revealed that the average size of the resulting Au NPs embedded within the polymer structure was about 10 ± 5 nm. This catalyst system was highly active for the reaction of selective oxidation of 2-hydroxybenzyl alcohol to salicylaldehyde, and 99% selectivity with 33% yield could be obtained using molecular oxygen in water at 90 °C after 24h. Additionally, after use the catalyst could be easily separated from the reaction medium and recycled for two consecutive runs without noticeable loss of activity.

Figure 35. Catalytic mechanism of the alcohol oxidation with Au NPs/IL/NHPI (Reprinted with permission from ref. (Hosseini-Monfared et al., 2013). Copyright 2013 Elsevier)

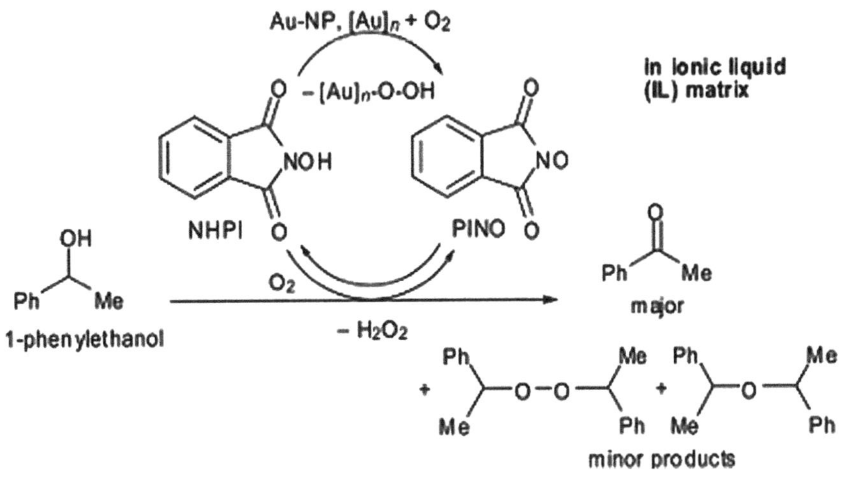

Figure 36. (a) Vesicles structure obtained by the self-assembly and polymerization of IL; (b) the formation of Au NPs inside the vesicles on irradiation, Compound 2 is photocuring agent namely 2-hydroxy-4'-(2-hydroxyethoxy)-2-methylpropiophenone (Reprinted with permission from ref. (Buaki et al., 2009). Copyright 2009 WILEY-VCH Verlag GmbH & Co. KGaA)

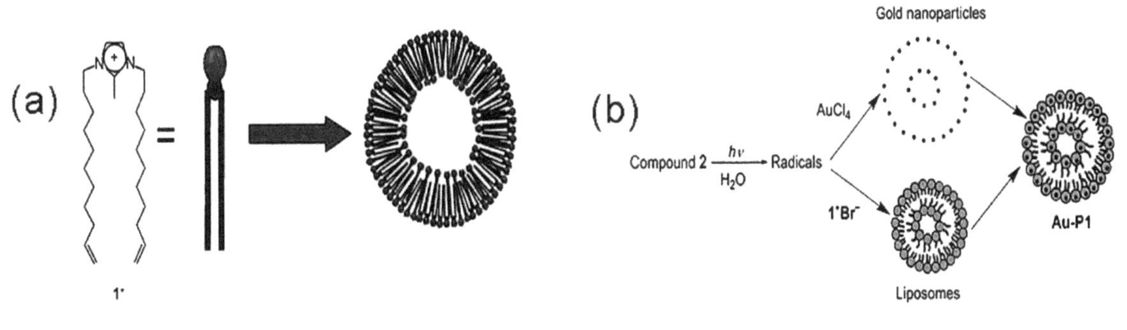

Other Reactions

Moreover, some other catalytic reactions catalyzed by IL-supported or mediated MNPs including A^3 coupling, hydroformylation, dehalogenation, *etc.* will be briefly introduced as follows.

A^3 Coupling

Generally, A^3 coupling reaction is one pot three-component reaction of an alkyne, an aldehyde and an amine to yield propargylamines. Au NPs supported on PMO-IL (Au@PMO-IL) with about 5 nm particle size were found to have excellent catalytic activity for the A^3 coupling reaction (B. Karimi, Gholinejad, & Khorasani, 2012), Figure 37. 88% yield of 1-(3-phenyl-1-p-tolylprop-2-ynyl) piperidine was achieved in the three-component reaction of 4-methylbenzaldehyde, piperidine and phenylacetylene at 60°C for 11 h in chloroform. Moreover, the Au@PMO-IL was stable and reusable up to three reaction runs, giving an average of 83% isolated yield under same conditions. More importantly, this catalyst applied well for a wide range of the substrates bearing various functional groups. In order to verify whether cationic gold species are functioning as actual catalysts in this system, a reduced Au@PMO-IL sample by treating the pristine catalyst with an excess of $NaBH_4$ was also prepared. As compared to the pristine Au@PMO-IL, only <15% conversion was obtained for the reduced Au@PMO-IL system. In combination with XPS analysis and catalytic results, it was suggested that ionic gold species in the form of Au(III) might be the active component in the active catalytic system.

One variation of A^3 reaction is called the oxidative A^3 coupling reaction, and the aldehyde is replaced by an alcohol in this case. Au NPs supported on graphene oxide (GO) with ionic liquid framework (Au@GO-IL) is an efficient catalyst for the oxidative A^3-coupling reactions of benzyl alcohols, terminal alkynes and secondary amines (Movahed, Lehi, & Dabiri, 2014)(Movahed, Lehi, & Dabiri, 2014). The Au@GO-IL nanocomposite displayed excellent selectivity and yields for a range of propargylamine through domino aerobic oxidation at very low catalyst loading (1 mol% of Au) under aerobic conditions in aqueous media. Furthermore, the catalyst could be reused 5 times without a significant decrease in its catalytic activity, and the TEM image of the recovered catalyst revealed the presence of well-distributed Au NPs on the GO-IL sheets without any aggregation.

Dehalogenation

The catalytic hydrodehalogenation of halogenated aromatic wastes has been recognized as an important chemical transformation in organic synthesis. The heteronuclear Au-Pd NPs were found to be effective catalysts in the dehalogenation of halogenated aromatic in $[C_2OHmim][NTf_2]$ (Yuan et al., 2013). The bimetallic NPs were obtained from the co-deposition of $Pd(OAc)_2$ and $Au(OAc)_3$ in ILs. All the bimetallic MNPs with different compositions are spherical with about 4 nm particle size. Analyses showed that Au

Figure 37. A^3 coupling of aldehydes, secondary amines and terminal alkynes over Au@PMO-IL

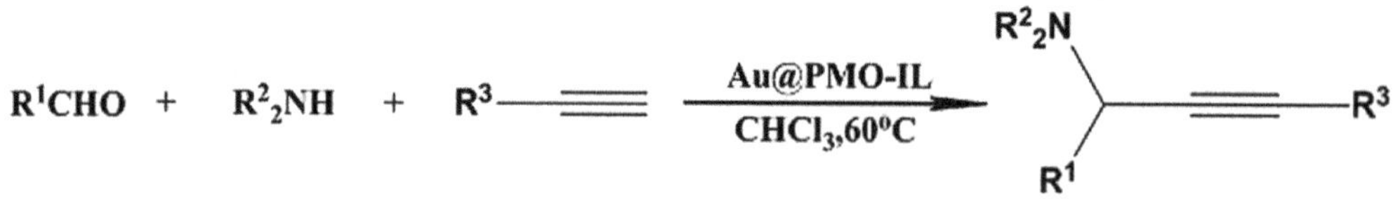

Figure 38. The structures of the Au-Pd NPs, partial coreshell structure (Pd-enriched surface layers). (Reprinted with permission from ref. (Yuan et al., 2013). Copyright 2013 WILEY-VCH Verlag GmbH & Co. KGaA)

atoms were mostly located in the NP core leaving the Pd atoms embedded on the surface, Figure 38, and the content of Pd species could be controlled by varying the ratio of Au and Pd salts. When the Au-Pd NPs were employed, the dehalogenation of 2-chloronitrobenzene, 4-bromotoluene and 3,5-dibromotoluene can be efficiently catalyzed. As compared to the one-component Pd or Au NPs, the bimetallic Au-Pd NPs showed much higher catalytic activity, in which the $Au_{0.8}Pd_{0.2}$ NPs were ca. ten-times more active than the Pd NPs. The activity of the bimetallic MNPs might be attributed to charge transfer from Pd to Au and consequently result in more efficient reductive elimination.

Hydroformylation

Hydroformylation is an important process for the production of aldehydes from alkenes. Rh is well-known as the highly active catalyst for this reaction. A new thermoregulated biphasic system involving ILs and cyclohexane with Rh NPs was developed and used for olefin hydroformylation (Xu, Wang, Zeng, Jiang, & Jin, 2012). Rh NPs with a diameter of 1.3 ± 0.2 nm were synthesized by simple H_2 reduction in the quaternary ammonium-based IL1, Figure 39. This IL/cyclohexane biphasic system with IL-stabilized Rh NPs exhibited the characteristics of thermoregulated phase-transition. By virtue of this property, the hydroformylation reaction can be performed under homogeneous conditions and subsequent efficient separation of catalyst from products can be realized by cooling the reaction mixture to room temperature. Under optimized conditions, various olefins could be converted completely with 98% selectivity for aldehydes without noticeable deactivation after five runs.

Figure 39. The structure of employed IL1-N,N-dimethyl-N-(2-(2-methoxyethoxy)ethyl)-N- (2-(2-octyloxyethoxy)ethyl)ammonium methanesulfonate

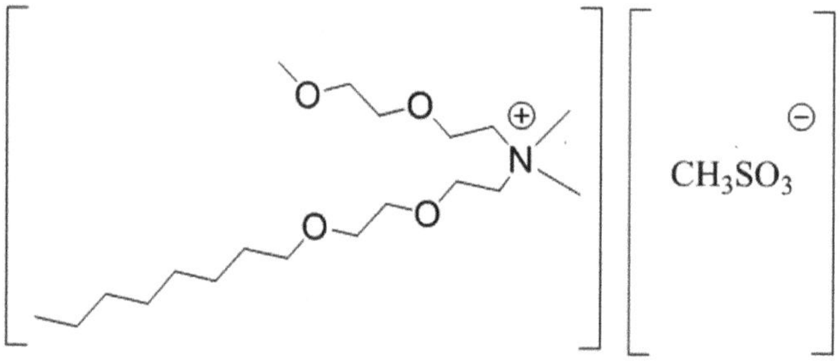

Figure 40. Biphasic hydrosilylation of phenylacetylene with triethylsilane by Pt-NP dispersions in [BMIm][BF$_4$] to yield distal triethyl(styryl)silane [triethyl(2-phenylvinyl)silane] and proximal triethyl(1-phenylvinyl)silane

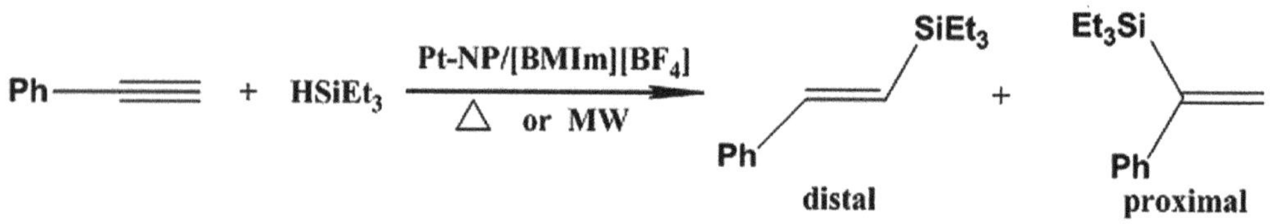

Hydrosilylation

A series of crystalline Pt NPs were prepared by decomposition of the organometallic precursor (MeCp) $PtMe_3$ in the presence of ILs and some other stabilizers to improve hydrosilylation (Marquardt, Barthel, Braun, Ganter, & Janiak, 2012), Figure 40. The resulting Pt-NPs showed highly crystalline structure with the small diameters of 1.0-2.3 nm and narrow diameter distribution ($\pm$ 0.7 nm). These Pt NPs/ILs as the catalyst has shown to be highly catalytically active to the hydrosilylation. Noteworthy, all the Pt-NPs were stable for several months without any detectable aggregation. These Pt-NPs/IL catalysts also showed highly activity for the biphasic hydrosilylation of phenylacetylene with triethylsilane, and quantitative conversion could be reached with a TOF of 96000 h^{-1} within 5 min MW irradiation.

SUMMARY AND OUTLOOK

It is no doubt that ILs have provided a new flexible liquid platform for the preparation and catalysis of MNPs, in which ILs can act as the solvent, stabilizer, ligand, and immobilizing agent, *etc.* Firstly, ILs are remarkable and excellent media for the synthesis and stabilization of MNPs. ILs can be regarded as a supramolecular three-dimensional electrostatic and hydrogenbonded network, which could prevent MNPs agglomeration and aggregation. At present, three usual methods could be carried out to synthesize the stable MNPs, *i.e.* chemical reduction method, direct decomposition of precursors and sputter deposition technique. In short, a wide range of MNPs with desired size could be prepared through the simple chemical reduction, however the stabilisers and reducing agents should be added in this process, which may introduce unwanted impurities; in general, no extra stabilizers or capping molecules are needed to achieve stable particle in ILs using direct decomposition from precursors, but the type of MNPs might be limited by the precursors; the sputter deposition technique as the innovative synthesis is considered as the extremely clean process, which combines the advantages of sputter deposition and the extremely low vapor pressure of ILs, besides this method could be extended for the generation of binary and multinary NPs. In this light, the sputter deposition technique may provide the promising platform to prepare the stable MNPs with tailored compositions and uniform sizes in ILs. Secondly, it is feasible to *in situ* characterize the MNPs either dispersed in the ILs or after isolation using modern techniques, especially vacuum techniques, due to the negligible vapor pressure and excellent stability of ILs. In this way, more data and information could be conveniently obtained to help deeply understanding the structures-property relationship as well as assembly mechanism for the ILs/MNPs catalytic system. Thirdly, it is now possible to design ILs/MNPs system for specific applications to improve the catalytic activity, selectivity stability and reusability of the catalysts, and even to change the reaction selectivity. More importantly, the soluble MNPs in ILs can behave as "homogeneous-like" heterogeneous catalysts in the catalytic reactions, and also, they can serve as a reservoir of soluble active catalytic species in some C-C coupling reactions (Christian Vollmer & Janiak, 2011; Yan, Xiao, & Kou, 2010). In this sense, the ILs-supported MNPs indeed provide a new approach for catalysis, *i.e.* how to design a "macroscopically homogeneous and microscopically heterogeneous" catalyst system.

Although progress has been achieved, there is still a tremendous room for further development of IL-supported or mediated MNPs catalysis. For example, (1) in most cases, ILs were thought to be capable of stabilizing the MNPs. It seems that, however, the stabilizing role of ILs has been exaggerated. To data, the preparation of very small-sized MNPs (< 2 nm) in ILs is still an interesting and enormous challenge. Therefore, to develop the very small-sized MNPs (<1 nm) in ILs would be highly desired. (2) Most of the previously reported works about ILs supported MNPs catalyst usually just showed the excellent activity of the whole system in specific reactions. The "true" influence of ILs on the morphology and size of MNPs, the intrinsic nature of the synergistic effect or promotion role of IL as the ligand or co-catalyst, as well as the catalytic mechanism is still unclear. (3) Although the ILs-supported MNPs is an ideal platform to explore that catalysis which is not typical homogeneous or heterogeneous, understanding the relationship between homogeneous and heterogeneous catalysis is not clear at this stage. Therefore, this is still in the early stage of investigation, much research room is existing, and it is expected that the intersection of homogeneous and heterogeneous catalysis involved in ILs and metal nanopaticles, though there would be a long way to go and much work are needed, would gestate new catalytic material, new catalytic phenomenon, new catalytic performance, and even new catalysis theory and application.

REFERENCES

Banerjee, A., & Scott, R. W. J. (2015). Optimization of transition metal nanoparticle-phosphonium ionic liquid composite catalytic systems for deep hydrogenation and hydrodeoxygenation reactions. *Green Chemistry*, *17*(3), 1597–1604. doi:10.1039/C4GC01716A

Banerjee, A., Theron, R., & Scott, R. W. (2013). Redispersion of transition metal nanoparticle catalysts in tetraalkylphosphonium ionic liquids. *Chemical Communications*, *49*(31), 3227–3229. doi:10.1039/c3cc40726h PMID:23482965

Beier, M. J., Andanson, J.-M., Mallat, T., Krumeich, F., & Baiker, A. (2012). Ionic Liquid-Supported Pt Nanoparticles as Catalysts for Enantioselective Hydrogenation. *ACS Catalysis*, *2*(3), 337–340. doi:10.1021/cs2006197

Biondi, I., Laporte, V., & Dyson, P. J. (2012). Application of a Versatile Nanoparticle Stabilizer in Phase Transfer and Catalysis. *ChemPlusChem*, *77*(8), 721–726. doi:10.1002/cplu.201200108

Boon, J. A., Levisky, J. A., Pflug, J. L., & Wilkes, J. S. (1986). Friedel Crafts Reactions in Ambient-Temperature Molten-Salts. *The Journal of Organic Chemistry*, *51*(4), 480–483. doi:10.1021/jo00354a013

Buaki, M., Aprile, C., Dhakshinamoorthy, A., Alvaro, M., & Garcia, H. (2009). Liposomes by polymerization of an imidazolium ionic liquid: Use as microreactors for gold-catalyzed alcohol oxidation. *Chemistry (Weinheim an der Bergstrasse, Germany)*, *15*(47), 13082–13089. doi:10.1002/chem.200902395 PMID:19856353

Calò, V., Nacci, A., Monopoli, A., & Cotugno, P. (2009). Palladium-nanoparticle-catalysed Ullmann reactions in ionic liquids with aldehydes as the reductants: Scope and mechanism. *Chemistry (Weinheim an der Bergstrasse, Germany)*, *15*(5), 1272–1279. doi:10.1002/chem.200801621 PMID:19072963

Calò, V., Nacci, A., Monopoli, A., Detomaso, A., & Iliade, P.Study on Factors Affecting the Regioselectivity of the Coupling Process. (2003). Pd Nanoparticle Catalyzed Heck Arylation of 1,1-Disubstituted Alkenes in Ionic Liquids. Study on Factors Affecting the Regioselectivity of the Coupling Process. *Organometallics*, *22*(21), 4193–4197. doi:10.1021/om034020w

Calò, V., Nacci, A., Monopoli, A., Fornaro, A., Sabbatini, L., Cioffi, N., & Ditaranto, N. (2004). Heck Reaction Catalyzed by Nanosized Palladium on Chitosan in Ionic Liquids. *Organometallics*, *23*(22), 5154–5158. doi:10.1021/om049586e

Calò, V., Nacci, A., Monopoli, A., & Montingelli, F. (2005). Pd nanoparticles as efficient catalysts for Suzuki and Stille coupling reactions of aryl halides in ionic liquids. *The Journal of Organic Chemistry*, *70*(15), 6040–6044. doi:10.1021/jo050801q PMID:16018701

Campbell, P. S., Santini, C. C., Bayard, F., Chauvin, Y., Collière, V., Podgoršek, A., & Sá, J. et al. (2010). Olefin hydrogenation by ruthenium nanoparticles in ionic liquid media: Does size matter? *Journal of Catalysis*, *275*(1), 99–107. doi:10.1016/j.jcat.2010.07.018

Carvalho, M. S., Lacerda, R. A., Leão, J. P. B., Scholten, J. D., Neto, B. A. D., & Suarez, P. A. Z. (2011). In situ generated palladium nanoparticles in imidazolium-based ionic liquids: A versatile medium for an efficient and selective partial biodiesel hydrogenation. *Catalysis Science & Technology*, *1*(3), 480–488. doi:10.1039/c0cy00028k

Cassol, C. C., Umpierre, A. P., Machado, G., Wolke, S. I., & Dupont, J. (2005). The role of Pd nanoparticles in ionic liquid in the Heck reaction. *Journal of the American Chemical Society*, *127*(10), 3298–3299. doi:10.1021/ja0430043 PMID:15755145

Cheng, J., Tang, L., & Xu, J. (2010). An Economical, Green Pathway to Biaryls: Palladium Nanoparticles Catalyzed Ullmann Reaction in Ionic Liquid/ Supercritical Carbon Dioxide System. *Advanced Synthesis & Catalysis*, *352*(18), 3275–3286. doi:10.1002/adsc.201000475

Chiappe, C., Pieraccini, D., Zhao, D., Fei, Z., & Dyson, P. J. (2006). Remarkable Anion and Cation Effects on Stille Reactions in Functionalised Ionic Liquids. *Advanced Synthesis & Catalysis*, *348*(1-2), 68–74. doi:10.1002/adsc.200505271

Dash, P., & Scott, R. W. (2009). 1-Methylimidazole stabilization of gold nanoparticles in imidazolium ionic liquids. *Chemical Communications*, (7): 812–814. doi:10.1039/b816446k PMID:19322450

Deshmukh, R. R., Rajagopal, R., & Srinivasan, K. V. (2001). Ultrasound promoted C–C bond formation: Heck reaction at ambient conditions in room temperature ionic liquids. *Chemical Communications*, (17): 1544–1545. doi:10.1039/b104532f PMID:12240373

Dewan, M., Kumar, A., Saxena, A., De, A., & Mozumdar, S. (2012). Using Hydrophilic Ionic Liquid, [bmim]BF4 - Ethylene Glycol System as a Novel Media for the Rapid Synthesis of Copper Nanoparticles. *PLoS ONE*, *7*(1), e29131. doi:10.1371/journal.pone.0029131 PMID:22238589

Dupont, J., Fonseca, G. S., Umpierre, A. P., Fichtner, P. F. P., & Teixeira, S. R. (2002). Transition-Metal Nanoparticles in Imidazolium Ionic Liquids: Recyclable Catalysts for Biphasic Hydrogenation Reactions. *Journal of the American Chemical Society*, *124*(16), 4228–4229. doi:10.1021/ja025818u PMID:11960449

Dupont, J., & Scholten, J. D. (2010). On the structural and surface properties of transition-metal nanoparticles in ionic liquids. *Chemical Society Reviews*, *39*(5), 1780–1804. doi:10.1039/b822551f PMID:20419219

Dykeman, R. R., Yan, N., Scopelliti, R., & Dyson, P. J. (2011). Enhanced rate of arene hydrogenation with imidazolium functionalized bipyridine stabilized rhodium nanoparticle catalysts. *Inorganic Chemistry*, *50*(3), 717–719. doi:10.1021/ic102041q PMID:21207990

Dykeman, R. R., Yuan, Y., Yan, N., Asakura, H., Teramura, K., Tanaka, T., & Dyson, P. J. (2012). Rational Design of a Molecular Nanocatalyst-Stabilizer that Enhances both Catalytic Activity and Nanoparticle Stability. *ChemCatChem*, *4*(12), 1907–1910. doi:10.1002/cctc.201200552

Fei, Z., Zhao, D., Pieraccini, D., Ang, W. H., Geldbach, T. J., Scopelliti, R., & Dyson, P. J. et al. (2007). Development of Nitrile-Functionalized Ionic Liquids for C–C Coupling Reactions: Implication of Carbene and Nanoparticle Catalysts. *Organometallics*, *26*(7), 1588–1598. doi:10.1021/om060950e

Fonseca, G. S., Machado, G., Teixeira, S. R., Fecher, G. H., Morais, J., Alves, M. C., & Dupont, J. (2006). Synthesis and characterization of catalytic iridium nanoparticles in imidazolium ionic liquids. *Journal of Colloid and Interface Science*, *301*(1), 193–204. doi:10.1016/j.jcis.2006.04.073 PMID:16780858

Fonseca, G. S., Scholten, J. D., & Dupont, J. (2004). Iridium Nanoparticles Prepared in Ionic Liquids: An Efficient Catalytic System for the Hydrogenation of Ketones. *Synlett*, (9): 1525–1528. doi:10.1055/s-2004-829061

Fonseca, G. S., Umpierre, A. P., Fichtner, P. F., Teixeira, S. R., & Dupont, J. (2003). The use of imidazolium ionic liquids for the formation and stabilization of ir0 and rh0 nanoparticles: Efficient catalysts for the hydrogenation of arenes. *Chemistry (Weinheim an der Bergstrasse, Germany)*, *9*(14), 3263–3269. doi:10.1002/chem.200304753 PMID:12866070

Gao, S., Zhang, H., Wang, X., Mai, W., Peng, C., & Ge, L. (2005). Palladium nanowires stabilized by thiol-functionalized ionic liquid: Seed-mediated synthesis and heterogeneous catalyst for Sonogashira coupling reaction. *Nanotechnology*, *16*(8), 1234–1237. doi:10.1088/0957-4484/16/8/042

Guo, S., Shi, F., Gu, Y., Yang, J., & Deng, Y. (2005). Size-controllable Synthesis of Gold Nanoparticles via Carbonylation and Reduction of Hydrochloroauric Acid with CO and H2O in Ionic Liquids. *Chemistry Letters*, *34*(6), 830–831. doi:10.1246/cl.2005.830

Hallett, J. P., & Welton, T. (2011). Room-temperature ionic liquids: Solvents for synthesis and catalysis. 2. *Chemical Reviews*, *111*(5), 3508–3576. doi:10.1021/cr1003248 PMID:21469639

Hatakeyama, Y., Onishi, K., & Nishikawa, K. (2011). Effects of sputtering conditions on formation of gold nanoparticles in sputter deposition technique. *RSC Advances*, *1*(9), 1815–1821. doi:10.1039/c1ra00688f

Hirano, M., Enokida, K., Okazaki, K., Kuwabata, S., Yoshida, H., & Torimoto, T. (2013). Composition-dependent electrocatalytic activity of AuPd alloy nanoparticles prepared via simultaneous sputter deposition into an ionic liquid. *Physical Chemistry Chemical Physics*, *15*(19), 7286–7294. doi:10.1039/c3cp50816a PMID:23575517

Hosseini-Monfared, H., Meyer, H., & Janiak, C. (2013). Dioxygen oxidation of 1-phenylethanol with gold nanoparticles and N-hydroxyphthalimide in ionic liquid. *Journal of Molecular Catalysis A Chemical*, *372*, 72–78. doi:10.1016/j.molcata.2013.02.007

Hu, Y., Yu, Y., Hou, Z., Li, H., Zhao, X., & Feng, B. (2008). Biphasic Hydrogenation of Olefins by Functionalized Ionic Liquid-Stabilized Palladium Nanoparticles. *Advanced Synthesis & Catalysis*, *350*(13), 2077–2085. doi:10.1002/adsc.200800290

Huang, J., Jiang, T., Han, B., Gao, H., Chang, Y., Zhao, G., & Wu, W. (2003). Hydrogenation of olefins using ligand-stabilized palladium nanoparticles in an ionic liquid. *Chemical Communications*, (14): 1654–1655. doi:10.1039/b302750c

Itoh, H., Naka, K., & Chujo, Y. (2004). Synthesis of gold nanoparticles modified with ionic liquid based on the imidazolium cation. *Journal of the American Chemical Society*, *126*(10), 3026–3027. doi:10.1021/ja039895g PMID:15012113

Jutz, F., Andanson, J.-M., & Baiker, A. (2009). A green pathway for hydrogenations on ionic liquid-stabilized nanoparticles. *Journal of Catalysis*, *268*(2), 356–366. doi:10.1016/j.jcat.2009.10.006

Karimi, B., Behzadnia, H., Bostina, M., & Vali, H. (2012). A nano-fibrillated mesoporous carbon as an effective support for palladium nanoparticles in the aerobic oxidation of alcohols "on pure water". *Chemistry (Weinheim an der Bergstrasse, Germany)*, *18*(28), 8634–8640. doi:10.1002/chem.201200373 PMID:22707318

Karimi, B., Elhamifar, D., Clark, J. H., & Hunt, A. J. (2011). Palladium containing periodic mesoporous organosilica with imidazolium framework (Pd@PMO-IL): An efficient and recyclable catalyst for the aerobic oxidation of alcohols. *Organic & Biomolecular Chemistry*, *9*(21), 7420–7426. doi:10.1039/c1ob05752a PMID:21904726

Karimi, B., Gholinejad, M., & Khorasani, M. (2012). Highly efficient three-component coupling reaction catalyzed by gold nanoparticles supported on periodic mesoporous organosilica with ionic liquid framework. *Chemical Communications*, *48*(71), 8961–8963. doi:10.1039/c2cc33320a PMID:22842770

Karimi, B., Mansouri, F., & Vali, H. (2014). A highly water-dispersible/magnetically separable palladium catalyst based on a Fe3O4@SiO2 anchored TEG-imidazolium ionic liquid for the Suzuki–Miyaura coupling reaction in water. *Green Chemistry*, *16*(5), 2587–2596. doi:10.1039/c3gc42311e

Kim, K.-S., Demberelnyamba, D., & Lee, H. (2004). Size-Selective Synthesis of Gold and Platinum Nanoparticles Using Novel Thiol-Functionalized Ionic Liquids. *Langmuir*, *20*(3), 556–560. doi:10.1021/la0355848 PMID:15773074

König, D., Richter, K., Siegel, A., Mudring, A.-V., & Ludwig, A. (2014). High-Throughput Fabrication of Au-Cu Nanoparticle Libraries by Combinatorial Sputtering in Ionic Liquids. *Advanced Functional Materials*, *24*(14), 2049–2056. doi:10.1002/adfm.201303140

Krämer, J., Redel, E., Thomann, R., & Janiak, C. (2008). Use of Ionic Liquids for the Synthesis of Iron, Ruthenium, and Osmium Nanoparticles from Their Metal Carbonyl Precursors. *Organometallics*, *27*(9), 1976–1978. doi:10.1021/om800056z

Leger, B., Denicourt-Nowicki, A., Olivier-Bourbigou, H., & Roucoux, A. (2008). Rhodium nanocatalysts stabilized by various bipyridine ligands in nonaqueous ionic liquids: Influence of the bipyridine coordination modes in arene catalytic hydrogenation. *Inorganic Chemistry*, *47*(19), 9090–9096. doi:10.1021/ic8010713 PMID:18781733

Li, Z., Liu, Z., Zhang, J., Han, B., Du, J., Gao, Y., & Jiang, T. (2005). Synthesis of single-crystal gold énanosheets of large size in ionic liquids. *The Journal of Physical Chemistry B*, *109*(30), 14445–14448. doi:10.1021/jp0520998 PMID:16852818

Luska, K. L., Migowski, P., & Leitner, W. (2015). Ionic liquid-stabilized nanoparticles as catalysts for the conversion of biomass. *Green Chemistry*, *17*(6), 3195–3206. doi:10.1039/C5GC00231A

Luska, K. L., & Moores, A. (2012). Functionalized Ionic Liquids for the Synthesis of Metal Nanoparticles and their Application in Catalysis. *Chemcatchem*, *4*(10), 1534–1546. doi:10.1002/cctc.201100366

Luska, K. L., & Moores, A. (2012). Ruthenium nanoparticle catalysts stabilized in phosphonium and imidazolium ionic liquids: Dependence of catalyst stability and activity on the ionicity of the ionic liquid. *Green Chemistry*, *14*(6), 1736–1742. doi:10.1039/c2gc35241a

Luza, L., Gual, A., Eberhardt, D., Teixeira, S. R., Chiaro, S. S. X., & Dupont, J. (2013). "Imprinting" Catalytically Active Pd Nanoparticles onto Ionic-Liquid-Modified Al2O3Supports. *ChemCatChem*, *5*(8), 2471–2478. doi:10.1002/cctc.201300123

Luza, L., Gual, A., Rambor, C. P., Eberhardt, D., Teixeira, S. R., Bernardi, F., & Dupont, J. et al. (2014). Hydrophobic effects on supported ionic liquid phase Pd nanoparticle hydrogenation catalysts. *Physical Chemistry Chemical Physics*, *16*(34), 18088–18091. doi:10.1039/C4CP03063J PMID:25057970

Marquardt, D., Barthel, J., Braun, M., Ganter, C., & Janiak, C. (2012). Weakly-coordinated stable platinum nanocrystals. *CrystEngComm*, *14*(22), 7607–7615. doi:10.1039/c2ce25904d

Migowski, P., & Dupont, J. (2007). Catalytic applications of metal nanoparticles in imidazolium ionic liquids. *Chemistry (Weinheim an der Bergstrasse, Germany)*, *13*(1), 32–39. doi:10.1002/chem.200601438 PMID:17115465

Migowski, P., Machado, G., Texeira, S. R., Alves, M. C., Morais, J., Traverse, A., & Dupont, J. (2007). Synthesis and characterization of nickel nanoparticles dispersed in imidazolium ionic liquids. *Physical Chemistry Chemical Physics*, *9*(34), 4814–4821. doi:10.1039/b703979d PMID:17712460

Movahed, S. K., Lehi, N. F., & Dabiri, M. (2014). Gold nanoparticle supported on ionic liquid-modified graphene oxide as an efficient and recyclable catalyst for one-pot oxidative A3-coupling reaction of benzyl alcohols. *RSC Advances*, *4*(79), 42155–42158. doi:10.1039/C4RA05190D

Okazaki, K.-i., Kiyama, T., Hirahara, K., Tanaka, N., Kuwabata, S., & Torimoto, T. (2008). Single-step synthesis of gold–silver alloy nanoparticles in ionic liquids by a sputter deposition technique. *Chemical Communications*(6), 691-693.

Ott, L. S., Campbell, S., Seddon, K. R., & Finke, R. G. (2007). Evidence that imidazolium-based ionic ligands can be metal(0)/nanocluster catalyst poisons in at least the test case of iridium(0)-catalyzed acetone hydrogenation. *Inorganic Chemistry*, *46*(24), 10335–10344. doi:10.1021/ic700976z PMID:17975891

Parvulescu, V. I., & Hardacre, C. (2007). Catalysis in ionic liquids. *Chemical Reviews*, *107*(6), 2615–2665. doi:10.1021/cr050948h PMID:17518502

Pryjomska-Ray, I., Gniewek, A., Trzeciak, A. M., Ziółkowski, J. J., & Tylus, W. (2006). Homogeneous/heterogeneous palladium based catalytic system for Heck reaction. The reversible transfer of palladium between solution and support. *Topics in Catalysis*, *40*(1-4), 173–184. doi:10.1007/s11244-006-0119-1

Qin, Y., Song, Y., Sun, N., Zhao, N., Li, M., & Qi, L. (2008). Ionic Liquid-Assisted Growth of Single-Crystalline Dendritic Gold Nanostructures with a Three-Fold Symmetry. *Chemistry of Materials*, *20*(12), 3965–3972. doi:10.1021/cm8002386

Redel, E., Thomann, R., & Janiak, C. (2008). Use of ionic liquids (ILs) for the IL-anion size-dependent formation of Cr, Mo and W nanoparticles from metal carbonyl M(CO)6 precursors. *Chemical Communications*, (15): 1789–1791. doi:10.1039/b718055a PMID:18379694

Ryu, H. J., Sanchez, L., Keul, H. A., Raj, A., & Bockstaller, M. R. (2008). Imidazolium-based ionic liquids as efficient shape-regulating solvents for the synthesis of gold nanorods. *Angewandte Chemie International Edition in English*, *47*(40), 7639–7643. doi:10.1002/anie.200802185 PMID:18712726

Scheeren, C. W., Machado, G., Dupont, J., Fichtner, P. F., & Texeira, S. R. (2003). Nanoscale Pt(0) particles prepared in imidazolium room temperature ionic liquids: Synthesis from an organometallic precursor, characterization, and catalytic properties in hydrogenation reactions. *Inorganic Chemistry*, *42*(15), 4738–4742. doi:10.1021/ic034453r PMID:12870966

Scheeren, C. W., Machado, G., Teixeira, S. R., Morais, J., Domingos, J. B., & Dupont, J. (2006). Synthesis and characterization of Pt0 nanoparticles in imidazolium ionic liquids. *The Journal of Physical Chemistry B*, *110*(26), 13011–13020. doi:10.1021/jp0623037 PMID:16805608

Scholten, J. D., Leal, B. C., & Dupont, J. (2012). Transition Metal Nanoparticle Catalysis in Ionic Liquids. *Acs Catalysis*, *2*(1), 184–200. doi:10.1021/cs200525e

Schrekker, H. S., Gelesky, M. A., Stracke, M. P., Schrekker, C. M., Machado, G., Teixeira, S. R., & Dupont, J. et al. (2007). Disclosure of the imidazolium cation coordination and stabilization mode in ionic liquid stabilized gold(0) nanoparticles. *Journal of Colloid and Interface Science*, *316*(1), 189–195. doi:10.1016/j.jcis.2007.08.018 PMID:17761189

Silveira, E. T., Umpierre, A. P., Rossi, L. M., Machado, G., Morais, J., Soares, G. V., & Dupont, J. et al. (2004). The partial hydrogenation of benzene to cyclohexene by nanoscale ruthenium catalysts in imidazolium ionic liquids. *Chemistry (Weinheim an der Bergstrasse, Germany)*, *10*(15), 3734–3740. doi:10.1002/chem.200305765 PMID:15281157

Sugden, S., & Wilkins, H. (1929). CLXVII.-The parachor and chemical constitution. Part XII. Fused metals and salts. *Journal of the Chemical Society*, 1291–1298. doi:10.1039/jr9290001291

Suzuki, S., Suzuki, T., Tomita, Y., Hirano, M., Okazaki, K.-, Kuwabata, S., & Torimoto, T. (2012). Compositional control of AuPt nanoparticles synthesized in ionic liquids by the sputter deposition technique. *CrystEngComm*, *14*(15), 4922–4926. doi:10.1039/c2ce25235j

Torimoto, T., Ohta, Y., Enokida, K., Sugioka, D., Kameyama, T., Yamamoto, T., & Kuwabata, S. et al. (2015). Ultrathin oxide shell coating of metal nanoparticles using ionic liquid/metal sputtering. *Journal of Materials Chemistry A*, *3*(11), 6177–6186. doi:10.1039/C4TA06643J

Torimoto, T., Okazaki, K.-, Kiyama, T., Hirahara, K., Tanaka, N., & Kuwabata, S. (2006). Sputter deposition onto ionic liquids: Simple and clean synthesis of highly dispersed ultrafine metal nanoparticles. *Applied Physics Letters*, *89*(24), 243117. doi:10.1063/1.2404975

Umpierre, A. P., Machado, G., Fecher, G. H., Morais, J., & Dupont, J. (2005). Selective Hydrogenation of 1,3-Butadiene to 1-Butene by Pd(0) Nanoparticles Embedded in Imidazolium Ionic Liquids. *Advanced Synthesis & Catalysis*, *347*(10), 1404–1412. doi:10.1002/adsc.200404313

Vanecht, E., Binnemans, K., Patskovsky, S., Meunier, M., Seo, J. W., Stappers, L., & Fransaer, J. (2012). Stability of sputter-deposited gold nanoparticles in imidazolium ionic liquids. *Physical Chemistry Chemical Physics*, *14*(16), 5662–5671. doi:10.1039/c2cp23677j PMID:22422275

Vollmer, C., & Janiak, C. (2011). Naked metal nanoparticles from metal carbonyls in ionic liquids: Easy synthesis and stabilization. *Coordination Chemistry Reviews*, *255*(17-18), 2039–2057. doi:10.1016/j.ccr.2011.03.005

Vollmer, C., Redel, E., Abu-Shandi, K., Thomann, R., Manyar, H., Hardacre, C., & Janiak, C. (2010). Microwave irradiation for the facile synthesis of transition-metal nanoparticles (NPs) in ionic liquids (ILs) from metal-carbonyl precursors and Ru-, Rh-, and Ir-NP/IL dispersions as biphasic liquid-liquid hydrogenation nanocatalysts for cyclohexene. *Chemistry (Weinheim an der Bergstrasse, Germany)*, *16*(12), 3849–3858. doi:10.1002/chem.200903214 PMID:20187043

Welton, T. (1999). Room-Temperature Ionic Liquids. Solvents for Synthesis and Catalysis. *Chemical Reviews*, *99*(8), 2071–2084. doi:10.1021/cr980032t PMID:11849019

Wilkes, J. S., & Zaworotko, M. J. (1992). Air and water stable 1-ethyl-3-methylimidazolium based ionic liquids. *Journal of the Chemical Society. Chemical Communications*, (13): 965–967. doi:10.1039/c39920000965

Xu, X., Li, Y., Gong, Y., Zhang, P., Li, H., & Wang, Y. (2012). Synthesis of palladium nanoparticles supported on mesoporous N-doped carbon and their catalytic ability for biofuel upgrade. *Journal of the American Chemical Society*, *134*(41), 16987–16990. doi:10.1021/ja308139s PMID:23030399

Xu, Y., Wang, Y., Zeng, Y., Jiang, J., & Jin, Z. (2012). A Novel Thermoregulated Ionic Liquid and Organic Biphasic System With Rh Nanoparticles for Olefin Hydroformylation. *Catalysis Letters*, *142*(7), 914–919. doi:10.1007/s10562-012-0841-1

Yan, N., Xiao, C., & Kou, Y. (2010). Transition metal nanoparticle catalysis in green solvents. *Coordination Chemistry Reviews*, *254*(9-10), 1179–1218. doi:10.1016/j.ccr.2010.02.015

Yuan, X., Sun, G., Asakura, H., Tanaka, T., Chen, X., Yuan, Y., & Yan, N. et al. (2013). Development of palladium surface-enriched heteronuclear Au-Pd nanoparticle dehalogenation catalysts in an ionic liquid. *Chemistry (Weinheim an der Bergstrasse, Germany)*, *19*(4), 1227–1234. doi:10.1002/chem.201203605 PMID:23293076

Zahmakiran, M., & Ozkar, S. (2011). Metal nanoparticles in liquid phase catalysis; from recent advances to future goals. *Nanoscale*, *3*(9), 3462–3481. doi:10.1039/c1nr10201j PMID:21833406

Zhang, B., Zhan, X., Zhao, P., & Li, Z. (2015). Ionic liquid-mediated synthesis of unique PtPd bimetallic particles with tiny subunits for efficient electrocatalytic and catalytic applications. *RSC Advances*, *5*(71), 57640–57646. doi:10.1039/C5RA09832G

Zhang, H., & Cui, H. (2009). Synthesis and characterization of functionalized ionic liquid-stabilized metal (gold and platinum) nanoparticles and metal nanoparticle/carbon nanotube hybrids. *Langmuir*, *25*(5), 2604–2612. doi:10.1021/la803347h PMID:19437685

Zhang, Q., Zhang, S., & Deng, Y. (2011). Recent advances in ionic liquid catalysis. *Green Chemistry*, *13*(10), 2619–2637. doi:10.1039/c1gc15334j

Zhao, D., Fei, Z., Geldbach, T. J., Scopelliti, R., & Dyson, P. J. (2004). Nitrile-functionalized pyridinium ionic liquids: Synthesis, characterization, and their application in carbon-carbon coupling reactions. *Journal of the American Chemical Society*, *126*(48), 15876–15882. doi:10.1021/ja0463482 PMID:15571412

Zhu, W., Yu, Y., Yang, H., Hua, L., Qiao, Y., Zhao, X., & Hou, Z. (2013). Cooperative effects in catalytic hydrogenation regulated by both the cation and anion of an ionic liquid. *Chemistry (Weinheim an der Bergstrasse, Germany)*, *19*(6), 2059–2066. doi:10.1002/chem.201202707 PMID:23255466

KEY TERMS AND DEFINITIONS

A^3 Coupling Reaction: In general, one pot three-component reaction of an alkyne, an aldehyde and an amine to yield propargylamines.

C-C Coupling: A reaction which link the carbon atoms of two hydrocarbon fragments.

Hydrogenation: A chemical reaction which transfer hydrogen from hydrogen source to a compound or element.

Ionic Liquid: Molten salts composing of bulky organic cations and inorganic or organic anions, which melt around 100 °C or below.

Metal Nanoparticle: The metal particles between 1 and 100 nanometers in size.

Oxidation of Alcohols: A reaction which remove hydrogen atoms of alcohols with the help of an oxidant generally forming aldehydes or ketones.

Turnover Frequency: The overall catalytic reaction takes place per catalytic site per unit time for a fixed set of reaction conditions.

Chapter 12
Kinetic Models for Complex Parallel–Consecutive Reactions Assessment of Reaction Network and Product Selectivity

Hamdy Farag
Kyushu University, Japan & Mansoura University, Egypt

Masahiro Kishida
Kyushu University, Japan

ABSTRACT

Kinetic models were developed to account for the partial contributions of intermediates in complex parallel–consecutive reactions. The models allow precise estimation of the apparent rate constants of all steps in such a reaction network. The hydrodesulfurization (HDS) of dibenzothiophene (DBT) over CoMo-based alumina and carbon catalysts, and over an unsupported molybdenum sulfide catalyst, were investigated in a batch reactor and used to represent this type of reaction. The HDS reactions proceeded through two parallel–consecutive reaction pathways, i.e., direct desulfurization (DDS) and hydrogenation (HYD), in which two main intermediates, namely biphenyl and partially hydrogenated DBT, were involved. Different selectivities in terms of yield fraction (percentage ratio of HYD/DDS) were observed for these catalysts. The results are discussed in the context of proposed HDS reaction networks. Use of these models enables more accurate assessment of differences among the performances of different catalysts.

1. INTRODUCTION

A reaction between molecules implies the breaking of existing bonds and/or the formation of new bonds. A catalyst is a substance that affects the rate or the direction of a chemical reaction, but is not consumed or altered during the process. This definition is, however, valid only theoretically in terms of chemistry, because the catalyst may undergo physical and sometimes chemical changes.) A catalyst therefore affects only the rate of a chemical reaction and not its equilibrium, i.e., a catalyst may increase or decrease the

DOI: 10.4018/978-1-4666-9975-5.ch012

reaction rate and it may influence the direction or the selectivity of a reaction, but it cannot change the ultimate thermodynamic equilibrium. Heterogeneous catalysis enables the development of large-scale continuous processes, whereas homogeneous catalysis is mainly used in fine chemicals manufacturing. Catalysis is mainly developed through advances in three areas: kinetics, synthesis, and characterization. Chemical kinetics is an important tool for studying catalytic reactions, and is concerned primarily with chemical changes and the energy and mass fluxes associated with them. Kinetics is simply the measurement and analysis of reaction rates. The rate of a chemical reaction is defined as

$$R = \frac{1}{v}\frac{dC}{dt}$$

where *V*, *C*, and *t* are the system volume, extent of reaction, and time, respectively.

Usually, integral methods based on integration of the reaction rate expressions and differential methods based on differentiation of experimental concentration versus time data are the techniques used to determine empirical reaction rate functions. The rate equations of simple reactions for which the concentrations of reactants, intermediates, and products depend on time can be solved analytically. However, for complex parallel–consecutive reactions, the rate expression becomes unmanageable and closed-form solutions cannot be easily obtained, especially for the time dependence of various species concentrations in the system. Parallel–consecutive reactions are dominant in complex petrochemical reactions (Vasudevan & Fierro, 1996; Whitehurst et al., 1998). There are also numerous chemical reactions that show this parallel–consecutive behavior; examples include HYD, oxidative dehydrogenation, hydrodesulfurization (HDS), and selective oxidation (Babich et al., 2003; Grabowski, 2006; Singh & Vannice, 2001). The term consecutive reactions refer to those reactions in which one or more of the products formed initially (primary product) undergoes a subsequent reaction to give another product. A secondary product may therefore be produced by various routes. In handling such reactions, it is necessary to deal with the problems of determining the orders and rate constants for each individual reaction. Many catalysts can stimulate one reaction with various activities and selectivities. Appropriate and reliable kinetic investigations are crucial for ranking catalysts in order of activities and selectivities. The product distribution as a function of contact time between catalyst and feed can help to identify the reaction network. The validity of the postulated reaction network is difficult to judge unless extensive kinetic studies are undertaken. Here, we focus on such reactions and discuss the key issues in selecting an appropriate reaction network by analyzing the reaction selectivities. The rate-law models described are valid for application to such reactions. Experimental results provide factual information on reactions. Although agreement between a rate law and the experimental data suggests that the proposed mechanism is valid, disagreement is enough to rule out the proposed mechanism. However, It should be noted that agreement between the predicted and measured kinetics is not always enough to assign a mechanism, unless the number of constants in the numerical rate formula is limited. Sophisticated reaction rate laws can be considered, but such rate laws usually include a high number of constants, which makes it difficult to identify certain mechanisms.

In this chapter, we provide some models to describe substrates that are transformed via parallel–consecutive reaction networks. We use the HDS of dibenzothiophene (DBT) as a representative reaction. Such aromatic sulfur-containing species accumulate mainly in high-boiling-point petroleum fractions such as middle distillate fuels (Mochida et al., 2003). These compounds have been proved to be the most refractory sulfur-containing compounds toward HDS. DBT undergoes HDS via reaction networks that involve both parallel and successive reactions, which are widely known as direct desulfurization (DDS); they involve cleavage of C–S bonds and lead to biphenyl (BP), and HYD pathways, in which the substrate molecule is hydrogenated to form partially hydrogenated DBT intermediates. Sulfur in these intermediates is removed by further sequential reactions. The activity of a given catalyst in the HDS of DBTs is evaluated from the overall activity. The HDS of most refractory sulfur-containing compounds occurs preferentially through the hydrogenation (HYD) pathway, therefore a catalyst with significant hydrogenating activity is important (Bej et al., 2004; Farag et al., 2003; Kagami et al., 2005). Precise determination of the catalytic selectivity in HDS reactions is therefore as important as determining the overall activity. The selectivity in these reactions, defined by the rate ratio (SR) of HYD to direct DDS reactions, depends on the relative reactivities of the intermediates, i.e., BP or similar species and partially hydrogenated DBT or similar species, in further sequential reactions. There is still uncertainty regarding the assessment of the contribution from each individual intermediate in the overall reaction network. For catalyst evaluation and ranking in deep HDS processes, a kinetic model that solves the rate for each reaction step in the proposed reaction network is important, because it may give insights into the HDS mechanism. Although a significant number of studies have been carried out in this area, there is as yet no consensus on a method for estimating the selectivity in terms of the apparent individual rate constant in the reaction network (Kim et al., 2005; Pawelec et al., 2008; Lewandowski et al., 2014; García-Gutiérrez et al., 2014; Castillo-Araiza et al., 2015). An approach using the initial selectivity in terms of yield fraction (percentage ratio, SF) to estimate the individual rate constants for the HYD and DDS pathways in the HDS of 4,6-dimethyldibenzothiophene has recently been reported by Kim et al. (2005). However, as the conversion level of the substrate changes, the selectivity in terms of product yield may also change. In addition, this method gives no explanation of the subsequent reactions of the intermediates. Other studies have simplified the reaction network by assuming that BP and similar species undergo no sequential reactions (Daage & Chianelli, 1994; Li et al., 2007; Michaud et al., 1998; Wang et al., 2008). In this chapter, we discuss the use of closed kinetic models to determine the reaction network for the HDS of DBT over various hydrotreatment catalysts. From the models, a reliable individual rate constant for each step in the reaction network can be estimated. The models also provide valuable quantitative data for intermediate contributions in the HDS reactions or any like reactions.

2. REACTIONS INVESTIGATED

We focus on the HDS of DBT as a representative complex parallel–consecutive reaction. Three catalysts were examined: a laboratory-synthesized CoMo-based carbon catalyst (CoMo/C), a commercial CoMo/Al_2O_3 catalyst, and an unsupported MoS_2 catalyst. We used the published experimental data of the aforementioned catalysts in the kinetic models. The HDS data for the CoMo/C and commercial CoMo/Al_2O_3 catalysts were obtained from the study by Farag et al. (2000), and the data for the unsupported MoS_2 catalyst were taken from the study by Daage & Chianelli (1994). The data for the HDS reaction of DBT are shown in Tables 1, 2, and 3. All the catalysts were sulfided with an H_2S/H_2 gas mixture. The experimental results show that as the reaction time changed, the conversion level of the substrate changed, and the product selectivity changed significantly. The product selectivity, SF, is expressed as *the yield ratio of a product/sum of yield ratios of all products*. The product selectivity curves showed that BP and 1,2,3,4-tetrahydrodibenzothiophene (H4-DBT) are the primary reaction products, and can

Table 1. Relative species concentrations in HDS of DBT over commercial $CoMo/Al_2O_3$ catalyst at different times[a,b]

Time, min	Fraction Converted	A	B	C	D
5	0.16	0.84	0.14	0.0034	0.01
15	0.34	0.66	0.30	0.0048	0.03
31	0.52	0.48	0.47	0.0044	0.05
63	0.79	0.21	0.71	0.0027	0.08
90	0.95	0.05	0.84	0.0000	0.11
122	0.98	0.02	0.85	0.0000	0.13

[a]Conditions: 3 MPa H_2 and 340 °C. [b]A, DBT; B, BP; C, H4-DBT; D, CB.

Table 2. Relative species concentrations in HDS of DBT over CoMo/C catalyst at different times.[a,b]

Time, min	Fraction Converted	A	B	C	D
0	0	1	0	0	0
2	0.289	0.711	0.276	0.005	0.008
10	0.469	0.531	0.447	0.005	0.016
15	0.589	0.411	0.562	0.005	0.022
29	0.780	0.220	0.741	0.003	0.036
40	0.903	0.097	0.853	0.002	0.048
59	0.979	0.021	0.912	0.001	0.066

[a]Conditions: 3 MPa H_2 and 340 °C. [b]A, DBT; B, BP; C, H4-DBT; D, CB

therefore be considered as intermediates in the reaction network. All reactions were conducted in a typical autoclave reactor. The reaction and product mixtures were analyzed using a gas chromatograph equipped with a flame-ionization detector, and gas chromatography-mass spectrometry. The reactions with the $CoMo/Al_2O_3$ and CoMo/C catalysts were performed at 340 °C under H_2 at 3 MPa. For the unsupported MoS_2 catalyst, the HDS reaction was performed at 350 °C under H_2 at 3 MPa. The HDS reaction was therefore performed under similar conditions over all the catalysts. The kinetics were established on the assumption that the products from the HDS reaction have no significant influence on the reaction rate. The following reaction models were proposed to describe the reaction systems.

3. KINETIC MODELS

To identify an appropriate relationship to determine the changes in species selectivity (SF) and product distribution as a function of time, it was assumed that all reaction steps follow pseudo-first-order kinetics. It is important to mention that the present model focuses on simultaneously simulating two correlations: the correlation between intermediate yield changes and reaction time, and the correlation between intermediate selectivity changes and reaction time. Studies have shown that treatments of the rates of such reactions based on the Langmuir–Hinshelwood mechanism eventually come to use the classic trends of gross pseudo-first-order reactions (Bataille et al., 2000; Farag, 2006; Hynaux et al., 2007).

Table 3. Relative species concentrations in HDS of DBT at different times over unsupported MoS_2 catalyst[a,b]

Time, h	Fraction Converted	A	B	C	D
1	0.037	0.963	0.013	0.023	0.001
1.5	0.092	0.908	0.036	0.051	0.005
2	0.138	0.862	0.053	0.069	0.016
3	0.222	0.778	0.096	0.083	0.043
3.5	0.258	0.742	0.118	0.088	0.052
4.5	0.327	0.673	0.158	0.083	0.086
5	0.361	0.639	0.173	0.082	0.106
6	0.424	0.576	0.215	0.073	0.136
7	0.491	0.509	0.256	0.068	0.167

[a]Conditions: 3 MPa H_2 and 350 °C. [b]A, DBT; B, BP; C, H4-DBT; D, CB.

Figure 1. Reaction network for HDS in model 1

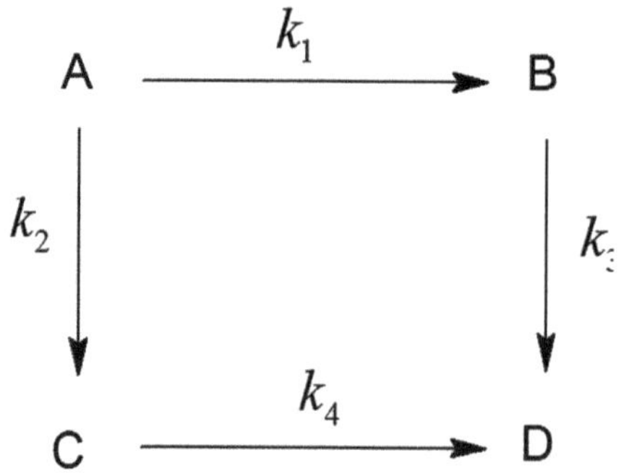

3.1. Model 1

We manipulated two complex reaction networks that can describe the parallel–consecutive reaction behavior in the HDS of DBT over different catalysts. The first network is displayed in Figure 1.

In Figure 1, A denotes the parent sulfur species, and B and C refer to the HDS intermediates, i.e., BP and H4-DBT in the case of DBT. Each step in Figure 1 is described by a set of differential equations based on first-order treatments. In this reaction network, there are two ways in which the intermediate can contribute to the overall reaction: both B and C contribute significantly to the reaction (case 1), or C is the sole reaction intermediate (case 2). In the following discussion, we will kinetically analyze these two cases.

Case 1

If B and C are significant intermediates (both contributing substantially to the reaction), then the disappearance rate of the parent molecule (e.g., DBT), R_{t1}, can be written as

$$R_{t1} = R_{HYD} + R_{DDS} \tag{1}$$

$$R_{t1} = k_2^2 \cdot C_A + k_1^1 \cdot C_A \tag{2}$$

$$R_{t1} = (k_1^1 + k_2^2) \cdot C_A = k_\circ \cdot C_A \tag{3}$$

where R_{HYD} and R_{DDS} are the rates for the HYD and DDS routes, respectively. The changes in the concentrations of B and C as a function of reaction time are described by the following model equations, and solved based on the material balance of each intermediate.

$$\frac{dC_A}{dt} = -(k_1K_1 + k_2K_2)C_A \tag{4}$$

$$\frac{dC_B}{dt} = k_1K_1C_A - k_3K_3C_B \tag{5}$$

$$\frac{dC_C}{dt} = k_2K_2C_A - k_4K_4C_C \tag{6}$$

$$\frac{dC_D}{dt} = k_3K_3C_B + k_4K_4C_C \tag{7}$$

where k1, k_2, k_3, and k4 are the intrinsic kinetic rate constants for the reactions in Figure 1 and K1, K2, K3, and K4 are the equilibrium adsorption constants of the corresponding species over the catalytic active sites. k_1^1, k_2^2 k_3^3, and k_4^4 replaced k_1K_1, k_2K_2, k_3K_3, and k_4K_4, respectively.

The change in the concentration of A as a function of time can readily be calculated. The above equations can be solved algebraically using C_B and $C_C = 0$ at $t = 0$ as boundary conditions. The results of integrating these equations give the following relationships between the integrated yields of the desired product with reaction time.

$$C_A = C_{A^\circ} \exp^{-k_\circ t} \tag{8}$$

$$C_B = \frac{C_{A^\circ}k_1^1}{k_3^3 - k_\circ}\left[\exp^{-k_\circ t} - \exp^{-k_3^3 t}\right] \tag{9}$$

$$C_C = \frac{C_{A^\circ} k_2^2}{k_4^4 - k_\circ}\left[\exp^{-k_\circ t} - \exp^{-k_4^4 t}\right] \quad (10)$$

$$\frac{C_B}{C_{A^\circ} - C_A} = \frac{-k_1^1}{k_3^3 - k_\circ}\left[\frac{\exp^{-k_\circ t} - \exp^{-k_3^3 t}}{\exp^{-k_\circ t} - 1}\right] \quad (11)$$

$$\frac{C_C}{C_{A^\circ} - C_A} = \frac{-k_2^2}{k_4^4 - k_\circ}\left[\frac{\exp^{-k_\circ t} - \exp^{-k_4^4 t}}{\exp^{-k_\circ t} - 1}\right] \quad (12)$$

Here, k_1^1, k_2^2, k_3^3, and k_4^4 are the individual apparent rate constants in the Figures, and C_{A°, C_C, and C_B are the initial concentration of species A (the mother component), and the concentrations of species C and B, respectively. The overall rate constant, $k_\circ$, can be directly determined by solving equation 8. Equations 9 and 10 show that only three constant parameters are required to solve the relationship between the intermediate concentration and reaction time kinetically. One of these is the gross apparent rate constant, $k_\circ$, which can be readily obtained by solving equation 8. The values of k_3^3 and/or k_4^4 can be estimated, and used to determine the contribution(s) of the intermediate(s) B and/or C to form D.

The behavior of the primary product formation is the same as that of an intermediate, and it shows an inflection point when the concentration reaches a maximum, at $t = t_{max}$.

$$t_{max.B} = \frac{k_B \cdot \ln\left(k_B\right)}{k_3^3 \cdot \left(k_B - 1\right)} \quad (13)$$

$$B_{t_{max}} = \frac{k_1^1 . k_B . A_\circ}{k_3^3 . (k_B - 1)} . \left[e^{-\left(\frac{k_3^3}{k_B}\right) . t_{max}} - e^{-k_3^3 . t_{max}} \right] \quad (14)$$

where $k_B = \frac{k_3^3}{k_\circ}$.

The model shows that the values of the individual rate constants, k_1^1 and k_2^2, are dependent on the estimated values for k_3^3 and k_4^4, respectively. The apparent individual rate constants for HYD and DDS, i.e., k_1^1 and k_2^2, can be determined independently. Equations 11 and 12 describe the reaction selectivity in terms of yield fraction of the species (SF). A product is considered to be primary when its selectivity at time zero is not zero. According to the present model, a simple plot of the experimental selectivity (SF) of the intermediate against reaction time shows whether or not components B and C are intermediates. Furthermore, the model can best be solved by matching the experimental data on product distribution and selectivity (SF) against reaction time with equations 9 and 11, in the case of species B, or equations 10 and 12, in the case of species C. Some observations can be made based on these equations. Although it is possible for the intermediate yield to show a maximum value, its selectivity (SF) will never exceed the initial value. The model therefore puts limits on the validity of matching the experimental data with the postulated reaction network.

Case 2

If C is the sole intermediate in the reaction network, then all the hydrogenated products should be produced exclusively from it. If the successive reactions in the HYD route are treated as pseudo-first-order reactions, then the following equations can be derived to describe such a system, after some algebraic manipulation. The results are as follows:

For the HYD route:

$$R_{HYD} = k_2^2 \cdot C_A \tag{15}$$

$$R_{HYD} = k_2^2 . C_{A^\circ} \exp^{-k_\circ t} \tag{16}$$

$$C_H = C_C + C_D \tag{17}$$

$$C_H = \frac{C_{A^\circ} k_2^2}{k_\circ}\left[1 - \exp^{-k_\circ t}\right] \tag{18}$$

$$C_H = \frac{k_2^2}{k_\circ}\left[C_{A^\circ} - C_A\right] \tag{19}$$

[illegible] (20)

For the DDS route:

$$R_{DDS} = k_1^1 \cdot C_A \tag{21}$$

$$R_{DDS} = k_1^1 \cdot C_{A^\circ} \exp^{-k_\circ t} \tag{22}$$

$$C_B = \frac{k_1^1}{k_\circ}\left[C_{A^\circ} - C_A\right] \tag{23}$$

The equations for C_H and C_B (equations 19 and 23) can be easily solved by plotting C_H or C_B versus $\left[C_{A^\circ} - C_A\right]$; the individual apparent rate constants, k_2^2 and k_1^1, can easily be determined from the slopes of the straight lines. Consequently, the ratio of the contributions of HYD and DDS can be determined. Equation 21 predicts that the selectivity (SF) for intermediate B is constant, irrespective of the conversion level (reaction time). In addition, it indicates that the yield of B as a function of reaction time will never have a maximum value. Furthermore, the selectivity (SF) for C will never increase with reaction time. These observations place clear limits when evaluating the proposed reaction network and mechanism. In all cases, the intermediate concentrations are governed by an equation containing only two dependent apparent rate constants. This approach can certainly assist in making more accurate estimates of the independent contributions from each route. In addition, this approach is applicable to similar catalytic reactions._

3.2. Model 2

There is a special case in which the partially hydrogenated intermediate may be partly transformed to produce BP or similar components in addition to the main transformation to PC or similar compounds. Such reaction mechanisms have been discussed in many reported studies (Bataille et al., 2000; Li et al., 2007; Rabarihoela-Rakotovao et al., 2007; Richard et al., 2007; Zuo et al., 2004). The reaction network is shown in Figure 2.

In this case, under the assumption that the reactions followed pseudo-first-order kinetics, the following kinetic equations were extracted from the material balance equations to describe the changes in the concentrations of intermediates with reaction time.

Figure 2. Reaction network for HDS in model 2

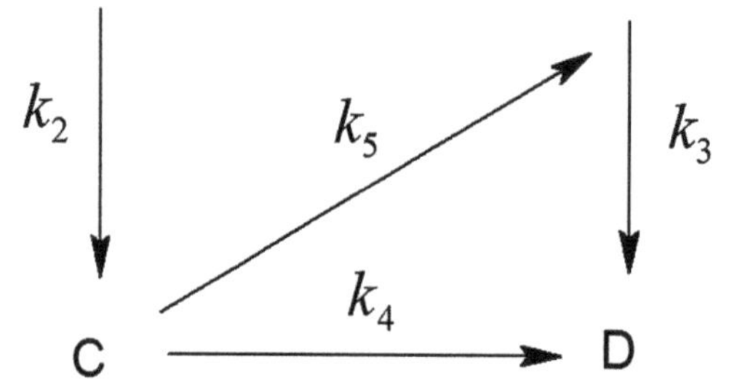

$$C_B = C_{A\circ} \cdot \left[\begin{array}{l} \left[\dfrac{k_5^5 \cdot k_2^2}{(k_4^{'})^2 - k_3^3 \cdot k_4^{'} - k_\circ \cdot k_4^{'} + k_3^3 \cdot k_\circ} \right] \cdot \left[\exp^{-k_4^{'} \cdot t} - \exp^{-k_3^3 \cdot t} \right] - \\ \left[\dfrac{k_1^1 \cdot k_\circ - k_4^{'} \cdot k_1^1 - k_5^5 \cdot k_2^2}{k_3^3 \cdot k_4^{'} - k_3^3 \cdot k_\circ + (k_\circ)^2 - k_4^{'} \cdot k_\circ} \right] \cdot \left[\exp^{-k_\circ \cdot t} - \exp^{-k_3^3 \cdot t} \right] \end{array} \right] \tag{24}$$

$$C_C = \frac{C_{A\circ} k_2^2}{k_4^{'} - k_\circ} \left[\exp^{-k_\circ t} - \exp^{-k_4^{'} t} \right] \tag{25}$$

where $k_4^{'}$ equals $k_4^4 + k_5^5$. Equation 24 shows, indirectly, that the selectivity (SF) for B may increase with reaction time. Such a model is important because many HDS studies have identified such subtle behavior experimentally. According to this model, component C is not only transformed to product D, but also partly reacts to form B. Some aspects of this model need to be discussed. First, each intermediate in the reaction network can be described by two independent plots: one showing changes in the intermediate fraction yield against reaction time and the other showing changes in selectivity (SF) as a function of reaction time. It is essential to match the experimental results with these two equations to describe the observed data adequately. Only two constants, which are easily obtainable through trial and error, are required to curve fit the experimental data with theoretically derived curves and to obtain accurate estimates of the contributions of all reaction intermediates.

Data of the present catalysts have also been manipulated in accordance with the initial rate method. In this method, the selectivity of products at close to zero conversion is estimated to determine the k_1^1 / k_2^2 ratio from the following equation:

$$\frac{k_1^1}{k_2^2} = \frac{S_B}{S_C} \tag{26}$$

where k_1^1 and k_2^2 are the apparent individual rate constants in Figures 1 or 2, and S_B and S_C are the initial selectivities for species B and C, respectively. A reaction product is considered to be the primary product if its selectivity at zero conversion is non-zero, whereas it considered to be a secondary product if its selectivity at zero conversion is zero. To obtain accurate estimates of the rates of all reaction steps in these reaction networks, we used Mathcad software (MathSoft, Inc.) to curve fit the experimental data with curves derived theoretically from models 1 and 2. This program enables curve fitting to be carried out easily with interactive graphics, and has been applied to all of the kinetic data reported here. The data obtained using the present catalysts were taken as examples to show the utility of these models in interpreting experimental observations in the HDS of DBT. Equations 9–12, 24, and 25 are the main governing equations of these models.

4. DATA APPLICATIONS

4.1. HDS of DBT Over CoMo-Based Catalysts

The observed products from the HDS of DBT over CoMo/Al_2O_3 and/or CoMo/C were BP, cyclohexylbenzene (CB), and partially hydrogenated DBT (mainly H4-DBT). Isomers of these compounds were also observed, but in small amounts. The product distribution trend differed significantly from that observed in the case of HDS of DBT over MoS_2 catalyst. The product distributions and selectivities in the HDS

Figure 3. Product distributions and selectivities (SF) versus reaction time in HDS of DBT over CoMo/ Al_2O_3 catalyst. Lines: correlations obtained by fitting experimental data to model 1

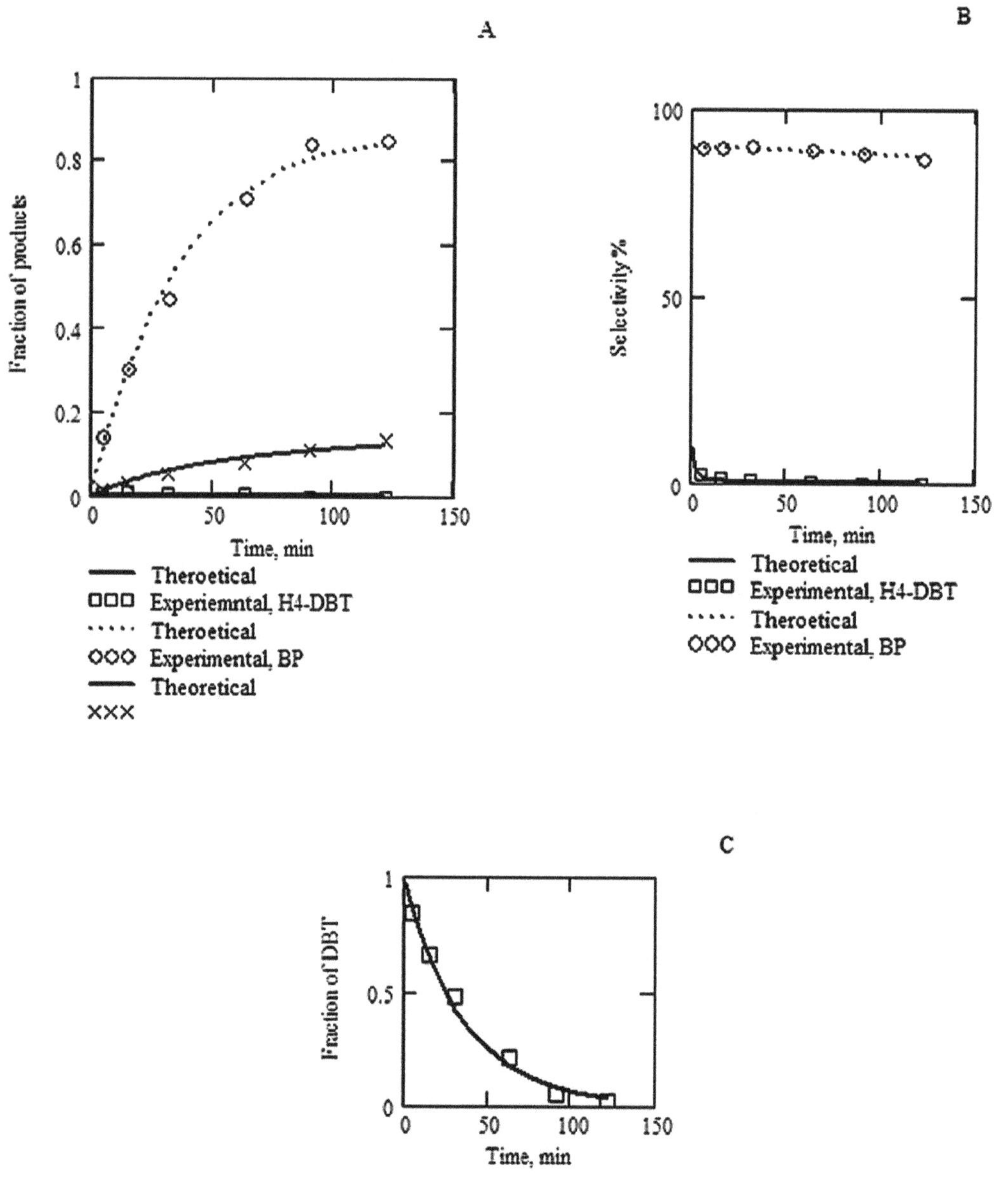

Figure 4. Time dependence of product concentrations in the HDS of DBT over CoMo/C catalyst

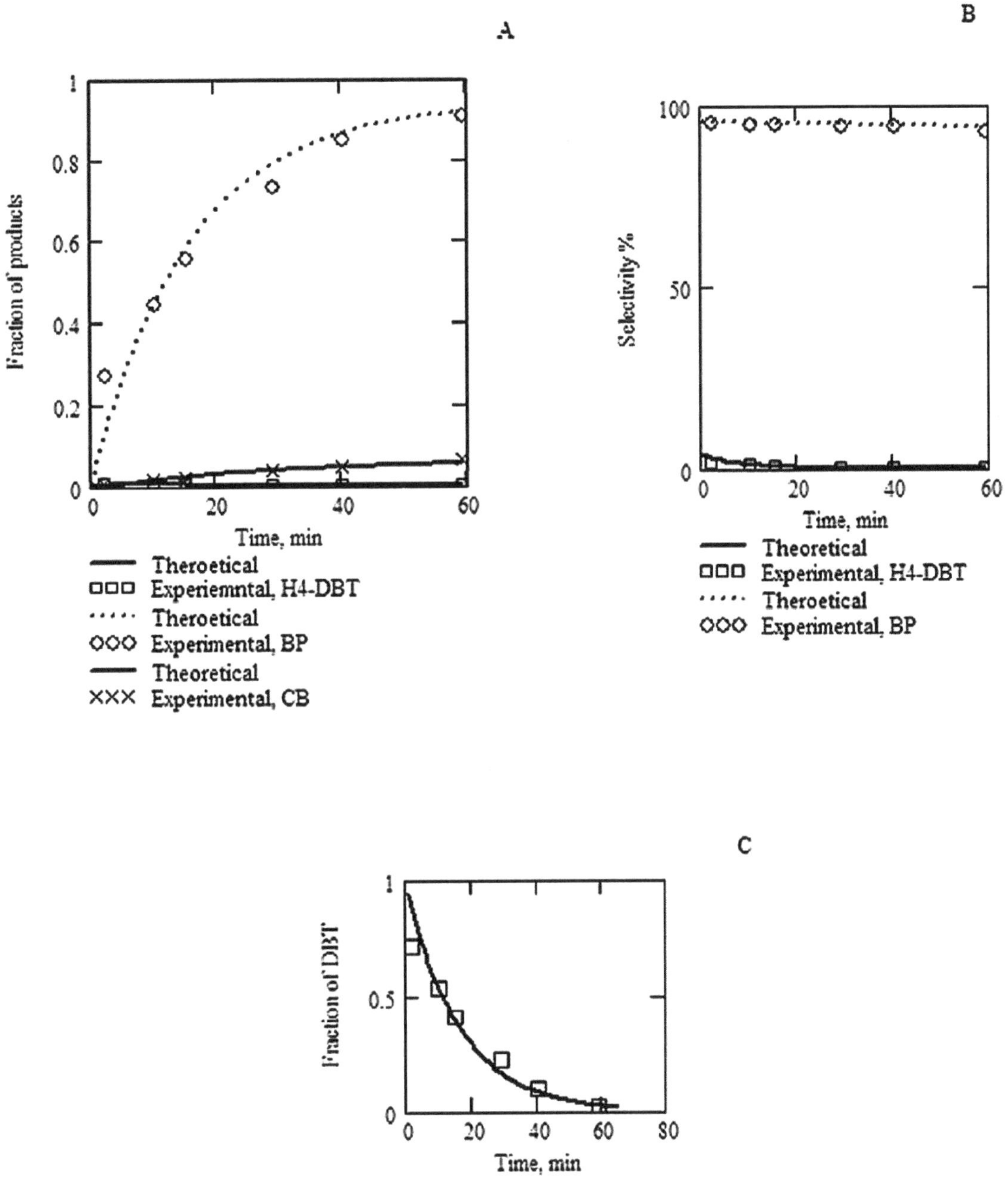

of DBT over $CoMo/Al_2O_3$ and CoMo/C catalysts are plotted as a function of reaction time in Figures 3 and 4, respectively. The raw experimental data are listed in Tables 1 and 2. The reactions in the HDS of DBT over $CoMo/Al_2O_3$ and CoMo/C reasonably fitted pseudo-first-order kinetics with respect to DBT, as shown in Figures 3 and 4(C).

4.2. HDS of DBT Over MoS_2 Catalyst

The data for HDS of DBT over the MoS_2 catalyst were manipulated. The main reported products were BP, partially hydrogenated DBT (mainly H4-DBT), and CB (Table 3). The product distributions and selectivities in the HDS of DBT over MoS_2 are shown in Figure 3. It can be observed that the reaction over MoS_2 proceeded via contributions from the DDS (as indicated by BP) and HYD (indicated by H4-DBT and CB) pathways. Figure 3 shows that the selectivity (SF) for BP formation was highest among

Figure 5. Product distributions and selectivities (SF) versus reaction time in HDS of DBT over unsupported MoS_2 catalyst. Lines: correlations obtained by fitting experimental data to model 2

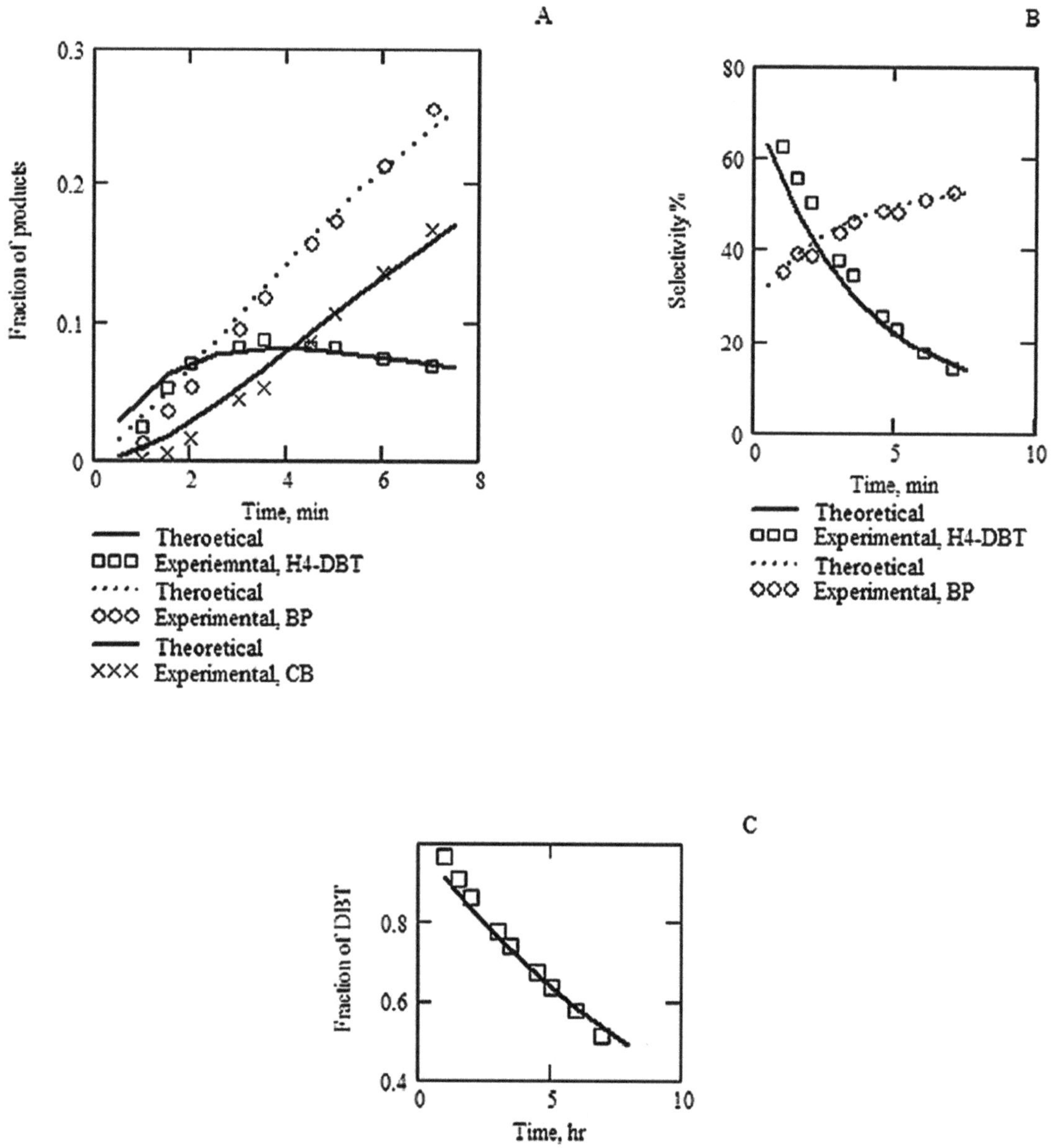

all the products, especially at high conversion levels. It is interesting to note that this selectivity for BP changed increasingly with reaction time. The selectivity for formation of H4-DBT decreased exponentially with reaction time. The HDS of DBT followed pseudo-first-order kinetics with respect to DBT, as shown in Figure 5(C).

5. DISCUSSION

5.1. Kinetic Analysis

This study is an attempt to establish reasonable kinetic models, to gain insights into the nature of the parallel and consecutive routes in such reactions. Detailed kinetic analyses are necessary to understand the subtle changes in product compositions that result from changes in catalyst formulations and/or reaction conditions in the HDS of DBT. They may also elucidate the influence of other feed components (inhibitors) on the progress of HDS. Experimental data on the HDS reaction and on the relative rates of the chemical processes involved are necessary. For a better overview of the kinetics of these reactions, it is important to carry out a complete analysis of the shortest possible HDS reaction time and of successively longer periods up to the point at which a near-complete conversion of the substrate is achieved. In this manner, primary products formed directly from the parent substrate can be distinguished from secondary products formed by subsequent reactions of the primary products (intermediates). Data simulation with the present models can provide valuable clues to the identities of the reaction intermediates involved and may assist in precise screening of catalyst activities and selectivities. The distribution of the reaction products enables a reaction figure for the catalytic transformation of DBT to be developed. The HDS of these refractory sulfur compounds often proceeds via two basic reaction routes, namely HYD and DDS, as already reported in the literature. The type of catalyst and the nature of the sulfur-containing compounds both play crucial roles in directing the HDS reaction toward a certain reaction network; for example, in the HDS of DBT, the DDS route is usually dominant. However, some recent classes of catalyst, particularly unsupported ones, have demonstrated the possibility of this reaction proceeding via a major contribution from the HYD route (Bej et al., 2004; Egorova et al., 2004; Eijsbouts et al., 2007; Kagami et al., 2005; Mendoza-Nieto et al., 2013; Topsoe et al., 2005). The nature of the associated active sites plays a role in this respect. Although many studies have been conducted in this field, only a few have reported comprehensive concentration–reaction time curves for HDS reactions (Egorova et al., 2004; Houalla et al., 1978; Li et al., 2007). Such kinetic data are needed to achieve better analyses of the reaction behavior.

A more precise kinetic model to describe the catalytic selectivity for such a relatively complex reaction network is essential, and crucial in any discussion of ranking catalysts according to their performances. Various approaches have been used to determine the selectivity (SR) and contribution ratios from intermediates in this type of reaction network. It has been suggested that an approximation can be obtained using the initial selectivities (SF) of intermediates, where the selectivity plots for all the resultant species can be extrapolated as a straight line to zero conversion (Kim et al., 2005; Oyama et al., 2009). If these lines intersect with the *y*-axis, the selectivity (SF) ratios can be used to determine the individual apparent rate constant ratios. However, this approach applies to specific cases, and it cannot provide any information on sequential reactions. Other researchers have performed numerical simulations to estimate the individual rate constants involved in such reaction networks (Daage et al., 1994;

Froment et al., 2008; Rodriguez et al., 2012; Vanrysselberghe et al., 1998). However, most of the proposed models contain a considerable number of constants. This makes it difficult to rely on the derived data. In addition, these treatments, from the kinetic point of view, provide no explanation of the relationship between the intermediate selectivity (SF) and reaction time. In other studies, the selectivity has

Figure 6. Mathcad sheet for estimating individual rate constants in parallel–consecutive reactions

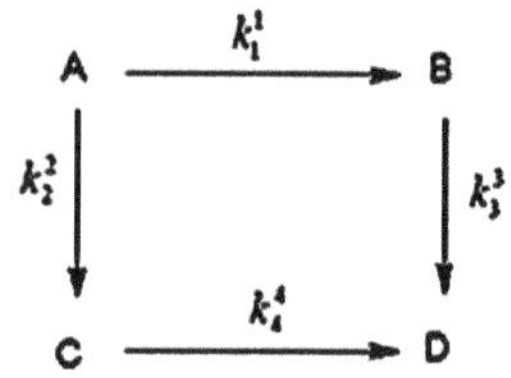

$$BP := \begin{pmatrix} 0.276 \\ 0.447 \\ 0.562 \\ 0.741 \\ 0.853 \\ 0.912 \end{pmatrix} \quad H := \begin{pmatrix} 0.0046 \\ 0.0053 \\ 0.005 \\ 0.003 \\ 0.0016 \\ 0.0005 \end{pmatrix} \quad tt := \begin{pmatrix} 2 \\ 10 \\ 15 \\ 29 \\ 40 \\ 59 \end{pmatrix} \quad seH := \begin{pmatrix} 1.594 \\ 1.132 \\ 0.849 \\ 0.385 \\ 0.177 \\ 0.051 \end{pmatrix} \quad conv := \begin{pmatrix} 0.289 \\ 0.4687 \\ 0.589 \\ 0.7798 \\ 0.903 \\ 0.9787 \end{pmatrix} \quad P := \begin{pmatrix} 0.008 \\ 0.016 \\ 0.0217 \\ 0.036 \\ 0.0475 \\ 0.066 \end{pmatrix} \quad seB := \begin{pmatrix} 95.63 \\ 95.45 \\ 95.46 \\ 95.0 \\ 94.56 \\ 93.20 \end{pmatrix}$$

$Ao := 1 \quad kt := .062 \quad k1 := .0593 \quad k3 := .00033 \quad k4 := .28$

$t := .1 .. 60 \quad k2 = 2.7 \times 10^{-3} \quad k2 := kt - k1$

$$B(t) = \frac{k1 \cdot Ao}{k3 - kt} \cdot \left(e^{-kt \cdot t} - e^{-k3 \cdot t}\right) \qquad B(t) := \frac{k1 \cdot Ao}{k3 - kt} \cdot \left(e^{-kt \cdot t} - e^{-k3 \cdot t}\right)$$

$$\frac{-k1}{k3 - kt} \cdot \left(\frac{e^{-kt \cdot t} - e^{-k3 \cdot t}}{e^{-kt \cdot t} - 1}\right)$$

$$\frac{-k1}{(k3 - kt)^2} \cdot \frac{(\exp(-kt \cdot t) - \exp(-k3 \cdot t))}{(\exp(-kt \cdot t) - 1)} + \frac{k1}{(k3 - kt)} \cdot t \cdot \frac{\exp(-kt \cdot t)}{(\exp(-kt \cdot t) - 1)} - \frac{k1}{(k3 - kt)} \cdot \frac{(\exp(-kt \cdot t) - \exp(-k3 \cdot t))}{(\exp(-kt \cdot t) - 1)^2} \cdot t \cdot \exp(-kt \cdot t)$$

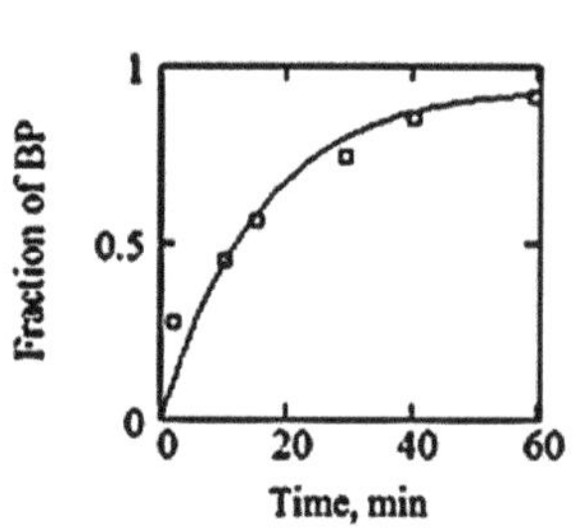

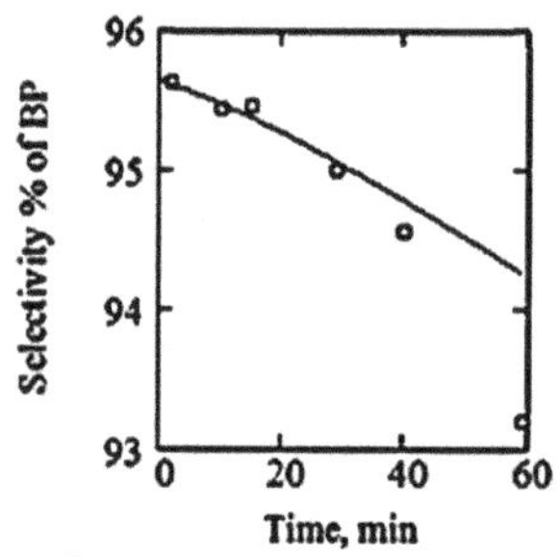

$ktt := \frac{kt}{60 \cdot .03} \quad ktt = 3.444 \times 10^{-2} \quad k11 := \frac{k1}{60 \cdot .03} \quad k22 := \frac{k2}{60 \cdot .03} \quad k22 = 1.5 \times 10^{-3}$

$k11 = 3.294 \times 10^{-2} \quad k33 := \frac{k3}{60 \cdot .1} \quad k33 = 5.5 \times 10^{-5} \quad k44 := \frac{k4}{60 \cdot .03} \quad k44 = 1.556 \times 10^{-1}$

$$Cc(t) = \frac{k2 \cdot Ao}{k4 - kt} \cdot \left(e^{-kt \cdot t} - e^{-k4 \cdot t}\right) \qquad se(t) := \frac{\frac{k2}{k4 - kt} \cdot \left(e^{-kt \cdot t} - e^{-k4 \cdot t}\right)}{1 - e^{-kt \cdot t}}$$

Figure 7. Mathcad sheet for estimating individual rate constants in parallel–consecutive reactions (cont.)

$$Cc(t) = \frac{k2\cdot Ao}{k4 - kt}\cdot\left(e^{-kt\cdot t} - e^{-k4\cdot t}\right) \qquad se(t) := \frac{\frac{k2}{k4 - kt}\cdot\left(e^{-kt\cdot t} - e^{-k4\cdot t}\right)}{1 - e^{-kt\cdot t}}$$

$$Cc(t) := \frac{k2\cdot Ao}{k4 - kt}\cdot\left(e^{-kt\cdot t} - e^{-k4\cdot t}\right)$$

$$tt := \begin{pmatrix} 2 \\ 10 \\ 15 \\ 29 \\ 40 \\ 59 \end{pmatrix} \qquad dd := \begin{pmatrix} 0.711 \\ 0.5313 \\ 0.411 \\ 0.2202 \\ 0.097 \\ 0.0213 \end{pmatrix}$$

$c := .062$

$A(n) := Ao\cdot e^{-c\cdot n}$

$Ao := 1 \qquad n := 1 .. 65$

$c = 6.2 \times 10^{-2}$

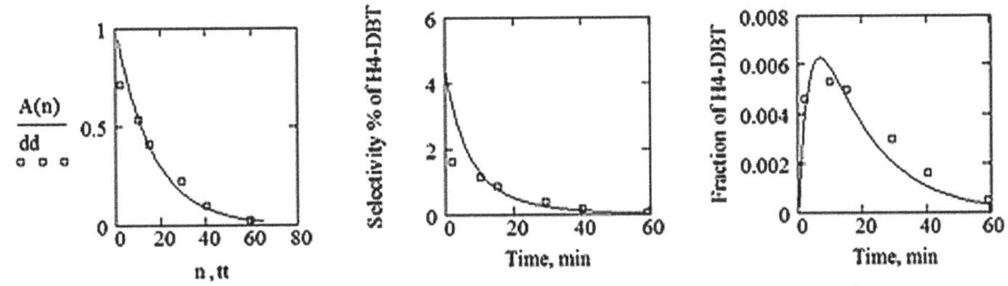

$$PC = \left(\frac{-1}{kt}\cdot\exp(-kt\cdot t) + \frac{1}{k3}\cdot\exp(-k3\cdot t)\right)\cdot k3\cdot k1\cdot\frac{Ao}{(k3 - kt)} + \left(\frac{-1}{kt}\cdot\exp(-kt\cdot t) + \frac{1}{k4}\cdot\exp(-k4\cdot t)\right)\cdot k4\cdot k2\cdot\frac{Ao}{(k4 - kt)} + Ao\cdot\frac{(k1 + k2)}{kt}$$

$$PC(t) := \left(\frac{-1}{kt}\cdot\exp(-kt\cdot t) + \frac{1}{k3}\cdot\exp(-k3\cdot t)\right)\cdot k3\cdot k1\cdot\frac{Ao}{(k3 - kt)} + \left(\frac{-1}{kt}\cdot\exp(-kt\cdot t) + \frac{1}{k4}\cdot\exp(-k4\cdot t)\right)\cdot k4\cdot k2\cdot\frac{Ao}{(k4 - kt)} + Ao\cdot\frac{(k1 + k2)}{kt}$$

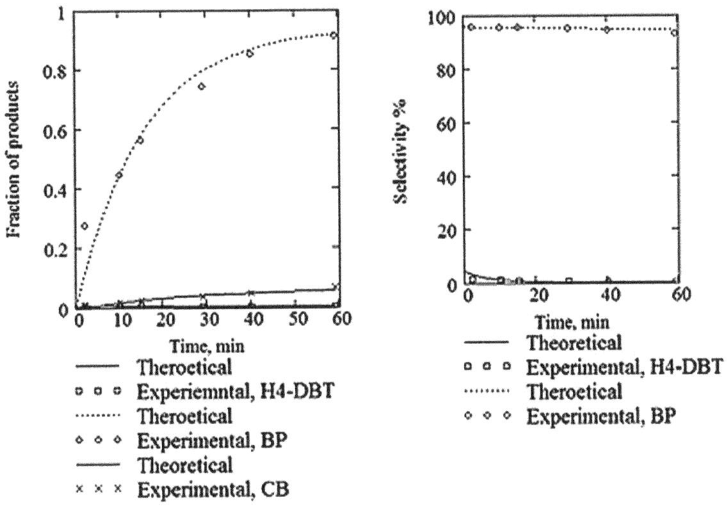

been simplified by ignoring further successive reactions of one intermediate, so the calculation of individual apparent rate constants becomes feasible. However, this simplification has not yet been successfully used to describe the changes in yield of, and selectivity for, an intermediate with reaction time. The kinetics of the HDS of DBT has been described in many studies as pseudo-first order. This is further supported by the present results (Figures 3, 4, and 5). The experimental results for the HDS of DBT over CoMo/C and commercial CoMo/Al_2O_3 catalysts fitted model 1, whereas model 2 was found to be more appropriate for the HDS of DBT over an MoS_2 catalyst. Figure 6 (continued Figure 7) shows an example of the Mathcad sheet used to estimate the individual rate constants in such parallel–consecutive reactions. The estimated individual apparent rate constants are presented in Table 4. Table 4 also shows the individual apparent rate constant values estimated from the present models, and the results determined using the initial selectivity approach. It can be seen that the estimated k_1^1 and k_2^2 values (individual apparent rate constants for the DDS and HYD reaction pathways, Figures 1 and 2) obtained using the initial selectivity method differ greatly from those derived using the present models. According to the governing equations used in models 1 and 2, the estimated values of k_2^2 and k_2^2 depend strongly on the values suggested for k_3^3 and k_4^4, respectively. The models describe, separately and independently, the yield changes of each intermediate as a function of reaction time. The conversion of each intermediate is governed by a single equation with only two adjustable parameters. Once these two parameters are established, they can be used in the equation that describes the other intermediate. The equation describing this intermediate will eventually contain one free parameter. This enables an optimum fit of the experimental results with these models. Valencia & Klimova (2012 recently used this model to estimate the kinetic parameters of the HDS of DBT.

5.2. Comparison of Network Reaction Rates in HDS of DBT over CoMo-Based Catalysts

Table 4 shows the overall apparent pseudo-first-order rate constants, calculated from the integrated conversion data for HDS of DBT, using the least-squares method, with 95% confidence limits on the linear regression parameter, over the present catalysts. In the HDS of DBT, the activity was higher over CoMo/C than over CoMo/Al_2O_3. The rate constants of the reaction steps in Figure 1 were calculated by fitting the experimental data at various conversion levels with model 1. The experimental and simulated results for the selectivities and distributions of BP and H4-DBT from the HDS of DBT (Figures 3 and 4) showed reasonable fits with model 1. The selectivity (SF) for H4-DBT was related to that for CB, i.e., as the selectivity for H4-DBT decreased, that for CB increased. This reveals that H4-DBT is an essential precursor for CB production. The kinetic results (Table 4) show that the conversion of H4-DBT to CB was very fast, whereas that of BP to CB was the slowest step in the network. This indicates that HYD was more difficult for the second phenyl ring than for the first phenyl ring. The apparent rate constant k_1^1 (BP formation) in the HDS of DBT over CoMo/C was about eight times higher than that over CoMo/Al_2O_3. However, the apparent rate constant for H4-DBT formation (k_2^2) was about three times higher over CoMo/C than over CoMo/Al_2O_3. The selectivity (SR) in the HDS of DBT over CoMo/Al_2O_3, according to model 1, was 0.11, whereas it was 0.027 using the initial selectivity approach developed by Kim et al. (2005). The corresponding selectivities were 0.046 (model 1) and 0.023 (initial selectivity approach) when the reaction was carried out over CoMo/C. Clearly, there are large differences between the selectivity (SR) estimated using the present model and that obtained from the initial selectivity approach. The differences between the estimated selectivities obtained for the same data may result in incorrect ranking of catalyst efficacies. It is worth noting that the selectivities (SF) for BP over CoMo/Al_2O_3 and CoMo/C changed only slightly during the reaction. This suggests that further HYD of BP to CB barely occurs in the HDS of DBT under the present reaction conditions. Because of the ge-

Table 4. Comparison of individual apparent rate constants of DBT HDS over CoMo-based catalysts estimated using models 1 and 2, and initial selectivity method

	$CoMo/Al_2O_3$		CoMo/C	
	Rate Constant × 10^{-4} s^{-1} (g cat)$^{-1}$			
	Model 1	**Initial Selectivity Method**[a]	**Model 1**	**Initial Selectivity Method**[a]
k_o [b]	45.0	45.0	344.4	344.4
k_1	42.1	43.8	329.4	336.7
k_2	4.6	1.2	15.0	7.7
k_3	0.6	-	0.6	-
k_4	667.0	-	1556.0	-
k_2 / k_1	0.11	0.027	0.046	0.023
MoS_2 Catalyst (Based on Data from Daage & Chianelli, 1994)				
	Rate Constant × 10^{-4} s^{-1} (g cat)$^{-1}$			
	Model 2[c]	**Initial Selectivity Method**[a]		
k_o [b]	0.251	0.251		
k_1	0.07	0.09		
k_2	0.18	0.16		
k_3	0.042	-		
k_4	0.86	-		
k_5	0.74	-		
k_2 / k_1	2.6	1.9		

[a]Calculation according to Kim et al., 2005.
[b]Pseudo-first-order overall rate constants.
[c]Fitting based on model 1 ignored increase in selectivity for BP with time. Model 2 best fitted the experimental data.

ometry of the partially hydrogenated sulfur substrate, i.e., H4-DBT, its reactivity is much higher than that of the parent compound. It is well known that the C–S bond can be broken by elimination, to form H_2S and a new π bond, or by hydrogenolysis, to form H_2S and a C–C single bond. H4-DBT can undergo double-bond-shift isomerization prior to elimination, which eventually leads to the formation of BP.

5.3. Selectivity and Reaction Network

The selectivity (SF) plot is important because it can help to decide which reaction network (figure) is suitable for describing a parallel–consecutive system. Models 1 and 2 were established for parallel–consecutive reactions that follow pseudo-first-order kinetics with respect to the substrate. Reasonable fits of the HDS experimental data with models 1 and 2 can easily be observed in the figures. According to the present models, the selectivity (SF) can be considered as a good predictor for successive intermediate reactions. The present kinetic analysis describes the overall behavior of the reaction and the selectivities (SF) for the intermediates. In the HDS of DBT, if the selectivity for an intermediate does not change with changing reaction time, then further successive reactions correlated with this intermediate can be safely ruled out. It is worth noting that although the initial selectivity approach can, in principle, be applied to such cases, generalizing it to all situations is questionable. The usefulness of the approximation obtained using the initial selectivity values to estimate the individual rate constants for the DDS and HYD routes in the HDS reaction of sulfur-containing compounds is limited, because the approximation does not represent the overall behavior of the reaction. This is because if the conversion level (substrate residence time) changes, the selectivity changes. Furthermore, it does not reveal much about the behavior of intermediates. In contrast, if the selectivity (SF) for an intermediate decreases with reaction time, then a substantial consecutive reaction of this intermediate must be considered (intermediate transformation). Model 1 can adequately describe the parallel–consecutive reaction system in this case. If the selectivity (SF) for the intermediate increases with reaction time, this suggests that the intermediate is formed by more than one route. This conclusion has to come after analyzing any possible experimental artifacts of the results. In such a case, a new reaction network should be considered. Model 2 provides a solution for this kind of network.

6. CONCLUSION

The HDS follows pseudo-first-order kinetics with respect to the substrate. Kinetic models that describe the reaction network were suggested. Each model uses a set of equations that simulate changes in both the product distribution and the selectivity as a function of reaction time. The models provide valuable quantitative data regarding the contributions of intermediates in the HDS reaction. In general, the catalytic HDS of DBT occurs via two pathways, i.e., HYD and DDS. The HYD pathway was dominant in the HDS of DBT over an MoS_2 catalyst, whereas DDS is favored in the case of CoMo-based catalysts. The kinetic results estimated by simulating the experimental data with models 1 and 2 indicate that the intermediate H4-DBT is mainly converted to CB. The models simultaneously track the changes in the yields of, and selectivities for, intermediates as a function of reaction time. The present models enable accurate assessments of the differences among the performances of different catalysts such as CoMo/C, CoMo/Al_2O_3, and MoS_2, the basic causes of selectivity changes, and the transformation mechanisms.

REFERENCES

Babich, I. V., & Moulijn, J. A. (2003). Science and technology of novel processes for deep desulfurization of oil refinery streams: A review. *Fuel*, *82*(6), 607–631. doi:10.1016/S0016-2361(02)00324-1

Bataille, F., Lemberton, J., Michaud, P., Perot, G., Vrinat, M., Lemaire, M., & Kasztelan, S. et al. (2000). Alkyldibenzothiophenes hydrodesulfurization–promoter effect, reactivity, and reaction mechanism. *Journal of Catalysis*, *191*(2), 409–422. doi:10.1006/jcat.1999.2790

Bej, S. K., Maity, S. K., & Turaga, U. T. (2004). Search for an efficient 4,6-DMDBT hydrodesulfurization catalyst: A Review of recent studies. *Energy & Fuels*, *18*(5), 1227–1237. doi:10.1021/ef030179+

Castillo-Araiza, C. O., Chávez, G., Dutta, A., de los Reyes, J. A., Nuñez, S., & García-Martínez, J. C. (2015). Role of Pt–Pd/γ-Al_2O_3 on the HDS of 4,6-DMBT: Kinetic modeling & contribution analysis. *Fuel Processing Technology*, *132*, 164–172. doi:10.1016/j.fuproc.2014.12.028

Daage, M., & Chianelli, R. R. (1994). Structure-function relations in molybdenum sulfide catalysts: The "Rim-Edge" model. *Journal of Catalysis*, *149*(2), 414–427. doi:10.1006/jcat.1994.1308

Egorova, M., & Prins, R. (2004). Hydrodesulfurization of dibenzothiophene and 4,6-dimethyldibenzothiophene over sulfided NiMo/γ-Al_2O_3, CoMo/γ-Al_2O_3, and Mo/γ- Al_2O_3 catalysts. *Journal of Catalysis*, *225*(2), 417–427. doi:10.1016/j.jcat.2004.05.002

Eijsbouts, S., Mayo, S. W., & Fujita, K. (2007). Unsupported transition metal sulfide catalysts: From fundamentals to industrial application. *Applied Catalysis A, General*, *322*, 58–66. doi:10.1016/j.apcata.2007.01.008

Farag, H. (2006). Kinetic analysis of the hydrodesulfurization of dibenzothiophene: Approach solution to the reaction network. *Energy & Fuels*, *20*(5), 1815–1821. doi:10.1021/ef060225g

Farag, H., Mochida, I., & Sakanishi, K. (2000). Fundamental comparison studies on hydrodesulfurization of dibenzothiophenes over CoMo-based carbon and alumina catalysts. *Applied Catalysis A, General*, *194*, 147–157. doi:10.1016/S0926-860X(99)00363-4

Farag, H., Sakanishi, K., Kouzu, M., Matsumura, A., Sugimoto, Y., & Saito, I. (2003). Dibenzothiophene hydrodesulfurization over synthesized MoS_2 catalysts. *Journal of Molecular Catalysis A Chemical*, *206*(1-2), 399–408. doi:10.1016/S1381-1169(03)00445-X

Froment, G., Castaneda-Lopez, L. C., & Marin-Rosas, C. (2008). Kinetic modeling of the hydrotreatment of light cycle oil and heavy gas oil using the structural contributions approach. *Catalysis Today*, *130*(2-4), 446–454. doi:10.1016/j.cattod.2007.10.108

García-Gutiérrez, J. L., Laredo, G. C., Fuentes, G. A., García-Gutiérrez, P., & Jiménez-Cruz, F. (2014). Effect of nitrogen compounds in the hydrodesulfurization of straight-run gas oil using a CoMoP/g-Al_2O_3 catalyst. *Fuel*, *138*, 98–103. doi:10.1016/j.fuel.2014.08.008

Grabowski, R. (2006). Kinetics of oxidative dehydrogenation of C-2–C-3 alkanes on oxide catalysts. *Catalysis Reviews. Science and Engineering*, *48*(2), 199–268. doi:10.1080/01614940600631413

Houalla, M., Nag, N., Sapre, A., Broderick, D., & Gates, B. (1978). Hydrodesulfurization of dibenzothiophene catalyzed by sulfided CoO-MoO_3γ-Al_2O_3: The reaction network. *AIChE Journal. American Institute of Chemical Engineers*, *24*(6), 1015–1021. doi:10.1002/aic.690240611

Hynaux, A., Sayag, C., Suppan, S., Trawczynski, J., Lewandowski, M., Szymanska-Kolasa, A., & Djega-Mariadassou, G. (2007). Kinetic study of the hydrodesulfurization of dibenzothiophene over molybdenum carbides supported on functionalized carbon black composite: Influence of indole. *Applied Catalysis B: Environmental*, *72*(1-2), 62–70. doi:10.1016/j.apcatb.2006.09.019

Kagami, N., Vogelaar, B. M., van Langeveld, A. D., & Moulijn, J. A. (2005). Reaction pathways on NiMo/Al_2O_3 catalysts for hydrodesulfurization of diesel fuel. *Applied Catalysis A, General*, *293*, 11–23. doi:10.1016/j.apcata.2005.05.042

Kim, J., Ma, X., Song, C., Lee, Y., & Oyama, T. (2005). Hydrodesulfurization of 4,6-dimethyldibenzothiophene over NiMo, CoMo sulfide and Ni phosphide catalysts: A kinetic study on individual reaction pathways. *Energy & Fuels*, *19*, 353–364. doi:10.1021/ef049804g

Lewandowski, M., Szymańska-Kolasa, A., Sayag, C., Beaunier, P., & Djéga-Mariadassou, G. (2014). Atomic level characterization and sulfur resistance of unsupported W2C during dibenzothiophene hydrodesulfurization. Classical kinetic simulation of the reaction. *Applied Catalysis B: Environmental*, *144*, 750–759. doi:10.1016/j.apcatb.2013.08.011

Li, X., Wang, A., Egorova, M., & Prins, R. (2007). Kinetics of the HDS of 4,6-dimethyldibenzothiophene and its hydrogenated intermediates over sulfided Mo and NiMo on γ-Al_2O_3. *Journal of Catalysis*, *250*(2), 283–293. doi:10.1016/j.jcat.2007.06.005

Mendoza-Nieto, J. A., Vera-Vallejo, O., Escobar-Alarcon, L., Solis-Casados, D., & Klimova, T. (2013). Development of new trimetallic NiMoW catalysts supported on SBA-15 for deep hydrodesulfurization. *Fuel*, *110*, 268–277. doi:10.1016/j.fuel.2012.07.057

Michaud, P., Lemberton, J., & Pérot, G. (1998). Hydrodesulfurization of dibenzothiophene and 4,6-dimethyldibenzothiophene: Effect of an acid component on the activity of a sulfided NiMo on alumina catalyst. *Applied Catalysis A, General*, *169*(2), 343–353. doi:10.1016/S0926-860X(98)00021-0

Mochida, I., & Choi, K. (2004). An overview of hydrodesulfurization and hydrodenitrogenation. *Journal of the Japan Petroleum Institute*, *47*(3), 145–163. doi:10.1627/jpi.47.145

Oyama, S. T., Gott, T., Zhao, H., & Lee, Y. (2009). Transition metal phosphide hydroprocessing catalysts: A review. *Catalysis Today*, *143*(1-2), 94–107. doi:10.1016/j.cattod.2008.09.019

Pawelec, B., Fierro, J., Montesinos, A., & Zepeda, T. (2008). Influence of the acidity of nanostructured CoMo/P/Ti-HMS catalysts on the HDS of 4,6-DMDBT reaction pathways. *Applied Catalysis B: Environmental*, *80*(1-2), 1–14. doi:10.1016/j.apcatb.2007.10.039

Rabarihoela-Rakotovao, V., Brunet, S., Perot, G., & Diehl, F. (2006). Effect of H_2S partial pressure on the HDS of dibenzothiophene and 4,6-dimethyldibenzothiophene over sulfided NiMoP/Al_2O_3 and CoMoP/Al_2O_3 catalysts. *Applied Catalysis A, General*, *306*, 34–44. doi:10.1016/j.apcata.2006.03.029

Richard, F., Boita, T., & Perot, G. (2007). Reaction mechanism of 4,6-dimethyldibenzothiophene desulfurization over sulfided NiMoP/Al_2O_3-zeolite catalysts. *Applied Catalysis A, General*, *320*, 69–79. doi:10.1016/j.apcata.2006.12.014

Rodriguez, M. A., Elizalde, I., & Ancheyta, J. (2012). Comparison of kinetic and reactor models to simulate a trickle-bed bench-scale reactor for hydrodesulfurization of VGO. *Fuel*, *100*, 91–99. doi:10.1016/j.fuel.2012.01.061

Singh, U. K., & Vannice, M. A. (2001). Kinetics of liquid-phase hydrogenation reactions over supported metal catalysts — a review. *Applied Catalysis A, General*, *213*(1), 1–24. doi:10.1016/S0926-860X(00)00885-1

Topsoe, T., Hinnemann, B., Norskov, J. K., Lauritsen, J. V., Besenbacher, F., Hansen, P. L., & Knudsen, K. G. et al. (2005). The role of reaction pathways and support interactions in the development of high activity hydrotreating catalysts. *Catalysis Today*, *107–108*, 12–22. doi:10.1016/j.cattod.2005.07.165

Valencia, D., & Klimova, T. (2012). Kinetic study of NiMo/SBA-15 catalysts prepared with citric acid in hydrodesulfurization of dibenzothiophene. *Catalysis Communications*, *21*, 77–81. doi:10.1016/j.catcom.2012.02.003

Vanrysselberghe, V., LeGall, R., & Froment, G. F. (1998). Hydrodesulfurization of 4-methyldibenzothiophene and 4,6-dimethyldibenzothiophene on a CoMo/Al_2O_3 catalyst: Reaction network and kinetic. *Industrial & Engineering Chemistry Research*, *37*(4), 1235–1242. doi:10.1021/ie970533p

Vasudevan, P. T., & Fierro, J. L. G. (1996). A review of deep hydrodesulfurization catalysis. *Catalysis Reviews. Science and Engineering*, *38*(2), 161–188. doi:10.1080/01614949608006457

Wang, H., & Prins, R. (2008). Hydrodesulfurization of dibenzothiophene and its hydrogenated intermediates over sulfided Mo/γ-Al_2O_3. *Journal of Catalysis*, *258*(1), 153–164. doi:10.1016/j.jcat.2008.06.005

Whitehurst, D. D., Isoda, T., & Mochida, I. (1998). Present state of the art and future challenges in the hydrodesulfurization of polyaromatic sulfur compounds. *Advances in Catalysis*, *42*, 345–471. doi:10.1016/S0360-0564(08)60631-8

Zuo, D., Li, D., Nie, H., Shi, Y., Lacroix, M., & Vrinat, M. (2004). Acid–base properties of NiW/Al_2O_3 sulfided catalysts: Relationship with hydrogenation, isomerization and hydrodesulfurization reactions. *Journal of Molecular Catalysis A Chemical*, *211*(1-2), 179–189. doi:10.1016/j.molcata.2003.10.018

Chapter 13
Valorisation of Glycerol to Fine Chemicals and Fuels

Nikolaos Dimitratos
Cardiff University, UK

Carine E. Chan-Thaw
Università degli Studi di Milano, Italy

Alberto Villa
Università degli Studi di Milano, Italy

Ceri Hammond
Cardiff University, UK

Laura Prati
Università degli Studi di Milano, Italy

ABSTRACT

The selective transformation of biomass-derived compounds to useful fine chemicals and fuels has developed rapidly in recent years, and biomass compounds will soon become one of the main resource contributors for the production of chemicals. In the near future, it is expected that biomass derived compounds will contribute substantially to global chemical production along with fossil-based analogues. Although, there is still debate about the sustainability of the usage of biomass-derived molecules, it is important to emphasise that effort has been made to use biomass in the most efficient way, and that the biomass resources used are not suitable for food purposes. In this review we will focus to present selected examples on the transformation of glycerol in three distinct areas; (i) glycerol oxidation, (ii) glycerol hydrogenolysis and (iii) glycerol aqueous reforming, using supported metal nanoparticles as the chosen catalysts.

1. INTRODUCTION

The selective transformation of biomass-derived compounds to useful fine chemicals and fuels has developed rapidly in recent years, and biomass compounds will soon become one of the main resource contributors for the production of chemicals, (Alonso, Wettstein, & Dumesic, 2012; Corma, Iborra, & Velty, 2007; Vennestrom, Osmundsen, Christensen, & Taarning, 2011). In the near future, it is expected that biomass derived compounds will contribute substantially to global chemical production along with fossil-based analogues. Although, there is still debate about the sustainability of the usage of biomass-

DOI: 10.4018/978-1-4666-9975-5.ch013

derived molecules, it is important to emphasise that effort has been made to use biomass in the most efficient way, and that the biomass resources used are not suitable for food purposes. A few important examples from the current literature that have shown effective ways of transforming biomass-derived molecules to useful chemicals and fuels are: (i) cyclodehydration of sugars, leading to furan-type products, with subsequent further transformation of these intermediates yielding more useful compounds such as levulinic acid and γ-valerolactone, which can be used as platform chemicals (Chheda, Roman-Leshkov, & Dumesic, 2007), (ii) hydrogenolysis of sugars to yield ethylene and propylene glycols, and (iii) glycerol valorisation to fine chemicals and fuels, such as acrylic acid, 1,2-propanediol, glyceric acid, hydrogen and methanol production. Figure 1 shows the common reactions used to obtain high value products from glycerol. In this review we will focus to present selected examples on the transformation of glycerol in three distinct areas; (i) glycerol oxidation, (ii) glycerol hydrogenolysis and (iii) glycerol aqueous reforming, using supported metal nanoparticles as the chosen catalysts.

2. SELECTIVE OXIDATION OF GLYCEROL

The valorisation of glycerol is an emerging area that has been extensively studied in recent years (Chheda, Huber, & Dumesic, 2007; Rahmat, Abdullah, & Mohamed, 2010; Zhou, Beltramini, Fan, & Lu, 2008), with a large body of this research focusing on the oxidation of glycerol to fine chemical products. The selective oxidation of glycerol can provide a range of useful fine chemical products used in pharmaceutical and perfume industry (Figure 2). A range of the presented catalysts developed for this oxidation are often also used to for the oxidation of interrelated polyols such as ethylene glycol, 1,2-propanediol (Dimitratos, Lopez-Sanchez, Meenakshisundaram, et al., 2009; Prati & Rossi, 1998) and 1,3-propanediol (Dimitratos, Lopez-Sanchez, Meenakshisundaram, et al., 2009; Biella, Castiglioni, Fumagalli, Prati, & Rossi, 2002; Taarning, Madsen, Marchetti, Egeblad, & Christensen, 2008).

Many research groups have focused on the optimisation of catalytic performance of materials that are able to tune activity and selectivity to the desired products. Therefore, glycerol oxidation is one of the most extensively studied reactions in recent years, and as such it can provide insights into the design and synthesis of new catalysts, covering a wide range of materials, such as nanomaterials, multicomponent oxides and heteropolyacids, (Beltran-Prieto, Kolomaznik, & Pecha, 2013; Sheldon, 2014). Glycerol is a by-product from biodiesel production, (Bharathiraja et al., 2014) and due to the fact that it is a highly functionalised molecule a large number of products can be formed from glycerol oxidation, (Rampino, Kavanagh, & Nord, 1943; Pagliaro, Ciriminna, Kimura, Rossi, & Della Pina, 2007). The first step for glycerol oxidation involves the formation of dihydroxyacetone under acidic conditions or glyceraldehyde under alkaline conditions. Sequential oxidation can subsequently lead to the formation of glyceric acid and tartronic acid as a product of the subsequent oxidation of glyceric acid. Formation of mesoxalic acid could occur due to the successive oxidation of tartronic acid. C-C cleavage products such as, glycolic, glyoxylic oxalic and formic acids could form depending on the experimental conditions. Therefore minimising C-C cleavage and enhancing the selectivity to desired products such as acids or to the corresponding aldehyde or ketone is a highly challenge. One of the main solutions for this challenge is the design of catalysts with specific properties, to allow control of activity and selectivity to the desired product. One of the main classes of catalysts used for the oxidation of glycerol are supported metal nanoparticles.

Figure 1. Processes of catalytic conversion of glycerol into useful chemicals

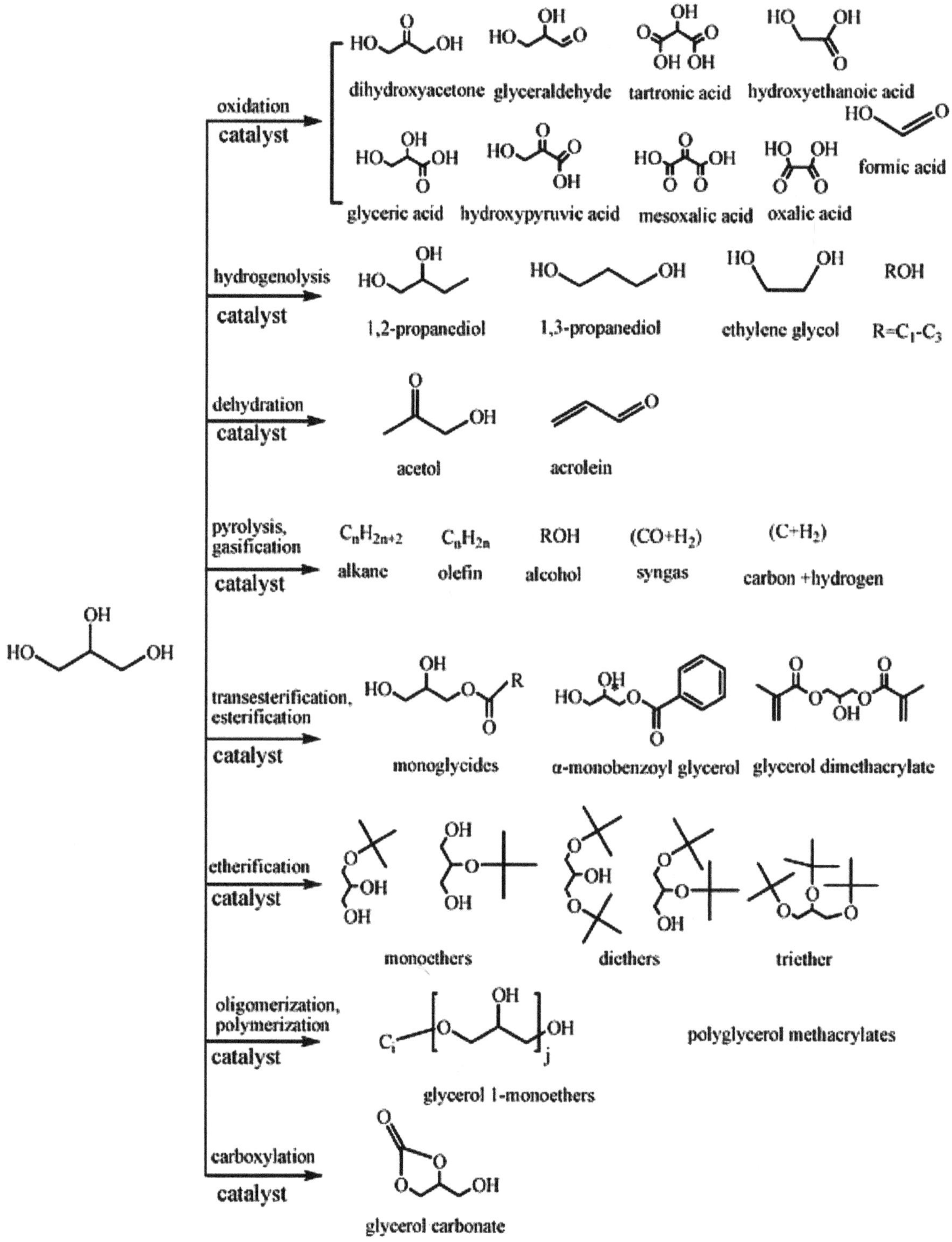

Figure 2. Oxidation products of glycerol

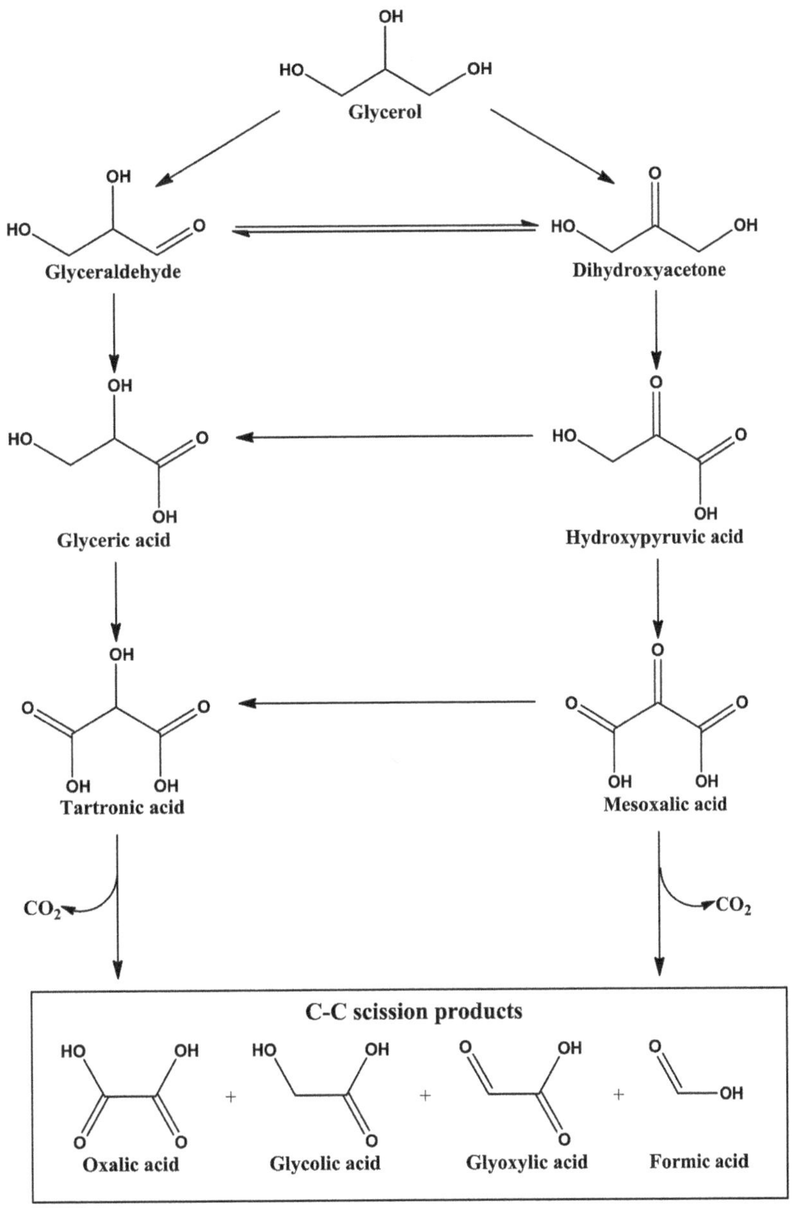

2.1 Oxidation of Glycerol Using Supported Monometallic Nanoparticles

In table 1 a summary of the published catalytic data for the oxidation of glycerol using supported monometallic nanoparticles is presented. One of the first examples showing the use of supported metal nanoparticles for the effective selective oxidation of glycerol under aerobic conditions was published by Kimura *et al.*, (Kimura, 1993; Kimura, Tsuto, Wakisawa, Kazumi, & Inaya, 1995). The addition of bismuth to supported Pt nanoparticles resulted in enhanced activity, stability and improved selectivity to dihydroxyacetone (80%), one of the primary products. The yield of the desired product was dependent on two factors; the reactor configuration and the catalyst composition. By optimising the Bi-Pt composition and using a batch reactor a 20% yield was achieved using a 1%Bi-3%Pt/C catalyst. In the case of using a fixed bed continuous flow reactor and using a 0.6%Bi-3%Pt catalyst, 30% yield was achieved demonstrating the continuous production of dihydroxyacetone under mild conditions (50 °C). The authors proposed that the high selectivity towards dihydroxyacetone was due to the formation of a bismuth submonolayer on Pt. The main role of bismuth atoms is to block the Pt sites that are responsible for unselective oxidation, and therefore to control the reaction selectivity towards dihydroxyacetone. To demonstrate the commercial potential of their catalyst, a comparison with the conventional fermentation process was carried out and the group demonstrated that the catalytic process could achieve higher productivity than other commercial approaches.

One of the first examples to show the catalytic formation of glyceric acid by glycerol oxidation at mild conditions was reported by Garcia *et al.*,(Garcia, Besson, & Gallezot, 1995). After performing a systematic study and optimising the experimental conditions, such as pH and temperature, glyceric acid was the main product formed using supported mono metallic Pd and Pt nanoparticles. The addition of Bi to the supported Pt nanoparticles at low Bi/Pt molar ratio, was found to improve the selectivity of the reaction towards dihydroxyacetone, and a maximum of 50% selectivity was achieved at 70% conversion (35% yield).(Garcia et al., 1995) This work was one of the first examples to show that by tuning the reaction conditions and the catalyst structure, the reaction selectivity could be directed to either the oxidation of the primary or secondary hydroxyl group. The formation of β-hydroxypuruvic acid (a flavour component) from glycerol was reported by Abbadi and van Bekkum using a Pt-Bi catalyst (Abbadi & vanBekkum, 1996).

Hutchings and co-workers reported for the first time the successful utilization of gold catalysts (Bond & Thompson, 1999; Della Pina, Falletta, Prati, & Rossi, 2008; Della Pina, Falletta, & Rossi, 2012; Hashmi & Hutchings, 2006) in the selective oxidation of glycerol under alkaline conditions (Carrettin, McMorn, Johnston, Griffin, & Hutchings, 2002). By synthesizing supported gold nanoparticles of around 25 nm mean particle size and varying parameters, such as pressure, sodium hydroxide and catalyst amount, high selectivity to glycerate (95% at 90% conversion) was reported (Figure 3). It was proposed that the role of sodium hydroxide was essential for the initial dehydrogenation pathway, since in the presence of base the hydrogen is readily abstracted from one of the primary hydroxyl groups of glycerol. In subsequent studies they compared the catalytic performance of Au, Pd and Pt based catalysts and the improved performance of the gold catalyst with respect to Pd and Pt based catalysts in terms of high selectivity to glycerate and higher catalytic activity and stability was reported (Carrettin et al., 2003).

At the same period, Prati and co-workers investigated the catalytic performance of supported gold nanoparticles for glycerol oxidation by using a range of preparation methods for the synthesis of gold catalysts. The effect of particle size was studied by using immobilisation (Prati & Villa, 2014), and impregnation methods. From their studies they demonstrated that high activity to glycerate could be

Figure 3. Glycerol oxidation using Au/graphite in presence of a base

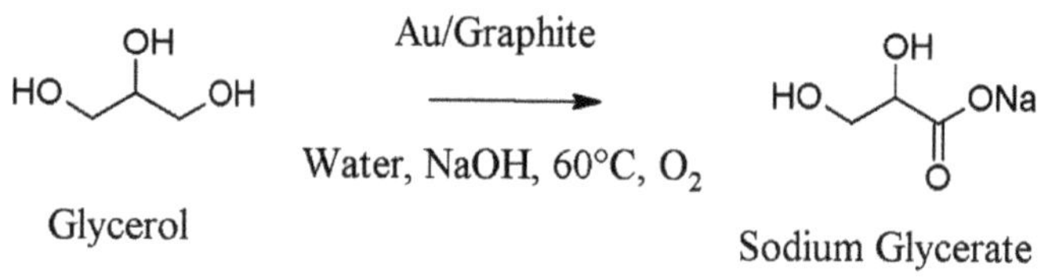

achieved (Porta & Prati, 2004), by using small well-dispersed gold nanoparticles of 6 nm mean diameter. Time on line analysis data showed that the initial selectivity could not be maintained due to the consecutive oxidation of glycerate to tartronate. Large nanoparticles of around 20 nm were responsible for high selectivity to glycerate without over-oxidation of glycerate although the activity was lower. By performing mechanistic studies they concluded that the overall selectivity of the reaction is affected from a combination of factors such as initial selectivity of the catalyst, base-catalysed inter-conversion and stability of the products.

Claus and co-workers investigated the effect of support and gold particle size using colloidal methods for the synthesis of Au catalysts. It was reported that at the same reaction conditions and with similar particle size, the carbon supported gold nanoparticles are more active than the oxide supported gold nanoparticles (Demirel-Gulen, Lucas, & Claus, 2005). By varying the gold particle size they concluded that that the reaction is structure-sensitive in agreement with previous observations by Hutchings and Prati (Carrettin et al., 2002; Porta & Prati, 2004);Dimitratos et al., 2006). In subsequent studies they investigated the effect of ceria as an alternative support. Ceria supported gold nanoparticles were active; however there was a significant decrease in catalytic activity during recycling tests. Leaching experiments were carried out and confirmed leaching of gold (Demirel, Kern, Lucas, & Claus, 2007).

Davis and co-workers studied the effect of gold particle size and it was demonstrated that small gold nanoparticles are more active than large gold nanoparticles and the selectivity to glycerate dropped with the decrease of gold particle size as previous studies by Prati and Hutchings. A new insight was discovered by mechanistic studies performed by Davis *et al.* They observed the formation of hydrogen peroxide over the series of gold catalysts investigated via the formation of hydroperoxo species. The concentration of hydrogen peroxide had a direct relationship with the selectivity towards C3 and C2 products; the presence of lower concentrations of hydrogen peroxide was associated with higher selectivity to glycerate and lower formation of glycolate. They concluded that the formation of glycolate is due to the C-C cleavage and the formation of hydrogen peroxide (Ketchie, Fang, Wong, Murayama, & Davis, 2007; Ketchie, Murayama, & Davis, 2007).

In subsequent studies, Davis and co-workers reported a significant breakthrough in the mechanism for glycerol oxidation to acids over various supported Au catalysts (Zope, Hibbitts, Neurock, & Davis, 2010; Ide & Davis, 2014). By using labelling experiments with $^{18}O_2$ and $H_2{}^{18}O$, they demonstrated that oxygen atoms originating from hydroxide ions are incorporated into the alcohol during the oxidation reaction, instead of molecular oxygen (Figure 4).

Their studies were also supported by density functional theory (DFT) calculations. According to the authors, the role of oxygen is to participate in the catalytic cycle not by dissociation to atomic oxygen, but by generating hydroxide ions via the catalytic decomposition of peroxide intermediate.

The utilization of active and selective supported Au nanoparticles immobilized on $MgAl_2O_4$ spinels as alternative basic supports was demonstrated by Prati and co-workers for glycerol oxidation (Villa, Gaiassi, et al., 2010). By using a variety of preparation methods, such as flame spray pyrolysis and co-precipitation for the synthesis of spinels, and colloidal, deposition-precipitation and magnetron sputtering methods for the deposition of Au, they explored the effect of particle size, surface area and Al/Mg surface ratio for the selective oxidation of glycerol under alkaline mild conditions. It was concluded that the Al/Mg surface ratio was affecting the selectivity to C2 and C3 products; Al-rich surface was promoting C-C bond cleavage with similar particle size of Au (2-3 nm). Selectivity to C3 products (glycerate and tartronate) was highest with Mg-rich support materials (61% selectivity at 50% conversion). Moreover, from an activity point of view, the most active catalysts were comprised of small supported gold particles and high

Figure 4. LC mass spectrum (electronegative ion mode) of glyceric acid (molecular weight 106) formed during glycerol oxidation over Au/C in $^{18}O_2$+$H_2{}^{16}O$ or $^{16}O_2$+$H_2{}^{18}O$. Glycerol oxidation in $^{16}O_2$ + $H_2{}^{18}O$ results in the incorporation of multiple ^{18}O atoms in the product glyceric acid.m/z, mass/charge ratio (Zope et al., 2010)

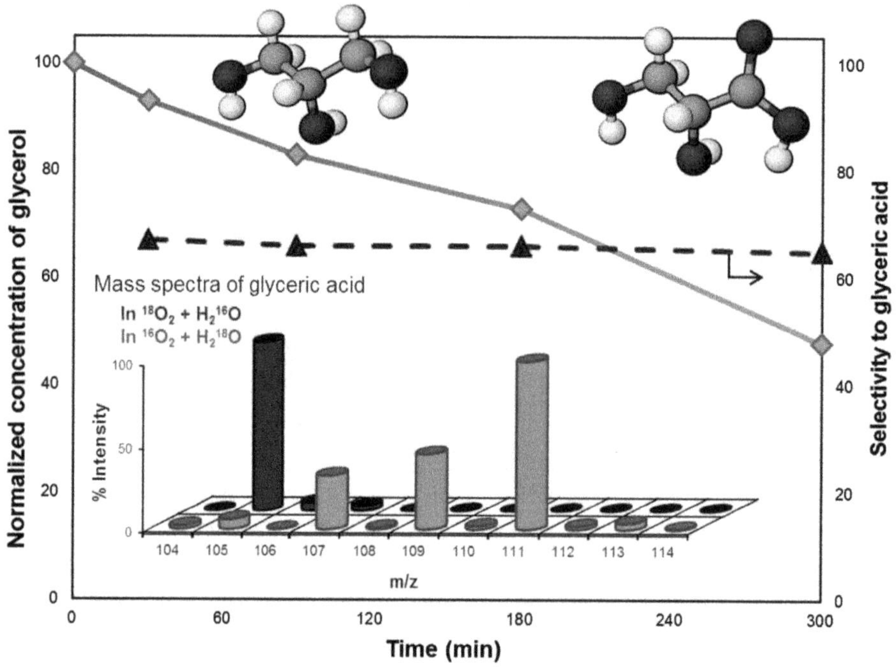

support surface areas. Another important parameter that influenced activity was the atomic percentage of Au, measured by XPS for spinels prepared by co-precipitation and flame pyrolysis methods. It was observed that with similar gold particle size (around 3 nm) for both samples the Au surface exposure was much higher on the flame-pyrolysis spinel. The higher activity was therefore ascribed to the higher exposure of active sites to the substrate due to the higher concentration of surface gold species. The utilization of functionalised carbon nanotubes and carbon nanofibers as supports for gold nanoparticles was also investigated by Prati and co-workers (Prati et al., 2011). Commercial carbon nanotubes were modified by chemical and physical methods. Oxidative treatment with nitric acid and with chemical treatment with ammonia at different temperatures was used to introduce carboxylic functionalities, and to produce N-containing groups for enhancing basic properties of the support respectively. Supported Au nanoparticles on the modified supports were prepared via a colloidal method, which ensures the formation of gold nanoparticles with a narrow particle size distribution and similar particle size. In this way, the basic and acidic properties of the supported nanoparticles on the functionalized materials were studied in the case of glycerol oxidation under alkaline conditions. The authors showed that increase of basic functionalities were accompanied with a constituent increase in the activity of the catalyst. These results are in agreement with the general accepted theory that basic reaction conditions can promote alkoxide formation and the subsequent C-H bond cleavage. Moreover, the important role of controlling the support's hydrophobicity/hydrophilicity was demonstrated. Increase of the hydrophobic surface was accompanied with increase in the selectivity to glycerate and tartronate as the main C3 products from

65% to 82% at 90% conversion. This result was explained due to the minimisation of H_2O_2 production as Davis *et al* has shown before (Ketchie, Fang, et al., 2007) and therefore an overall decrease in C-C bond cleavage. Finally, the authors demonstrated that by using functionalised carbon materials it is possible to control both activity and selectivity in glycerol oxidation by varying at the same time not only the basic properties of the support, but also hydrophobicity/hydrophilicity ratio. In subsequent studies Prati and co-workers investigated the role of NiO as a novel support for depositing Au nanoparticles, and explored the catalytic performance of Au/NiO and Au/NiO/TiO_2 catalysts in the case of glycerol and ethylene-1,2-diol oxidation (Villa, Veith, et al., 2013). In this way, the influence of the support (NiO) and specifically the electronic effect of NiO on Au could be systematically studied. It was demonstrated that the dispersion of NiO nanoparticles on TiO_2 could significantly influence the activity and selectivity of the supported Au nanoparticles, depending on the relative ratio of NiO to TiO_2. Au/NiO catalysts showed high activity but lower selectivity to glycerate (55% selectivity at 90% conversion). By varying the NiO/TiO_2 ratio and optimising the desired ratio (NiO_{10}-$TiO_{2\,90}$), the selectivity towards glycerate increased to 75% at 90% conversion. To explain the improvement in activity by using NiO as support, the authors performed thorough FTIR, HRTEM and XPS studies, and demonstrated that the most important parameter was the interaction of the support with CO as a probe molecule. XPS analysis showed that there was no change of oxide electronic properties. On the contrary, CO adsorption studies revealed that the presence of NiO was strongly influencing the CO adsorption which was in agreement by HRTEM analysis performed. HRTEM analysis showed that Au nanoparticles were preferentially associated with NiO either in the pure Au/NiO or Au/NiO/TiO_2 materials, with this interaction improving the stability of the Au nanoparticles. From these studies, the authors concluded that the interaction and interface between NiO and Au nanoparticles is the essential factor for affecting activity and selectivity for Au/NiO and Au/NiO/TiO_2 catalysts.

The effect of gold particle shape at similar gold particle size was studied by Prati and co-workers using as a reaction model the oxidation of glycerol (Wang, Villa, Su, Prati, & Schlogl, 2013). Despite the high number of studies dealing with the effect of metal nanoparticle shape, the majority of these studies concern gas-phase reactions, and the effect of shape has not been studied at similar particle sizes. Prati and co-workers showed that the variation of particle shape at similar gold particle size has a significant effect in liquid phase oxidation of glycerol. For the synthesis of Au nanoparticles with different shapes, the author's approach was to utilise commercial carbon nanofibers with different properties. The impregnation of preformed colloidal gold particles onto two different types of carbon nanofibers produced supported Au nanoparticles with similar particle size (3.5-3.7 nm). However, the Au particle shape was related to the surface structure of the support and therefore to the different interaction with the support. It was noted that a significant higher frequency of [111] planes of Au particles was observed to be parallel and highly favoured to the one type of carbon nanofiber (Figure 5).

In terms of activity and selectivity, performing glycerol oxidation at 50 °C at 3 atm of O_2 under alkaline conditions, it was observed that similar activities were obtained with the different particle morphologies. However, in terms of selectivity, it was found that the supported Au catalysts with the highest frequency of [111] planes of Au nanoparticles in contact to the support, therefore with the lowest exposure of [111] planes of Au nanoparticles, promoted the formation of C1 and C2 products. The authors concluded that increasing the number of Au (111) sites would favour the selectivity towards C3 products as it was hypothesised in previous studies (Dimitratos et al., 2006). The role of the capping agents in terms of affecting activity and selectivity has attracted considerable investigation the recent years (Kumar & Anthony, 2014; Rossi, Vono, Garcia, Faria, & Lopez-Sanchez, 2013; Shen, Ziaei-Azad, & Semagina,

Figure 5. Aberration-corrected HRTEM images of Au particles supported on PR24-LHT. a) A representative 2–4 nm particle with (1 1 1) surface epitaxially parallel to the graphitic layer of CNF. b) Structure model derived from the image in part a. c) A larger particle with spherical shape showing carbon binding to the CNF surface (Wang et al., 2013)

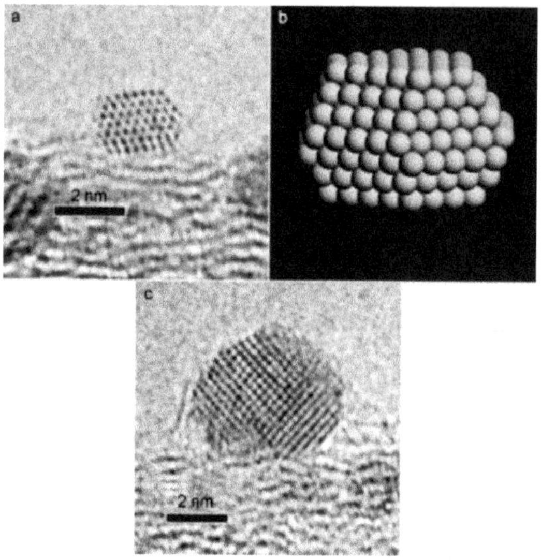

2014; Zhang & Xu, 2010). Prati and co-workers investigated the influence of capping agent (polyvinylalcohol) in terms of activity and selectivity in glycerol oxidation (Villa, Wang, Veith, Vindigni, & Prati, 2013). The presence of the protective layer around the metal nanoparticle and the surface of the support are expected to influence catalyst properties by affecting metal-support and reactant-metal interactions. By varying the polyvinylalcohol/metal weight ratio and synthesizing Au/TiO_2 catalysts with and without the presence of the capping agent the role of the capping agent was explored. It was observed that the presence of PVA lowered the catalytic activity of Au/TiO_2 catalyst but increase the stability of the catalyst during recycling tests in agreement with recent results by Hutchings and co-workers (Jose A.

Table 1. Oxidation of glycerol using Au-based supported catalysts

Substrate	Catalyst	T,°C	Solvent	Conv.%	Sel.% (Glycerate)	Ref
Glycerol	Au/Graphite	60	H_2O/NaOH	56	100	Carrettin et al., (2002)
Glycerol	Au/Graphite	60	H_2O/NaOH	72	86	Carrettin et al., (2003)
Glycerol	Au/Graphite	60	H_2O/NaOH	90	83	Porta, & Prati, (2004)
Glycerol	Au/CeO_2	60	H_2O/NaOH	40	45	Demirel-Gulen et al, (2005)
	Au/Calgon	60	H_2O/NaOH	50	62	Ketchie et al, (2007)
Glycerol	Au/$MgAl_2O_4$	50	H_2O/NaOH	50	61	Villa et al, (2010)
Glycerol	Au/N-CNF 873K	50	H_2O/NaOH	90	68	Prati et al, (2011)
Glycerol	Au/$NiO_{10}TiO_{2-90}$	50	H_2O/NaOH	90	70	Villa et al, (2013)
Glycerol	Au/CNFs-LHT	50	H_2O/NaOH	90	22	Wang et al, (2013)

Lopez-Sanchez et al., 2011). Moreover, by optimizing the ratio of capping agent to metal it was found that a small amount of capping agent is beneficial for increasing activity by a factor of 2. Finally, by studying the variation of capping agent amount on a preformed Au/TiO_2 catalysts prepared by deposition-precipitation method selectivity towards C3 products, such as glycerate and tartonate, was increased by approximately 7% at similar conversion levels. For elucidating the scientific reasons for this observation, FTIR adsorption studies reveal a direct interaction between capping agent and the glycerol molecule, thus affecting the adsorption of glycerol on the active sites of the catalyst. The authors concluded that the different adsorption mode of glycerol on the active sites in the presence or absence of capping agent can vary activity, selectivity and stability. Therefore a new way of designing metal supported catalysts by the careful choice of capping agent and amount has been demonstrated.

2.2 Oxidation of Glycerol using Supported Bimetallic Nanoparticles

Table 2 summarises key catalytic data in the oxidation of glycerol using supported bimetallic nanoparticles. Supported bimetallic nanoparticles have been shown to offer significant improvements in terms of catalytic reactivity and stability for a range of reactions, such as the oxidation of alkanes, alcohols and polyols (Dimitratos et al., 2006; Enache et al., 2006; Hou, Dehm, & Scott, 2008), and hydrogen peroxide synthesis (Edwards, Ntainjua, et al., 2009; Edwards et al., 2005; Edwards, Solsona, et al., 2009; Lopez-Sanchez et al., 2008; Jose Antonio Lopez-Sanchez et al., 2011). In the case of glycerol oxidation one of the first reports to demonstrate the successful utilisation of supported Au-based bimetallic catalysts for the oxidation of glycerol was reported by Prati and co-workers (Dimitratos, Porta, & Prati, 2005; Bianchi, Canton, Dimitratos, Porta, & Prati, 2005). Alloying Au with Pt or Pd could lead to an enhancement in the activity for glycerol oxidation but also it was shown to determine the product selectivity. The utilization of supported Au-Pd nanoparticles showed to improve the selectivity towards to the products of the oxidation of the terminal hydroxyl groups (glycerate and tartronate) with selectivities over 90% at high conversion levels (90%). In the case of supported Au-Pt catalysts, similar TOFs values were obtained with respect to the supported Au-Pd catalysts, and an enhancement in the formation of glycolic acid was observed (Bianchi et al., 2005; Dimitratos, Messi, Porta, Prati, & Villa, 2006). The monometallic Pt and Pd catalysts were shown to deactivate on reuse and it was reported that alloying gold with platinum not only led to an enhancement in the activity of the catalysts but the deactivation of the catalysts on reuse was significantly decreased.

By using a two-step experimental approach for the synthesis of bimetallic nanoparticles, which relies on the formation of preformed particles of the first metal as nucleation centres for the second metal, the synthesis of bimetallic nanoparticles with single phase and random alloy structure was achieved (Wang, Villa, Porta, Su, & Prati, 2006). Detailed characterisation of the synthesised bimetallic nanoparticles by means of HRTEM/EDX analysis showed the presence of similar Au:Pd ratio among the single particles regardless particle size, without Au or Pd segregation and uniform structure among within each particle, confirming that particles consist of random Au-Pd alloy with the Au-Pd ratio close to 6:4 (Figure 6). The catalytic performance of the supported single-phase Au-Pd nanoparticles showed a significant increase in catalytic activity with respect to monometallic Au and Pd supported nanoparticles and Au-Pd bimetallic nanoparticles prepared in one step approach. In following studies, thorough recycling tests were carried out to show the high resistance of the catalyst to poisoning and the high stability of the catalyst that retain similar catalytic activity after 11 cycles (Prati, Villa, Porta, Wang, & Su, 2007). The used catalyst was characterised by means of SEM and TEM and showed minimum degree of segregated particles, low degree of metal leaching without significant reconstruction of metal particle on the support surface.

Figure 6. (a) Representative HRTEM image of single particles of Au-Pd/AC and (b) Overall EDX spectrum and the representative one for individual single particles for Au-Pd/AC (Wang et al., 2006)

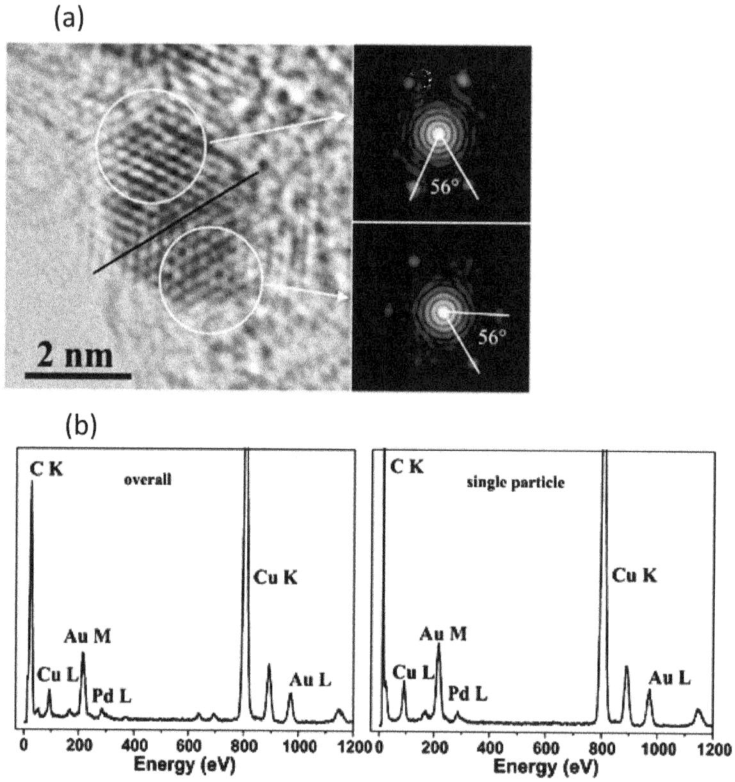

In subsequent studies Prati and co-workers investigated the ratio of the metals in supported carbon bimetallic Au-Pd catalysts in terms of activity and selectivity. The variation of the Au:Pd atomic ratio from 9:1 to 1:4 was carried out by using the sol-immobilisation method (Villa, Campione, & Prati, 2007). The particle size distribution of the supported bimetallic nanoparticles was in the range of 2.9 to 3.7 nm and supported bimetallic nanoparticles were observed to be homogeneous alloys (Wang, Villa, Porta, Prati, & Su, 2008). The variation of Au:Pd atomic ratio led to a significant change in the catalytic performance with the gold-rich catalysts showing higher catalytic activity and stability. The highest levels of activity were found with Au:Pd ratio of 9:1 and the selectivity towards glyceric acid at 90% iso-conversion was similar (70-80%) for all compositions tested. From their studies the authors concluded that the improvement in terms of activity and lifetime of the catalyst is due to the presence of surface Pd monomers in contact with Au particles. Ketchie *et al.* expanded on their earlier studies on monometallic Au catalysts to study at the effect of using bimetallic catalysts and also physical mixtures of the monometallic Au and Pd catalysts (Ketchie, Murayama, et al., 2007). It was noted that the bimetallic catalysts had higher TOF for the oxidation reaction and the selectivity towards glyceric acid at 50% iso-conversion was increased with the usage of Au-Pd bimetallic catalysts and the physically mixed monometallic catalysts with selectivities values in the range 80-85% in agreement with previous studies (Bianchi et al., 2005). Moreover the Au-Pd bimetallic catalysts produced significantly less hydrogen peroxide than the gold analogue catalyst, which indicates that the presence of palladium in the reaction can lead to the reduction in hydrogen peroxide production. This would suggest that the palladium is facilitating the decomposition of the hydrogen peroxide. Luque *et al.* further probed the effect of hydrogen peroxide on the oxidation process by using hydrogen peroxide as the oxidant under microwave irradiation and showed almost total selectivity to C-C bond cleavage products, such as glycolic, oxalic and formic acids) (Luque, Budarin, Clark, & Macquarrie, 2008).

Hutchings and co-workers synthesised supported bimetallic Au-Pd nanoparticles using two preparation methods (impregnation and sol-immobilisation methods) and studied the catalytic performance of the synthesised materials in the liquid phase oxidation of glycerol under aerobic and alkaline conditions (Dimitratos, Lopez-Sanchez, & Hutchings, 2009; Nikolaos Dimitratos, Lopez-Sanchez, Anthonykutty, et al., 2009). Higher activity with moderate selectivity to the desired product (glycerate) (60-70%) was observed with the bimetallic catalysts synthesised by the colloidal method due to the fact that high activity can be achieved by gold-rich surface bimetallic nanoparticles with mean particle size of 3-5 nm and metallic oxidation state. On the contrary, larger particles (over 6 nm) with surfaces enriched in Pd and prepared by the impregnation method led to a significant lower catalytic activity.

Prati and co-workers showed the beneficial role of Bi as a promoter for Au-Pd catalysts in the case of liquid phase oxidation of alcohols (Villa, Wang, Veith, & Prati, 2012). By using two different protocols Au-Pd-Bi catalysts were prepared, where Bi was either deposited prior or after the immobilization of Au-Pd colloidal nanoparticles onto the support (Figure 7). Characterization studies showed that the prior deposition of Bi and the Bi loading onto the support did not affect the particle size of the Au-Pd nanoparticles, but did not allow a homogenous composition of the alloyed Au-Pd phase to be obtained. Conversely, the deposition of Bi on preformed Au-Pd supported nanoparticles did not affect the alloy composition. In terms of catalysis, in the case of liquid phase oxidation of benzyl alcohol, a suppression of the parallel reaction pathway that forms toluene was observed. In the case of glycerol oxidation, a promotion of the consecutive reaction leading to tartronate was found with yield of 78%. The authors attributed the described catalytic performance due to the selective inhibition of active AuPd sites by Bi.

While the use of bimetallic catalysts has facilitated the increase catalytic activity, and has allowed selectivity to the desired products for the oxidation of glycerol to be tuned, all the initial studies were performed under alkaline conditions for facilitating conversion. The disadvantage of using alkaline conditions is that the products formed are present in the form of organic salts, therefore increasing the financial cost of the chemical process, due to the additional steps required for purification of the organic products. The first reported studies using base free reactions showed the successful transformation of glycerol under aerobic liquid phase conditions, however the level of conversion was low and the catalysts suffered from deactivation phenomena (Carrettin et al., 2003; Porta & Prati, 2004). To make the process truly catalytic the amount of base used must be decreased or removed from the reaction mixture. This has required a revision of the catalyst design and initial studies have investigated monometallic catalysts. Liang *et al.* prepared a series of platinum catalysts supported on carbon and also on carbon nanotubes, significant conversion was achieved with these catalysts with oxygen as the oxidant, however the selectivity to glyceric acid was low (Liang et al., 2009; Liang et al., 2011). The first example of bimetallic catalysts for base-free oxidations was reported by Prati and co-workers who showed that the choice of support was a crucial parameter (Villa, Veith, & Prati, 2010). Au-Pt nanoparticles supported on H-mordenite showed high selectivity to glyceric acid than the equivalent nanoparticles supported on neutral or basic supports such as TiO_2 and $MgAl_2O_4$. The higher selectivity to glyceric acid was attributed to the lower amount of H_2O_2 formed using H-mordenite as support, which is responsible for the C-C cleavage. This study demonstrates the synergistic effect of alloying metals with gold and in this case the catalyst design can ensure that the activation of glycerol can be achieved using base-free aerobic liquid phase conditions.

Figure 7. (a) STEM image of 1%Pd@Au/0.1% Bi on ACand (b) 1%Pd@Au/3% Bi on AC; (c) EDX spectra corresponding to element mapping on) 1%Pd@Au/3% Bi on AC with the integrated EDX spectrum from the mapping area (Villa et al., 2012)

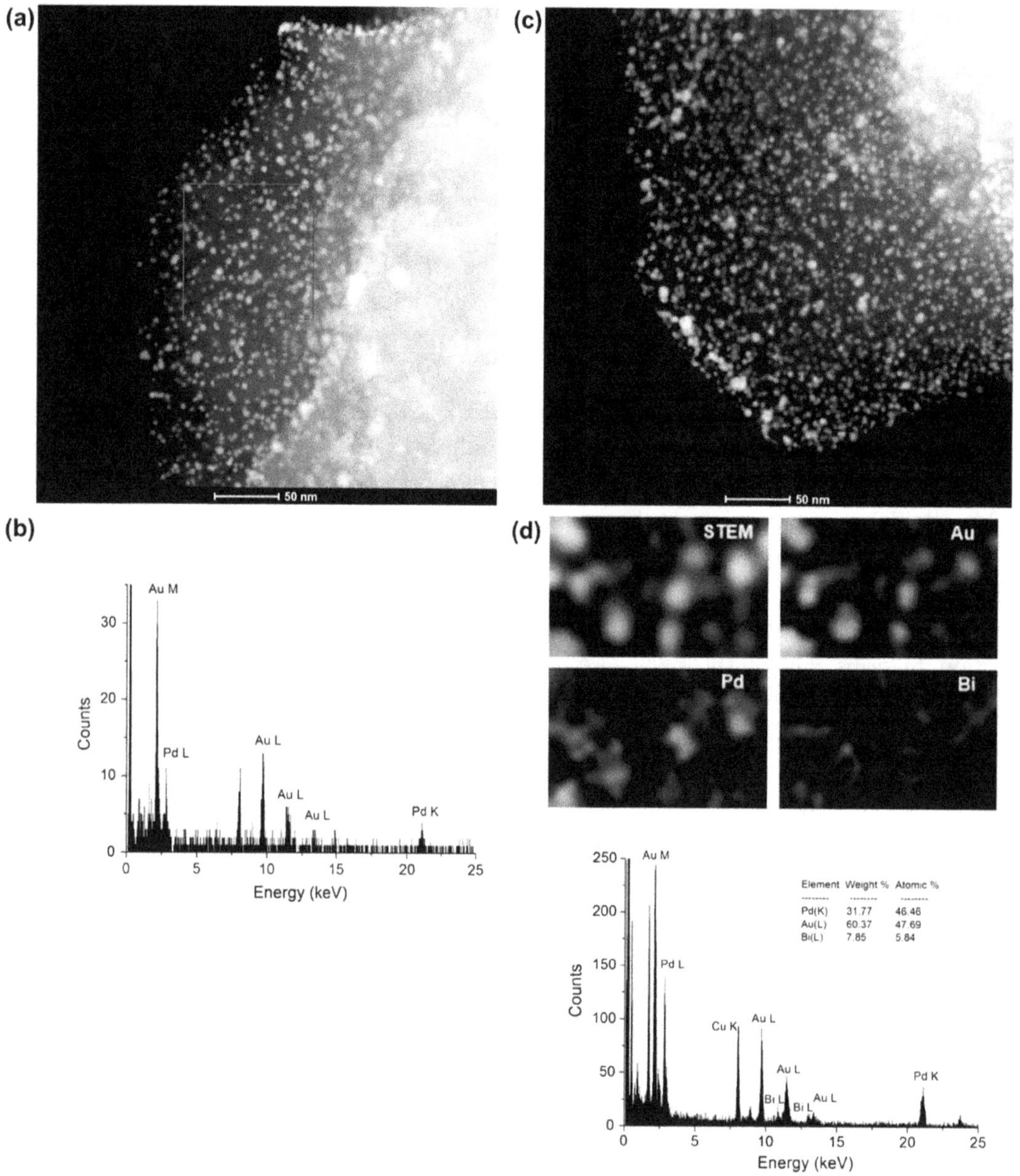

Hutchings and co-workers showed that by using MgO as a basic support and by fine tuning the temperature, and the metal ratio between Au-Pd and Au-Pt, significant conversion of glycerol could be achieved using mild base-free conditions (Brett et al., 2011). The microstructure of the bimetallic catalysts was investigated by means of STEM/HAADF/EDX analysis and it was shown that the individual nanoparticles contained both elements (Figure 8). While the support is nominally referred to as MgO, $Mg(OH)_2$ was formed during the preparation step, and the use of this solid base enabled the reaction to

Figure 8. HAADF image of the AuPt(1:1)/MgO catalyst (dried at 120°C), along with the corresponding $Au\text{-}L_\alpha$, $Pt\text{-}L_\alpha$, $Mg\text{-}K_\alpha$, and ($Au\text{-}M_\alpha + Pt\text{-}M_\alpha$) elemental maps from the same area (Brett et al., 2011)

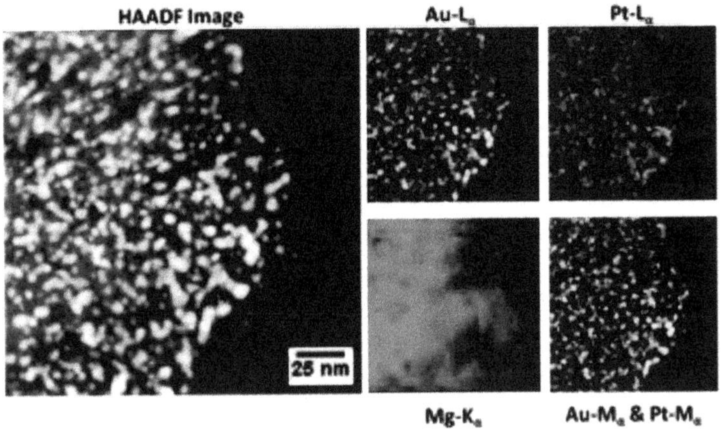

proceed smoothly. Loss of Mg^{2+} into the reaction mixture by leaching was observed to be negligible. The choice of metals for the bimetallic catalysts was also found to play a significant role as platinum led to an increase in both the activity and selectivity to C_3 products (85%) when compared with palladium. AuPt/MgO was active for this reaction at ambient temperatures. The catalysts were observed to be reusable, however, while there was no decrease in activity with the used catalyst the selectivity trend shifted significantly from glyceric acid towards tartronic acid indicating that there is some change in the catalyst during reusability tests.

Overall it is evident that significant progress in the oxidation of glycerol under acidic, alkaline and free base liquid phase conditions has been achieved in recent years. The main focus has been in the catalyst design taking into account the choice of support, noble metals and the molar ratio of the metals. In particular, the design and use of bimetallic catalysts has provided significant increases in the activity and stability of the catalysts using base-free conditions. However, at this stage the conversion

Table 2. Oxidation of glycerol using bimetallic Au, Pd and Pt based supported catalysts

Substrate	Catalyst	T,°C	Solvent	Conv.%	Sel.% (Glycerate)	Ref
Glycerol	Au-Pd/C	30	H_2O/NaOH	90	70	Dimitratos et al, (2005)
Glycerol	Au-Pt/C	30	H_2O/NaOH	90	58	Dimitratos et al, (2005)
Glycerol	Pd@Au/AC	50	H_2O/NaOH	90	77	Wang et al, (2006)
Glycerol	Au90-Pd10/AC	50	H_2O/NaOH	90	75	Villa et al, (2008)
Glycerol	Au-Pd/AC	60	H_2O/NaOH	50	84	Ketchie et al, (2007)
Glycerol	Au-Pd/AC	60	H_2O/NaOH	90	72	Dimitratos et al, (2009)
Glycerol	3%Bi@1%AuPd/AC	50	H_2O/NaOH	90	71	Villa et al, (2012)
Glycerol	Pt-Cu/C	40	H_2O	86	71	Liang et al, (2011)
Glycerol	Au-Pt/Mordenite	100	H_2O	70	83	Villa et al, (2010)
Glycerol	Au-Pt/MgO	40	H_2O	43	85	Brett et al, (2011)

of these catalysts is still lower than when sacrificial base is used and the opportunity for future catalyst design will focus in improving activity for the glycerol oxidation under base free conditions with high substrate/metal molar ratio. The reusability of these catalysts is also an area which needs attention in terms of catalyst development; whilst the activity of these catalysts for glycerol conversion is recovered for re-used catalysts, the selectivity profile of the major products is not retained. This suggests that the catalysts are undergoing a level of restructuring during the reaction and while this presents difficulties in terms of producing a truly reusable catalyst it also presents opportunities. An improved understanding of the structural changes during reaction will facilitate the development of better catalysts. Thus, the use of in situ/operando techniques for exploring active site speciation and chemistry under reaction conditions is essential. Furthermore, the current literature clearly demonstrates that the choice of metals in bimetallic catalysts can have a significant effect of on the product distribution and further investigation into the field of bimetallic catalysts could lead to the optimisation of these processes leading to designer catalysts for each target molecule.

3. HYDROGENOLYSIS OF GLYCEROL

Hydrogenolysis of glycerol involves the cleavage of C-C or C-O bond using H_2, and can lead to the production of a range of important bulk/commodity chemicals such as 1,3- and 1,2-propanediols, methanol and ethylene glycol as they are in high demand in industry (Figure 9) (Nakagawa & Tomishige, 2010; Nakagawa & Tomishige, 2011).

For converting glycerol, the oxygen content should be reduced by breaking C-O bonds in order to be more useful as a chemical fuel product. However, due to the presence of more than one hydroxyl groups the synthesis of desired products with high selectivity is a high technical challenge. 1,3-propane diol is an important monomer for the production of several polymers *e.g.* polyesters, polyethers and poly-

Figure 9. Pathway of glycerol hydrogenolysis (Nakagawa & Tomishige, 2011)

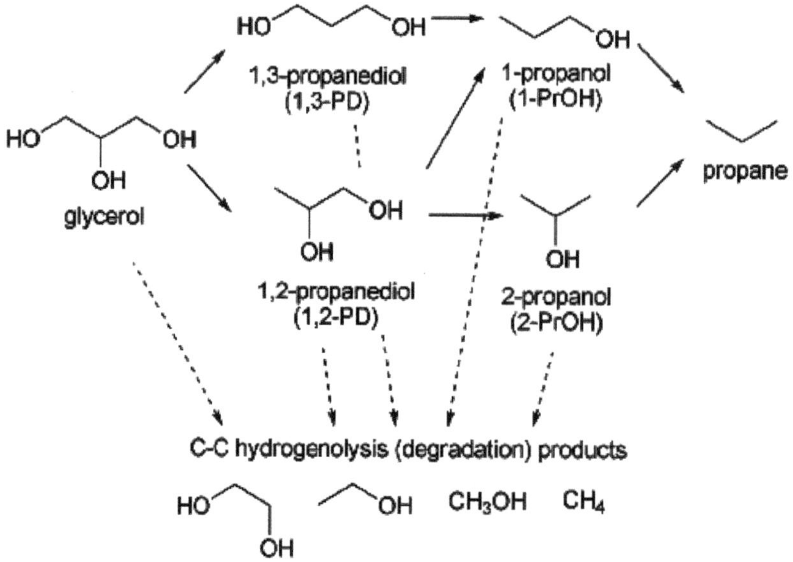

urethanes. Dupont produces a polymer (SORONA), where the 1,3-propane diol is produced chemically by the hydration of acrolein or by the fermentation of glucose and annual production of 1 million tons. Shell produces a similar polymer, under the brand name CORTERRA, but in this case the 1,3-propane diol is obtained *via* the hydroformylation of ethylene oxide followed by hydrogenation (Nakagawa & Tomishige, 2011). 1,2-propanediol is also used as a monomer in the preparation of polyester resin, as an antifreeze agent, used in paints, detergents, cosmetics and food, with an annual global production close to 1.4 million tons. Currently, 1,2-propanediol is produced industrially by the hydration of propylene oxide. Therefore, alternative processes via the chemical route from glycerol to propanediols are from academic and industrial point of view of highly interest (Jerome, Pouilloux, & Barrault, 2008). However, to be cost effective and competitive to the current industrial processes highly active, selective and stable catalysts need to be discovered. A crucial prerequisite for an active catalyst for this transformation is the selective breaking of C-O bond in preference to a C-C bond, in cases where propanediols are the target products. Over recent years, a number of catalytic systems have been reported for this reaction and these have been extensively reviewed, but most of the reported catalysts to date are based on supported monometallic catalysts (Nakagawa & Tomishige, 2011).

3.1 Hydrogenolysis of Glycerol using Supported Bimetallic Nanoparticles

For other catalytic systems including the use of oxides and multicomponet oxides and zeolites for the hydrogenolysis of glycerol to propanediols, readers are referred to some of the previous reviews on this topic (Perosa & Tundo, 2005; Ruppert, Weinberg, & Palkovits, 2012; ten Dam & Hanefeld, 2011). Among the reported catalysts, supported Cu, Ni and noble metal (Au, Pd, Pt) monometallic nanoparticles are the most studied catalytic systems for the hydrogenolysis of glcerol. However, the exploitation of supported bimetallic nanoparticles is a new area of research for the hydrogenolysis of glycerol with fascinating results in terms of tuning activity and selectivity to the desired products.

Ru-based bimetallic catalysts were reported for the hydrogenolysis of glycerol by Maris and co-workers. The strategy they followed for the synthesis of a bimetallic catalyst was based on a seed-growth technique; the addition of either Pt or Au metal to a commercially preformed Ru nanoparticles supported on carbon (5.0%Ru/C) was carried out (Maris, Ketchie, Murayama, & Davis, 2007). The hydrogenolysis activities of the bimetallic catalysts and physical mixtures were compared with the monometallic analogue catalysts. It is interesting to note that Pt is partially miscible with Ru whereas Au is not miscible in the bulk (Antolini, Giorgi, Cardellini, & Passalacqua, 2001). Among the catalysts evaluated for glycerol hydrogenolysis, the catalytic activity reported was in the following order: Ru/C ≈ Ru +Au/C (Physical mixture) > Ru+Pt/C (physical mixture) > Pt-Ru/C (bimetallic) > Au-Ru/C (bimetallic) >> Pt/C > Au/C (Antolini et al., 2001). It is evident from the catalytic trend that the supported monometallic nanoparticles were more active than their combined supported bimetallic nanoparticles. Characterisation studies were carried out for revealing the structure and morphology of the supported bimetallic nanoparticles. In the case of the used Ru-Au/C catalyst the Au particles migrates out of the Ru particles during the reaction, and forms large gold nanoparticles on the surface of the support, therefore during the reaction there is a dealloying process and segregation of particles. In subsequent studies Ma *et al.* modified supported Ru nanoparticles with Re metal were synthesised and the catalytic performance was investigated (Ma & He, 2009, 2010; Ma, He, & Li, 2008; Ma, Li, & He, 2011). Initially, Re was added using $Re_2(CO)_{10}$ organometallic complex as the chosen metal precursor during the reaction along with the supported Ru catalyst. A promotion effect was observed for all the supported modified Ru catalysts. In the case of

Ru/Al_2O_3 catalyst, the conversion increased from 19% to 53% by the addition of $Re_2(CO)_{10}$, for Ru/ZrO_2 the increase was from 20% to 27% and for Ru/C the effect of Re addition was an increase in conversion from 30% to 60%. Ru was found to be in metallic state in fresh and spent catalysts whereas the oxidation state of Re could not be determined thoroughly by XPS. For further investigating the promotional effect by Re the same group synthesised a series of supported Ru-Re nanoparticles, using a typical impregnation method, and evaluated the catalysts for the hydrogenolysis of glycerol (Ma & He, 2009). For all the supported monometallic and bimetallic nanoparticles evaluated, bimetallic Ru-Re catalysts exhibited superior catalytic activities compared to their monometallic analogues. The use of H-ZSM-5 as support was beneficial and supported Ru-Re bimetallic nanoparticles on H-ZSM-5 showed similar activity with $Ru-Re/Al_2O_3$ catalysts. In both cases, the addition of Re component to the catalyst was beneficial in terms of activity and the selectivity to ethylene glycol was substantially reduced, minimising C-C cleavage. The higher activity and improved selectivity was reported to be due the following reasons; (i) improved dispersion of Ru by Re and prevention of Ru aggregation on the surface of the catalysts and (ii) the enhanced acidity of the ReO_x particles. The effect of pre-treatment was studied and it was found that heat pre-treatment of the catalyst under reducing conditions and high temperatures resulted in forming mostly Re^0 particles that are found to be catalytically inactive. Based on the pre-treatment studies at several experimental conditions, the most optimised pre-treatment was found to be either at low temperature gas phase reduction at 350 °C or an *in situ* reduction at 200 °C (Ma & He, 2010). These conditions resulted in the synthesis of a mixture of metallic Ru particles and oxidised ReO_3 particles, which resulted in an active catalyst. As stated earlier, the acidity of the ReO_3 component of the bimetallic catalyst was found to be one of the main reasons for the enhanced catalytic activity. To confirm that the active phase of Re component in the catalyst is ReO_3 addition of Re metallic nanoparticles to the solid Ru/SiO_2 catalyst was carried out. The glycerol conversion was similar to that of the monometallic Ru/SiO_2 catalyst. On the contrary, when ReO_3 particles were added to the Ru/SiO_2 catalyst, the conversion increased from 17% to 34%, which is similar to the value obtained for the bimetallic catalyst. The role of acidic ReO_3 according to the authors is to facilitate the dehydration of glycerol, which is the first step of hydrogenolysis followed by the hydrogenation of the intermediate (Ma et al., 2011). This hypothesis was further confirmed by adding equivalent amount of acid instead of Re component in the catalyst resulting in almost the same trend in the catalytic activity (Figure 10).

Cu based catalysts were investigated for the hydrogenolysis of glycerol, albeit at a higher temperature was needed for the hydrogenolysis of glycerol (Balaraju, Rekha, Prasad, Prasad, & Lingaiah, 2008; Huang et al., 2008; Vasiliadou & Lemonidou, 2011). Jiang *et al.* synthesised Ru-Cu bimetallic nanoparticles supported on bentonite clay and using 1,1,3,3-tetramethylguanidinium lactate (TGML) ionic liquid as solvent, for the hydrogenolysis of glycerol (Jiang, Zhou, Liang, Liu, & Han, 2009).

A series of catalysts with different Ru/Cu atomic ratio were synthesised and the catalytic performance was carried out for glycerol hydrogenolysis. Supported monometallic Ru catalyst exhibited the highest activity with 91% conversion among the series of monometallic and bimetallic catalysts and the selectivity to 1,2-propanediol was 63%. The addition of Cu to Ru catalyst improved substantially the selectivity towards to 1,2-propanediol but the conversion dropped with the increase in Cu amount (Jiang et al., 2009). The optimum atomic ratio of Ru/Cu was found to be 3/1 and further optimisation of the catalytic performance was carried out by varying reaction parameters. With this optimised Ru-Cu catalyst, a maximum of 86% selectivity to 1,2-propanediol was achieved. Reusability studies were carried out for at least of five cycles and minimum deactivation was observed. In subsequent studies Later Wu *et al.* synthesised Cu-Ru bimetallic nanoparticles supported on multiwalled carbon nanotubes (MWCNTs)

Figure 10. Schematic representation of the two steps involved in the hydrogenolysis of glycerol using Ru-Re/SiO$_2$ (Ma et al., 2011)

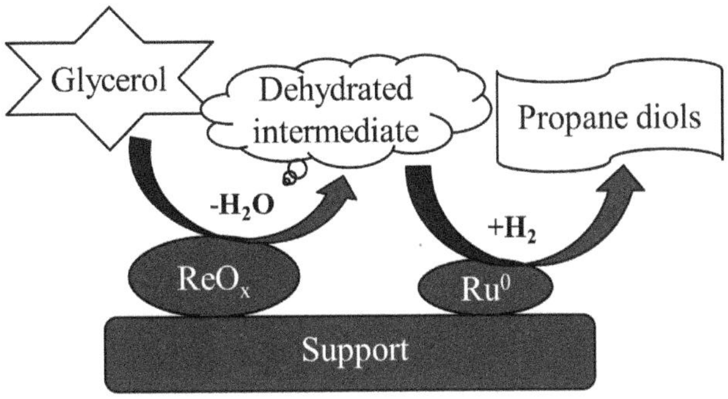

for the hydrogenolysis of glycerol (Z. J. Wu, Mao, Wang, & Zhang, 2011). These supported bimetallic nanoparticles were prepared by chemical replacement method resulting in the formation of core-shell structures, where Ru particles were deposited on the external surfaces of Cu. The authors reported that the role of Ru nanoparticles was primarily the activation of H_2 and therefore the increase availability of active hydrogen species for hydrogenolysis at the Cu site through hydrogen spill over mechanism (J. B. Wu, Gross, & Yang, 2011). In view of this co-operative mechanism, the activity of the bimetallic catalyst was found to be much higher than that of the corresponding monometallic catalysts. Liu *et al.* subsequently studied the effect of the support on the same Ru-Cu bimetallic nanoparticle system and found that ZrO_2 was the preferred support among those studied, such as SiO_2, TiO_2, Al_2O_3, HY and NaY. Using the ZrO_2-supported Cu-Ru bimetallic catalyst an overall yield of 79% towards 1,2-propanediol was achieved (Liu, Liang, Jiang, Han, & Zhou, 2012). Li *et al.* used Fe as a possible candidate for modifying modified Ru catalysts. Ru-Fe nanoparticles supported on carbon nanotubes (CNT) were synthesised and comparison of the catalytic performance with the monometallic counterparts resulted in a more active, selective and stable catalyst for the conversion of glycerol to 1,2-propanediol with an overall conversion of 86% and 53% selectivity for 1,2-propanediol (Li et al., 2011). Monometallic Fe catalyst was inactive, however the addition of Fe to Ru increased the activity and more importantly the selectivity towards 1,2-propanediol. Similar to previous cases, the addition of Fe aided to the formation of smaller particles and better dispersion. The ratio between the two metals appears to be crucial for the activity; for example, $Ru_{10}Fe_1$/CNT (average particle size 5.3 nm) has a similar activity (conversion and selectivity) to that of the monometallic Ru/CNT (average particle size 6.9 nm), whereas Ru_1Fe_1/CNT (average particle size 3 nm) outperformed both the catalysts in conversion as well as selectivity to 1,2-propanediol. The clear advantage of the Ru-Fe/CNT catalysts with respect to the monometallic Ru/CNT was an improvement in reusability and lowering the financial cost of the catalyst. Ru/CNT catalyst was deactivated on recovery and reuse, whereas the Ru-Fe/CNT catalyst was stable for up to four reuses. Detailed characterisation of the catalyst was carried out and revealed that two kinds of iron species have been found on the surface of the catalyst after reduction: (a) metallic Fe alloyed to Ru to form Ru-Fe alloy on the surface of the catalyst and (b) FeO and FeO_{1-x} ($0 < x < 0.5$). The observed higher catalytic active for the bimetallic catalyst has been attributed to both alloy and oxidic species being active and working in a synergistic manner.

Table 3. Hydrogenolysis of glycerol over supported Ru-, Re- and Cu-based bimetallic nanoparticles

Catalyst	Reaction Conditions	Conv. (%)	Main Product/ Selectivity (%)[c]	Ref
0.85%Au-5.0%Ru / C	200 °C/4 MPa/5 h 0.8 M NaOH	100	FA/44	Maris et al, (2007)
4.92% Ru / C + $Re_2(CO)_{10}$	160 °C/8 MPa/8h	59	1,2-PG/57	Ma&He, (2009)
Ru-Cu / TMG-BEN[a] Ru/Cu = 3/2	195 °C/10 MPa/18h	64	1,2-PG/70	Huang et al, (2008)
Ru-Cu / TMG-BEN[a] Ru/Cu = 0/3	195 °C/10 MPa/18h	27	1,2-PG/83	Balaraju et al, (2008)
Cu-Ru / MWCNTs[b]	200 °C/4 MPa/6h	100	1,2-PD/87	Wu et al, (2011)
Pt-Re / C	170 °C/4 MPa/8 h 0.8 M NaOH	93	1,2-PG/29	Daniel et al, (2010)
Ir – ReO_x / SiO_2 Re/Ir = 3	120 °C/8 MPa/12h H^+ / Re = 1 (H_2SO_4)	58	1,3-PD/45	Shinmi et al, (2010)

[a] BEN = bentonite; TMG = 1,1,3,3- tetraguanidynum lactate.
[b] MWCNT = Multi-walled carbon nanotubes
[c] FA=Formic acid; 1,2-PG=1,2-propanediol; 1,3-PG=1,3-propanediol; EG=Ethylene glycol

Bimetallic Pt-Re nanoparticles supported on activated C were synthesised by using standard incipient wetness method by Daniel *et al.*, extensively characterised, and their catalytic performance was carried out for the hydrogenolysis of glycerol to propanediols (Daniel et al., 2010). The bimetallic catalyst with Pt:Re molar ratio of 1 shower higher catalytic activity when compared to their monometallic analogues. For the hydrogenolysis of glycerol to 1,3-propanediol using water as solvent a selectivity of 34% at 170 °C was observed. The Pt-Re bimetallic catalysts were calcined at 700 °C and it was found to be slightly more active and selective towards 1,3-propanediol compared to the uncalcined catalyst using identical reaction conditions. The heat-treatment at high temperature was beneficial for improving the formation of alloyed particles and therefore the improved activity of the bimetallic system, compared to the monometallic systems. The role of Re component is to directly activate the C-OH bond of glycerol using its oxophilic character since it exists as Re-OH in the reaction medium (Daniel et al., 2010). An similar model has been proposed by Tomishige *et al.* for the hydrogenolysis of glycerol using bimetallic catalysts contained Re as a component. In this case the role of Re component in the catalyst is to anchor the glycerol through its OH bond, using its oxophilic nature, and the role of the other metal component, in this case, Rh to deoxygenate the glycerol with hydrogen to form the final propanediols (Shinmi, Koso, Kubota, Nakagawa, & Tomishige, 2010).

4. AQUEOUS-PHASE REFORMING OF GLYCEROL USING SUPPORTED BIMETALLIC NANOPARTICLES

A range of established and developing technologies for hydrogen production have been presented the last 10 years, (Chheda, Huber, et al., 2007; Davda, Shabaker, Huber, Cortright, & Dumesic, 2005; Huber & Dumesic, 2006; Huber, Iborra, & Corma, 2006; Nishiyama, Kubota, Kimura, Tsuruya, & Masai, 1997; Tanksale, Beltramini, & Lu, 2010) utilising various processes, including (i) partial oxidation and autothermal reforming of hydrocarbons using natural gas,(Rostrup-Nielsen, 2001) (ii) steam reforming (iii) coal gasification,(Lin, Harada, Suzuki, & Hatano, 2002) (iv) methane decomposition (Poirier & Sapundzhiev, 1997; Steinberg, 1998) and (v) methane aromatisation (Solymosi, Cserenyi, Szoke, Bansagi, & Oszko, 1997; Xu, Bao, & Lin, 2003). Processes (i-iii) produce H_2 along with emissions of CO and CO_2, whereas processes (iv,v) result in CO_2 free emissions. However, a truthfully sustainable and clean process for producing H_2 requires it to be carbon neutral. CO_2 neutral H_2 production can be explored by means of the following processes; (i) pyrolysis of bio-oils (Taralas & Kontominas, 2006), (ii) conversion of biomass *via* gasification (Chen, Andries, Spliethoff, & Fang, 2004), (iii) steam reforming of biomass derived alkanes and alcohols (Rostrup-Nielsen, 2001) and (iv) aqueous phase reforming of oxygenated biomass-derived hydrocarbons (Cortright, Davda, & Dumesic, 2002). Aqueous phase reforming (APR) has been shown to be a promising route to generate H_2 and alkanes as value-added chemicals, from aqueous phase carbohydrates found in waste-water from biomass processing.

Carbohydrates are oxygenated hydrocarbons with a C:O ratio of 1:1 where hydrogen can be generated in the following reaction (1):

$$C_nH_{2y}O_n \leftrightarrows nCO + yH_2 \qquad (1)$$

Dumesic and co-workers, pioneering research in the area of aqueous phase reforming, have shown the efficient conversion sugars and sugar alcohols with water into H_2 and CO_2 at temperatures below 230 °C using a Pt/Al_2O_3 catalyst (Cortright et al., 2002). High H_2 yields (50-70%) were obtained using methanol, ethylene glycol, glycerol, sorbitol and glucose as feed molecules, with the highest selectivity obtained using methanol (99%) and the lowest selectivity (50%) using glucose. Moreover, they demonstrated that by varying reaction conditions such as temperature and concentration a significant variation of rates of H_2 production can be observed. They concluded that, although Pt-based catalysts showed high activities together with high selectivity for the production of H_2 from sugars and alcohols, improvements are necessary in the case of design materials with a lower cost and higher activity. In subsequent studies, Dumesic and co-workers evaluated the catalytic performance of a series of metal supported catalysts (Ni, Pd, Pt, Ru and Rh nanoparticles supported on silica) in the aqueous reforming of ethylene glycol as a model reaction. It was reported that the highest selectivities were exhibited by the utilisation of Pt and Pd nanoparticles (78% and 99% respectively at 225 °C) whereas Ru and Rh nanoparticles showed the lowest selectivities (7% and 0.2% respectively at 225 °C). It is important to emphasise that silica-supported Ru and Rh catalysts showed very high selectivity to the formation of alkane products (over 60%) and Ni-silica supported catalysts showed significant deactivation at 225 °C. From these studies the authors concluded that it may be possible to obtain better catalysts by using metal alloys such as Pt-Pd, Ni-Pd or a combination of precious/non-precious metal alloys (Davda, Shabaker, Huber, Cortright, & Dumesic, 2003). Dumesic and co-workers developed a strategy for developing less expensive catalysts based on non-precious metals. They developed a high-throughput, 48-well parallel batch reactor for catalyst screening and evaluated the catalytic behaviour of more than 300 catalysts for aqueous-phase reforming of ethylene glycol. This including three series of catalysts; (i) monometallic, (ii) bimetallic and (iii) metallic catalysts promoted with various metal oxides and synthesized by wetness impregnation. The synthesized catalysts were treated using different heat treatment procedures for developing the final morphology. From these studies they concluded that the most effective catalyst was a Sn-promoted Raney-Ni catalyst (Huber, Shabaker, & Dumesic, 2003). A Sn-Ni catalyst with bulk Sn/Ni atomic ratio of 0.075 performed better than the Raney-Ni and Pt catalysts, with normalised TOFs to be higher by a factor of 6. Moreover, higher selectivity for production of H_2 and lower selectivity for production of alkanes was achieved by the Raney-Ni-Sn catalyst, the catalyst stability was also improved considerably due to the reduced level of irreversible CO chemisorption and therefore lower potential of CO poisoning. They reported that the beneficial effect of Sn on the selectivity to H_2 relates to the presence of Sn at Ni-defect sites and by the formation of Ni-Sn alloy surfaces, such as a Ni_3Sn alloy, that can suppress side reactions such as methanation. In subsequent studies Dumesic and co-workers investigated bimetallic Ni-based catalysts using aqueous reforming of ethylene glycol as the model reaction. They found that addition of Sn, Au and Zn can improve the hydrogen selectivity of the catalyst, with Ni-Sn/Al_2O_3 catalysts being the most active and selective. However, Ni-Sn/Al_2O_3 catalysts were not stable with respect to time on line during aqueous-phase reforming, so they concentrated on effective procedures for adding Sn to Ni with a significant focus on lowering the level of deactivation. They developed a method of adding Sn to Ni by using Raney-Ni catalysts, varying the Ni:Sn atomic ratio. They observed that only a very low Ni:Sn atomic ratio of 400:1 was necessary to significantly increase selectivity to H_2 by markedly decreasing side reactions, such as methane formation. Indeed methane formation was nearly switched off by decreasing the Ni:Sn atomic ratio to 18:1. Moreover, a pronounced effect in the distribution of liquid products was observed, with addition of Sn significantly enhancing the selectivity towards methanol in preference over ethanol, acetaldehyde and acetic acid. The major role of Sn was to suppress the over-oxidation of alcohols to acids. Shabaker, Huber, & Dumesic, 2004). The catalytic deactivation of bimetallic Raney-Ni-Sn catalysts was explored during hydrogen production by aqueous-phase reforming of ethylene glycol by means of X-ray photoelectron spectroscopy, CO and H_2 adsorption microcalorimetry, X-ray diffraction and thermogravimetric analysis (Shabaker, Simonetti, Cortright, & Dumesic, 2005). It was reported that during the first 48 h on stream, the deactivation proceeded rapidly on the monometallic Ni catalysts, a consequence of sintering of small Ni particles in parallel with substantial leaching. The deactivation of the bimetallic Ni-Sn was much slower as a result of alloying, which suppressed sintering and dissolution of Ni by improving the oxidation resistance of Ni surfaces. Qiao and co-workers reported the synthesis of Sn-modified Ni and subsequent catalytic performance in aqueous-phase reforming of ethylene glycol (Xie et al., 2006). The Sn-Ni catalysts were prepared in a two-step process, where the first step involved the synthesis of $Ni_{50}Al_{50}$ alloy by using a single-roller melt-spinning technique, followed by the second step of addition of Sn. This preparative route yielded a series of Sn-Ni catalysts with a range of Sn:Ni molar ratios. SEM-EDX experiments were employed to confirm the homogeneity of Sn distribution associated with Ni. XRD analysis was exploited to identify the absence of any Sn-related phases, implying the dispersion was homogeneous, with only small crystallites of deposited Sn. XRD was also used to confirm the presence and formation of hexagonal phase Ni_3Sn. XPS studies revealed surface Sn-enrichment, with 70% of Sn present in the metallic state and the other 30% as oxidized Sn species (Sn^{2+} and Sn^{4+}). XPS studies also identified metallic state Ni species. The effect of Sn addition on the Ni catalyst was highly beneficial, with increased levels of activity as well as selectivity. Selectivity to H_2 increased with Sn content, where Sn:Ni ratios over 4.4% gave rise to a selectivity towards H_2 in excess of 90%with reduced levels of selectivity towards alkanes. They attributed the beneficial addition of Sn to Ni as (i) an electronic effect that weakened the bonding of CO, consequently suppressing the methanation reaction, and (ii) due to the formation of the Pt_3Sn alloy that was able to act as a centre for H_2O activation. The authors proposed that with activation of H_2O, surface OH groups could form on Sn and could facilitate the oxidation of adsorbed CO molecules on neighbouring Ni sites, therefore leading to higher H_2 selectivity.

Dumesic and co-workers studied the effect of addition of Ni, Co, and Fe on Pd and Pt monometallic catalysts in the liquid phase reforming of ethylene glycol and sorbitol, where the incipient impregnation technique was used for the synthesis of bimetallic catalysts (Huber, Shabaker, Evans, & Dumesic, 2006). Bimetallic Pt-Ni and Pt-Co had enhanced turnover frequencies for H_2 production in comparison to monometallic Pt catalysts by a factor of 1.5-2.8. Variations in Pt-Ni and Pt-Co atomic ratio showed that the most active catalysts were those containing a Pt:Ni atomic ratio of 1:1 and Pt:Co atomic ratio of 1:5. Only slight promotional effects were observed for the case of mixing Pd with Ni or Co throughout the course of the high-throughput studies. However, by adding Fe, Pd-Fe bimetallic catalysts showed a strong promotional effect with regards to the production of H_2, where the optimum atomic ratio of Pd:Fe was found to be 1:1.18. Although there was no detailed characterization to prove the existence of alloy formation the authors suggest that alloying Pt with Ni, Co or Fe could improve the activity for H_2 production. This was ascribed to a decrease in the enthalpy of adsorption of CO and H_2, thereby increasing the fraction of surface available for reaction with ethylene glycol. In subsequent studies Dumesic and

Figure 11. (a, b) TEM micrographs of Ni supported on alumina nanofibres. (a) STEM micrograph of a Ni-Pt/alumina nanofibre sample (c) and; EDX spectra of point 1 and 2 (d) (Tanksale et al., 2008)

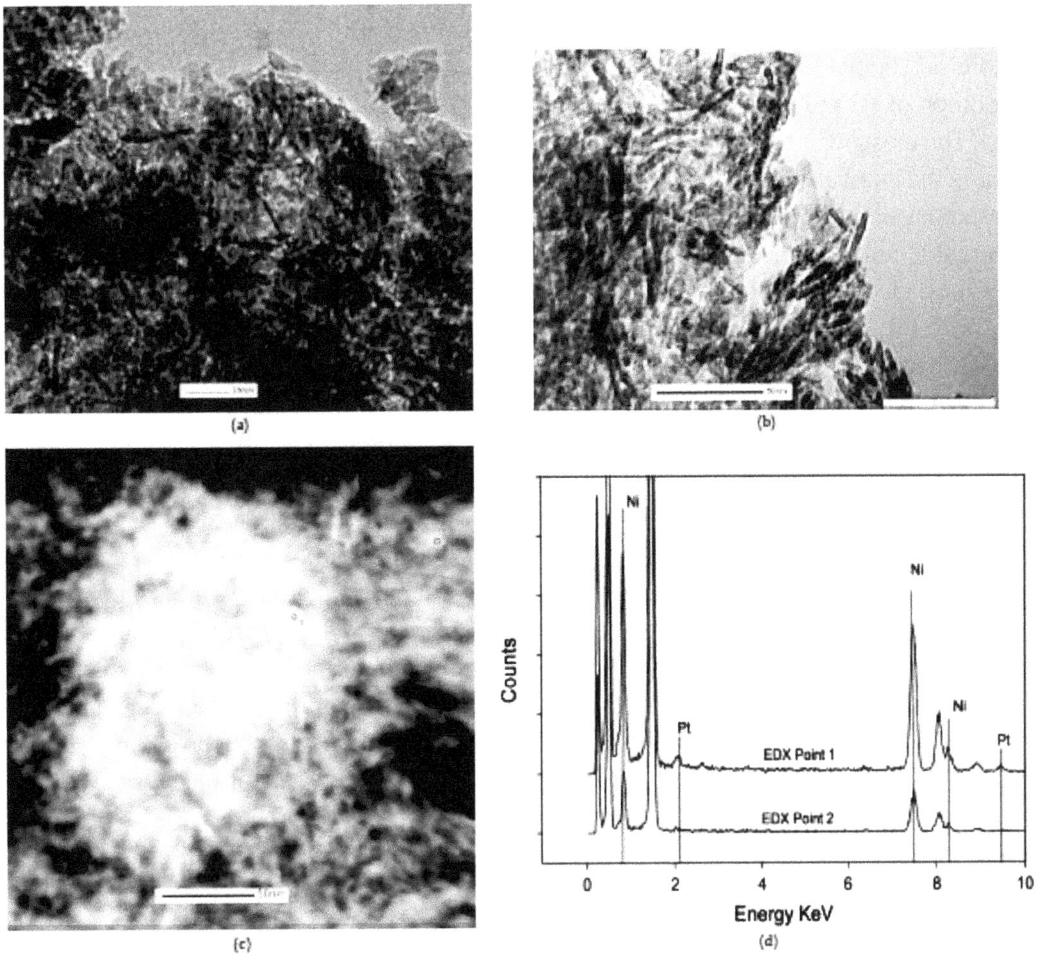

Figure 12. Microstructure of Pt-Ru catalysts distributed onto the surface of carbon fibers (Chang et al., 2011)

co-workers evaluated the catalytic performance of Ni-based catalysts synthesized by wet impregnation methods (Tanksale, Beltramini, Dumesic, & Lu, 2008). The support chosen for these studies was nano-fibred alumina due to its high surface area. The addition of Ni to Pt or Pd resulted in a 1.5-2 fold increase in catalytic activity, with the selectivity towards hydrogen showing a slight decrease to 76% when using a Ni-Pt catalyst, whereas a monometallic Pt catalyst has a selectivity of 79%. The drop of selectivity to H_2 was more substantial with Ni-Pd systems but still exhibited high levels (67%). TPR studies showed that the addition of Pd and Pt to Ni aided the reducibility of Ni catalysts, thus increasing the number of active sites. The most important findings where achieved when performing TPD studies, which indicated a reduction in the number of strong CO-adsorption sites by the formation of Ni-Pd or Ni-Pt alloys, leading to substantial improvement in catalyst stability. STEM studies showed the presence of rod shaped particles around 3nm in diameter and 10-20 nm in length as well as larger agglomerated particles. EDX analysis of individual nanoparticles (Figure 11) showed the presence of both metals, suggesting alloy formation, with the Pt:Ni atomic ratio to be in close agreement with the ICP analysis.

Low cost Pt-Ru bimetallic catalysts were developed by Chang and co-workers for converting biomass to bio-hydrogen fuel (Chang, Louh, Wong, Tseng, & Lee, 2011). The synthesis involved the coating of commercial Pt-Ru/C catalysts on carbon textile by using electrophoretic deposition or a dip-coating process. SEM analysis showed that the Pt-Ru/C catalysts with mean particle size of 3 nm were evenly distributed on the surface of the carbon fibers (Figure 12). The bimetallic Ru-Pt catalysts showed an improvement in the yield of hydrogen compared to the monometallic Ru and Pt catalysts at lower temperatures.

It is clear from the results presented that the utilization of supported bimetallic nanoparticles based mainly on Ni-modified nanoparticles have shown promising results in terms of enhanced activity, selectivity and stability of the catalyst for the aqueous reforming of polyols and the production of H_2. However, there are still important challenges and opportunities for improving the yield and the stability of the materials, and to lower the cost of the synthesised catalysts. It will be worthwhile to investigate the effects of pure alloy and core-shell structures alongside systematic variations in particle morphology, size and shape to aid the development of bimetallic materials for these processes.

5. CONCLUSION AND FUTURE PERSPECTIVES

A variety of supported monometallic, bimetallic and trimetallic nanoparticles have been explored for the valorisation of glycerol to fine chemicals and fuels. The majority of the recent supported metal nanoparticles are based on the utilisation of bimetallic nanoparticles with controlled morphology (size, shape) and in a less extent few examples have been shown the potential of using supported trimetallic nanoparticles. In many cases it is clear the advantages of using multimetallic nanoparticles due to the fact that (i) improvement in terms of activity, (ii) control of selectivity and (iii) improvement in the lifetime of the catalyst. The majority of the scientific reports have attributed the improvement due to the alteration of the electronic and geometric properties of the synthesised nanoparticles. Moreover, in many cases it has been shown that the metal-support interaction is an important parameter that affects activity, selectivity and durability of the catalyst. In the majority of the investigated reactions, it is general accepted that the synthesis of nanoparticles of small particle size (2-5 nm) seems to be the "key" for enhancing catalytic activity. However, the selectivity to the desired product is affected by the particle size, shape and oxidation state of the metal. Therefore, it is important to study these parameters in a systematic manner and also perform *in situ/operando* studies that will allow the formation of direct structure-activity relationships, and will unravel the reaction pathways. Therefore, in order to design and synthesise better and more efficient materials it is essential to increase our level of mechanistic understanding in the case of the formation of nanomaterials as well as in the case of understanding the interaction of the metal with the substrate/product. From the current literature it is evident that a significant percentage of the scientific research is focused in the development of cheaper catalysts using more abundant metals. Finally, it is evident that the synthesis of multimetallic nanoparticles is a growing and promising research area with new materials to be discovered.

REFERENCES

Abbadi, A., & vanBekkum, H. (1996). Selective chemo-catalytic routes for the preparation of beta-hydroxypyruvic acid. *Applied Catalysis A, General*, *148*(1), 113–122. doi:10.1016/S0926-860X(96)00229-3

Alonso, D. M., Wettstein, S. G., & Dumesic, J. A. (2012). Bimetallic catalysts for upgrading of biomass to fuels and chemicals. *Chemical Society Reviews*, *41*(24), 8075–8098. doi:10.1039/c2cs35188a PMID:22872312

Antolini, E., Giorgi, L., Cardellini, F., & Passalacqua, E. (2001). Physical and morphological characteristics and electrochemical behaviour in PEM fuel cells of PtRu/C catalysts. *Journal of Solid State Electrochemistry*, *5*(2), 131–140. doi:10.1007/s100080000116

Balaraju, M., Rekha, V., Sai Prasad, P. S., Prasad, R. B. N., & Lingaiah, N. (2008). Selective Hydrogenolysis of Glycerol to 1, 2 Propanediol Over Cu-ZnO Catalysts. *Catalysis Letters*, *126*(1-2), 119–124. doi:10.1007/s10562-008-9590-6

Beltran-Prieto, J. C., Kolomaznik, K., & Pecha, J. (2013). A Review of Catalytic Systems for Glycerol Oxidation: Alternatives for Waste Valorization. *Australian Journal of Chemistry*, *66*(5), 511–521. doi:10.1071/ch12514

Bharathiraja, B., Chakravarthy, M., Kumar, R. R., Yuvaraj, D., Jayamuthunagai, J., Kumar, R. P., & Palani, S. (2014). Biodiesel production using chemical and biological methods - A review of process, catalyst, acyl acceptor, source and process variables. *Renewable & Sustainable Energy Reviews*, *38*, 368–382. doi:10.1016/j.rser.2014.05.084

Bianchi, C. L., Canton, P., Dimitratos, N., Porta, F., & Prati, L. (2005). Selective oxidation of glycerol with oxygen using mono and bimetallic catalysts based on Au, Pd and Pt metals. *Catalysis Today*, *102*, 203–212. doi:10.1016/j.cattod.2005.02.003

Biella, S., Castiglioni, G. L., Fumagalli, C., Prati, L., & Rossi, M. (2002). Application of gold catalysts to selective liquid phase oxidation. *Catalysis Today*, *72*(1-2), 43–49. doi:10.1016/S0920-5861(01)00476-X

Bond, G. C., & Thompson, D. T. (1999). Catalysis by gold. *Catalysis Reviews. Science and Engineering*, *41*(3-4), 319–388. doi:10.1081/CR-100101171

Brett, G. L., He, Q., Hammond, C., Miedziak, P. J., Dimitratos, N., Sankar, M., & Hutchings, G. J. et al. (2011). Selective Oxidation of Glycerol by Highly Active Bimetallic Catalysts at Ambient Temperature under Base-Free Conditions. *Angewandte Chemie International Edition*, *50*(43), 10136–10139. doi:10.1002/anie.201101772 PMID:21990249

Carrettin, S., McMorn, P., Johnston, P., Griffin, K., & Hutchings, G. J. (2002). Selective oxidation of glycerol to glyceric acid using a gold catalyst in aqueous sodium hydroxide. *Chemical Communications*, (7): 696–697. doi:10.1039/b201112n PMID:12119680

Carrettin, S., McMorn, P., Johnston, P., Griffin, K., Kiely, C. J., & Hutchings, G. J. (2003). Oxidation of glycerol using supported Pt, Pd and Au catalysts. *Physical Chemistry Chemical Physics*, *5*(6), 1329–1336. doi:10.1039/b212047j

Chang, A. C. C., Louh, R. F., Wong, D., Tseng, J., & Lee, Y. S. (2011). Hydrogen production by aqueous-phase biomass reforming over carbon textile supported Pt-Ru bimetallic catalysts. *International Journal of Hydrogen Energy*, *36*(14), 8794–8799. doi:10.1016/j.ijhydene.2010.09.094

Chen, G., Andries, J., Spliethoff, H., Fang, M., & van de Enden, P. J. (2004). Biomass gasification integrated with pyrolysis in a circulating fluidised bed. *Solar Energy*, *76*(1-3), 345–349. doi:10.1016/j.solener.2003.08.021

Chheda, J. N., Huber, G. W., & Dumesic, J. A. (2007). Liquid-phase catalytic processing of biomass-derived oxygenated hydrocarbons to fuels and chemicals. *Angewandte Chemie International Edition*, *46*(38), 7164–7183. doi:10.1002/anie.200604274 PMID:17659519

Chheda, J. N., Roman-Leshkov, Y., & Dumesic, J. A. (2007). Production of 5-hydroxymethylfurfural and furfural by dehydration of biomass-derived mono- and poly-saccharides. *Green Chemistry*, *9*(4), 342–350. doi:10.1039/b611568c

Corma, A., Iborra, S., & Velty, A. (2007). Chemical routes for the transformation of biomass into chemicals. *Chemical Reviews*, *107*(6), 2411–2502. doi:10.1021/cr050989d PMID:17535020

Cortright, R. D., Davda, R. R., & Dumesic, J. A. (2002). Hydrogen from catalytic reforming of biomass-derived hydrocarbons in liquid water. *Nature*, *418*(6901), 964–967. doi:10.1038/nature01009 PMID:12198544

Daniel, O. M., DeLaRiva, A., Kunkes, E. L., Datye, A. K., Dumesic, J. A., & Davis, R. J. (2010). X-ray Absorption Spectroscopy of Bimetallic Pt-Re Catalysts for Hydrogenolysis of Glycerol to Propanediols. *Chemcatchem*, *2*(9), 1107–1114. doi:10.1002/cctc.201000093

Davda, R. R., Shabaker, J. W., Huber, G. W., Cortright, R. D., & Dumesic, J. A. (2003). Aqueous-phase reforming of ethylene glycol on silica-supported metal catalysts. *Applied Catalysis B: Environmental*, *43*(1), 13–26. doi:10.1016/S0926-3373(02)00277-1

Davda, R. R., Shabaker, J. W., Huber, G. W., Cortright, R. D., & Dumesic, J. A. (2005). A review of catalytic issues and process conditions for renewable hydrogen and alkanes by aqueous-phase reforming of oxygenated hydrocarbons over supported metal catalysts. *Applied Catalysis B: Environmental*, *56*(1-2), 171–186. doi:10.1016/j.apcatb.2004.04.027

Della Pina, C., Falletta, E., Prati, L., & Rossi, M. (2008). Selective oxidation using gold. *Chemical Society Reviews*, *37*(9), 2077–2095. doi:10.1039/b707319b PMID:18762847

Della Pina, C., Falletta, E., & Rossi, M. (2012). Update on selective oxidation using gold. *Chemical Society Reviews*, *41*(1), 350–369. doi:10.1039/C1CS15089H PMID:21727977

Demirel, S., Kern, P., Lucas, M., & Claus, P. (2007). Oxidation of mono- and polyalcohols with gold: Comparison of carbon and ceria supported catalysts. *Catalysis Today*, *122*(3-4), 292–300. doi:10.1016/j.cattod.2006.12.002

Demirel-Gulen, S., Lucas, M., & Claus, P. (2005). Liquid phase oxidation of glycerol over carbon supported gold catalysts. *Catalysis Today*, *102*, 166–172. doi:10.1016/j.cattod.2005.02.033

Dimitratos, N., Lopez-Sanchez, J. A., Anthonykutty, J. M., Brett, G., Carley, A. F., Tiruvalam, R. C., & Hutchings, G. J. et al. (2009). Oxidation of glycerol using gold-palladium alloy-supported nanocrystals. *Physical Chemistry Chemical Physics*, *11*(25), 4952–4961. doi:10.1039/b904317a PMID:19562125

Dimitratos, N., Lopez-Sanchez, J. A., & Hutchings, G. J. (2009). Green Catalysis with Alternative Feedstocks. *Topics in Catalysis*, *52*(3), 258–268. doi:10.1007/s11244-008-9162-4

Dimitratos, N., Lopez-Sanchez, J. A., Lennon, D., Porta, F., Prati, L., & Villa, A. (2006). Effect of particle size on monometallic and bimetallic (Au,Pd)/C on the liquid phase oxidation of glycerol. *Catalysis Letters*, *108*(3-4), 147–153. doi:10.1007/s10562-006-0036-8

Dimitratos, N., Lopez-Sanchez, J. A., Meenakshisundaram, S., Anthonykutty, J. M., Brett, G., Carley, A. F., & Hutchings, G. J. et al. (2009). Selective formation of lactate by oxidation of 1,2-propanediol using gold palladium alloy supported nanocrystals. *Green Chemistry*, *11*(8), 1209–1216. doi:10.1039/b823285g

Dimitratos, N., Messi, C., Porta, F., Prati, L., & Villa, A. (2006). Investigation on the behaviour of Pt(0)/carbon and Pt(0),Au(0)/carbon catalysts employed in the oxidation of glycerol with molecular oxygen in water. *Journal of Molecular Catalysis a-Chemical, 256*(1-2), 21-28. doi:10.1016/j.molcata.2006.04.019

Dimitratos, N., Porta, F., & Prati, L. (2005). Au, Pd (mono and bimetallic) catalysts supported on graphite using the immobilisation method - Synthesis and catalytic testing for liquid phase oxidation of glycerol. *Applied Catalysis A, General*, *291*(1-2), 210–214. doi:10.1016/j.apcata.2005.01.044

Dimitratos, N., Villa, A., Wang, D., Porta, F., Su, D., & Prati, L. (2006). Pd and Pt catalysts modified by alloying with Au in the selective oxidation of alcohols. *Journal of Catalysis*, *244*(1), 113–121. doi:10.1016/j.jcat.2006.08.019

Edwards, J. K., Ntainjua, E., Carley, A. F., Herzing, A. A., Kiely, C. J., & Hutchings, G. J. (2009). Direct Synthesis of H2O2 from H-2 and O-2 over Gold, Palladium, and Gold-Palladium Catalysts Supported on Acid-Pretreated TiO2. *Angewandte Chemie International Edition*, *48*(45), 8512–8515. doi:10.1002/anie.200904115 PMID:19802864

Edwards, J. K., Solsona, B., N, E. N., Carley, A. F., Herzing, A. A., Kiely, C. J., & Hutchings, G. J. (2009). Switching Off Hydrogen Peroxide Hydrogenation in the Direct Synthesis Process. *Science*, *323*(5917), 1037–1041. doi:10.1126/science.1168980 PMID:19229032

Edwards, J. K., Solsona, B. E., Landon, P., Carley, A. F., Herzing, A., Kiely, C. J., & Hutchings, G. J. (2005). Direct synthesis of hydrogen peroxide from H-2 and O-2 using TiO2-supported Au-Pd catalysts. *Journal of Catalysis*, *236*(1), 69–79. doi:10.1016/j.jcat.2005.09.015

Enache, D. I., Edwards, J. K., Landon, P., Solsona-Espriu, B., Carley, A. F., Herzing, A. A., & Hutchings, G. J. et al. (2006). Solvent-free oxidation of primary alcohols to aldehydes using Au-Pd/TiO2 catalysts. *Science*, *311*(5759), 362–365. doi:10.1126/science.1120560 PMID:16424335

Garcia, R., Besson, M., & Gallezot, P. (1995). Chemoselective catalytic-oxidation of glycerol with air on platinum metals. *Applied Catalysis A, General*, *127*(1-2), 165–176. doi:10.1016/0926-860X(95)00048-8

Hashmi, A. S. K., & Hutchings, G. J. (2006). Gold catalysis. *Angewandte Chemie International Edition*, *45*(47), 7896–7936. doi:10.1002/anie.200602454 PMID:17131371

Hou, W. B., Dehm, N. A., & Scott, R. W. J. (2008). Alcohol oxidations in aqueous solutions using Au, Pd, and bimetallic AuPd nanoparticle catalysts. *Journal of Catalysis*, *253*(1), 22–27. doi:10.1016/j.jcat.2007.10.025

Huang, Z. W., Cui, F., Kang, H. X., Chen, J., Zhang, X. Z., & Xia, C. G. (2008). Highly dispersed silica-supported copper nanoparticles prepared by precipitation-gel method: A simple but efficient and stable catalyst for glycerol hydrogenolysis. *Chemistry of Materials*, *20*(15), 5090–5099. doi:10.1021/cm8006233

Huber, G. W., & Dumesic, J. A. (2006). An overview of aqueous-phase catalytic processes for production of hydrogen and alkanes in a biorefinery. *Catalysis Today*, *111*(1-2), 119–132. doi:10.1016/j.cattod.2005.10.010

Huber, G. W., Iborra, S., & Corma, A. (2006). Synthesis of transportation fuels from biomass: Chemistry, catalysts, and engineering. *Chemical Reviews*, *106*(9), 4044–4098. doi:10.1021/cr068360d PMID:16967928

Huber, G. W., Shabaker, J. W., & Dumesic, J. A. (2003). Raney Ni-Sn catalyst for H-2 production from biomass-derived hydrocarbons. *Science*, *300*(5628), 2075–2077. doi:10.1126/science.1085597 PMID:12829777

Huber, G. W., Shabaker, J. W., Evans, S. T., & Dumesic, J. A. (2006). Aqueous-phase reforming of ethylene glycol over supported Pt and Pd bimetallic catalysts. *Applied Catalysis B: Environmental*, *62*(3-4), 226–235. doi:10.1016/j.apcatb.2005.07.010

Ide, M. S., & Davis, R. J. (2014). The Important Role of Hydroxyl on Oxidation Catalysis by Gold Nanoparticles. *Accounts of Chemical Research*, *47*(3), 825–833. doi:10.1021/ar4001907 PMID:24261465

Jerome, F., Pouilloux, Y., & Barrault, J. (2008). Rational Design of Solid Catalysts for the Selective Use of Glycerol as a Natural Organic Building Block. *ChemSusChem*, *1*(7), 586–613. doi:10.1002/cssc.200800069 PMID:18702160

Jiang, T., Zhou, Y. X., Liang, S. G., Liu, H. Z., & Han, B. X. (2009). Hydrogenolysis of glycerol catalyzed by Ru-Cu bimetallic catalysts supported on clay with the aid of ionic liquids. *Green Chemistry*, *11*(7), 1000–1006. doi:10.1039/b901425j

Ketchie, W. C., Fang, Y. L., Wong, M. S., Murayama, M., & Davis, R. J. (2007). Influence of gold particle size on the aqueous-phase oxidation of carbon monoxide and glycerol. *Journal of Catalysis*, *250*(1), 94–101. doi:10.1016/j.jcat.2007.06.001

Ketchie, W. C., Murayama, M., & Davis, R. J. (2007). Selective oxidation of glycerol over carbon-supported AuPd catalysts. *Journal of Catalysis*, *250*(2), 264–273. doi:10.1016/j.jcat.2007.06.011

Kimura, H. (1993). Selective oxidation of glycerol on a platinum-bismuth catalyst by using a fixed-bed reactor. *Applied Catalysis A, General*, *105*(2), 147–158. doi:10.1016/0926-860X(93)80245-L

Kimura, H., Tsuto, K., Wakisawa, T., Kazumi, Y., & Inaya, Y. (1995). Selective oxidation of glycerol on a platinum bismuth catalyst. *Applied Catalysis A, General*, *123*(2), 323–323. doi:10.1016/0926-860X(95)00008-9

Kumar, V. V., & Anthony, S. P. (2014). AuNP based selective colorimetric sensor for cysteine at a wide pH range: Investigation of capping molecule structure on the colorimetric sensing and catalytic properties. *RSC Advances*, *4*(35), 18467–18472. doi:10.1039/c4ra00345d

Li, B. D., Wang, J., Yuan, Y. Z., Ariga, H., Takakusagi, S., & Asakura, K. (2011). Carbon Nanotube-Supported RuFe Bimetallic Nanoparticles as Efficient and Robust Catalysts for Aqueous-Phase Selective Hydrogenolysis of Glycerol to Glycols. *Acs Catalysis*, *1*(11), 1521–1528. doi:10.1021/cs200386q

Liang, D., Gao, J., Sun, H., Chen, P., Hou, Z. Y., & Zheng, X. M. (2011). Selective oxidation of glycerol with oxygen in a base-free aqueous solution over MWNTs supported Pt catalysts. *Applied Catalysis B: Environmental*, *106*(3-4), 423–432. doi:10.1016/j.apcatb.2011.05.050

Liang, D., Gao, J., Wang, J. H., Chen, P., Hou, Z. Y., & Zheng, X. M. (2009). Selective oxidation of glycerol in a base-free aqueous solution over different sized Pt catalysts. *Catalysis Communications*, *10*(12), 1586–1590. doi:10.1016/j.catcom.2009.04.023

Lin, S. Y., Harada, M., Suzuki, Y., & Hatano, H. (2002). Hydrogen production from coal by separating carbon dioxide during gasification. *Fuel*, *81*(16), 2079–2085. doi:10.1016/S0016-2361(02)00187-4

Liu, H. Z., Liang, S. G., Jiang, T., Han, B. X., & Zhou, Y. X. (2012). Hydrogenolysis of Glycerol to 1,2-Propanediol over Ru-Cu Bimetals Supported on Different Supports. *Clean-Soil Air Water*, *40*(3), 318–324. doi:10.1002/clen.201000227

Lopez-Sanchez, J. A., Dimitratos, N., Glanville, N., Kesavan, L., Hammond, C., Edwards, J. K., & Hutchings, G. J. et al. (2011). Reactivity studies of Au-Pd supported nanoparticles for catalytic applications. *Applied Catalysis A, General*, *391*(1-2), 400–406. doi:10.1016/j.apcata.2010.05.010

Lopez-Sanchez, J. A., Dimitratos, N., Hammond, C., Brett, G. L., Kesavan, L., White, S., & Hutchings, G. J. et al. (2011). Facile removal of stabilizer-ligands from supported gold nanoparticles. *Nature Chemistry*, *3*(7), 551–556. doi:10.1038/nchem.1066 PMID:21697877

Lopez-Sanchez, J. A., Dimitratos, N., Miedziak, P., Ntainjua, E., Edwards, J. K., Morgan, D., & Hutchings, G. J. et al. (2008). Au-Pd supported nanocrystals prepared by a sol immobilisation technique as catalysts for selective chemical synthesis. *Physical Chemistry Chemical Physics*, *10*(14), 1921–1930. doi:10.1039/b719345a PMID:18368185

Luque, R., Budarin, V., Clark, J. H., & Macquarrie, D. J. (2008). Glycerol transformations on polysaccharide derived mesoporous materials. *Applied Catalysis B: Environmental*, *82*(3-4), 157–162. doi:10.1016/j.apcatb.2008.01.015

Ma, L., & He, D. H. (2009). Hydrogenolysis of glycerol to propanediols over highly active Ru-Re bimetallic catalysts. *Topics in Catalysis*, *52*(6-7), 834–844. doi:10.1007/s11244-009-9231-3

Ma, L., & He, D. H. (2010). Influence of catalyst pretreatment on catalytic properties and performances of Ru-Re/SiO(2) in glycerol hydrogenolysis to propanediols. *Catalysis Today*, *149*(1-2), 148–156. doi:10.1016/j.cattod.2009.03.015

Ma, L., He, D. H., & Li, Z. P. (2008). Promoting effect of rhenium on catalytic performance of Ru catalysts in hydrogenolysis of glycerol to propanediol. *Catalysis Communications*, *9*(15), 2489–2495. doi:10.1016/j.catcom.2008.07.009

Ma, L., Li, Y. M., & He, D. H. (2011). Glycerol Hydrogenolysis to Propanediols over Ru-Re/SiO2: Acidity of Catalyst and Role of Re. *Chinese Journal of Catalysis*, *32*(5), 872–876. doi:10.1016/S1872-2067(10)60198-2

Maris, E. P., Ketchie, W. C., Murayama, M., & Davis, R. J. (2007). Glycerol hydrogenolysis on carbon-supported PtRu and AuRu bimetallic catalysts. *Journal of Catalysis*, *251*(2), 281–294. doi:10.1016/j.jcat.2007.08.007

Nakagawa, Y., & Tomishige, K. (2010). Total hydrogenation of furan derivatives over silica-supported Ni-Pd alloy catalyst. *Catalysis Communications*, *12*(3), 154–156. doi:10.1016/j.catcom.2010.09.003

Nakagawa, Y., & Tomishige, K. (2011). Heterogeneous catalysis of the glycerol hydrogenolysis. *Catalysis Science & Technology*, *1*(2), 179–190. doi:10.1039/c0cy00054j

Nishiyama, S., Kubota, T., Kimura, K., Tsuruya, S., & Masai, M. (1997). Unique hydrogenation activity of supported tin catalyst: Selective hydrogenation catalyst for unsaturated aldehydes. *Journal of Molecular Catalysis a-Chemical, 120*(1-3), L17-L22. doi:10.1016/s1381-1169(97)00049-6

Pagliaro, M., Ciriminna, R., Kimura, H., Rossi, M., & Della Pina, C. (2007). From glycerol to value-added products. *Angewandte Chemie International Edition, 46*(24), 4434–4440. doi:10.1002/anie.200604694 PMID:17471485

Perosa, A., & Tundo, P. (2005). Selective hydrogenolysis of glycerol with Raney nickel. *Industrial & Engineering Chemistry Research, 44*(23), 8535–8537. doi:10.1021/ie0489251

Poirier, M. G., & Sapundzhiev, C. (1997). Catalytic decomposition of natural gas to hydrogen for fuel cell applications. *International Journal of Hydrogen Energy, 22*(4), 429–433. doi:10.1016/S0360-3199(96)00101-2

Porta, F., & Prati, L. (2004). Selective oxidation of glycerol to sodium glycerate with gold-on-carbon catalyst: An insight into reaction selectivity. *Journal of Catalysis, 224*(2), 397–403. doi:10.1016/j.jcat.2004.03.009

Prati, L., & Rossi, M. (1998). Gold on carbon as a new catalyst for selective liquid phase oxidation of diols. *Journal of Catalysis, 176*(2), 552–560. doi:10.1006/jcat.1998.2078

Prati, L., & Villa, A. (2014). Gold Colloids: From Quasi-Homogeneous to Heterogeneous Catalytic Systems. *Accounts of Chemical Research, 47*(3), 855–863. doi:10.1021/ar400170j PMID:24266851

Prati, L., Villa, A., Chan-Thaw, C. E., Arrigo, R., Wang, D., & Su, D. S. (2011). Gold catalyzed liquid phase oxidation of alcohol: The issue of selectivity. *Faraday Discussions, 152*, 353–365. doi:10.1039/c1fd00016k PMID:22455055

Prati, L., Villa, A., Porta, F., Wang, D., & Su, D. S. (2007). Single-phase gold/palladium catalyst: The nature of synergistic effect. *Catalysis Today, 122*(3-4), 386–390. doi:10.1016/j.cattod.2006.11.003

Rahmat, N., Abdullah, A. Z., & Mohamed, A. R. (2010). Recent progress on innovative and potential technologies for glycerol transformation into fuel additives: A critical review. *Renewable & Sustainable Energy Reviews, 14*(3), 987–1000. doi:10.1016/j.rser.2009.11.010

Rampino, L. D., Kavanagh, K. E., & Nord, F. F. (1943). Relationship between particle size and efficiency of palladium polyvinyl alcohol (Pd PVA) catalysts. *Proceedings of the National Academy of Sciences of the United States of America, 29*(8), 246–256. doi:10.1073/pnas.29.8.246 PMID:16578085

Rossi, L. M., Vono, L. L. R., Garcia, M. A. S., Faria, T. L. T., & Lopez-Sanchez, J. A. (2013). Screening of Soluble Rhodium Nanoparticles as Precursor for Highly Active Hydrogenation Catalysts: The Effect of the Stabilizing Agents. *Topics in Catalysis, 56*(13-14), 1228–1238. doi:10.1007/s11244-013-0089-z

Rostrup-Nielsen, J. R. (2001). Conversion of hydrocarbons and alcohols for fuel cells. *Physical Chemistry Chemical Physics, 3*(3), 283–288. doi:10.1039/b004660o

Ruppert, A. M., Weinberg, K., & Palkovits, R. (2012). Hydrogenolysis Goes Bio: From Carbohydrates and Sugar Alcohols to Platform Chemicals. *Angewandte Chemie International Edition, 51*(11), 2564–2601. doi:10.1002/anie.201105125 PMID:22374680

Shabaker, J. W., Huber, G. W., & Dumesic, J. A. (2004). Aqueous-phase reforming of oxygenated hydrocarbons over Sn-modified Ni catalysts. *Journal of Catalysis*, *222*(1), 180–191. doi:10.1016/j.jcat.2003.10.022

Shabaker, J. W., Simonetti, D. A., Cortright, R. D., & Dumesic, J. A. (2005). Sn-modified Ni catalysts for aqueous-phase reforming: Characterization and deactivation studies. *Journal of Catalysis*, *231*(1), 67–76. doi:10.1016/j.jcat.2005.01.019

Sheldon, R. A. (2014). Green and sustainable manufacture of chemicals from biomass: State of the art. *Green Chemistry*, *16*(3), 950–963. doi:10.1039/C3GC41935E

Shen, J., Ziaei-Azad, H., & Semagina, N. (2014). Is it always necessary to remove a metal nanoparticle stabilizer before catalysis? *Journal of Molecular Catalysis a-Chemical, 391*, 36-40. doi:10.1016/j.molcata.2014.03.027

Shinmi, Y., Koso, S., Kubota, T., Nakagawa, Y., & Tomishige, K. (2010). Modification of Rh/SiO2 catalyst for the hydrogenolysis of glycerol in water. *Applied Catalysis B: Environmental*, *94*(3-4), 318–326. doi:10.1016/j.apcatb.2009.11.021

Solymosi, F., Cserenyi, J., Szoke, A., Bansagi, T., & Oszko, A. (1997). Aromatization of methane over supported and unsupported Mo-based catalysts. *Journal of Catalysis*, *165*(2), 150–161. doi:10.1006/jcat.1997.1478

Steinberg, M. (1998). Production of hydrogen and methanol from natural gas with reduced CO2 emission. *International Journal of Hydrogen Energy*, *23*(6), 419–425. doi:10.1016/S0360-3199(97)00092-X

Taarning, E., Madsen, A. T., Marchetti, J. M., Egeblad, K., & Christensen, C. H. (2008). Oxidation of glycerol and propanediols in methanol over heterogeneous gold catalysts. *Green Chemistry*, *10*(4), 408–414. doi:10.1039/b714292g

Tanksale, A., Beltramini, J. N., Dumesic, J. A., & Lu, G. Q. (2008). Effect of Pt and Pd promoter on Ni supported catalysts - A TPR/TPO/TPD and microcalorimetry study. *Journal of Catalysis*, *258*(2), 366–377. doi:10.1016/j.jcat.2008.06.024

Tanksale, A., Beltramini, J. N., & Lu, G. M. (2010). A review of catalytic hydrogen production processes from biomass. *Renewable & Sustainable Energy Reviews*, *14*(1), 166–182. doi:10.1016/j.rser.2009.08.010

Taralas, G., & Kontominas, M. G. (2006). Pyrolysis of solid residues commencing from the olive oil food industry for potential hydrogen production. *Journal of Analytical and Applied Pyrolysis*, *76*(1-2), 109–116. doi:10.1016/j.jaap.2005.08.004

ten Dam, J., & Hanefeld, U. (2011). Renewable Chemicals: Dehydroxylation of Glycerol and Polyols. *ChemSusChem*, *4*(8), 1017–1034. doi:10.1002/cssc.201100162 PMID:21887771

Vasiliadou, E. S., & Lemonidou, A. A. (2011). Investigating the performance and deactivation behaviour of silica-supported copper catalysts in glycerol hydrogenolysis. *Applied Catalysis A, General*, *396*(1-2), 177–185. doi:10.1016/j.apcata.2011.02.014

Vennestr, øm, P. N. R., Osmundsen, C. M., Christensen, C. H., & Taarning, E. (2011). Beyond Petrochemicals: The Renewable Chemicals Industry. *Angewandte Chemie (International ed. in English), 50*(45), 10502–10509. doi:10.1002/anie.201102117 PMID:22045621

Villa, A., Campione, C., & Prati, L. (2007). Bimetallic gold/palladium catalysts for the selective liquid phase oxidation of glycerol. *Catalysis Letters, 115*(3-4), 133–136. doi:10.1007/s10562-007-9077-x

Villa, A., Gaiassi, A., Rossetti, I., Bianchi, C. L., van Benthem, K., Veith, G. M., & Prati, L. (2010). Au on MgAl2O4 spinels: The effect of support surface properties in glycerol oxidation. *Journal of Catalysis, 275*(1), 108–116. doi:10.1016/j.jcat.2010.07.022

Villa, A., Veith, G. M., Ferri, D., Weidenkaff, A., Perry, K. A., Campisi, S., & Prati, L. (2013). NiO as a peculiar support for metal nanoparticles in polyols oxidation. *Catalysis Science & Technology, 3*(2), 394–399. doi:10.1039/C2CY20370G

Villa, A., Veith, G. M., & Prati, L. (2010). Selective Oxidation of Glycerol under Acidic Conditions Using Gold Catalysts. *Angewandte Chemie (International ed. in English), 49*(26), 4499–4502. doi:10.1002/anie.201000762 PMID:20461742

Villa, A., Wang, D., Veith, G. M., & Prati, L. (2012). Bismuth as a modifier of Au-Pd catalyst: Enhancing selectivity in alcohol oxidation by suppressing parallel reaction. *Journal of Catalysis, 292*, 73–80. doi:10.1016/j.jcat.2012.04.021

Villa, A., Wang, D., Veith, G. M., Vindigni, F., & Prati, L. (2013). Sol immobilization technique: A delicate balance between activity, selectivity and stability of gold catalysts. *Catalysis Science & Technology, 3*(11), 3036–3041. doi:10.1039/c3cy00260h

Wang, D., Villa, A., Porta, F., Prati, L., & Su, D. S. (2008). Bimetallic gold/palladium catalysts: Correlation between nanostructure and synergistic effects. *The Journal of Physical Chemistry C, 112*(23), 8617–8622. doi:10.1021/jp800805e

Wang, D., Villa, A., Porta, F., Su, D. S., & Prati, L. (2006). Single-phase bimetallic system for the selective oxidation of glycerol to glycerate. *Chemical Communications*, (18): 1956–1958. doi:10.1039/b518069d PMID:16767248

Wang, D., Villa, A., Su, D. S., Prati, L., & Schlogl, R. (2013). Carbon-Supported Gold Nanocatalysts: Shape Effect in the Selective Glycerol Oxidation. *Chemcatchem, 5*(9), 2717–2723. doi:10.1002/cctc.201200535

Wu, J. B., Gross, A., & Yang, H. (2011). Shape and Composition-Controlled Platinum Alloy Nanocrystals Using Carbon Monoxide as Reducing Agent. *Nano Letters, 11*(2), 798–802. doi:10.1021/nl104094p PMID:21204581

Wu, Z. J., Mao, Y. Z., Wang, X. X., & Zhang, M. H. (2011). Preparation of a Cu-Ru/carbon nanotube catalyst for hydrogenolysis of glycerol to 1,2-propanediol via hydrogen spillover. *Green Chemistry, 13*(5), 1311–1316. doi:10.1039/c0gc00809e

Xie, F. Z., Chu, X. W., Hu, H. R., Qiao, M. H., Yan, S. R., Zhu, Y. L., & Zhang, X. X. et al. (2006). Characterization and catalytic properties of Sn-modified rapidly quenched skeletal Ni catalysts in aqueous-phase reforming of ethylene glycol. *Journal of Catalysis, 241*(1), 211–220. doi:10.1016/j.jcat.2006.05.001

Xu, Y. D., Bao, X. H., & Lin, L. W. (2003). Direct conversion of methane under nonoxidative conditions. *Journal of Catalysis*, *216*(1-2), 386–395. doi:10.1016/S0021-9517(02)00124-0

Zhang, G. R., & Xu, B. Q. (2010). Surprisingly strong effect of stabilizer on the properties of Au nanoparticles and Pt boolean AND Au nanostructures in electrocatalysis. *Nanoscale*, *2*(12), 2798–2804. doi:10.1039/c0nr00295j PMID:20938521

Zhou, C. H. C., Beltramini, J. N., Fan, Y. X., & Lu, G. Q. M. (2008). Chemoselective catalytic conversion of glycerol as a biorenewable source to valuable commodity chemicals. *Chemical Society Reviews*, *37*(3), 527–549. doi:10.1039/B707343G PMID:18224262

Zope, B. N., Hibbitts, D. D., Neurock, M., & Davis, R. J. (2010). Reactivity of the Gold/Water Interface During Selective Oxidation Catalysis. *Science*, *330*(6000), 74–78. doi:10.1126/science.1195055 PMID:20929807

Chapter 14
Dual Role of Perovskite Hollow Fiber Membrane in the Methane Oxidation Reactions

Serbia M. Rodulfo-Baechler
University of Newcastle, UK

ABSTRACT

The Mixed Ionic and Electronic Conducting (MIEC) membrane reactors are of interest because they have the potential to produce high purity oxygen from air at lower costs and provide a continuous oxygen supply to reactions or/and industrial processes. The study of the dual role oxygen flux and catalytic performance of the unmodified and Ni-coated La0.6Sr0.4Co0.2Fe0.8O3-δ hollow fibre membranes (LSCF6428 HFM) in the methane oxidation reactions (i.e., partial oxidation of methane and methane combustion) by using air on lumen side and methane on shell side are shown in this chapter. The LSCF6428 HFM participates not only in the oxygen flux but also in the methane conversion to C2. A Ni-coated LSCF6428 HFM under lean O2/CH4 gradient (i.e., 0.5) showed the production of syngas, carbon dioxide and C2 products in agreement with the thermodynamic calculation. At rich O2/CH4 gradient (i.e., 1.0), the formation of carbon dioxide was facilitated. The main catalytic pathway at lean O2/CH4 gradient and H2 reduction treatment was partial oxidation of methane to C2 and syngas.

INTRODUCTION

The partial oxidation of methane (POM) to synthesis gas or syngas (CO/H_2) has attracted much attention in recent years. Syngas is a versatile feedstock for many industrial synthetic processes and can be easily transformed to methanol or hydrocarbons via the Fischer-Tröpsch reaction (Tang et al., 2007). These catalytic routes could transform natural gas reserves, whose principal component is methane, into more valuable chemicals and fuels. However, in order to apply the POM process at industrial level, cost effective methods for the supply of pure oxygen are required (Kondratenko et al. 2009; Ten Elshof, Bouwmeester & Verweij, 1995; Xu & Thomson, 1997). The current method for obtaining pure oxygen on a large scale is by cryogenic air separation (fractional distillation of liquefied air), PSA (pressure swing

DOI: 10.4018/978-1-4666-9975-5.ch014

Figure 1. Reaction mechanisms for partial oxidation of methane through MIEC membrane. Adapted from (Bouwmeester, 2003; Hamakawa, et al., 2006; Thursfield & Metcalfe, 2006)

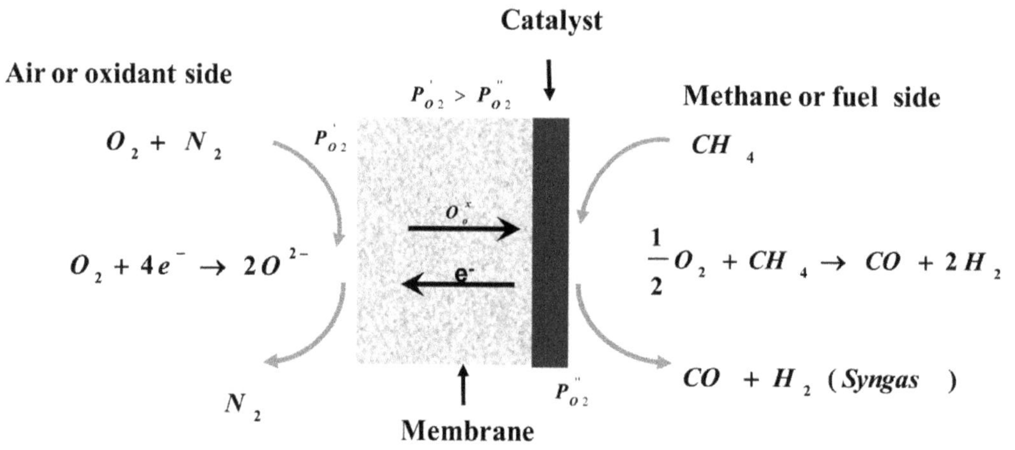

adsorption) and VSA (vacuum swing adsorption). The first process, cryogenic air separation, utilizes low temperatures and elevated pressures that make it very expensive (Baker, 2002), because it is energy intensive, although it produces 99.5% pure oxygen. The other processes, PSA and VSA, require less energy but produce lower purity oxygen (90-93%) (Bartholomew & Farrauto, 2006).

An alternative and more energy efficient method is the use of mixed ionic and electronic conducting (MIEC) metal oxide ceramic membranes (Tan & Li, 2006). The process is very promising because this enhances the performance of methane conversion processes due to simultaneous production of pure oxygen and syngas by combining oxygen separation and high temperature catalytic partial oxidation into a single unit, simplifying the process and reducing the capital costs (Bouwmeester, 2003; Caro et al., 2006; Hamakawa et al., 2006; Jin et al., 2000; Lui et al., 2008; Schiestel et al., 2005; Tan, Liu & Li, 2005; Thursfield & Metcalfe, 2007).

Figure 1 presents the simultaneous production of pure oxygen and syngas in MIEC membrane. At high temperature (800 to 1000) °C, in air or oxidant side, gaseous oxygen is adsorbed, dissociated and reduced to lattice oxygen .. on the membrane surface and diffuse through the combination of mobile oxygen vacancies and electronic defects (holes) since the oxygen partial pressure on the air side is lower than that on the methane side.

The driving for is the oxygen partial pressure difference across the membrane. On the fuel or methane side, the permeated oxygen is consumed by methane (CH_4) for their partial oxidation to syngas $(CO \text{ and } H_2)$ or the complete oxidation to carbon dioxide and water. The MIEC membrane allow separation of oxygen from air supply at elevated temperature (>700 °C). By combining air separation and catalytic partial oxidation of methane to syngas into membrane reactor, it is expected that this technology will reduces the capital costs of natural gas to liquid (GTL) (Bouwmeester, 2003) and improves the selectivity and yields of the products (Au & Wang, 1997).

The overall aim of this chapter is to illustrate the potential application of LSCF6428 hollow fibre membranes in the methane oxidation reactions (i.e., partial oxidation and combustion), in which the LSCF6428 HFM can be used not only for oxygen flux but also catalyzing the methane oxidation reactions. This dual functionality can be tuned by the deposition of nickel through 5%Ni-LSCF6428 onto

LSCF6428 HFM because of its significant influence on the methane conversion. Also comparison can be made of the apparent overall rate constant for both reactors under different gradients (i.e., Air/He and Air/CH_4) at 850 °C. In order to gain insight into the dual performance of the HFM-based system, characterization of the samples obtained from both reactors (i.e., LSCF6428HFM and 5%Ni-LSCF6428/LSCF6428-HFM) after methane oxidation has been done. Various techniques were used such as X-ray powder diffraction (XRPD), scanning electron microscopy (SEM), energy-dispersive x-rays spectroscopy (EDXS) and hydrogen temperature-programmed reduction (H_2-TPR).

LITERATURE REVIEW

Mechanism for Partial Oxidation of Methane to Syngas through Mixed Ionic-Electronic Conductor Membrane (MIEC)

Sierra *et al.* proposed a mechanism for the partial oxidation of methane over $LaNiO_{3-\delta}$ perovskite through pulse studies (Sierra Gallego, Batiot-Dupeyrat & Mondragón, 2010). They concluded that the product distribution of the POM reaction is dependent upon the concentration of surface oxygen species (i.e. O_2^-, O^- and $O_{2(ads)}$). Initially, the O_2 is incorporated into the lattice through different intermediate oxygen species, which are represented in eqns. (1) where O^{2-} corresponds to the lattice oxygen, O_2^- is superoxide, O_2^{2-} is peroxide, O^- is oxido, $h^\bullet$ electron hole and e' is an electron (Gellings and Bouwmeester, 2000; Shao et al., 2001; Tan et al., 2005; Van Hassel et al., 1993).

$$O_2(g) \leftrightarrow O_2(ads) \xleftrightarrow{+e'} O_2^- + h^\bullet \xleftrightarrow{+e'} O_2^{2-} + h^\bullet \leftrightarrow 2O^- \xleftrightarrow{+e'} 2O_{lattice}^{2-} + h^\bullet \tag{1}$$

Sierra Gallegos and co-workers did not distinguish the different surface oxygen species and used the Kröger-Vink notation O^* to represent the oxygen species, O_O^x a regular lattice oxygen ion and $V_O^{\bullet\bullet}$ denotes an oxygen vacancy.

At high concentration of reactive lattice oxygen, CH_4 is completely oxidized to form CO_2 and H_2O.

$$CH_4 + 4O_O^x \rightarrow CO_2 + 2H_2O + 4V_O^{\bullet\bullet} \tag{2}$$

or

$$CH_4 + 4O^* \rightarrow CO_2 + 2H_2O \tag{3}$$

At low concentration of lattice oxygen, on the other hand, It is favored the partial oxidation of methane as can be observed using the reaction mechanism given below. Note that the subscripts s and b refer to surface and bulk, respectively.

$$CH_4 \rightarrow CH_{4-x} + xH_s \tag{4}$$

$$CH_{4-x} \rightarrow C_S + (4-x) H_S \tag{5}$$

$$H_S + H_S \rightarrow H_2 \rightarrow H_{2,g} \tag{6}$$

$$C_S + O^* \rightarrow CO_S \tag{7}$$

$$CO_S \rightarrow CO_g \tag{8}$$

$$CO + O^* \rightarrow CO_{2,S} \tag{9}$$

$$CO_{2,S} \rightarrow CO_{2,g} \tag{10}$$

$$2H_S + O^* \rightarrow H_2O_S \tag{11}$$

$$H_2O_S \rightarrow H_2O_g \tag{12}$$

The CH_4 can be also be dissociated by the O^* according to the following equations (Sierra Gallego, 2010):

$$CH_{4,S} + O^* \rightarrow CH_{3,S} + OH_S \tag{13}$$

$$CH_{3,S} + 3O^* \rightarrow C_S + 3OH_S \tag{14}$$

In this step, the absorbed H atom may combine to form hydrogen which easily desorbs. The chemisorbed carbon species are either oxidized to CO or CO_2 by lattice oxygen.

Finally, the re-oxidation process can be written as a function of an incorporation and migration process:

$$O^- + V_{O,S}^{\bullet\bullet} + e' \leftrightarrow O_{O,S}^x \text{ (Incorporation)} \tag{15}$$

Figure 2. Scheme of surface mechanism for methane conversion through MIEC membrane. Adapted from (Dittmeyer & Caro, 2008)

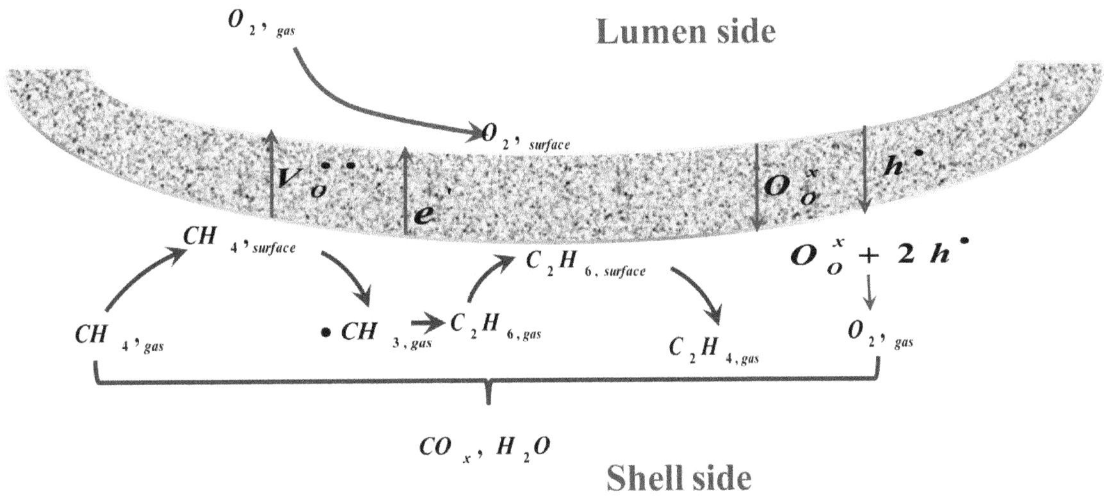

$$O^x_{O,S} + V^{\bullet\bullet}_{O,b} \leftrightarrow O^x_{O,b} + V^{\bullet\bullet}_{O,S} \text{ (Migration)} \tag{16}$$

Dittmeyer and Caro proposed a mechanism for oxidative coupling of methane to C_2 hydrocarbons (ethylene and ethane) using different catalysts on MIEC perovskite membrane at 700-900°C, which is displayed in Figure 2, (Dittmeyer & Caro, 2008).

Initially, the reaction of methane with the lattice oxygen and electron holes produces $\bullet CH_3$ radical, water and an oxygen vacancy:

$$CH_4 + O^x_O + 2h^\bullet \leftrightarrow 2\left(\bullet CH_3\right)_{ads} + H_2O + V^{\bullet\bullet}_O \tag{17}$$

Two $\bullet CH_3$ radicals form ethane which can react again with lattice oxygen and electron holes giving, ethylene, water and oxygen vacancy.

$$2\left(\bullet CH_3\right)_{ads} \rightarrow C_2H_6 \tag{18}$$

$$C_2H_6 + O^x_O + 2h^\bullet \rightarrow C_2H_4 + H_2O + V^{\bullet\bullet}_O \tag{19}$$

Deeper oxidation of C_1 and C_2 to CO_x can be achieved when the lattice oxygen is not completely consumed:

$$2O^x_O + 4h^\bullet \rightarrow O_{s,g} + 2V^{\bullet\bullet}_O \tag{20}$$

Therefore, the methane conversion is increased but the C_2 selectivity decreases when the oxygen transported across the membrane is too high.

Stability of Mixed Ionic Electronic Conductor Membranes

A review of the literature related to the stability of the $La_{0.6}Sr_{0.4}Co_{0.2}Fe_{0.8}O_{3-\delta}$ (LSCF6428) membrane and other perovskite membranes is given below.

The mechanical properties and structural stability of mixed ionic electronic conductor membrane have been studied by Lee and co-workers (Lee et al., 2006). They considered the mechanical properties as relevant factors for applications of ion conducting oxides since stresses chemically induced can produce component degradation. The reduction of the membrane thickness can also improve the oxygen permeation performance. The mechanical properties such as fracture strength and toughness of the membrane could be explained with micro structural features and relative density.

Fundamental mechanical properties of LSCF6428 and $La_{0.6}Sr_{0.3}Ga_{0.6}Fe_{0.4}O_{3-\delta}$ (LSGF6364) membranes were evaluated. The membrane failures during the permeation experiment were attributed to the following factors:

1. Dimensional discord, by the difference in non-stoichiometry between the surfaces of the membrane.
2. Surface weakening or phase decomposition due to the instability of the perovskite structure in a severe reducing environment for a long time.
3. The mismatch in thermal expansion between the membrane and the sustaining tube. It should be mentioned that each mechanism could be responsible for cracking under specific experimental conditions.

In the case of the LSGF7364 membrane, even though there were decomposed phases present after the exposure to a highly reducing environment, $CH_4 + He$ at 950°C for 600 h, the syngas yield and the methane conversion were unaffected.

The stability of oxygen permeable membranes of perovskite $La_{0.3}Sr_{0.7}CoO_{3-\delta}$ under partial oxygen pressure was studied by Van Doorn and coworkers (Van Doorn, Bouwmeester & Burggraaf, 1998). They observed phase separation in the oxygen lean side when the membrane was exposed to streams of air and inert gas at opposite sides. This process was detected when the (homogenous) oxide is annealed in either flowing air or nitrogen for several days, indicating that the phase separation was induced by the dynamic forces acting during oxygen permeation.

The chemical stability of the LSCF6428 membrane in conjunction with its potential use as a catalytic membrane for the oxidative coupling of methane (OCM) were also investigated (Xu & Thomson, 1998). They reported that the module reached steady state oxygen-defect gradient in (15-20 h) and high stability under air/nitrogen gradients at temperatures up to 960 °C. Also, they found that the apparent activation energy was 159 kJ mol^{-1}. However, they observed that severe near-surface etching occurs when one side of the membrane is exposed to 100% CH_4 at 850 °C under atmospheric pressure, which was reflected on a 5-fold increase in oxygen flux. In addition, this also produced an increase in the OCM reaction rate, the selectivity of C_{2+} hydrocarbons was cut down from 40 to 10%. These results indicated that OCM is possible if CH_4 pressures are reduced to 0.1 atm.

Lein and coworkers reported that oxygen permeation through $La_{0.5}Sr_{0.5}Fe_{1-x}Co_xO_{3-\delta}$ (x = 0, 0.5 and 1) membranes under oxygen partial pressure gradients, temperature and time (Lein, Wiik & Grande, 2006). They found that at low pO_2 gradients the flux was controlled by diffusion, while at larger pO_2 gradients the flux was controlled by a surface exchange. The activation energy varied in the range 67–105 kJ mol^{-1}. The membranes after about 720 hours at 1150°C under O_2/N_2 gradient exhibited only the original perovskite phase on the reducing side, while on the oxidizing side various secondary crystalline phases were observed dependent on the composition at the Fe/Co-site and the (Sr + La)/(Fe + Co) ratio of the materials. Moreover, kinetic demixing of the main perovskite phase was also observed, particularly near the surfaces. Grain growth and pore coalescence resulting in membrane expansion were also observed in some cases.

Wang and coworkers investigated the effects of kinetic demixing/decomposition and impurity segregation on stabilities of the LSCF6428 hollow fibre membranes under oxygen separation condition (Wang et al., 2009). The membranes were operated under 0.21/0.008 oxygen partial pressure (bar) difference at 950°C for 1128 h. They found that cobalt oxide grains were separated on the air side of the membrane surface after long-term operation, whilst Sr and Fe enriched the surface layer of air side at 50 nm of depth. The oxygen permeation performance was stable after operating 1128 h, indicating that the demixing/decomposition and impurity segregation would not affect the LSCF6428 HF membranes at the operating conditions applied.

A stable permeation flux of 0.02 µmol cm^{-2} sec^{-1} oxygen during the long-term operation of a $La_{0.58}Sr_{0.4}Co_{0.2}Fe_{0.8}O_{3-\delta}$ membrane at 800°C for nearly 3000 h was recently reported (Schlehuber et al., 2010). They found degradation of the membrane surfaces due to kinetic demixing of mobile cations and different oxide layers on the feed and the permeate sides of the membrane.

The oxygen-separation on three LSCF6428HFM modules under a 0.21/0.009bar oxygen partial pressure difference at 950°C for 1128, 3672 and 5512 h, respectively, were studied by Li and co-workers (Wang et al., 2011). They found that the microstructure of the membranes exhibited kinetic demixing and decomposition as was determined through different techniques of characterization (i.e., Secondary Ion Mass Spectrometry, SEM, EDXS and XRD). Also, they found secondary crystalline phase grains on the air side surface of the membranes after the long-term operation. In addition, they also observed Co and Fe enrichment as well as La depletion on the surface and in the bulk at the air side. The order of cation diffusivities was Co>Fe>Sr>La. Kinetic demixing and decomposition rates of the membranes at the air side were observed to be self-accelerating with time. The role of A-site deficiency in the perovskite lattice in the bulk near the air side surface was considered in the mechanism. The oxygen permeation of the LSCF6428HFM modules was not affected by the kinetic demixing and decomposition during long-term operation (up to 5512 h). However, they supposed that oxygen permeability may be affected by secondary phase formation on the air-side surface at even longer operation times.

Sulfur Contamination of Mixed Ionic Electronic Conductor Membrane

Li *et al.* measured the oxygen permeation through tubular $La_{0.6}Sr_{0.4}Co_{0.2}Fe_{0.8}O_{3-\delta}$ membranes prepared by isostatic pressing (Li et al., 2000). The oxygen permeation fluxes were studied at various downstream oxygen partial pressures and temperatures. The oxygen permeation obtained was about 0.07 µmol cm^{-2} sec^{-1}(STP) at 900°C, when the oxygen partial pressure on the feed and permeate sides are 0.21 and 0.001 atm, respectively. New crystalline phases such as $SrSO_4$, $CoSO_4$, SrO, Co_2O_3, and La_2O_3 were identified by XRD and EDXS. The formation of sulphur-containing phases was attributed to surface interaction of the tubular membrane with SO_2 trace presence in the air and helium flows, whereas metal oxides are due to phase segregation of the elements at the surface.

Xu *et al.* prepared tubular LSCF6428 membranes and developed a mathematical model for simulating the oxygen permeation performance (Xu et al., 1999). They reported that the experimental oxygen permeation increased with decreasing downstream oxygen partial pressure and increasing helium flow rate, which coincided with the results of the oxygen permeation modelling study. Also, they have found that oxygen permeation slightly decreased after long-term operation (i.e., 110 h). Furthermore, new crystalline phase such as $SrSO_4$, $CoSO_4$, SrO, Co_2O_3, and La_2O_3 were observed on the surfaces of the tubular membrane, sulphur phases are also associated with the interaction between SO_2 trace present in air and helium.

The catalytic combustion of methane and the effects of operating methane and air feed flow rates on the performance of the LSCF6428 HFM was investigated by Tan and coworkers (Tan et al., 2008). They also developed a mathematical model to predict the performance of the membrane reactor for methane combustion, which combines the local oxygen permeation rate with approximate catalytic reaction kinetics. They have reported that both the methane conversion and oxygen permeation rate can be improved by coating platinum on the air side of the hollow fibre membranes. Also, they found that the membranes changes in the metal cation stoichiometry and microstructure together with contamination with sulphur might have their origins in the CH_4 gas supply and SO_2 in the atmosphere. According to the authors, the LSCF6428-catalyst system developed a second phase in lumen side, which did not adversely affect the methane combustion process and performance.

Subsequently, Thursfield and Metcalfe studied oxygen permeation through LSCF6428 HFM unmodified and Pt modified under air/helium gradient at 650–1000°C (Thursfield & Metcalfe, 2007). They reported the absence of mass transfer limitations at lower temperatures and also that the rate-determining step for oxygen permeation significantly enhanced by deposition of a porous platinum catalyst film onto the air-side of the membranes. It was also found that oxygen permeation (i.e., 0.56 $\mu mol\ cm^{-2}s^{-1}$) was limited by mass transfer at 1000°C whilst the non-mass transfer limited permeation at 800 °C was 0.14 $\mu mol\ cm^{-2}\ s^{-1}$. The module was operated during 500 h before the onset of significant leakage. The characterization of the post operation membranes by XRD, SEM and EDS revealed localised changes in the stoichiometry of the membranes and the presence of sulphur in the form of sulphates of strontium, cobalt and calcium. Furthermore, changes of surface microstructure for both shell side and lumen sides of the membrane surfaces were also observed. They associated the sulphur contaminations to the glass-ceramic sealing material employed and sulphur dioxide presents in the helium sweep gas and atmosphere. Sulphur contamination may contribute to the decrease in the long-term oxygen permeation rates.

Metcalfe and coworkers reported that LSCF6428 micro tubular membranes tested for oxygen permeation and hydrogen production by membrane-based steam reforming at 900°C (Franca, Thursfield & Metcalfe, 2012). The characterization of the post operation membrane revealed the presence of a strontium enriched dense layer on the water exposed membrane surface and crystallites enriched with cobalt and sulphur on the methane feed side surface.

A comparative study of the oxygen transport mechanism through sulphur-free and -containing $Ba_{0.5}Sr_{0.5}Co_{0.8}Fe_{0.2}O_{3-\delta}$ (BSCF) membranes was carried out by Yaremchenko *et al.* (Yaremchenko et al., 2013). They have found that oxygen permeation rate decreases due to sulphates partial blocking of the surface oxygen exchange, whilst the bulk ambipolar conductivity remains essentially unchanged. Also, they observed segregation of $BaSO_4$ at the grain boundaries by analysis of SEM/EDXS. The negative impact of sulphur contamination on oxygen permeation was remarked at temperatures below 850°C. The authors demonstrated that a surface activation process can increase the oxygen flux through sulphur-containing BSCF membranes to the level of that of sulphur-free membranes.

Liu *et al.* have found an important degradation mechanism of BSCF perovskite membranes to the operation-induced surface segregation of trace impurities of sulphur that cause a considerable impact on oxygen permeation at low-temperature (Liu et al., 2013). Also, they reported an effective method that enables stabilization of the permeation flux at low-temperature in BSCF perovskite membranes. It was recently reported that the sulphate ions can emerge into the lattice of perovskite oxides (Porras-Vazquez & Slater, 2012; Shin at al., 2011), and they occupy the B-site of the perovskite (ABO_3). However, the detailed structure state of the sulphate ions in the perovskite lattice is still unclear.

According to Liu et al. (Liu et al., 2013) for a mixed electronic–ionic conductor, the equilibrium of interest is the following:

$$0.5O_2 + 2e^- \leftrightarrow O^{2-} \tag{21}$$

In the mixed conducting materials, the possible species involved in the charge transfer are electrons, anions and cations, consequently the current balance can be written as:

$$I_e + I_{anions} + I_{cations} = 0 \tag{22}$$

At low temperatures, the motion of metallic ions is very slow, and can be neglected; moreover, there is no proton transport under the conditions for oxygen permeation. Eqn. (22) can be written as:

$$I_e + I_{anions} = 0 \tag{23}$$

Here, the anions are oxygen and sulphate, so eqn. (23) is re-written as:

$$I_e + I_{O^{2-}} + I_{SO_4^{2-}} = 0 \tag{24}$$

When the membranes were treated in air or helium, there is no oxygen chemical potential difference across the membranes, so the electronic current and ions current in the membrane bulk is zero at a balance state, as shown in Figure 3a.

However, on the membrane interface zone the reaction (eqn. 22) still takes place. The thickness of the interface zone varies with membranes and operation conditions, since the dipole electrical conductivity of this zone is difficult to measure, but it is usually in micron dimension and can be estimated from the z Once the sulphate anion transport from the interface zone to the outside surface, they react with barium ions to produce $BaSO_4$, in which sulphate anions have lower chemical potential than occupying B-site of the perovskite. As a result, there is $BaSO_4$ grains found on the membrane surfaces. Only the interface zones are involved in this process, so there are few $BaSO_4$ grains formed. When the membranes were operated for oxygen permeation, there is an oxygen chemical potential difference across the membranes,

Figure 3. Diagram of the transport of electrons and ions when the membrane are (a) treated by air or helium, and (b) operated for oxygen separation. Adapted from (Liu et al., 2013)

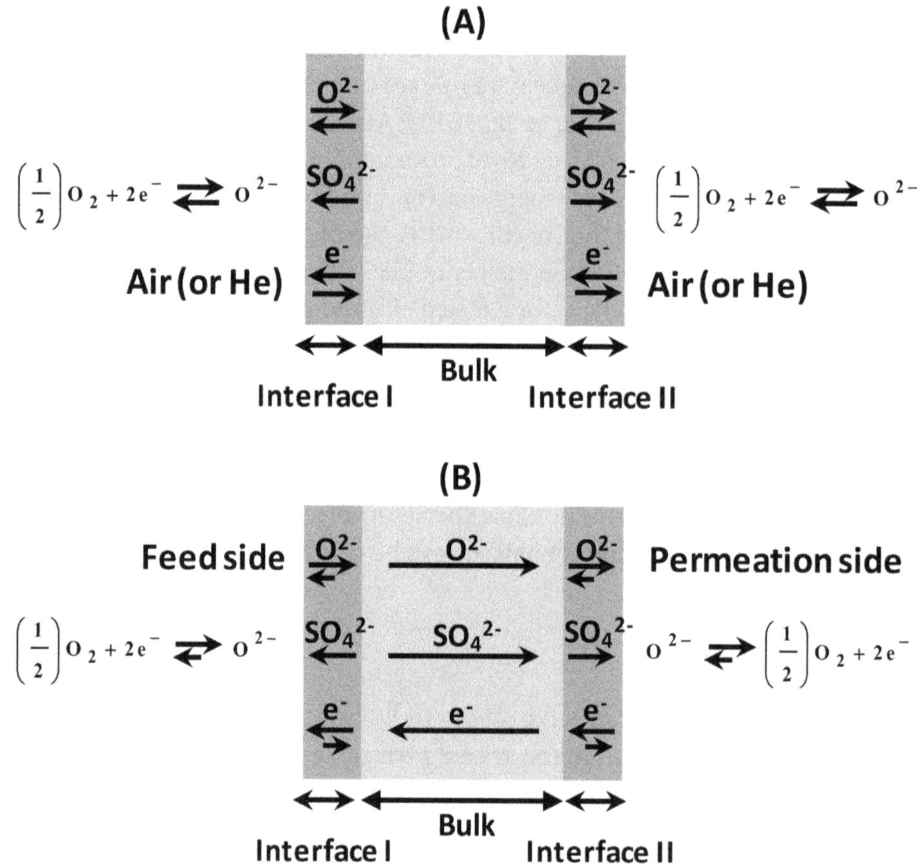

so the electronic current and ions current in the membrane bulk zone is not equal to zero at a steady state, as shown in Figure 3b.

Under this condition, both interface zones and bulk zone of the membrane are involved in the transport process. After a long time, the sulphate anions in the interface zone of the permeation side and bulk zone transport to the outside surface and react with barium ions to produce $BaSO_4$; and the sulphate anions in the interface zone of the feed side partially transport to the outside surface and react with barium ions to produce $BaSO_4$ and partially transport to the bulk zone.

EXPERIMENTAL SETUP

Preparation and Modification of Mixed Ionic Electronic Conductor Membrane

The LSCF6428-hollow fibre membrane (HFM) was prepared using a combined phase-inversion/sintering technique as was previously described by Li and co-workers (Deshmukh & Li, 1998; Tan et al., 2005). This membrane had a wall thickness of 250 μm and outer and inner radiuses of 0.07 and 0.05 cm, respectively.

Nickel nanopowder of average particle size 90 nm (Sigma-Aldrich) was used. The nanopowder was weighed out (0.05 g) and mixed with a 0.95 g of LSCF6428 powder to obtain 5 wt% Ni dispersed on LSCF6428. The mixture was stirred in ethylene glycol for 24 hours to obtain an even paste, the liquid was evaporated and the resultant powder was ground. Four hollow fibres were coated on the shell side surface with 5 wt. % Ni-LSCF6428 HFM catalyst. This was applied to the shell-side surface of the central 5 cm of each hollow fibre using a small fine brush. The membranes were thermally treated to remove the organic resin using a programmed temperature ramp of 1 °C min^{-1} from room temperature up to 400°C and held for 1 hour. Then the temperature was increased up to 850°C at 1°C min^{-1} and held for 1 hour. Subsequently, the membrane reactor was cooled down to 25°C at 1°C min^{-1}.

The average surface loading of the catalytically modified hollow fibres was approximately 57 mg cm^{-2}, and the layer thickness was 26 μm ± 2.

Methane Conversion Module Assembly

The reactor was assembled using four hollow fibre membranes and a pair of engineered MACOR tubes in order to create a hollow fibre (HF) bundle. Both ends of the bundle were attached to the gas inlet and outlet by a length of flexible silicone tubing, in order to accommodate the thermal expansion of the membranes. The temperature was controlled by a furnace (Vecstar Furnaces, UK) of total length 15 cm equipped with an eight-segment Eurothermcontroller. The heated zone (13 cm in length) had a central linear heating zone of 5 cm. The sealant used was a commercial high temperature water-based glass-ceramic (Heavy Grade Fortafix, UK). The same sealant was also used to partially cover the four hollow fibre membranes so that only the centre zone (5 cm) was uncovered, as shown in Figure 4, which coincided with the isothermal zone of the furnace. A heating or cooling rate of 0.1 or 0.2°C min^{-1} was used throughout this investigation. The HF bundle was housed inside a quartz tube (outer diameter of 3 cm) that allowed control of the gas composition and flow rate at the external surface of the HF membranes. The internal side of the fibres is referred to as the lumen side (helium-side), and the outside of the hollow fibres is referred to as the shell side (air-side in these experiments). A schematic representation of the hollow fibre membrane reactor is given in Figure 4.

Figure 4. Diagram of the LSCF6428 hollow fibre membrane reactor module

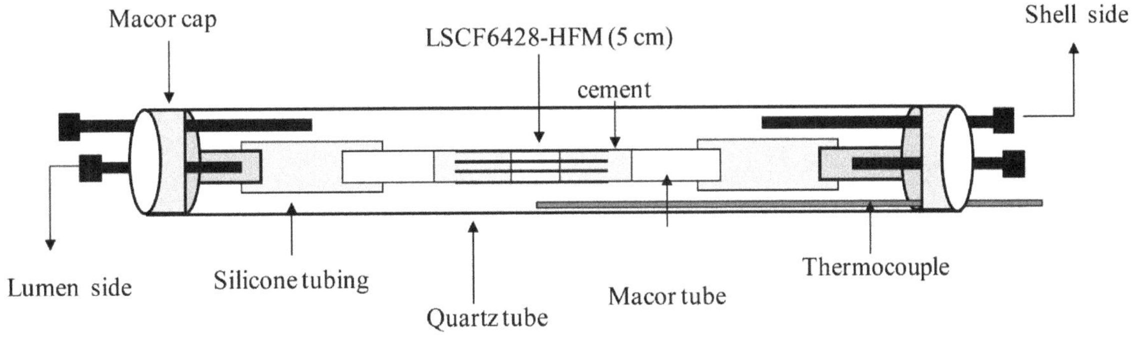

GAS CHROMATOGRAPHY CALIBRATIONS

Two Varian 3900 chromatographs were used (denoted GCA and GCB) each equipped with a thermal conductivity detector (TCD) and Molecular sieve 5A column. Helium was used as carrier gas and set to 20 ml min^{-1}.

The GCB calibration was performed using a gas mixture containing 2% oxygen and 2% nitrogen balance gas helium. Calibration of GCA was carried out using air. GCA was used to analyse the oxygen permeate at the lumen side and GCB the depleted oxygen at the shell side. The gas analysis was carried out at least six times to test repeatability and the maximum relative error was below 5% in all cases.

The experimental setup for the POM experiments is described below, under the following flow configurations:

1. 4% CH_4/He flow fed at 25 ml (STP) min^{-1} to the shell side of the reactor, while air flow rate at 25 ml (STP) min^{-1} was co-currently introduced into the lumen side (MSAL).
2. 4% CH_4/He flow supply in the lumen side of the reactor at 25 ml (STP) min^{-1}, whilst air flow rate at 25 ml (STP) min^{-1} was sweep into the shell side co-currently (MLAS).

The experiments were carried out at 850 °C and the products were analyzed on-line using two Varian 3900 chromatographs. The oxygen permeation flux was calculated from oxygen atoms of all the oxygen containing products eqn. (25), except the water, which was not measured from the shell side. The conversion of methane was determined from eqn. (25). Also, the selectivities to carbon monoxide, carbon dioxide, ethane and ethylene products were calculated using the equations shown below [i.e., eqns. (27) – (31)] and the carbon balances was calculated from eqn. (32).

$$O_{2\ permeation} = \left(n_{O_2\ out} + \frac{1}{2} n_{CO\ out} + n_{CO_2\ out} + \frac{1}{2} n_{H_2O} \right) \times 1/S_A \tag{25}$$

$$X_{CH_4} = \frac{n_{CH_4\ \text{inlet}} - n_{CH_4\ \text{outlet}}}{n_{CH_4\ \text{inlet}}} \times 100 \tag{26}$$

Selectivity of each product was calculated using the following equations:

$$S_{CO} = \frac{n_{CO\ \text{out}}}{n_{CO} + n_{CO_2} + 2\left(n_{C_2H_2}\right) + 2\left(n_{C_2H_6}\right)} \times 100 \tag{27}$$

$$S_{CO_2} = \frac{n_{CO_2\ \text{out}}}{n_{CO} + n_{CO_2} + 2\left(n_{C_2H_4}\right) + 2\left(n_{C_2H_6}\right)} \times 100 \tag{28}$$

$$S_{C_2H_4} = \frac{2\left(n_{C_2H_4\ \text{out}}\right)}{n_{CO} + n_{CO_2} + 2\left(n_{C_2H_4}\right) + 2\left(n_{C_2H_6}\right)} \times 100 \tag{29}$$

$$S_{C_2H_6} = \frac{2\left(n_{C_2H_6\ \text{out}}\right)}{n_{CO} + n_{CO_2} + 2\left(n_{C_2H_4}\right) + 2\left(n_{C_2H_6}\right)} \times 100 \tag{30}$$

$$S_{H_2} = \frac{1}{2} \times \frac{n_{H_2\ \text{out}}}{n_{CO} + n_{CO_2} + 2\left(n_{C_2H_4}\right) + 3\left(n_{C_2H_6}\right)} \times 100 \tag{31}$$

The carbon balance was calculated from the following relation:

$$\text{Carbon balance} = \frac{n_{CO} + n_{CO_2} + 2\left(n_{C_2H_4}\right) + 2\left(n_{C_2H_6}\right)}{n_{CH_4\ \text{inlet}} - n_{CH_4\ \text{outlet}}} \times 100 \tag{32}$$

Characterization of the Unmodified and Ni-LSCF6428 HFM Modified-LSCF6428 Hollow Fibre Membranes

X-Ray Powder Diffraction

The characterisation of the samples by X-ray diffraction (XRD) was carried out using a PANalytical XPert- Pro diffractometer fitted with a X' Celerator and equipped with an X-ray tube ($\boldsymbol{CuK_\alpha}$ radiation λ = 1.54060 Å; 40 kV, 30 mA). The specimens were prepared by grinding a small amount of each sample (ca. 100-150 mg) using an agate mortar and pestle and then loading into a flat sample holder. In studies on membrane the sample was cut into sections approximately 1 cm in length and 0.2 cm wide and arrayed on the amorphous plasticene attached to the sample holder. The data were collected at room temperature in θ/2θ reflection mode, from 3° to 100° 2θ, using steps of 0.033°, with time per step of 100s in continuous mode. The instrument was interfaced with a computer for collected data.

Scanning Electron Microscopy (SEM) and Energy-Dispersive X-Rays Spectroscopy (EDXS)

The microstructure characterisation of the fresh and spent samples was carried out using a Quanta 200 and 30XL (FEI) instruments. The SEM images were simulated using the software Scanning Probe Image Processor (SPIP 6.0.9) in order to determine the diameter of grain boundary and pore areas. Semi-quantitative elemental compositions were verified by energy-dispersive X-rays spectroscopy using an Emission Scanning Electron Microscope Röntec Quanta 1, 200 30XL FEI which was equipped with a light-element energy-dispersive X-ray spectrometer (EDXS). To investigate structural changes at the membrane surfaces, membrane specimens were prepared by carefully breaking the hollow fibre membrane into sections of length 1 cm and 0.2 cm wide. The instrument was interfaced with a computer to collect data.

Hydrogen Temperature-Programmed Reduction (H_2-TPR)

The TPR profiles for the oxide precursors were carried out on a Micromeritics AutoChem II 2920 instrument. Between 110 and 130 mg of the solid samples were analysed using 10% H_2 in Ar (50 $cm^3.min^{-1}$) and 5 °C min^{-1} heating rate from room temperature up to 950-1000 °C and hold this temperature for approximately 30 min. The amount of consumed H_2 was quantified using a thermal conductivity detector whose response was calibrated with different H_2 concentrations.

RESULTS AND DISCUSSION

Oxidation of Methane over Unmodified and Ni-Modified LSCF6428 Hollow Fibre Membranes

In Figures 5 and 6 is shown the conversion data obtained with unmodified hollow fibre membrane of LSCF6428, which was operated at 850°C for ca. 16 hours under two flow configurations (i.e., MSAL and MLAS).

When 4% CH_4/He is supplied into the shell side, the oxygen permeation slightly increased during the initial 45 min to obtain a constant value of 0.00033 μmol. $s^{-1}.cm^{-2}$. Meanwhile, the methane conversion was below 5% upon operation time (Figure 5a). Also, it can be appreciated that C_2 product selectivity (i.e., C_2H_4 and C_2H_6 selectivities were 23% and 60%, respectively) and this was much higher than

Figure 5. Catalytic performance of the POM on the LSCF6428-HFM reactor, absent catalyst, where 4% CH_4/He flow supply in shell side while air flow rate sweep into to the lumen side (MSAL), co-current

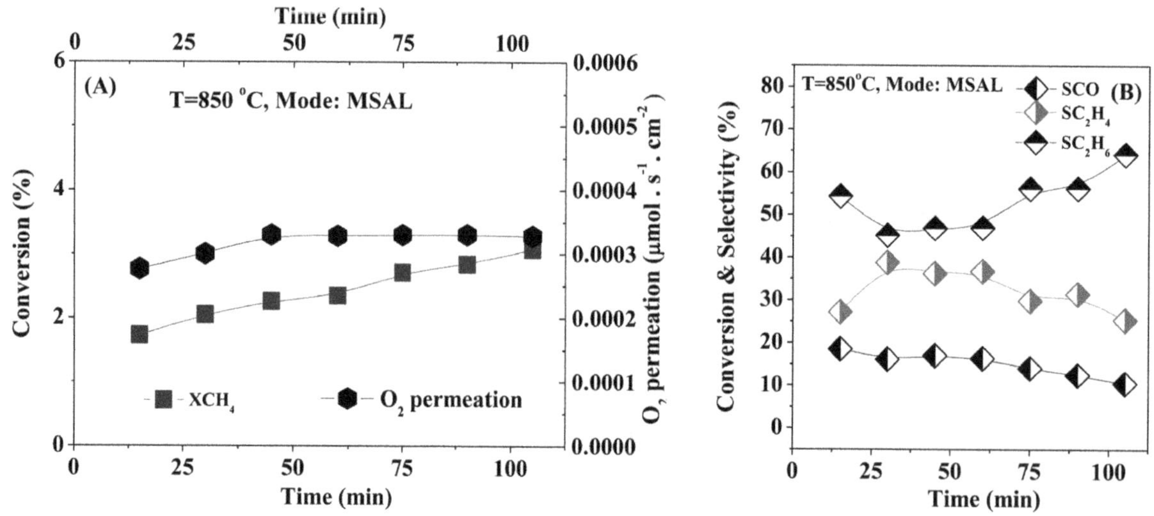

Figure 6. Catalytic performance of the methane oxidation reaction on the unmodified LSCF6428-HFM reactor, where 4% CH_4/He flow was supplied in the lumen side and air flow fed into the shell side (MLAS) at 850 °C

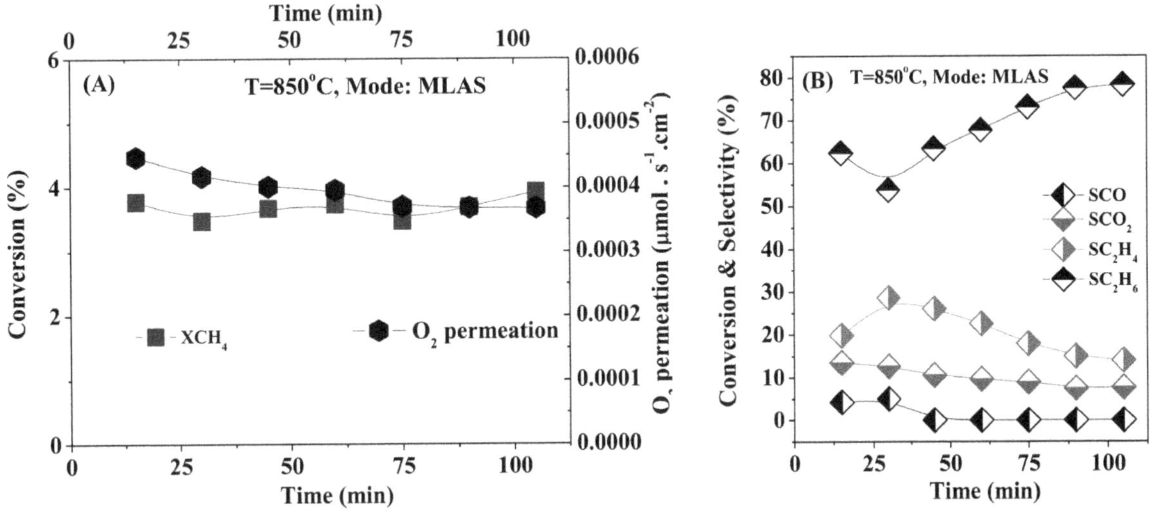

CO selectivity (i.e., 10%) (Figure 5b). Comparable catalytic data have been obtained elsewhere over LSCF6428 (Tang et al., 2007; Xu & Thomson, 1997). Ten Elshof *et al.* reported C_2 selectivities of up to 70% for methane conversion between 1 and 3% over LSCF6482 (Ten Elshof et al., 1995).

As LSCF6428-HFM reactor remained from the last experiment was supplied with 4% CH_4/He flow in the shell side while air flow rate swept into the lumen side (MSAL), co-current. It was decided to change the gas flow configuration (i.e., 4% CH_4/He flow was fed into the lumen side while air flow was introduced into the shell side (MLAS), co-current mode) on the HFM reactor and then it was tested for the methane oxidation reaction. It can be noted that oxygen permeation decreased slightly from 0.00045 to 0.00037 µmol s^{-1} cm^{-2} during the operation time. Methane conversion was approximately constant (i.e., 4%) during operation time (Figure 6a).

Furthermore, it was found that CO and CO_2 selectivity decreases gradually from 5% to 0% and 13% to 8%, respectively during the time on stream. Note that SCO_2 is higher than SCO, this is tentatively attributed to the enrichment of oxygen in the lumen side because the air flow was supplied in this side during the previous experiment (mode: MSAL).

In addition, during the time on stream, C_2H_6 selectivity rose slightly from 65% to 78%, while C_2H_4 selectivity declined slightly from 20% to 14% (Figure 6b). The relatively low CO_x selectivity compared to C_2 selectivity would indicate that the amount of oxygen permeated was not sufficient for oxidizing methane to CO and/or CO_2. Furthermore, the trend observed for C_2H_6 and C_2H_4 suggests that C_2H_4 formation occurred through consecutive oxidation reactions, eqns. (33) and (34), in agreement with a previous report (Kondratenko et al., 2009):

$$2CH_4 + O \rightarrow C_2H_6 + H_2O \tag{33}$$

$$C_2H_6 + O \rightarrow C_2H_4 + H_2O \tag{34}$$

In addition, the carbon deposited on the membrane surface was apparently between 10% and 20%, according to the carbon balance. This finding is in line with previous reports (Tan & Li, 2006; Thursfield & Metcalfe, 2006).

In Figure 7 are given the dependence of the methane conversion and product selectivity dependence with oxygen permeation under two flow operation modes (i.e. MSAL and MLAS) at 850 °C. Comparing the methane conversion versus oxygen permeation for both flow operation modes (Figure 7a), shows that methane conversion for the MSAL operation mode increased slightly with rising oxygen permeation. On the other hand, methane conversion for the other operation mode (i.e. MLAS) did not vary markedly with change of the oxygen permeation. One should remark that methane conversions for MLAS operation mode were slightly higher than these obtained in MSAL mode as a consequence of the larger oxygen permeation values.

It is well established that different oxygen species exist at the membrane surface (Gellings & Bouwmeester, 2000; Shao et al. 2001; Tan et al., 2005), eqn. (1). However, not all these species are able to generate selective oxidation products from methane activation (González-Cortés et al., 2007). According to Bielanski and Haber, only the nucleophilic lattice oxygen ions (i.e., O^{2-}) conduce to selective oxidation products formation whereas that electrophilic oxygen species (i.e. O_2^-, O^- and $O_{2\,(ads)}$) generate total oxidation products (Bielanski & Haber, 1991). One can envisage that the activity-selectivity performance of the LSCF6428-HFM in methane conversion is likely associated with the concentrations of electrophilic and nucleophilic oxygen species and also that different oxygen species produced depend upon the abundance of the oxygen vacancies (i.e., $V_0^{\bullet\bullet}$) in the membrane, see eqn. (35).

$$\frac{1}{2}O_2 + V_0^{\bullet\bullet} + 2e' \rightarrow O^{2-} + 2h^{\bullet} \tag{35}$$

The Figure 7b are illustrated the C_2H_4, C_2H_6 selectivity against oxygen permeation for both modes of flow operation. It can be noted that the C_2H_6 and C_2H_4 selectivity exhibit the same behaviour for both flow operation modes, that is, C_2H_6 selectivity decreased with increasing of oxygen permeation whereas C_2H_4 selectivity showed an opposite trend. However, C_2H_6 selectivity was markedly higher than C_2H_4 selectivity, particularly for MLAS mode. Again, this trend can be rationalized assuming that C_2H_4 formation occurred via consecutive oxidation reactions of C_2H_6 initially formed, see eqn. (34).

In addition, in Figure 7c are presented the CO and CO_2 selectivity for both flow operation modes. It can be noted that the CO selectivity decreased while the oxygen permeation increased for MSAL mode, while no CO_2 was detected. In contrast, for MLAS mode, the CO_2 and CO selectivity increased when there was rising the oxygen permeation values.

It is worth mentioning that the oxygen permeation values for MLAS mode were higher than those values obtained in MSAL mode. This is likely due to high oxygen concentration in the lumen side, owing to the previous experiment under MSAL mode.

Different oxygen species exist in the lumen side of the membrane likely due to different pathways and selectivity for the total methane oxidation [i.e., eqns. (36) and (39), respectively] or/and consecutive oxidation from primary products (CO, C_2H_6 and C_2H_4) to CO_2, see eqns. (37), (38), (40) and (41).

Figure 7. Dependence of methane conversion and product selectivity with oxygen permeation under two different flow operation modes at 850 °C. (a) Methane conversion, (b) C_2H_6 and C_2H_4 selectivity; (c) CO and CO_2 selectivity

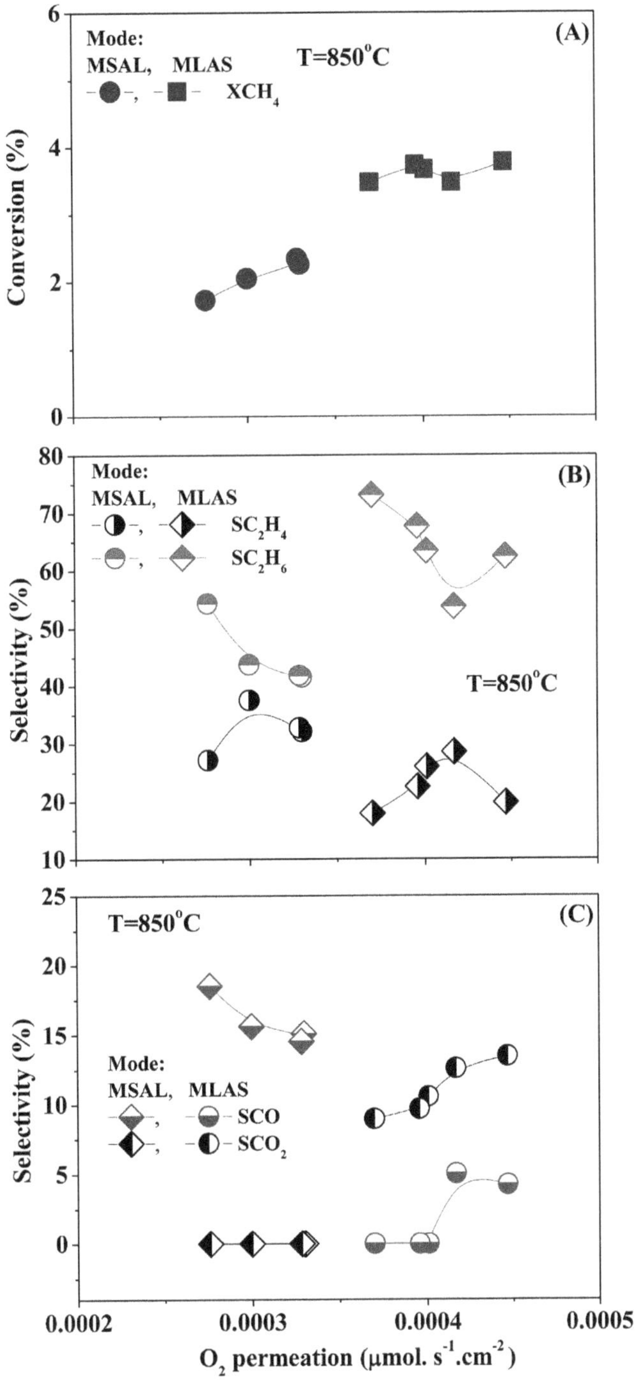

$$CH_4 + 3O \rightarrow CO + 2H_2O \quad (36)$$

$$CO + H_2O \rightarrow CO_2 + H_2 \quad (37)$$

$$CO + O \rightarrow CO_2 \quad (38)$$

$$CH_4 + 4O \rightarrow CO_2 + 2H_2O \quad (39)$$

$$C_2H_6 + 7O \rightarrow 2CO_2 + 3H_2O \quad (40)$$

$$C_2H_4 + 6O \rightarrow 2CO_2 + 2H_2O \quad (41)$$

The methane conversion results and oxygen permeation for the 5%Ni-LSCF6428/LSCF6428-HFM reactor are shown in Figure 8, when 4% CH_4/He was supplied on the shell side and air flow swept in the lumen side, with a flow rate of 25 ml (STP) min^{-1} for both sides.

Figure 8. Catalytic performance of the methane conversion on LSCF6428-HFM-coated 5%Ni-LSCF6428 reactor at 850 °C. Air flow fed on the lumen side and 4% CH_4/He on the shell side with flow of 25ml (STP) min^{-1} in both sides

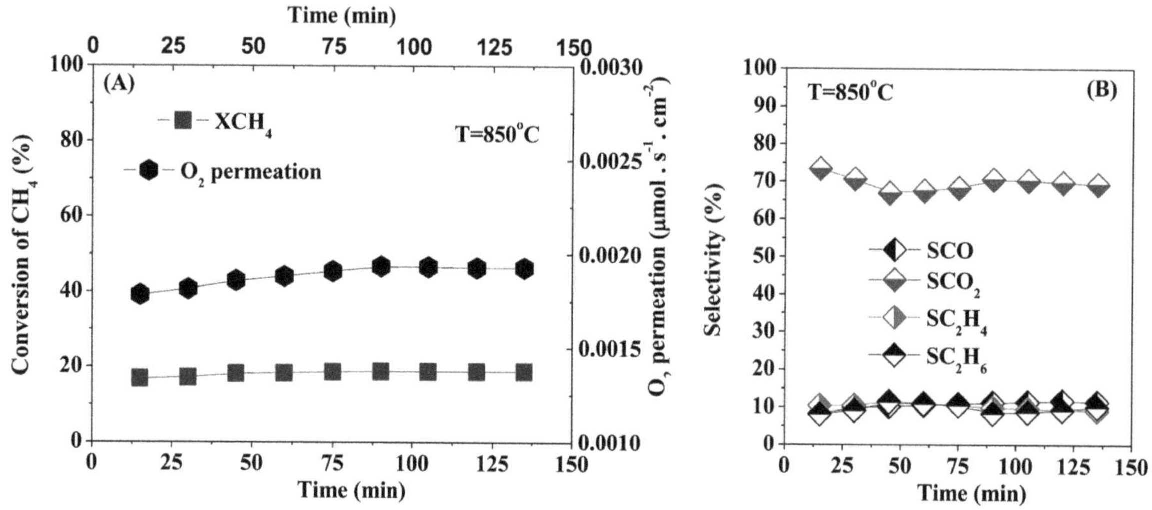

It was found that both oxygen permeation and methane conversion hold constant during operation time (i.e. 0.0019 μmol $s^{-1}cm^{-2}$ and 19%, respectively), see Figure 8a. Furthermore, the CO_2 selectivity is slightly decreased from 73% to 69% and CO, C_2H_4 and C_2H_6 selectivity presented a similar trend with values between 8% and 11% for each product (Figure 8b). Carbon balance was between 97 and 100 wt. %, indicating that only small amount of carbon was deposited in the module.

The modification effect of the catalyst (i.e., 5% Ni-LSCF6428) coated on the hollow fibre membrane over methane conversion is shown in Figure 9. Three effects of the 5%Ni-LSCF6428 catalyst on the membrane reactor performance were noted: (i) Oxygen permeation increased from ca. 0.0003 μmol s^{-1} cm^{-2} to ca. 0.0019 μmol s^{-1} cm^{-2} as was expected (ii) Methane conversion increased from approximately 3% to 19% (iii) CO_x becomes the main reaction product, as a consequence of high oxygen permeation. Note that the unmodified HFM reactor favoured C_2 (C_2H_4 and C_2H_6) selectivity (i.e., oxidative coupling of methane). Both the products distribution and the high methane conversions suggest that CH_4 reacts with oxygen species that diffused over the catalyst to produce CO_x and H_2O. This would indicate that methane conversion occurs over the surface of 5%Ni-LSCF6428 catalyst and facilitates the total oxidation catalytic pathway (Kondratenko et al., 2009; Tang et al., 2007).

Figure 9. Dependence of the oxygen permeation, methane conversion and products selectivity against time on stream for unmodified and modified LSCF6428HFM at 850 °C. Air flow fed on the lumen side and 4% CH_4/He on the shell side with flow of 25ml (STP) min^{-1} in both sides

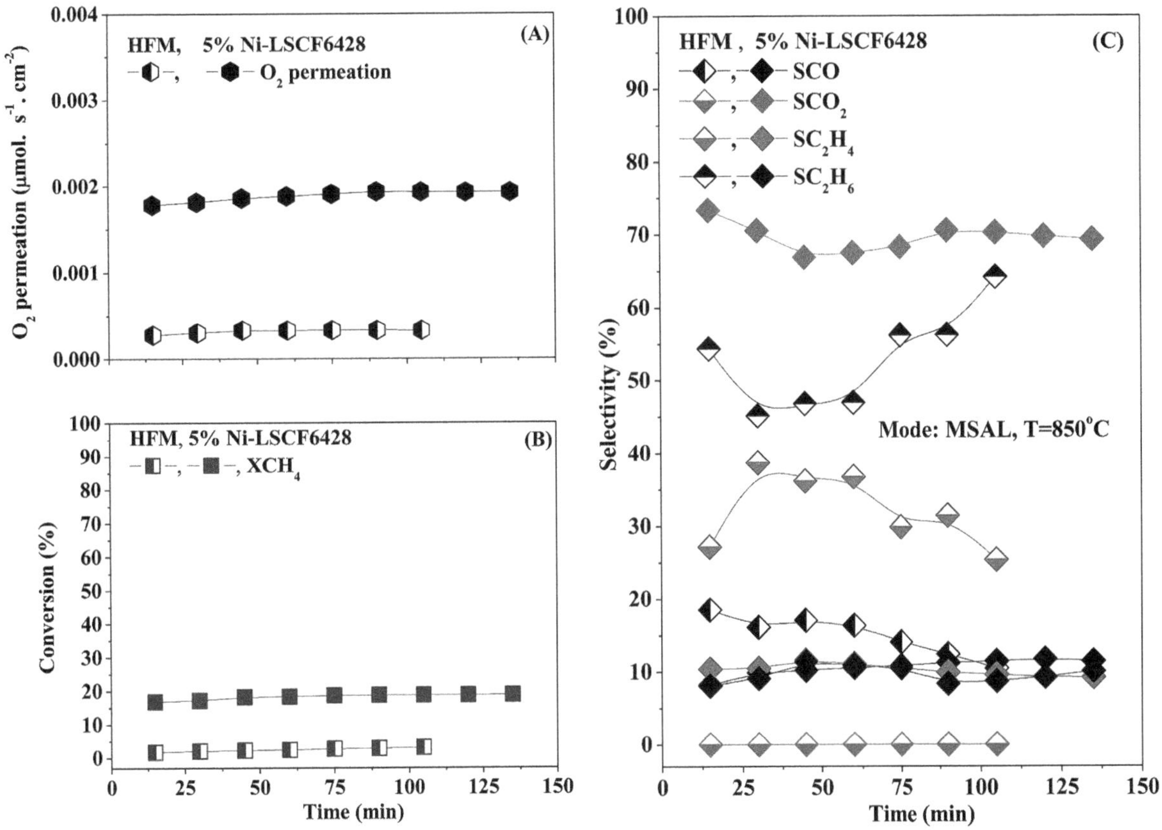

Table 1 .Apparent overall rate constant for LSCF6428-HFM and 5%Ni-LSCF6428/LSCF6428-HFM at 850 °C, where CH_4 or He flow supplied in the shell side and air sweep in the lumen side

Catalyst	He Flow (ml. min^{-1})	4%CH_4 Flow (ml. min^{-1})	Percentage of Oxygen (%)	O_2 Flow (ml. min^{-1})	Rate Constant (μmol.sec^{-1}. bar^{-1}.cm^{-2})
Unmodified	25	-	20	25	0.16
Unmodified	-	25	20	25	0.26
5%Ni/LSCF64268	25	-	20	25	0.50
5%Ni/LSCF64268	-	10	2	10	nd
5%Ni/LSCF64268	-	25	2	25	nd
5%Ni/LSCF64268	-	25	20	25	0.76

In Table 1 are given the apparent overall rate constants for LSCF6428-HFM and 5%Ni-LSCF6428/LSCF6428-HFM under different reducing gas atmospheres (i.e., He and CH_4) at 850 °C. Comparison of the apparent overall rate constant for LSCF6428-HFM without catalyst obtained with different reducing atmospheres (CH_4 and He) under MSAL or ALHS mode are given in Table 1. It is found that the apparent overall rate constant for air/CH_4 gradient increased 1.5 fold (0.26 μmol s^{-1} bar^{-1} cm^{-2}) compared to air/He gradient (0.16 μmol s^{-1} bar^{-1} cm^{-2}). This finding is likely due to the reducing character of CH_4 relative to He; the air/CH_4 gradient is larger than air/He gradient, therefore the oxygen concentration as driving force of the oxygen transport across the membrane is higher compared to air/He gradient (Tan et al., 2008). Moreover, the methane combustion is an exothermic reaction and releases a great amount of energy and thereby produces hot spots on the surface of the hollow fibre membrane reactor (Caro et al., 2007). Helium, on the other hand, is inert on the HFM surface, facilitating a uniform temperature profile in the reactor (Tan et al., 2008). A similar tendency was also observed in 5%Ni-LSCF6428/LSCF6428-HFM reactor under an air/CH_4 gradient (i.e. 0.76 μmol s^{-1} bar^{-1} cm^{-2}) and air/ He gradient (i.e. 0.50 μmol s^{-1} bar^{-1}cm^{-2}), see Table 1, in agreement with previous reports (Jin et al., 2000; Xu et al., 1998; Zeng, Lin & Swartz, 1998).

On the other hand, comparison of the apparent overall rate constant for the unmodified LSCF6428-HFM with the modified LSCF6428-HFM by 5%Ni-LSCF6428 catalyst under an air/He gradient shows that this increased from 0.16 μmol. s^{-1} bar^{-1} cm^{-2} to 0.50 μmol s^{-1} bar^{-1}cm^{-2}. This is tentatively attributed to the nickel catalyst that facilities the surface reaction of molecular oxygen and the possible enlargement of the surface area of the membrane by catalyst coated on the shell side.

Similar behaviour was observed under an air/CH_4 gradient for both the reactors unmodified LSCF6428-HFM (i.e. 0.26 μmol s^{-1} bar^{-1} cm^{-2}) and 5%Ni-LSCF6428-HFM (i.e. 0.76 μmol s^{-1} bar^{-1} cm^{-2}). This can be explained considering that the catalyst facilitates the dissociation of methane into methyl radical $(\bullet CH_3)$ at the active sites on the catalyst surface. These active sites are usually reactive nucleophilic lattice oxygen ions (i.e. O^{2-}) and electrophilic oxygen species (i.e. O_2^-, O^- and $O_{2(ads)}$), which react with $\bullet CH_3$ to form gas products (Bielanski & Haber, 1991; Lin & Zeng, 1996; Tan et al., 2008; Tang et al., 2007; Yacou et al., 2011). It is worth mentioning that in the experiments carried out at low oxygen concentration (i.e., 2%) it was not possible to determine the apparent rate constant due to the low amount of oxygen permeated.

Figure 10. Catalytic performance and oxygen flux versus time on stream for 5%Ni-LSCF6428/LSCF6428-HFM reactor at 850 °C. 4% CH_4/He flow on the shell side and 2% O_2/He flow rate on the lumen side, both flows were 10 ml (STP) min^{-1}

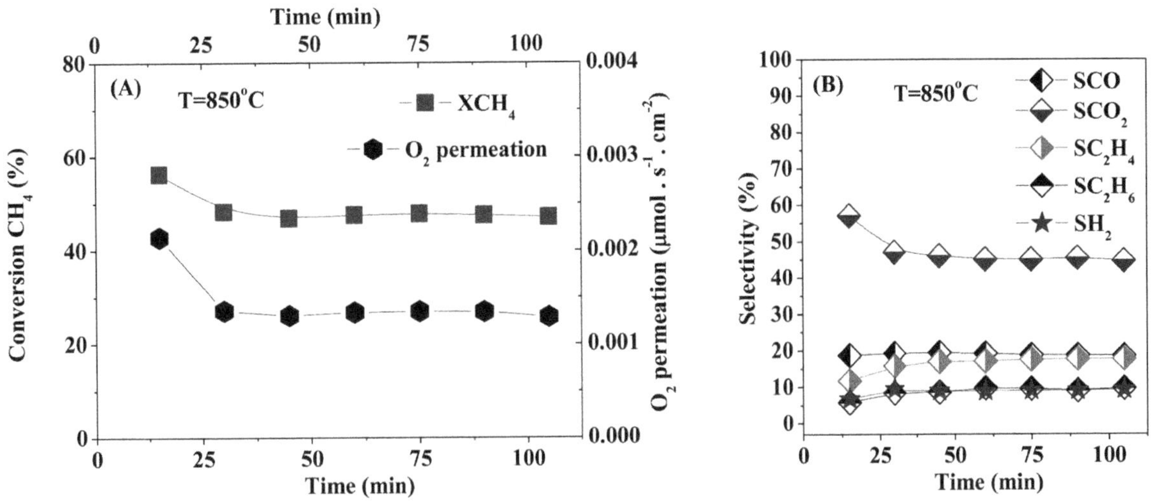

Figure 10. shows the catalytic performance and oxygen flux for 5%Ni-LSCF6428/LSCF6428-HFM reactor as a function of operation time under MSAL mode (i.e., 10 ml (STP) min^{-1} of 4% CH_4/He flow was supplied in the shell side while in the lumen side a sweep 10 ml (STP) min^{-1} of 2% O_2/He flow), co current at 850 °C. It should be remarked that this module was previously tested for 1968 hours in an oxygen permeation experiment and was then treated with 30 ml (STP) min^{-1} of 5% H_2/He. As can be seen in Figure 10A, the oxygen permeation flux decreased slightly from 0.0021 to 0.0013 μmol s^{-1} cm^{-2} and also methane conversion gradually decreased from 56% to 47%. Nevertheless, a considerable improvement compared to a blank of LSCF6428-HFM reactor (Figure 10A) was obtained.

This is probably due to changes in the catalytic performance of the membrane reactor as the presence of 5%Ni-LSCF6428 catalyst coated on the surface of LSCF6428-HFM, which plays the following role: increases the oxygen transport through the hollow fibre membrane reactor and speeds up the methane oxidation reaction. In addition, this trend suggests that the methane conversion occurs mainly over the 5%Ni-LSCF6428 catalyst rather than the hollow fibre membrane (Kondratenko et al., 2009).

In Figure 10B it can be observed that CO and H_2 selectivity was held constant (19% and 9%, respectively) during operation time. In contrast, CO_2 selectivity decreased initially from 57% to 45% during 30 minutes of reaction, but became constant during the remaining operation time. C_2 selectivity (C_2H_4 and C_2H_6), on the other hand, increased from 24% up to 27% during the methane conversion experiment. It was estimated that around 60-70 wt. % of carbon was deposited on the module surface (i.e., 5%Ni-LSCF6428-HFM) under MSAL operation mode.

The above-mentioned results clearly suggest that the three catalytic pathways can take place on 5%Ni-LSCF6428/LSCF6428-HFM: 1) Partial oxidation of methane to syngas (CO + H_2) on metallic nickel particles. 2) Oxidative coupling of methane to C_2 and 3) total oxidation of methane to CO_2 products on LSCF6248 HFM and NiO particles (Li & Smith, 2008; Lu et al., 1996; Matsuma & Nakamori, 2004).

Figure 11. XRD patterns of the unmodified $La_{0.6}Sr_{0.4}Co_{0.2}Fe_{0.8}O_{3-\delta}$ hollow fibre membrane. LSCF6428 powder (a), Lumen side (b) and Shell side (c) before the oxygen flux measurements. Centre lumen side (B) and centre shell side (C) after methane conversion experiments

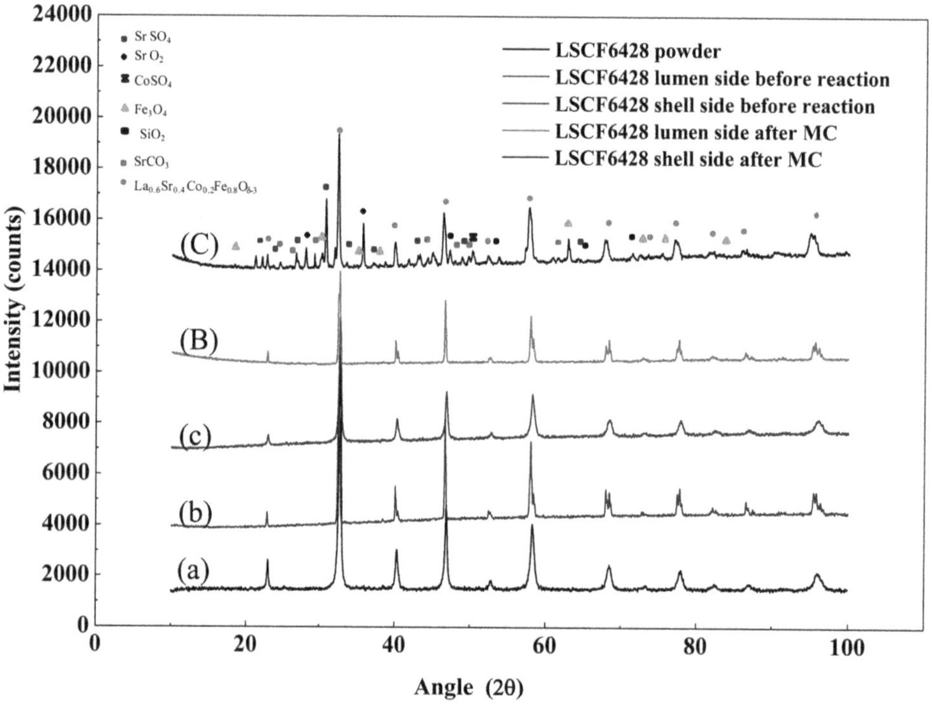

Characterization of the Unmodified LSCF6428-HFM Reactor Before and After Methane Oxidation Tests:

X Ray-Diffraction (XRD)

The diffraction profiles for the unmodified LSCF 6428 hollow fibre membrane system before and after methane conversion are shown in Figure 11. The XRD patterns for lumen side and shell side of unmodified LSCF 6428 hollow fibre membrane before reaction are shown in Figs. 11b and 11c, respectively. Both XRD profiles display the characteristic sign of the hollow fibre membrane (i.e., La0.6Sr0.4Co0.8Fe0.2O3-δ) as the main crystalline phase (Sathe et al., 1998).

The diffraction profiles for centre lumen side and centre shell side for unmodified LSCF6428-HFM reactors after methane conversion experiments are illustrated in Figures 11B and 11C. It can be clearly observed that the principal crystalline phase (i.e. LSCF6428) was detected in both sides (i.e., centre lumen and centre shell sides). However, other crystalline phases such as SrO2 (00-005-0490), SrSO4 (00-005-5935), CoSO4 (10124-43-3), Fe3O4 (01-088-0315), SiO2 (00-005-0490) and SrCO3 (1633-05-2) were also observed for the centre shell side.

Scanning Electron Microscopy (SEM)

The SEM micrographs of the centre shell side of the La0.6Sr0.4Co0.2Fe0.8O3-δ hollow fibre membrane before and after methane conversion are shown in Figures 12a and 12A, respectively. Comparing the panoramic images of the centre shell side before and after methane conversion experiment, it is clearly

Figure 12. SEM images of the shell side (a) and simulated image (b) before reaction. SEM micrograph centre shell side (A), image simulated by SPIP6.0.9 (B). Distribution of the grain boundary perimeter of shell side before and after methane conversion measurements (C)

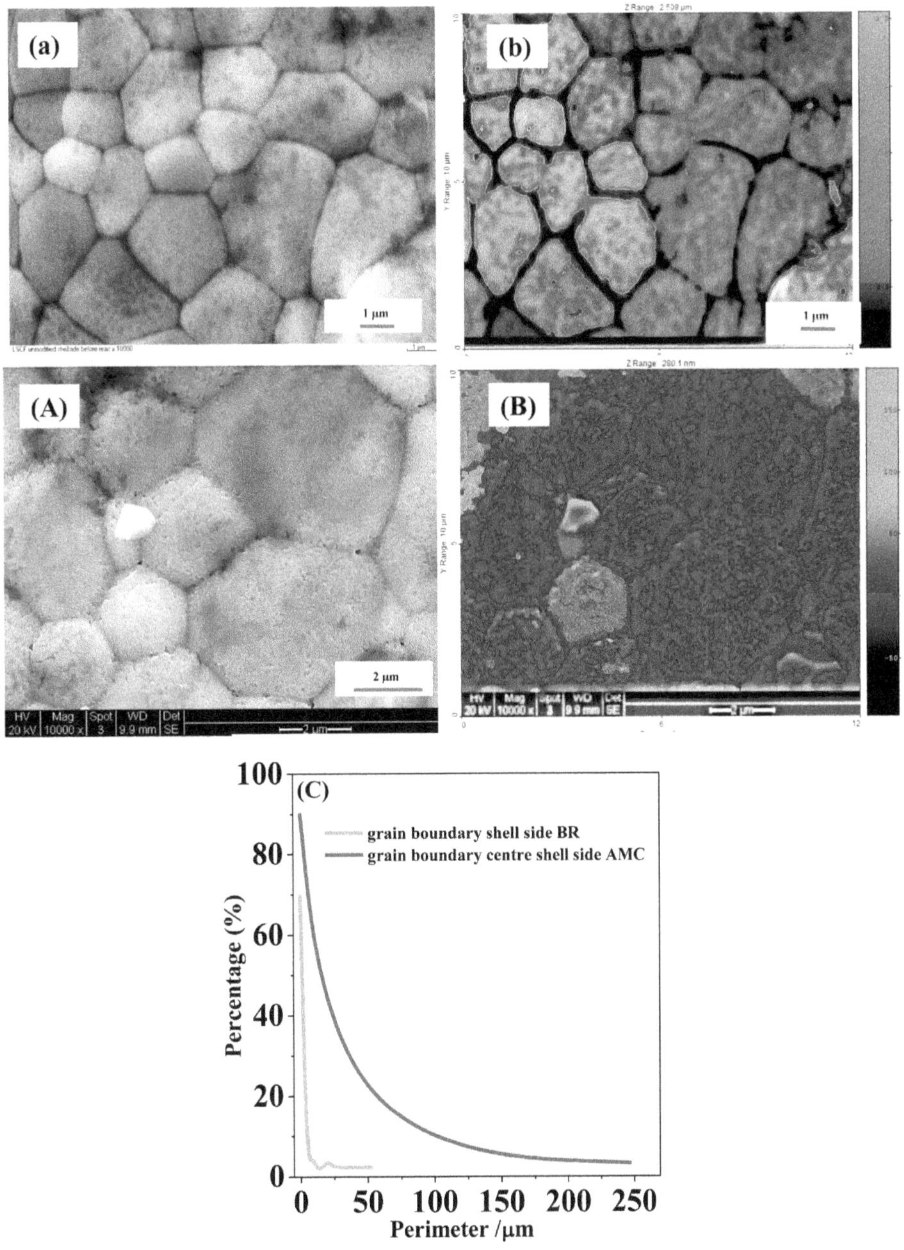

Figure 13. SEM images of the lumen side (a) and image simulated (b) before reaction and centre lumen side after methane conversion measurements (C) and image simulated (D). Distribution of the grain boundaries perimeter obtained by simulation with SPIP 6.0.9. (E)

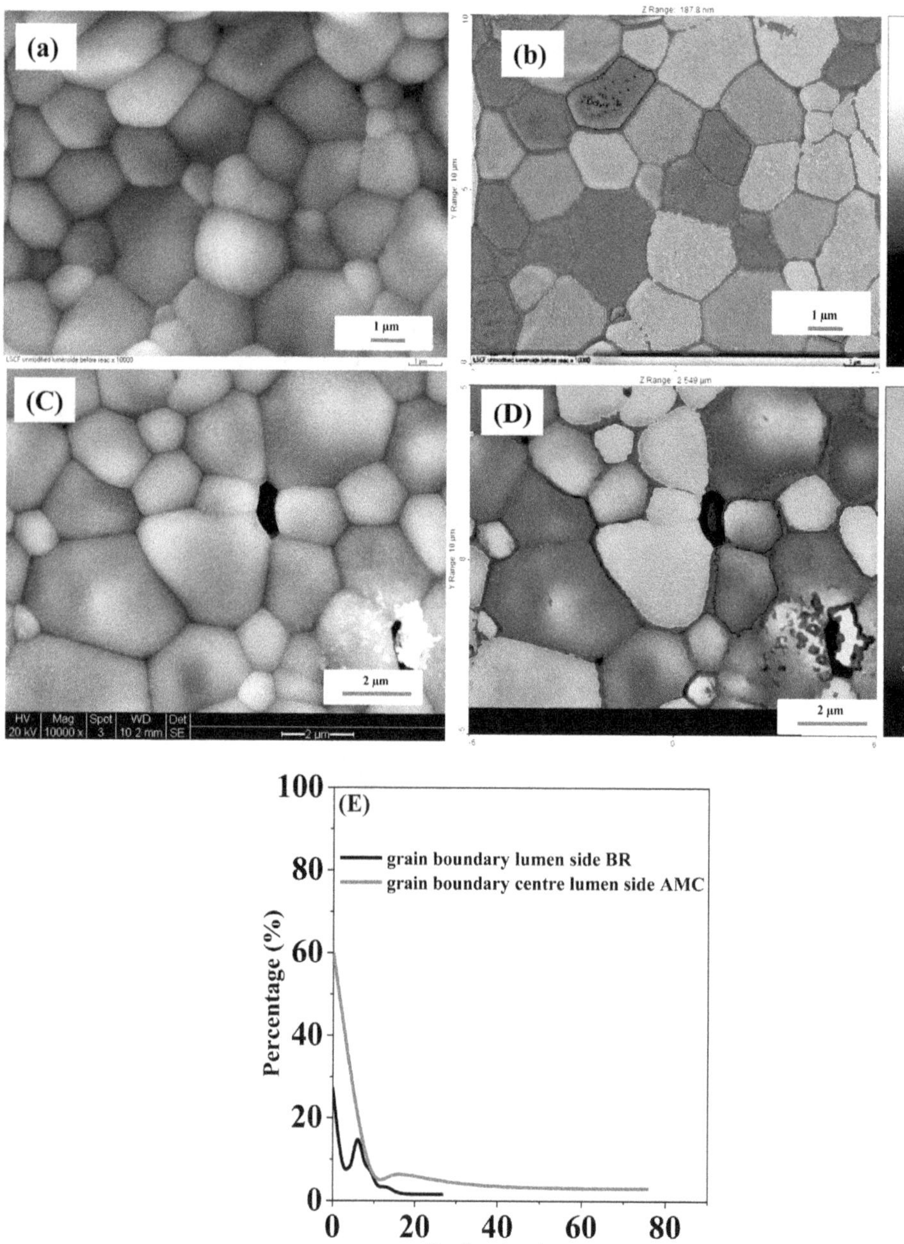

illustrated that the smooth surface of the grain boundaries changed to rough surface due to the influence of the reducing gas atmosphere used (CH4) and produced (CO) during the experiments (Figures 12a and 12A). Furthermore, the grain boundary perimeters slightly increased from (0.02 - 20 µm) before reaction to (0.03-175 µm) after methane conversion (Figures 12b, 12B and 12C). These ranges of perimeters represent between 93% and 96% of the total grains boundaries analysed. The perimeter increase of the grain boundaries could be associated to the possible presence of hot spots during the methane conversion reaction.

The microstructures of the centre lumen side for the unmodified LSCF6428-HFM before and after methane conversion are given in Figures 13a and 13C, respectively. It can be noted that the grain boundary perimeter after partial oxidation of methane slightly increased up to (0.03 to 34 μm), which represent 97% of the total grain boundaries, compared to the grain boundary perimeter from the lumen side (0.02-16 μm) before reaction that correspond to 91% of the total grain boundaries, according to the simulation of the images by SPIP 6.0.9 (Figure 13E). Furthermore, it can be noted that small cavities or pores partially decorated with small particles appear at the interface of the grain boundaries after reaction (Figure 13C).

Comparison of the microstructure from the centre shell side and the centre lumen side after methane conversion revealed that the external membrane side exhibits larger changes in the surface and morphology of the grain boundary than the centre lumen side surface. This is probably due to the fact that the first experiment of methane conversion was under MSAL mode (i.e. 4%CH4/He supplied in the shell side and air sweep in the lumen side). Then, it was changed the operation in MLAS mode (i.e. 4%CH4/He supplied in the lumen side and air sweep in the shell side), where CH4 reacted with relative high oxygen concentration in the lumen side because air was supplied to this side in previous experiments.

In Figure 14 is given the SEM photographs of the end shell side and end lumen side of the LSCF6428-HFM after methane conversion. The magnification of the end shell side after reaction is shown in Figure 14B. It can be noted that the surface and morphology of the grain boundary severely changed as a consequence of the formation of small particles because of the relative high methane concentration in the module inlet.

On the other hand, the microstructure of end lumen side after methane conversion is shown in Figures 14C and 14D. It can be observed that the grain boundary perimeter slightly increased (i.e., 0.03 to 39 μm) compared to the grain boundary perimeter from the lumen side (i.e., 0.02-16 μm) before reaction. The former represents 97% whilst the later corresponds to 91% of the total grain boundaries simulated by SPIP 6.0.9, Figure 14E. In addition, small cavities or pores partially decorated with small particles appear at the interface of the grain boundaries after reaction (Figure 14C). Note that the surface morphologies for the end lumen side compared to end shell side after methane conversion did not vary markedly (Figures 14C and 14B). Comparing the centre shell side (Figure 12C) with end shell side (Figures 14A and 14B) one can infer that the later side underwent further deterioration. This finding can be rationalised considering that in the module extreme, the oxygen permeation is lower than that in the centre shell side because of lower temperature, according to internal temperature profiles of the membrane module. This is reflected on a relatively high methane concentration in the end shell side of the module and hence higher reducing character.

Energy Dispersive X-Ray Spectroscopy (EDXS)

In Table 2 is given the EDXS results for different zones of the LSCF6428 hollow fibre membrane before and after methane conversion. Comparison of the centre shell side of unmodified LSCF6428 post methane conversion with shell side before reaction indicates that the elemental compositions of all metals decreased from (7.2% to 5%) for La; (5.6% to 2.5%) for Sr, (2.2% to 1.2%) for Co and (9.8% to 6.1%) for Fe. It is worth remarking that sulphur (0.3%) and carbon (14.7%) were also observed on this side of the membrane; see Table 2.

Figure 14. SEM micrograph of the end shell side (A) and image magnified (B). Microstructure of the end lumen side (C) and image simulated (D) after methane conversion measurements. Distribution of the grain boundary perimeter before and after methane conversion (E)

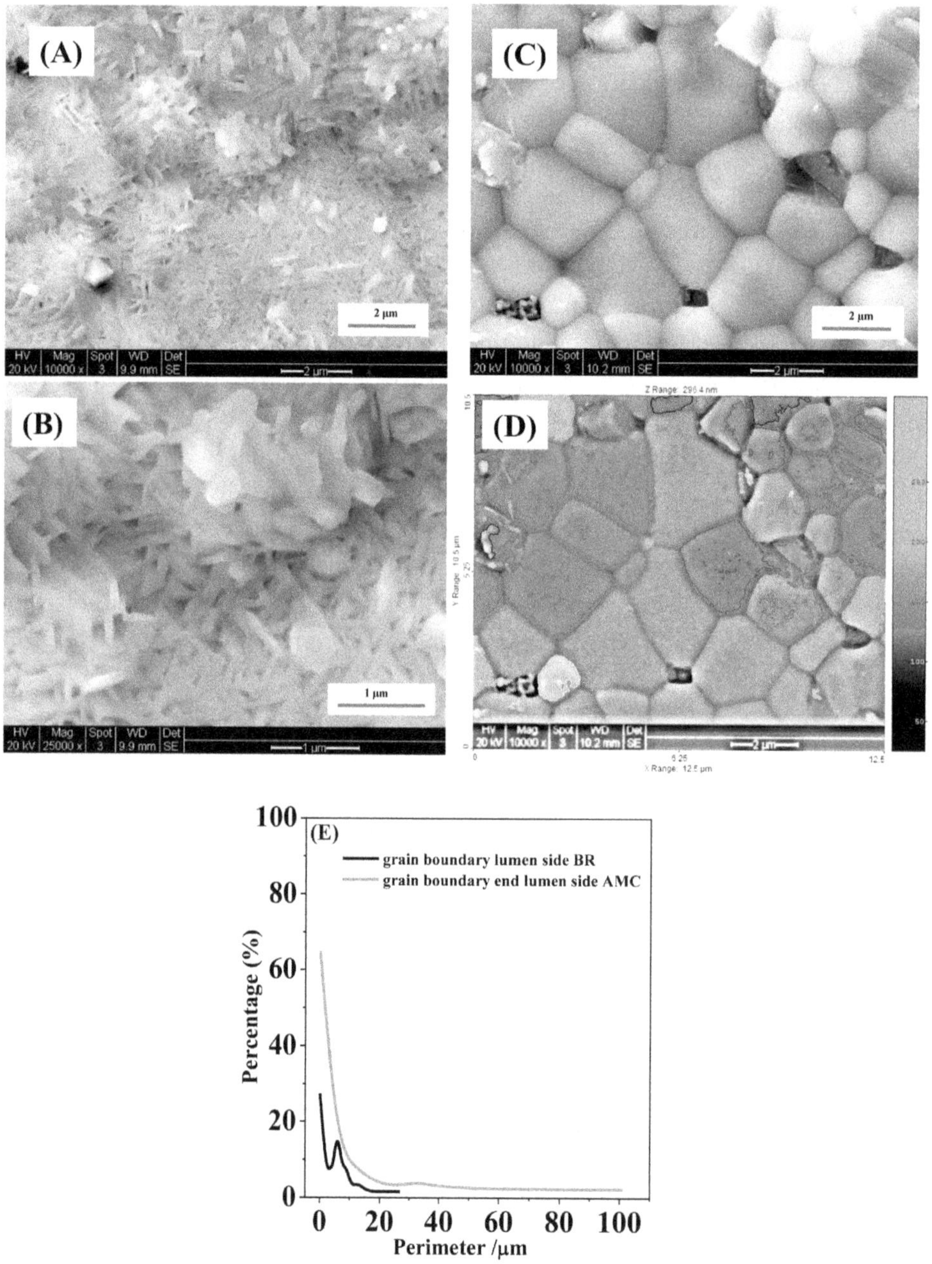

Similar metal composition trends were displayed by the end shell side after methane conversion. These changes in the elemental composition on the surface of HFM are attributed to the reducing gases used or generated during methane conversion and sulphur impurity. Similar tendency was reported elsewhere (Tan et al., 2008).

Table 2. Summary of EDXS results of different zones of the LSCF6428 hollow fibre membrane before and after partial oxidation of methane

Membrane	Condition	Section (Side)	Absolute Atomic %								Experimental Stoichiometry
			La	Sr	Co	Fe	O	C	S	Al	
LSCF6428	**BR**	**Lumen**	**8**	**5.6**	**2.3**	**10.3**	**73.8**				**La0.6:Sr0.4:Co0.2:Fe0.8**
LSCF6428	AR	Centre lumen	7.6±0.92	3.5±1.13	1.2±0.50	8.8±1.11	74.6±9.52	2.3	0.92		---
LSCF6428	AR	End lumen	7.5±0.6	3.41±0.63	1.2±0.5	9.02±1.05	64.8±12.9	13.0	0.98		---
LSCF6428	**BR**	**Shell**	**7.2**	**5.6**	**2.2**	**9.8**	**75.4**				**La0.6:Sr0.4:Co0.2:Fe0.8**
LSCF6428	AR	Centre shell	5.0±0.78	2.5±1.1	1.2±0.59	6.1±1.2	70.0±11.6	14.7	0.3		---
LSCF6428	AR	End shell	4.6±0.66	3.2±0.81	1.3±0.55	6.1±1.03	71.3±13.1	13.4	0.1		---

BR=Before Reaction, AR=After Reaction

Considering the above mentioned XRD, SEM and EDXS results from unmodified LSCF6428-HFM after methane conversion, it can be concluded that crystalline phases, morphology, perimeter and metal composition of the grain boundary changed on the centre shell side and the end shell side compared to the membrane composition before reaction. This is attributed to kinetic demixing of mobile cations and decomposition of the membrane due to the axial temperature generated by the exothermic reaction of methane combustion (Stephens et al., 2000). Note that large amount of residual carbon remains on the membrane after the catalytic reaction, affecting the efficiency of the catalyst and/or membrane performance.

Moreover, the membrane was contaminated by sulphur impurity which comes from air and helium gases employed during the experiments, polyethersulfone used during fabrication of the LSCF6428 hollow fibre and the cement used to assemble the reactor. Also observed was the segregation of strontium, cobalt and iron to form different crystalline oxide and sulphate phases such as SrO_2, $SrSO_4$, $CoSO_4$ and Fe_3O_4. Additionally, the formation of $SrCO_3$ was observed, owing to the reaction of CO_2 and/or CO with segregated strontium oxide, according to eqns. (22) and (23) (Balachandran et al., 1997; Tan et al., 2008; Ten Elshof et al., 1995). According to the ΔG calculation at 850°C, the eqn. (42) would have a minor contribution toward the formation of $SrCO_3$. (-52.3 kJ mol-1) that the eqn. (43) (-258 kJ mol^{-1}).

$$SrO + CO_2 \rightarrow SrCO_3 \qquad (42)$$

$$SrO_2 + CO \rightarrow SrCO_3 \qquad (43)$$

Characterization of the 5%Ni-LSCF6428/LSCF6428-HFM Reactor Before and After Methane Oxidation Reactions

X Ray-Diffraction (XRD)

The XRD patterns of the 5%Ni-LSCF6428/LSCF 6428 hollow fibre membrane module before and after methane conversion are shown in Figure 15.

Figure 15. XRD patterns of the 5%Ni-LSCF6428/LSCF6428 hollow fibre membrane. LSCF6428 powder (a), 5%Ni/LSCF6428 powder (b), shell side (c) and lumen side (d) before the oxygen flux measurements. Centre lumen side (C) and centre shell side (D) after methane conversion experiments

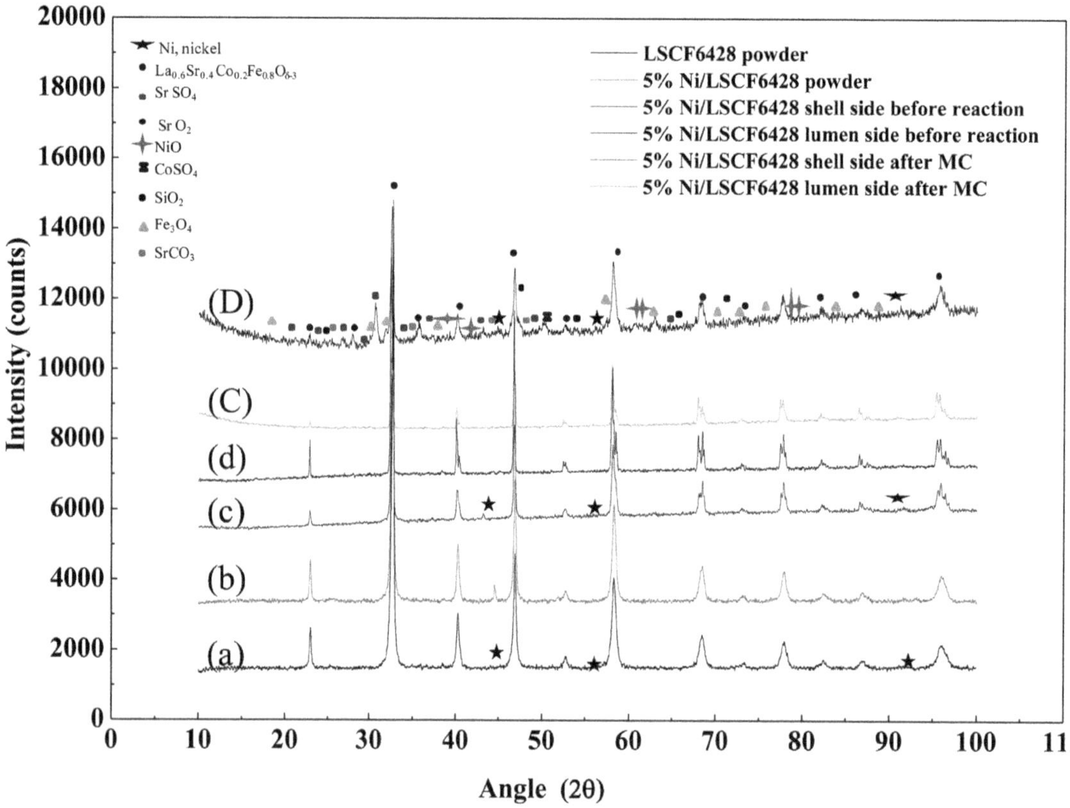

Both the shell side and lumen side samples exhibit the characteristic peaks of LSCF6428 as the major phase and only the 5%Ni-LSCF6428/LSCF6428-HFM powder and shell side before reaction presented a second crystalline phase that corresponds to metallic nickel, Ni (00-004-0850). Furthermore, the diffraction profiles for centre lumen side and centre shell side post methane conversion experiments showed LSCF6428 as the main crystalline phase and other minor crystalline phases such as SrO_2 (00-007-0234), $SrSO_4$ (00-005-5935) and only for the centre shell side were observed $CoSO_4$ (10124-43-3), Fe_3O_4 (01-088-0315), NiO (1313-99-1), SiO_2 (00-005-0490) and $SrCO_3$ (1633-05-2), see Figure 15D. It is worth mentioning that only one XRD profile from the lumen side and the shell side after methane oxidation are shown in Figures 15C and 15D, respectively.

The above-mentioned XRD results show that the shell side of 5%Ni-LSCF6428-HFM after methane conversion exhibited certain impurities such as sulphur and silica detected as $SrSO_4$, $CoSO_4$ and SiO_2 crystalline phases, Figures 15D. Similar sulphur-containing crystalline phases have been previously reported (Li et al., 2000; Thursfield & Metcalfe, 2006; Thursfield & Metcalfe, 2007; Xu et al., 1999). The formation of these phases was attributed to the presence of sulphur in the air and helium gases employed during the experiments and the polyethersulfone used during the fabrication of the LSCF6428 hollow or/and the cement used in the reactor assemble. Additionally, $SrCO_3$ was observed as a consequence of the CO_2 interaction with segregated strontium oxide (Balachandran et al., 1997; Tan et al., 2008; Ten Elshof et al., 1995).

Figure 16. SEM images of the shell side before reaction (a), image simulated by SPIP 6.0.9 (b). SEM images of the grain boundary in lumen side before reaction (c), image simulated by SPIP 6.0.9 of the lumen side (d) and distribution of particle (5%Ni-LSCF6428 catalyst) and grain boundary perimeter (e)

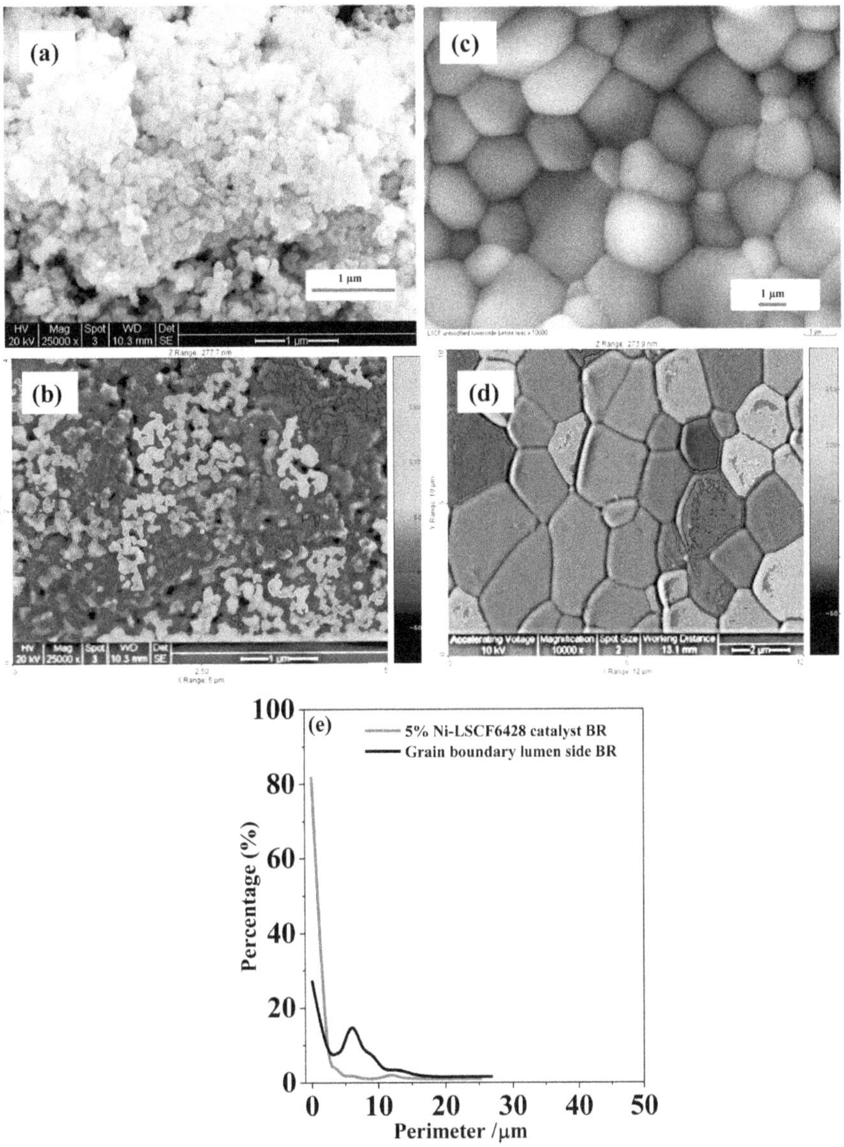

Scanning Electron Microscopy (SEM)

The SEM photographs of the shell side and the lumen side of the modified LSCF6428-HFM pre operation are given in Figure 16.

The magnification image for shell side before reaction is shown in Figure 16a; it can be observed that the catalyst was well-distributed on the shell side of the LSCF6428-HFM. The particle perimeter range was between 0.01 and 4.5 µm, which represent around 92% of the total grain analysed in Figure 16e.

Furthermore, note that the grain boundary perimeter for lumen side exhibits ranges between 0.02 to 16 µm, representing 91% of the overall grain boundary perimeter, see Figure 16e. The deposition of 5%Ni/LSCF6428-HFM catalyst fully covered the grain boundary of the HFM. . The particle perimeter range was between 0.01 and 4.5 µm, which represent around 92% of the total grain analysed in Figure 16e.

The SEM photographs of the centre shell side and the centre lumen side of the modified LSCF6428-HFM post methane conversion are illustrated in Figure 17. It can be clearly observed in the microstructure of the centre shell side after methane conversion (Figures 17A and 17B) that the grain size of 5%Ni/LSCF6428 catalyst increased compared to the grain size range before reaction (Figure 16a). This was attributed to the agglomeration of catalyst particle and oxidation of Ni metal to NiO during the operation time. Also, note that the microstructure of the centre shell side after methane conversion was markedly affected by the reaction and product atmosphere (Figures 17A and 17B).

SEM pictures of the centre lumen side of 5%Ni-LSCF6428/LSCF6428-HFM post methane oxidation are shown in Figures 17C and 17D. It can be noted that the grain boundaries perimeter show significant changes with respect to the grain boundaries perimeter before reaction. In addition, the surface morphology of grain boundaries was modified with small particles and cavities or pores among the grain boundaries due to high temperature and the reaction atmosphere (i.e., CH_4, CO and CO_2).

The microstructures of the end shell side and the end lumen side after methane oxidation are given in Figure 18. These photographs show similar changes in the morphology both in the catalyst surface (i.e., end shell side) and in cavities or pores at the interfaces of the grain boundaries of the end lumen side compared to shell side and lumen side (see Figures 16a and 16c) before methane oxidation.

Figure 17. SEM microstructure of the centre shell side of 5%Ni-LSCF6428/LSCF6428-HFM (A), image magnified (B), the centre lumen side (C), Image magnified (D) after methane oxidation

Figure 18. SEM microstructure of the end shell side (A), images magnified (B), the end lumen side (C), Image simulated (D) after methane conversion and distribution of the grain boundary perimeter (E)

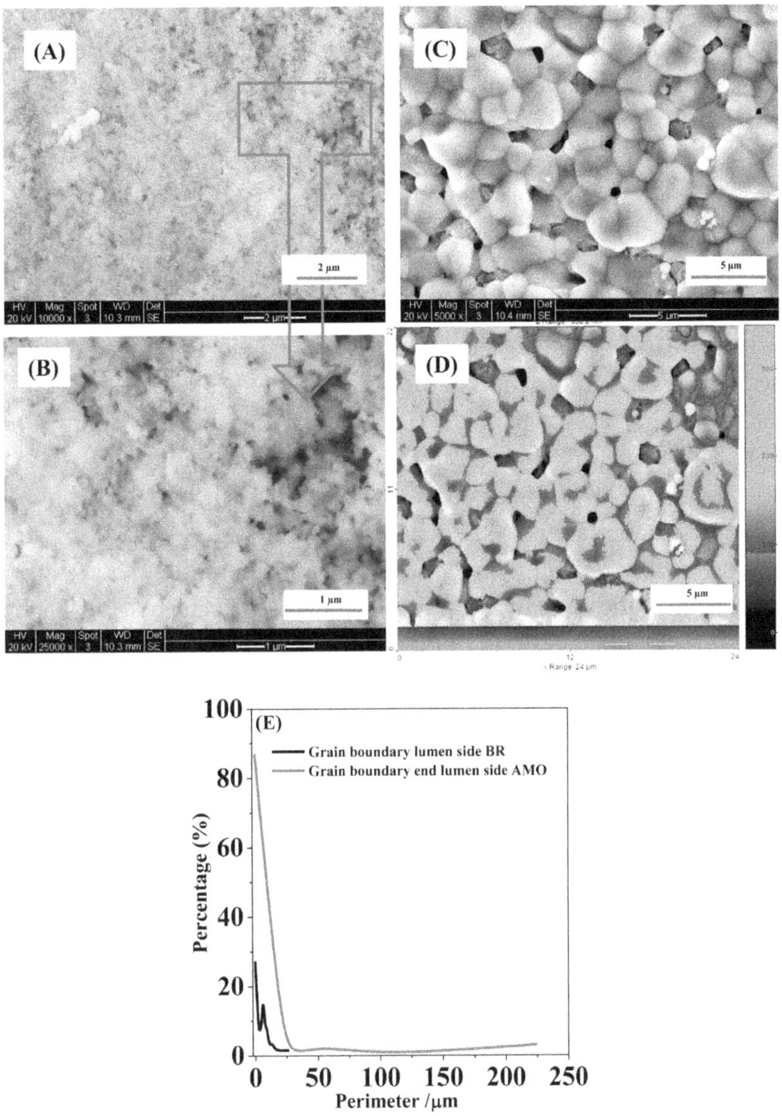

In addition, the grain boundary perimeter for end lumen side (i.e., 0.05 to 50 μm) with 97% of the overall grain boundary perimeter analysed, see Figure 18E, slightly increased compared to lumen side before reaction, which presented grain boundary perimeter within 0.02 and 16 μm that correspond to 91% of the overall grain boundary perimeter simulated, see Figure 18E.

Taking into account the above discussed XRD and SEM results for the centre shell side of the 5%Ni/LSCF6428-HFM after methane conversion. It can be noted that the morphology of the catalyst changes as a consequence of an oxidation process of metallic nickel to NiO and to the agglomeration of the particles due to the high thermal treatment (Figures 17A and 17B). Furthermore, re-dispersion of the catalyst during the methane conversion produced small particles on the catalyst surface (Figure 17A).

Energy Dispersive X-Rays Spectroscopy (EDXS)

The Table 3 shows the EDXS results for different zones of the 5%Ni-LSCF6428 hollow fibre membrane before and after methane conversion reaction. Comparison of the centre shell side of 5%Ni-LSCF6428-HFM before reaction with the shell side after methane conversion, it can be noted that the elemental compositions decreased from (7.4% to 6.3%) for La; (4.7% to 3.3) for Sr, (9.1% to 7.8%) for Fe, (2.2%to 1.6%) for Co, whilst Ni (ca.1.6%) contents did not change. In addition, it can be noted the presence of carbon (14.1%) and sulphur (1.0%) on this membrane section, see Table 3.

On the other hand, an overall diminution of the metal composition for La, Sr, Co and Fe (i.e., 5.29%, 3.4%, 1.3% and 8.3%, respectively) in the centre lumen side post methane conversion compared to lumen side before reaction of 5%Ni-LSCF6428/LSCF6428-HFM was observed. Also, it can be observed that sulphur content decreases (0.1%) compared to the fresh lumen side (0.3%), see Table 3. A high carbon content was detected (i.e. 13.6%) in this area of the HFM.

Meanwhile, the end shell side showed a clear diminution of the metal compositions for Ni (1.7% to 1.1%), La (7.4% to 5.5%), Co (2.2% to 1.3%), Fe (9.1% to 6.4%), and Sr (4.7% to 3%). In addition, sulphur (1%) and carbon (14.8%) were detected on this membrane section; see Table 3. Furthermore, at the end lumen side La, Co and Fe cations content decreased from 8% to 5.2%; 2.3% to 0.98% and 10.3% to 6.4%, respectively, whilst the remaining metal (i.e., Sr) concentration increased from 5.5% to 6.7%, see Table 3. These changes in the elemental composition on the surface of HFM are a consequence of several factors such as the reducing gas mixture used (4%CH_4/He) or generated (CO) during methane conversion, impurities and the kinetic demixing of mobile cations of the membrane.

Table 3. Summary of EDXS results of different zones of the 5%Ni-LSCF6428/LSCF6428-HFM before and after methane conversion

Membrane	Condition	Section (Side)	Absolute Atomic %									Experimental Stoichiometry
			La	Sr	Co	Fe	O	Ni	S	Ca	C	
LSCF6428	**BR**	**Lumen**	**8.0±0.9**	**5.5±0.8**	**2.3±0.6**	**10.3±1.2**	**73.8±13.5**		**0.3**	**0.5**		**La0.6:Sr0.4:Co0.2:Fe0.8**
LSCF6428	**BR**	**Shell**	**7.2**	**5.6**	**2.2**	**9.8**	**75.4**					**La0.6:Sr0.4:Co0.2:Fe0.8**
5%Ni-LSCF6428	**BR**	**Shell**	**7.4±0.8**	**4.7±0.9**	**2.2±0.5**	**9.1±1.1**	**74.9±14.0**	**1.7±0.6**				**La0.6:Sr0.4:Co0.2:Fe0.8**
5%Ni-LSCF6428	**BR**	**Cross**	**7.4±1.8**	**5.5±1.7**	**2.3±0.6**	**10.2±1.4**	**72.8±15.5**		**0.3**			**La0.6:Sr0.4:Co0.2:Fe0.8**
5%Ni-LSCF6428	AR	Centre lumen	5.2	3.4	1.3	8.3	68.9		0.14	0.5	13.6	
5%Ni-LSCF6428	AR	Centre shell	6.3	3.3	1.6	7.8	64.4	1.6	0.98		14.1	
5%Ni-LSCF6428	AR	Cross	4.7	2.8	1.4	6.2	70.4		1.26		13.2	
5%Ni-LSCF6428	AR	End shell	5.5	3	1.3	6.4	66.9	1.1	1		14.8	
5%Ni-LSCF6428	AR	End lumen	5.2	6.7	0.98	6.4	65.5		3.8		11.4	

BR=Before Reaction, AR=After Reaction

According to the results obtained from XRD, SEM and EDXS analysis on the shell side of the 5%Ni-LSCF6428/LSCF6428-HFM module after methane conversion suggest that the 5%Ni-LSCF6428 catalyst coated on the shell side of the LSCF6428 hollow fibre membrane displayed morphology changes likely due to the oxidation of metallic nickel to NiO by oxygen chemisorption on active sites of the catalyst. In addition, the particles were agglomerated by the high thermal treatment; furthermore carbon deposition on the HF membrane surface occurred during the experiments. These processes could affect the number of active sites in the catalyst, as reflected in the products selectivity during methane oxidation experiments.

XRD characterization of the module reactor after methane conversion showed different crystalline oxide/sulphate/carbonate phases (i.e., SrO_2, Fe_3O_4, $SrSO_4$, $CoSO_4$ and $SrCO_3$). The presence of sulphate and carbonate phases are attributed to sulphur impurities and reducing gases used and generated (i.e., CH_4 and CO), which could cause deterioration of the 5%Ni/LSCF6428-HFM performance. Also, it can be observed that the module was contaminated by silica (SiO_2), which can likely come from the sealant or from the quartz.

H_2-TPR Analysis of the Unmodified and Ni-LSCF6428-HFM Modified LSCF6428-HFM Reactor Before and After Methane Oxidation Reactions

The analysis of $La_{0.6}Sr_{0.4}Co_{0.2}Fe_{0.8}O_{3-\delta}$ hollow fibre membrane by H_2-TPR can give further details of the influence of the reaction atmosphere on its thermal stability and partial reduction of Co^{3+} and Fe^{3+} cations. The H_2-TPR profiles of the unmodified (A) and modified (B) LSCF6428 HFM before and after the methane oxidation reactions are shown in Figure 19. The TPR profiles display three major reductions steps denoted as α (between 200 and 550 °C), β (between 550 and 875 °C) and γ (above 875 °C). Note that the deconvolution of the HFM TPR profile (Figure 19A) clearly reveals that these reduction processes involve a complex number of species and reduction steps, whose maximum reduction temperature and overall hydrogen consumption are strongly dependent upon the HFM stage (i.e., before or after the methane oxidation reactions) and the Ni modification of the $La_{0.6}Sr_{0.4}Co_{0.2}Fe_{0.8}O_{3-\delta}$ membrane, see Table 4.

Table 4. Maximum reduction temperatures, moles of H_2 consumed per gram of sample or mole of metal and assignment of overall reduction process for unmodified and modified $La_{0.6}Sr_{0.4}Co_{0.2}Fe_{0.8}O_{3-\delta}$ hollow fibre membrane

Sample	A (°C)	B (°C)	Γ (°C)	mmol H_2/ Sample-g	(H_2/Metals) Experimental	(H_2/Metals) Theoretical	[a]Reduction (H_2 mol. %)	Assignment
HFM	380	710	900	3.46	0.76	0.70	50.7	$Co^{3+} \rightarrow Co^0$ $Fe^{3+} \rightarrow Fe^{2+}$
Spent HFM	408	760	990	5.13	1.13	1.10	75.1	$Co^3 \rightarrow Co^0$ $Fe^{3+} \rightarrow Fe^+$
0.65 wt% Ni-HFM	365	850	960	4.18	0.90	0.90	61.2	$CO^{3+} \rightarrow Co^0$ $Fe^{3+} \rightarrow Fe^+$ $Ni^{2+} \rightarrow Ni^0$
Spent 0.65 wt% Ni-HFM	455	720	975	3.60	0.78	0.70	52.8	$Co^3 \rightarrow Co^0$ $Fe^{3+} \rightarrow Fe^{2+}$ $Ni^{2+} \rightarrow Ni^0$

[a]Relative to the total reduction of HFM (i.e., 6.83 mmol H_2/ HFM-g)

Figure 19. H_2-Temperature-programmed reduction of the unmodified (A) and modified (B) $La_{0.6}Sr_{0.4}Co_{0.2}Fe_{0.8}O_{3-\delta}$ hollow fibre membrane before (dark line) and after (red line) the methane oxidation reactions

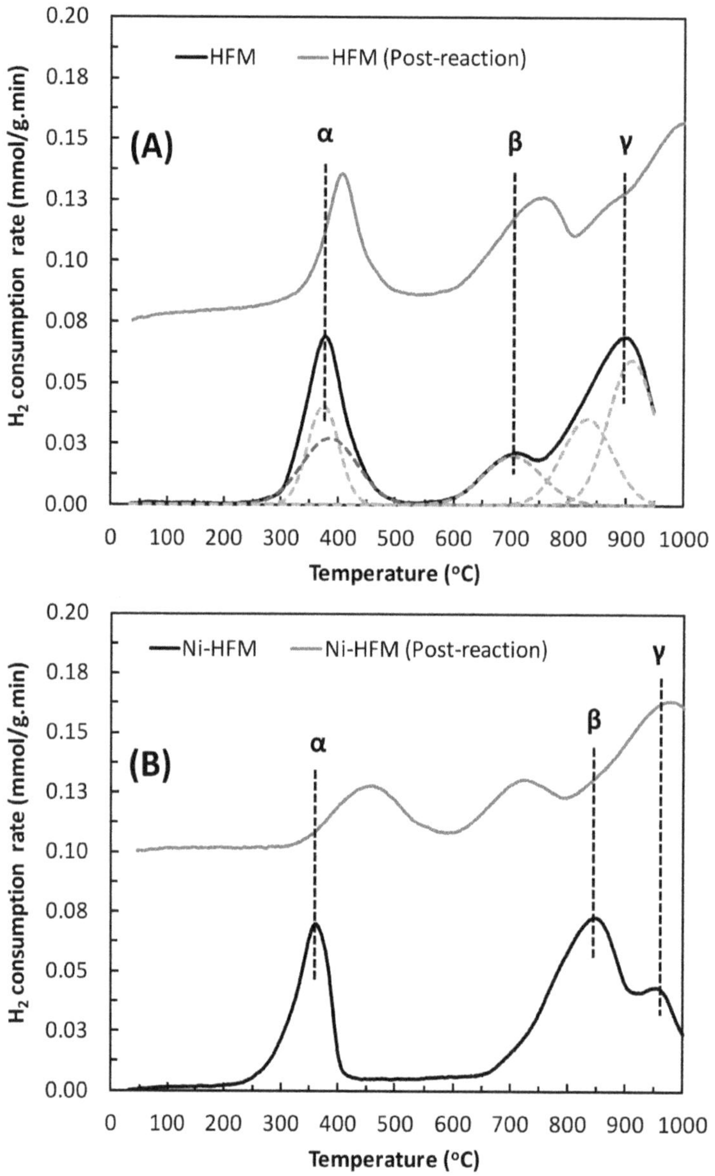

The theoretical and experimental moles of H_2 consumed per moles of metals (i.e., Fe, Co and Ni), the overall perceptual reduction and the assignment of the overall reduction process are also listed in Table 4. The unmodified HFM before the methane oxidation reactions showed three well-defined reduction steps at 380, 710 and 900 °C. The measured H_2 consumption in the peaks at 380 and 710 °C suggests that Co^{3+} is reduced to Co^0 based on the peak deconvolution and the Co^{3+} content in the HFM, whilst Fe^{3+} is reduced to Fe^{2+} considering the measured H_2 consumption from the wide peak at 900 °C. It is worth remarking that $SrLaFeO_4$ (i.e., K_2NiF_4-type structure) was identified after temperature-programmed reduction of $La_{0.5}Sr_{0.5}Co_{0.2}Fe_{0.8}O_{3-\delta}$ (Berry et al., 2005) and $La_{0.8}Sr_{0.2}Co_{0.5(0.4)}Fe_{0.5(0.6)}O_{3-\delta}$ perovskites (Petkovic, Utgikar & Rashkeev, 2011) because of the reduction of Co^{3+} cation. Note that ca. 51 mole % of H_2 was consumed relative to the overall hydrogen consumption of the stoichiometric reduction reaction (eqn. 44) as a consequence of its relatively high thermal stability and high calcination temperature (i.e., 1100-1280 °C).

$$La_{0.6}Sr_{0.4}Co_{0.2}Fe_{0.8}O_{2.8} + 1.5H_2 \rightarrow 0.3La_2O_3 + 0.4SrO + 0.2Co + 0.8Fe + 1.5H_2O \quad (44)$$

Indeed, an equivalent composition of perovskite phase calcined at 900 °C showed an overall H_2 consumption of 4.76 mmol H_2/sample-g (Huang, Shen & Chou, 2009), which represents ca. 70 mole % of H_2 of the total hydrogen uptake. On the other hand, the maximum temperature of the major H_2 consumption peaks of the HFM after the methane oxidation reactions were shifted toward higher temperatures compared to the $La_{0.6}Sr_{0.4}Co_{0.2}Fe_{0.8}O_{3-\delta}$ HFM before the catalytic reaction. Nevertheless, the overall hydrogen consumption is significantly higher than that of HFM, owing to the subsequent reduction of Fe^{2+} to Fe^{+}. This finding could be attributed to the segregation of Fe_3O_4 from the HFM structure upon the methane oxidation reactions as it was confirmed by XRD (see Figures 11 and 15).

The H_2-TPR profiles of the Ni- LSCF6428-HFM modified LSCF6428 HFM before the methane oxidation reactions displays an enhancement in the reducibility of the low temperature reduction α peak compared with the unmodified HFM data, Table 4, likely due to the autocatalytic effect of metallic Ni on the H_2 dissociation reaction. This tendency to higher reducibility was also reflected in the overall percentage of H_2 consumption. Apparently, Co^{3+} is partially reduced to Co^{2+} whilst Fe^{3+} was reduced to Fe^{+}. The post-reaction Ni-LSCF6428-HFM modified HFM showed a markedly higher initial reduction temperature than that for Ni-modified $La_{0.6}Sr_{0.4}Co_{0.2}Fe_{0.8}O_{3-\delta}$ HFM before the catalytic reaction. A similar tendency showed the unmodified HFM, this is attributed to the presence of $CoSO_4$, whose reduction temperature is considerable higher than that of cobalt oxides (Wang et al., 2013). Furthermore, the segregation of Sr^{2+} cation from the $La_{0.6}Sr_{0.4}Co_{0.2}Fe_{0.8}O_{3-\delta}$ lattice could decrease the reducibility of the perovskite structure based on its effect over the H_2-TPR profiles of $LaCoO_3$ and $La_{0.6}Sr_{0.4}CoO_{3-\delta}$ perovskite (Niu et al., 2007). It is worth remarking that the catalyst reducibility and overall H_2 consumption can also be affected by the presence of residual carbon in the post-reaction samples.

Thermodynamic Calculations

Figure 20 shows the standard Gibbs free energy for the oxidation of methane, ethane and ethylene at different temperatures. The thermodynamic calculations were carried out using HSC Chemistry 5 software.

It can be observed that the total oxidation reactions of methane and C_2 are thermodynamically favourable because of their lower ΔG. Furthermore, the total oxidation of methane is more favourable than the partial oxidation of methane at high oxygen concentration, according to thermodynamic calculation, Figure 20. These findings are in agreement with experimental results obtained in this work and previous reports (Al-Sayari, 2013).

The thermodynamic stability of perovskite phases under O_2/CH_4 atmosphere was evaluated. The standard Gibbs free energy for single perovskite phases, rather than LSCF6428 perovskite due to absence of thermodynamic data, were calculated and given in Figure 21. It can be appreciated that the reaction of $LaCoO_3$ perovskite with O_2/CH_4 atmosphere exhibits lower ΔG compared to $LaFeO_3$ and $Sr_2Fe_2O_5$

Figure 20. Dependence of standard Gibbs free energy with temperature for partial and total oxidation of methane, ethane and ethylene

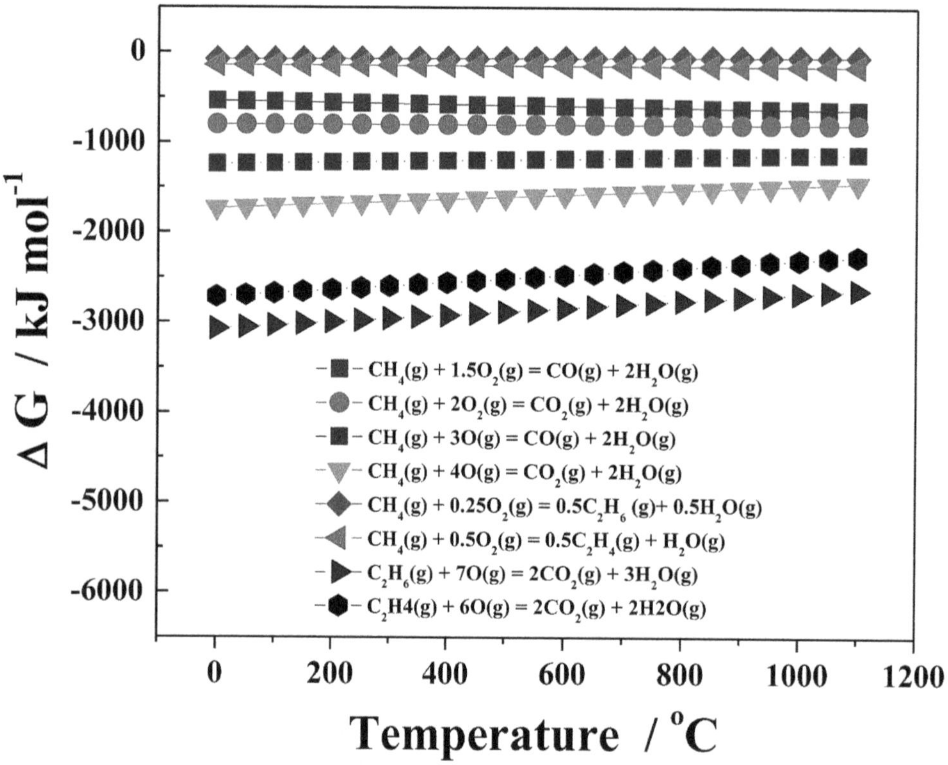

perovskites. It is envisaged that $La_{0.6}Sr_{0.4}Co_{0.2}Fe_{0.8}O_{3-\delta}$ could follow a similar thermodynamic tendency than $LaCoO_3$ and $LaFeO_3$ perovskite. According to the results given in Figure 21, the $LaCoO_3$ perovskite was selected to find out its stability under O_2/CH_4 atmosphere. Below are shown the standard Gibbs free energy for various oxidations of methane reactions at $LaCoO_3$ perovskite model (Figure 22).

It can be noted that the reactions that exhibit an O_2/CH_4 ratio of 0.33 are thermodynamically more favourable because of its lower ΔG, see Figure 22A and eqn. (45).

$$6LaCoO_3 + 3CH_4(g) + O_2(g) \rightarrow 3La_2O_3 + 2Co_3O_4 + 3CO(g) + 6H_2(g) \quad (45)$$

The dependence of ΔG at 850°C with the O_2/CH_4 ratio indicates that the $LaCoO_3$ perovskite is less stable at O_2/CH_4 ratio below 0.5, see Figure 22B. At the operation condition carried out the experiments in this work, it was observed $CoSO_4$ rather than Co_3O_4 and/or CoO because of its sulphurization with sulphur impurity. Also, note that syngas formation is thermodynamically favourable. On the other hand, the total oxidation of methane to CO_2 are thermodynamically favourable when the ratio is between 0.5 and 2 see Figure 22B. This finding suggests that high O_2/CH_4 ratio would improve the $LaCoO_3$ perovskite stability relative to low O_2/CH_4 ratio (i.e., below 0.5).

Figure 21. Dependence of standard Gibbs free energy with temperature for total and partial oxidations of methane reactions over $LaCoO_3$, $LaFeO_3$ and $Sr_2Fe_2O_5$ perovskites

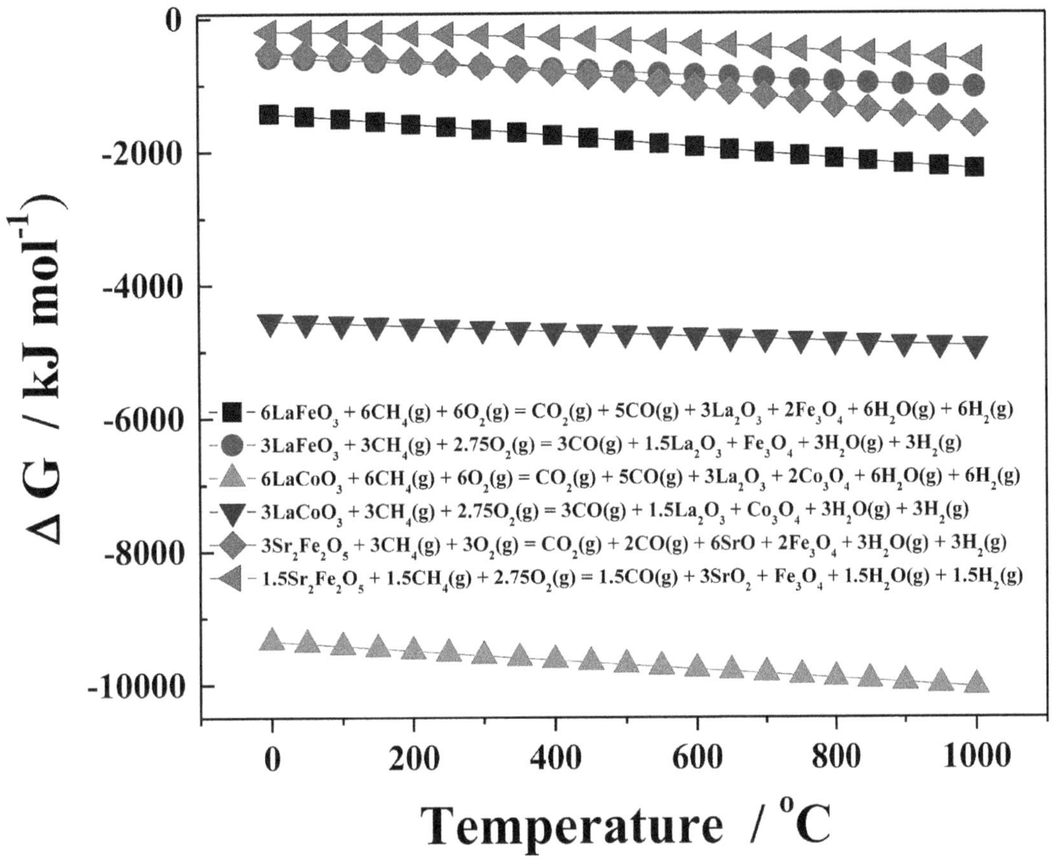

Additional evidence that support the above-discussed thermodynamic data is given in the phase diagram for La-Co-O system (see Figure 23). It can be observed that Co_3O_4 and La_2O_3 are thermodynamically favourable under different partial pressure of oxygen at 850°C.

Similar thermodynamic studies were carried out for $LaFeO_3$ and $Sr_2Fe_2O_5$ perovskites. It can be noted that the phase more thermodynamic stable are La_2O_3 and Fe_3O_4 for La-Fe-O system while SrO_2, Fe_2O_3 and Fe_3O_4 were the most stable phases for Sr-Fe-O system at 850°C. These findings are in agreement with experimental results obtained in this work, which suggests that the thermodynamic calculation for single perovskite phases can give some insight into the phase evolution of $La_{0.6}Sr_{0.4}Co_{0.2}Fe_{0.8}O_{3-\delta}$ perovskite.

FUTURE RESEARCH DIRECTIONS

The use of unmodified and modified LSCF6428 reactors in the reforming of methane could bring some benefit not only in the oxygen separation/purification process but also in the catalytic reaction. This dual role of the perovskite hollow fibre membrane is a key factor to control not only the performance

Figure 22. Dependence of standard Gibbs free energy with temperature for various oxidations of methane reactions of the $LaCoO_3$ perovskite (A), dependence of the standard Gibbs free energy versus O_2/CH_4 ratio for $LaCoO_3$ perovskite at 850 °C (B)

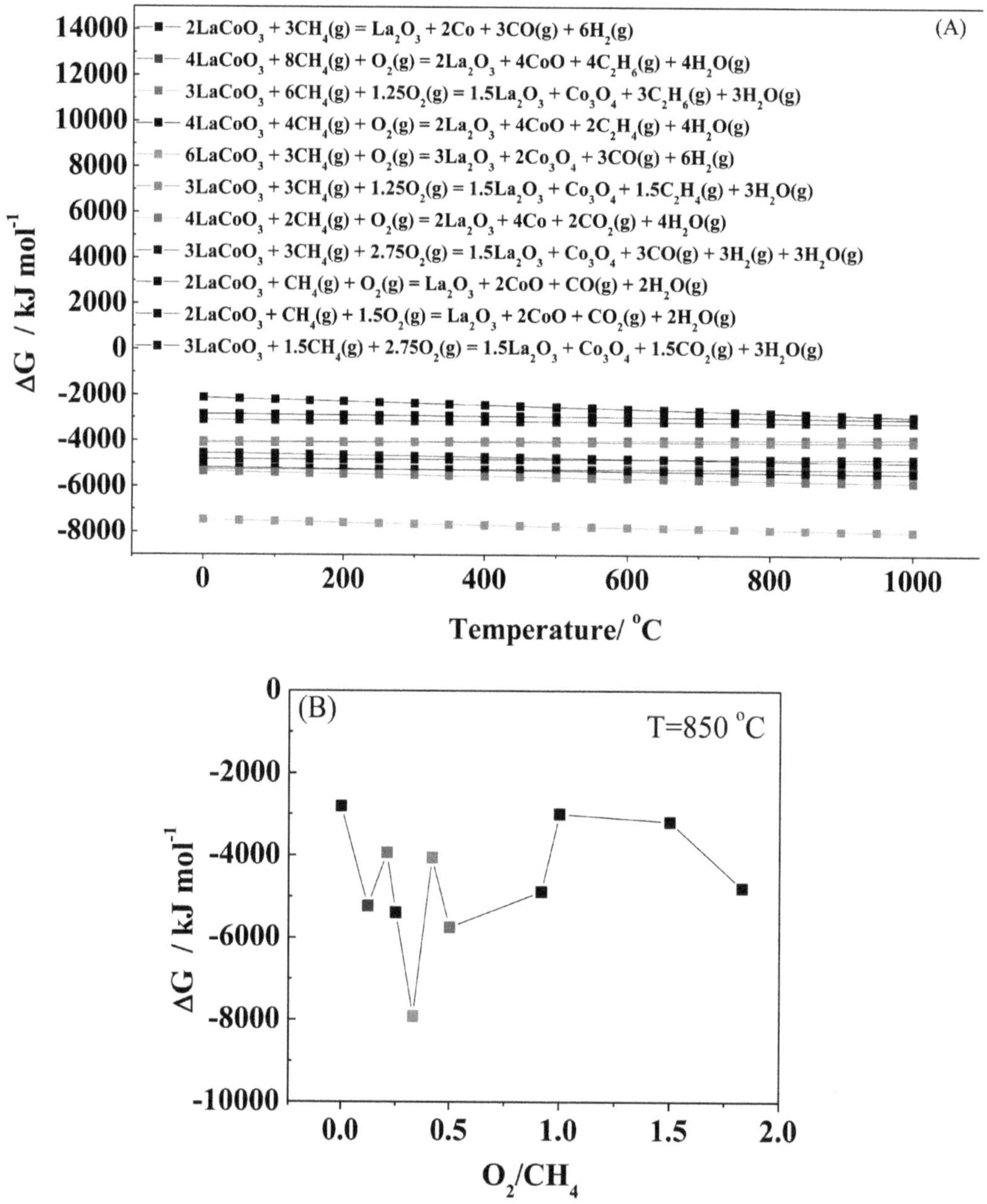

of the catalyst-membrane system but also the membrane integrity. Furthermore, the impurities such as sulphur, steam and carbon dioxide present in the feed or as reaction products may cause membrane degradations. It is either needed ceramic membranes further resistant to this stream or the capture of these gases before feeding the rector membrane.

Figure 23. Phase diagram for $LaCoO_3$ system in function of $logpO_2$ against temperature

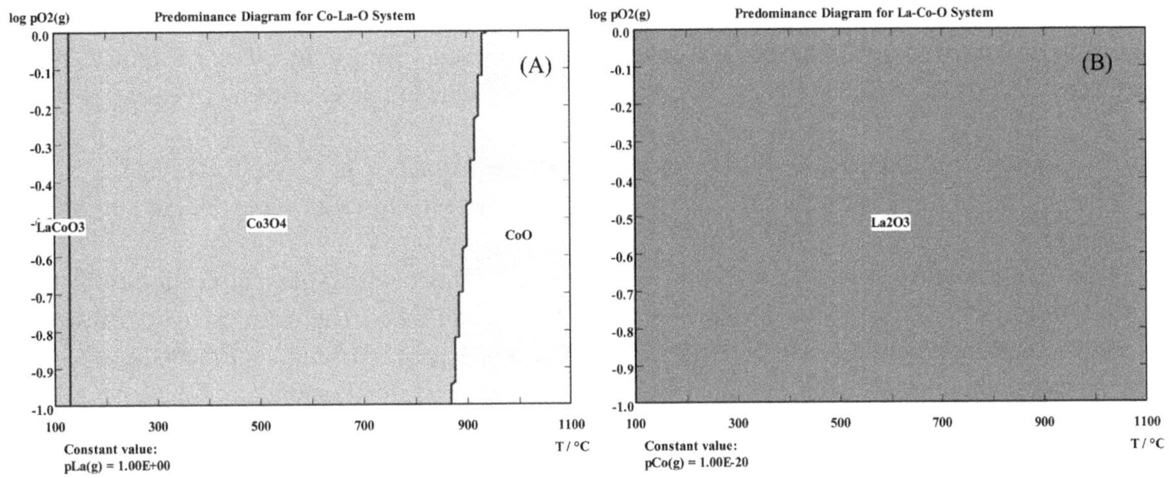

The assemble LSCF6428 hollow fibre membrane reactor quite often presents problem of sealing between the two ends of the membrane with cement, owing to the expansion at high temperatures that produce cement/membrane fracture and thereby generate gas leaks in the module. We recommend using a high temperature resistant silicone sealant to seal the ends of the membrane. Wei *et al.* recently reported sealing problems in the hollow fiber membrane due to the expansion at high temperature (Wei et al., 2011). They proposed the use of U-shaped hollow fiber membranes, whose two ends can be fixed on the same side by ceramic sealant. The hollow fiber can expand and shrink freely when varying temperature.

In a membrane reactor, the change in oxygen partial pressure over the membrane is relatively large (Den Exter, Haije & Vente, 2009). This will put significant constraints on the membrane materials as they have to be stable under a wide range of reducing conditions. However, the requirements on the intrinsic transport properties of the materials could be less demanding because of the large generated driving force during the operation of the dense ceramic membrane. In the case of highly exothermic partial oxidation reactions the formation of hot spots and runaway condition need to be mitigated.

The scientific and engineering community have done a great effort to promote the commercialisation of perovskite-type ceramic membranes. A number of issues still need to be addressed such as the membrane reliability, large production of the membrane at acceptable cost, and high temperature sealing. Nevertheless, there has been an important progress by Air Products and Chemicals, Inc. by developing flat membranes in a wafer configuration, in which the membrane wafers are assembled in stacks with a central oxygen collector tube which conducts the permeate stream (Anderson et al., 2011; Bose et al., 2009). This design reduces the sealing area per membrane area, whilst also providing a large membrane packing density, which is reflected in high oxygen fluxes per unit volume. A significant step forward toward the commercialisation of this technology based on perovskite membrane design was given by Air products by commissioning a new oxygen production facility able to produce 100 tons per day (Mak et al., 2013). This progress gives us a great confidence that the challenges for the commercial applications of dense hollow fibre ceramic membranes could be a reality in the near future.

CONCLUSION

The results discussed above for the methane conversion experiments using unmodified LSCF6428-HFM and 5%Ni-LSCF6428/LSCF6428-HFM reactors lead to the following conclusions:

1. The unmodified LSCF6428-HFM reactor favoured the formation of C_2 products (C_2H_6 + C_2H_4) at methane conversion below 5% when Air/4% CH_4 gradients operated under MSAL and MLAS modes.
2. The increase of oxygen concentration from 2% to 20% improves oxygen permeation ca. 2 fold, but this was not reflected on methane conversion (15% and 20%). The main catalytic pathway at low oxygen concentration and H_2 reduction treatment was partial oxidation of methane to C_2 and syngas. At high oxygen concentration and no H_2 treatment the total oxidation of methane to CO_2 was the main pathway in agreement with the thermodynamic calculation data.
3. The modification of the LSCF6428-HFM with 5% Ni-LSCF6428 catalyst increases the performance of the reactor and oxygen permeation 6 fold compared to the unmodified LSCF6428-HFM. In addition, CO_x become the main reaction product, instead the unmodified HFM reactor favoured the C_2 (C_2H_4 and C_2H_6) selectivity (i.e., oxidative coupling of methane).
4. According to the products distribution and the high methane conversions it is proposed that CH_4 reacts with oxygen species that diffused or dissociated over the catalyst to produce CO_x and H_2O. This would indicate that methane conversion mainly occurs over the surface of 5%Ni-LSCF6428 catalyst and facilitates the total oxidation catalytic pathway.
5. The presence of 5%Ni-LSCF6428 catalyst improved 3 fold the apparent overall rate constant under Air/CH_4 gradient relative to unmodified LSCF6428-HFM. A similar trend was obtained for Air/He gradient, which reflects the positive effect of the catalyst on the oxygen flux.
6. The morphology of the catalyst (centre shell side) and metal composition of the grain boundary (centre lumen side and end lumen side) of the 5%Ni-LSCF6428-HFM after methane conversion significantly changed relative to the modified membrane before reaction. The centre shell side and the end shell side of the unmodified LSCF6428-HFM reactor also showed changes in crystalline phases, morphology and metal composition of the grain boundary compared to the fresh unmodified membrane. This is tentatively attributed to kinetic demixing of mobile cations and decomposition of the membrane due to the axial temperature generated by the exothermic reaction of methane combustion.
7. Both the spent LSCF6428-HFM and 5%Ni/LSCF6428-HFM reactors showed the presence of $SrSO_4$ and $CoSO_4$, which affected markedly their reducibility. The formation of those phases is tentatively attributed to the air gas employed during the experiments. Sulphur impurity could also be produced by polyethersulfone employed during fabrication of the LSCF6428 hollow and the cement used to assemble the reactor.
8. The segregation of strontium, cobalt and iron to form different crystalline oxide/carbonate/sulphate phases such as Fe_3O_4, SrO_2, $SrCO_3$, $SrSO_4$ and $CoSO_4$ occurred on LSCF6428-HFM and 5%Ni/LSCF6428-HFM reactors. Thermodynamic calculation and membrane phase diagrams supported the formation of these species under Air/CH_4 gradient and sulphur impurity.

ACKNOWLEDGMENT

SMRB is indebted to the University of Newcastle upon Tyne and EPSRC for funding and my supervisor, Professor Ian Metcalfe for his help, encouragement and valuable advice on the development of this work and also for giving me the opportunity to work in his Group. Also to Professor M.L.H. Green (Oxford University) and Professor B. Fontal (University of Los Andes) for his valuable and relevant technical comments that significantly improved the writing of this manuscript and Dr. S. Gonzalez-Cortes (Oxford University) for H_2-temperature-programmed reduction analysis and interpretation.

REFERENCES

Al-Sayari, S. A. (2013), Recent Developments in the Partial Oxidation of Methane to Syngas. The Open Catal. J., *6,* 17-28.

Anderson, L. L., Armstrong, P. A., Repasky, J. M., & Stein, V. E. (2011). Enabling clean coal power generation: ITM oxygen technology. *Paper presented at the International Pittsburgh Coal Conference*, Pittsburgh, USA.

Au, C. T., & Wang, H. Y. (1997). Mechanistic studies of methane partial oxidation to syngas over SiO2-supported rhodium catalysts. *Journal of Catalysis*, *167*(2), 337–345. doi:10.1006/jcat.1997.1609

Baker, R. W. (2002). Future Directions of Membrane Gas Separation Technology. *Industrial & Engineering Chemistry Research*, *41*(6), 1393–1411. doi:10.1021/ie0108088

Balachandran, U., Dusek, J. T., Maiya, P. S., Ma, B., Mieville, R. L., Kleefisch, M. S., & Udovich, C. A. (1997). Ceramic membrane reactor for converting methane to syngas. *Catalysis Today*, *36*(3), 265–272. doi:10.1016/S0920-5861(96)00229-5

Bartholomew, C. H., & Farrauto, R. J. (2006). *Fundamentals of industrial catalytic processes* (2nd ed.). New Jersey: John Wiley & Sons, Inc.

Berry, F. J., Gancedo, J. R., Marco, J. F., & Ren, X. (2005). Characterisation of the reduction properties of La0.5Sr0.5Fe0.5Co0.5O3. *Hyperfine Interactions*, *116*(1-4), 449–453.

Bielanski, A., & Haber, J. (1991). *Oxygen in Catalysis*. New York, NY: Marcel Dekker Inc.

Bose, A. C., Stiegel, G. J., Armstrong, P. A., Halper, B. J., & Foster, E. P. (2009). Progress in ion transport membranes for gas separation applications. In A. C. Bose (Ed.), Inorganic membranes for energy and environmental applications (pp. 3-25). New York, NY: Springer Science+Business Media. doi:10.1007/978-0-387-34526-0_1

Bouwmeester, H. J. M. (2003). Dense ceramic membranes for methane conversion. *Catalysis Today*, *82*(1-4), 141–150. doi:10.1016/S0920-5861(03)00222-0

Caro, J., Caspary, K. J., Hamel, C., Hoting, B., Kölsch, P., Langanke, B., & Wolf, A. et al. (2007). Catalytic membrane reactors for partial oxidation using perovskite hollow fiber membranes and for partial hydrogenation using a catalytic membrane contactor. *Industrial & Engineering Chemistry Research*, *46*(8), 2286–2294. doi:10.1021/ie0609620

Caro, J., Wang, H. H., Tablet, C., Kleinert, A., Feldhoff, A., Schiestel, T., & Werth, S. et al. (2006). Evaluation of perovskites in hollow fibre and disk geometry in catalytic membrane reactors and in oxygen separators. *Catalysis Today*, *118*(1-2), 128–135. doi:10.1016/j.cattod.2005.12.015

Den Exter, M. J., Haije, W. G., & Vente, J. F. (2009). Viability of ITM technology for oxygen production and oxidation processes: Material, system, and process Aspects. In A. C. Bose (Ed.), Inorganic membranes for energy and environmental applications (pp. 27-51). New York, NY: Springer Science+Business Media.

Deshmukh, S. P., & Li, K. (1998). Effect of ethanol composition in water coagulation bath on morphology of PVDF hollow fibre membranes. *Journal of Membrane Science*, *150*(1), 75–85. doi:10.1016/S0376-7388(98)00196-3

Dittmeyer, R., & Caro, J. (2008). Catalytic membrane reactors. In G. Ertl, H. Knözinger, F. Schüth, & J. Weitkamp (Eds.), *Handbook of Heterogeneous Catalysis* (2nd ed., pp. 2198–2248). Weinheim: Wiley-VCH. doi:10.1002/9783527610044.hetcat0117

Franca, R. V., Thursfield, A., & Metcalfe, I. S. (2012). La0.6Sr0.4Co0.2Fe0.8O3-δ microtubular membranes for hydrogen production from water splitting. *Journal of Membrane Science*, *389*, 173–181. doi:10.1016/j.memsci.2011.10.027

Gellings, P. J., & Bouwmeester, H. J. M. (2000). Solid state aspects of oxidation catalysis. *Catalysis Today*, *58*(1), 1–53. doi:10.1016/S0920-5861(00)00240-6

González-Cortés, S. L., Aray, I., Rodulfo-Baechler, S. M. A., Lugo, C. A., Del Castillo, H. L., Loaiza-Gil, A., & Rueda, F. et al. (2007). On the structure and surface properties of NiO/MgO-La2O3 catalyst: Influence of the support composition and preparation method. *Journal of Materials Science*, *42*(16), 6532–6540.

Hamakawa, S., Sato, K., Inoue, T., Nishioka, M., Kobayashi, K., & Mizukami, F. (2006). Design of one-component ceramic membrane-reactor for natural gas conversion. *Catalysis Today*, *117*(1-3), 297–303. doi:10.1016/j.cattod.2006.05.042

Huang, T.-J., Shen, X.-D., & Chou, C.-L. (2009). Characterization of Cu, Ag and Pt added La0.6Sr0.4Co0.2Fe0.8O3−δ and gadolinia-doped ceria as solid oxide fuel cell electrodes by temperature-programmed techniques. *Journal of Power Sources*, *187*(2), 348–355. doi:10.1016/j.jpowsour.2008.11.019

Jin, W., Li, S., Huang, P., Xu, N., Shi, J., & Lin, Y. S. (2000). Tubular lanthanum cobaltite perovskite-type membrane reactors for partial oxidation of methane to syngas. *Journal of Membrane Science*, *166*(1), 13–22. doi:10.1016/S0376-7388(99)00245-8

Kondratenko, E. V., Wang, H., Kondratenko, V. A., & Caro, J. (2009). Selective oxidation of CH4 and C2H6 over a mixed oxygen ion and electron conducting perovskite—A TAP and membrane reactors study. *Journal of Molecular Catalysis A Chemical*, *297*(2), 142–149. doi:10.1016/j.molcata.2008.09.015

Lee, S., Woo, S. K., Lee, K. S., & Kim, D. K. (2006). Mechanical properties and structural stability of perovskite-type, oxygen-permeable, dense membranes. *Desalination*, *193*(1-3), 236–243. doi:10.1016/j.desal.2005.06.061

Lein, H. L., Wiik, K., & Grande, T. (2006). Kinetic demixing and decomposition of oxygen permeable membranes. *Solid State Ionics, 177*(19-25), 1587-1590.

Li, J., & Smith, K. J. (2008). Methane decomposition and catalyst regeneration in a cyclic mode over supported Co and Ni catalysts. *Applied Catalysis A, General*, *349*(1-2), 116–124. doi:10.1016/j.apcata.2008.07.011

Li, S., Jin, W., Huang, P., Xu, N., Shi, J., & Lin, Y. S. (2000). Tubular lanthanum cobaltite perovskite type membrane for oxygen permeation. *Journal of Membrane Science*, *166*(1), 51–61. doi:10.1016/S0376-7388(99)00244-6

Lin, Y. S., & Zeng, Y. (1996). Catalytic properties of oxygen semipermeable perovskite-type ceramic membrane materials for oxidative coupling of methane. *Journal of Catalysis*, *164*(1), 220–231. doi:10.1006/jcat.1996.0377

Liu, Y., Zhu, X., Li, M., Liu, H., Cong, Y., & Yang, W. (2013). Stabilization of low-temperature degradation in mixed ionic and electronic conducting perovskite oxygen permeation membranes. *Angewandte Chemie International Edition*, *52*(11), 3232–3236. doi:10.1002/anie.201209077 PMID:23401278

Lu, Y., Xue, J.-Z., Ding, X. J., & Shen, S. K. (1996). TPR and CTPR studies on Ni/Al2O3 catalyst for catalytic partial oxidation of methane to syngas. *Journal of Natural Gas Chemistry*, *5*(1), 45–50.

Lui, H., Pang, Z., Tan, X., Shao, Z., Sunarso, J., Ding, R., & Liu, S. (2008). Effects of preparation methods on the oxygen nonstoichiometry, B-site cation valences and catalytic efficiency of perovskite La0.6Sr0.4Co0.2Fe0.8O3-δ. *Ceramics International*, *35*(8), 3201–3206.

Mak, A., Foster, E. P., Armstrong, P. A., Repasky, J. M., & Stein, V. E. (2013). *ITM Oxygen Technology for Clean Energy Applications*. Paper presented at the International Pittsburgh Coal Conference, Beijing,China.

Matsuma, Y., & Nakamori, T. (2004). Steam reforming of methane over nickel catalysts at low reaction temperature. *Applied Catalysis A, General*, *258*(1), 107–114. doi:10.1016/j.apcata.2003.08.009

Niu, J., Deng, J., Liu, W., Zhang, L., Wang, G., Dai, H., & Zi, X. et al. (2007). Nanosized perovskite-type oxides La1-xSrxMO3-δ (M = Co, Mn; x = 0, 0.4) for the catalytic removal of ethylacetate. *Catalysis Today*, *126*(3-4), 420–429. doi:10.1016/j.cattod.2007.06.027

Petkovic, L. M., Utgikar, V., & Rashkeev, S. N. (2011). Structural stability and catalytic activity of lanthanum-based perovskites. *The Journal of Physical Chemistry C*, *115*(17), 8709–8715. doi:10.1021/jp112422m

Porras-Vazquez, J. M., & Slater, P. R. (2012). Synthesis of oxyanion-doped barium strontium cobalt ferrites: Stabilization of the cubic perovskite and enhancement in conductivity. *Journal of Power Sources*, *209*, 180–183. doi:10.1016/j.jpowsour.2012.02.105

Sathe, V. G., Paranjpe, S. K., Siruguri, V., & Pimpale, A. V. (1998). Novel magnetic phases in La0.7Sr0.3Co1−yFeyO3 (0:0 6 y 6 1:0): a neutron diffraction study. *Journal of Physics Condensed Matter*, *10*, 4045–4055. doi:10.1088/0953-8984/10/18/014

Schiestel, T., Kilgus, M., Peter, S., Caspary, K., Wang, H., & Caro, J. (2005). Hollow fibre perovskite membranes for oxygen separation. *Journal of Membrane Science*, *258*(1-2), 1–4. doi:10.1016/j.memsci.2005.03.035

Schlehuber, D., Wessel, E., Singheiser, L., & Markus, T. (2010). Long-term operation of a La0.58Sr0.4Co0.2Fe0.8O3−δ-membrane for oxygen separation. *Journal of Membrane Science*, *351*(1-2), 16–20. doi:10.1016/j.memsci.2010.01.022

Shao, Z., Xiong, G., Dong, H., Yang, W., & Lin, L. (2001). Synthesis, oxygen permeation study and membrane performance of a Ba0.5Sr0.5Co0.8Fe0.2O3−δ oxygen-permeable dense ceramic reactor for partial oxidation of methane to syngas. *Separation and Purification Technology*, *25*(1-3), 97–116. doi:10.1016/S1383-5866(01)00095-8

Shin, J. F., Orera, A., Apperley, D. C., & Slater, P. R. (2011). Oxyanion doping strategies to enhance the ionic conductivity in Ba2In2O5. *Journal of Materials Chemistry*, *21*(3), 874–879. doi:10.1039/C0JM01978J

Sierra Gallego, G., & Batiot-Dupeyrat, C. & Mondragón. (2010). Methane partial oxidation by the lattice oxygen of the LaNiO3-δ perovskite. A pulse study. *Dyna*, *77*(163), 141–150.

Stephens, W. T., Mazanec, T. J., & Anderson, H. U. (2000). Influence of gas flow rate on oxygen flux measurements for dense oxygen conducting ceramic membranes. *Solid State Ionics*, *129*(1-4), 271–284. doi:10.1016/S0167-2738(99)00332-X

Tan, X., & Li, K. (2006). Oxidative coupling of methane in a perovskite hollow-fiber membrane reactor. *Industrial & Engineering Chemistry Research*, *45*(1), 142–149. doi:10.1021/ie0506320

Tan, X., Li, K., Thursfield, A., & Metcalfe, I. S. (2008). Oxyfuel combustion using a catalytic ceramic membrane reactor. *Catalysis Today*, *131*(1-4), 292–304. doi:10.1016/j.cattod.2007.10.081

Tan, X., Liu, Y., & Li, K. (2005). Mixed conducting ceramic hollow-fiber membranes for air separation. *AIChE Journal. American Institute of Chemical Engineers*, *51*(7), 1991–2000. doi:10.1002/aic.10475

Tang, X., Pang, Z., Gu, Z., & Liu, S. (2007). Catalytic perovskite hollow fibre membrane reactors for methane oxidative coupling. *Journal of Membrane Science*, *302*(1-2), 109–114. doi:10.1016/j.memsci.2007.06.033

Ten Elshof, J. E., Bouwmeester, H. J. M., & Verweij, H. (1995). Oxidative coupling of methane in a mixed-conducting perovskite membrane reactor. *Applied Catalysis A, General*, *130*(2), 195–212. doi:10.1016/0926-860X(95)00098-4

Thursfield, A., & Metcalfe, I. S. (2006). Methane oxidation in a mixed ionic–electronic conducting ceramic hollow fibre reactor module. *Journal of Solid State Electrochemistry*, *10*(8), 604–616. doi:10.1007/s10008-006-0139-6

Thursfield, A., & Metcalfe, I. S. (2007). Air separation using a catalytically modified mixed conducting ceramic hollow fibre membrane module. *Journal of Membrane Science*, *288*(1-2), 175–187. doi:10.1016/j.memsci.2006.11.015

Van Doorn, R. H. E., Bouwmeester, H. J. M., & Burggraaf, A. J. (1998). Kinetic decomposition of La0.3Sr0.7CoO3-δ perovskite membranes during oxygen permeation. *Solid State Ionics*, *111*(3-4), 263–272. doi:10.1016/S0167-2738(98)00191-X

Van Hassel, B. A., Kawada, A., Sakai, T., Yokokawa, N., Dokiya, H., & Bouwmeester, H. J. M. (1993). Oxygen permeation modelling of perovskites. *Solid State Ionics*, *66*(3-4), 295–305. doi:10.1016/0167-2738(93)90419-4

Wang, B., Zydorczak, B., Poulidi, D., Metcalfe, I. S., & Li, K. (2011). A further investigation of the kinetic demixing/decomposition of La0.6Sr0.4Co0.2Fe0.8O3−δ oxygen separation membranes. *Journal of Membrane Science*, *369*(1-2), 526–535. doi:10.1016/j.memsci.2010.12.040

Wang, B., Zydorczak, B., Wu, Z., & Li, K. (2009). Stabilities of La0.6Sr0.4Co0.2Fe0.8O3-δ oxygen separation membranes. Effects of kinetic demixing/decomposition and impurity segregation. *Journal of Membrane Science*, *344*(1-2), 101–106. doi:10.1016/j.memsci.2009.07.040

Wang, H., Ren, F., Liu, C., Si, R., Yu, D., Pfefferle, L. D., & Chen, Y. et al. (2013). CoSO4/SiO2 catalyst for selective synthesis of (9, 8) single-walled carbon nanotubes: Effect of catalyst calcination. *Journal of Catalysis*, *300*, 91–101. doi:10.1016/j.jcat.2012.12.032

Wei, Y., Liu, H., Xue, J., Li, Z., & Wang, H. (2011). Preparation and oxygen permeation of U-shaped perovskite hollow-fiber membranes. *AIChE Journal. American Institute of Chemical Engineers*, *57*(4), 975–984. doi:10.1002/aic.12321

Xu, N., Li, S., Jin, W., Shi, J., & Lin, Y. S. (1999). Experimental and modeling study on tubular dense membranes for oxygen permeation. *AIChE Journal. American Institute of Chemical Engineers*, *45*(12), 2519–2526. doi:10.1002/aic.690451207

Xu, S., & Thomson, W. (1997). Perovskite-type oxide membranes for the oxidative coupling of methane. *AIChE Journal. American Institute of Chemical Engineers*, *43*(S11), 2731–2740. doi:10.1002/aic.690431319

Xu, S., & Thomson, W. (1998). Stability of La0.6Sr0.4Co0.2Fe0.8O3-δ perovskite membranes in reducing and nonreducing environments. *Industrial & Engineering Chemistry Research*, *37*(4), 1290–1299. doi:10.1021/ie970761j

Yacou, C., Sunarso, J., Lin, C. X. C., Smart, S., Liu, S., & Diniz da Costa, J. C. (2011). Palladium surface modified La0.6Sr0.4Co0.2Fe0.8O3−δ hollow fibres for oxygen separation. *Journal of Membrane Science*, *380*(1-2), 223–231. doi:10.1016/j.memsci.2011.07.008

Yaremchenko, A. A., Buysse, C., Middelkoop, V., Snijkers, F., Buekenhoudt, A., Frade, J. R., & Kovalevsky, A. V. (2013). Impact of sulphur contamination on the oxygen transport mechanism through Ba0.5Sr0.5Co0.8Fe0.2O3-δ: Relevant issues in the development of capillary and hollow fibre membrane geometry. *Journal of Membrane Science*, *428*, 123–130. doi:10.1016/j.memsci.2012.10.033

Zeng, Y., Lin, Y. S., & Swartz, S. L. (1998). Perovskite-type ceramic membrane: Synthesis, oxygen permeation and membrane reactor performance for oxidative coupling of methane. *Journal of Membrane Science*, *150*(1), 87–98. doi:10.1016/S0376-7388(98)00182-3

Zhu, X. F., Liu, H. Y., Cong, Y., & Yang, W. S. (2012). Permeation model and experimental investigation of mixed conducting membranes. *AIChE Journal. American Institute of Chemical Engineers*, *58*(6), 1744–1754. doi:10.1002/aic.12710

KEY TERMS AND DEFINITIONS

Grain Boundary: It is the interface between two grains, particles or crystallites in a material. They are crystal structure defects that hinder the conductivity and electrical properties of a crystallite material.

Kinetic Demixing of the Membrane: It is a process in which cations in the membrane material with high diffusion coefficients will be enriched at the high oxygen partial pressure side, and slower cations will be enriched at the low oxygen partial pressure side. Finally a steady state will be achieved and continuous gradients in cation concentrations across the membrane will be formed.

Methane Oxidation Reactions: Chemical reactions of methane and oxygen that include methane (dry) reforming, partial oxidation of methane, oxidative coupling of methane and total oxidation of methane.

Mixed Ionic and Electronic Conductor (MIEC) Ceramic Membrane: It is a dense mixed-metal oxide (e.g., perovskite, brownmillerite, K_2NiF_4-type structure) with crystalline defects that shows highly mobile electrons and mobile oxygen ions with good chemical stability.

Oxidative Coupling of Methane: It is a type of chemical reaction to upgrade methane (or natural gas) to C_2 products (ethane + ethylene) using oxygen as oxidizing agent. The reaction is exothermic and occurs at high temperatures (750–950°C).

Oxygen Permeation: It is a complex process that involves a series of steps such as mass transfer of gaseous oxygen, reaction between the molecular oxygen and oxygen vacancies, diffusion of oxygen species through the combination of mobile oxygen vacancies and electronic defects (holes), reaction between the lattice and electron hole at the membrane surface and finally, the mass transfer of oxygen from the membrane surface to the gas stream.

Partial Oxidation of Methane: It is a chemical reaction in which the methane (or natural gas) reacts exothermically with oxygen to produce syngas as main product. This catalytic reaction is usually carried out at temperature between 750 and 900 °C.

Steam Methane Reforming: Chemical reaction to convert methane (or natural gas) with steam into hydrogen and carbon monoxide (syngas). A Ni-based catalyst operating into a relatively wide gradient of reaction temperatures (400-1000 °C) is usually employed.

Total Oxidation of Methane: Chemical reaction between methane (or natural gas) and oxygen (or air) to produce carbon dioxide and water. This reaction is carried out at high temperatures and in the presence of a metal or metal oxide catalyst.

Chapter 15
Preparation, Characterization and Desulfurization of the Supported Nickel Phosphide Catalysts

Zhoujun Wang
Beijing University of Chemical Technology, China

Pingyi Wu
Beijing University of Chemical Technology, China

Ling Lan
PetroChina, China

Shengfu Ji
Beijing University of Chemical Technology, China

ABSTRACT

Hydrodesulfurization (HDS) is an important technology to produce clean fuels, in which the nickel phosphide catalysts exhibit excellent catalytic performances. In this work, a series of Ni_xP/SBA-15 catalysts with various P/Ni molar ratios were prepared using the mesoporous molecular sieve SBA-15 as the support. The structure and surface properties of the catalysts were characterized by X-ray diffraction (XRD), N_2 sorption analysis, transmission electron microscopy (TEM), Fourier transform infrared spectrometry (FTIR), and in situ diffuse reflectance infrared Fourier transfer spectroscopy (DRIFTS). The catalytic performances for the HDS of dibenzothiophene (DBT) were evaluated. The results demonstrated that the Ni_xP/SBA-15 catalysts possessed high specific surface area and the mesoporous structures, which benefited the elimination of the internal diffusion limitation in the HDS reactions. Both Ni_2P and $Ni_{12}P_5$ phases showed catalytic activity in HDS reactions while Ni_2P was more active. The optimal P/Ni molar ratio was about 0.75. The DBT conversion can reach 95.8% under the reaction condition of pressure at 3.0 MPa, H_2:oil=600, WHSV at 26.7 h^{-1}, and temperature at 340°C. In situ DRIFTS spectra indicated that the coordinative unsaturated $Ni^{\delta+}$ ($0<\delta<1$) species on the catalysts surface were the active sites for the HDS of DBT.

DOI: 10.4018/978-1-4666-9975-5.ch015

INTRODUCTION

The exhaust emission, especially the SO_2 emission, has attracted more and more attentions by the governments all over the world due to the developments of automobile industry and increasingly grievous situation in environmental protections. Accordingly, more and more stringent environmental regulations have been enacted to limit the S content in the fuels. To meet the regulations, most of the S should be eliminated by the deep hydrodesulfurization (HDS) during the petroleum refining processes. It is difficult to remove the S content to ultra-low levels using a conventional Mo(W)Co(Ni)/Al_2O_3 HDS catalyst. The transition metal phosphides are reported as a novel catalytic material with exceptional HDS performances, showing great prospect as the new generation of catalysts in deep HDS. Among the transition metal phosphides, nickel phosphide showed the best activity. Previously the nickel phosphides were usually supported on Al_2O_3 or SiO_2. The specific surface areas of Al_2O_3 or SiO_2 supported catalysts were relatively low, which was unfavorable for the dispersion of the active phases, leading to an unsatisfactory activity or stability. The mesoporous molecular sieve SBA-15 was regarded as a superior support due to the higher specific surface area, higher pore volume and narrower pore size distribution.

In the present work the mesoporous molecular sieve SBA-15 was synthesized via hydrothermal method using the cheap sodium silicate as the silicon source, the triblock copolymer (P123, $EO_{20}PO_{70}EO_{20}$) and cetyltrimethyl ammonium bromide (CTAB) as the co-template reagents. The Ni_xP/SBA-15 catalysts with a fixed Ni content of 20 wt% and a varying initial P/Ni molar ratio (0.25, 0.5, 0.75, 1.0, 1.25, 1.5, 2.0) were prepared by a co-impregnation method using $Ni(NO_3)_2 \cdot 6H_2O$ as the nickel source, $(NH_4)_2HPO_4$ as the phosphorus source and SBA-15 as the support, and following by the H_2 temperature-programmed reduction (H_2-TPR) treatment.

The phase formation, specific surface area and pore structure of the prepared catalysts were characterized by X-ray diffraction (XRD), transmission electron microscopy (TEM), N_2 sorption analysis and Fourier transform infrared spectrometry (FTIR) techniques. The nature of the active sites on the surface of the catalysts was investigated by CO adsorption using *in situ* diffuse reflectance infrared Fourier transfer spectroscopy (DRIFTS). The HDS of dibenzothiophene (DBT) was employed as a model reaction to test the HDS performances of the catalyst. The effects of the external diffusion, the internal diffusion, the total pressure, the volume ratio of H_2 to liquid feed, and the reaction temperature on the activity of the catalyst was examined systematically. The stability of the catalyst and the evolution of the structure during the stability test were investigated as well.

The obtained results showed that with the cheap sodium silicate as the silicon source, the prepared mesoporous molecular sieve SBA-15 had the specific surface area (S_{BET}) of 674.0 m^2/g, the pore volume (V_p) of 1.12 m^3/g and average pore diameter (D_{BJH}) of 6.7 nm. For the prepared Ni_xP/SBA-15, Ni_2P and $Ni_{12}P_5$ phases existed at P/Ni = 0.5. Only Ni_2P phase was detected at P/Ni ≥ 0.75. Both Ni_2P and $Ni_{12}P_5$ phases showed catalytic activity in HDS reactions while Ni_2P was more active. The optimal P/Ni molar ratio was found at 0.75. Either too high or too low P/Ni molar ratio would lead to a diminished HDS activity. The mesoporous structure was mainly maintained with the P/Ni molar ratio ranging from 0.25 to 2.0. All of the Ni_xP/SBA-15 catalysts had high S_{BET} and V_p as well as uniform pore size distribution, which would benefit the dispersion of the active phases on the catalysts and the diffusion of reactants during HDS reactions.

The prepared Ni_xP/SBA-15 catalysts exhibited good activity in the HDS of DBT, among which the Ni_xP/SBA-15 (P/Ni=0.75) catalyst possessed the highest DBT conversion. The DBT conversion of the Ni_xP/SBA-15 (P/Ni=0.75) catalyst reached 95.8% at 340 °C, under 3.0 MPa, with the volume ratio of

H_2 to liquid feed at 600 and WHSV at 26.7 h^{-1}. Stability test showed that the Ni_xP/SBA-15 (P/Ni=0.75) catalyst exhibited excellent activity and selectivity up to 300 h. Characterization results demonstrated that the structure of the catalyst was well maintained after the stability test. *In situ* DRIFTS spectra of the Ni_xP/SBA-15 (P/Ni=0.75) catalyst suggested the coordinatively unsaturated $Ni^{\delta+}$ ($0<\delta<1$) sites as the active sites in the HDS reactions. In summary, the Ni_xP/SBA-15 catalysts supported on the mesoporous molecular sieves showed great prospect as a promising catalyst in the deep HDS reactions.

BACKGROUND

HDS is an important process in the petroleum refining industry. The research and development of HDS catalysts have attracted significant attention during the past decades. Recently, environmental regulations have been enacted in many countries to limit the S content in the light oils such as gasoline and diesel. On the other hand, the quality of the crude oil declined continuously as the depletion of the petroleum. The content of the impurities such as S and N in the crude oil became higher and higher, and the molecular structure of the S- or N-contained compounds became more and more complicated. Therefore, it is critical to develop new catalyst and technology for HDS process under the stringent regulations.

HDS is the process that removes the S impurity from oils such as gasoline and diesel by converting the organosulfur compounds (thiophene and the deviates) to hydrogen sulfide (H_2S) over a catalyst under H_2 partial pressure. HDS technology has been widely used in the industry of desulfurization processes. Compared with other desulfurization technologies, HDS technology is a relatively mature one. As a traditional technology for the desulfurization processes, HDS technology can reduce the S content under the required limit after developing new catalyst with exceptional catalytic performances and optimizing the process operating conditions. To merely eliminate the S from the oils is not difficult, but to yield the aimed product with satisfactory quality is still challenging. For example, if the aimed product is gasoline, it is required to reduce the amount of the olefins and obtain the specified octane number besides the eliminating of S. If the aimed product is diesel, the polycyclic aromatics have to be hydrocracked and saturated besides the removal of organosulfur and organonitrogen compounds to improve the cetane number.

The HDS catalysts mainly consist of active component, support and promoters. The active components are usually the transition metal (Co, Mo, Ni, W) sulfides, between which the synergism effect is generally observed. Therefore, most of the active components for HDS catalysts consist of binary or ternary transition metal sulfides. The most widely used commercial HDS catalysts employ binary transition metal sulfides such as Co-Mo, Ni-W and Ni-Mo sulfides as the active components. Sometimes ternary metal sulfides such as Ni-W-Mo or Co-Ni-Mo sulfides were also used as the active component of HDS catalyst (Coulier et al., 2002; Dhas et al., 2001; Faro Jr et al., 2006; Fujikawa et al., 2006; Hatanaka et al., 1997; Zhang et al., 2010).

In 1973, Levy and Boudart reported that the tungsten carbide (WC) displayed similar catalytic properties as the noble metal Pt catalyst in a series of catalytic reactions (Levv et al., 1973). After that, the transition metal nitrides and carbides have received great attention as a new family of catalytic materials (Costa et al., 2004; Furimsky et al., 2003; Gong et al., 2006; Korlann et al., 2002; Liu et al., 2002; Schwartz et al., 2000; Wang et al., 2007). A number of studies demonstrated that the transition metal nitrides and carbides exhibit comparable catalytic performances as the noble metal catalysts do in many cases. In 1998, Li and Oyama co-workers (Li et al., 1998) firstly reported the preparation of transition metal phosphides with the phosphate as the precursor using TPR procedure. Both bulk and supported transition metal phosphides can be prepared by this method in a simple way. Up to now, a series of supported and un-supported binary transition metal phosphides such as MoP, WP, Ni_2P, CoP, Co_2P and Fe_2P as well as ternary metal phosphides such as CoMoP and NiMoP have been successfully prepared (Burns et al., 2007; Clark et al., 2001; Clark et al., 2003; Li et al., 2008; Oyama et al., 2002; Wang et al., 2002;) Burn et al. (Burns et al., 2007) obtained MoP and Ni_2P crystals with higher specific surface area by using 80% CH_4/N_2 as the reducing reagent. Guan et al. (Guan et al., 2009) prepared Ni_2P crystal by the thermal decomposition method using $NiCl_2{\cdot}6H_2O$ as the nickel source and $NaH_2PO_2{\cdot}H_2O$ as the phosphorous source. Such prepared Ni_2P catalyst exhibited high HDS activity. Guan et al. (Guan et al., 2009) used hydrogen plasma to prepare a series of transition metal phosphides under low temperature conditions, which reduced the energy consumption by avoiding the use of the high temperature procedures. Cecilia et al. (Cecilia et al., 2009; Cecilia et al., 2009) obtained highly active Ni_2P and CoP catalysts with $Ni(HPO_3H)_2$ as the nickel source and $Co(HPO_3H)_2$ as the cobalt source, which were reduced at 375 and 475 °C respectively without the calcination processes. Yang et al. (Yang et al., 2006) prepared bulk Ni_2P catalyst with high surface area by adding a polymer surfactant (Triton X-114) and ethylene glycol to an aqueous solution of $Ni(NO_3)_2$ and $(NH_4)_2HPO_4$ prior to drying and calcination. Cheng et al. (Cheng et al., 2007) and Ibeh et al. (Ibeh et al., 2009) used citric acid (CA) as a structure template or chelating agent and prepared the well dispersed Ni_2P and WP crystals.

The application of phosphide catalysts in HDS has been reported in recent ten years. In 1998 Oyama and co-workers firstly prepared pure MoP phase with the TPR method. They applied such prepared catalyst in the hydrodenitrogenation (HDN) of quinoline and observed a good activity (Li et al., 1998). After that, the catalytic properties of transition metal phosphides in HDS reactions were extensively investigated. In recent years, the transition metal phosphides, especially the nickel phosphides have been the hot research subject in HDS catalysts due to their exceptional catalytic performances. The transition metal phosphides are regarded as a new catalytic material with high activity, good stability and excellent sulfur resistance (Clark et al., 2001; Clark et al., 2003; Li et al., 1998; Li et al., 2008; Oyama et al., 2002; Wang et al., 2002) .

The support of HDS catalysts is the substance to which the active components are affixed and dispersed, which is an important constituents of the catalyst. While γ-Al_2O_3 has been the most commonly used support for HDS catalysts in industry, its main drawback for transition metal phosphide catalysts is its strong interaction with phosphorus, which is found to negatively affect the HDS activity. Due to the negative interaction between γ-Al_2O_3 and transition metal precursors, more and more studies employ the mesoporous molecular sieves, such as MCM-41, SBA-15, MCM-48 and HMS, as the support. The mesoporous molecular sieves show superior HDS performances than other supports due to the high specific surface areas, high pore volumes and uniform pore size distributions. For example, Wang et al. (Wang et al., 2005) prepared a series of MCM-41 supported nickel phosphide catalysts with various P/Ni molar ratios and found that the *in situ* reduced catalysts show superior catalytic performances than the conventionally *ex situ* reduced catalysts in the HDS of DBT. Oyama et al. (Oyama et al., 2008) loaded Ni_2P catalyst on MCM-41, high surface area and low surface area SiO_2, founding that the Ni_2P/MCM-41 catalyst exhibit the highest activity in the HDS of 4,6-DMDBT, which is substantially more active than the commercial Ni-Mo-S/Al_2O_3 catalyst. Guan et al. (Guan et al., 2009) prepared Ni_2P/MCM-41 catalyst with the thermal decomposition method instead of the conventional TPR method and the obtained catalyst showed excellent activity in the HDS of DBT. Cecilia et al. (Cecilia et al., 2009) prepared MCM-41 supported Ni_2P catalysts with various amount of Ni using a new TPR method procedure and investigated the activity and stability of such prepared catalysts in the HDS of DBT. Korányi et al. (Korányi et al.,

2008) prepared SBA-15 and CMK-5 supported nickel phosphide catalysts with various initial P/Ni molar ratios. It was found the SBA-15 supported catalyst exhibited higher activity than the CMK-5 supported one. Besides, the Ni_2P phase was catalytically more active than the $Ni_{12}P_5$ phase. Therefore, SBA-15 was employed as the support of HDS catalysts in the present work.

Our group has prepared a bunch of HDS catalyst with SBA-15 as the support and Ni_2P as the active component (Gao et al., 2013; Guan et al., 2012; Guo et al., 2010; Huang et al., 2008; Li et al., 2014; Liu et al., 2009; Ma et al., 2011; Wei et al., 2009; Wei et al., 2011; Wu et al., 2012; Wu et al., 2014; Zhao et al., 2010; Zhao et al., 2011). However, some fundamental issues such as the structure-activity relationship and the active sites of the catalysts are still unclear while these issues are critical for the scale-up and industrialization of the catalyst. Herein, the Ni_2P/SBA-15 catalyst was studied from the perspective of industrialization. The mesoporous molecular sieve SBA-15 support was synthesized via hydrothermal method using the cheap sodium silicate as the silicon source, triblock copolymer (P123, $EO_{20}PO_{70}EO_{20}$) and cetyltrimethyl ammonium bromide (CTAB) as the co-template reagents. Subsequently, the Ni_xP/SBA-15 catalysts with various P/Ni molar ratios (0.25, 0.5, 0.75, 1.0, 1.25, 1.5, 2.0) were prepared by the H_2-TPR treatment following a co-impregnation method using $Ni(NO_3)_2 \cdot 6H_2O$ as the nickel source and $(NH_4)_2HPO_4$ as the phosphorus source. The phase formation, specific surface area and pore structure of the prepared catalysts were characterized by XRD, TEM, N_2 sorption and FTIR techniques. The nature of the active sites on the surface of the catalysts was investigated by CO adsorption using *in situ* DRIFTS. The outcome of this study is expected to provide some valuable insight into the scale-up and industrialization of the Ni_xP/SBA-15 catalyst in HDS reactions.

1. EXPERIMENTAL

1.1. Catalyst Preparation

1.1.1. Synthesis of SBA-15

The mesoporous molecular sieve SBA-15 was hydrothermally synthesized using sodium silicate as the silicon source, sulphuric acid as acid source, P123 and CTAB as co-template agents. Specifically, P123 (4 g), CTAB(0.6 g) and $Na_2SiO_3 \cdot 9H_2O$ (11.6 g) were dissolved in deionized water (120 ml) followed by a stirring at 40 °C for 2 h. 12 ml of concentrated sulphuric acid was then quickly added to the obtained aqueous mixture and subsequently stirred at 40 °C for 22 h. After that, the gel mixture was transferred into a Teflon bottle and aged at 100 °C for 24 h under static condition. The obtained solid sample was filtered, washed with deionized water for five times until neutral pH, and dried at room temperature. To remove the organic template, the sample was finally heated at 2 °C/min and calcined at 550 °C for 6 h. The mesoporous molecular sieve SBA-15 was collected after the white powder was cooled down to room temperature.

1.1.2. Preparation of Ni_xP/SBA-15 Catalysts

The Ni_xP/SBA-15 catalysts with a fixed Ni content of 20 wt% and a varying initial P/Ni molar ratio (0.25, 0.5, 0.75, 1.0, 1.25, 1.5, 2.0) were prepared by a co-impregnation method using $Ni(NO_3)_2 \cdot 6H_2O$ as the nickel source, $(NH_4)_2HPO_4$ as the phosphorus source and SBA-15 as the support. SBA-15 was impregnated with the required amount of $Ni(NO_3)_2·6H_2O$ and $(NH_4)_2HPO_4$ aqueous solution at room temperature for 24 h, followed by drying at 100 °C for another 24 h. Then the precursor was calcined at 500 °C for 4 h. After that, the precursor was pelletized to 40 - 60 mesh (250 - 420 μm) and treated by the H_2-TPR in a fixed-bed quartz reactor under atmospheric pressure in flowing H_2 (100 ml/min). The temperature was increased from room temperature to 300 °C at a rate of 10 °C/min, then to 650 °C at 1 °C/min, and held at 650 °C for 2 h, followed by cooling to room temperature in the H_2 flow. Finally, the samples were passivated at room temperature with a 1.0 vol% O_2/Ar flow (50 ml/min) for 2 h. Such passivated samples were used for structure characterization or performance investigations. The sample was denoted as Ni_xP/SBA-15 (P/Ni=n) catalyst, where n was the P/Ni molar ratio.

1.2. Catalyst Characterizations

1.2.1. XRD

The XRD patterns were collected on a Rigaku D/MAX 2500VB 2+ /PC system using Cu Kα radiation (λ=0.154056 nm) at 40 kV and 200 mA. The low-angle XRD patterns were collected at a scanning speed of 1°/min in the 2θ range of 0.5 - 8° while the wide-angle XRD patterns at a scanning speed of 10°/min in the 2θ range of 10 - 85°.

1.2.2. N_2 Adsorption-Desorption Isotherms

The surface areas and porosity of samples were characterized with N_2 sorption analysis. N_2 adsorption-desorption isotherms were measured at -196 °C on a Sorptomatic 1990 analyzer (Thermo Corp.). Before the measurements, the samples were outgassed at 300 °C for 4.0 h. The Brumauer-Emmett-Teller (BET) method was employed to calculate the specific surface areas. The pore volumes and pore size distributions were derived from the desorption branches of the isotherms using the Barrett-Joyner-Halanda (BJH) model.

1.2.3. FTIR

The surface hydroxyl and framework of the mesoporous molecular sieve were studied by FTIR spectroscopy. FTIR spectra were recorded on a Bruker Tensor 27 FTIR spectrometer equipped with a MCT detector at room temperature. The sample diluted in KBr was pressed into a wafer under 8 MPa. All of the spectra were recorded at a resolution of 4 cm^{-1} for 64 scans from 4000 to 600 cm^{-1}.

1.2.4. TEM

The TEM images were recorded on a JOEL JEM-100CXII system operated at 100 kV. The powder sample was sonicated in ethanol for 10 min, dropped and dried on a carbon-coated copper grid before test.

1.2.5. *In Situ* DRIFTS

In situ DRIFTS spectra were recorded on a Bruker Tensor 27 FTIR spectrometer equipped with a Harrick diffuse reflectance accessory and a high temperature sample cup. The catalyst powder (approximately 20 mg) was finely ground and placed into the sample cup. The catalyst sample was reduced *in situ* at 500

°C for 4.0 h using flowing H_2 (20 ml/min) and then flushed with flowing He (30 ml/min) for 0.5 h. The sample was subsequently cooled down in the He flow and the background spectrum was measured at 30 °C. After that, CO (99.9%) was introduced and adsorbed for 1.0 h. Then the sample was purged with flowing He (30 ml/min) for 0.5 h to remove the weakly physisorbed gaseous CO. The DRIFTS spectra were recorded after the sample was cooled down to 10 °C. The DRIFTS spectra at higher temperature were acquired after the sample was heated at 10 °C/min and held at each temperature for 5 min. All of the spectra were recorded with background spectrum subtracted at a resolution of 4 cm^{-1} for 128 scans from 4000 to 600 cm^{-1} and converted to Kubelka-Munk form directly.

1.3. Catalytic Reaction

The catalytic performance of the catalysts was evaluated by the HDS of DBT, which was carried out in a high pressure fixed-bed continuous flow stainless steel reactor (i.d. 9.0 mm) with a central thermocouple to monitor the temperature of the catalyst bed. For each test, the catalyst (40 - 60 mesh) was diluted with quartz sands to fill the reactor. H_2 flow was regulated by a mass flow controller and the liquid feed consisting of 1.0 wt% DBT in decalin was introduced into the reactor by a piston pump. The S content in the feed corresponded to 1740 ppm. The catalyst was pre-reduced *in situ* by flowing H_2 at 40 ml/min at 500 °C for 2 h. The catalytic activities were investigated under the total pressure in the range of 2.0 - 3.5 MPa, at thc temperature from 280 to 360 °C, with the volume ratio of H_2 to liquid feed from 300 to 1000. The activity data were collected after 0.5 h duration at each temperature.

The liquid products were collected at 1.0 h intervals and analyzed off-line by a gas chromatography (SP 2100, Beijing Beifenruili Analytic Instrument (Group) Co., Ltd) equipped with a flame ionization detector (FID) and a capillary column (HJ. PONA, 50 m × 0.20 mm × 0.50 mm). The main products of the reaction were biphenyl (BP) and cyclohexylbenzene (CHB) with the S content being converted to H_2S. Therefore, the HDS performance of the catalysts can be evaluated by the conversion of DBT.

2. RESULTS AND DISCUSSIONS

2.1. Structure Characterization

2.1.1. XRD Characterization

XRD is a rapid and widely used analytical technique to identify the bulk phase of a crystalline material based on constructive interference between monochromatic X-rays and a crystalline sample. XRD patterns provide information of crystalline phases, crystal sizes, lattice constants, etc. The low-angle XRD was employed to investigate the impact of the loading of the phosphides on the long-range order of the molecular sieve materials. As shown in Figure 1, all of the Ni_xP/SBA-15 catalysts with various P/Ni molar ratios presented a major peak at about 0.92° due to the (100) reflection, characteristic of the well-ordered hexagonal (*p6mm*) structure in SBA-15. However, another two peaks at around 1.6° and 1.9° corresponding to (110) and (200) reflections of SBA-15 were severely decayed, suggesting that the long-range regularity of mesoporous molecular sieve was partially destroyed at the Ni content of 20 wt%.

Figure 1. Low-angle XRD patterns of Ni_xP/SBA-15 catalysts with different P/Ni molar ratio
(a) Ni_xP/SBA-15 (P/Ni=0.25); (b) Ni_xP/SBA-15 (P/Ni=0.5); (c) Ni_xP/SBA-15 (P/Ni=0.75); (d) Ni_xP/SBA-15 (P/Ni=1.0); (e) Ni_xP/SBA-15 (P/Ni=1.25); (f) Ni_xP/SBA-15 (P/Ni=1.5); (g) Ni_xP/SBA-15 (P/Ni=2.0)

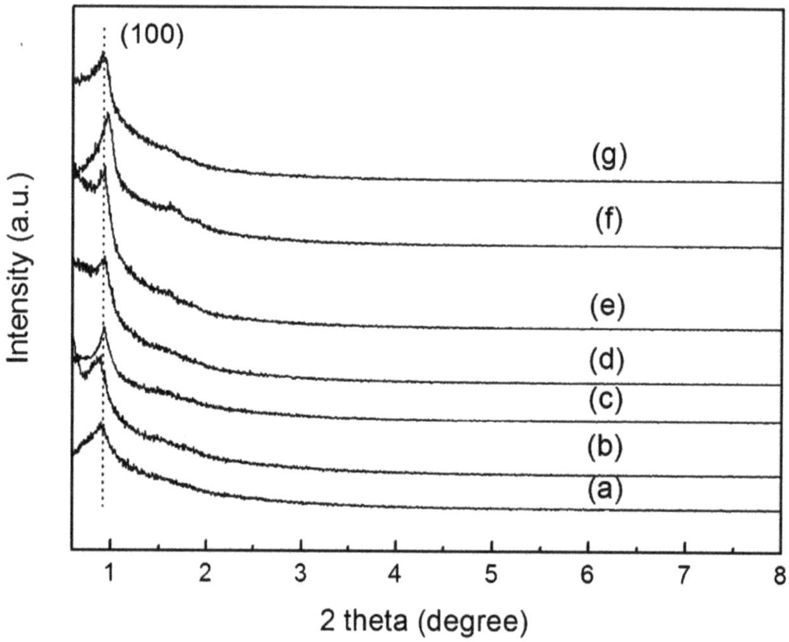

The evolution of the phase of Ni_xP as a function of the P/Ni molar ratio was investigated by the wide-angle XRD. As shown in Figure 2a, the Ni_xP/SBA-15 (P/Ni=0.25) catalyst displayed two series of diffraction peaks. The peaks at 2θ of 44.4°, 51.8° and 76.4° were due to metallic Ni phase (JCPDS card, File No. 65-2865), while the peaks at 41.7°, 43.6°, 46.5° and 52.6° were assigned to phosphide Ni_3P phase (JCPDS card, File No. 34-0501). That is, both metallic Ni and phosphide Ni_3P phases existed in the Ni_xP/SBA-15 catalyst at the lowest P content (Guan et al., 2012). As the P/Ni molar ratio increased to 0.5 (Figure 2b), Ni and Ni_3P phases disappeared. In the meantime, another two phosphide phases developed. The peaks at 38.3°, 41.6°, 46.8° and 48.8 ° matched the features of $Ni_{12}P_5$ phase (JCPDS card, File No. 65-1623) while the peaks at 40.6°, 44.5°, 47.2°, 54.1°, 54.9°, 74.8° and 80.1° were identified as the Ni_2P phase (JCPDS card, File No. 03-0953) (Gao et al., 2013). For the Ni_xP/SBA-15 catalysts with P/Ni≥0.75 (Figure 2(c-g)), only N_2P phase was detected. Moreover, the intensity and the full width at half maximum (FWHM) of the Ni_2P peaks were nearly constant. Our XRD results were consistent with the literature data . Oyama et al. and Korányi et al. (Oyama et al., 2008; Korányi et al., 2008) observed a mixture of $Ni_{12}P_5$ and Ni_2P phases at P/Ni=0.5 and pure Ni_2P phase at P/Ni>0.5, which may be related to the loss of P during the TPR procedure. It is generally accepted that the Ni_2P phase is catalytically more active than the $Ni_{12}P_5$ phase in HDS reactions. Therefore, an excess of P in the precursors is needed to obtain the Ni_2P phase which possessing higher HDS activity.

2.1.2. Surface Areas and Porosity

Surface area and porosity are important characteristics of catalysts that highly determine the catalytic performances. Gas adsorption isotherms are usually measured to analyze the surface area and pore size distribution of catalysts based on the physical adsorption of a vapor or gas onto the surface of a solid. The

Figure 2. High-angle XRD patterns of Ni_xP/SBA-15 catalysts with different P/Ni molar ratio
(a) Ni_xP/SBA-15 (P/Ni=0.25); (b) Ni_xP/SBA-15 (P/Ni=0.5); (c) Ni_xP/SBA-15 (P/Ni=0.75); (d) Ni_xP/SBA-15 (P/Ni=1.0); (e) Ni_xP/SBA-15 (P/Ni=1.25); (f) Ni_xP/SBA-15 (P/Ni=1.5); (g) Ni_xP/SBA-15 (P/Ni=2.0)

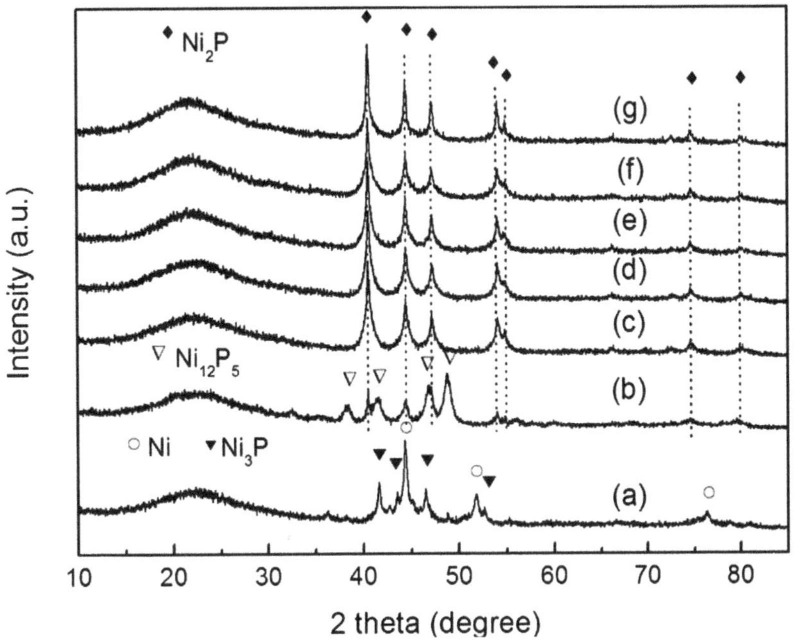

Figure 3. Nitrogen adsorption-desorption isotherms of SBA-15 and Ni_xP/SBA-15 catalysts
(a) SBA-15; (b) Ni_xP/SBA-15 (P/Ni=0.25); (c) Ni_xP/SBA-15 (P/Ni=0.75); (d) Ni_xP/SBA-15 (P/Ni=1.25); (e) Ni_xP/SBA-15 (P/Ni=2.0)

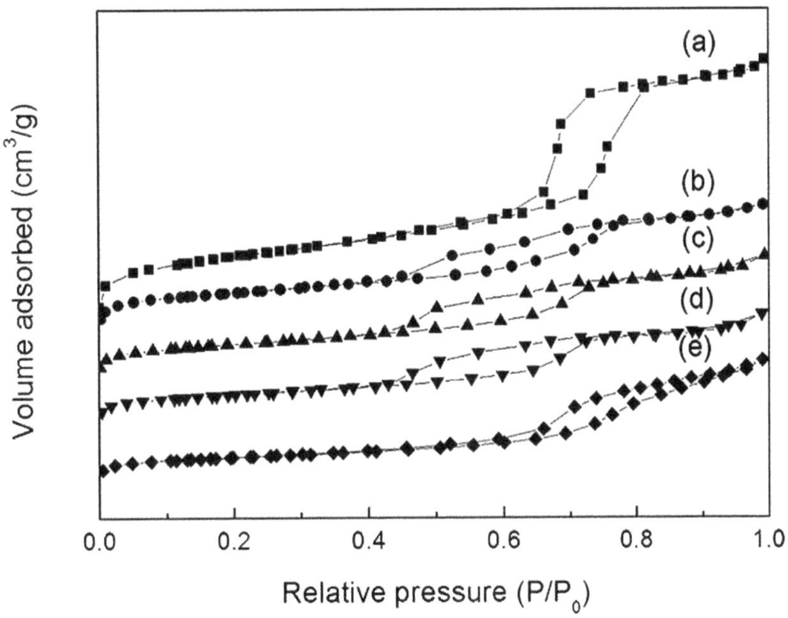

Figure 4. Pore size distributions of SBA-15 and Ni_xP/SBA-15 catalysts with different P/Ni
(a) SBA-15; (b) Ni_xP/SBA-15 (P/Ni=0.25); (c) Ni_xP/SBA-15 (P/Ni=0.75); (d) Ni_xP/SBA-15 (P/Ni=1.25); (e) Ni_xP/SBA-15 (P/Ni=2.0)

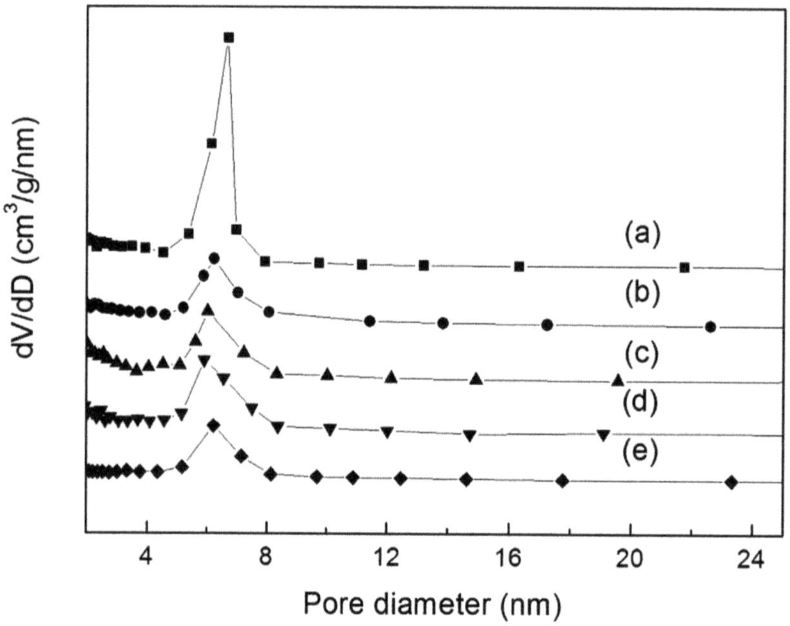

Ni_xP/SBA-15 catalysts with various P/Ni molar ratios, together with the ordered mesoporous molecular sieve SBA-15, were characterized by N_2 adsorption-desorption measurements. Figure 3 illustrated that the isotherms for SBA-15 and Ni_xP/SBA-15 samples were of type IV with a hysteresis loop, which were typical for the regular mesoporous materials. The deformation of the hysteresis loop for Ni_xP/SBA-15 samples was noticed, which may be caused by some particles situated inside the mesoporous structure, partially blocking the cross-section of the mesopores. The pore size distribution in Figure 4 further confirmed the impact of the loaded nickel phosphides. That is, the pore size distribution of Ni_2P/SBA-15 catalysts was broader than that of SBA-15 support.

Table 1. Physico-chemical properties of SBA-15 and Ni_xP/SBA-15 catalysts with different P/Ni

Sample	P/Ni (mol/mol)	Ni (wt%)	S_{BET} [a] ($m^2 g^{-1}$)	V_P [b] ($cm^3 g^{-1}$)	D_{BJH} [c] (nm)
SBA-15	—	—	674.0	1.12	6.7
Ni_xP/SBA-15 (P/Ni=0.25)	0.25	20	368.2	0.53	6.2
Ni_xP/SBA-15 (P/Ni=0.75)	0.75	20	334.1	0.50	6.0
Ni_xP/SBA-15 (P/Ni=1.25)	1.25	20	255.7	0.47	5.9
Ni_xP/SBA-15 (P/Ni=2.0)	2.0	20	222.8	0.48	6.2

[a]Specific surface area determined by Brunauer-Emmett-Teller (BET) method.
[b]Total pore volume recorded at $P/P_0 = 0.99$.
[c]Pore diameter calculated by Barrett-Joyner-Halenda (BJH) method.

The pore structure parameters for the characterized samples were summarized in Table 1. Compared with the SBA-15 support, a significant decrease in both specific surface area and pore volume was observed and a slight decrease in the average pore diameter occurred on the Ni_xP/SBA-15 catalysts, which suggested that the mesopores were partially blocked after the incorporation of Ni_xP. Moreover, the specific surface area of the Ni_xP/SBA-15 catalyst decreased as the P/Ni molar ratio increased, which indicated

Figure 5. TEM images of Ni_xP/SBA-15 catalysts with different P/Ni
(A) Ni_xP/SBA-15 (P/Ni=0.25); (B) Ni_xP/SBA-15 (P/Ni=0.5); (C) Ni_xP/SBA-15 (P/Ni=0.75); (D) Ni_xP/SBA-15 (P/Ni=1.0); (E) Ni_xP/SBA-15 (P/Ni=1.25); (F) Ni_xP/SBA-15 (P/Ni=1.5); (G) Ni_xP/SBA-15 (P/Ni=2.0)

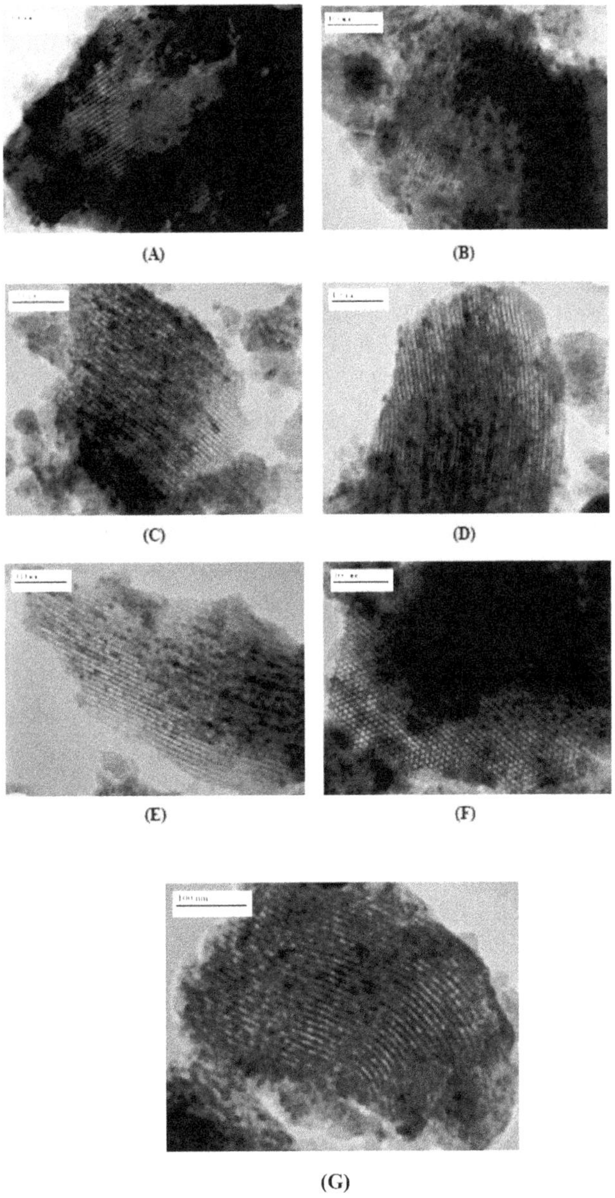

that the excess of P accumulated in the mesopores. Generally, the N_2 sorption results were in excellent consistence with the aforementioned low-angle XRD patterns. Both of the characterizations demonstrated that the ordered mesoporous structure was mainly maintained after the incorporation of Ni_xP.

2.1.3. TEM

TEM is a microscopy technique in which an ultra-thin specimen (<100 nm) is exposed to a high-energy (20-300 keV) electron beam. TEM images contain contrast due to diffraction, differences in atomic mass and/or thickness or crystalline structure, providing size and morphology information for the catalysts. TEM images were taken to investigate the dispersion and morphology of the Ni_xP/SBA-15 catalysts with various P/Ni molar ratios. As displayed in Figure 5, all of the catalysts possessed a well-ordered mesoporous channel structure, which supported the low-angle XRD and N_2 sorption results. Dark particles were observed in Figure 5a, which were assigned as metallic Ni nanoparticles based on the XRD analysis of Ni_xP/SBA-15 (P/Ni=0.25) catalyst (Figure 2a). As the P/Ni molar ratio increased, the dark particles became much vaguer and homogenously distributed in the mesopores (Figure 5 (b-g)). Combined with the XRD results (Figure 2 (b-g)), it was speculated that the $Ni_{12}P_5$ and Ni_2P phosphide phases exhibited a good dispersion on the Ni_xP/SBA-15 catalysts with P/Ni=0.5, 0.75, 1.0, 1.25, 1.5, 2.0.

2.1.4. FTIR

FTIR is a method to measure an infrared absorption spectrum using a Fourier transform (a mathematical process) to convert the raw data into the actual spectrum. An FTIR spectrum represents a fingerprint of a sample with absorption peaks corresponding to the vibration frequencies from the bonds making up the material. The FTIR spectra of SBA-15 and Ni_xP/SBA-15 catalysts are shown in Figure 6 in the range of 4000-600 cm^{-1}. The absorption peaks at 1082 and 802 cm^{-1} corresponded to the asymmetric and symmetric stretching vibration of Si-O-Si bonds in the framework, respectively. The absorption peak at 965 cm^{-1} was due to the stretching of nonbridging oxygen atoms of Si-OH bonds. The absorption peak at 1082 cm^{-1} and the shoulder peak in the range of 1100-1300 cm^{-1} were assigned to the synergistic stretching of the Si-O-Si bonds, which was caused by the partial ordering of the silicon framework on the surface of the mesoporous. It is noticed that after the introduction of nickel phosphides, the absorption peak at 965 cm^{-1} was decayed. Moreover, the intensity of this peak continues to decrease as the increase of the P/Ni molar ratio, which suggested that the excess of P would consume the Si-OH bonds on the surface of the mesopores.

2.2. Catalytic Performances

Based on our previous study (Huang et al., 2008) which examined the effect of the Ni content of the Ni_2P/SBA-15 catalysts on the activity in the HDS of thiophene and DBT as well as the literature results (Ma et al., 2011), the HDS catalytic performances of Ni_xP/SBA-15 catalysts were evaluated with the liquid feed consisting of 1.0 wt% DBT in decalin as the model compound. The external and internal diffusion limitations were eliminated by changing the space velocity and particle size. The effects of the reaction parameters such as the total pressure and the volume ratio of H_2 to liquid feed on the HDS activity were investigated as well. The HDS activity of Ni_xP/SBA-15 catalysts with various P/Ni molar ratios (P/Ni=0.25, 0.5, 0.75, 1.0, 1.25, 1.5 and 2.0) was evaluated after eliminating the external and internal diffusion limitations.

Figure 6. FTIR spectra of SBA-15 and Ni_xP/SBA-15 catalysts with different P/Ni
(a) SBA-15; (b) Ni_xP/SBA-15 (P/Ni=0.25); (c) Ni_xP/SBA-15 (P/Ni=0.75); (d) Ni_xP/SBA-15 (P/Ni=1.25); (e) Ni_xP/SBA-15 (P/Ni=2.0)

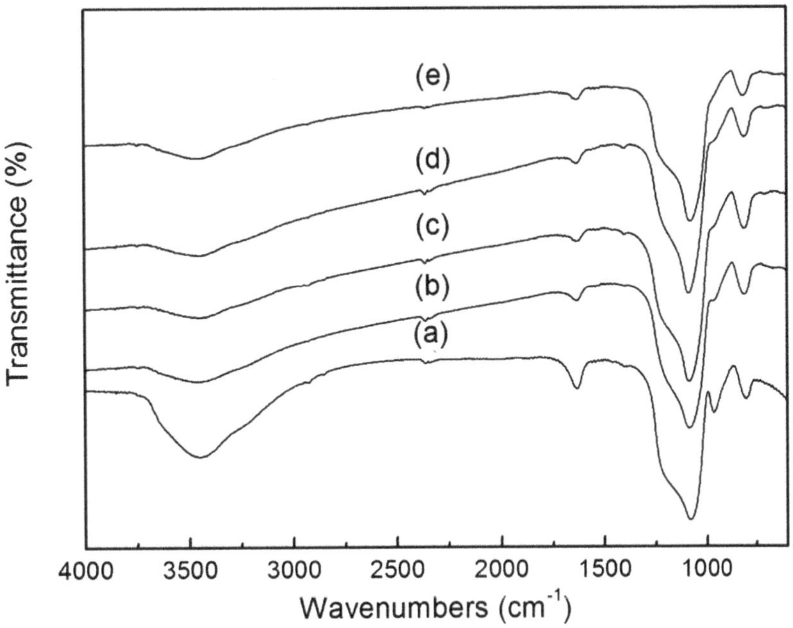

2.2.1. Elimination of External and Internal Diffusion Limitation

It is necessary to eliminate both the external and internal diffusion limitations before evaluating the catalytic activity of the Ni_xP/SBA-15 catalysts with various P/Ni molar ratios. The influence of the external diffusion was examined by comparing the HDS activity between two catalysts with different loadings while other reaction conditions were kept constant. In the present work, the reaction was carried out at 300 °C under 3.0 MPa with the volume ratio of H_2 to liquid feed at 600 over 0.3 g and 0.5 g of Ni_xP/SBA-15 (P/Ni=0.75) catalysts, respectively. The DBT conversion as a function of contact time was acquired by changing the amount of feeds. As shown in Figure 7, the curves became coincident when the contact time was lower than 0.04 h, which indicated the complete elimination of the external diffusion limitation. Therefore the contact time should be set at lower than 0.04 h, corresponding to WHSV higher than 25 h^{-1}. The WHSV in the following studies was set at 26.7 h^{-1}.

After eliminating the external diffusion limitation, the influence of the internal diffusion was examined by comparing the HDS activity between catalysts with various particle sizes. The reaction was carried out at 300 °C under 3.0 MPa with the volume ratio of H_2 to liquid feed at 600 and WHSV of 26.7 h^{-1}. As shown in Figure 8, the DBT conversion was 77.6% when the catalyst particle size was in the range of 60-80 mesh. The DBT conversion show little change when the catalyst particle size increased to 40-60 mesh. The DBT conversion decreased to 76.2% when the catalysts particle size further increased to 20- 40 mesh. Therefore, internal diffusion limitation was completely eliminated when the particle size was in the range of 40-60 mesh.

Figure 7. External diffusion test of Ni_xP/SBA-15 (P/Ni=0.75) catalyst

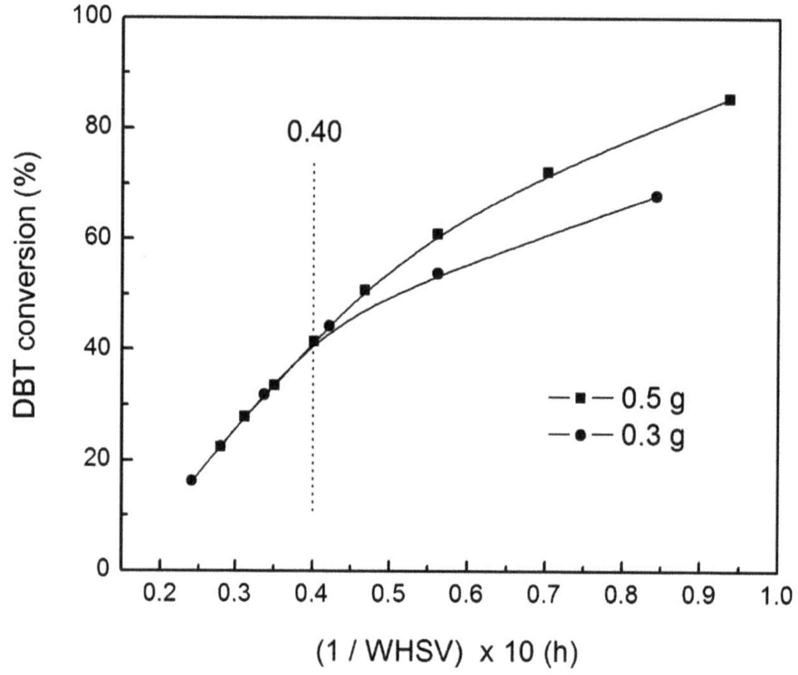

Figure 8. Effect of particle size of Ni_xP/SBA-15 (P/Ni=0.75) catalyst on the DBT conversion

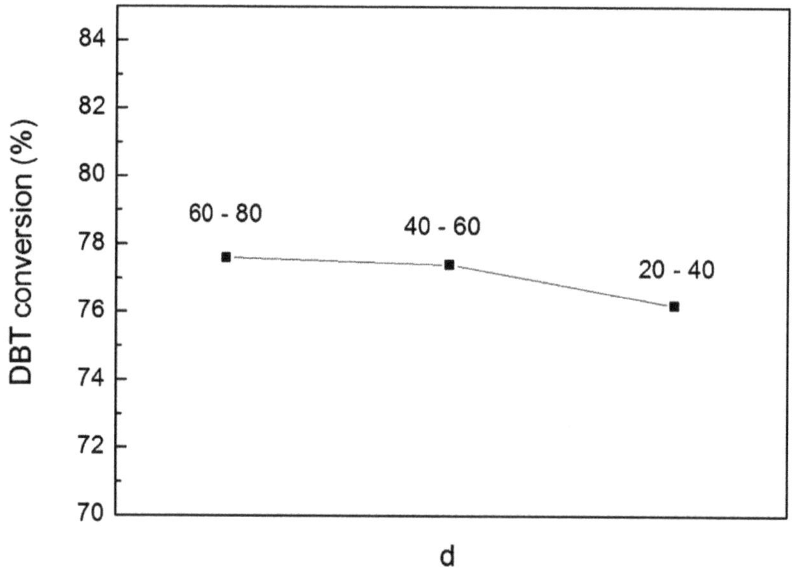

Figure 9. Effect of system pressure on the DBT conversion

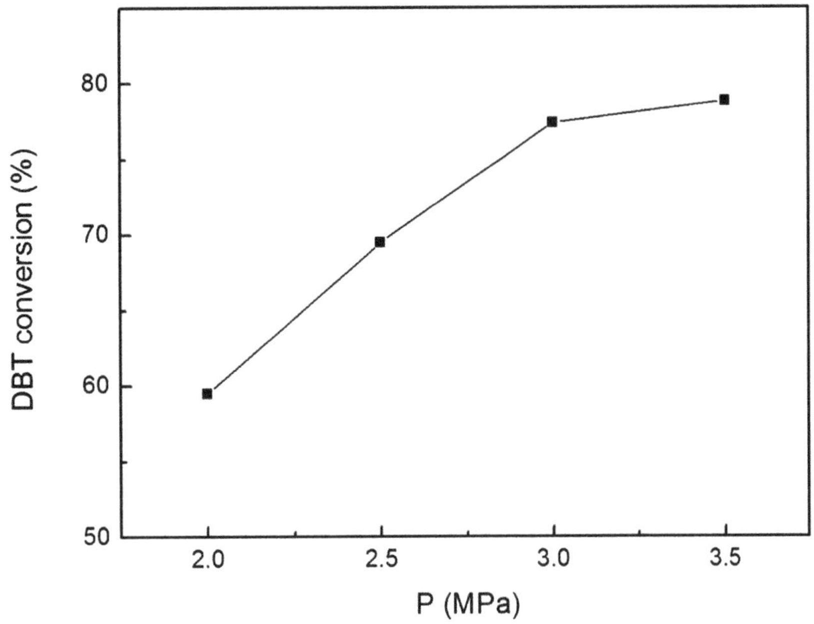

2.2.2. Effect of the Total Pressure

Since the model feed contains only 1 wt% of DBT in the present work, the total pressure of the reaction system mainly consists of the partial pressure of H_2 and the saturated pressure of decalin. We examined the effect of the total pressure on the DBT conversion with the Ni_xP/SBA-15 (P/Ni=0.75) catalyst at 320 °C with the volume ratio of H_2 to liquid feed at 600. As shown in Figure 9, the DBT conversion increased significantly from 59.5% to 77.4% as the total pressure increased from 2.0 to 3.0 MPa. A further increase of the total pressure up to 3.5 MPa only led to a slight increase for the DBT conversion. In other words, the DBT conversion was influenced significantly by the total pressure below 3.0 MPa, which may be related to the limited H_2 dissolution under low pressure. When the total pressure was higher than 3.0 MPa, the dissolved H_2 was enough for the HDS of DBT. Therefore, 3.0 MPa of the total pressure was employed in the following work.

2.2.3. Effect of the Volume Ratio of H_2 to Liquid Feed

H_2 consumption is a key parameter for the energy cost in the HDS process; therefore, the volume ratio of H_2 to liquid feed should be set as low as possible given that accepted HDS reactivity can be reached. In the present work, the effect of the volume ratio of H_2 to liquid feed on the DBT conversion was examined with the Ni_xP/SBA-15 (P/Ni=0.75) catalyst at 320 °C under 3.0 MPa. As demonstrated in Figure 10, the DBT conversion increased dramatically from 61.5% to 77.4% as the volume ratio of H_2/liquid feed increased from 300 to 600. Only slight increase on the DBT conversion was observed when the volume ratio of H_2/liquid feed further increased from 600 to 1000. Therefore, the volume ratio of H_2/liquid feed should be set at 600.

Figure 10. Effect of H_2/oil ratio on the DBT conversion

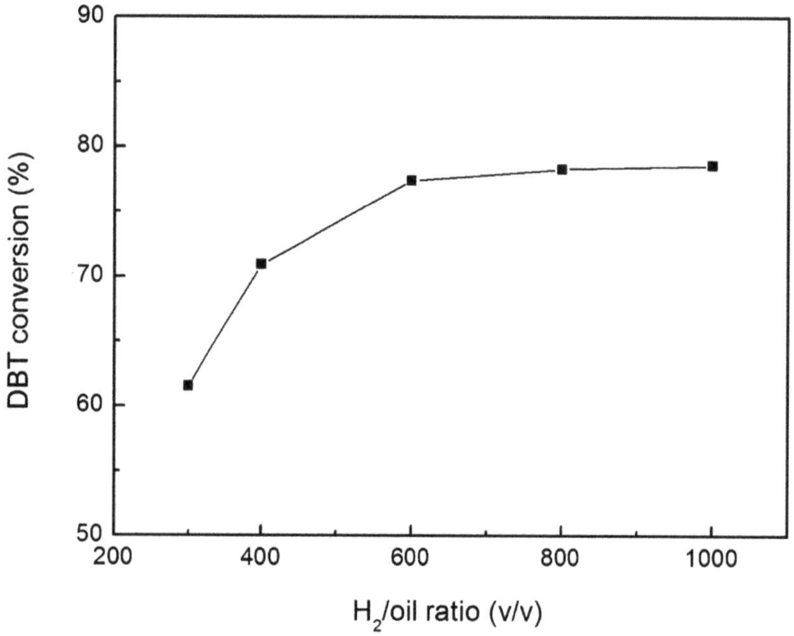

2.2.4. Effect of the P/Ni Molar Ratio

Wide-angle XRD patterns have clearly documented that the P/Ni molar ratio has significant influence on the phase formation of Ni_xP/SBA-15 catalysts, which may impact the catalytic performances in HDS reactions. The effect of P/Ni molar ratio on the DBT conversion was investigated at reaction temperature from 280 to 360 °C. As demonstrated in Figure 11, Ni_xP/SBA-15 (P/Ni=0.25) catalyst exhibited the lowest activity, suggesting that the metallic Ni and phosphide Ni_3P were not active in the HDS of DBT. The DBT conversion increased for the Ni_xP/SBA-15 (P/Ni=0.5) catalyst and reached the highest value for the Ni_xP/SBA-15 (P/Ni=0.75) catalyst, which indicated Ni_2P phase was the most active phase and $Ni_{12}P_5$ was more active than Ni or Ni_3P. Compared with the Ni_xP/SBA-15 (P/Ni=0.75) catalyst, the Ni_xP/SBA-15 catalyst exhibited a close activity at P/Ni=1.0, and a lower activity as the P/Ni molar ratio further increased. The decrease of activity for Ni_xP/SBA-15 catalysts at P/Ni >0.75 may be related to the partial blockage of the active sites by the excess P as the aforementioned N_2 sorption data suggested. Similar phenomena have been reported in the literature (Korányi et al., 2008).

In addition, the DBT conversion increased with the temperature for all of the Ni_xP/SBA-15 catalysts. Except the Ni_xP/SBA-15 (P/Ni=0.25) catalyst, all of the other Ni_xP/SBA-15 catalysts exhibited a DBT conversion higher than 95% at 360 °C. Besides, the Ni_xP/SBA-15 (P/Ni=0.75) catalyst reached a DBT conversion of 95.8% at 340 °C.

It is well established that the HDS of DBT proceeds mainly via two parallel reaction pathways. One is the direct desulfurization (DDS) with BP as the product. The other is the hydrogenation followed by the desulfurization (HYD) with CHB as the final product. Figure 12 illustrates the selectivity profiles for the HDS of DBT on the Ni_xP/SBA-15 catalysts with various P/Ni molar ratios at reaction temperature from 280 to 360 °C. For the Ni_xP/SBA-15 catalysts at P/Ni=0.25 and 0.5, the BP selectivity was much higher

Figure 11. HDS of DBT over Ni_xP/SBA-15 catalysts with different P/Ni
(a) Ni_xP/SBA-15 (P/Ni=0.25); (b) Ni_xP/SBA-15 (P/Ni=0.5); (c) Ni_xP/SBA-15 (P/Ni=0.75); (d) Ni_xP/SBA-15 (P/Ni=1.0); (e) Ni_xP/SBA-15 (P/Ni=1.25); (f) Ni_xP/SBA-15 (P/Ni=1.5); (g) Ni_xP/SBA-15 (P/Ni=2.0)

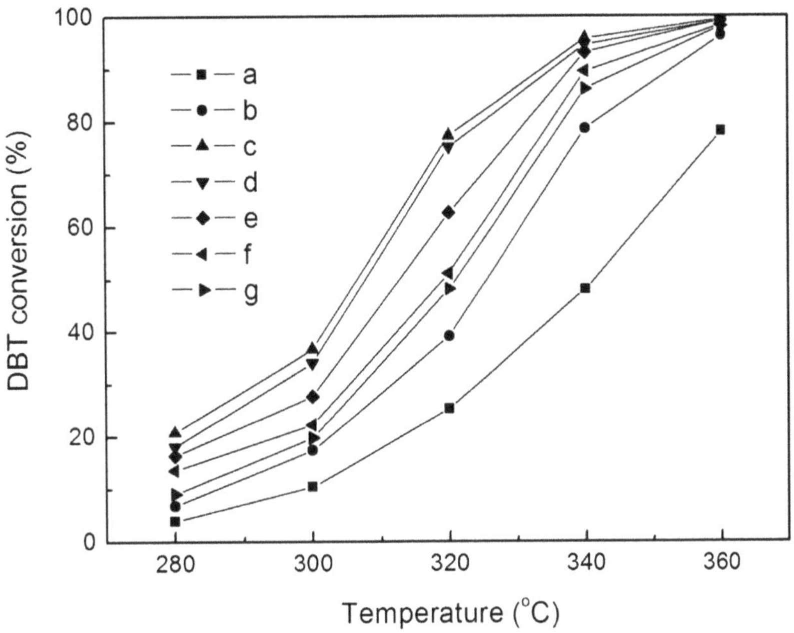

than the CHB selectivity and increased with the temperature in the investigated range, indicating that DBT was mainly converted via the DDS pathway. For the Ni_xP/SBA-15 catalysts at P/Ni≥0.75, the BP selectivity increased with the temperature from 260 to 320 °C and then decreased with the temperature from 320 to 360 °C. Specifically, the BP selectivity was close to 70% at temperature between 260 and 340 °C and lower than 60% at 360 °C, which indicated that the DDS pathway was favored at temperature between 260 and 340 °C while both the DDS and HYD pathways were important at 360 °C.

2.2.5. Investigations of Stability

Since the Ni_xP/SBA-15 (P/Ni=0.75) catalyst showed the best activity, it was employed to investigate the stability. The stability test was carried out with liquid feed consisting of 1.0 wt% DBT in decalin as the model compound at 340 °C under 3.0 MPa with the volume ratio of H_2 to liquid feed at 600 and WHSV of 26.7 h^{-1} over the 40 - 60 mesh catalysts. Characterizations were also performed to compare the structure of the Ni_xP/SBA-15 (P/Ni=0.75) catalyst before and after the stability test.

2.2.5.1. Stability Test

Shown in Figure 13 is the evolution of DBT conversion as well as BP and CHB selectivity with time on stream. The DBT conversion increased from 90.4% to 98.0% during the initial 12 h and then maintained in the range of 96.8% - 99.4% up to 300 h. The BP selectivity of the Ni_xP/SBA-15 (P/Ni=0.75) catalyst increased slowly from around 60% to around 70% during the initial 100 h and then maintained stable. The selectivity profiles confirmed that the DBT was mainly converted via the DDS pathway on the Ni_xP/

SBA-15 (P/Ni=0.75) catalyst. Both the DBT conversion and BP/CHB selectivity profiles demonstrated that the Ni_xP/SBA-15 (P/Ni=0.75) catalyst possessed an excellent stability in the HDS of DBT under the current conditions.

Figure 12. BP and CHB selectivity over Ni_xP/SBA-15 catalysts with different P/Ni for DBT HDS
(a) Ni_xP/SBA-15 (P/Ni=0.25); (b) Ni_xP/SBA-15 (P/Ni=0.5); (c) Ni_xP/SBA-15 (P/Ni=0.75); (d) Ni_xP/SBA-15 (P/Ni=1.0); (e) Ni_xP/SBA-15 (P/Ni=1.25); (f) Ni_xP/SBA-15 (P/Ni=1.5); (g) Ni_xP/SBA-15 (P/Ni=2.0)

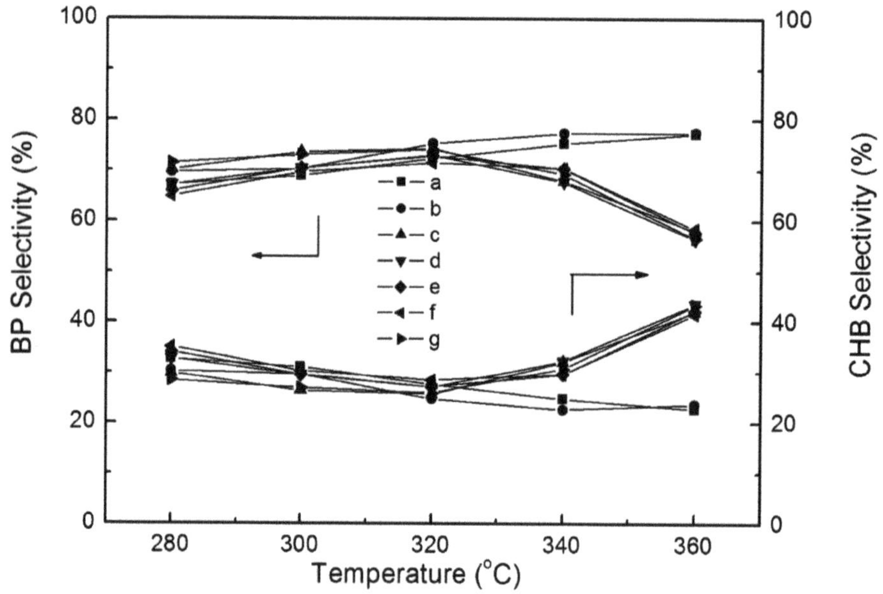

Figure 13. Stability test of Ni_xP/SBA-15 (P/Ni=0.75) catalyst for HDS of DBT

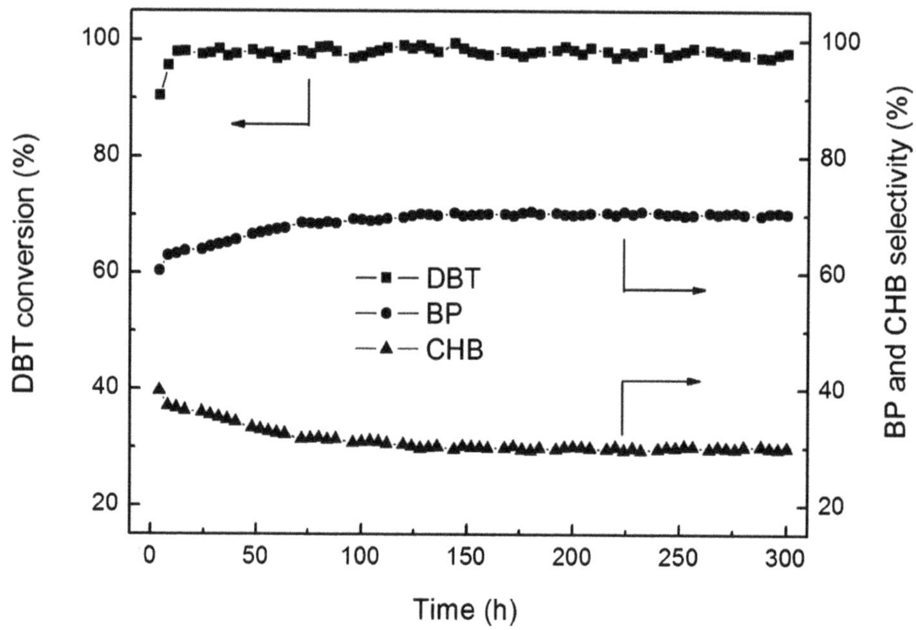

Figure 14. Low-angle XRD patterns of Ni_xP/SBA-15 (P/Ni=0.75) catalyst before (a) and after (b) stability test

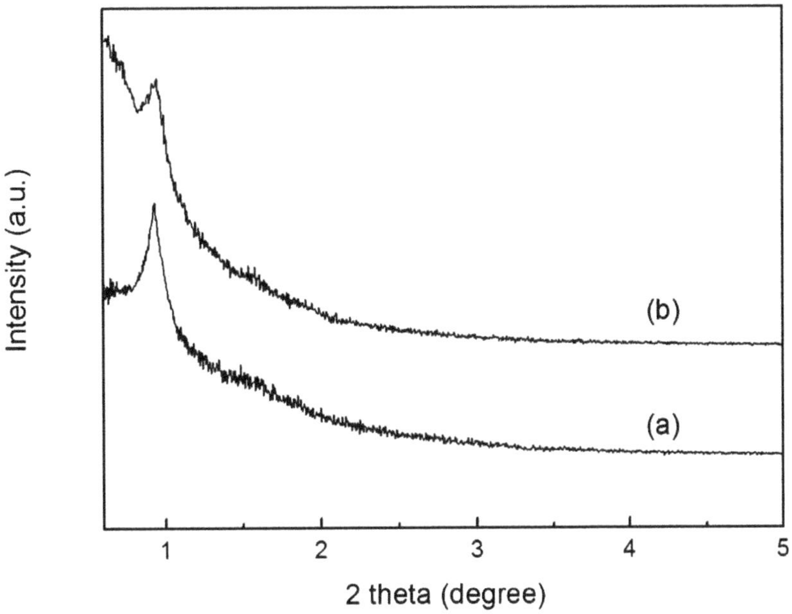

2.2.5.2. Structure Characterization

After 300 h of stability test, the Ni_xP/SBA-15 (P/Ni=0.75) catalyst was cooled down in flowing N_2 to room temperature and then passivated with 50 ml/min of 1.0 vol.% O_2/Ar for 2h. The passivated sample was characterized with XRD, N_2 sorption and TEM techniques to investigate the structure properties of the Ni_xP/SBA-15 (P/Ni=0.75) catalyst after the stability test.

Figure 14 and Figure 15 compare the low-angle and wide-angle XRD patterns of the Ni_xP/SBA-15 (P/Ni=0.75) catalyst before and after the stability test. No significant change occurred for the characteristic features in the low-angle XRD patterns, which meant that the ordered mesoprorous structure was well maintained. Excellent consistence was observed in the wide-angle XRD patterns, demonstrating that the active Ni_2P phosphide phase was stable.

Figure 16 shows the isotherm of the Ni_xP/SBA-15 (P/Ni=0.75) catalyst before and after the stability test. The similarity observed suggested that the pore structure was well maintained. After the stability test, the S_{BET} decreased slightly from 334 m^2/g to 328 m^2/g while both the V_p and D_{BJH} kept almost constant at 0.5 cm^3/g and 6.0 nm, respectively.

Figure 17 presents the TEM images of the Ni_xP/SBA-15 (P/Ni=0.75) catalyst after the stability test. A well-ordered mesoporous channel structure can still be observed and the phosphide phase still exhibits a good dispersion, which is in agreement with that in Figure 5c.

To summarize, XRD, N_2 sorption and TEM characterizations demonstrated that the pore structure as well as the phase and the dispersion of Ni_2P phosphide in the Ni_xP/SBA-15 (P/Ni=0.75) catalyst were stable after 300 h of HDS reactions, which may explain the excellent activity/selectivity observed in the stability test.

Figure 15. High-angle XRD patterns of Ni_xP/SBA-15 (P/Ni=0.75) catalyst before (a) and after (b) stability test

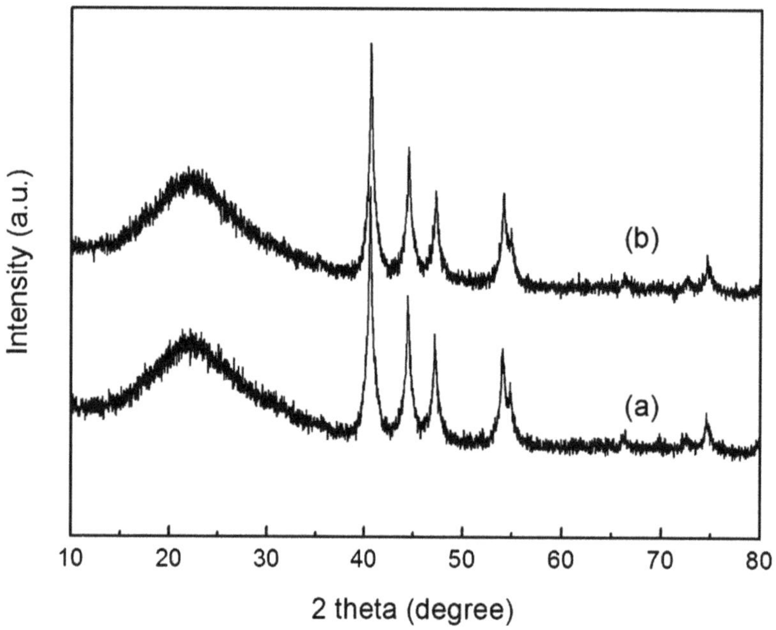

Figure 16. Nitrogen adsorption-desorption isotherms of Ni_xP/SBA-15 (P/Ni=0.75) catalyst before (a) and after (b) stability test

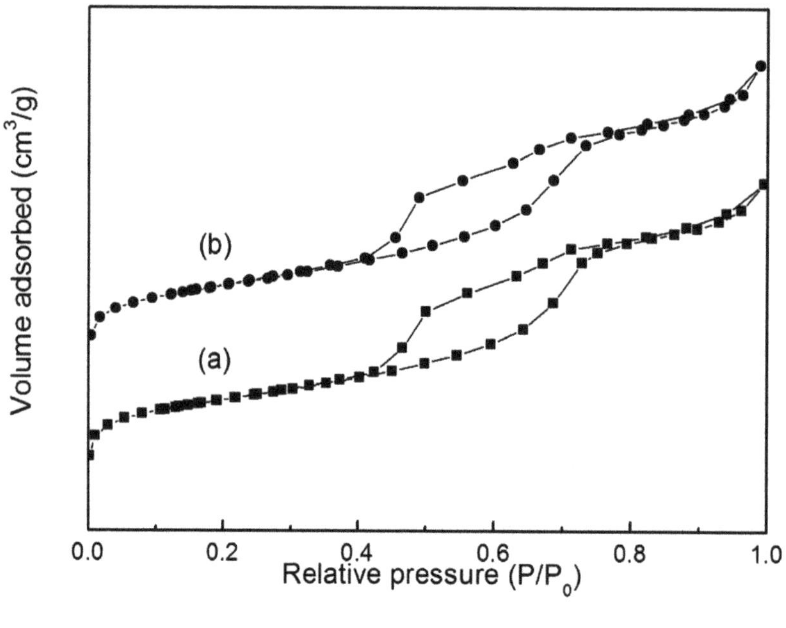

Figure 17. TEM image of Ni_xP/SBA-15 (P/Ni=0.75) catalyst after stability test

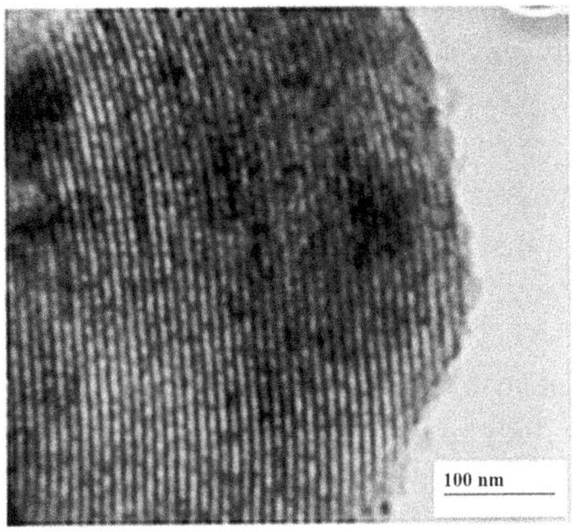

2.3. *In Situ* DRIFTS Spectra

To further understand the nature of the active sites of the Ni_2P catalyst in HDS, the surface properties of the Ni_xP/SBA-15 (P/Ni=0.75) catalyst were studied by CO adsorption using *in situ* DRIFTS measurements. *In situ* DRIFTS of CO adsorbed to solid surfaces is an accepted and widely used approach to probe the surface structure of the catalysts. CO adsorption bands provide structure information of the catalyst surface due to the high sensitivity of ν(CO) towards the coordinated surface. Shown in Figure 18 are *in situ* DRIFTS spectra of CO adsorption on the Ni_xP/SBA-15 (P/Ni=0.75) catalyst at various

Figure 18. In situ DRIFT spectra of CO adsorption on Ni_xP/SBA-15 (P/Ni=0.75)
(a) 10 °C; (b) 20 °C; (c) 40 °C; (d) 80 °C; (e) 100 °C; (f) 120 °C; (g) 200 °C; (h) 300 °C; (i) 450 °C

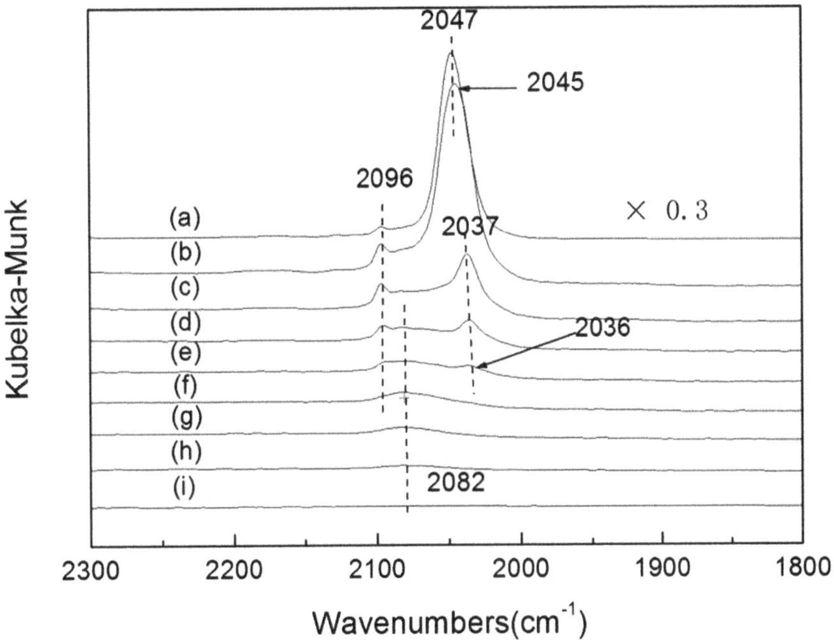

temperature. Intense signals and two characteristic bands were observed at 10 °C. One strong band was centered at 2047 cm^{-1} and another weak band appeared as a shoulder peak at 2096 cm^{-1}. The strong band at 2047 cm^{-1} decayed significantly and shifted slightly to 2045 cm^{-1} after the sample was heated to 20 °C. When the sample temperature was raised to 40 °C, this band shifted negatively to 2037 cm^{-1}. Upon further heating, the band at 2037 cm^{-1} decayed gradually and shifted slightly to 2036 cm^{-1}. The complete desorption of this band was observed at 120 °C. In contrast, the band at 2096 cm^{-1} exhibited much better stability, the intensity of which only decayed slightly as the sample temperature was raised up to 100 °C. This band shifted negatively to 2082 cm^{-1} and the peak became significantly broadened after the sample was further heated to 120 °C. The band at 2082 cm^{-1} remained stable up to 200 °C and decayed substantially at 300 °C. The complete desorption of this band was observed at 450 °C.

Our previous *in situ* DRIFTS study of CO adsorption on SBA-15 has documented that a characteristic band at 2047 cm^{-1} was observed due to the physisorbed CO on SBA-15 at 20 °C, which decayed significantly as the sample temperature increased to 40 °C. In addition, it is reported in the literature (Liu et al., 2009) that nickel tetracarbonyl ($Ni(CO)_4$) can be formed upon CO adsorption showing the characteristic band around 2050 cm^{-1}. Therefore, the bands around 2047 cm^{-1} between 10 and 20 °C were assigned to the combination of physisorbed CO on SBA-15 and $Ni(CO)_4$ species formed upon CO adsorption. As the sample temperature increased to 40 °C, most of the physisorbed CO was removed from SBA-15 and the remained bands were mainly due to the stretching of $Ni(CO)_4$ species. The remaining $Ni(CO)_4$ species desorbed gradually as the sample was heated up to 100 °C, with the complete desorption observed at 120 °C. The bands between 2096 and 2082 cm^{-1} have been well documented in the literature (Liu et al., 2009), which corresponded to CO terminally bonded to the coordinatively unsaturated $Ni^{\delta+}$ ($0<\delta<1$) sites. The wavenumbers of CO adsorption on the $Ni^{\delta+}$ sites were sensitive to the temperature of the sample, which explained the shift observed at 120 °C in the current work. The high stability of CO adsorbed on the $Ni^{\delta+}$ sites demonstrated that these $Ni^{\delta+}$ cations had a high d-electron density and were able to form π-back-bonded CO species. A previous theoretic study by Rodriguez et al. [Rodriguez et al., 2003] has found a correlation between the electronic properties of the metal cations and their HDS activities for Ni_2P, MoP, and MoS_2 catalysts supported on silica. The higher electron density of nickel cations in Ni_2P was suggested to enhance HDS activity by facilitating the dissociation of thiophene molecules. Similarly, Layman and Bussell [Layman et al., 2004] concluded from the infrared spectroscopic study of CO adsorption on reduced and sulfided Ni_2P/SiO_2 catalysts that the active site for thiophene HDS was most likely a $Ni^{\delta+}$ atop site, which may be associated with a surface NiP_xS_y phase. Therefore, the coordinatively unsaturated $Ni^{\delta+}$ ($0<\delta<1$) sites are proposed as the active sites for the Ni_2P/SBA-15 catalyst in the HDS of DBT reactions.

FUTURE RESEARCH DIRECTIONS

The prsent work reported the preparation, structure characterization and catalytic performance of the Ni_xP/SBA-15 catalysts in the HDS of DBT. The active sites of the Ni_2P/SBA-15 catalyst was discussed as well. Although some insightful results have been acquired, the following issues still need to be further addressed for the future industializaitions.

1. It is generally accepted that the Ni_2P phase is catalytically more active than the $Ni_{12}P_5$ phase. Therefore, the initial P/Ni molar ratio would critically influence the HDS activity of the nickel phosphide catalyts. Since some of the P content would be lost during the TPR procedure, an excess of phosphorus in the precursors was needed to obtain the desired stoichiometry Ni_2P phase. However, the excessive P may also partially block the active phase on the surface and thus lower the HDS activity. Therefore, it is critical to adopt an optimal P/Ni molar ratio during the preparation of the nickel phosphide catalysts.
2. The active phase of Ni_xP plays a critical role on the Ni_xP/SBA-15 catalysts. However, little information has been acquired about the distribution status of Ni_xP inside the mesopores of SBA-15. Therefore, further studies are still needed to clarify such issues.
3. The preparation parameters are still needed to be optimized to obtain improved HDS activity. The relationship between the structure and the activity on the Ni_xP/SBA-15 catalysts is also needed to be further clarified.
4. The essence of determining the proper catalytic material for the process lie in understanding the catalytic mechanisms and thus identifying the active site for a specific reaction. In HDS reaction, the complexity of hetergenous systems as well as the structural evolution of catalysts under reaction conditions make understanding of the catalytic mechanisms difficult. To gain more information towards catalytic mechanisms, it is of prime importance to carry out *in situ* characterizations of the working catalysts with morden techniques.

CONCLUSION

A series of Ni_xP/SBA-15 catalysts with a fixed Ni content of 20 wt% and a varying P/Ni molar ratio (0.25, 0.5, 0.75, 1.0, 1.25, 1.5, 2.0) have been prepared by the H_2-TPR treatment following co-impregnation method. The mesoporous molecular sieve SBA-15 support was synthesized using the cheap sodium silicate as the silicon source. The Ni_xP/SBA-15 catalysts was prepared using $Ni(NO_3)_2 \cdot 6H_2O$ as the nickel source and $(NH_4)_2HPO_4$ as the phosphorus source. The prepared catalysts were characterized by XRD, TEM, N_2 sorption analysis and FTIR techniques. The obtained results demonstrated that Ni_2P and $Ni_{12}P_5$ phases existed at P/Ni = 0.5 while Pure Ni_2P phase was detected at P/Ni $\geq$ 0.75. Both Ni_2P and $Ni_{12}P_5$ phases showed activity in HDS reactions while Ni_2P was more active. The optimal P/Ni was found at 0.75. The catalytic performances would be deteriorated at either too high or too low P/Ni molar ratios. The mesoporous structure was mainly maintained with high S_{BET} and V_P as well as uniformly distributed mesopores, which would benefit the dispersion of the active phases and the diffusion of the reactants during the HDS reactions.

The prepared Ni_xP/SBA-15 catalysts show good activity in the HDS of DBT, among which the Ni_xP/SBA-15 (P/Ni=0.75) catalyst exhibited the highest DBT conversion. The DBT conversion on the Ni_xP/SBA-15 (P/Ni=0.75) catalyst reached 95.8% at 340 °C, under 3.0 MPa, with the volume ratio of H_2 to liquid feed at 600 and WHSV at 26.7 h^{-1}. The HDS of DBT on the Ni_xP/SBA-15 (P/Ni=0.75) catalyst proceeded mainly via the DDS pathway. Besides, stability test showed that the Ni_xP/SBA-15 (P/Ni=0.75) catalyst possessed excellent activity/selectivity up to 300 h. Characterization results demonstrated that the pore structure and the phosphide phase of the catalyst were well maintained after the stability test.

In situ DRIFTS spectra of CO adsorption on the Ni_xP/SBA-15 (P/Ni=0.75) catalyst suggested the coordinatively unsaturated $Ni^{\delta+}$ ($0<\delta<1$) sites as the active sites for the Ni_xP/SBA-15 (P/Ni=0.75) catalyst during HDS reactions. The mesoporous molecular sieve supported Ni_xP/SBA-15 catalyst is expected to be very promising in the industrialization of HDS catalysts. The present work provides valuable insight into the scale-up and industrialization of the Ni_xP/SBA-15 catalyst in HDS reactions.

REFERENCES

Burns, S., Hargreaves, J. S., & Hunter, S. M. (2007). On the use of methane as a reductant in the synthesis of transition metal phosphides. *Catalysis Communications*, *8*(6), 931–935. doi:10.1016/j.catcom.2006.10.001

Cecilia, J. A., Infantes-Molina, A., Rodríguez-Castellón, E., & Jiménez-López, A. (2009). A novel method for preparing an active nickel phosphide catalyst for HDS of dibenzothiophene. *Journal of Catalysis*, *263*(1), 4–15. doi:10.1016/j.jcat.2009.02.013

Cecilia, J. A., Infantes-Molina, A., Rodríguez-Castellón, E., & Jiménez-López, A. (2009). Dibenzothiophene hydrodesulfurization over cobalt phosphide catalysts prepared through a new synthetic approach: Effect of the support. *Applied Catalysis B: Environmental*, *92*(1-2), 100–113. doi:10.1016/j.apcatb.2009.07.017

Cheng, R., Shu, Y., Li, L., Zheng, M., Wang, X., Wang, A., & Zhang, T. (2007). Synthesis and characterization of high surface area molybdenum phosphide. *Applied Catalysis A, General*, *316*(2), 160–168. doi:10.1016/j.apcata.2006.08.036

Clark, P., Li, W., & Oyama, S. T. (2001). Synthesis and activity of a new catalyst for hydroprocessing: Tungsten phosphide. *Journal of Catalysis*, *200*(1), 140–147. doi:10.1006/jcat.2001.3189

Clark, P., & Oyama, S. T. (2003). Alumina-supported molybdenum phosphide hydroprocessing catalysts. *Journal of Catalysis*, *218*(1), 78–87. doi:10.1016/S0021-9517(03)00086-1

Costa, P. D., Potvin, C., Manoli, J. M., Genin, B., & Djéga-Mariadassou, G. (2004). Deep hydrodesulphurization and hydrogenation of diesel fuels on alumina-supported and bulk molybdenum carbide catalysts. *Energy & Fuels*, *83*(13), 1717–1726. doi:10.1016/j.fuel.2004.03.007

Coulier, L., Kishan, G., van Veen, J. A. R., & Niemantsverdriet, J. W. (2002). Influence of support-interaction on the sulfidation behavior and hydrodesulfurization activity of Al_2O_3-supported W, CoW, and NiW model catalysts. *The Journal of Physical Chemistry B*, *106*(23), 5897–5906. doi:10.1021/jp0136821

Dhas, N. A., Ekhtiarzadeh, A., & Suslick, K. S. (2001). Sonochemical preparation of supported hydrodesulfurization catalysts. *Journal of the American Chemical Society*, *123*(34), 8310–8316. doi:10.1021/ja010516y PMID:11516280

Faro, A. Jr, & Dossantos, A. (2006). Cumene hydrocracking and thiophene HDS on niobia-supported Ni, Mo and Ni–Mo catalysts. *Catalysis Today*, *118*(3-4), 402–409. doi:10.1016/j.cattod.2006.07.027

Fujikawa, T., Kimura, H., Kiriyama, K., & Hagiwara, K. (2006). Development of ultra-deep HDS catalyst for production of clean diesel fuels. *Catalysis Today*, *111*(3-4), 188–193. doi:10.1016/j.cattod.2005.10.024

Furimsky, E. (2003). Metal carbides and nitrides as potential catalysts for hydroprocessing. *Applied Catalysis A, General*, *240*(1-2), 1–28. doi:10.1016/S0926-860X(02)00428-3

Gao, L., Wu, P., Lan, L., Guan, Y., & Ji, S. (2013). Influence of sulfurization behavior on structure and hydrodesulfurization performance of Ni_2p/SBA-15 catalysts. *Acta Petrolei Sinica*, *29*, 198–204.

Gong, S. W., Chen, H. K., Li, W., & Li, B. Q. (2006). Catalytic behaviors of β-$Mo_2N_{0.78}$ as a hydrodesulfurization catalyst. *Energy & Fuels*, *20*(4), 1372–1376. doi:10.1021/ef050208b

Guan, J., Wang, Y., Qin, M., Yang, Y., Li, X., & Wang, A. (2009). Synthesis of transition-metal phosphides from xidic precursors by reduction in hydrogen plasma. *Journal of Solid State Chemistry*, *182*(6), 1550–1555. doi:10.1016/j.jssc.2009.03.026

Guan, Q., Li, W., Zhang, M., & Tao, K. (2009). Alternative synthesis of bulk and supported nickel phosphide from the thermal decomposition of hypophosphites. *Journal of Catalysis*, *263*(1), 1–3. doi:10.1016/j.jcat.2009.02.008

Guan, Y., Wu, P., Lan, L., Gao, L., & Ji, S. (2012). Preparation of Ni_2P/SBA-15 catalysts using citrate acid method and its hydrodesulfurization performance. *J. Mol. Catal (China)*, *26*, 197–203.

Guo, Y., Zeng, P., Ji, S., Wei, N., Liu, H., & Li, C. (2010). Effect of Mo promoter content on performance of Mo-Ni_2P/SBA-15/cordierite monolithic catalyst for hydrodesulfurization. *Chinese Journal of Catalysis*, *31*, 329–334.

Hatanaka, S., Yamada, M., & Sadakane, O. (1997). Hydrodesulfurization of catalytic cracked gasoline. 2. The difference between HDS active site and olefin hydrogenation active site. *Industrial & Engineering Chemistry Research*, *36*(12), 5110–5117. doi:10.1021/ie9703494

Huang, X., Ji, S., Wu, P., Liu, Q., Liu, H., Zhu, J., & Li, C. (2008). Structure and hydrodesulfurization performances of Ni_2P/SBA-15 catalysts. *Acta Physico-Chimica Sinica*, *24*, 1773–1779.

Ibeh, B., Zhang, S., & Hill, J. M. (2009). Effect of citric acid on the synthesis of tungsten phosphide hydrotreating catalysts. *Applied Catalysis A, General*, *368*(1-2), 127–131. doi:10.1016/j.apcata.2009.08.017

Korányi, T. I., Vít, Z., & Nagy, J. B. (2008). Support and pretreatment effects on the hydrotreating activity of SBA-15 and CMK-5 supported nickel phosphide catalysts. *Catalysis Today*, *130*(1), 80–85. doi:10.1016/j.cattod.2007.09.004

Korányi, T. I., Vít, Z., Poduval, D. G., Ryoo, R., Kim, H., & Hensen, E. (2008). SBA-15-supported nickel phosphide hydrotreating catalysts. *Journal of Catalysis*, *253*(1), 119–131. doi:10.1016/j.jcat.2007.10.012

Korlann, S., Diaz, B., & Bussell, M. E. (2002). Synthesis of bulk and alumina-supported bimetallic carbide and nitride catalysts. *Chemistry of Materials*, *14*(10), 4049–4058. doi:10.1021/cm011508j

Layman, K. A., & Bussell, M. E. (2004). Infrared spectroscopic investigation of CO adsorption on silica-supported nickel phosphide catalysts. *The Journal of Physical Chemistry B*, *108*(30), 10930–10941. doi:10.1021/jp037101e

Levv, R. B., & Boudart, M. (1973). Platinum-like behavior of tungsten carbide in surface catalysis. *Science*, *181*(4099), 547–549. doi:10.1126/science.181.4099.547 PMID:17777803

Li, Q., Wu, P., Lan, L., Liu, H., & Ji, S. (2013). The effect of Ca on structure and performance of hydrodesulfurization of Ca-Ni_2P/SBA-15/Cord monolithic catalysts. *Catalysis Today*, *216*, 38–43. doi:10.1016/j.cattod.2013.06.016

Li, Q., Wu, P., Lan, L., Wei, N., & Ji, S. (2014). Preparation of novel mesoporous Ce-Ni_2P/SBA-15 catalysts and their catalytic performance for hydrodesulfurization of dibenzothiophene. *Applied Petrochemical Research*, *4*(2), 209–216. doi:10.1007/s13203-013-0039-5

Li, W., Dhandapani, B., & Oyama, S. T. (1998). Molybdenum phosphide: A novel catalyst for hydrogenitrogenation. *Chemistry Letters*, *27*(3), 207–208. doi:10.1246/cl.1998.207

Li, X., Lu, M., Wang, A., Song, C., & Hu, Y. (2008). Promoting effect of TiO_2 on the hydrodenitrogenation performance of nickel phosphide. *The Journal of Physical Chemistry C*, *112*(42), 16584–16592. doi:10.1021/jp803676v

Liu, Q., Ji, S., Wu, P., Hu, L., Huang, X., Zhu, J., & Li, C. (2009). *In situ* diffuse reflectance FTIR spectroscopy study of CO adsorption on Ni_2P/Mesoporous molecule sieve catalysts. *Spectroscopy and Spectral Analysis*, *29*, 1227–1231. PMID:19650459

Liu, Y., Liu, C., & Que, G. (2002). Hydrodesulfurization of dibenzothiophene over cobalt-molybdenum nitride catalysts. *Energy & Fuels*, *16*(3), 531–535. doi:10.1021/ef000272n

Ma, Q., Zeng, P., Ji, S., Wei, N., Liu, H., & Li, C. (2011). Effect of metal promoters on the structure and performance of the Ni_2P/SBA-15 catalyst for hydrodesulfurization. *Acta Petrolei Sinica*, *27*, 181–186.

Oyama, S. T., Clark, P., Wang, X., Shido, T., Iwasawa, Y., Hayashi, S., & Requejo, F. G. et al. (2002). Structural characterization of tungsten Phosphide (WP) hydrotreating catalysts by X-ray absorption spectroscopy and nuclear magnetic resonance spectroscopy. *The Journal of Physical Chemistry B*, *106*(8), 1913–1920. doi:10.1021/jp0136056

Oyama, S. T., & Lee, Y. K. (2008). The active site of nickel phosphide catalysts for the hydrodesulfurization of 4,6-DMDBT. *Journal of Catalysis*, *258*(2), 393–400. doi:10.1016/j.jcat.2008.06.023

Rodriguez, J. A., Kim, J.-Y., Hanson, J. C., Sawhill, S. J., & Bussell, M. E. (2003). Physical and chemical properties of MoP, Ni_2P, and MoNiP hydrodesulfurization catalysts: Time-resolved X-ray diffraction, density functional, and hydrodesulfurization activity studies. *The Journal of Physical Chemistry B*, *107*(26), 6276–6285. doi:10.1021/jp022639q

Schwartz, V., Oyama, S. T., & Chen, J. G. (2000). Supported bimetallic Nb-Mo carbide: Synthesis, characterization, and reactivity. *The Journal of Physical Chemistry B*, *104*(37), 8800–8806. doi:10.1021/jp0010233

Wang, A., J., Ruan, L., F., Teng, Y., et al. (2005). Hydrodesulfurization of dibenzothiophene over siliceous MCM-41 -supported nickel phosphide catalysts. *J. Catal.*, *229*, 314-321.

Wang, H., Wang, X., Zhang, M., Du, X.-Y., Li, W., & Tao, K.-Y. (2007). Synthesis of bulk and supported molybdenum carbide by a single-step thermal carburization method. *Chemistry of Materials*, *19*(7), 1801–1807. doi:10.1021/cm0615471

Wang, X., Clark, P., & Oyama, S. T. (2002). Synthesis, characterization, and hydrotreating activity of several iron group transition metal phosphides. *Journal of Catalysis*, *208*(2), 321–331. doi:10.1006/jcat.2002.3604

Wei, N., Ji, S., Guan, Y., Liu, H., & Li, C. (2011). Influence of W on structure and hydrodesulfurization performance of W-Ni_2P/SBA-15 catalysts. [Petroleum Processing Section]. *Acta Petrolei Sinica*, *27*, 852–858.

Wei, N., Ji, S., Wu, P., Guo, Y., Liu, H., Zhu, J., & Li, C. (2009). Preparation of nickel phosphide/SBA-15/cordierite monolithic catalysts and catalytic activity for hydrodesulfurization of dibenzothiophene. *Catalysis Today*, *147S*, S66–S70. doi:10.1016/j.cattod.2009.07.053

Wei, N., Zeng, P., Ji, S., Zhao, P., Liu, H., & Li, C. (2011). Effect of Lanthanum promoter on structure and hydrodesulfurization performance of La-Ni_2P/SBA-15 catalysts. *Journal of Rare Earths*, *29*, 310–315.

Wu, P., Lan, L., Ji, S., Wei, N., Lv, Z., & Liu, H. (2012). Co-Ni2P/SBA-15/cordierite monolithic catalysts: Preparation and performance for hydrodesulfurization of dibenzothiophene. *Wuji Huaxue Xuebao*, *28*, 565–571.

Wu, P., Li, Q., Lan, L., Liu, H., Ju, N., Piao, J., & Ji, S. (2014). Preparation, structure characterization and hydrodesulfurization performance of B-Ni_2P/SBA-15/cord monolithic catalysts. *Chinese Journal of Chemical Engineering*, *22*(8), 903–908. doi:10.1016/j.cjche.2014.06.014

Wu, P., Wang, J., Lan, L., Ju, Y., Piao, J., Jin, C., & Ji, S. et al. (2014). Preparation of La-Ni_2P/SBA-15/cordierite monolithic catalysts and their catalytic performance of hydrodesulfurization. *Industrial Catalysis*, *22*, 272–276.

Yang, S., Liang, C., & Prins, R. (2006). Preparation and hydrotreating activity of unsupported nickel phosphide with high surface area. *Journal of Catalysis*, *241*(2), 465–469. doi:10.1016/j.jcat.2006.05.014

Zhang, X., Guo, A., Wang, F., & Duan, X. (2010). Directly producing clean and low softening point diesel using integrated hydrotreating and hydroisomerizing catalysts. *Energy & Fuels*, *24*(7), 3772–3777. doi:10.1021/ef901373w

Zhao, P., Ji, S., Wei, N., Liu, H., & Li, C. (2010). Preparation, characterization and hydrodesulphurization catalytic performance of Al-Ni_2P/SBA-15 catalysts. *Wuji Huaxue Xuebao*, *26*, 2273–2279.

Zhao, P., Ji, S., Wei, N., Ma, Q., Liu, H., & Li, C. (2011). Effect of Boron promoter on the structure and hydrodesulfurization activity of Ni_2P/SBA-15 catalysts. *Acta Physico-Chimica Sinica*, *27*, 1737–1742.

KEY TERMS AND DEFINITIONS

Catalyst: A substance which changes the velocity of the chemical reactions while keeps the change of the standard Gibbs free enthalpy constant.

Dibenzothiophene: An organosulfur compound difficult to remove in the oils which can be used as the drug intermediate.

Diesel: A transparent mixture of liquid consisting of C_{10}-C_{22} organic compounds that is mainly used as a fuel in diesel engines.

Gasoline: A transparent mixture of liquid consisting of C_4-C_{10} organic compounds that is mainly used as a fuel in gasoline engines.

Hydrodesulfurization: A process that removes the S impurity from oils such as gasoline and diesel by converting the organosulfur compounds to hydrogen sulfide over a catalyst under H_2 partial pressure.

Nickel Phosphide: A compound consisting of phosphorous and transition metal nickel which is an exceptional hydrodesulfurization catalyst.

SBA-15: A mesoporous molecular sieve which has a high specific surface area and pore volume as well as uniformly distributed mesoporous.

Support: A material which can improve the efficiency of active component on the weight basis by dispersing the active component.

Compilation of References

Abbadi, A., & vanBekkum, H. (1996). Selective chemo-catalytic routes for the preparation of beta-hydroxypyruvic acid. *Applied Catalysis A, General*, *148*(1), 113–122. doi:10.1016/S0926-860X(96)00229-3

Abdelhady, F., Bamufleh, H., El-Halwagi, N. M., & Ponce-Ortega, J. M. (2015). Optimal Design and Integration of Solar Thermal Collection, Storage, and Dispatch with Process Cogeneration Systems. *Chemical Engineering Science*, *136*, 158–167. doi:10.1016/j.ces.2015.04.043

Abdelsayed, V., Shekhawat, D., Poston, J. A. Jr, & Spivey, J. J. (2013). Synthesis, characterization, and catalytic activity of Rh-based lanthanum zirconate pyrochlores for higher alcohol synthesis. *Catalysis Today*, *207*(0), 65–73. doi:10.1016/j.cattod.2012.07.011

Abro, R., Abdeltawab, A. A., Al-Deyab, S. S., Yu, G., Qazi, A. B., Gao, S., & Chen, X. (2014). A review of extractive desulfurization of fuel oils using ionic liquids. *RSC Adv.*, 4(67), 35302-35317. doi:10.1039/C4RA03478C

Abro, R., Gao, S., Chen, X., Yu, G., Abdeltawabb, A. A., & Al-Deyabb, S. S. (2015). Oxidative Desulfurization of Gasoline by Ionic Liquids Coupled with Extraction by Organic Solvents. *Journal of the Brazilian Chemical Society*. doi:10.5935/0103-5053.20150355

ABSG, C. I. (2004). *Consequence assessment methods for incidents involving releases from liquefied natural gas carriers.* Federal Energy Regulatory Commision.

Ahmed, S., Alasiri, H. S., Rahman, F., & Al-Amer, A. M. (2011). *US Patent No. US8623781(B2).*

Al Zarooni, M., & Elshorbagy, W. (2006). Characterization and assessment of Al Ruwais refinery wastewater. *Journal of Hazardous Materials*, *136*(3), 398–405. doi:10.1016/j.jhazmat.2005.09.060 PMID:16859828

Albright, L. F., Crynes, B. L., & Nowak, S. (1992). *Novel production methods for ethylene, light hydrocarbons, and aromatics*. New York, NY: Marcel Dekker Inc.

Al-Dalama, K., & Stanislaus, A. (2006). A Comparative Study of the Influence of Chelating Agents on the Hydrodesulfurization (HDS) Activity of Alumina and Silica-Alumina-Supported CoMo Catalysts. *Energy Fuels*, 20(5), 1777-1783. doi:10.1021/ef060125a

Al-Dalama, K., Aravind, B., & Stanislaus, A. (2005). Influence of complexing agents on the adsorption of molybdate and nickel ions on alumina. *Appl. Catal.*, A, 296(1), 49-53. doi:10.1016/j.apcata.2005.07.041

Ali, S. A., Ahmed, S., Ahmed, K. W., & Al-Saleh, M. A. (2012). Simultaneous hydrodesulfurization of dibenzothiophene and substituted dibenzothiophene over phosphorus modified CoMo/Al2O3 catalysts. *Fuel Process. Technol.*, 98, 39-44. doi: 10.1016/j.fuproc.2012.01.027

Aliabadi, M., Aroujalian, A., & Raisi, A. (2012). Removal of styrene from petrochemical wastewater using pervaporation process. *Desalination, 284*, 116–121. doi:10.1016/j.desal.2011.08.044

Ali, M. F., Al-Malki, A., El-Ali, B., Martinie, G., & Siddiqui, M. N. (2006). Deep desulphurization of gasoline and diesel fuels using non-hydrogen consuming techniques. *Fuel, 85*(10-11), 1354–1363. doi:10.1016/j.fuel.2005.12.006

Ali, S. A. (2014). *Development of improved catalysts for deep HDS of diesel fuels*. Appl. Petrochem. Res. Doi:10.1007/s13203-014-0082-x

Al-Malack, M. H. (2013). Treatment of petroleum refinery wastewater using crossflow and immersed membrane processes. *Desalination and Water Treatment, 51*(37-39), 6985–6993. doi:10.1080/19443994.2013.778798

Al-Megren, H. A. (2009). Hydrodesulfurization of thiophene over bimetallic Ni-Mo sulfide catalysts prepared by different methods. *Arabian J. Sci. Eng. Sect. A, 34*(1A), 55–66.

Al-Megren, H. A., Barbieri, G., Mirabelli, I., Brunetti, A., Drioli, E., & Al-Kinany, M. C. (2013b). Direct conversion of *n*-butane to isobutene in a membrane reactor: A thermodynamic analysis. *Industrial & Engineering Chemistry Research, 52*(31), 10380–10386. doi:10.1021/ie400006c

Al-Megren, H. A., Poerio, T., Brunetti, A., Barbieri, G., Drioli, E., Al-Hebaid, B. S. A., & Al-Kinany, M. C. et al. (2013a). Liquid phase benzene hydroxylation to phenol using semi batch and continuous membrane reactors. *Separation and Purification Technology, 107*, 195–203. doi:10.1016/j.seppur.2013.01.019

Alonso, D. M., Wettstein, S. G., & Dumesic, J. A. (2012). Bimetallic catalysts for upgrading of biomass to fuels and chemicals. *Chemical Society Reviews, 41*(24), 8075–8098. doi:10.1039/c2cs35188a PMID:22872312

Al-Sayari, S. A. (2013), Recent Developments in the Partial Oxidation of Methane to Syngas. The Open Catal. J., *6*, 17-28.

Al-Zeghayer, Y. S., Soliman, M. A., & Al-Mayman, S. & S, A.-A. A. (2013). *US Patent No. US2014235893A1.*

Al-Zeghayer, Y. (2014). Partial Oxidation of Ethane to Acetic Acid Catalyzed by MoVNbPd Catalyst Supported on Titania. *International Journal of Chemical Engineering and Applications, 5*(1), 50–57. doi:10.7763/IJCEA.2014.V5.350

Al-Zeghayer, Y. S., & Al-Awadi, A. S. (2013). Partial Oxidation of Ethane to Acetic Acid on Titania Supported MoVNbPd Catalyst. *Asian Journal of Chemistry, 25*(14), 7979–7985. doi:10.14233/ajchem.2013.14886

Amarasekara, S. A., & Owereh, O. S. (2010). Synthesis of a sulfonic acid functionalized acidic ionic liquid modified silica catalyst and applications in the hydrolysis of cellulose. *Catalysis Communications, 11*(13), 1072–1075. doi:10.1016/j.catcom.2010.05.012

Anderson, L. L., Armstrong, P. A., Repasky, J. M., & Stein, V. E. (2011). Enabling clean coal power generation: ITM oxygen technology. *Paper presented at theInternational Pittsburgh Coal Conference*, Pittsburgh, USA.

Ania, C. O., Parra, J. B., Arenillas, A., Rubiera, F., Bandosz, T. J., & Pis, J. J. (2007). On the mechanism of reactive adsorption of dibenzothiophene on organic waste derived carbons. *Applied Surface Science, 253*(13), 5899–5903. doi:10.1016/j.apsusc.2006.12.065

Antolini, E., Giorgi, L., Cardellini, F., & Passalacqua, E. (2001). Physical and morphological characteristics and electrochemical behaviour in PEM fuel cells of PtRu/C catalysts. *Journal of Solid State Electrochemistry, 5*(2), 131–140. doi:10.1007/s100080000116

Anugwom, I., Salmi, T., & Mikkola, J.-P. (2011). Ionic liquid assisted extraction of nitrogen and sulphur-containing air pollutants from model oil and regeneration of the spent ionic liquid. *Journal of Environmental Protection, 2*(06), 796–802. doi:10.4236/jep.2011.26091

Argyle, M. D., Chen, K., Bell, A. T., & Iglesia, E. (2002a). Effect of catalyst structure on oxidative dehydrogenation of ethane and propane on alumina-supported vanadia. *Journal of Catalysis*, *208*(1), 139–149. doi:10.1006/jcat.2002.3570

Argyle, M. D., Chen, K., Bell, A. T., & Iglesia, E. (2002b). Ethane oxidative dehydrogenation pathways on vanadium oxide catalysts. *The Journal of Physical Chemistry B*, *106*(21), 5421–5427. doi:10.1021/jp0144552

Arstad, B., & Kolboe, S. (2001). The Reactivity of Molecules Trapped within the SAPO-34 Cavities in the Methanol-to-Hydrocarbons Reaction. *Journal of the American Chemical Society*, *123*(33), 8137–8138. doi:10.1021/ja010668t PMID:11506579

Asumana, C., Haque, M. R., Yu, L., Wu, X., Chen, X., & Yu, G. (2013). Desulphurization of real fuel oils by extraction with ILs. *Separation Science and Technology*, *48*(17), 2582–2588. doi:10.1080/01496395.2013.804559

Asumana, C., Yu, G., Li, X., Zhao, J., Liu, G., & Chen, X. (2010). Extractive desulfurization of fuel oils with low-viscosity dicyanamide-based ionic liquids. *Green Chemistry*, *12*(11), 2030–2037. doi:10.1039/c0gc00118j

Atallah, P., Wagener, K. B., & Schulz, M. D. (2013). ADMET: The Future Revealed. *Macromolecules*, *46*(12), 4735–4741. doi:10.1021/ma400067b

Au, C. T., & Wang, H. Y. (1997). Mechanistic studies of methane partial oxidation to syngas over SiO2-supported rhodium catalysts. *Journal of Catalysis*, *167*(2), 337–345. doi:10.1006/jcat.1997.1609

Babak, V. N., Babak, T. B., Zakiev, S. E., & Kholpanov, L. P. (2009). Theoretical study of hydrocarbon dehydrogenation at high temperatures. *Theoretical Foundations of Chemical Engineering*, *43*(1), 74–87. doi:10.1134/S0040579509010102

Babich, I. V., & Moulijn, J. A. (2003). Science and technology of novel processes for deep desulphurization of oil refinery streams: A review. *Fuel*, *82*(6), 607–631. doi:10.1016/S0016-2361(02)00324-1

Baeza, P., Aguila, G., Gracia, F., & Araya, P. (2008). Desulfurization by adsorption with copper supported on zirconia. *Catalysis communications,* 9(5), 751-755.

Baker, R. W. (2006). *Membranes for vapor/gas separation*. Retrieved from http://www.mtrinc.com/publications/MT01%20Fane%20Memb%20for%20VaporGas_Sep%202006%20Book%20Ch.pdf

Baker, R. W., Hofmann, T., Kaschemekat, J., & Lokhandwala, K. A. (2001). *Field Demonstration of a Membrane Process to Recover Heavy Hydrocarbons and to Remove Water from Natural Gas* (Annual Report). Contract Number DE-FC26-99FT40723.

Baker, J. E., Burch, R., Hibble, S. J., & Loader, P. K. (1990). Properties of silica-supported Cu/Co bimetallic catalysts in the synthesis of higher alcohols. *Applied Catalysis*, *65*(2), 281–292. doi:10.1016/S0166-9834(00)81603-6

Baker, R. W. (2002). Future directions of membrane gas separation technology. *Industrial & Engineering Chemistry Research*, *41*(6), 1393–1404. doi:10.1021/ie0108088

Baker, R. W., & Lokhandwala, K. (2008). Natural Gas Processing with Membranes: An Overview. *Industrial & Engineering Chemistry Research*, *47*(7), 2109–2121. doi:10.1021/ie071083w

Balachandran, U., Dusek, J. T., Maiya, P. S., Ma, B., Mieville, R. L., Kleefisch, M. S., & Udovich, C. A. (1997). Ceramic membrane reactor for converting methane to syngas. *Catalysis Today*, *36*(3), 265–272. doi:10.1016/S0920-5861(96)00229-5

Balaraju, M., Rekha, V., Sai Prasad, P. S., Prasad, R. B. N., & Lingaiah, N. (2008). Selective Hydrogenolysis of Glycerol to 1, 2 Propanediol Over Cu-ZnO Catalysts. *Catalysis Letters*, *126*(1-2), 119–124. doi:10.1007/s10562-008-9590-6

Bamufleh, H. S., Ponce-Ortega, J. M., & El-Halwagi, M. M. (2013). Multi-Objective Optimization of Process Cogeneration Systems with Economic, Environmental, and Social Tradeoffs. *Clean Technologies and Environmental Policy*, *15*(1), 185–197. doi:10.1007/s10098-012-0497-y

Banerjee, S., Mani, K., Leonard, L., & Kokayeff, P. (2011). Distillate hydrotreating to ultra-low-sulfur diesel - the impact of aromatics. *Indian Chem. Eng.*, 53(3), 152-169. doi:10.1080/00194506.2011.696372

Banerjee, A., & Scott, R. W. J. (2015). Optimization of transition metal nanoparticle-phosphonium ionic liquid composite catalytic systems for deep hydrogenation and hydrodeoxygenation reactions. *Green Chemistry*, *17*(3), 1597–1604. doi:10.1039/C4GC01716A

Banerjee, A., Theron, R., & Scott, R. W. (2013). Redispersion of transition metal nanoparticle catalysts in tetraalkylphosphonium ionic liquids. *Chemical Communications*, *49*(31), 3227–3229. doi:10.1039/c3cc40726h PMID:23482965

Bao, J., Sun, Z.-H., Fu, Y.-L., Bian, G.-Z., Zhang, Y., & Tsubaki, N. (2009). Mixed Alcohol Synthesis from Syngas on K-Co-Mo/C Catalyst Prepared by a Sol-Gel Method. *Topics in Catalysis*, *52*(6-7), 789–794. doi:10.1007/s11244-009-9200-x

Barbara, I., Flourat, A. L., & Allais, F. (2015). Renewable polymers derived from ferulic acid and biobased diols via ADMET. *European Polymer Journal*, *62*, 236–243. doi:10.1016/j.eurpolymj.2014.11.035

Barbieri, G., Brunetti, A., Caravella, A., & Drioli, E. (2011). Pd-based Membrane Reactors for one-stage Process of Water Gas Shift. *RSC Advances*, *1*(4), 651–661. doi:10.1039/c1ra00375e

Bardin, B. B., & Hammond, R. R. (2001). *WO Patent No. WO2003028881(A1).*

Bartholomew, C. H., & Farrauto, R. J. (2006). *Fundamentals of industrial catalytic processes* (2nd ed.). New Jersey: John Wiley & Sons, Inc.

Bataille, F., Lemberton, J., Michaud, P., Perot, G., Vrinat, M., Lemaire, M., & Kasztelan, S. et al. (2000). Alkyldibenzothiophenes hydrodesulfurization–promoter effect, reactivity, and reaction mechanism. *Journal of Catalysis*, *191*(2), 409–422. doi:10.1006/jcat.1999.2790

Baudot, A. (2011). Applications of membrane gas separation in hydrocarbon industries. In E. Drioli & G. Barbieri (Eds.), *Membrane Engineering for the treatment of gases* (pp. 150–195). Cambridge, The United Kingdom: The Royal Society of Chemistry.

Bauer, K., Garbe, D., & Surburg, H. . (1988). Common Fragrance and Flavor Materials. In Ullmann's Encyclopedia of Industrial Chemistry (Vol. 5, 92-117). Weinheim: VCH.

Baughman, T. W., Chan, C. D., Winey, K. I., & Wagener, K. B. (2007). Synthesis and Morphology of Well-Defined Poly(ethylene-co-acrylic acid) Copolymers. *Macromolecules (Washington, DC, U. S.)*, *40*(18), 6564–6571. doi:10.1021/ma070841r

Baxendale, J. H., Evans, M. G., & Kilham, J. K. (1946). The kinetics of polymerisation reactions in aqueous solution. *Transactions of the Faraday Society*, *42*, 668–675. doi:10.1039/tf9464200668

Behrens, M., Studt, F., Kasatkin, I., Kühl, S., Hävecker, M., Abild-Pedersen, F., & Schlögl, R. (2012). The Active Site of Methanol Synthesis over Cu/ZnO/Al_2O_3 Industrial Catalysts. *Science*, *336*(6083), 893–897. doi:10.1126/science.1219831 PMID:22517324

Beier, M. J., Andanson, J.-M., Mallat, T., Krumeich, F., & Baiker, A. (2012). Ionic Liquid-Supported Pt Nanoparticles as Catalysts for Enantioselective Hydrogenation. *ACS Catalysis*, *2*(3), 337–340. doi:10.1021/cs2006197

Beiramar, J. M., Griboval-Constant, A., & Khodakov, A. Y. (2014). Effects of Metal Promotion on the Performance of CuZnAl Catalysts for Alcohol Synthesis. *ChemCatChem*, *6*(6), 1788–1793. doi:10.1002/cctc.201402037

Bej, S. K., Maity, S. K., & Turaga, U. T. (2004). Search for an Efficient 4,6-DMDBT Hydrodesulfurization Catalyst: A Review of Recent Studies. *Energy Fuels*, 18(5), 1227-1237. doi:10.1021/ef030179+

Beletskaya, I. P., & Cheprakov, A. V. (2000). The Heck Reaction as a Sharpening Stone of Palladium Catalysis. *Chemical Reviews*, *100*(8), 3009–3066. doi:10.1021/cr9903048 PMID:11749313

Beltran-Prieto, J. C., Kolomaznik, K., & Pecha, J. (2013). A Review of Catalytic Systems for Glycerol Oxidation: Alternatives for Waste Valorization. *Australian Journal of Chemistry*, *66*(5), 511–521. doi:10.1071/ch12514

Benhmid, A., Kalevaru, V. N., Martin, A., & Lücke, B. (2004). One-step synthesis of benzyl acetate by gas phase acetoxylation of toluene over highly active and selective Pd-Sb-TiO_2 catalysts. *Chemical Communications*, 2004, 2118–2119. doi:10.1039/b406396a PMID:15368003

Benichou, L., & Mayr, S. (2014). Rogeaulito: a world energy scenario modeling tool for transparent energy system thinking. *Frontiers in Energy Research, 1.*

Benkalowycz, N. C., Blum, P. R., Cirjak, L. M., Lemanski, M. F., Paparizos, C., Pepera, M. A., & Wagner, D. R. (1996). *U.S. Patent No. 5,550,281.* Washington, DC: U.S. Patent and Trademark Office.

Bentaleb, F., Che, M., Dubreuil, A.-C., Thomazeau, C., & Marceau, E. (2014). Influence of organic additives on the properties of impregnation solutions and on nickel oxide particle size for Ni/Al2O3 catalysts. *Catal. Today*, 235, 250-255. doi:10.1016/j.cattod.2014.02.020

Beretta, A., Sun, Q., Herman, R. G., & Klier, K. (1996). Production of Methanol and Isobutyl Alcohol Mixtures over Double-Bed Cesium-Promoted Cu/ZnO/Cr_2O_3 and ZnO/Cr_2O_3 Catalysts. *Industrial & Engineering Chemistry Research*, *35*(5), 1534–1542. doi:10.1021/ie9505219

Bernardo, P., & Drioli, E. (2010). Membrane technology: latest applications in the refinery and petrochemical field. In E. Drioli & L. Giorno (Eds.), Comprehensive Membrane Science and Engineering - Volume 4: Membrane Contactors and Integrated Membrane Operations (pp. 211-239). Kidlington, United Kingdom: Elsevier B.V. doi:10.1016/B978-0-08-093250-7.00054-2

Bernardo, P., Barbieri, G., & Drioli, E. (2002). An exergetic analysis of membrane unit operations integrated in the ethylene production cycle. *Chemical Engineering Research & Design*, *84*(5), 405–411. doi:10.1205/cherd05014

Bernardo, P., Criscuoli, A., Clarizia, G., Barbieri, G., Drioli, E., Fleres, G., & Picciotti, M. (2004). Applications of membrane unit operations in ethylene process. *Clean Technologies and Environmental Policy*, *6*(2), 78–95. doi:10.1007/s10098-003-0225-8

Berry, F. J., Gancedo, J. R., Marco, J. F., & Ren, X. (2005). Characterisation of the reduction properties of La0.5Sr0.5Fe0.5Co0.5O3. *Hyperfine Interactions*, *116*(1-4), 449–453.

Bessell, S. (1993). Support effects in cobalt-based fischer-tropsch catalysis. *Applied Catalysis A, General*, *96*(2), 253–268. doi:10.1016/0926-860X(90)80014-6

Bezergianni, S., Dimitriadis, A., & Meletidis, G. (2014). Effectiveness of CoMo and NiMo catalysts on co-hydroprocessing of heavy atmospheric gas oil-waste cooking oil mixtures. *Fuel*, 125, 129-136. doi:10.1016/j.fuel.2014.02.010

Bharathiraja, B., Chakravarthy, M., Kumar, R. R., Yuvaraj, D., Jayamuthunagai, J., Kumar, R. P., & Palani, S. (2014). Biodiesel production using chemical and biological methods - A review of process, catalyst, acyl acceptor, source and process variables. *Renewable & Sustainable Energy Reviews*, *38*, 368–382. doi:10.1016/j.rser.2014.05.084

Bhasin, M., McCain, J., Vora, B., Imai, T., & Pujado, P. (2001). Dehydrogenation and oxydehydrogenation of paraffins to olefins. *Applied Catalysis A, General*, *221*(1-2), 397–419. doi:10.1016/S0926-860X(01)00816-X

Bhore, N. A., Gould, R. M., Jacob, S. M., Staffeld, P. O., McNally, D., Smiley, P. H., & Wildemuth, C. R. (1999). New membrane process debottlenecks solvent dewaxing unit. *Oil & Gas Journal*, *97*(46), 67–74.

Bianchi, C. L., Canton, P., Dimitratos, N., Porta, F., & Prati, L. (2005). Selective oxidation of glycerol with oxygen using mono and bimetallic catalysts based on Au, Pd and Pt metals. *Catalysis Today*, *102*, 203–212. doi:10.1016/j.cattod.2005.02.003

Bielanski, A., & Haber, J. (1991). *Oxygen in Catalysis*. New York, NY: Marcel Dekker Inc.

Biella, S., Castiglioni, G. L., Fumagalli, C., Prati, L., & Rossi, M. (2002). Application of gold catalysts to selective liquid phase oxidation. *Catalysis Today*, *72*(1-2), 43–49. doi:10.1016/S0920-5861(01)00476-X

Bielmeier, E., Haeberle, T., Siegert, H.-J., & Gruber, W. (1993). *US Patent No. US5380932.*

Bi, J., Guo, X., Liu, M., & Wang, X. (2010). High effective dehydration of bio-ethanol into ethylene over nanoscale HZSM-5 zeolite catalysts. *Catalysis Today*, *149*(1–2), 143–147. doi:10.1016/j.cattod.2009.04.016

Bilstad, T., & Espedal, E. (1996). Membrane separation of produced water. *Water Science and Technology*, *34*(9), 239–246. doi:10.1016/S0273-1223(96)00810-4

Biondi, I., Laporte, V., & Dyson, P. J. (2012). Application of a Versatile Nanoparticle Stabilizer in Phase Transfer and Catalysis. *ChemPlusChem*, *77*(8), 721–726. doi:10.1002/cplu.201200108

Bissot, T. C. (1977). *U.S. Patent No. 4,048,096.* Washington, DC: U.S. Patent and Trademark Office.

Blank, J. H., Hembre, R., Ponasik, J., & Cole-Hamilton, D. J. (2013). A Tertiary Phosphonium Salt as a Promoter for the Hydrogenation of CO. *ChemCatChem*, *5*(5), 1075–1078. doi:10.1002/cctc.201200176

Blank, J. H., Hembre, R., Ponasik, J., & Cole-Hamilton, D. J. (2014). Alternative pathways in the ruthenium catalysed hydrogenation of CO to alcohols. *Catalysis Science & Technology*, *4*(1), 218–223. doi:10.1039/C3CY00423F

Blaser, H.-U., Indolese, A., Schnyder, A., Steiner, H., & Studer, M. (2001). Supported palladium catalysts for fine chemicals synthesis. *Journal of Molecular Catalysis A Chemical*, *173*(1-2), 3–18. doi:10.1016/S1381-1169(01)00143-1

Boam, A., & Nozari, A. (2006). Fine chemical: OSN – a lower energy alternative. *Filtration & Separation*, *43*(3), 46–48. doi:10.1016/S0015-1882(06)70819-9

Bond, G. C., & Thompson, D. T. (1999). Catalysis by gold. *Catalysis Reviews. Science and Engineering*, *41*(3-4), 319–388. doi:10.1081/CR-100101171

Boon, J. A., Levisky, J. A., Pflug, J. L., & Wilkes, J. S. (1986). Friedel Crafts Reactions in Ambient-Temperature Molten-Salts. *The Journal of Organic Chemistry*, *51*(4), 480–483. doi:10.1021/jo00354a013

Bortolotto, L., & Dittmeyer, R. (2010). Direct hydroxylation of benzene to phenol in a novel microstructured membrane reactor with distributed dosing of hydrogen and oxygen. *Separation and Purification Technology*, *73*(1), 51–58. doi:10.1016/j.seppur.2009.10.019

Bos, A. N., Schoonebeek, R., Spies, F., & Verhaak, M. J. (2015). Alkane oxidative dehydrogenation and/or alkene oxidation using mixed metal oxide catalyst (WO 2015082602 A1).

Bose, A. C., Stiegel, G. J., Armstrong, P. A., Halper, B. J., & Foster, E. P. (2009). Progress in ion transport membranes for gas separation applications. In A. C. Bose (Ed.), Inorganic membranes for energy and environmental applications (pp. 3-25). New York, NY: Springer Science+Business Media. doi:10.1007/978-0-387-34526-0_1

Bosmann, A., Datsevich, L., Jess, A., Lauter, A., Schmitz, C., & Wasserscheid, P. (2001). Deep desulphurization of diesel fuel by extraction with ILs. Chemica*l Communications* (23), 2494-2495. doi:10.1039/B108411A

Botella, P., García-González, E., Dejoz, A., López Nieto, J. M., Vázquez, M. I., & González-Calbett, J. (2004). Selective oxidative dehydrogenation of ethane on MoVTeNbO mixed metal oxide catalysts. *Journal of Catalysis*, *225*(2), 428–438. doi:10.1016/j.jcat.2004.04.024

Bounaceur, R., Lape, N., Roizard, D., Vallieres, C., & Favre, E. (2006). Membrane processes for post-combustion carbon dioxide capture: A parametric study. *Energy*, *31*(14), 2556–2570. doi:10.1016/j.energy.2005.10.038

Bourlinos, B. A., Herrera, R., Chalkias, N., Jiang, D. D., Zhang, Q., Archer, L. A., & Giannelis, E. P. (2005). Surface-Functionalized Nanoparticles with Liquid-Like Behavior. *Advanced Materials*, *17*(2), 234–237. doi:10.1002/adma.200401060

Bouwmeester, H. J. M. (2003). Dense ceramic membranes for methane conversion. *Catalysis Today*, *82*(1-4), 141–150. doi:10.1016/S0920-5861(03)00222-0

Boz, E., Ghosal, A., Alamo, R. G., & Wagener, K. B. (2006). Polyolefins with precisely placed halogens. *Polym. Prepr. (Am. Chem. Soc., Div. Polym. Chem.)*, *47*(1), 219.

Boz, I., Sahibzada, M., & Metcalfe, I. S. (1994). Kinetics of the Higher Alcohol Synthesis over a K-promoted CuO/ZnO/Al_2O_3 Catalyst. *Industrial & Engineering Chemistry Research*, *33*(9), 2021–2028. doi:10.1021/ie00033a001

Bradow, C. W., Grenoble, D. C., Foley, R. M., Murray, B. D., Winquist, B. H. C., & Milham, S. N. (2001). *U.S. Patent No. 6,210,561*. Washington, DC: U.S. Patent and Trademark Office.

Brazdil, J. F. (2006). Strategies for the selective catalytic oxidation of alkanes. *Topics in Catalysis*, *38*(4), 289–294. doi:10.1007/s11244-006-0027-4

Breman, B. B., Beenackers, A. A. C. M., Schuurman, H. A., & Oesterholt, E. (1995). Kinetics of the gas-slurry methanol-higher alcohol synthesis from CO/CO_2/H_2 over a Cs-Cu/ZnO/Al_2O_3 catalyst, including simultaneous formation of methyl esters and hydrocarbons. *Catalysis Today*, *24*(1–2), 5–14. doi:10.1016/0920-5861(95)00097-Y

Brett, G. L., He, Q., Hammond, C., Miedziak, P. J., Dimitratos, N., Sankar, M., & Hutchings, G. J. et al. (2011). Selective Oxidation of Glycerol by Highly Active Bimetallic Catalysts at Ambient Temperature under Base-Free Conditions. *Angewandte Chemie International Edition*, *50*(43), 10136–10139. doi:10.1002/anie.201101772 PMID:21990249

Brown Bourzutschky, J. A., Homs, N., & Bell, A. T. (1990a). Conversion of synthesis gas over $LaMn_{1-x}Cu_xO_{3+\lambda}$ perovskites and related copper catalysts. *Journal of Catalysis*, *124*(1), 52–72. doi:10.1016/0021-9517(90)90103-Q

Brown Bourzutschky, J. A., Homs, N., & Bell, A. T. (1990b). Hydrogenation of CO_2 and CO_2/CO mixtures over copper-containing catalysts. *Journal of Catalysis*, *124*(1), 73–85. doi:10.1016/0021-9517(90)90104-R

Brunetti, A., Barbieri, G., & Drioli, E. (2011b). Membrane reactors in hydrogen production. In E. Drioli & G. Barbieri (Eds.), Membrane engineering for the treatment of gases (Vol. 2, pp. 87-109). Cambridge, The United Kingdom: The Royal Society of Chemistry.

Brunetti, A., Barbieri, G., & Drioli, E. (2015). Membranes applications in oil refining and petrochemical industry. In A. K. Pabby, S. S. H. Rizvi, & A. M. Sastre (Eds.), *Handbook of Membrane Separations: Chemical, Pharmaceutical, Food and Biotechnological Applications* (2nd ed., pp. 77–99). London, United Kingdom: CRC – press, Taylor and Francis Group. doi:10.1201/b18319-7

Brunetti, A., Bernardo, P., Drioli, E., & Barbieri, G. (2011a). Membrane engineering progresses and potentialities in gas separations. In Y. Yampolskii & B. Freeman (Eds.), *Membrane Gas Separation* (pp. 281–312). Chichester, United Kingdom: Wiley & Sons.

Brunetti, A., Santoro, S., Macedonio, F., Figoli, A., Drioli, E., & Barbieri, G. (2013). Waste Gaseous Streams: From Environmental Issue to Source of Water by Using Membrane Condensers. *Clean – Soil, Air. Water (Basel)*, *41*(9999), 1–9.

Brunetti, A., Scura, F., Barbieri, G., & Drioli, E. (2010). Membrane technologies for CO_2 separation. *Journal of Membrane Science*, *359*(1-2), 115–125. doi:10.1016/j.memsci.2009.11.040

Buaki, M., Aprile, C., Dhakshinamoorthy, A., Alvaro, M., & Garcia, H. (2009). Liposomes by polymerization of an imidazolium ionic liquid: Use as microreactors for gold-catalyzed alcohol oxidation. *Chemistry (Weinheim an der Bergstrasse, Germany)*, *15*(47), 13082–13089. doi:10.1002/chem.200902395 PMID:19856353

Buitrago, C. F., Opper, K. L., Wagener, K. B., & Winey, K. I. (2012). Precise Acid Copolymer Exhibits a Face-Centered Cubic Structure. *ACS Macro Lett.*, *1*(1), 71–74. doi:10.1021/mz2000237

Buonomo, F., Sanfilippo, D., & Trifirò, F. (1997). Dehydrogenation reactions. In G. Ertl, H. Knozinger, & J. Weitkamp (Eds.), *Handbook of Heterogeneous Catalysis* (Vol. 5, pp. 2140–2151). Weinheim, Germany: Wiley-VCH.

Burcham, M. M., Herman, R. G., & Klier, K. (1998). Higher Alcohol Synthesis over Double Bed Cs–Cu/ZnO/Cr_2O_3 Catalysts: Optimizing the Yields of 2-Methyl-1-propanol (Isobutanol). *Industrial & Engineering Chemistry Research*, *37*(12), 4657–4668. doi:10.1021/ie9705620

Burch, R., & Hayes, M. J. (1997). The Preparation and Characterisation of Fe-Promoted Al_2O_3-Supported Rh Catalysts for the Selective Production of Ethanol from Syngas. *Journal of Catalysis*, *165*(2), 249–261. doi:10.1006/jcat.1997.1482

Burch, R., Loader, P. K., & Cruise, N. A. (1996). An investigation of the deactivation of Rh/alumina catalysts under strong oxidising conditions. *Applied Catalysis A, General*, *147*(2), 375–394. doi:10.1016/S0926-860X(96)00212-8

Burch, R., & Swarnakar, R. (1991). Oxidative dehydrogenation of ethane on vanadium-molybdenum oxide and vanadium-niobium-molybdenum oxide catalysts. *Applied Catalysis*, *70*(1), 129–148. doi:10.1016/S0166-9834(00)84159-7

Burns, S., Hargreaves, J. S., & Hunter, S. M. (2007). On the use of methane as a reductant in the synthesis of transition metal phosphides. *Catalysis Communications*, *8*(6), 931–935. doi:10.1016/j.catcom.2006.10.001

Calaza, F., Stacchiola, D., Neurock, M., & Tysoe, W. T. (2010). Coverage Effects on the Palladium-Catalyzed Synthesis of Vinyl Acetate: Comparison between Theory and Experiment. *Journal of the American Chemical Society*, *132*(7), 2202–2207. doi:10.1021/ja907061m PMID:20121234

Calderon, N., Ofstead, E. A., & Judy, W. A. (1967). Ring-opening polymerization of unsaturated alicyclic compounds. *J. Polym. Sci., Part A-1 Polym. Chem.*, *5*(9), 2209–2217. doi:10.1002/pol.1967.150050901

Calò, V., Nacci, A., Monopoli, A., & Cotugno, P. (2009). Palladium-nanoparticle-catalysed Ullmann reactions in ionic liquids with aldehydes as the reductants: Scope and mechanism. *Chemistry (Weinheim an der Bergstrasse, Germany)*, *15*(5), 1272–1279. doi:10.1002/chem.200801621 PMID:19072963

Calò, V., Nacci, A., Monopoli, A., Detomaso, A., & Iliade, P.Study on Factors Affecting the Regioselectivity of the Coupling Process. (2003). Pd Nanoparticle Catalyzed Heck Arylation of 1,1-Disubstituted Alkenes in Ionic Liquids. Study on Factors Affecting the Regioselectivity of the Coupling Process. *Organometallics*, *22*(21), 4193–4197. doi:10.1021/om034020w

Calò, V., Nacci, A., Monopoli, A., Fornaro, A., Sabbatini, L., Cioffi, N., & Ditaranto, N. (2004). Heck Reaction Catalyzed by Nanosized Palladium on Chitosan in Ionic Liquids. *Organometallics*, *23*(22), 5154–5158. doi:10.1021/om049586e

Calò, V., Nacci, A., Monopoli, A., & Montingelli, F. (2005). Pd nanoparticles as efficient catalysts for Suzuki and Stille coupling reactions of aryl halides in ionic liquids. *The Journal of Organic Chemistry*, *70*(15), 6040–6044. doi:10.1021/jo050801q PMID:16018701

Calverley, E. M., & Smith, K. T. (1991). The effects of carbon dioxide, methanol, and alkali promoter concentration on the higher alcohol synthesis over a $Cu/ZnO/Cr_2O_3$ catalyst. *Journal of Catalysis*, *130*(2), 616–626. doi:10.1016/0021-9517(91)90141-P

Campbell, P. S., Santini, C. C., Bayard, F., Chauvin, Y., Collière, V., Podgoršek, A., & Sá, J. et al. (2010). Olefin hydrogenation by ruthenium nanoparticles in ionic liquid media: Does size matter? *Journal of Catalysis*, *275*(1), 99–107. doi:10.1016/j.jcat.2010.07.018

Campos, M. (1996). Promoter Effect of Cesium on C–C Bond Formation during Alcohol Synthesis from CO/H_2 over $Cu/ZnO/Cr_2O_3$Catalysts. *Journal of Catalysis*, *163*(2), 418–428. doi:10.1006/jcat.1996.0343

Campos-Martin, J. M., Capel-Sanchez, M. C., Perez-Presas, P., & Fierro, J. L. G. (2010). Oxidative processes of desulphurization of liquid fuels. *Journal of Chemical Technology and Biotechnology (Oxford, Oxfordshire)*, *85*(7), 879–890. doi:10.1002/jctb.2371

Camposmartin, J. M., Guerreroruiz, A., & Fierro, J. L. G. (1995). Structural and Surface Properties of $CuO-ZnO-Cr_2O_3$ Catalysts and Their Relationship with Selectivity to Higher Alcohol Synthesis. *Journal of Catalysis*, *156*(2), 208–218. doi:10.1006/jcat.1995.1248

Cao, C., Hu, J., Li, S., Wilcox, W., & Wang, Y. (2009). Intensified Fischer–Tropsch synthesis process with microchannel catalytic reactors. *Catalysis Today*, *140*(3–4), 149–156. doi:10.1016/j.cattod.2008.10.016

Carenco, S., Tuxen, A., Chintapalli, M., Pach, E., Escudero, C., Ewers, T. D., & Salmeron, M. (2013). Dealloying of Cobalt from CuCo Nanoparticles under Syngas Exposure. *The Journal of Physical Chemistry C*, *117*(12), 6259–6266. doi:10.1021/jp4000297

Caro, J., Caspary, K. J., Hamel, C., Hoting, B., Kölsch, P., Langanke, B., & Wolf, A. et al. (2007). Catalytic membrane reactors for partial oxidation using perovskite hollow fiber membranes and for partial hydrogenation using a catalytic membrane contactor. *Industrial & Engineering Chemistry Research*, *46*(8), 2286–2294. doi:10.1021/ie0609620

Caro, J., Wang, H. H., Tablet, C., Kleinert, A., Feldhoff, A., Schiestel, T., & Werth, S. et al. (2006). Evaluation of perovskites in hollow fibre and disk geometry in catalytic membrane reactors and in oxygen separators. *Catalysis Today*, *118*(1-2), 128–135. doi:10.1016/j.cattod.2005.12.015

Carrettin, S., McMorn, P., Johnston, P., Griffin, K., & Hutchings, G. J. (2002). Selective oxidation of glycerol to glyceric acid using a gold catalyst in aqueous sodium hydroxide. *Chemical Communications*, (7): 696–697. doi:10.1039/b201112n PMID:12119680

Carrettin, S., McMorn, P., Johnston, P., Griffin, K., Kiely, C. J., & Hutchings, G. J. (2003). Oxidation of glycerol using supported Pt, Pd and Au catalysts. *Physical Chemistry Chemical Physics*, *5*(6), 1329–1336. doi:10.1039/b212047j

Carvalho, M. S., Lacerda, R. A., Leão, J. P. B., Scholten, J. D., Neto, B. A. D., & Suarez, P. A. Z. (2011). In situ generated palladium nanoparticles in imidazolium-based ionic liquids: A versatile medium for an efficient and selective partial biodiesel hydrogenation. *Catalysis Science & Technology*, *1*(3), 480–488. doi:10.1039/c0cy00028k

Casanave, D., Giroir-Fendler, A., Sanchez, J., Loutaty, R., & Dalmon, J. A. (1995). Control of transport properties with a microporous membrane reactor to enhance yields in dehydrogenation reactions. *Catalysis Today*, *25*(3-4), 309–314. doi:10.1016/0920-5861(95)00129-4

Cassol, C. C., Umpierre, A. P., Machado, G., Wolke, S. I., & Dupont, J. (2005). The role of Pd nanoparticles in ionic liquid in the Heck reaction. *Journal of the American Chemical Society*, *127*(10), 3298–3299. doi:10.1021/ja0430043 PMID:15755145

Castillo-Araiza, C. O., Chávez, G., Dutta, A., de los Reyes, J. A., Nuñez, S., & García-Martínez, J. C. (2015). Role of Pt–Pd/γ-Al_2O_3 on the HDS of 4,6-DMBT: Kinetic modeling & contribution analysis. *Fuel Processing Technology*, *132*, 164–172. doi:10.1016/j.fuproc.2014.12.028

Cattaneo, S., Marciano, F., Masotti, L., Vecchiato, G., Verlicchi, P., & Zaffaroni, C. (2008). Improvement in the removal of micropollutants at Porto Marghera industrial wastewater treatment plant by MBR technology. *Water Science and Technology*, *58*(9), 1789–1796. doi:10.2166/wst.2008.531 PMID:19029720

Cavalcanti, F. A., Chaturvedi, S., Gaffney, A. M., Han, S., Song, R., & Vickery, E. M. (2004). *US Patent No. US7304014B2.*

Cavani, F., & Trifiro, F. (2004). Partial Oxidation of C2 to C4 Paraffins. In M. Baerns, Basic Principles in Applied Catalysis (pp. 19-84). doi:10.1007/978-3-662-05981-4_2

Cavani, F. (2010). Catalytic selective oxidation: The forefront in the challenge for a more sustainable chemical industry. *Catalysis Today*, *157*(1-4), 8–15. doi:10.1016/j.cattod.2010.02.072

Cavani, F., Ballarini, N., & Cericola, A. (2007). Oxidative dehydrogenation of ethane and propane: How far from commercial implementation? *Catalysis Today*, *127*(1-4), 113–131. doi:10.1016/j.cattod.2007.05.009

Cavani, F., Ferroni, L., Frattini, A., Lucarelli, C., Mazzini, A., Raabova, K., & Babini, P. et al. (2011). Evidence for the presence of alternative mechanisms in the oxidation of cyclohexanone to adipic acid with oxygen, catalysed by Keggin polyoxometalates. *Applied Catalysis A, General*, *391*(1-2), 118–124. doi:10.1016/j.apcata.2010.04.032

Cavani, F., & Trifirò, F. (1995). The oxidative dehydrogenation of ethane and propane as an alternative way for the production of light olefins. *Catalysis Today*, *24*(3), 307–313. doi:10.1016/0920-5861(95)00051-G

Cecilia, J. A., Infantes-Molina, A., Rodríguez-Castellón, E., & Jiménez-López, A. (2009). A novel method for preparing an active nickel phosphide catalyst for HDS of dibenzothiophene. *Journal of Catalysis*, *263*(1), 4–15. doi:10.1016/j.jcat.2009.02.013

Cecilia, J. A., Infantes-Molina, A., Rodríguez-Castellón, E., & Jiménez-López, A. (2009). Dibenzothiophene hydrodesulfurization over cobalt phosphide catalysts prepared through a new synthetic approach: Effect of the support. *Applied Catalysis B: Environmental*, *92*(1-2), 100–113. doi:10.1016/j.apcatb.2009.07.017

Celaya Sanfiz, A., Timpe, O., Trunschke, A., & Schlogl, R. (2008). *EP Patent No. EP 2179793.*

Celaya Sanfiz, A., Timpe, O., Trunschke, A., & Schlogl, R. (2010). *European Patent No. EP2179790.*

Celaya Sanfiz, A., Hansen, T. W., Sakthivel, A., Trunschke, A., Schlogl, R., Knoester, A., & Hamid, S. B. et al. (2008). How important is the (001) plane of M1 for selective oxidation of propane to acrylic acid? *Journal of Catalysis*, *258*(1), 35–43. doi:10.1016/j.jcat.2008.05.028

Centi, G., Cavani, F., & Trifirò, F. (2001). *Selective Oxidation by Heterogeneous Catalysis*. New York, NY: Kluwer Academic Press/Plenum Publishers. doi:10.1007/978-1-4615-4175-2

Cevasco, G., & Chiappe, C. (2014). Are ionic liquids a proper solution to current environmental challenges? *Green Chemistry*, *16*(5), 2375–2385.

Chaker, M., Meher-Homji, C. B., Pillai, P., Bhattacharya, D., & Messersmith, D. (2011). Application of Boil Off Gas Compressors in Liquefied Natural Gas Plants. *Journal of Engineering for Gas Turbines and Power*.

Chalakov, L., Rihko-Struckmann, L. K., Munder, B., & Sundmacher, K. (2007). Feasibility study of the oxidative dehydrogenation of ethane in an electrochemical packed-bed membrane reactor. *Industrial & Engineering Chemistry Research*, *46*(25), 8665–8673. doi:10.1021/ie070089i

Chalakov, L., Rihko-Struckmann, L. K., Munder, B., & Sundmacher, K. (2009). Oxidative dehydrogenation of ethane in an electrochemical packed-bed membrane reactor: Model and experimental validation. *Chemical Engineering Journal*, *145*(3), 385–392. doi:10.1016/j.cej.2008.08.017

Chalavekilian, B., & Dumon, R. (1976). New route to ethylene-pyrolysis. *Chemical and Engineering News*, 1976, 176–178.

Chang, A. C. C., Louh, R. F., Wong, D., Tseng, J., & Lee, Y. S. (2011). Hydrogen production by aqueous-phase biomass reforming over carbon textile supported Pt-Ru bimetallic catalysts. *International Journal of Hydrogen Energy*, *36*(14), 8794–8799. doi:10.1016/j.ijhydene.2010.09.094

Chang, C. D. (1983). Hydrocarbons from methanol. *Catalysis Reviews. Science and Engineering*, *25*(1), 1–118. doi:10.1080/01614948308078874

Chang, C. D. (1992). The New Zealand gas-to-gasoline plant: An engineering tour de force. *Catalysis Today*, *13*(1), 103–111. doi:10.1016/0920-5861(92)80190-X

Che-Galicia, G., Quintana-Solorzano, R., Ruiz-Martinez, R. S., Valente, J. S., & Castillo-Araiza, C. O. (2014). Kinetic modelling of the oxidative dehydrogenation of ethane to ethylene over a MoVTeNbO catalytic system. *Chemical Engineering Journal*, *252*, 75–88. doi:10.1016/j.cej.2014.04.042

Che-Galicia, G., Ruiz-Martínez, R. S., López-Isunza, F., & Castillo-Araiza, C. O. (2015). Modeling of oxidative dehydrogenation of ethane to ethylene on a MoVTeNbO/TiO2 catalyst in an industrial-scale packed bed catalytic reactor. *Chemical Engineering Journal*, *280*, 682–694.

Chen, M., Gath, K. K., Wei, T., Kumar, D., & Goodman, D. W. (2007, August). Ensemble effects of Pd-Au and Pd-Sn bimetals for vinyl acetate and benzene synthesis. *Proceedings of the234th ACS National Meeting*, Boston, MA, United States.

Chen, T.-M., Wang, C.-M., Wang, I., & Tsai, T.-C. (2010). Promoter effect of vanadia on Co/Mo/Al2O3 catalyst for deep hydrodesulfurization via the hydrogenation reaction pathway. *J. Catal.*, 272(1), 28-36. doi:10.1016/j.jcat.2010.03.015

Chen, Y., Wang, L., Zhang, Y., Liu, T., Liu, X., Jiang, Z., & Li, C. (2014). A new multi-metallic bulk catalyst with high hydrodesulfurization activity of 4,6-DMDBT prepared using layered hydroxide salts as structural templates. *Appl. Catal.*, A, 474, 69-77. doi: 10.1016/j.apcata.2013.09.002

Chen, G., Andries, J., Spliethoff, H., Fang, M., & van de Enden, P. J. (2004). Biomass gasification integrated with pyrolysis in a circulating fluidised bed. *Solar Energy*, *76*(1-3), 345–349. doi:10.1016/j.solener.2003.08.021

Chen, G., Li, S., Jiao, F., & Yuan, Q. (2007). Catalytic dehydration of bioethanol to ethylene over $TiO_2/\gamma\text{-}Al_2O_3$ catalysts in microchannel reactors. *Catalysis Today*, *125*(1–2), 111–119. doi:10.1016/j.cattod.2007.01.071

Cheng, J., Tang, L., & Xu, J. (2010). An Economical, Green Pathway to Biaryls: Palladium Nanoparticles Catalyzed Ullmann Reaction in Ionic Liquid/ Supercritical Carbon Dioxide System. *Advanced Synthesis & Catalysis*, *352*(18), 3275–3286. doi:10.1002/adsc.201000475

Cheng, R., Shu, Y., Li, L., Zheng, M., Wang, X., Wang, A., & Zhang, T. (2007). Synthesis and characterization of high surface area molybdenum phosphide. *Applied Catalysis A, General*, *316*(2), 160–168. doi:10.1016/j.apcata.2006.08.036

Chen, M., & Goodman, D. W. (2008). Promotional effects of Au in Pd-Au catalysts for vinyl acetate synthesis. *Cuihua Xuebao*, *29*, 1178–1186.

Chen, M., Kumar, D., Yi, C. W., & Goodman, D. W. (2005). The Promotional Effect of Gold in Catalysis by Palladium-Gold. *Science*, *310*(5746), 291–293. doi:10.1126/science.1115800 PMID:16224016

Chen, W., Pan, X. L., Willinger, M. G., Su, D. S., & Bao, X. H. (2006). Facile autoreduction of iron oxide/carbon nanotube encapsulates. *Journal of the American Chemical Society*, *128*(10), 3136–3137. doi:10.1021/ja0567211 PMID:16522081

Chen, X., Guan, Y., Abdeltawab, A. A., Al-Deyab, S. S., Yuan, X., Wang, C., & Yu, G. (2015). Using functional acidic ionic liquids as both extractant and catalyst in oxidative desulfurization of diesel fuel: An investigation of real feedstock. *Fuel*, *146*, 6–12. doi:10.1016/j.fuel.2014.12.091

Chen, X., Song, D., Asumana, C., & Yu, G. (2012). Deep oxidative Desulphurization of diesel fuels by Lewis acidic ILs based on 1-*n*-butyl-3-methylimidazolium metal chloride. *Journal of Molecular Catalysis A Chemical*, *359*, 8–13. doi:10.1016/j.molcata.2012.03.014

Chheda, J. N., Huber, G. W., & Dumesic, J. A. (2007). Liquid-phase catalytic processing of biomass-derived oxygenated hydrocarbons to fuels and chemicals. *Angewandte Chemie International Edition*, *46*(38), 7164–7183. doi:10.1002/anie.200604274 PMID:17659519

Chheda, J. N., Roman-Leshkov, Y., & Dumesic, J. A. (2007). Production of 5-hydroxymethylfurfural and furfural by dehydration of biomass-derived mono- and poly-saccharides. *Green Chemistry*, *9*(4), 342–350. doi:10.1039/b611568c

Chiappe, C., Pieraccini, D., Zhao, D., Fei, Z., & Dyson, P. J. (2006). Remarkable Anion and Cation Effects on Stille Reactions in Functionalised Ionic Liquids. *Advanced Synthesis & Catalysis*, *348*(1-2), 68–74. doi:10.1002/adsc.200505271

Choi, K.-M., Mizugaki, T., Ebitani, K., & Kaneda, K. (2002, April 7-11). Novel preparation of monodispersed palladium nanoclusters and their catalysis for oxidative acetoxylation of toluene using molecular oxygen. *Proceedings of the223rd ACS National Meeting*, Orlando, FL, USA.

Choi, Y., & Liu, P. (2009). Mechanism of Ethanol Synthesis from Syngas on Rh(111). *Journal of the American Chemical Society*, *131*(36), 13054–13061. doi:10.1021/ja903013x PMID:19702298

Christodoulakis, A., Heracleous, E., Lemonidou, A. A., & Boghosian, S. (2006). An operando Raman study of molecular structure and reactivity of MoO_3/Al_2O_3 catalysts for ethane ODH. *Journal of Catalysis*, *242*, 16–25. doi:10.1016/j.jcat.2006.05.024

Ciferno, J. P., Fout, T. E., Jones, A. P., & Murphy, J. T. (2009). Capturing Carbon from Existing Coal-Fired Power Plants. *Chemical Engineering Progress*, *105*(4), 33–41.

Clark, P., Li, W., & Oyama, S. T. (2001). Synthesis and activity of a new catalyst for hydroprocessing: Tungsten phosphide. *Journal of Catalysis*, *200*(1), 140–147. doi:10.1006/jcat.2001.3189

Clark, P., & Oyama, S. T. (2003). Alumina-supported molybdenum phosphide hydroprocessing catalysts. *Journal of Catalysis*, *218*(1), 78–87. doi:10.1016/S0021-9517(03)00086-1

ConocoPhillips and EPA's Natural Gas STAR Program. (2006, May 25). Retrieved from Environment Protection Agency website: http://www.epa.gov/gasstar/documents/lng_opportunities.pdf

Contreras, J. F. (2005, May). Dynamic simulation: a case study. *Hydrocarbon Engineering*. Retrieved from Aspentech: www.aspentech.com

Cook, J., Ellis, B., Howard, P., Jones, M. D., & Kitchen, S. J. (n.d.). *US Patent No. US6350716(B1).*

Cooper, B., & Knudsen, K. (2006). Ultra deep desulfurization of diesel: How an understanding of the underlying kinetics can reduce investment costs. *Practical Advances in Petroleum Processing*: 297-316.

Cooper, M.S., & Heaney, , HNewbold, A.J.,, & Sanderson, , W.R. (1991). Oxidation Reactions Using Urea-Hydrogen Peroxide. A Safe Alternative to Anhydrous Hydrogen Peroxide. *ChemInform*, *22*(5).

Coq, B., & Figueras, F. (2001). Bimetallic palladium catalysts: Influence of the co-metal on the catalyst performance. *Journal of Molecular Catalysis A Chemical*, *173*(1-2), 117–134. doi:10.1016/S1381-1169(01)00148-0

Corma, A. (2014). Introduction: Chemicals from Coal, Alkynes, and Biofuels. *Chemical Reviews*, *114*(3), 1545–1546. doi:10.1021/cr400719e PMID:24517085

Corma, A., Iborra, S., & Velty, A. (2007). Chemical routes for the transformation of biomass into chemicals. *Chemical Reviews*, *107*(6), 2411–2502. doi:10.1021/cr050989d PMID:17535020

Cortes-Jácome, M. A., Morales, M., Angeles Chavez, C., Ramírez-Verduzco, L. F., López-Salinas, E., & Toledo-Antonio, J. A. (2007). WO x/TiO2 catalysts via titania nanotubes for the oxidation of dibenzothiophene. *Chemistry of Materials*, *19*(26), 6605–6614. doi:10.1021/cm702010k

Cortright, R. D., Davda, R. R., & Dumesic, J. A. (2002). Hydrogen from catalytic reforming of biomass-derived hydrocarbons in liquid water. *Nature*, *418*(6901), 964–967. doi:10.1038/nature01009 PMID:12198544

Costa, P. D., Potvin, C., Manoli, J. M., Genin, B., & Djéga-Mariadassou, G. (2004). Deep hydrodesulphurization and hydrogenation of diesel fuels on alumina-supported and bulk molybdenum carbide catalysts. *Energy & Fuels*, *83*(13), 1717–1726. doi:10.1016/j.fuel.2004.03.007

Coulier, L., Kishan, G., van Veen, J. A. R., & Niemantsverdriet, J. W. (2002). Influence of support-interaction on the sulfidation behavior and hydrodesulfurization activity of Al_2O_3-supported W, CoW, and NiW model catalysts. *The Journal of Physical Chemistry B*, *106*(23), 5897–5906. doi:10.1021/jp0136821

Courty, P., Durand, D., Freund, E., & Sugier, A. (1982). C_1-C_6 alcohols from synthesis gas on copper-cobalt catalysts. *Journal of Molecular Catalysis*, *17*(2–3), 241–254. doi:10.1016/0304-5102(82)85035-9

Cramer, H. R. (1991). *Risk assessment and risk management for the chemical process industry*. New York: Van Nostrand Reinhold.

Crathorne, E., MacGowan, D., Morris, S. R., & Rawlinson, A. P. (1994). Application of Isotopic Transient Kinetics to Vinyl Acetate Catalysis. *Journal of Catalysis*, *149*(2), 254–267. doi:10.1006/jcat.1994.1294

Criscuoli, A., & Drioli, E. (2007). New metrics for evaluating the performance of membrane operations in the logic of process intensification. *Industrial & Engineering Chemistry Research*, *46*(8), 2268–2271. doi:10.1021/ie0610952

Cruz-Campa, H. J., & Cruz-Gomez, M. J. (2010, March). Determine SIS and SIL using HAZOPS. *Process Safety Progress*, 22-31.

Czuprat, O., Werth, S., Schirrmeister, S., Schiestel, T., & Caro, J. (2009). Olefin production by a multistep oxidative dehydrogenation in a perovskite hollow-fiber membrane reactor. *ChemCatChem*, *1*(3), 401–405. doi:10.1002/cctc.200900176

Daage, M., & Chianelli, R. R. (1994). Structure-function relations in molybdenum sulfide catalysts: The "Rim-Edge" model. *Journal of Catalysis*, *149*(2), 414–427. doi:10.1006/jcat.1994.1308

Dahl, I. M., & Kolboe, S. (1993). On the reaction mechanism for propene formation in the MTO reaction over SAPO-34. *Catalysis Letters*, *20*(3-4), 329–336. doi:10.1007/BF00769305

Dahl, I. M., & Kolboe, S. (1994). On the Reaction mechanism for hydrocarbon formation from methanol over SAPO-34: I. Isotopic labeling studies of the co-reaction of ethene and methanol. *Journal of Catalysis*, *149*(2), 458–464. doi:10.1006/jcat.1994.1312

Dalmon, J. A., Chaumette, P., & Mirodatos, C. (1992). Higher alcohols synthesis on cobalt based model catalysts. *Catalysis Today*, *15*(1), 101–127.

Daniel, O. M., DeLaRiva, A., Kunkes, E. L., Datye, A. K., Dumesic, J. A., & Davis, R. J. (2010). X-ray Absorption Spectroscopy of Bimetallic Pt-Re Catalysts for Hydrogenolysis of Glycerol to Propanediols. *Chemcatchem*, *2*(9), 1107–1114. doi:10.1002/cctc.201000093

Dash, P., & Scott, R. W. (2009). 1-Methylimidazole stabilization of gold nanoparticles in imidazolium ionic liquids. *Chemical Communications*, (7): 812–814. doi:10.1039/b816446k PMID:19322450

Datta, A. K., & Sen, P. K. (2006). Optimization of membrane unit for removing carbon dioxide from natural gas. *Journal of Membrane Science*, *283*(1-2), 291–300. doi:10.1016/j.memsci.2006.06.043

Dauber, F., & Span, R. (2012). Achieving higher accuracies for process simulations by implementing the new reference equation for natural gases. *Computers & Chemical Engineering*, *37*, 15–21. doi:10.1016/j.compchemeng.2011.09.009

Davda, R. R., Shabaker, J. W., Huber, G. W., Cortright, R. D., & Dumesic, J. A. (2003). Aqueous-phase reforming of ethylene glycol on silica-supported metal catalysts. *Applied Catalysis B: Environmental*, *43*(1), 13–26. doi:10.1016/S0926-3373(02)00277-1

Davda, R. R., Shabaker, J. W., Huber, G. W., Cortright, R. D., & Dumesic, J. A. (2005). A review of catalytic issues and process conditions for renewable hydrogen and alkanes by aqueous-phase reforming of oxygenated hydrocarbons over supported metal catalysts. *Applied Catalysis B: Environmental*, *56*(1-2), 171–186. doi:10.1016/j.apcatb.2004.04.027

Davidson, S. D., Sun, J., Hong, Y., Karim, A. M., Datye, A. K., & Wang, Y. (2014). The effect of ZnO addition on Co/C catalyst for vapor and aqueous phase reforming of ethanol. *Catalysis Today*, *233*, 38–45. doi:10.1016/j.cattod.2013.12.044

De Aquino, A. D., & Gomez Cobo, A. J. (2001). Synthesis of higher alcohols with cobalt and copper based model catalysts: Effect of the alkaline metals. *Catalysis Today*, *65*(2–4), 209–216. doi:10.1016/S0920-5861(00)00575-7

De León, M. A., Santos, C., Latronica, L., Cesio, A. M., Volzone, C., Castiglioni, J., & Sergio, M. (2014). High catalytic activity at low temperature in oxidative dehydrogenation of propane with Cr–Al pillared clay. *Chemical Engineering Journal*, *241*, 336–343. doi:10.1016/j.cej.2013.10.086

de Rooij, F., & Schoonebeek, R. J. J, S. J., & Verhaak, M. J. (2012). *WO Patent No. WO2013164418(A1).*

Della Pina, C., Falletta, E., Prati, L., & Rossi, M. (2008). Selective oxidation using gold. *Chemical Society Reviews*, *37*(9), 2077–2095. doi:10.1039/b707319b PMID:18762847

Della Pina, C., Falletta, E., & Rossi, M. (2012). Update on selective oxidation using gold. *Chemical Society Reviews*, *41*(1), 350–369. doi:10.1039/C1CS15089H PMID:21727977

Deluzarche, A., Hindermann, J. P., Kiennemann, A., & Kieffer, R. (1985). Application of chemical trapping to determination of surface species and to the study of their evolution under reaction conditions in heterogeneous catalysis. *Journal of Molecular Catalysis*, *31*(2), 225–250. doi:10.1016/0304-5102(85)85066-5

Demirel-Gulen, S., Lucas, M., & Claus, P. (2005). Liquid phase oxidation of glycerol over carbon supported gold catalysts. *Catalysis Today*, *102*, 166–172. doi:10.1016/j.cattod.2005.02.033

Demirel, S., Kern, P., Lucas, M., & Claus, P. (2007). Oxidation of mono- and polyalcohols with gold: Comparison of carbon and ceria supported catalysts. *Catalysis Today*, *122*(3-4), 292–300. doi:10.1016/j.cattod.2006.12.002

Den Exter, M. J., Haije, W. G., & Vente, J. F. (2009). Viability of ITM technology for oxygen production and oxidation processes: Material, system, and process Aspects. In A. C. Bose (Ed.), Inorganic membranes for energy and environmental applications (pp. 27-51). New York, NY: Springer Science+Business Media.

Deng, S., Li, S., Li, H., & Zhang, Y. (2009). Oxidative dehydrogenation of ethane to ethylene with CO2 over Fe-Cr/ZrO_2 catalysts. *Industrial & Engineering Chemistry Research*, *48*(16), 7561–7566. doi:10.1021/ie9007387

Denisov, A. A., Makarenko, V. A., & Shamrai, A. A. (1996). Impregnation as a method for control of the distribution of active components along the depth of a solid ceramic catalyst support, 3. Solutions and catalysts. Ekotekhnol. Resursosberezhenie(5-6), 55-61.

Deshmukh, R. R., Rajagopal, R., & Srinivasan, K. V. (2001). Ultrasound promoted C–C bond formation: Heck reaction at ambient conditions in room temperature ionic liquids. *Chemical Communications*, (17): 1544–1545. doi:10.1039/b104532f PMID:12240373

Deshmukh, S. P., & Li, K. (1998). Effect of ethanol composition in water coagulation bath on morphology of PVDF hollow fibre membranes. *Journal of Membrane Science*, *150*(1), 75–85. doi:10.1016/S0376-7388(98)00196-3

Dewan, M., Kumar, A., Saxena, A., De, A., & Mozumdar, S. (2012). Using Hydrophilic Ionic Liquid, [bmim]BF4 - Ethylene Glycol System as a Novel Media for the Rapid Synthesis of Copper Nanoparticles. *PLoS ONE*, *7*(1), e29131. doi:10.1371/journal.pone.0029131 PMID:22238589

Dhas, N. A., Ekhtiarzadeh, A., & Suslick, K. S. (2001). Sonochemical preparation of supported hydrodesulfurization catalysts. *Journal of the American Chemical Society*, *123*(34), 8310–8316. doi:10.1021/ja010516y PMID:11516280

Di Fabio, S., Malamis, S., Katsou, E., Vecchiato, G., Cecchi, F., & Fatone, F. (2013). Are centralized MBRs coping with the current transition of large petrochemical areas? A pilot study in Porto-Marghera (Venice). *Chemical Engineering Journal*, *214*, 68–77. doi:10.1016/j.cej.2012.10.015

Dimitratos, N., Messi, C., Porta, F., Prati, L., & Villa, A. (2006). Investigation on the behaviour of Pt(0)/carbon and Pt(0),Au(0)/carbon catalysts employed in the oxidation of glycerol with molecular oxygen in water. *Journal of Molecular Catalysis a-Chemical, 256*(1-2), 21-28. doi:10.1016/j.molcata.2006.04.019

Dimitratos, N., Lopez-Sanchez, J. A., Anthonykutty, J. M., Brett, G., Carley, A. F., Tiruvalam, R. C., & Hutchings, G. J. et al. (2009). Oxidation of glycerol using gold-palladium alloy-supported nanocrystals. *Physical Chemistry Chemical Physics*, *11*(25), 4952–4961. doi:10.1039/b904317a PMID:19562125

Dimitratos, N., Lopez-Sanchez, J. A., & Hutchings, G. J. (2009). Green Catalysis with Alternative Feedstocks. *Topics in Catalysis*, *52*(3), 258–268. doi:10.1007/s11244-008-9162-4

Dimitratos, N., Lopez-Sanchez, J. A., Lennon, D., Porta, F., Prati, L., & Villa, A. (2006). Effect of particle size on monometallic and bimetallic (Au,Pd)/C on the liquid phase oxidation of glycerol. *Catalysis Letters*, *108*(3-4), 147–153. doi:10.1007/s10562-006-0036-8

Dimitratos, N., Lopez-Sanchez, J. A., Meenakshisundaram, S., Anthonykutty, J. M., Brett, G., Carley, A. F., & Hutchings, G. J. et al. (2009). Selective formation of lactate by oxidation of 1,2-propanediol using gold palladium alloy supported nanocrystals. *Green Chemistry*, *11*(8), 1209–1216. doi:10.1039/b823285g

Dimitratos, N., Porta, F., & Prati, L. (2005). Au, Pd (mono and bimetallic) catalysts supported on graphite using the immobilisation method - Synthesis and catalytic testing for liquid phase oxidation of glycerol. *Applied Catalysis A, General*, *291*(1-2), 210–214. doi:10.1016/j.apcata.2005.01.044

Dimitratos, N., Villa, A., Wang, D., Porta, F., Su, D., & Prati, L. (2006). Pd and Pt catalysts modified by alloying with Au in the selective oxidation of alcohols. *Journal of Catalysis*, *244*(1), 113–121. doi:10.1016/j.jcat.2006.08.019

Ding, L., Qiu, J., Wei, J., & Zhu, Z. (2014). Convenient divergent synthesis of linear-dendron block polyphosphoesters via acyclic diene metathesis polymerization. *Polym. Chem.*, *5*(14), 4285–4292. doi:10.1039/c4py00327f

Ding, L., Wei, J., Qiu, J., Wang, J., & Zhu, Z. (2014). Star-shaped polyphosphoesters with reactive end groups synthesized via acyclic diene metathesis polymerization and their transformation to nanostructures. *RSC Advances*, *4*(43), 22342–22349. doi:10.1039/c4ra01546k

Ding, W., Kan, Q., Liu, H., Liu, L., & Guan, J. (2012). Catalytic activity of V-substituted Cs2Te0.2H0.6 + x PMo12 – x V x O n heteropoly compounds in selective oxidation of isobutane. *Kinetics and Catalysis*, *53*(3), 404–408. doi:10.1134/S0023158412030068

Dittmeyer, R., & Caro, J. (2008). Catalytic membrane reactors. In G. Ertl, H. Knözinger, F. Schüth, & J. Weitkamp (Eds.), *Handbook of Heterogeneous Catalysis* (2nd ed., pp. 2198–2248). Weinheim: Wiley-VCH. doi:10.1002/9783527610044.hetcat0117

Diya'uddeen, B. H., Wan Daud, W. M. A., & Abdul Aziz, A. R. (2011). Treatment technologies for petroleum refinery effluents: A review. *Process Safety and Environmental Protection*, *89*(2), 95–105. doi:10.1016/j.psep.2010.11.003

Docherty, K. M., & Kulpa, C. F. Jr. (2005). Toxicity and antimicrobial activity of imidazolium and pyridinium ionic liquids. *Green Chemistry*, *7*(4), 185–189. doi:10.1039/b419172b

Dong, Y., Nie, Y., & Zhou, Q. (2013). Highly efficient oxidative desulfurization of fuels by Lewis acidic ionic liquids based on iron chloride. *Chemical Engineering & Technology*, *36*(3), 435–442. doi:10.1002/ceat.201200570

Donsı`, F., Williams, K. A., & Schmidt, L. D. (2005). Multistep mechanism for ethane oxidative dehydrogenation on Pt and Pt-Sn coated monoliths. *Industrial & Engineering Chemistry Research*, *44*(10), 3453–3470. doi:10.1021/ie0493356

Drioli, E., Brunetti, A., Di Profio, G., & Barbieri, G. (2012). Process intensification strategies and membrane engineering. *Green Chemistry*, *14*(6), 1561–1572. doi:10.1039/c2gc16668b

Duan, A., Fan, H., Zhao, Z., Jiang, G., & Liu, J. (2011). Hydrodesulfurization performances of NiMo supported on micro/mesoporous beta/MCM-48 composite catalyst. *Adv. Mater. Res.* (Durnten-Zurich, Switz.), 233-235(Pt. 1, Fundamental of Chemical Engineering), 827-832. doi:10.4028/www.scientific.net/AMR.233-235.827

Duan, A., Wan, G., Zhang, Y., Zhao, Z., Jiang, G., & Liu, J. (2011). Optimal synthesis of micro/mesoporous beta zeolite from kaolin clay and catalytic performance for hydrodesulfurization of diesel. *Catal. Today*, 175(1), 485-493. doi:10.1016/j.cattod.2011.03.044

Duguid, I. (2001, July). Take this safety database to heart.

Dunn, R. F., & El-Halwagi, M. M. (2003). Process Integration Technology Review: Background and Applications in the Chemical Process Industry. *Journal of Chemical Technology and Biotechnology (Oxford, Oxfordshire)*, *78*(9), 1011–1021. doi:10.1002/jctb.738

Dupont, J., Fonseca, G. S., Umpierre, A. P., Fichtner, P. F. P., & Teixeira, S. R. (2002). Transition-Metal Nanoparticles in Imidazolium Ionic Liquids: Recyclable Catalysts for Biphasic Hydrogenation Reactions. *Journal of the American Chemical Society*, *124*(16), 4228–4229. doi:10.1021/ja025818u PMID:11960449

Dupont, J., & Scholten, J. D. (2010). On the structural and surface properties of transition-metal nanoparticles in ionic liquids. *Chemical Society Reviews*, *39*(5), 1780–1804. doi:10.1039/b822551f PMID:20419219

Durasul highly cross linked polyethylene. (n. d.). *Zimmer*. Retrieved from http://www.zimmernederland.nl/producten/heup/alternative-bearing-solutions/durasulr-highly-cross-linked-polyethylene/

Dykeman, R. R., Yan, N., Scopelliti, R., & Dyson, P. J. (2011). Enhanced rate of arene hydrogenation with imidazolium functionalized bipyridine stabilized rhodium nanoparticle catalysts. *Inorganic Chemistry*, *50*(3), 717–719. doi:10.1021/ic102041q PMID:21207990

Dykeman, R. R., Yuan, Y., Yan, N., Asakura, H., Teramura, K., Tanaka, T., & Dyson, P. J. (2012). Rational Design of a Molecular Nanocatalyst-Stabilizer that Enhances both Catalytic Activity and Nanoparticle Stability. *ChemCatChem*, *4*(12), 1907–1910. doi:10.1002/cctc.201200552

Eberle, J., Jung, H., Heidenreich, R., & Roscher, A. (2013). *DE-OS 102011081786*. München, Germany: Deutsches Patent und Markenamt.

Ebitani, K., Choi, K.-M., Mizugaki, T., & Kaneda, K. (2002). Novel Preparation of Palladium Nanoclusters Using Metal Nitrates and Their Catalysis for Oxidative Acetoxylation of Toluene in the Presence of Molecular Oxygen. *Langmuir*, *18*(5), 1849–1855. doi:10.1021/la011359j

Edwards, J. K., Ntainjua, E., Carley, A. F., Herzing, A. A., Kiely, C. J., & Hutchings, G. J. (2009). Direct Synthesis of H2O2 from H-2 and O-2 over Gold, Palladium, and Gold-Palladium Catalysts Supported on Acid-Pretreated TiO2. *Angewandte Chemie International Edition*, *48*(45), 8512–8515. doi:10.1002/anie.200904115 PMID:19802864

Edwards, J. K., Solsona, B. E., Landon, P., Carley, A. F., Herzing, A., Kiely, C. J., & Hutchings, G. J. (2005). Direct synthesis of hydrogen peroxide from H-2 and O-2 using TiO2-supported Au-Pd catalysts. *Journal of Catalysis*, *236*(1), 69–79. doi:10.1016/j.jcat.2005.09.015

Edwards, J. K., Solsona, B., N, E. N., Carley, A. F., Herzing, A. A., Kiely, C. J., & Hutchings, G. J. (2009). Switching Off Hydrogen Peroxide Hydrogenation in the Direct Synthesis Process. *Science*, *323*(5917), 1037–1041. doi:10.1126/science.1168980 PMID:19229032

Egbebi, A., Schwartz, V., Overbury, S. H., & Spivey, J. J. (2010). Effect of Li Promoter on titania-supported Rh catalyst for ethanol formation from CO hydrogenation. *Catalysis Today*, *149*(1–2), 91–97. doi:10.1016/j.cattod.2009.07.104

Egorova, M., & Prins, R. (2004). Hydrodesulfurization of dibenzothiophene and 4,6-dimethyldibenzothiophene over sulfided NiMo/γ-Al_2O_3, CoMo/γ-Al_2O_3, and Mo/γ- Al_2O_3 catalysts. *Journal of Catalysis*, *225*(2), 417–427. doi:10.1016/j.jcat.2004.05.002

Eijsbouts, S., Mayo, S. W., & Fujita, K. (2007). Unsupported transition metal sulfide catalysts: From fundamentals to industrial application. *Applied Catalysis A, General*, *322*, 58–66. doi:10.1016/j.apcata.2007.01.008

Ekiner, O. M., Hayes, R. A., & Manos, P. (1992). Novel multicomponent fluid separation membranes, *U.S. Patent 5,085,676*.

El-Halwagi, M. M., & D. C. Y. Foo . (2012). Process Synthesis and Integration. In Kirk and Othmer Encyclopedia of Chemical Technology. Wiley.

El-Halwagi, M. M. (1997). *Pollution Prevention through Process Integration: Systematic Design Tools*. San Diego: Academic Press.

El-Halwagi, M. M. (2006). *Process Integration*. Amsterdam: Elsevier.

El-Halwagi, M. M. (2012). *Sustainable Design through Process Integration: Fundamentals and Applications to Industrial Pollution Prevention, Resource Conservation, and Profitability Enhancement*. Butterworth-Heinemann/Elsevier.

El-Halwagi, M. M., & Spriggs, H. D. (1998). Solve design Puzzles with Mass Integration. *Chemical Engineering Progress*, *94*(August), 25–44.

Elliott, D. J., & Pennella, F. (1988). Mechanism of ethanol formation from synthesis gas over $CuO/ZnO/Al_2O_3$. *Journal of Catalysis*, *114*(1), 90–99. doi:10.1016/0021-9517(88)90011-5

Enache, D. I., Edwards, J. K., Landon, P., Solsona-Espriu, B., Carley, A. F., Herzing, A. A., & Hutchings, G. J. et al. (2006). Solvent-free oxidation of primary alcohols to aldehydes using Au-Pd/TiO2 catalysts. *Science*, *311*(5759), 362–365. doi:10.1126/science.1120560 PMID:16424335

Endres, F. (2004). Ionic Liquids: Promising solvents for electrochemistry. *Impact Factor*, 255-284. DOI:10.1524/zpch.218.2.255.25920

Enger, B. C., Fossan, Å.-L., Borg, Ø., Rytter, E., & Holmen, A. (2011). Modified alumina as catalyst support for cobalt in the Fischer–Tropsch synthesis. *Journal of Catalysis*, *284*(1), 9–22. doi:10.1016/j.jcat.2011.08.008

Escobar, J., Barrera, M. C., Toledo, J. A., Cortes-Jacome, M. A., Angeles-Chavez, C., Nunez, S., . . . Pacheco, J. G. (2009). Effect of ethyleneglycol addition on the properties of P-doped NiMo/Al2O3 HDS catalysts: Part I. Materials preparation and characterization. *Appl. Catal.*, B, 88(3-4), 564-575. doi:10.1016/j.apcatb.2008.10.005

Eßer, J., Wasserscheid, P., & Jess, A. (2004). Deep desulfurization of oil refinery streams by extraction with ionic liquids. *Green Chemistry*, *6*(7), 316–322. doi:10.1039/b407028c

Faiz, R., & Li, K. (2012). Polymeric membranes for light olefin/paraffin separation. *Desalination*, *287*, 82–97. doi:10.1016/j.desal.2011.11.019

Fan, Y., Xiao, H., Shi, G., Liu, H., Qian, Y., Wang, T., . . . Bao, X. (2011). Citric acid-assisted hydrothermal method for preparing NiW/USY-Al2O3 ultradeep hydrodesulfurization catalysts. *J. Catal.*, 279(1), 27-35. doi:10.1016/j.jcat.2010.12.014

Fang, K., Li, D., Lin, M., Xiang, M., Wei, W., & Sun, Y. (2009). A short review of heterogeneous catalytic process for mixed alcohols synthesis via syngas. *Catalysis Today*, *147*(2), 133–138. doi:10.1016/j.cattod.2009.01.038

Fang, Y. Z., Liu, Y., & Zhang, L. H. (2011). $LaFeO_3$-supported nano Co-Cu catalysts for higher alcohol synthesis from syngas. *Applied Catalysis A, General*, *397*(1–2), 183–191. doi:10.1016/j.apcata.2011.02.032

Fan, Z., Chen, W., Pan, X., & Bao, X. (2009). Catalytic conversion of syngas into C_2 oxygenates over Rh-based catalysts—Effect of carbon supports. *Catalysis Today*, *147*(2), 86–93. doi:10.1016/j.cattod.2009.03.004

Farag, H. (2006). Kinetic analysis of the hydrodesulfurization of dibenzothiophene: Approach solution to the reaction network. *Energy & Fuels*, *20*(5), 1815–1821. doi:10.1021/ef060225g

Farag, H., Mochida, I., & Sakanishi, K. (2000). Fundamental comparison studies on hydrodesulfurization of dibenzothiophenes over CoMo-based carbon and alumina catalysts. *Applied Catalysis A, General*, *194*, 147–157. doi:10.1016/S0926-860X(99)00363-4

Farag, H., Sakanishi, K., Kouzu, M., Matsumura, A., Sugimoto, Y., & Saito, I. (2003). Dibenzothiophene hydrodesulfurization over synthesized MoS_2 catalysts. *Journal of Molecular Catalysis A Chemical*, *206*(1-2), 399–408. doi:10.1016/S1381-1169(03)00445-X

Faro, A. Jr, & Dossantos, A. (2006). Cumene hydrocracking and thiophene HDS on niobia-supported Ni, Mo and Ni–Mo catalysts. *Catalysis Today*, *118*(3-4), 402–409. doi:10.1016/j.cattod.2006.07.027

Favre, E. (2007). Carbon dioxide recovery from post-combustion processes: Can gas permeation membranes compete with absorption? *Journal of Membrane Science*, *294*(1-2), 50–59. doi:10.1016/j.memsci.2007.02.007

Federal Energy Regulatory Commission. (2015). *Energy Primer: A Handbook of Energy Market Basics*. The Division of Energy Market Oversight.

Fei, Z., Zhao, D., Pieraccini, D., Ang, W. H., Geldbach, T. J., Scopelliti, R., & Dyson, P. J. et al. (2007). Development of Nitrile-Functionalized Ionic Liquids for C–C Coupling Reactions: Implication of Carbene and Nanoparticle Catalysts. *Organometallics*, *26*(7), 1588–1598. doi:10.1021/om060950e

Ferencz, Z., Erdőhelyi, A., Baán, K., Oszkó, A., Óvári, L., Kónya, Z., & Kiss, J. (2014). Effects of Support and Rh Additive on Co-Based Catalysts in the Ethanol Steam Reforming Reaction. *ACS Catalysis*, *4*(4), 1205–1218. doi:10.1021/cs500045z

Few, C. S., Wagener, K. B., & Thompson, D. L. (2014). Systematic Studies of Morphological Changes of Precision Polyethylene. *Macromolecular Rapid Communications*, *35*(2), 123–132. doi:10.1002/marc.201300672 PMID:24317938

Finn, A., Johnson, G., & Tomlinson, T. (1999). Developments in natural gas liquefaction. *Hydrocarbon Process*, 47-59.

Fonseca, G. S., Machado, G., Teixeira, S. R., Fecher, G. H., Morais, J., Alves, M. C., & Dupont, J. (2006). Synthesis and characterization of catalytic iridium nanoparticles in imidazolium ionic liquids. *Journal of Colloid and Interface Science*, *301*(1), 193–204. doi:10.1016/j.jcis.2006.04.073 PMID:16780858

Fonseca, G. S., Scholten, J. D., & Dupont, J. (2004). Iridium Nanoparticles Prepared in Ionic Liquids: An Efficient Catalytic System for the Hydrogenation of Ketones. *Synlett*, (9): 1525–1528. doi:10.1055/s-2004-829061

Fonseca, G. S., Umpierre, A. P., Fichtncr, P. F., Tcixcira, S. R., & Dupont, J. (2003). The use of imidazolium ionic liquids for the formation and stabilization of ir0 and rh0 nanoparticles: Efficient catalysts for the hydrogenation of arenes. *Chemistry (Weinheim an der Bergstrasse, Germany)*, *9*(14), 3263–3269. doi:10.1002/chem.200304753 PMID:12866070

Foo, D. C. Y. (2009). State-of-the-Art Review of Pinch Analysis Techniques for Water Network Synthesis. *Industrial & Engineering Chemistry Research*, *48*(11), 5125–5159. doi:10.1021/ie801264c

Foo, D. C. Y., El-Halwagi, M. M., & Tan, R. R. (Eds.). (2014). *Recent Advances in Sustainable Process Design and Optimization. Series on Advances in Process Systems Engineering*. World Scientific Publishing Company.

Forzatti, P., Tronconi, E., & Pasquon, I. (1991). Higher Alcohol Synthesis. *Catalysis Reviews*, *33*(1-2), 109–168. doi:10.1080/01614949108020298

Fraga, M. A., & Jordao, E. (1998). Alcohol synthesis over bulk aluminium-nickel-copper based catalysts from syngas. *Reaction Kinetics and Catalysis Letters*, *64*(2), 331–336. doi:10.1007/BF02475353

Franca, R. V., Thursfield, A., & Metcalfe, I. S. (2012). La0.6Sr0.4Co0.2Fe0.8O3-δ microtubular membranes for hydrogen production from water splitting. *Journal of Membrane Science*, *389*, 173–181. doi:10.1016/j.memsci.2011.10.027

Frank, B., Morassutto, M., Schomäcker, R., Schlögl, R., & Su, D. S. (2010). Oxidative dehydrogenation of ethane over multiwalled carbon nanotubes. *ChemCatChem*, *2*(6), 644–648. doi:10.1002/cctc.201000035

Frolich, P. K. (1931). Pressure as a Tool in the Chemical Industry of the Future. *Industrial & Engineering Chemistry*, *23*(1), 111–115. doi:10.1021/ie50253a039

Froment, G., Castaneda-Lopez, L. C., & Marin-Rosas, C. (2008). Kinetic modeling of the hydrotreatment of light cycle oil and heavy gas oil using the structural contributions approach. *Catalysis Today*, *130*(2-4), 446–454. doi:10.1016/j.cattod.2007.10.108

Fujikawa, T. (2006). Highly active CoMo hydrodesulfurization catalyst for the production of clean diesel fuels. *Catal. Surv. Asia*, 10(2), 89-97. doi:10.1007/s10563-006-9006-9

Fujikawa, T. (2007). Development of new CoMo HDS catalyst for ultra-low sulfur diesel fuel production. *J. Jpn. Pet. Inst.*, 50(5), 249-261. doi:10.1627/jpi.50.249

Fujikawa, T. (2009). Highly Active HDS Catalyst for Producing Ultra-low Sulfur Diesel Fuels. *Top. Catal.*, 52(6-7), 872-879. doi:10.1007/s11244-009-9228-y

Fujikawa, T., Chiyoda, O., Tsukagoshi, M., Idei, K., & Takehara, S. (1998). Development of a high activity HDS catalyst for diesel fuel: from basic research to commercial experience. *Catal. Today*, 45(1-4), 307-312. doi:10.1016/S0920-5861(98)00235-1

Fujikawa, T., Kato, M., Ebihara, T., Hagiwara, K., Kubota, T., & Okamoto, Y. (2005). Development of highly active Co-Mo catalysts with phosphorus and citric acid for ultra-deep desulfurization of diesel fractions (part 2) characterization of active sites. *J. Jpn. Pet. Inst.*, 48(2), 114-120. doi:10.1627/jpi.48.114

Fujikawa, T., Kimura, H., Kiriyama, K., & Hagiwara, K. (2006). Development of ultra-deep HDS catalyst for production of clean diesel fuels. *Catal. Today*, 111(3-4), 188-193. doi:10.1016/j.cattod.2005.10.024

Furimsky, E. (2003). Metal carbides and nitrides as potential catalysts for hydroprocessing. *Applied Catalysis A, General*, *240*(1-2), 1–28. doi:10.1016/S0926-860X(02)00428-3

Gaab, S., Machli, M., Find, J., Grasselli, R. K., & Lercher, J. A. (2003). Oxidative dehydrogenation of ethane over novel Li/Dy/Mg mixed oxides: Structure–activity study. *Topics in Catalysis*, *23*(1/4), 151–158. doi:10.1023/A:1024836707308

Gadewar, S. B., Vicente, B. C., Stoimenov, P. K., & Julka, V. (2015). *Production of higher alcohols*. Google Patents.

Gaffney, A. M., & Song, R. (2006). *US Patent No. US7718568(B2).*

Gaffney, A. M., Ghosh, R., Song, R., Yeh, C. C., & Langner, T. (2009). *US Patent No. US8105972(B2).*

Gaffney, A. M., Ghosh, R., Song, R., Yeh, C. Y., & Langner, T. (2009). *US Patent No. US8105971(B2).*

Gaffney, A. M., Heffner, M. D., & Song, R. (2007). *US Patent No. US7396955(B2).*

Gallezot, P. (1979). The State and Catalytic Properties of Platinum and Palladium in Faujasite-type Zeolites. *Catalysis Reviews. Science and Engineering*, *20*(1), 121–154. doi:10.1080/03602457908065108

Galownia, J. M., Wight, A. P., Blanc, A., Labinger, J. A., & Davis, M. E. (2005). Partially reduced heteropolyanions for the oxidative dehydrogenation of ethane to ethylene and acetic acid at atmospheric pressure. *Journal of Catalysis*, *236*(2), 356–365. doi:10.1016/j.jcat.2005.10.010

Gangadharan, P., Singh, R., Cheng, F., & Lou, H. (2013). Novel methodology for inherent safety assessment in the process design stage. *Industrial & Engineering Chemistry Research*, *52*(17), 5921–5933. doi:10.1021/ie303163y

Gao, Q., Ofosu, T. N. K., Ma, S.-G., Komvokis, V. G., Williams, C. T., & Segawa, K. (2011). Catalyst development for ultra-deep hydrodesulfurization (HDS) of dibenzothiophenes. I: Effects of Ni promotion in molybdenum-based catalysts. *Catal. Today*, 164(1), 538-543. doi:10.1016/j.cattod.2010.10.016

Gao, J., Mo, X., Chien, A. C.-Y., Torres, W., & Goodwin, J. G. Jr. (2009). CO hydrogenation on lanthana and vanadia doubly promoted Rh/SiO_2 catalysts. *Journal of Catalysis*, *262*(1), 119–126. doi:10.1016/j.jcat.2008.12.006

Gao, J., Mo, X., & Goodwin, J. G. Jr. (2009). La, V, and Fe promotion of Rh/SiO_2 for CO hydrogenation: Detailed analysis of kinetics and mechanism. *Journal of Catalysis*, *268*(1), 142–149. doi:10.1016/j.jcat.2009.09.012

Gao, L., Wu, P., Lan, L., Guan, Y., & Ji, S. (2013). Influence of sulfurization behavior on structure and hydrodesulfurization performance of Ni_2p/SBA-15 catalysts. *Acta Petrolei Sinica*, *29*, 198–204.

Gao, S., Zhang, H., Wang, X., Mai, W., Peng, C., & Ge, L. (2005). Palladium nanowires stabilized by thiol-functionalized ionic liquid: Seed-mediated synthesis and heterogeneous catalyst for Sonogashira coupling reaction. *Nanotechnology*, *16*(8), 1234–1237. doi:10.1088/0957-4484/16/8/042

García-Gutiérrez, J. L., Laredo, G. C., Fuentes, G. A., García-Gutiérrez, P., & Jiménez-Cruz, F. (2014). Effect of nitrogen compounds in the hydrodesulfurization of straight-run gas oil using a CoMoP/g-Al_2O_3 catalyst. *Fuel*, *138*, 98–103. doi:10.1016/j.fuel.2014.08.008

García-Mota, M., & Lopez, N. J. (2008). Template Effects in Vinyl Acetate Synthesis on PdAu Surface Alloys: A Density Functional Theory Study. *Journal of the American Chemical Society*, *130*(44), 14406–14407. doi:10.1021/ja805519v PMID:18839957

Garcia, R., Besson, M., & Gallezot, P. (1995). Chemoselective catalytic-oxidation of glycerol with air on platinum metals. *Applied Catalysis A, General*, *127*(1-2), 165–176. doi:10.1016/0926-860X(95)00048-8

Gartner, C. A., van Veen, A. C., & Lercher, J. A. (2013). Oxidative Dehydrogenation of Ethane: Common Principles and Mechanistic Aspects. *ChemCatChem*, *5*(11), 3196–3217. doi:10.1002/cctc.201200966

Gary, J. H., Handwerk, G. E., & Kaiser, M. J. (2007). *Petroleum refining technology and economics* (5th ed.). Boca Raton, FL: CRC Press.

Gaspar, N. J., & Pasternak, I. S. (1971). H_2S promoted oxidative dehydrogenation of ethane. *Canadian Journal of Chemical Engineering*, *49*(2), 248–251. doi:10.1002/cjce.5450490213

Gatla, S., Madaan, N., Radnik, J., Kalevaru, V. N., Lücke, B., Martin, A., & Brückner, A. et al. (2011a). Impact of co-components on the state of Pd and the performance of supported Pd/TiO_2 catalysts in the gas-phase acetoxylation of toluene. *ChemCatChem*, *3*(12), 1893–1901. doi:10.1002/cctc.201100225

Gatla, S., Madaan, N., Radnik, J., Kalevaru, V. N., Pohl, M.-M., Lücke, B., & Brückner, A. et al. (2011b). Key properties promoting high activity and stability of supported PdSb/TiO_2 catalysts in the acetoxylation of toluene to benzyl acetate. *Applied Catalysis A, General*, *398*(1-2), 104–112. doi:10.1016/j.apcata.2011.03.025

Gatla, S., Madaan, N., Radnik, J., Kalevaru, V. N., Pohl, M.-M., Lücke, B., & Brückner, A. et al.(2013). Rutile – a superior support for highly selective and stable Pd-based catalysts in the gas phase acetoxylation of toluene. *Journal of Catalysis*, *297*, 256–263. doi:10.1016/j.jcat.2012.10.016

Gatla, S., Radnik, J., Madaan, N., Pohl, M.-M., Mathon, O., Rogalev, A., & Brückner, A. et al. (2015). New insights into the nature of co-components and their impact on Pd structure: X-ray Absorption studies on toluene acetoxylation catalysts. *Chemistry (Weinheim an der Bergstrasse, Germany)*, *21*(43), 15280–15289. doi:10.1002/chem.201502162 PMID:26471442

Ge, J., Zhou, Y., Yang, Y., & Xue, M. (2011). Catalytic oxidative desulphurization of gasoline using ionic liquid emulsion system. *Industrial & Engineering Chemistry Research*, *50*(24), 13686–13692. doi:10.1021/ie201325e

Gellings, P. J., & Bouwmeester, H. J. M. (2000). Solid state aspects of oxidation catalysis. *Catalysis Today*, *58*(1), 1–53. doi:10.1016/S0920-5861(00)00240-6

Giernoth, R. (2010). Task-Specific Ionic Liquids. *Angewandte Chemie International Edition*, *49*(16), 2834–2839. doi:10.1002/anie.200905981 PMID:20229544

Girgsdies, F., Schlögl, R., & Trunschke, A. (2012). In-situ X-ray diffraction study of phase crystallization from an amorphous MoVTeNb oxide catalyst precursor. *Catalysis Communications*, *18*, 60–62. doi:10.1016/j.catcom.2011.11.009

Global Vinyl Acetate Production to Post 4.5% Annual Growth Through 2017, According to In-demand Report by Merchant Research & Consulting. (2014). *PRWeb*. Retrieved from http://www.prweb.com/releases/2014/06/prweb11958764.htm

Gogate, M. R., & Davis, R. J. (2009). X-ray Absorption Spectroscopy of an Fe-Promoted Rh/TiO_2 Catalyst for Synthesis of Ethanol from Synthesis Gas. *ChemCatChem*, *1*(2), 295–303. doi:10.1002/cctc.200900104

Gong, S. W., Chen, H. K., Li, W., & Li, B. Q. (2006). Catalytic behaviors of β-$Mo_2N_{0.78}$ as a hydrodesulfurization catalyst. *Energy & Fuels*, *20*(4), 1372–1376. doi:10.1021/ef050208b

Gonzalez-Cortes, S. L., Rugmini, S., Xiao, T., Green, M. L. H., Rodulfo-Baechler, S. M., & Imbert, F. E. (2014). Deep hydrotreating of different feedstocks over a highly active Al2O3-supported NiMoW sulfide catalyst. *Appl. Catal.*, A, 475, 270-281. doi:10.1016/j.apcata.2014.01.045

Gonzalez-Cortes, S. L., Xiao, T.-C., Costa, P. M. F. J., Fontal, B., & Green, M. L. H. (2004). Urea-organic matrix method: an alternative approach to prepare Co-MoS2/γ-Al2O3 HDS catalyst. Appl. *Catal., A*, 270(1-2), 209-222. doi:10.1016/j.apcata.2004.05.006

González-Cortés, S. L., Aray, I., Rodulfo-Baechler, S. M. A., Lugo, C. A., Del Castillo, H. L., Loaiza-Gil, A., & Rueda, F. et al. (2007). On the structure and surface properties of NiO/MgO-La2O3 catalyst: Influence of the support composition and preparation method. *Journal of Materials Science*, *42*(16), 6532–6540.

Górak, A., & Stankiewicz, A. (Eds.). (2011). *Research Agenda for Process Intensification: Towards a Sustainable World of 2050. Creative Energy - Energy Transition*. Retrieved from http://3me.tudelft.nl/fileadmin/Faculteit/3mE/Actueel/Nieuws/2011/docs/DSD_Research_Agenda.pdf

Göransson, K., Söderlind, U., He, J., & Zhang, W. (2011). Review of syngas production via biomass DFBGs. *Renewable & Sustainable Energy Reviews*, *15*(1), 482–492. doi:10.1016/j.rser.2010.09.032

Gould, R. M., White, L. S., & Wildemuth, C. R. (2001). Membrane separation in solvent lube dewaxing. *Environment and Progress*, *20*(1), 12–16. doi:10.1002/ep.670200110

Grabowski, R. (2006). Kinetics of Oxidative Dehydrogenation of C2-C3 Alkanes on Oxide Catalysts. *Catalysis Reviews*, *48*(2), 199–268. doi:10.1080/01614940600631413

Graham, U. M., Jacobs, G., Gnanamani, M. K., Lipka, S. M., Shafer, W. D., Swartz, C. R., & Davis, B. H. (2014). Fischer–Tropsch Synthesis: Higher Oxygenate Selectivity of Cobalt Catalysts Supported on Hydrothermal Carbons. *ACS Catalysis*, *4*(6), 1662–1672. doi:10.1021/cs400965t

Grasselli, R. K., Lugmair, C. G., & Volpe, A. F. Jr. (2008). Doping of MoVNbTeO (M1) and MoVTeO (M2) Phases for Selective Oxidation of Propane and Propylene to Acrylic Acid. *Topics in Catalysis*, *50*(1-4), 66–73. doi:10.1007/s11244-008-9097-9

Grasselli, R., Buttrey, D., DeSanto, P. Jr, Burrington, J., Lugmair, C., Volpe, A. Jr, & Weingand, T. (2004). Active centers in Mo-V-Nb-Te-Ox (amm)oxidation catalysts. *Catalysis Today*, *91-92*, 251–258. doi:10.1016/j.cattod.2004.03.060

Graves, G. D. (1931). Higher Alcohols Formed from Carbon Monoxide and Hydrogen1,2. *Industrial & Engineering Chemistry*, *23*(12), 1381–1385. doi:10.1021/ie50264a016

Gronchi, P., Tempesti, E., & Mazzocchia, C. (1994). Metal dispersion dependent selectivities for syngas conversion to ethanol on V_2O_3 supported rhodium. *Applied Catalysis A, General*, *120*(1), 115–126. doi:10.1016/0926-860X(94)80337-4

Grubbs, R.H. (2004). Handbook of Metathesis. Wiley VCH: Weinheim.

Guan, J., Wang, Y., Qin, M., Yang, Y., Li, X., & Wang, A. (2009). Synthesis of transition-metal phosphides from xidic precursors by reduction in hydrogen plasma. *Journal of Solid State Chemistry*, *182*(6), 1550–1555. doi:10.1016/j.jssc.2009.03.026

Guan, Q., Li, W., Zhang, M., & Tao, K. (2009). Alternative synthesis of bulk and supported nickel phosphide from the thermal decomposition of hypophosphites. *Journal of Catalysis*, *263*(1), 1–3. doi:10.1016/j.jcat.2009.02.008

Guan, Y., Wu, P., Lan, L., Gao, L., & Ji, S. (2012). Preparation of Ni_2P/SBA-15 catalysts using citrate acid method and its hydrodesulfurization performance. *J. Mol. Catal (China)*, *26*, 197–203.

Gudgila, R., & Leclerc, C. A. (2011). Support effects on the oxidative dehydrogenation of ethane to ethylene over platinum catalysts. *Industrial & Engineering Chemistry Research*, *50*(14), 8438–8443. doi:10.1021/ie200035q

Guerrero-Perez, O. M., Lopez-Medina Ricardo, B. M., M, R. J., & Rodriguez-Mirasol Jose, C. T. (2010). A New and Highly Selective Vanadium Carbon-based catalysts. *American Carbon Society Proceedings.*

Gui, J., Liu, D., Sun, Z., Liu, D., Min, D., Song, B., & Peng, X. (2010). Deep oxidative desulfurization with task-specific ionic liquids: An experimental and computational study. *Journal of Molecular Catalysis A Chemical*, *331*(1), 64–70. doi:10.1016/j.molcata.2010.08.003

Guo, H., Li, S., Zhang, H., Peng, F., Xiong, L., Yang, J., & Chen, Y. (2013). Reaction Condition Optimization and Lumped Kinetics Study for Lower Alcohols Synthesis from Syngas Using a Two-Stage Bed Catalyst Combination System. *Industrial & Engineering Chemistry Research*, *53*(1), 123–131. doi:10.1021/ie402422p

Guo, S., Shi, F., Gu, Y., Yang, J., & Deng, Y. (2005). Size-controllable Synthesis of Gold Nanoparticles via Carbonylation and Reduction of Hydrochloroauric Acid with CO and H2O in Ionic Liquids. *Chemistry Letters*, *34*(6), 830–831. doi:10.1246/cl.2005.830

Guo, Y., Zeng, P., Ji, S., Wei, N., Liu, H., & Li, C. (2010). Effect of Mo promoter content on performance of Mo-Ni_2P/SBA-15/cordierite monolithic catalyst for hydrodesulfurization. *Chinese Journal of Catalysis*, *31*, 329–334.

Gupta, M., Smith, M. L., & Spivey, J. J. (2011). Heterogeneous Catalytic Conversion of Dry Syngas to Ethanol and Higher Alcohols on Cu-Based Catalysts. *ACS Catalysis, 1*(6), 641-656.

H, M. J., W, K. S., & L, O. G. (1987). *US Patent No. US5162578(A).*

Hagemeyer, A., & Jahn, U. (2012). *Vinylacetate*. Retrieved from https://roempp.thieme.de/roempp4.0/do/data/RD-22-00794

Hagemeyer, A., Dingerdissen, U., Kühlein, K., Manz, A., & Fischer, R. (2003). *U.S. Patent No. 6,603,038.* Washington, DC: U.S. Patent and Trademark Office.

Hagemeyer, A., Mestl, G., Neumann, S., Margitfalvi, J., Tompos, A., & Vegvari, I. L. (2012). *WO Patent No. WO2013021020(A1).*

Hagemeyer, A., Mestl, G., Neumann, S., Tompos, A., Margitfalvi, J., & Vegvari, I. L. (2012). *WO Patent No. WO2013021034(A1).*

Haider, M. A., Gogate, M. R., & Davis, R. J. (2009). Fe-promotion of supported Rh catalysts for direct conversion of syngas to ethanol. Journal of Catalysis, 261(1), 9–16. doi:10.1016/j.jcat.2008.10.013 doi:10.1016/j.jcat.2008.10.013

Håkonsen, S. F., Walmsley, J. C., & Holmen, A. (2010). Ethene production by oxidative dehydrogenation of ethane at short contact times over Pt-Sn coated monoliths. *Applied Catalysis A, General*, *378*(1), 1–10. doi:10.1016/j.apcata.2010.01.032

Hallale, N. (2001). Burning Bright: Trends in Process Integration. *Chemical Engineering Progress*, *97*(7), 30–41.

Hallett, J. P., & Welton, T. (2011). Room-temperature ILs: Solvents for synthesis and catalysis. 2. *Chemical Reviews*, *111*(5), 3508–3576. doi:10.1021/cr1003248 PMID:21469639

Hamakawa, S., Sato, K., Inoue, T., Nishioka, M., Kobayashi, K., & Mizukami, F. (2006). Design of one-component ceramic membrane-reactor for natural gas conversion. *Catalysis Today*, *117*(1-3), 297–303. doi:10.1016/j.cattod.2006.05.042

Hamel, C., Wolff, T., & Seidel-Morgenstern, A. (2011). Compatibility of transport and reaction in membrane reactors used for the oxidative dehydrogenation of short-chain hydrocarbons. *International Journal of Chemical Reactor Engineering,* 9(1), 1542-6580.

Hamza, A., Towe, G., & Keyvani, M. (2013). Stable Facilitated Transport Membrane for Olefin/Paraffin Separation. *Presented at the Fuel and Petrochemical Division AIChE Spring Meeting and Global Congress on Process Safety.* Retrieved from http://www.cepmagazine.net/resources/chemeondemand/conference-presentations/stable-facilitated-transport-membrane-olefinparaffin-separation

Han, L., Mao, D., Yu, J., Guo, Q., & Lu, G. (2013). C2-oxygenates synthesis through CO hydrogenation on SiO2-ZrO2 supported Rh-based catalyst: The effect of support. Applied Catalysis A, General, 454(0), 81–87. doi:10.1016/j.apcata.2013.01.008 doi:10.1016/j.apcata.2013.01.008

Han, S., & Frick, C. D. & J, M. D. (2012). *Patent No. WO2013148006(A1).*

Han, S., & Frick, C. D. & J, M. D. (2012). *WO Patent No. WO2013148006(A1).*

Han, S., Jin, L.; Martenak, D. J., Suib, S. L., (2010). Improved process for oxidative dehydrogenation of ethane and related reactions (EP 2165997 A1).

Han, P., Axnanda, S., Lyubinetsky, I., & Goodman, D. W. (2007). Atomic-Scale Assembly of a Heterogeneous Catalytic Site. *Journal of the American Chemical Society*, *129*(46), 14355–14361. doi:10.1021/ja074891n PMID:17967014

Hanrieder, E.-K., Jentys, A., & Lercher, J.-A. (2014, August 10-14). Interaction of the alkaline acetate promoter with bimetallic PdAu catalysts for the gas phase vinyl acetate synthesis. *Proceedings of the248th ACS National Meeting & Exposition*, San Francisco, CA, USA.

Han, Y. F., Kumar, D., & Goodman, D. W. (2005). Particle size effects in vinyl acetate synthesis over Pd/SiO_2. *Journal of Catalysis*, *230*(2), 353–358. doi:10.1016/j.jcat.2004.12.018

Hara, Y., Watanabe, E., Wada, K., & Onoda, T. (1989). Effective hydrogenation of carbon monoxide to ethylene glycol using rhodium-trialkylphosphine-nitrogen base catalyst: Critical role of nitrogen base promoter in controlling selectivity. Journal of Organometallic Chemistry, 359(1), 97–104. doi:10.1016/0022-328X(89)85255-6 doi:10.1016/0022-328X(89)85255-6

Harries-Rees, K., Chauvin, Y., Grubbs, R., & Schrock, R. (2005). Three share prize for metathesis work. *Chem. World*, *2*(11), 42–44.

Hartmann-Schreier, J. (2012). *Benzylacetate.* Römpp Online. Retrieved from https://roempp.thieme.de/roempp4.0/do/data/RD-02-00911

Hashmi, A. S. K., & Hutchings, G. J. (2006). Gold catalysis. *Angewandte Chemie International Edition*, *45*(47), 7896–7936. doi:10.1002/anie.200602454 PMID:17131371

Hatakeyama, Y., Onishi, K., & Nishikawa, K. (2011). Effects of sputtering conditions on formation of gold nanoparticles in sputter deposition technique. *RSC Advances*, *1*(9), 1815–1821. doi:10.1039/c1ra00688f

Hatanaka, S., Yamada, M., & Sadakane, O. (1997). Hydrodesulfurization of catalytic cracked gasoline. 2. The difference between HDS active site and olefin hydrogenation active site. *Industrial & Engineering Chemistry Research*, *36*(12), 5110–5117. doi:10.1021/ie9703494

Hävecker, M., Wrabetz, S., Kröhnert, J., Csepei, L.-I., d'Alnoncourt, R. N., Kolen'ko, Y. V., & Trunschke, A. et al. (2012). Surface chemistry of phase-pure M1 MoVTeNb oxide during operation in selective oxidation of propane to acrylic acid. *Journal of Catalysis*, *285*, 48–60.

He, M., Xiao, B., Hu, Z., Liu, S., Guo, X., & Luo, S. (2009). Syngas production from catalytic gasification of waste polyethylene: Influence of temperature on gas yield and composition. International Journal of Hydrogen Energy, 34(3), 1342–1348. doi:10.1016/j.ijhydene.2008.12.023 doi:10.1016/j.ijhydene.2008.12.023

Heck, R. F. (1982). Palladium-catalyzed vinylation of organic halides. *Organic Reactions*, *27*, 345–390.

Heracleous, E., & Lemonidou, A. A. (2006). Ni-Nb-O mixed oxides as highly active and selective catalysts for ethene production via ethane oxidative dehydrogenation. II. Mechanistic aspects and kinetic modelling. *Journal of Catalysis*, *237*(1), 175–189. doi:10.1016/j.jcat.2005.11.003

Heracleous, E., & Lemonidou, A. A. (2006). Ni–Nb–O mixed oxides as highly active and selective catalysts for ethene production via ethane oxidative dehydrogenation. Part I: Characterization and catalytic performance. *Journal of Catalysis*, *237*(1), 162–174. doi:10.1016/j.jcat.2005.11.002

Heracleous, E., & Lemonidou, A. A. (2010). Ni-Me-O mixed metal oxides for the effective oxidative dehydrogenation of ethane to ethylene - Effect of promoting metal Me. *Journal of Catalysis*, *270*(1), 67–75. doi:10.1016/j.jcat.2009.12.004

Herisson, J. L., & Chauvin, Y. (1971). Catalysis of olefin transformations by tungsten complexes. II. Telomerization of cyclic olefins in the presence of acyclic olefins. *Die Makromolekulare Chemie*, *141*(1), 161–176. doi:10.1002/macp.1971.021410112

Herman, R. G. (2000). Advances in catalytic synthesis and utilization of higher alcohols. Catalysis Today, 55(3), 233–245. doi:10.1016/S0920-5861(99)00246-1 doi:10.1016/S0920-5861(99)00246-1

Higman, C., & Tam, S. (2013). Advances in Coal Gasification, Hydrogenation, and Gas Treating for the Production of Chemicals and Fuels. Chemical Reviews, 114(3), 1673–1708. doi:10.1021/cr400202m PubMed doi:10.1021/cr400202m PMID:24144161

Hilmen, A.-M., Xu, M., Gines, M. J. L., & Iglesia, E. (1998). Synthesis of higher alcohols on copper catalysts supported on alkali-promoted basic oxides. Applied Catalysis A, General, 169(2), 355–372. doi:10.1016/S0926-860X(98)00025-8 doi:10.1016/S0926-860X(98)00025-8

Hirano, M., Enokida, K., Okazaki, K., Kuwabata, S., Yoshida, H., & Torimoto, T. (2013). Composition-dependent electrocatalytic activity of AuPd alloy nanoparticles prepared via simultaneous sputter deposition into an ionic liquid. *Physical Chemistry Chemical Physics*, *15*(19), 7286–7294. doi:10.1039/c3cp50816a PMID:23575517

Hkonsen, S. F., & Holmen, A. (2007). Oxidative dehydrogenation of ethane at short contact times. *Studies in Surface Science and Catalysis*, *167*, 337–342. doi:10.1016/S0167-2991(07)80154-6

Ho, T. C. (2004). Deep HDS of diesel fuel: chemistry and catalysis. *Catal. Today*, 98(1-2), 3-18. doi:10.1016/j.cattod.2004.07.048

Hofstadt, C. E., Schneider, M., Bock, O., & Kochloefl, K. (1983). Effect of preparation methods and promoters on activity and selectivity of cu-zno-al2o3-k catalysts in aliphatic alcohols synthesis from CO and H2. In P. G. G. Poncelet & P. A. Jacobs (Eds.), Studies in Surface Science and Catalysis (Vol. Volume 16, pp. 709-721): Elsevier.

Hornig, L., Fernholz, H., Schmidt, H.-J., Wunder, F., & Quadflieg, T. (1972). *U.S. Patent No. 3,658,888*. Washington, DC: U.S. Patent and Trademark Office.

Hosseini-Monfared, H., Meyer, H., & Janiak, C. (2013). Dioxygen oxidation of 1-phenylethanol with gold nanoparticles and N-hydroxyphthalimide in ionic liquid. *Journal of Molecular Catalysis A Chemical*, *372*, 72–78. doi:10.1016/j.molcata.2013.02.007

Houalla, M., Nag, N., Sapre, A., Broderick, D., & Gates, B. (1978). Hydrodesulfurization of dibenzothiophene catalyzed by sulfided CoO-$MoO_3\gamma$-Al_2O_3: The reaction network. *AIChE Journal. American Institute of Chemical Engineers*, *24*(6), 1015–1021. doi:10.1002/aic.690240611

Hou, W. B., Dehm, N. A., & Scott, R. W. J. (2008). Alcohol oxidations in aqueous solutions using Au, Pd, and bimetallic AuPd nanoparticle catalysts. *Journal of Catalysis*, *253*(1), 22–27. doi:10.1016/j.jcat.2007.10.025

http://www.iea-amf.org/content/fuel_information/diesel_gasoline, http://ec.europa.eu/, (Essar et al, 2004).

Hu, H., Wachs, I. E., & Bare, S. R. (1995). Surface Structures of Supported Molybdenum Oxide Catalysts: Characterization by Raman and Mo L3-Edge XANES. *J. Phys. Chem.*, 99(27), 10897-10910. doi:10.1021/j100027a034

Hu, J., Wang, Y., Cao, C., Elliott, D. C., Stevens, D. J., & White, J. F. (2005). Conversion of Biomass Syngas to DME Using a Microchannel Reactor. Industrial & Engineering Chemistry Research, 44(6), 1722–1727. doi:10.1021/ie0492707 doi:10.1021/ie0492707

Hu, J., Wang, Y., Cao, C., Elliott, D. C., Stevens, D. J., & White, J. F. (2007). Conversion of biomass-derived syngas to alcohols and C2 oxygenates using supported Rh catalysts in a microchannel reactor. Catalysis Today, 120(1), 90–95. doi:10.1016/j.cattod.2006.07.006 doi:10.1016/j.cattod.2006.07.006

Huang, M., & Cho, K. (2009). Density Functional Theory Study of CO Hydrogenation on a MoS2 Surface. The Journal of Physical Chemistry C, 113(13), 5238–5243. doi:10.1021/jp807705y doi:10.1021/jp807705y

Huang, Y., Deng, W., Guo, E., Chung, P.-W., Chen, S., Trewyn, B. G., & Lin, V. S. Y. (2012). Mesoporous Silica Nanoparticle-Stabilized and Manganese-Modified Rhodium Nanoparticles as Catalysts for Highly Selective Synthesis of Ethanol and Acetaldehyde from Syngas. ChemCatChem, 4(5), 674–680. doi:10.1002/cctc.201100460 doi:10.1002/cctc.201100460

Huang, C.-P., Liu, Z.-C., Xu, C.-M., Chen, B.-H., & Liu, Y.-F. (2004). Effects of additives on the properties of chloroaluminate ionic liquids catalyst for alkylation of isobutane and butene. *Applied Catalysis A, General*, *277*(1), 41–43. doi:10.1016/j.apcata.2004.08.019

Huang, D., Lu, Y. C., Wang, Y. J., Yang, L., & Luo, G. S. (2008). Intensification of catalytic oxidation with a T-junction microchannel reactor for deep desulfurization. *Industrial & Engineering Chemistry Research*, *47*(11), 3870–3875. doi:10.1021/ie701781r

Huang, J., Jiang, T., Han, B., Gao, H., Chang, Y., Zhao, G., & Wu, W. (2003). Hydrogenation of olefins using ligand-stabilized palladium nanoparticles in an ionic liquid. *Chemical Communications*, (14): 1654–1655. doi:10.1039/b302750c

Huang, T.-J., Shen, X.-D., & Chou, C.-L. (2009). Characterization of Cu, Ag and Pt added La0.6Sr0.4Co0.2Fe0.8O3−δ and gadolinia-doped ceria as solid oxide fuel cell electrodes by temperature-programmed techniques. *Journal of Power Sources*, *187*(2), 348–355. doi:10.1016/j.jpowsour.2008.11.019

Huang, X., Ji, S., Wu, P., Liu, Q., Liu, H., Zhu, J., & Li, C. (2008). Structure and hydrodesulfurization performances of Ni_2P/SBA-15 catalysts. *Acta Physico-Chimica Sinica*, *24*, 1773–1779.

Huang, Z. W., Cui, F., Kang, H. X., Chen, J., Zhang, X. Z., & Xia, C. G. (2008). Highly dispersed silica-supported copper nanoparticles prepared by precipitation-gel method: A simple but efficient and stable catalyst for glycerol hydrogenolysis. *Chemistry of Materials*, *20*(15), 5090–5099. doi:10.1021/cm8006233

Huber, G. W., & Dumesic, J. A. (2006). An overview of aqueous-phase catalytic processes for production of hydrogen and alkanes in a biorefinery. *Catalysis Today*, *111*(1-2), 119–132. doi:10.1016/j.cattod.2005.10.010

Huber, G. W., Iborra, S., & Corma, A. (2006). Synthesis of transportation fuels from biomass: Chemistry, catalysts, and engineering. *Chemical Reviews*, *106*(9), 4044–4098. doi:10.1021/cr068360d PMID:16967928

Huber, G. W., Shabaker, J. W., & Dumesic, J. A. (2003). Raney Ni-Sn catalyst for H-2 production from biomass-derived hydrocarbons. *Science*, *300*(5628), 2075–2077. doi:10.1126/science.1085597 PMID:12829777

Huber, G. W., Shabaker, J. W., Evans, S. T., & Dumesic, J. A. (2006). Aqueous-phase reforming of ethylene glycol over supported Pt and Pd bimetallic catalysts. *Applied Catalysis B: Environmental*, *62*(3-4), 226–235. doi:10.1016/j.apcatb.2005.07.010

Huff, M. C., & Schmidt, L. D. (1993). Ethylene formation by oxidative dehydrogenation of ethane over monoliths at very short contact times. *Journal of Physical Chemistry*, *97*(45), 11815–11815. doi:10.1021/j100147a040

Huff, M. C., & Schmidt, L. D. (1996). Elementary step model of ethane oxidative dehydrogenation on Pt-coated monoliths. *AIChE Journal. American Institute of Chemical Engineers*, *42*(12), 3484–3497. doi:10.1002/aic.690421218

Hunter, D. (2007). Air Liquide to install ASU for Dongbei Special Steel Group. *China Chemical Reporter*, *18*(14), 10.

Hu, Y., Yu, Y., Hou, Z., Li, H., Zhao, X., & Feng, B. (2008). Biphasic Hydrogenation of Olefins by Functionalized Ionic Liquid-Stabilized Palladium Nanoparticles. *Advanced Synthesis & Catalysis*, *350*(13), 2077–2085. doi:10.1002/adsc.200800290

Huynh, Q., Selmi, A., Corbel, G., Lacorre, P., & Millet, J. (2009). Atypical synergetic effect between Te- and V-substituted phosphomolybdic cesium salt and LAMOX-type phases for the oxidation of isobutane into methacrylic acid. *Journal of Catalysis 266 (2009) 64–70*, *266*, 64-70.

Huynh, Q., Schuurman, Y., Delichere, P., Loridant, S., & Millet, J. (2009). Study of Te and V as counter-cations in Keggin type phosphomolybdic polyoxometalate catalysts for isobutane oxidation. *Journal of Catalysis*, *261*(2), 166–176. doi:10.1016/j.jcat.2008.11.013

Hynaux, A., Sayag, C., Suppan, S., Trawczynski, J., Lewandowski, M., Szymanska-Kolasa, A., & Djega-Mariadassou, G. (2007). Kinetic study of the hydrodesulfurization of dibenzothiophene over molybdenum carbides supported on functionalized carbon black composite: Influence of indole. *Applied Catalysis B: Environmental*, *72*(1-2), 62–70. doi:10.1016/j.apcatb.2006.09.019

Ibeh, B., Zhang, S., & Hill, J. M. (2009). Effect of citric acid on the synthesis of tungsten phosphide hydrotreating catalysts. *Applied Catalysis A, General*, *368*(1-2), 127–131. doi:10.1016/j.apcata.2009.08.017

Iddir, H., Skavysh, V., Öğüt, S., Browning, N. D., & Disko, M. M. (2006). Preferential growth of Pt on rutile TiO_2. *Physical Review B: Condensed Matter and Materials Physics*, *73*(4), 041403. doi:10.1103/PhysRevB.73.041403

Ide, M. S., & Davis, R. J. (2014). The Important Role of Hydroxyl on Oxidation Catalysis by Gold Nanoparticles. *Accounts of Chemical Research*, *47*(3), 825–833. doi:10.1021/ar4001907 PMID:24261465

IEA. (2004). *Energy statistics of OECD countries 2001/2002 and energy statistics of non-OECD countries 2001/2002*. Paris: International Energy Agency.

Inci, B., & Wagener, K. B. (2011). Decreasing the Alkyl Branch Frequency in Precision Polyethylene: Pushing the Limits toward Longer Run Lengths. *Journal of the American Chemical Society*, *133*(31), 11872–11875. doi:10.1021/ja2040046 PMID:21766883

Independent Statistics and Analysis. (2014, May 7). *U.S Energy Information Administration*. Retrieved from http://www.eia.gov/forecasts/aeo/

Infantes-Molina, A., Romero-Perez, A., Sanchez-Gonzalez, V., Pawelec, B., Fierro, J. L. G., Jimenez-Lopez, A., & Rodriguez-Castellon, E. (2011). Role of Cs on Hydrodesulfurization Activity of RuS2 Catalysts Supported on a Mesoporous SBA-15 Type Material. *ACS Catal.*, 1(3), 175-186. doi:10.1021/cs100053e

Iranmahboob, J., Toghiani, H., & Hill, D. O. (2003). Dispersion of alkali on the surface of CO-MoS2/clay catalyst: A comparison of K and Cs as a promoter for synthesis of alcohol. Applied Catalysis A, General, 247(2), 207–218. doi:10.1016/S0926-860X(03)00092-9 doi:10.1016/S0926-860X(03)00092-9

Ishikawa, S., Yi, X., Murayama, T., & Ueda, W. (2014). Catalysis field in orthorhombic Mo3VOx oxide catalyst for the selective oxidation of ethane, propane and acrolein. *Catalysis Today*, *238*, 35–40. doi:10.1016/j.cattod.2013.12.054

Ishikawa, S., Yi, X., Murayama, T., & Ueda, W. (2014). Heptagonal channel micropore of orthorhombic Mo3VOx as catalysis field for the selective oxidation of ethane. *Applied Catalysis A, General*, *474*, 10–17. doi:10.1016/j.apcata.2013.07.050

Ismagilov, Z. R., Yashnik, S. A., Startsev, A. N., Boronin, A. I., Stadnichenko, A. I., Kriventsov, V. V., . . . Moulijn, J. A. (2009). Deep desulphurization of diesel fuels on bifunctional monolithic nanostructured Pt-zeolite catalysts. *Catal. Today*, 144(3-4), 235-250. doi:10.1016/j.cattod.2009.01.030

Ito, E., & van Veen, J. A. R. (2006). On novel processes for removing sulphur from refinery streams. *Catalysis Today*, *116*(4), 446–460. doi:10.1016/j.cattod.2006.06.040

Itoh, H., Naka, K., & Chujo, Y. (2004). Synthesis of gold nanoparticles modified with ionic liquid based on the imidazolium cation. *Journal of the American Chemical Society*, *126*(10), 3026–3027. doi:10.1021/ja039895g PMID:15012113

Ivars, F., Solsona, B., Hernandez, S., & Lopez Nieto, J. (2010). Influence of gel composition in the synthesis of MoVTeNb catalysts over their catalytic performance in partial propane and propylene oxidation. *Catalysis Today*, *149*(3-4), 260–266. doi:10.1016/j.cattod.2009.09.018

Ivars, F., Solsona, B., Soriano, M., & Lopez Nieto, J. (2008). Selective Oxidation of Propane Over AMoVSbO Catalysts (A = Li, Na, K, Rb or Cs). *Topics in Catalysis*, *50*(1-4), 74–81. doi:10.1007/s11244-008-9103-2

Iwamoto, M., Yoda, Y., Egashira, M., & Seiyama, T. (1976). Study of metal oxide catalysts by temperature programmed desorption. 1. Chemisorption of oxygen on nickel oxide. *Journal of Physical Chemistry*, *80*(18), 1989–1994. doi:10.1021/j100559a008

Jahangiri, H., Bennett, J., Mahjoubi, P., Wilson, K., & Gu, S. (2014). A review of advanced catalyst development for Fischer-Tropsch synthesis of hydrocarbons from biomass derived syn-gas. Catalysis Science & Technology, 4(8), 2210–2229. doi:10.1039/C4CY00327F doi:10.1039/C4CY00327F

Jasper, S., & El-Halwagi, M. M. (2015). A Techno-Economic Comparison of Two Methanol-to-Propylene Processes. *Processes*, *3*(3), 684–698. doi:10.3390/pr3030684

Jerome, F., Pouilloux, Y., & Barrault, J. (2008). Rational Design of Solid Catalysts for the Selective Use of Glycerol as a Natural Organic Building Block. *ChemSusChem*, *1*(7), 586–613. doi:10.1002/cssc.200800069 PMID:18702160

Jess, A., Wasserscheid, P., & Esser, J. (2003). Hydrogen-free deep desulfurization of diesel oil by extraction with ionic liquids. *Proceedings of the DGMK-Conference on Innovation in the Manufacture and Use of Hydrogen* (pp. 313-320).

Jiang, D., Ding, Y., Pan, Z., Li, X., Jiao, G., Li, J., & Luo, H. (2007). Roles of chlorine in the CO hydrogenation to C2-oxygenates over Rh-Mn-Li/SiO2 catalysts. Applied Catalysis A, General, 331(0), 70–77. doi:10.1016/j.apcata.2007.06.038 doi:10.1016/j.apcata.2007.06.038

Jiang, T., Niu, Y., & Zhong, B. (2001). Synthesis of higher alcohols from syngas over Zn–Cr–K catalyst in supercritical fluids. Fuel Processing Technology, 73(3), 175–183. doi:10.1016/S0378-3820(01)00178-3 doi:10.1016/S0378-3820(01)00178-3

Jiang, T., Zhou, Y. X., Liang, S. G., Liu, H. Z., & Han, B. X. (2009). Hydrogenolysis of glycerol catalyzed by Ru-Cu bimetallic catalysts supported on clay with the aid of ionic liquids. *Green Chemistry*, *11*(7), 1000–1006. doi:10.1039/b901425j

Jing, F., Katryniok, B., Dumeignil, F., Bordes-Richard, E., & Paul, S. (2014). Catalytic selective oxidation of isobutane to methacrylic acid on supported (NH4)3HPMo11VO40 catalysts. *Journal of Catalysis*, *309*, 121–135. doi:10.1016/j.jcat.2013.09.014

Jin, W., Li, S., Huang, P., Xu, N., Shi, J., & Lin, Y. S. (2000). Tubular lanthanum cobaltite perovskite-type membrane reactors for partial oxidation of methane to syngas. *Journal of Membrane Science*, *166*(1), 13–22. doi:10.1016/S0376-7388(99)00245-8

Jira, R. (2009). Acetaldehyde from Ethylene - A Retrospective on the Discovery of the Wacker Process. *Angewandte Chemie International Edition*, *48*(48), 9034–9037. doi:10.1002/anie.200903992 PMID:19834923

Johnson Matthey plc, (2014). *Palladium-Market Data*, Retrieved from http://www.platinum.matthey.com/services/market-research/market-data-charts/palladium

Jones, C. W., Agrawal, P. K., & Nguyen, T. T. (2013). Catalyst compositions for converting syngas to produce higher alcohols. Google Patents.

Juan, A., & Damiani, D. E. (1992). Alumina-supported Ru-Mo Catalysts for the syngas reaction. Catalysis Today, 15(3–4), 469–480. doi:10.1016/0920-5861(92)85012-B doi:10.1016/0920-5861(92)85012-B

Judd, S., & Jefferson, B. (Eds.). (2003). *Membrane for Industrial Wastewater Recovery and Re-Use*. Oxford, United Kingdom: Elsevier B.V.

Jutz, F., Andanson, J.-M., & Baiker, A. (2009). A green pathway for hydrogenations on ionic liquid-stabilized nanoparticles. *Journal of Catalysis*, *268*(2), 356–366. doi:10.1016/j.jcat.2009.10.006

Kabe, T., Qian, W. H., Ogawa, S., & Ishihara, A. (1993). Mechanism of Hydro Desulphurization of Dibenzothiophene on Co-Mo/Al2O 3 and Co/Al 2O3 Catalyst by the Use of Radioisotope sup S Tracer. *Journal of Catalysis*, *143*(1), 239–248. doi:10.1006/jcat.1993.1269

Kagami, N., Vogelaar, B. M., van Langeveld, A. D., & Moulijn, J. A. (2005). Reaction pathways on NiMo/Al_2O_3 catalysts for hydrodesulfurization of diesel fuel. *Applied Catalysis A, General*, *293*, 11–23. doi:10.1016/j.apcata.2005.05.042

Kalevaru, V. N., Benhmid, A., Radnik, J., Pohl, M.-M., Bentrup, U., & Martin, A. (2007). Marked influence of support on the catalytic performance of PdSb acetoxylation catalysts: Effects of Pd particle size, valence states, and acidity characteristics. *Journal of Catalysis*, *246*(2), 399–412. doi:10.1016/j.jcat.2007.01.002

Kaliya, M. L., Kogan, S. B., Froumin, N., & Herskowitz, M. (2002). Novel nitrogen containing heterogeneous catalysts for oxidative dehydrogenation of light paraffins. *Catalysis Communications*, *3*(8), 327–333. doi:10.1016/S1566-7367(02)00141-3

Kanoglu, M. (2001). Cryogenic turbine efficiencies. *Exergy Int. J*, 238-253.

Karim, K. S., Al-Hazmi, M. H., & Al-Andis, N. (1999). Catalysts for Oxidative Dehydrogenation of Ethane. *The Arabian Journal for Science and Engineering*, 41-48.

Karim, K., & Sheikh, K. (1999). *US Patent No. US6274764B1.*

Karim, K., Al-Hazmi, M. H., Khan, A. A., & Zaheer, S. I. (n.d.). *US Patent No. US6531631B1.*

Karim, K., Mamedov, E., Al-Hazmi, M. H., Fakeeha, A. H., Soliman, M. A., Al-Zeghayer, Y. S., . . . Al-Arify, A. A. (n.d.). *US Patent No. US6310241B1.*

Karimi, B., Behzadnia, H., Bostina, M., & Vali, H. (2012). A nano-fibrillated mesoporous carbon as an effective support for palladium nanoparticles in the aerobic oxidation of alcohols "on pure water". *Chemistry (Weinheim an der Bergstrasse, Germany)*, *18*(28), 8634–8640. doi:10.1002/chem.201200373 PMID:22707318

Karimi, B., Elhamifar, D., Clark, J. H., & Hunt, A. J. (2011). Palladium containing periodic mesoporous organosilica with imidazolium framework (Pd@PMO-IL): An efficient and recyclable catalyst for the aerobic oxidation of alcohols. *Organic & Biomolecular Chemistry*, *9*(21), 7420–7426. doi:10.1039/c1ob05752a PMID:21904726

Karimi, B., Gholinejad, M., & Khorasani, M. (2012). Highly efficient three-component coupling reaction catalyzed by gold nanoparticles supported on periodic mesoporous organosilica with ionic liquid framework. *Chemical Communications*, *48*(71), 8961–8963. doi:10.1039/c2cc33320a PMID:22842770

Karimi, B., Mansouri, F., & Vali, H. (2014). A highly water-dispersible/magnetically separable palladium catalyst based on a Fe3O4@SiO2 anchored TEG-imidazolium ionic liquid for the Suzuki–Miyaura coupling reaction in water. *Green Chemistry*, *16*(5), 2587–2596. doi:10.1039/c3gc42311e

Katzer, J. R., Sleight, A. W., Gajardo, P., Michel, J. B., Gleason, E. F., & McMillan, S. (1981). The Role of The Support in CO Hydrogenation Selectivity of Supported Rhodium. Faraday Discussions, 72, 121–133. doi:10.1039/dc9817200121 doi:10.1039/dc9817200121

Keim, W. (1987). Synthesis Gas Industrial Chemicals via C1 Processes (Vol. 328, pp. 1–16). American Chemical Society. doi:10.1021/bk-1987-0328.ch001 doi:10.1021/bk-1987-0328.ch001

Keith, J. A., & Henry, P. M. (2009). The Mechanism of the Wacker Reaction: A Tale of Two Hydroxypalladations. *Angewandte Chemie International Edition*, *48*(48), 9038–9049. doi:10.1002/anie.200902194 PMID:19834921

Kemp, I. (2009). *Pinch Analysis and Process Integration - A User Guide on Process Integration for the Efficient Use of Energy* (2nd ed.). Butterworth-Heinemann.

Ketchie, W. C., Fang, Y. L., Wong, M. S., Murayama, M., & Davis, R. J. (2007). Influence of gold particle size on the aqueous-phase oxidation of carbon monoxide and glycerol. *Journal of Catalysis*, *250*(1), 94–101. doi:10.1016/j.jcat.2007.06.001

Ketchie, W. C., Murayama, M., & Davis, R. J. (2007). Selective oxidation of glycerol over carbon-supported AuPd catalysts. *Journal of Catalysis*, *250*(2), 264–273. doi:10.1016/j.jcat.2007.06.011

Keyvanloo, K., Mardkhe, M. K., Alam, T. M., Bartholomew, C. H., Woodfield, B. F., & Hecker, W. C. (2014). Supported Iron Fischer–Tropsch Catalyst: Superior Activity and Stability Using a Thermally Stable Silica-Doped Alumina Support. ACS Catalysis, 4(4), 1071–1077. doi:10.1021/cs401242d doi:10.1021/cs401242d

Kim, J., Ma, X., Song, C., Lee, Y., & Oyama, T. (2005). Hydrodesulfurization of 4,6-dimethyldibenzothiophene over NiMo, CoMo sulfide and Ni phosphide catalysts: A kinetic study on individual reaction pathways. *Energy & Fuels*, *19*, 353–364. doi:10.1021/ef049804g

Kim, K.-S., Demberelnyamba, D., & Lee, H. (2004). Size-Selective Synthesis of Gold and Platinum Nanoparticles Using Novel Thiol-Functionalized Ionic Liquids. *Langmuir*, *20*(3), 556–560. doi:10.1021/la0355848 PMID:15773074

Kimura, H. (1993). Selective oxidation of glycerol on a platinum-bismuth catalyst by using a fixed-bed reactor. *Applied Catalysis A, General*, *105*(2), 147–158. doi:10.1016/0926-860X(93)80245-L

Kimura, H., Tsuto, K., Wakisawa, T., Kazumi, Y., & Inaya, Y. (1995). Selective oxidation of glycerol on a platinum bismuth catalyst. *Applied Catalysis A, General*, *123*(2), 323–323. doi:10.1016/0926-860X(95)00008-9

Kitson, M. (1993). *US Patent No. US5210293(A).*

Klier, K., Herman, R. G., Nunan, J. G., Smith, K. J., Bogdan, C. E., Young, C. W., & Santiesteban, J. G. (1988). Mechanism of Methanol and Higher Oxygenate Synthesis. In Studies in Surface Science and Catalysis (Vol. 36, pp. 109-125). Elsevier. doi:10.1016/S0167-2991(09)60506-1 doi:10.1016/S0167-2991(09)60506-1

Klimov, O. V., Pashigreva, A. V., Fedotov, M. A., Kochubey, D. I., Chesalov, Y. A., Bukhtiyarova, G. A., & Noskov, A. S. (2010). Co-Mo catalysts for ultra-deep HDS of diesel fuels prepared via synthesis of bimetallic surface compounds. *J. Mol. Catal. A: Chem.*, 322(1-2), 80-89. doi:10.1016/j.molcata.2010.02.020

Klose, F., Joshi, M., Hamel, C., & Seidel-Morgenstern, A. (2004). Selective oxidation of ethane over a VOx/g-Al_2O_3 catalyst-investigation of the reaction network. *Applied Catalysis A, General*, *260*(1), 101–110. doi:10.1016/j.apcata.2003.10.005

Knifton, J. F., Grigsby, R. A., & Lin, J. J. (1984). Syngas reactions. 6. Aliphatic alcohols and esters from synthesis gas. Organometallics, 3(1), 62–69. doi:10.1021/om00079a012 doi:10.1021/om00079a012

Knifton, J. F., Lin, J. J., Storm, D. A., & Wong, S. F. (1993). New synthesis gas chemistry. Catalysis Today, 18(4), 355-384.

Koh, T., Koo, H. M., Yu, T., Lim, B., & Bae, J. W. (2014). Roles of Ruthenium–Support Interactions of Size-Controlled Ruthenium Nanoparticles for the Product Distribution of Fischer–Tropsch Synthesis. *ACS Catalysis*, *4*(4), 1054–1060.

Koizumi, N., Murai, K., Ozaki, T., & Yamada, M. (2004). Development of sulfur tolerant catalysts for the synthesis of high quality transportation fuels. Catalysis Today, 89(4), 465–478. doi:10.1016/j.cattod.2004.02.002 doi:10.1016/j.cattod.2004.02.002

Komatsu, T., & Onda, A. (2008). Catalytic properties of single phase intermetallic compounds. *Catalysis Surveys from Asia*, *12*(1), 6–15. doi:10.1007/s10563-007-9031-3

Komintarachat, C., & Trakarnpruk, W. (2006). Oxidative desulfurization using polyoxometalates. *Industrial & Engineering Chemistry Research*, *45*(6), 1853–1856. doi:10.1021/ie051199x

Kondratenko, E. V., Wang, H., Kondratenko, V. A., & Caro, J. (2009). Selective oxidation of CH4 and C2H6 over a mixed oxygen ion and electron conducting perovskite—A TAP and membrane reactors study. *Journal of Molecular Catalysis A Chemical*, *297*(2), 142–149. doi:10.1016/j.molcata.2008.09.015

König, D., Richter, K., Siegel, A., Mudring, A.-V., & Ludwig, A. (2014). High-Throughput Fabrication of Au-Cu Nanoparticle Libraries by Combinatorial Sputtering in Ionic Liquids. *Advanced Functional Materials*, *24*(14), 2049–2056. doi:10.1002/adfm.201303140

Konya, T., Katou, T., Murayama, T., Ishikawa, S., Sadakane, M., Buttrey, D., & Ueda, W. (2013). An orthorhombic Mo_3VOx catalyst most active for oxidative dehydrogenation of ethane among related complex metal oxides. *Catalysis Science & Technology*, *3*(2), 380–387. doi:10.1039/C2CY20444D

Korányi, T. I., Vít, Z., & Nagy, J. B. (2008). Support and pretreatment effects on the hydrotreating activity of SBA-15 and CMK-5 supported nickel phosphide catalysts. *Catalysis Today*, *130*(1), 80–85. doi:10.1016/j.cattod.2007.09.004

Korányi, T. I., Vít, Z., Poduval, D. G., Ryoo, R., Kim, H., & Hensen, E. (2008). SBA-15-supported nickel phosphide hydrotreating catalysts. *Journal of Catalysis*, *253*(1), 119–131. doi:10.1016/j.jcat.2007.10.012

Korlann, S., Diaz, B., & Bussell, M. E. (2002). Synthesis of bulk and alumina-supported bimetallic carbide and nitride catalysts. *Chemistry of Materials*, *14*(10), 4049–4058. doi:10.1021/cm011508j

Kosaric, N., Duvnjak, Z., Farkas, A., Sahm, H., Bringer-Meyer, S., Goebel, O., & Mayer, D. (2000). Ethanol. In Ullmann's Encyclopedia of Industrial Chemistry. Wiley-VCH Verlag GmbH & Co. KGaA.

Kozhevnikov, I. V., Taraban'ko, V. E., Matveev, K. I., & Vardanyan, V. D. (1977). Acetoxylation of olefins catalyzed by a palladium(II)-heteropolyacid system. *Reaction Kinetics and Catalysis Letters*, *7*(3), 297–302. doi:10.1007/BF02062459

Kragten, D. D., van Santen, R. A., Neurock, M., & Lerou, J. J. (1999). A Density Functional Study of the Acetoxylation of Ethylene to Vinyl Acetate Catalyzed by Palladium Acetate. *The Journal of Physical Chemistry A*, *103*(15), 2756–2765. doi:10.1021/jp982956q

Krämer, J., Redel, E., Thomann, R., & Janiak, C. (2008). Use of Ionic Liquids for the Synthesis of Iron, Ruthenium, and Osmium Nanoparticles from Their Metal Carbonyl Precursors. *Organometallics*, *27*(9), 1976–1978. doi:10.1021/om800056z

Krause, T. R., Rathke, J. W., & Chen, M. J. (2013). Process for producing ethanol from syngas: Google Patents.

Krekel, H., & Kroenig, W. (1969). Ethylene and its Industrial Derivatives. In S. A. Miller (Ed.), *Proc. 7th World Petrol. Congr.*, London, (pp. 24-108). Ernest Benn Limited.

Kreye, O., Trefzger, C., Sehlinger, A., & Meier, M. A. R. (2014). Multicomponent Reactions with a Convertible Isocyanide: Efficient and Versatile Grafting of ADMET-Derived Polymers. *Macromolecular Chemistry and Physics*, *215*(22), 2207–2220. doi:10.1002/macp.201400187

Krishnan, G. (2005). What HAZOP studies cannot do. *Hydrocarbon Processing*, 93-95.

Kroenig, W., & Scharfe, G. (1974). Process for the production of vinyl esters. *U.S. Patent No. 3,822,308.* Washington, DC: U.S. Patent and Trademark Office.

Ku, A. Y. C., Acharya, H. R., Kulkarni, P. P., Miller, S. M., & Cook Livingood, W., III. (2013). Syngas cleanup section with carbon capture and hydrogen-selective membrane. *U.S. Patent 8,495,882 B2.*

Kulawska, M., & Skrzypek, J. (2001). Kinetics of the synthesis of higher aliphatic alcohols from syngas over a modified methanol synthesis catalyst. Chemical Engineering and Processing: Process Intensification, 40(1), 33-40.

Kulkarni, Prashant S., & Afonso, Carlos A. M. (2010). Deep desulphurization of diesel fuel using ILs: Current status and future challenges. *Green Chemistry,* 12(7), 1139-1149. doi:10.1039/C002113J

Kumar, D., & Schoonebeek, R. J. (2015). Process for the oxidative dehydrogenation of ethane to ethylene (WO 2015059275 A1).

Kumar, D., Chen, M. S., & Goodman, D. W. (2007). Synthesis of vinyl acetate on Pd-based catalysts. *Catalysis Today*, *123*(1-4), 77–85. doi:10.1016/j.cattod.2007.01.050

Kumar, V. V., & Anthony, S. P. (2014). AuNP based selective colorimetric sensor for cysteine at a wide pH range: Investigation of capping molecule structure on the colorimetric sensing and catalytic properties. *RSC Advances*, *4*(35), 18467–18472. doi:10.1039/c4ra00345d

Kurtz, M., Wilmer, H., Genger, T., Hinrichsen, O., & Muhler, M. (2003). Deactivation of Supported Copper Catalysts for Methanol Synthesis. *Catalysis Letters*, *86*(1-3), 77–80.

Kustov, L. M., Kucherov, A. V., Finashina, E. D., Kucherova, T. N., Isaeva, V. I., Krzywicki, A., & Cai, H. (2010). *US Patent No. US20120016171(A1).*

Kustov, L. M., Kucherov, A. V., Finashina, E. D., Shimanzhenkov, V., & Krzywicki, A. (2012). *US Patent No. US8642825(B2).*

Kustov, L. M., Kucherov, A. V., Finashina, E. D., Stakheev, A. Y., Sinev, I. M., & Krzywicki, A. (2011). CA Patent 2701089 A1.

Lahiere, R. J., Hellums, M. W., Wijmans, J. G., & Kaschemekat, J. (1993). Membrane Vapor Separation: Recovery of Vinyl Chloride Monomer from PVC Reactor Vents. *Industrial & Engineering Chemistry Research*, *32*(10), 2236–2241. doi:10.1021/ie00022a006

Lai, X., Clair, T. P. S., Valden, M., & Goodman, D. W. (1998). Scanning Tunneling Microscopy studies of metal clusters supported on TiO_2 (110): Morphology and electronic structure. *Progress in Surface Science*, *59*(1-4), 25–52. doi:10.1016/S0079-6816(98)00034-3

Lapidus, A. L., Mikhailov, M. N., & Kustov, L. M. (2010). Interaction of a Co6 cluster with oxides of different nature: a quantum chemical study. *Russ. Chem. Bull.*, 58(2), 273-279. doi:10.1007/s11172-010-0001-y

Larrubia, M. A., Gutièrrez-Alejandre, A., Ramìrez, J., & Busca, G. (2002). A FT-IR study of the adsorption of indole, carbazole, benzothiophene, dibenzothiophene and 4, 6-dibenzothiophene over solid adsorbents and catalysts. *Applied Catalysis A, General*, *224*(1-2), 167–178. doi:10.1016/S0926-860X(01)00769-4

Layman, K. A., & Bussell, M. E. (2004). Infrared spectroscopic investigation of CO adsorption on silica-supported nickel phosphide catalysts. *The Journal of Physical Chemistry B*, *108*(30), 10930–10941. doi:10.1021/jp037101e

Lebarbé, T., More, A. S., Sane, P. S., Grau, E., Alfos, C., & Cramail, H. (2014). Bio-Based Aliphatic Polyurethanes through ADMET Polymerization in Bulk and Green Solvent. *Macromolecular Rapid Communications*, *35*(4), 479–483. doi:10.1002/marc.201300695 PMID:24339353

Lebarbier, V. M., Mei, D., Kim, D. H., Andersen, A., Male, J. L., Holladay, J. E., & Wang, Y. (2011). Effects of La2O3 on the mixed higher alcohols synthesis from syngas over co catalysts: a combined theoretical and experimental study. The Journal of Physical Chemistry C, 115(35), 17440–17451. doi:10.1021/jp204003q doi:10.1021/jp204003q

Lee, J. H., Reddy, K. H., Jung, J. S., Yang, E.-H., & Moon, D. J. (2014). Role of support on higher alcohol synthesis from syngas. Applied Catalysis A, General, 480(0), 128–133. doi:10.1016/j.apcata.2014.04.026 doi:10.1016/j.apcata.2014.04.026

Lee, J. M., & Frankiewicz, T. (2005). Treatment of Produced Water with an Ultrafiltration (UF) Membrane - A Field Trial. *Presented at theSPE Annual Technical Conference and Exhibition*, Dallas, Texas. doi:10.2118/95735-MS

Lee, S.H.D., Kumar, R., & Krumpelt, M. (2002). Sulfur removal from diesel fuel-contaminated methanol. *Separation and Purification Technology*, *26*(2), 247–258.

Lee, S., Woo, S. K., Lee, K. S., & Kim, D. K. (2006). Mechanical properties and structural stability of perovskite-type, oxygen-permeable, dense membranes. *Desalination*, *193*(1-3), 236–243. doi:10.1016/j.desal.2005.06.061

Leger, B., Denicourt-Nowicki, A., Olivier-Bourbigou, H., & Roucoux, A. (2008). Rhodium nanocatalysts stabilized by various bipyridine ligands in nonaqueous ionic liquids: Influence of the bipyridine coordination modes in arene catalytic hydrogenation. *Inorganic Chemistry*, *47*(19), 9090–9096. doi:10.1021/ic8010713 PMID:18781733

Lein, H. L., Wiik, K., & Grande, T. (2006). Kinetic demixing and decomposition of oxygen permeable membranes. *Solid State Ionics, 177*(19-25), 1587-1590.

Leonard, J. K., & Wagener, K. B. (2008). Polyethylene prodrugs using precisely placed pharmaceutical agents. *Polym. Prepr. (Am. Chem. Soc., Div. Polym. Chem.), 49*(1), 874.

Leonard, J. K., Hopkins, T. E., Chaffin, K., & Wagener, K. B. (2008). Semicrystalline lysine functionalized precision polyolefins. *Macromolecular Chemistry and Physics, 209*(14), 1485–1494. doi:10.1002/macp.200700645

Lerou, J. J., Tonkovich, A. L., Silva, L., Perry, S., & McDaniel, J. (2010). Microchannel reactor architecture enables greener processes. Chemical Engineering Science, 65(1), 380-385.

Leveles, L., Seshan, K., Lercher, J., & Lefferts, A. L. (2003). Oxidative conversion of propane over lithium promoted magnesia catalyst I. Kinetics and mechanism. *Journal of Catalysis, 218*(2), 296–306. doi:10.1016/S0021-9517(03)00112-X

Levv, R. B., & Boudart, M. (1973). Platinum-like behavior of tungsten carbide in surface catalysis. *Science, 181*(4099), 547–549. doi:10.1126/science.181.4099.547 PMID:17777803

Lewandowski, M., Szymańska-Kolasa, A., Sayag, C., Beaunier, P., & Djéga-Mariadassou, G. (2014). Atomic level characterization and sulfur resistance of unsupported W2C during dibenzothiophene hydrodesulfurization. Classical kinetic simulation of the reaction. *Applied Catalysis B: Environmental, 144*, 750–759. doi:10.1016/j.apcatb.2013.08.011

Lewin, M., (Ed.), (2007). *Handbook of Fiber Chemistry* (Third ed.). CRC Press.

Liang, D., Gao, J., Sun, H., Chen, P., Hou, Z. Y., & Zheng, X. M. (2011). Selective oxidation of glycerol with oxygen in a base-free aqueous solution over MWNTs supported Pt catalysts. *Applied Catalysis B: Environmental, 106*(3-4), 423–432. doi:10.1016/j.apcatb.2011.05.050

Liang, D., Gao, J., Wang, J. H., Chen, P., Hou, Z. Y., & Zheng, X. M. (2009). Selective oxidation of glycerol in a base-free aqueous solution over different sized Pt catalysts. *Catalysis Communications, 10*(12), 1586–1590. doi:10.1016/j.catcom.2009.04.023

Liang, W. Q., & Hughes, R. (2005). The catalytic dehydrogenation of isobutane to isobutene in a palladium/silver composite membrane reactor. *Catalysis Today, 104*(2-4), 238–243. doi:10.1016/j.cattod.2005.03.045

Liang, W., Zhang, S., Li, H., & Zhang, G. (2013). Oxidative desulphurization of simulated gasoline catalyzed by acetic acid-based ILs at room temperature. *Fuel Processing Technology, 109*(0), 27–31. doi:10.1016/j.fuproc.2012.09.034

Li, B. D., Wang, J., Yuan, Y. Z., Ariga, H., Takakusagi, S., & Asakura, K. (2011). Carbon Nanotube-Supported RuFe Bimetallic Nanoparticles as Efficient and Robust Catalysts for Aqueous-Phase Selective Hydrogenolysis of Glycerol to Glycols. *Acs Catalysis, 1*(11), 1521–1528. doi:10.1021/cs200386q

Li, C., Jiang, Z., Gao, J., Yang, Y., Wang, S., Tian, F., & Han, C. et al. (2004). Ultra-Deep Desulfurization of Diesel: Oxidation with a Recoverable Catalyst Assembled in Emulsion. *Chemistry (Weinheim an der Bergstrasse, Germany), 10*(9), 2277–2280. PMID:15112217

Li, C., Li, D., Zou, S., Li, Z., Yin, J., Wang, A., & Zhao, Q. et al. (2013). Extraction desulfurization process of fuels with ammonium-based deep eutectic solvents. *Green Chemistry, 15*(10), 2793–2799. doi:10.1039/c3gc41067f

Li, J., & Smith, K. J. (2008). Methane decomposition and catalyst regeneration in a cyclic mode over supported Co and Ni catalysts. *Applied Catalysis A, General, 349*(1-2), 116–124. doi:10.1016/j.apcata.2008.07.011

Lin, W. M., Gan, S. F., Huang, C. R., Pan, D. R., & Noller, H. (1988). Synthesis of light alcohols from carbon monoxide and hydrogen over copper-iron catalysts. Reaction Kinetics and Catalysis Letters, 37(2), 487–492. doi:10.1007/BF02062104 doi:10.1007/BF02062104

Lindlar, H., & Dubuis, R. (1973). Palladium catalyst for partial reduction of acetylenes. *Organic Syntheses. Collect.*, *5*, 880–883.

Lin, H., Thompson, S. M., Serbanescu-Martin, A., Wijmans, J. G., Amo, K. D., Lokhandwala, K. A., & Merkel, T. C. et al. (2013). Dehydration of natural gas using membranes. Part II: Sweep/countercurrent design and field test. *Journal of Membrane Science*, *432*, 106–114. doi:10.1016/j.memsci.2012.12.049

Linke, D., Wolf, D., Baerns, M., Timpe, O., Schlogl, R., Zeyb, S., & Dingerdissen, U. (2002). Catalytic Partial Oxidation of Ethane to Acetic Acid over Mo1V0.25Nb0.12Pd0.0005Ox. I. Catalyst Performance and Reaction Mechanism. *Journal of Catalysis*, *205*, 16–31. doi:10.1006/jcat.2001.3367

Lin, L., Zhang, Y., & Kong, Y. (2009). Recent advances in sulfur removal from gasoline by pervaporation. *Fuel*, *88*(10), 1799–1809. doi:10.1016/j.fuel.2009.03.031

Lin, M. M. (2001). Selective oxidation of propane to acrylic acid with molecular oxygen. *Applied Catalysis A, General*, *207*(1-2), 1–16. doi:10.1016/S0926-860X(00)00609-8

Lin, S. Y., Harada, M., Suzuki, Y., & Hatano, H. (2002). Hydrogen production from coal by separating carbon dioxide during gasification. *Fuel*, *81*(16), 2079–2085. doi:10.1016/S0016-2361(02)00187-4

Lintz, H.-G., & Muller, S. P. (2009). The partial oxidation of propane on mixed metal oxides - A short overview. *Applied Catalysis A, General*, *357*(2), 178–183. doi:10.1016/j.apcata.2009.01.018

Lin, Y. S., & Zeng, Y. (1996). Catalytic properties of oxygen semipermeable perovskite-type ceramic membrane materials for oxidative coupling of methane. *Journal of Catalysis*, *164*(1), 220–231. doi:10.1006/jcat.1996.0377

Li, Q., Wu, P., Lan, L., Liu, H., & Ji, S. (2013). The effect of Ca on structure and performance of hydrodesulfurization of Ca-Ni_2P/SBA-15/Cord monolithic catalysts. *Catalysis Today*, *216*, 38–43. doi:10.1016/j.cattod.2013.06.016

Li, Q., Wu, P., Lan, L., Wei, N., & Ji, S. (2014). Preparation of novel mesoporous Ce-Ni_2P/SBA-15 catalysts and their catalytic performance for hydrodesulfurization of dibenzothiophene. *Applied Petrochemical Research*, *4*(2), 209–216. doi:10.1007/s13203-013-0039-5

Li, S., Jin, W., Huang, P., Xu, N., Shi, J., & Lin, Y. S. (2000). Tubular lanthanum cobaltite perovskite type membrane for oxygen permeation. *Journal of Membrane Science*, *166*(1), 51–61. doi:10.1016/S0376-7388(99)00244-6

Lissner, E., de Souza, W. F., Ferrera, B., & Dupont, J. (2009). Oxidative desulphurization of fuels with task-specific ILs. *ChemSusChem*, *2*(10), 962–964. doi:10.1002/cssc.200900171 PMID:19795435

Liu, J., Guo, Z., Childers, D., Schweitzer, N., Marshall, C. L., Klie, R. F., & Meyer, R. J. (2014). Correlating the degree of metal–promoter interaction to ethanol selectivity over MnRh/CNTs CO hydrogenation catalysts. Journal of Catalysis, 313(0), 149–158.

Liu, J., Tao, R., Guo, Z., Regalbuto, J. R., Marshall, C. L., Klie, R. F., & Meyer, R. J. (2013). Selective Adsorption of Manganese onto Rhodium for Optimized Mn/Rh/SiO2 Alcohol Synthesis Catalysts. ChemCatChem, 5(12), 3665–3672. doi:10.1002/cctc.201300479 doi:10.1002/cctc.201300479

Liu, Y. (2001). *US Patent No. US6891075(B2).*

Liu, Y. (2007). *US Patent No. US7674944(B2).*

Liu, Y. Y., Murata, K., Inaba, M., Takahara, I., & Okabe, K. (2013). Mixed alcohols synthesis from syngas over Cs- and Ni-modified Cu/CeO2 catalysts. Fuel, 104, 62–69. doi:10.1016/j.fuel.2010.08.014 doi:10.1016/j.fuel.2010.08.014

Liu, Z., Li, X., Close, M. R., Kugler, E. L., Petersen, J. L., & Dadyburjor, D. B. (1997). Screening of Alkali-Promoted Vapor-Phase-Synthesized Molybdenum Sulfide Catalysts for the Production of Alcohols from Synthesis Gas. Industrial & Engineering Chemistry Research, 36(8), 3085–3093. doi:10.1021/ie960648d doi:10.1021/ie960648d

Liu, D., Gui, J., Song, L., Zhang, X., & Sun, Z. (2008). Deep desulfurization of diesel fuel by extraction with task-specific ionic liquids. *Petroleum Science and Technology*, *26*(9), 973–982. doi:10.1080/10916460600695496

Liu, H. Z., Liang, S. G., Jiang, T., Han, B. X., & Zhou, Y. X. (2012). Hydrogenolysis of Glycerol to 1,2-Propanediol over Ru-Cu Bimetals Supported on Different Supports. *Clean-Soil Air Water*, *40*(3), 318–324. doi:10.1002/clen.201000227

Liu, Q., Ji, S., Wu, P., Hu, L., Huang, X., Zhu, J., & Li, C. (2009). *In situ* diffuse reflectance FTIR spectroscopy study of CO adsorption on Ni_2P/Mesoporous molecule sieve catalysts. *Spectroscopy and Spectral Analysis*, *29*, 1227–1231. PMID:19650459

Liu, Y., Liu, C., & Que, G. (2002). Hydrodesulfurization of dibenzothiophene over cobalt-molybdenum nitride catalysts. *Energy & Fuels*, *16*(3), 531–535. doi:10.1021/ef000272n

Liu, Y.-M., Wang, L.-C., Chen, M., Xu, J., Cao, Y., He, Y.-H., & Fan, K.-N. (2009). Highly Selective Ce–Ni–O Catalysts for Efficient Low Temperature Oxidative Dehydrogenation of Propane. *Catalysis Letters*, *130*(3-4), 350–354. doi:10.1007/s10562-009-9977-z

Liu, Y., Zhang, Y., Liu, X., Xue, J., & Li, S. (1998). The excellent performance of TiO2-supported Li-Mn-W oxide catalyst for oxidative dehydrogenation of ethane to ethylene. *Chemistry Letters*, 1998(10), 1057–1058. doi:10.1246/cl.1998.1057

Liu, Y., Zhu, X., Li, M., Liu, H., Cong, Y., & Yang, W. (2013). Stabilization of low-temperature degradation in mixed ionic and electronic conducting perovskite oxygen permeation membranes. *Angewandte Chemie International Edition*, *52*(11), 3232–3236. doi:10.1002/anie.201209077 PMID:23401278

Li, W., Dhandapani, B., & Oyama, S. T. (1998). Molybdenum phosphide: A novel catalyst for hydrogenitrogenation. *Chemistry Letters*, *27*(3), 207–208. doi:10.1246/cl.1998.207

Li, X., & Iglesia, E. (2008). Catalytic dehydroisomerization of *n*-alkanes to isoalkenes. *Journal of Catalysis*, *255*(1), 134–137. doi:10.1016/j.jcat.2008.01.021

Li, X., & Iglesia, E. (2008). Kinetics and Mechanism of Ethane Oxidation to Acetic Acid on Catalysts Based on Mo-V-Nb Oxides. *The Journal of Physical Chemistry C*, *112*(38), 15001–15008. doi:10.1021/jp801488y

Li, X., & Iglesia, E. (2008). Support and promoter effects in the selective oxidation of ethane to acetic acid catalyzed by Mo-V-Nb oxides. *Applied Catalysis A, General*, *334*(1-2), 339–347. doi:10.1016/j.apcata.2007.10.021

Li, X., Lu, M., Wang, A., Song, C., & Hu, Y. (2008). Promoting effect of TiO_2 on the hydrodenitrogenation performance of nickel phosphide. *The Journal of Physical Chemistry C*, *112*(42), 16584–16592. doi:10.1021/jp803676v

Li, X., Wang, A., Egorova, M., & Prins, R. (2007). Kinetics of the HDS of 4,6-dimethyldibenzothiophene and its hydrogenated intermediates over sulfided Mo and NiMo on γ-Al_2O_3. *Journal of Catalysis*, *250*(2), 283–293. doi:10.1016/j.jcat.2007.06.005

Li, X., Xu, B., Feng, N., Qiu, A., He, J. M. J., Yang, H., & Chen, Y. (2004). The effect of titania polymorph on the strong metal support interaction of Pd/TiO_2 catalysts and their application in the liquid phase selective hydrogenation of long chain alkadienes. *Journal of Molecular Catalysis A Chemical*, *216*(1), 107–114. doi:10.1016/j.molcata.2004.02.007

Li, Y., Chen, X., & Chein, M.-H. (2012). Flexible and cost-effective optimization of BOG (boil-off gas) recondensation process at LNG terminals. *Chemical Engineering Research & Design*, *90*(10), 1500–1505. doi:10.1016/j.cherd.2012.01.013

Li, Z.-L., Lv, A., Du, F.-S., & Li, Z.-C. (2014). Intrachain Cyclization via Postmodification of the Internal Alkenes of Periodic ADMET Copolymers: The Sequence Matters. *Macromolecules (Washington, DC, U. S.)*, *47*(17), 5942–5951. doi:10.1021/ma5013732

Li, Z., Liu, Z., Zhang, J., Han, B., Du, J., Gao, Y., & Jiang, T. (2005). Synthesis of single-crystal gold énanosheets of large size in ionic liquids. *The Journal of Physical Chemistry B*, *109*(30), 14445–14448. doi:10.1021/jp0520998 PMID:16852818

Li, Z.-r., Fu, Y.-l., & Jiang, M. (1999). Structures and performance of Rh–Mo–K/Al2O3 catalysts used for mixed alcohol synthesis from synthesis gas. *Applied Catalysis A, General*, *187*(2), 187–198.

Llevot, A., Grau, E., Carlotti, S., Grelier, S., & Cramail, H. (2014). Dimerization of abietic acid for the design of renewable polymers by ADMET. *Eur. Polym. J.* doi:10.1016/j.eurpolymj.2014.10.021

Lokhandwala, K. (1997). Membrane-Augmented Cryogenic Methane/Nitrogen Separation. *U.S. Patent 5,647,227.*

Lopez Nieto, J. M., Solsona, B., Concepcion, P., Ivars, F., Dejoz, A., & Vazquez, M. I. (2010). Reaction products and pathways in the selective oxidation of C2–C4 alkanes on MoVTeNb mixed oxide catalysts. *Catalysis Today*, *157*(1-4), 291–296. doi:10.1016/j.cattod.2010.01.046

López, N., García-Mota, M., & Gómez-Diaz, J. (2008). NH_3 Oxidation on Oxygen-Precovered Au(111): A Density Functional Theory Study on Selectivity. *The Journal of Physical Chemistry C*, *112*(1), 247–252. doi:10.1021/jp077205f

Lopez-Sanchez, J. A., Dimitratos, N., Glanville, N., Kesavan, L., Hammond, C., Edwards, J. K., & Hutchings, G. J. et al. (2011). Reactivity studies of Au-Pd supported nanoparticles for catalytic applications. *Applied Catalysis A, General*, *391*(1-2), 400–406. doi:10.1016/j.apcata.2010.05.010

Lopez-Sanchez, J. A., Dimitratos, N., Hammond, C., Brett, G. L., Kesavan, L., White, S., & Hutchings, G. J. et al. (2011). Facile removal of stabilizer-ligands from supported gold nanoparticles. *Nature Chemistry*, *3*(7), 551–556. doi:10.1038/nchem.1066 PMID:21697877

Lopez-Sanchez, J. A., Dimitratos, N., Miedziak, P., Ntainjua, E., Edwards, J. K., Morgan, D., & Hutchings, G. J. et al. (2008). Au-Pd supported nanocrystals prepared by a sol immobilisation technique as catalysts for selective chemical synthesis. *Physical Chemistry Chemical Physics*, *10*(14), 1921–1930. doi:10.1039/b719345a PMID:18368185

Lü, H., Gao, J., Jiang, Z., Jing, F., Yang, Y., Wang, G., & Li, C. (2006). Ultra-deep desulfurization of diesel by selective oxidation with $[C_{18}H_{37}N(CH_3)_3]_4[H_2NaPW_{10}O_{36}]$ catalyst assembled in emulsion droplets. *Journal of Catalysis*, *239*(2), 369–375. doi:10.1016/j.jcat.2006.01.025

Lui, H., Pang, Z., Tan, X., Shao, Z., Sunarso, J., Ding, R., & Liu, S. (2008). Effects of preparation methods on the oxygen nonstoichiometry, B-site cation valences and catalytic efficiency of perovskite La0.6Sr0.4Co0.2Fe0.8O3-δ. *Ceramics International*, *35*(8), 3201–3206.

Luis, P., Van der Bruggen, B., & Van der Gerven, T. (2011). Non-dispersive absorption for CO_2 capture: From the laboratory to industry. *Journal of Chemical Technology and Biotechnology (Oxford, Oxfordshire)*, *86*(6), 769–775. doi:10.1002/jctb.2614

Lu, L., Cheng, S., Gao, J., Gao, G., & He, M.-y, . (2007). Deep oxidative desulfurization of fuels catalyzed by ionic liquid in the presence of H2O2. *Energy & Fuels*, *21*(1), 383–384. doi:10.1021/ef060345o

Luo, H.-Y., Lin, P.-Z., Xie, S.-B., Zhou, H.-W., Xu, C.-H., Huang, S.-Y., & Xin, Q. (1997). The role of Mn and Li promoters in supported rhodium catalysts in the formation of acetic acid and acetaldehyde. Journal of Molecular Catalysis A Chemical, 122(2–3), 115–123. doi:10.1016/S1381-1169(97)00012-5 doi:10.1016/S1381-1169(97)00012-5

Luo, J. Z., & Wan, H. L. (1997). Oxidative dehydrogenation of ethane over LaF_3-CeO_2 catalysts. *Applied Catalysis A, General*, *158*(1-2), 137–144. doi:10.1016/S0926-860X(97)00039-2

Luo, J. Z., Zhou, X. P., Chao, Z. S., & Wan, H. L. (1997). Oxidative dehydrogenation of ethane over BaF_2 promoted SmO_3-LaF_3 catalysts. *Applied Catalysis A, General*, *159*(1-2), 9–19. doi:10.1016/S0926-860X(97)00046-X

Luque, R., Budarin, V., Clark, J. H., & Macquarrie, D. J. (2008). Glycerol transformations on polysaccharide derived mesoporous materials. *Applied Catalysis B: Environmental*, *82*(3-4), 157–162. doi:10.1016/j.apcatb.2008.01.015

Luska, K. L., Migowski, P., & Leitner, W. (2015). Ionic liquid-stabilized nanoparticles as catalysts for the conversion of biomass. *Green Chemistry*, *17*(6), 3195–3206. doi:10.1039/C5GC00231A

Luska, K. L., & Moores, A. (2012). Functionalized Ionic Liquids for the Synthesis of Metal Nanoparticles and their Application in Catalysis. *Chemcatchem*, *4*(10), 1534–1546. doi:10.1002/cctc.201100366

Luska, K. L., & Moores, A. (2012). Ruthenium nanoparticle catalysts stabilized in phosphonium and imidazolium ionic liquids: Dependence of catalyst stability and activity on the ionicity of the ionic liquid. *Green Chemistry*, *14*(6), 1736–1742. doi:10.1039/c2gc35241a

Luthra, N. P., & Cheng, W. C. (1987). Molybdenum-95 NMR study of the adsorption of molybdates on alumina. *J. Catal.*, 107(1), 154-160. doi:10.1016/0021-9517(87)90280-6

Lu, W., Fadeev, A. G., Qi, B. H., Smela, E., Mattes, B. R., & Ding, J. et al.. (2002). Use of ILs for pi-conjugated polymer electrochemical devices. *Science*, *297*(5583), 983–987. doi:10.1126/science.1072651 PMID:12098704

Lu, Y., Xue, J.-Z., Ding, X. J., & Shen, S. K. (1996). TPR and CTPR studies on Ni/Al2O3 catalyst for catalytic partial oxidation of methane to syngas. *Journal of Natural Gas Chemistry*, *5*(1), 45–50.

Luza, L., Gual, A., Eberhardt, D., Teixeira, S. R., Chiaro, S. S. X., & Dupont, J. (2013). "Imprinting" Catalytically Active Pd Nanoparticles onto Ionic-Liquid-Modified Al2O3Supports. *ChemCatChem*, *5*(8), 2471–2478. doi:10.1002/cctc.201300123

Luza, L., Gual, A., Rambor, C. P., Eberhardt, D., Teixeira, S. R., Bernardi, F., & Dupont, J. et al. (2014). Hydrophobic effects on supported ionic liquid phase Pd nanoparticle hydrogenation catalysts. *Physical Chemistry Chemical Physics*, *16*(34), 18088–18091. doi:10.1039/C4CP03063J PMID:25057970

Ma, C.-H., Li, H.-Y., Lin, G.-D., & Zhang, H.-B. (2010). MWCNT-Supported Ni-Mo-K Catalyst for Higher Alcohol Synthesis from Syngas. Catalysis Letters, 137(3-4), 171–179. doi:10.1007/s10562-010-0343-y doi:10.1007/s10562-010-0343-y

Macedonio, F., Ali, A., Poerio, T., El-Sayed, E., Drioli, E., & Abdel-Jawad, M. (2014). Direct contact membrane distillation for treatment of oilfield produced water. *Separation and Purification Technology*, *126*, 69–81. doi:10.1016/j.seppur.2014.02.004

Macias Hernandez, M. J., Morales, R. D., & Ramirez-Lopez, A. (2009). Simulation of the effectiveness factor for a trilobular catalyst on the hydrodesulfurization of diesel. *Int. J. Chem. React. Eng.*, 7. doi:10.2202/1542-6580.1806

Macleod, N., Keel, J. M., & Lambert, R. M. (2004). The effects of ageing a bimetallic catalyst under industrial conditions: A study of fresh and used Pd-Au-K/silica vinyl acetate synthesis catalysts. *Applied Catalysis A, General*, *261*(1), 37–46. doi:10.1016/j.apcata.2003.10.032

Madaan, N., Gatla, S., Kalevaru, V. N., Radnik, J., Lücke, B., Brückner, A., & Martin, A. (2011a). Optimization of reaction conditions and regeneration procedure of the PdSb/TiO_2 catalyst for acetoxylation of toluene. *Topics in Catalysis*, *54*(16-18), 1197–1205. doi:10.1007/s11244-011-9743-5

Madaan, N., Gatla, S., Kalevaru, V. N., Radnik, J., Lücke, B., Brückner, A., & Martin, A. (2011b). Deactivation and regeneration studies of a PdSb/TiO_2 catalyst used in the gas phase acetoxylation of toluene. *Journal of Catalysis*, *282*(1), 103–111. doi:10.1016/j.jcat.2011.06.005

Madaan, N., Gatla, S., Kalevaru, V. N., Radnik, J., Pohl, M.-M., Lücke, B., & Martin, A. et al. (2013). The Impact of Reaction Pressure on the Catalytic Performance of the Pd-Sb/TiO2 Catalyst in the Acetoxylation of Toluene into Benzyl Acetate. *ChemCatChem*, *5*(1), 185–191. doi:10.1002/cctc.201200522

Mader, J., Xiao, L., Schmidt, T. J., & Benicewicz, B. C. (2008). Polybenzimidazole/acid complexes as high-temperature membranes. *Adv. Polym. Sci.*, *216*(Fuel Cells II), 63-124. doi:10.1007/12_2007_129

Mahata, N., & Vishwanathan, V. (2000). Influence of Palladium Precursors on Structural Properties and Phenol Hydrogenation Characteristics of Supported Palladium Catalysts. *Journal of Catalysis*, *196*(2), 262–270. doi:10.1006/jcat.2000.3041

Maitlis, P. M. (2003). Metal catalysed CO hydrogenation: hetero- or homo-, what is the difference? Journal of Molecular Catalysis A: Chemical, 204–205, 54-61.

Majocchi, L., Lietti, L., Beretta, A., Forzatti, P., Micheli, E., & Tagliabue, L. (1998). Synthesis of short chain alcohols over a Cs-promoted Cu/ZnO/Cr2O3 catalyst. *Applied Catalysis A, General*, *166*(2), 393–405.

Majozi, T. (2010). *Batch Chemical Process Integration: Analysis, Synthesis, and Optimization*. Heidelberg: Springer. doi:10.1007/978-90-481-2588-3

Mak, A., Foster, E. P., Armstrong, P. A., Repasky, J. M., & Stein, V. E. (2013). *ITM Oxygen Technology for Clean Energy Applications*. Paper presented at the International Pittsburgh Coal Conference, Beijing,China.

Ma, L., & He, D. H. (2009). Hydrogenolysis of glycerol to propanediols over highly active Ru-Re bimetallic catalysts. *Topics in Catalysis*, *52*(6-7), 834–844. doi:10.1007/s11244-009-9231-3

Ma, L., & He, D. H. (2010). Influence of catalyst pretreatment on catalytic properties and performances of Ru-Re/SiO(2) in glycerol hydrogenolysis to propanediols. *Catalysis Today*, *149*(1-2), 148–156. doi:10.1016/j.cattod.2009.03.015

Ma, L., He, D. H., & Li, Z. P. (2008). Promoting effect of rhenium on catalytic performance of Ru catalysts in hydrogenolysis of glycerol to propanediol. *Catalysis Communications*, *9*(15), 2489–2495. doi:10.1016/j.catcom.2008.07.009

Ma, L., Li, Y. M., & He, D. H. (2011). Glycerol Hydrogenolysis to Propanediols over Ru-Re/SiO2: Acidity of Catalyst and Role of Re. *Chinese Journal of Catalysis*, *32*(5), 872–876. doi:10.1016/S1872-2067(10)60198-2

Malamis, S., Katsou, E., Di Fabio, S., Frison, N., Cecchi, F., & Fatone, F. (2015). Treatment of petrochemical wastewater by employing membrane bioreactors: A case study of effluents discharged to a sensitive water recipient. *Desalination and Water Treatment*, *53*(12), 3397–3406. doi:10.1080/19443994.2014.934112

Mamedov, E. A., & Shaikh, S. N. (1999). *US Patent No. US6235678(B1).*

Mamedov, E., & Karim, K. (2014). Selective Oxidation at SABIC: Innovative Catalysts and Technologies. In D. Duprez, & F. Cavani (Eds.), Handbook of Advanced Methods and Processes in Oxidation Catalysis: From Laboratory to Industry (pp. 291-301). Imperial College Press.

Mamedov, E. A., & Cortés Corberán, V. (1995). Oxidative dehydrogenation of lower alkanes on vanadium oxide-based catalysts. The present state of the art and outlooks. *Applied Catalysis A, General*, *127*(1-2), 1–40. doi:10.1016/0926-860X(95)00056-9

Ma, Q., Zeng, P., Ji, S., Wei, N., Liu, H., & Li, C. (2011). Effect of metal promoters on the structure and performance of the Ni_2P/SBA-15 catalyst for hydrodesulfurization. *Acta Petrolei Sinica*, *27*, 181–186.

Maris, E. P., Ketchie, W. C., Murayama, M., & Davis, R. J. (2007). Glycerol hydrogenolysis on carbon-supported PtRu and AuRu bimetallic catalysts. *Journal of Catalysis*, *251*(2), 281–294. doi:10.1016/j.jcat.2007.08.007

Marquardt, D., Barthel, J., Braun, M., Ganter, C., & Janiak, C. (2012). Weakly-coordinated stable platinum nanocrystals. *CrystEngComm*, *14*(22), 7607–7615. doi:10.1039/c2ce25904d

Mars, P., & van Krevelen, D. W. (1954). Oxidations carried out by means of vanadium oxide catalysts. *Chemical Engineering Science*, *3*(1), 41–59. doi:10.1016/S0009-2509(54)80005-4

Martin, A., & Kalevaru, V. N. (2010). Heterogeneously Catalyzed Ammoxidation: A Valuable Tool for One-Step Synthesis of Nitriles. *ChemCatChem*, *2*(12), 1504–1522. doi:10.1002/cctc.201000173

Matar, S., & Hatch, L. F. (2001). *Chemistry of petrochemical processes* (2nd ed.). Woburn: Gulf Professional Publishing.

Matsui, K., Seno, S., Nozue, Y., Shinohara, Y., Amemiya, Y., Berda, E. B., & Wagener, K. B. et al. (2013). Influence of Branch Incorporation into the Lamella Crystal on the Crystallization Behavior of Polyethylene with Precisely Spaced Branches. *Macromolecules (Washington, DC, U. S.)*, *46*(11), 4438–4446. doi:10.1021/ma400608q

Matsuma, Y., & Nakamori, T. (2004). Steam reforming of methane over nickel catalysts at low reaction temperature. *Applied Catalysis A, General*, *258*(1), 107–114. doi:10.1016/j.apcata.2003.08.009

Matsumoto, T., Kim, W. Y., Kishida, M., Nagata, H., & Wakabayashi, K. (1994). Catalytic Hydrogenation of Carbon Monoxide over Silica-Supported Ir-MO-Rh Catalyst. Catalysis Letters, 24(3-4), 391–394. doi:10.1007/BF00811812 doi:10.1007/BF00811812

Mazanec, T. J. (1986). On the mechanism of higher alcohol formation over metal oxide catalysts: I. A rationale for branching in the synthesis of higher alcohols from syngas. Journal of Catalysis, 98(1), 115-125.

Mazzei, R., Drioli, E., & Giorno, L. (2010). Biocatalytic membranes and membrane bioreactors. In E. Drioli & L. Giorno (Eds.), Comprehensive Membrane Science and Engineering (Vol 3., pp. 195-212). Oxford, United Kingdom: Elsevier B.V. doi:10.1016/B978-0-08-093250-7.00058-X

Mazzone, G., Rivalta, I., Russo, N., & Sicilia, E. (2009). The geometric effect in palladium-gold catalysis. Is the coupling the rate-determining step in the vinyl-acetate synthesis? *Chemical Communications*, *14*(14), 1852–1854. doi:10.1039/b823207e PMID:19319423

McCain, J. H. (1986). Oxydehydrogenation of ethane to ethylene. EP 166438 A2.

Mei, D., Rousseau, R., Kathmann, S. M., Glezakou, V.-A., Engelhard, M. H., Jiang, W., & Stevens, D. J. et al. (2010). Ethanol synthesis from syngas over Rh-based/SiO2 catalysts: A combined experimental and theoretical modeling study. *Journal of Catalysis*, *271*(2), 325–342.

Mei, H., Mei, B. W., & Yen, T. F. (2003). A new method for obtaining ultra-low sulphur diesel fuel via ultrasound assisted oxidative Desulphurization. *Fuel*, *82*(4), 405–414. doi:10.1016/S0016-2361(02)00318-6

Melin, T., Jefferson, B., Bixio, D., Thoeye, C., De Wilde, W., De Koning, J., & Wintgens, T. et al. (2006). Membrane bioreactor technology for wastewater treatment and reuse. *Desalination*, *187*(1-3), 271–282. doi:10.1016/j.desal.2005.04.086

Mendoza-Nieto, J. A., Vera-Vallejo, O., Escobar-Alarcon, L., Solis-Casados, D., & Klimova, T. (2013). Development of new trimetallic NiMoW catalysts supported on SBA-15 for deep hydrodesulfurization. *Fuel*, *110*, 268–277. doi:10.1016/j.fuel.2012.07.057

Merzouki, M., Taouk, B., Monceaux, L., Bordes, E., & Courtine, P. (1992). Catalytic properties of promoted vanadium oxide in the oxidation of ethane to acetic acid. *Studies in Surface Science and Catalysis*, *72*, 165–179. doi:10.1016/S0167-2991(08)61669-9

Mestl, G., Margitfalvi, J. L., Végvári, L., Szijjártó, G. P., & Tompos, A. (2014). Combinatorial design and preparation of transition metal dopedMoVTe catalysts for oxidation of propane to acrylic acid. *Applied Catalysis A, General*, *474*, 3–9. doi:10.1016/j.apcata.2013.08.022

Michaud, P., Lemberton, J., & Pérot, G. (1998). Hydrodesulfurization of dibenzothiophene and 4,6-dimethyldibenzothiophene: Effect of an acid component on the activity of a sulfided NiMo on alumina catalyst. *Applied Catalysis A, General*, *169*(2), 343–353. doi:10.1016/S0926-860X(98)00021-0

Migowski, P., & Dupont, J. (2007). Catalytic applications of metal nanoparticles in imidazolium ionic liquids. *Chemistry (Weinheim an der Bergstrasse, Germany)*, *13*(1), 32–39. doi:10.1002/chem.200601438 PMID:17115465

Migowski, P., Machado, G., Texeira, S. R., Alves, M. C., Morais, J., Traverse, A., & Dupont, J. (2007). Synthesis and characterization of nickel nanoparticles dispersed in imidazolium ionic liquids. *Physical Chemistry Chemical Physics*, *9*(34), 4814–4821. doi:10.1039/b703979d PMID:17712460

Miller, G. Q., & Stöcker, J. (1989). Selection of a Hydrogen Separation Process. *Presented at the1989 NPRA Annual Meeting*, San Francisco, California (USA).

Miller, S. A. (1969). *Ethylene and its industrial derivatives* (pp. 24–108). Ernest Benn Limited London.

Millet, J.-M., Lacorre, P., & Huynh, Q. (2007). *US Patent No. US8207080(B2).*

Mimura, N., Takahara, I., Inaba, M., Okamoto, M., & Murata, K. (2002). High-performance Cr/H-ZSM-5 catalysts for oxidative dehydrogenation of ethane to ethylene with CO_2 as an oxidant. *Catalysis Communications*, *3*(6), 257–262. doi:10.1016/S1566-7367(02)00117-6

Mirodatos, C., Brum Pereira, E., Gomez Cobo, A., Dalmon, J. A., & Martin, G. A. (1995). CO hydrogenation over Ni- and Co-based catalysts: influence of alkali addition on morphological and catalytic properties. *Top. Catal.*, 2(1-4, Fischer-Tropsch and Methanol Synthesis), 183-192. doi:10.1007/BF01491966

Mizuno, N., Suh, D.-J., Han, W., & Kudo, T. (1996). Catalytic performance of $Cs_{2.5}Fe_{0.08}H_{1.26}PVMo_{11}O_{40}$ for direct oxidation of lower alkanes. *Journal of Molecular Catalysis A Chemical*, *114*(1-3), 309–317. doi:10.1016/S1381-1169(96)00332-9

Mo, X., Gao, J., Umnajkaseam, N., & Goodwin, J. G., Jr. (2009). La, V, and Fe promotion of Rh/SiO2 for CO hydrogenation: Effect on adsorption and reaction. Journal of Catalysis, 267(2), 167–176. doi:10.1016/j.jcat.2009.08.007 doi:10.1016/j.jcat.2009.08.007

Mochida, I., & Choi, K. (2004). An overview of hydrodesulfurization and hydrodenitrogenation. *Journal of the Japan Petroleum Institute*, *47*(3), 145–163. doi:10.1627/jpi.47.145

Mochizuki, Y., & Sugawara, K. (2008). Removal of organic sulfur from hydrocarbon resources using ionic liquids. *Energy & Fuels*, *22*(5), 3303–3307. doi:10.1021/ef800400k

Moiseev, I. I., Vargaftic, M. N., & Syrkin, Y. K. (1960). Mechanism of the reaction of palladium salts with olefins in hydroxyl-containing solvents. *Doklady Akademii Nauk SSSR*, *133*, 377–380.

Molinari, R., Poerio, T., & Argurio, P. (2006). One-step production of phenol by selective oxidation of benzene in a biphasic system. *Catalysis Today*, *118*(1-2), 52–56. doi:10.1016/j.cattod.2005.11.089

Mondelli, C., Ferri, D., Grunwaldt, J.-D., Krumeich, F., Mangold, S., Psaro, R., & Baiker, A. (2007). Combined liquid-phase ATR-IR and XAS study of the Bi-promotion in the aerobic oxidation of benzyl alcohol over Pd/Al_2O_3. *Journal of Catalysis*, *252*(1), 77–87. doi:10.1016/j.jcat.2007.09.013

Morales, E., & Lunsford, J. H. (1989). Oxidative dehydrogenation of ethane over a lithium-promoted magnesium oxide catalyst. *Journal of Catalysis*, *118*(1), 255–265. doi:10.1016/0021-9517(89)90315-1

Mortazavi, A., Somers, C., Hwang, Y., Radermacher, R., Rodgers, P., & Al-Hashimi, S. (2012). Performance enhancement of propane pre-cooled mixed refrigerant LNG plant. *Applied Energy*, *93*, 125–131. doi:10.1016/j.apenergy.2011.05.009

Mouaddib, N., Perrichon, V., & Martin, G. A. (1994). Characterization of copper-cobalt catalysts for alcohol synthesis from syngas. Applied Catalysis A, General, 118(1), 63–72. doi:10.1016/0926-860X(94)80089-8 doi:10.1016/0926-860X(94)80089-8

Movahed, S. K., Lehi, N. F., & Dabiri, M. (2014). Gold nanoparticle supported on ionic liquid-modified graphene oxide as an efficient and recyclable catalyst for one-pot oxidative A3-coupling reaction of benzyl alcohols. *RSC Advances*, *4*(79), 42155–42158. doi:10.1039/C4RA05190D

Muramatsu, A., Tatsumi, T., & Tominaga, H.-. (1987). Mixed Alcohol Synthesis from CO/H2 by Use of KCl-Promoted Mo/SiO2 Catalysts. Bulletin of the Chemical Society of Japan, 60(9), 3157–3161. doi:10.1246/bcsj.60.3157 doi:10.1246/bcsj.60.3157

Murrell, J. N. L., & Laidler, K. J. (1968). Symmetries of activated complexes. Faraday Transactions, 64, 371. doi:10.1039/tf9686400371 doi:10.1039/tf9686400371

Nakagawa, Y., & Tomishige, K. (2010). Total hydrogenation of furan derivatives over silica-supported Ni-Pd alloy catalyst. *Catalysis Communications*, *12*(3), 154–156. doi:10.1016/j.catcom.2010.09.003

Nakagawa, Y., & Tomishige, K. (2011). Heterogeneous catalysis of the glycerol hydrogenolysis. *Catalysis Science & Technology*, *1*(2), 179–190. doi:10.1039/c0cy00054j

Nakamura, S., & Yasui, T. (1970). Mechanism of the palladium-catalyzed synthesis of vinyl acetate from ethylene in a heterogeneous gas reaction. *Journal of Catalysis*, *17*(3), 366–374. doi:10.1016/0021-9517(70)90113-2

Naraschewski, F. N., Praveen Kumar, C., Jentys, A., & Lercher, J. A. (2011). Phase formation and selective oxidation of propane over MoVTeNbOx catalysts with varying compositions. *Applied Catalysis A, General*, *391*(1-2), 63–69. doi:10.1016/j.apcata.2010.07.005

Naumann d'Alnoncourt, R., Csepei, L.-I., Hävecker, M., Girgsdies, F., Schuster, M. E., Schlögl, R., & Trunschke, A. (2014). The reaction network in propane oxidation over phase-pure MoVTeNb M1 oxide catalysts. *Journal of Catalysis*, *311*, 369–385. doi:10.1016/j.jcat.2013.12.008

Nava, J. A. O., & Krishna, R. (2004). In-situ stripping of H2S in gasoil hydrodesulphurization reactor design considerations. *Chem. Eng. Res. Des.*, 82(A2), 208-214. doi:10.1205/026387604772992783

Nava, R., Ortega, R. A., Alonso, G., Ornelas, C., Pawelec, B., & Fierro, J. L. G. (2007). CoMo/Ti-SBA-15 catalysts for dibenzothiophene desulfurization. *Catal. Today*, 127(1-4), 70-84. doi:10.1016/j.cattod.2007.02.034

Navarrete, C., Garcia, R., Sepulveda, C., Gil-Llambias, F. J., Fierro, J. L. G., & Escalona, N. (2011). Deep Hydrodesulphurization via Hydrogen Spillover. *Catal. Lett.*, 141(12), 1796-1802. doi:10.1007/s10562-011-0717-9

Ngamcharussrivichai, C., Chatratananon, C., Nuntang, S., & Prasassarakich, P. (2008). Adsorptive removal of thiophene and benzothiophene over zeolites from Mae Moh coal fly ash. *Fuel*, *87*(10-11), 2347–2351. doi:10.1016/j.fuel.2007.10.003

Nibbelke, R., Kauffman, S., & Pek, B. (2002, July8). Double mixed refrigerant LNG process provides viable alternative for tropocal conditions. *Oil & Gas Journal*, *100*(27), 64.

Nicolau, I., Broussard, J. A., & Colling, P. M. (1999). *U.S. Patent No. 5,948,724.* Washington, DC: U.S. Patent and Trademark Office.

Nielsen, P. E. H., Temel, B., & Beato, P. (2014). Process for the preparation of ethanol and higher alcohols. Google Patents.

Nieto, J. M. L., Botella, P., Vázquezb, M. I., & Dejoz, A. (2002). The selective oxidative dehydrogenation of ethane over hydrothermally synthesized MoVTeNb catalysts. *Chemical Communications*, 2002, 1906–1907. doi:10.1039/b204037a

Nikulshin, P. A., Mozhaev, A. V., Pimerzin, A. A., Konovalov, V. V., & Pimerzin, A. A. (2012). CoMo/Al2O3 catalysts prepared on the basis of Co2Mo10-heteropolyacid and cobalt citrate: Effect of Co/Mo ratio. *Fuel*, 100, 24-33. doi:10.1016/j.fuel.2011.11.028

Nishiyama, S., Kubota, T., Kimura, K., Tsuruya, S., & Masai, M. (1997). Unique hydrogenation activity of supported tin catalyst: Selective hydrogenation catalyst for unsaturated aldehydes. *Journal of Molecular Catalysis a-Chemical, 120*(1-3), L17-L22. doi:10.1016/s1381-1169(97)00049-6

Niu, J., Deng, J., Liu, W., Zhang, L., Wang, G., Dai, H., & Zi, X. et al. (2007). Nanosized perovskite-type oxides La1-xSrxMO3-δ (M = Co, Mn; x = 0, 0.4) for the catalytic removal of ethylacetate. *Catalysis Today*, *126*(3-4), 420–429. doi:10.1016/j.cattod.2007.06.027

Niwa, S., Eswaramoorthy, M., Jalajakumari, N., Anuj, R., Naotsugu, I., Hiroshi, S., & Fuji, M. et al. (2002). A one-step conversion of benzene to phenol with a palladium membrane. *Science*, *295*(5552), 105–107. doi:10.1126/science.1066527 PMID:11778042

Nobuhiro, M. (2006). BOG reliquefaction with cold energy storage and utilization of LNG cold energy using cascade process. *Energy Resource*, 334-336.

Noureldin, M. B. (2011). *Pinch Technology and Beyond: New Vistas on Energy Efficiency Optimization*. Nova Science Publishers.

Noureldin, M. M. B., & El-Halwagi, M. M. (2015). Synthesis of C-H-O Symbiosis Networks. *AIChE Journal. American Institute of Chemical Engineers*, *61*(4), 1242–1262. doi:10.1002/aic.14714

Novomer. (2014). Retrieved from http://www.novomer.com/novomer-co-process

Nunan, J. G., Bogdan, C. E., Klier, K., Smith, K. J., Young, C.-W., & Herman, R. G. (1988). Methanol and C2 oxygenate synthesis over cesium doped CuZnO and Cu/ZnO/Al2O3 catalysts: A study of selectivity and 13C incorporation patterns. Journal of Catalysis, 113(2), 410–433. doi:10.1016/0021-9517(88)90268-0 doi:10.1016/0021-9517(88)90268-0

Nunan, J. G., Herman, R. G., & Klier, K. (1989). Higher alcohol and oxygenate synthesis over Cs/Cu/ZnO/M2O3 (M = Al, Cr) catalysts. Journal of Catalysis, 116(1), 222–229. doi:10.1016/0021-9517(89)90087-0 doi:10.1016/0021-9517(89)90087-0

Nunan, J., Bogdan, C., Herman, R., & Klier, K. (1989). Efficient carbon-carbon bond formation in ethanol homologation by CO/H2 with specific C1 oxygen retention over Cs/Cu/ZnO catalysts. Catalysis Letters, 2(1), 49–55. doi:10.1007/BF00765330 doi:10.1007/BF00765330

Nunan, J., Klier, K., Young, C.-W., Himelfarb, P. B., & Herman, R. G. (1986). Promotion of methanol synthesis over Cu/ZnO catalysts by doping with caesium. Journal of the Chemical Society, Chemical Communications, 1986(3), 193–195. doi:10.1039/c39860000193 doi:10.1039/c39860000193

Ockwig, N. W., & Nenoff, T. M. (2007). Membranes for hydrogen separation. *Chemical Reviews, 107*(10), 4078–4110. doi:10.1021/cr0501792 PMID:17927157

O'Donnell, P. M., & Wagener, K. B. (2003). Graft Copolymers by Acyclic Diene Metathesis and Atom Transfer Radical Polymerization Techniques. *Journal of Polymer Science. Part A, Polymer Chemistry, 41*, 2816–2827.

Ojeda, M., Granados, M. L., Rojas, S., Terreros, P., García-García, F. J., & Fierro, J. L. G. (2004). Manganese-promoted Rh/Al2O3 for C2-oxygenates synthesis from syngas: Effect of manganese loading. Applied Catalysis A, General, 261(1), 47–55. doi:10.1016/j.apcata.2003.10.033 doi:10.1016/j.apcata.2003.10.033

Ojeda, M., Rojas, S., Boutonnet, M., Pérez-Alonso, F. J., Javier García-García, F., & Fierro, J. L. G. (2004). Synthesis of Rh nano-particles by the microemulsion technology: Particle size effect on the CO + H2 reaction. Applied Catalysis A, General, 274(1–2), 33–41. doi:10.1016/j.apcata.2004.05.014 doi:10.1016/j.apcata.2004.05.014

Okazaki, K.-i., Kiyama, T., Hirahara, K., Tanaka, N., Kuwabata, S., & Torimoto, T. (2008). Single-step synthesis of gold–silver alloy nanoparticles in ionic liquids by a sputter deposition technique. *Chemical Communications*(6), 691-693.

Old, D. W., Wolfe, J. P., & Buchwald, S. L. (1998). Highly Active Catalyst for Palladium-Catalyzed Cross-Coupling Reactions: Room-Temperature Suzuki Couplings and Amination of Unactivated Aryl Chlorides. *Journal of the American Chemical Society, 120*(37), 9722–9723. doi:10.1021/ja982250+

Onda, A., Ogo, S., & Yanagisawa, K. (2011). Catalyst and Alcohol Synthesis Method. Google Patents.

Opper, K. L., Fassbender, B., Brunklaus, G., Spiess, H. W., & Wagener, K. B. (2009). Polyethylene Functionalized with Precisely Spaced Phosphonic Acid Groups. *Macromolecules (Washington, DC, U. S.), 42*(13), 4407–4409. doi:10.1021/ma900884f

Opper, K. L., Markova, D., Klapper, M., Mullen, K., & Wagener, K. B. (2010). Precision Phosphonic Acid Functionalized Polyolefin Architectures. *Macromolecules (Washington, DC, U. S.), 43*(8), 3690–3698. doi:10.1021/ma902659y

Opper, K. L., & Wagener, K. B. (2011). ADMET: Metathesis polycondensation. *Journal of Polymer Science. Part A, Polymer Chemistry, 49*(4), 821–831. doi:10.1002/pola.24491

Ortmann, P., Heckler, I., & Mecking, S. (2014). Physical properties and hydrolytic degradability of polyethylene-like polyacetals and polycarbonates. *Green Chemistry, 16*(4), 1816–1827. doi:10.1039/c3gc42592d

Ortmann, P., Trzaskowski, J., Krumova, M., & Mecking, S. (2013). Precise Microstructure Self-Stabilized Polymer Nanocrystals. *ACS Macro Lett., 2*(2), 125–127. doi:10.1021/mz300606n

Otsuki, S., Nonaka, T., Takashima, N., Qian, W., Ishihara, A., Imai, T., & Kabe, T. (2000). Oxidative desulfurization of light gas oil and vacuum gas oil by oxidation and solvent extraction. *Energy & Fuels, 14*(6), 1232–1239. doi:10.1021/ef000096i

Ott, L. S., Campbell, S., Seddon, K. R., & Finke, R. G. (2007). Evidence that imidazolium-based ionic ligands can be metal(0)/nanocluster catalyst poisons in at least the test case of iridium(0)-catalyzed acetone hydrogenation. *Inorganic Chemistry, 46*(24), 10335–10344. doi:10.1021/ic700976z PMID:17975891

Ovsitser, O., & Kondratenko, E. V. (2009). Similarity and differences in the oxidative dehydrogenation of C2-C4 alkanes over nano-sized VOx species using N_2O and O_2. *Catalysis Today, 142*(3-4), 138–142. doi:10.1016/j.cattod.2008.09.012

Oyama, S. T., Clark, P., Wang, X., Shido, T., Iwasawa, Y., Hayashi, S., & Requejo, F. G. et al. (2002). Structural characterization of tungsten Phosphide (WP) hydrotreating catalysts by X-ray absorption spectroscopy and nuclear magnetic resonance spectroscopy. *The Journal of Physical Chemistry B, 106*(8), 1913–1920. doi:10.1021/jp0136056

Oyama, S. T., Gott, T., Zhao, H., & Lee, Y. (2009). Transition metal phosphide hydroprocessing catalysts: A review. *Catalysis Today*, *143*(1-2), 94–107. doi:10.1016/j.cattod.2008.09.019

Oyama, S. T., & Lee, Y. K. (2008). The active site of nickel phosphide catalysts for the hydrodesulfurization of 4,6-DMDBT. *Journal of Catalysis*, *258*(2), 393–400. doi:10.1016/j.jcat.2008.06.023

Ozkan, U. S., & Watson, R. B. (2000). *US Patent No. US6521808(B1).*

Pagliaro, M., Ciriminna, R., Kimura, H., Rossi, M., & Della Pina, C. (2007). From glycerol to value-added products. *Angewandte Chemie International Edition*, *46*(24), 4434–4440. doi:10.1002/anie.200604694 PMID:17471485

Palladium Alliance International, (2012). *Stillwater Palladium.* Retrieved from http://stillwaterpalladium.com/

Pan, X. L., Fan, Z. L., Chen, W., Ding, Y. J., Luo, H. Y., & Bao, X. H. (2007). Enhanced ethanol production inside carbon-nanotube reactors containing catalytic particles. Nature Materials, 6(7), 507–511. doi:10.1038/nmat1916 PubMed doi:10.1038/nmat1916 PMID:17515914

Pan, X., & Bao, X. (2011). The Effects of Confinement inside Carbon Nanotubes on Catalysis. Accounts of Chemical Research, 44(8), 553–562. doi:10.1021/ar100160t PubMed doi:10.1021/ar100160t PMID:21707038

París, R. S., Lopez, L., Barrientos, J., Pardo, F., Boutonneta, M., & Järås, S. G. (2015). Catalytic conversion of biomass-derived synthesis gas to fuels. Catalysis, 62-143.

Parkhurst, R. R., Balog, S., Weder, C., & Simon, Y. C. (2014). Synthesis of poly(sulfonate ester)s by ADMET polymerization. *RSC Advances*, *4*(96), 53967–53974. doi:10.1039/C4RA08788G

Parvulescu, V. I., & Hardacre, C. (2007). Catalysis in ionic liquids. *Chemical Reviews*, *107*(6), 2615–2665. doi:10.1021/cr050948h PMID:17518502

Pawelec, B., Fierro, J., Montesinos, A., & Zepeda, T. (2008). Influence of the acidity of nanostructured CoMo/P/Ti-HMS catalysts on the HDS of 4,6-DMDBT reaction pathways. *Applied Catalysis B: Environmental*, *80*(1-2), 1–14. doi:10.1016/j.apcatb.2007.10.039

Pedernera, E., Reimert, R., Nguyen, N. L., & van Buren, V. (2003). Deep Desulphurization of middle distillates: Process adaptation to oil fractions' compositions. *Catalysis Today*, *79*, 371–381. doi:10.1016/S0920-5861(03)00066-X

Perosa, A., & Tundo, P. (2005). Selective hydrogenolysis of glycerol with Raney nickel. *Industrial & Engineering Chemistry Research*, *44*(23), 8535–8537. doi:10.1021/ie0489251

Petkovic, M., Seddon, K. R., Rebelo, L. P. N., & Silva Pereira, C. (2011). Ionic liquids: A pathway to environmental acceptability. *Chemical Society Reviews*, *40*(3), 1383–1403. doi:10.1039/C004968A PMID:21116514

Petkovic, L. M., Utgikar, V., & Rashkeev, S. N. (2011). Structural stability and catalytic activity of lanthanum-based perovskites. *The Journal of Physical Chemistry C*, *115*(17), 8709–8715. doi:10.1021/jp112422m

Phillips, M. J., & Ternan, M., & Editors. (1988). Catalysis: Theory to Practice. Proceedings of the 9th International Congress on Catalysis, Calgary.

Piringer, O. G., Baner, A. L., (Eds.), (2008). *Plastic Packaging; Interactions with Food and Pharmaceuticals.*

Pirngruber, G. D., Seshan, K., & Lercher, J. A. (2000). Dehydroisomerization of *n*-butane over Pt-ZSM5: II. Kinetic and Thermodynamic Aspects. *Journal of Catalysis*, *190*(2), 338–351. doi:10.1006/jcat.1999.2753

Piroux, F., Espuche, E., Mercier, R., Pinéri, M., & Gebel, G. (2002). Gas transport mechanism in sulfonated polyimides. Consequences on gas selectivity. *Journal of Membrane Science*, *209*(1), 241–253. doi:10.1016/S0376-7388(02)00350-2

Plata, J. J., García-Mota, M., Braga, A. A. C., López, N., & Maseras, F. (2009). Vinyl Acetate Synthesis on Homogeneous and Heterogeneous Pd-Based Catalysts: A Theoretical Analysis on the Reaction Mechanisms. *The Journal of Physical Chemistry A*, *113*(43), 11758–11762. doi:10.1021/jp902929x PMID:19769362

Poirier, M. G., & Sapundzhiev, C. (1997). Catalytic decomposition of natural gas to hydrogen for fuel cell applications. *International Journal of Hydrogen Energy*, *22*(4), 429–433. doi:10.1016/S0360-3199(96)00101-2

Porras-Vazquez, J. M., & Slater, P. R. (2012). Synthesis of oxyanion-doped barium strontium cobalt ferrites: Stabilization of the cubic perovskite and enhancement in conductivity. *Journal of Power Sources*, *209*, 180–183. doi:10.1016/j.jpowsour.2012.02.105

Porta, F., & Prati, L. (2004). Selective oxidation of glycerol to sodium glycerate with gold-on-carbon catalyst: An insight into reaction selectivity. *Journal of Catalysis*, *224*(2), 397–403. doi:10.1016/j.jcat.2004.03.009

Prati, L., & Rossi, M. (1998). Gold on carbon as a new catalyst for selective liquid phase oxidation of diols. *Journal of Catalysis*, *176*(2), 552–560. doi:10.1006/jcat.1998.2078

Prati, L., & Villa, A. (2014). Gold Colloids: From Quasi-Homogeneous to Heterogeneous Catalytic Systems. *Accounts of Chemical Research*, *47*(3), 855–863. doi:10.1021/ar400170j PMID:24266851

Prati, L., Villa, A., Chan-Thaw, C. E., Arrigo, R., Wang, D., & Su, D. S. (2011). Gold catalyzed liquid phase oxidation of alcohol: The issue of selectivity. *Faraday Discussions*, *152*, 353–365. doi:10.1039/c1fd00016k PMID:22455055

Prati, L., Villa, A., Porta, F., Wang, D., & Su, D. S. (2007). Single-phase gold/palladium catalyst: The nature of synergistic effect. *Catalysis Today*, *122*(3-4), 386–390. doi:10.1016/j.cattod.2006.11.003

Prieto, G., Beijer, S., Smith, M. L., He, M., Au, Y., Wang, Z., & de Jongh, P. E. (2014). Design and Synthesis of Copper–Cobalt Catalysts for the Selective Conversion of Synthesis Gas to Ethanol and Higher Alcohols. Angewandte Chemie International Edition, 53(25), 6397–6401. doi:10.1002/anie.201402680 PubMed doi:10.1002/anie.201402680 PMID:24827541

Pryjomska-Ray, I., Gniewek, A., Trzeciak, A. M., Ziółkowski, J. J., & Tylus, W. (2006). Homogeneous/heterogeneous palladium based catalytic system for Heck reaction. The reversible transfer of palladium between solution and support. *Topics in Catalysis*, *40*(1-4), 173–184. doi:10.1007/s11244-006-0119-1

Pushpinder, P. (2011). Commercial Applications of Membranes in Gas Separations. In E. Drioli & G. Barbieri (Eds.), *Membrane Engineering for the treatment of gases* (pp. 215–244). Cambridge, United Kingdom: The Royal Society of Chemistry.

Qian, K., & Huang, W. (2011). Au–Pd alloying-promoted thermal decomposition of PdO supported on SiO_2 and its effect on the catalytic performance in CO oxidation. *Catalysis Today*, *164*(1), 320–324. doi:10.1016/j.cattod.2010.10.018

Qiao, A., Kalevaru, V. N., Sri Hari Kumar, A., Lingaiah, N., Sai Prasad, P. S., & Martin, A. (2012). Effect of CO2-admixture on the catalytic performance of Ni-Nb-M-O catalysts in oxidative dehydrogenation of ethane to ethylene. DGMK Tagungsbericht, 3, 97-103.

Qiao, A., Kalevaru, V. N., Radnik, J., Duevel, A., Heitjans, P., Kumar, A. S. H., & Martin, A. et al. (2014). Oxidative Dehydrogenation of Ethane to Ethylene over V_2O_5/Al_2O_3 Catalysts: Effect of Source of Alumina on the Catalytic Performance. *Industrial & Engineering Chemistry Research*, *53*(49), 18711–18721. doi:10.1021/ie5008344

Qiao, A., Kalevaru, V. N., Radnik, J., Srihari Kumar, A., Lingaiah, N., Sai Prasad, P. S., & Martin, A. (2013a). Oxidative dehydrogenation of ethane to ethylene over V_2O_5/Nb_2O_5 catalysts. *Catalysis Communications*, *30*, 45–50. doi:10.1016/j.catcom.2012.10.018

Qin, Y., Song, Y., Sun, N., Zhao, N., Li, M., & Qi, L. (2008). Ionic Liquid-Assisted Growth of Single-Crystalline Dendritic Gold Nanostructures with a Three-Fold Symmetry. *Chemistry of Materials*, *20*(12), 3965–3972. doi:10.1021/cm8002386

Qu, X. Y., Dong, H., Zhou, Z. J., Zhang, L., & Chen, H. L. (2010). Pervaporation Separation of Xylene Isomers by Hybrid Membranes of PAAS Filled with Silane-Modified Zeolite. *Industrial & Engineering Chemistry Research*, *49*(16), 7504–7514. doi:10.1021/ie100256u

Rabarihoela-Rakotovao, V., Brunet, S., Perot, G., & Diehl, F. (2006). Effect of H_2S partial pressure on the HDS of dibenzothiophene and 4,6-dimethyldibenzothiophene over sulfided NiMoP/Al_2O_3 and CoMoP/Al_2O_3 catalysts. *Applied Catalysis A, General*, *306*, 34–44. doi:10.1016/j.apcata.2006.03.029

Radnik, J., Benhmid, A., Kalevaru, V. N., Pohl, M.-M., Martin, A., Lücke, B., & Dingerdissen, U. (2005). Deactivation of Pd acetoxylation catalysts: Direct observations by XPS investigations. *Angewandte Chemie International Edition*, *44*(41), 6771–6774. doi:10.1002/anie.200501650 PMID:16187392

Radnik, J., Pohl, M.-M., Kalevaru, V. N., & Martin, A. (2007). First knowledge on the formation of novel core-shell structures in PdCu catalysts and their Influence on the prevention of catalyst deactivation. *The Journal of Physical Chemistry C*, *111*(28), 10166–10169. doi:10.1021/jp073184d

Rahman, F., Loughlin, K. F., Al-Saleh, M. A., Saeed, M. R., Tukur, N. M., Hossain, M. M., & Mamedov, A. et al. (2010). Kinetics and mechanism of partial oxidation of ethane to ethylene and acetic acid over MoV type catalysts. *Applied Catalysis A, General*, *375*(1), 17–25. doi:10.1016/j.apcata.2009.11.026

Rahmat, N., Abdullah, A. Z., & Mohamed, A. R. (2010). Recent progress on innovative and potential technologies for glycerol transformation into fuel additives: A critical review. *Renewable & Sustainable Energy Reviews*, *14*(3), 987–1000. doi:10.1016/j.rser.2009.11.010

Ramayya, S., Brittain, A., DeAlmeida, C., Mok, W., & Antal, M. J., Jr. (1987). Acid-catalysed dehydration of alcohols in supercritical water. Fuel, 66(10), 1364–1371. doi:10.1016/0016-2361(87)90183-9 doi:10.1016/0016-2361(87)90183-9

Ramirez, J., Macias, G., Cedeno, L., Gutierrez-Alejandre, A., Cuevas, R., & Castillo, P. (2004). The role of titania in supported Mo, CoMo, NiMo, and NiW hydrodesulfurization catalysts: analysis of past and new evidences. *Catal. Today*, 98(1-2), 19-30. doi:10.1016/j.cattod.2004.07.050

Rampino, L. D., Kavanagh, K. E., & Nord, F. F. (1943). Relationship between particle size and efficiency of palladium polyvinyl alcohol (Pd PVA) catalysts. *Proceedings of the National Academy of Sciences of the United States of America*, *29*(8), 246–256. doi:10.1073/pnas.29.8.246 PMID:16578085

Ranhotra, G. S., Bell, A. T., & Reimer, J. A. (1987). Catalysis over molybdenum carbides and nitrides: II. Studies of CO hydrogenation and C2H6 hydrogenolysis. Journal of Catalysis, 108(1), 40–49. doi:10.1016/0021-9517(87)90153-9 doi:10.1016/0021-9517(87)90153-9

Rashidi, F., Sasaki, T., Rashidi, A. M., Nemati Kharat, A., & Jozani, K. J. (2013). Ultradeep hydrodesulfurization of diesel fuels using highly efficient nanoalumina-supported catalysts: Impact of support, phosphorus, and/or boron on the structure and catalytic activity. *J. Catal.*, 299, 321-335. doi:10.1016/j.jcat.2012.11.012

Ravanchi, M. T., Kaghazchi, T., & Kargari, A. (2009). Application of membrane separation processes in petrochemical industry: A review. *Desalination*, *235*(1-3), 199–244. doi:10.1016/j.desal.2007.10.042

Redel, E., Thomann, R., & Janiak, C. (2008). Use of ionic liquids (ILs) for the IL-anion size-dependent formation of Cr, Mo and W nanoparticles from metal carbonyl M(CO)6 precursors. *Chemical Communications*, (15): 1789–1791. doi:10.1039/b718055a PMID:18379694

Reining, S., Kondratenko, E. V., Kalevaru, V. N., & Martin, A. (2014). Reaction pathways in the palladium-catalyzed acetoxylation of toluene. DGMK Tagungsbericht, 3, 51-56.

Renaudin, G. (1995). Improvement of natural gas liquefaction processes by using liquid turbines.*Proceedings of the Eleventh International Conference on Liquefied Natural Gas*. Chicago, Institute of Gas Technology.

Reubroycharoen, P., Yoneyama, Y., Vitidsant, T., & Tsubaki, N. (2003). Catalytically active supercritical fluid to accelerate methanol synthesis. Fuel, 82(18), 2255–2257. doi:10.1016/S0016-2361(03)00208-4 doi:10.1016/S0016-2361(03)00208-4

Richard, F., Boita, T., & Perot, G. (2007). Reaction mechanism of 4,6-dimethyldibenzothiophene desulfurization over sulfided NiMoP/Al_2O_3-zeolite catalysts. *Applied Catalysis A, General*, *320*, 69–79. doi:10.1016/j.apcata.2006.12.014

Rivalta, I., Mazzone, G., Russo, N., & Sicilia, E. (2010). The influence of surface oxygen and hydroxyl groups on the dehydrogenation of ethylene on PdAu surface alloys. A theoretical cluster model study. *Chemical Physics Letters*, *493*(1-3), 87–93. doi:10.1016/j.cplett.2010.05.033

Robinson, W. R. A. M., van Veen, J. A. R., de Beer, V. H. J., & van Santen, R. A. (1999). Development of deep hydrodesulfurization catalysts I. CoMo and NiMo catalysts tested with (substituted) dibenzothiophene. *Fuel Process. Technol.*, 61(1-2), 89-101. doi:10.1016/S0378-3820(99)00033-8

Rodríguez-Cabo, B., Rodríguez, H., Rodil, E., Arce, A., & Soto, A. (2014). Extractive and oxidative-extractive desulfurization of fuels with ionic liquids. *Fuel*, *117*, 882–889. doi:10.1016/j.fuel.2013.10.012

Rodriguez, J. A., Kim, J.-Y., Hanson, J. C., Sawhill, S. J., & Bussell, M. E. (2003). Physical and chemical properties of MoP, Ni_2P, and MoNiP hydrodesulfurization catalysts: Time-resolved X-ray diffraction, density functional, and hydrodesulfurization activity studies. *The Journal of Physical Chemistry B*, *107*(26), 6276–6285. doi:10.1021/jp022639q

Rodriguez, M. A., Elizalde, I., & Ancheyta, J. (2012). Comparison of kinetic and reactor models to simulate a trickle-bed bench-scale reactor for hydrodesulfurization of VGO. *Fuel*, *100*, 91–99. doi:10.1016/j.fuel.2012.01.061

Rodriguez, M. L., Ardissone, D. E., Heracleous, E., Lemonidou, A. A., Lopez, E., Pedernera, M. N., & Borio, D. O. (2010). Oxidative dehydrogenation of ethane to ethylene in a membrane reactor: A theoretical study. *Catalysis Today*, *157*(1-4), 303–309. doi:10.1016/j.cattod.2010.01.053

Rodriguez, M. L., Ardissone, D. E., Lemonidou, A. A., Heracleous, E., Lopez, E., Pedernera, M. N., & Borio, D. O. (2009). Simulation of a membrane reactor for the catalytic oxydehydrogenation of ethane. *Industrial & Engineering Chemistry Research*, *48*(3), 1090–1095. doi:10.1021/ie800564v

Rojas, G., Inci, B., Wei, Y., & Wagener, K. B. (2009). Precision Polyethylene: Changes in Morphology as a Function of Alkyl Branch Size. *Journal of the American Chemical Society*, *131*(47), 17376–17386. doi:10.1021/ja907521p PMID:19891434

Romanow-Garcia, S., & Hoffman, H. L. (2007). Petroleum and its products. In J.A. Kent & Riegel (Eds.), Handbook of industrial chemistry and biotechnology (pp. 801-842). New York: Springer. doi:10.1007/978-0-387-27843-8_18

Rossi, L. M., Vono, L. L. R., Garcia, M. A. S., Faria, T. L. T., & Lopez-Sanchez, J. A. (2013). Screening of Soluble Rhodium Nanoparticles as Precursor for Highly Active Hydrogenation Catalysts: The Effect of the Stabilizing Agents. *Topics in Catalysis*, *56*(13-14), 1228–1238. doi:10.1007/s11244-013-0089-z

Rossiter, A. P. (2010). Improve Energy Efficiency via Heat Integration. *Chemical Engineering Progress*, *106*(12), 33–42.

Rostrup-Nielsen, J. R. (2001). Conversion of hydrocarbons and alcohols for fuel cells. *Physical Chemistry Chemical Physics*, *3*(3), 283–288. doi:10.1039/b004660o

Roussel, M., Barama, S., Lofberg, A., Al-Sayari, S., Karim, K., & Bordes-Richard, E. (2009). MoV-based catalysts in ethane oxidation to acetic acid: Influence of additives on redox chemistry. *Catalysis Today*, *141*(3-4), 288–293. doi:10.1016/j.cattod.2008.04.046

Roussel, M., Bouchard, M., Karim, K., Al-Sayari, S., & Bordes-Richard, E. (2006). MoVO-based catalysts for the oxidation of ethane to ethylene and acetic acid. Influence of niobium and/or palladium on physicochemical and catalytic properties. *Applied Catalysis A, General*, *308*, 62–74. doi:10.1016/j.apcata.2006.04.017

Ruettinger, W., Benderly, A., Han, S., Shen, X., Ding, Y., & Suib, S. L. (2011). Influence of Support Surface Area and Niobium Addition on the Reactivity of Vanadium Catalysts for Propane Oxidative Dehydrogenation. *Catalysis Letters*, *141*(1), 15–21. doi:10.1007/s10562-010-0459-0

Ruppert, A. M., Weinberg, K., & Palkovits, R. (2012). Hydrogenolysis Goes Bio: From Carbohydrates and Sugar Alcohols to Platform Chemicals. *Angewandte Chemie International Edition*, *51*(11), 2564–2601. doi:10.1002/anie.201105125 PMID:22374680

Ryu, H. J., Sanchez, L., Keul, H. A., Raj, A., & Bockstaller, M. R. (2008). Imidazolium-based ionic liquids as efficient shape-regulating solvents for the synthesis of gold nanorods. *Angewandte Chemie International Edition in English*, *47*(40), 7639–7643. doi:10.1002/anie.200802185 PMID:18712726

Safety History. (2012, February 8). CH. IV International. Retrieved from [REMOVED HYPERLINK FIELD]http://www.ch-iv.com/assets/documents/safety-history.pdf

Samanos, B., Boutry, P., & Montarnal, R. (1971). The mechanism of vinyl acetate formation by gas-phase catalytic ethylene acetoxidation. *Journal of Catalysis*, *23*(1), 19–30. doi:10.1016/0021-9517(71)90019-4

Sanders, E., Clark, D. O., Jensvold, J. A., Beck, H. N., Lipscomb, G. G., & Coan, F. L. (1988). Process for preparing powadir membranes from Tetrahalobisphenol A polycarbonate, *U.S. Patent 4,772,392*.

Sandru, M., Kim, T. J., Capala, W., Huijbers, M., & Hägg, M. B. (2013). Pilot scale testing of polymeric membranes for CO_2 capture from coal fired power plants. *Energy Procedia*, *37*, 6473–6480. doi:10.1016/j.egypro.2013.06.577

Sankaranarayanan, T. M., Ingle, R. H., Gaikwad, T. B., Lokhande, S. K., Raja, T., Devi, R. N., & Manikandan, P. et al. (2008). Selective Oxidation of Ethane Over Mo–V–Al–O Oxide Catalysts: Insight to the Factors Affecting the Selectivity of Ethylene and Acetic Acid and Structure-activity Correlation Studies. *Catalysis Letters*, *121*(1-2), 39–51. doi:10.1007/s10562-007-9289-0

Santander, J. A., Lopez, E., Tonetto, G. M., & Pedernera, M. N. (2014). Preparation of NiNbO/AISI 430 ferritic stainless steel monoliths for catalytic applications. *Industrial & Engineering Chemistry Research*, *53*(28), 11312–11319. doi:10.1021/ie501884g

Santhanam, N., Conforti, T. A., Spieker, W., & Regalbuto, J. R. (1994). Nature of metal catalyst precursors adsorbed onto oxide supports. *Catal. Today*, 21(1), 141-156. doi:10.1016/0920-5861(94)80040-5

Santos, V. P., van der Linden, B., Chojecki, A., Budroni, G., Corthals, S., Shibata, H., & Gascon, J. (2013). Mechanistic Insight into the Synthesis of Higher Alcohols from Syngas: The Role of K Promotion on MoS2 Catalysts. ACS Catalysis, 3(7), 1634–1637. doi:10.1021/cs4003518 doi:10.1021/cs4003518

Sarrazin, P., Mouchel, B., & Kasztelan, S. (1989). Molybdenum-95 NMR study of the interaction of heptamolybdate solutions with alumina and silica. *J. Phys. Chem.*, 93(2), 904-908. doi:10.1021/j100339a069

Sathe, V. G., Paranjpe, S. K., Siruguri, V., & Pimpale, A. V. (1998). Novel magnetic phases in La0.7Sr0.3Co1−yFeyO3 (0:0 6 y 6 1:0): a neutron diffraction study. *Journal of Physics Condensed Matter*, *10*, 4045–4055. doi:10.1088/0953-8984/10/18/014

Sato, K., Hamakawa, S., Natsui, M., Nishioka, M., Inoue, T., & Mizukami, F. (2010). Palladium-based bifunctional membrane reactor for one-step conversion of benzene to phenol and cyclohexanone. *Catalysis Today*, *156*(3-4), 276–281. doi:10.1016/j.cattod.2010.04.045

Savova, B., Loridant, S., Filkova, D., & Millet, J. M. M. (2010). Ni-Nb-O catalysts for ethane oxidative dehydrogenation. *Applied Catalysis A, General*, *390*(1-2), 148–157. doi:10.1016/j.apcata.2010.10.004

Scheeren, C. W., Machado, G., Dupont, J., Fichtner, P. F., & Texeira, S. R. (2003). Nanoscale Pt(0) particles prepared in imidazolium room temperature ionic liquids: Synthesis from an organometallic precursor, characterization, and catalytic properties in hydrogenation reactions. *Inorganic Chemistry*, *42*(15), 4738–4742. doi:10.1021/ic034453r PMID:12870966

Scheeren, C. W., Machado, G., Teixeira, S. R., Morais, J., Domingos, J. B., & Dupont, J. (2006). Synthesis and characterization of Pt0 nanoparticles in imidazolium ionic liquids. *The Journal of Physical Chemistry B*, *110*(26), 13011–13020. doi:10.1021/jp0623037 PMID:16805608

Schiestel, T., Kilgus, M., Peter, S., Caspary, K., Wang, H., & Caro, J. (2005). Hollow fibre perovskite membranes for oxygen separation. *Journal of Membrane Science*, *258*(1-2), 1–4. doi:10.1016/j.memsci.2005.03.035

Schlehuber, D., Wessel, E., Singheiser, L., & Markus, T. (2010). Long-term operation of a La0.58Sr0.4Co0.2Fe0.8O3−δ-membrane for oxygen separation. *Journal of Membrane Science*, *351*(1-2), 16–20. doi:10.1016/j.memsci.2010.01.022

Schlogl, R., & Trunschke, A. (2006). *WO Patent No. WO2008068332(A1).*

Schmidt, B. (2004). Catalysis at the interface of ruthenium carbene and ruthenium hydride chemistry: Organometallic aspects and applications to organic synthesis. *European Journal of Organic Chemistry*, *2004*(9), 1865–1880. doi:10.1002/ejoc.200300714

Scholten, J. D., Leal, B. C., & Dupont, J. (2012). Transition Metal Nanoparticle Catalysis in Ionic Liquids. *Acs Catalysis*, *2*(1), 184–200. doi:10.1021/cs200525e

Schrekker, H. S., Gelesky, M. A., Stracke, M. P., Schrekker, C. M., Machado, G., Teixeira, S. R., & Dupont, J. et al. (2007). Disclosure of the imidazolium cation coordination and stabilization mode in ionic liquid stabilized gold(0) nanoparticles. *Journal of Colloid and Interface Science*, *316*(1), 189–195. doi:10.1016/j.jcis.2007.08.018 PMID:17761189

Schrock, R. R., & Hoveyda, A. H. (2003). Molybdenum and tungsten imido alkylidene complexes as efficient olefin-metathesis catalysts. *Angewandte Chemie International Edition*, *42*(38), 4592–4633. doi:10.1002/anie.200300576 PMID:14533149

Schulz, M. D., & Wagener, K. B. (2012, March 25-29). Exploring solvent effects with alternating acyclic diene metathesis (ADMET) polymerization. *Abstracts of Papers, Proceedings of the243rd ACS National Meeting & Exposition, San Diego, CA, USA.*

Schulz, H. (1999). Short history and present trends of Fischer–Tropsch synthesis. *Applied Catalysis A, General*, *186*(1-2), 3–12. doi:10.1016/S0926-860X(99)00160-X

Schulz, M. D., & Wagener, K. B. (2014). Precision Polymers through ADMET Polymerization. *Macromolecular Chemistry and Physics*, *215*(20), 1936–1945. doi:10.1002/macp.201400268

Schunk, S. A., & de Oliveira, A.L. (2008). Acetoxylation of Ethylene. In G. Ertl, H. Knözinger, F. Schüth, J. Weitkamp (Eds.), Handbook of Heterogeneous Catalysis (pp. 3464-3479). Weinheim, Germany: Wiley-VCH Verlag GmbH & Co. KGaA. doi:10.1002/9783527610044.hetcat0176

Schwartz, V., Campos, A., Egbebi, A., Spivey, J. J., & Overbury, S. H. (2011). EXAFS and FT-IR Characterization of Mn and Li Promoted Titania-Supported Rh Catalysts for CO Hydrogenation. ACS Catalysis, 1(10), 1298–1306. doi:10.1021/cs200281g doi:10.1021/cs200281g

Schwartz, V., Oyama, S. T., & Chen, J. G. (2000). Supported bimetallic Nb-Mo carbide: Synthesis, characterization, and reactivity. *The Journal of Physical Chemistry B*, *104*(37), 8800–8806. doi:10.1021/jp0010233

Schwendeman, J. E., & Wagener, K. B. (2002, April 7-11). Precisely placed gem-dimethyl branching in ADMET polyethylene. *Abstracts of Papers, Proceedings of the223rd ACS National Meeting, Orlando, FL, USA.*

Schwerdtel, W. (1968). Vinyl acetate via ethylene. *Hydrocarbon Processing*, *47*, 187–191.

Scirè, S., Burgio, G., Crisafulli, C., & Minicò, S. (2004). Dehydroisomerization of *n*-butane over H-Y zeolite supported Pt and Pt,Sn catalysts. *Applied Catalysis A, General*, *274*(1-2), 151–157. doi:10.1016/j.apcata.2004.06.019

Seitz, M. E., Chan, C. D., Opper, K. L., Baughman, T. W., Wagener, K. B., & Winey, K. I. (2010). Nanoscale Morphology in Precisely Sequenced Poly(ethylene-co-acrylic acid) Zinc Ionomers. *Journal of the American Chemical Society*, *132*(23), 8165–8174. doi:10.1021/ja101991d PMID:20496894

Shabaker, J. W., Huber, G. W., & Dumesic, J. A. (2004). Aqueous-phase reforming of oxygenated hydrocarbons over Sn-modified Ni catalysts. *Journal of Catalysis*, *222*(1), 180–191. doi:10.1016/j.jcat.2003.10.022

Shabaker, J. W., Simonetti, D. A., Cortright, R. D., & Dumesic, J. A. (2005). Sn-modified Ni catalysts for aqueous-phase reforming: Characterization and deactivation studies. *Journal of Catalysis*, *231*(1), 67–76. doi:10.1016/j.jcat.2005.01.019

Shao, Z., Xiong, G., Dong, H., Yang, W., & Lin, L. (2001). Synthesis, oxygen permeation study and membrane performance of a Ba0.5Sr0.5Co0.8Fe0.2O3−δ oxygen-permeable dense ceramic reactor for partial oxidation of methane to syngas. *Separation and Purification Technology*, *25*(1-3), 97–116. doi:10.1016/S1383-5866(01)00095-8

Shearouse, W. C., Lillie, L. M., Reineke, T. M., & Tolman, W. B. (2015). Sustainable Polyesters Derived from Glucose and Castor Oil: Building Block Structure Impacts Properties. *ACS Macro Lett.*, Ahead of Print. doi:10.1021/acsmacrolett.5b00099

Sheffer, G. R., & King, T. S. (1989). Differences in the promotional effect of the group IA elements on unsupported copper catalysts for carbon monoxide hydrogenation. Journal of Catalysis, 116(2), 488–497. doi:10.1016/0021-9517(89)90115-2 doi:10.1016/0021-9517(89)90115-2

Sheffer, G. R., Jacobson, R. A., & King, T. S. (1989). Chemical nature of alkali-promoted copper-cobalt-chromium oxide higher alcohol catalysts. Journal of Catalysis, 116(1), 95–107. doi:10.1016/0021-9517(89)90078-X doi:10.1016/0021-9517(89)90078-X

Sheldon, R. A. (2014). Green and sustainable manufacture of chemicals from biomass: State of the art. *Green Chemistry*, *16*(3), 950–963. doi:10.1039/C3GC41935E

Shen, J., Ziaei-Azad, H., & Semagina, N. (2014). Is it always necessary to remove a metal nanoparticle stabilizer before catalysis? *Journal of Molecular Catalysis a-Chemical, 391*, 36-40. doi:10.1016/j.molcata.2014.03.027

Shin, J. F., Orera, A., Apperley, D. C., & Slater, P. R. (2011). Oxyanion doping strategies to enhance the ionic conductivity in Ba2In2O5. *Journal of Materials Chemistry*, *21*(3), 874–879. doi:10.1039/C0JM01978J

Shinmi, Y., Koso, S., Kubota, T., Nakagawa, Y., & Tomishige, K. (2010). Modification of Rh/SiO2 catalyst for the hydrogenolysis of glycerol in water. *Applied Catalysis B: Environmental*, *94*(3-4), 318–326. doi:10.1016/j.apcatb.2009.11.021

Shi, X., Ji, S., Wang, K., & Li, C. (2008). Oxidative dehydrogenation of ethane with CO_2 over novel Cr/SBA-15/Al_2O_3/FeCrAl monolithic catalysts. *Energy & Fuels*, *22*(6), 3631–3638. doi:10.1021/ef800567v

Shu, Q., Wang, X., Ren, S., & Shi, W. (2005). Gas-phase oxidative acetoxylation of toluene over Pd-Sn-K/SiO_2 catalyst. *Cuihua Xuebao*, *26*, 869–873.

Sierra Gallego, G., & Batiot-Dupeyrat, C. & Mondragón. (2010). Methane partial oxidation by the lattice oxygen of the LaNiO3-δ perovskite. A pulse study. *Dyna*, *77*(163), 141–150.

Sijpkes, A. H., Wilhelmus, R. H., van der Puil, N., & Abdul Hamid, S. B. (2003). *US Patent No. US20070149390(A1).*

Sikdar, S. K. (2003). Sustainable development and sustainability metrics. *AIChE Journal. American Institute of Chemical Engineers*, *49*(8), 1928–1932. doi:10.1002/aic.690490802

Silveira, E. T., Umpierre, A. P., Rossi, L. M., Machado, G., Morais, J., Soares, G. V., & Dupont, J. et al. (2004). The partial hydrogenation of benzene to cyclohexene by nanoscale ruthenium catalysts in imidazolium ionic liquids. *Chemistry (Weinheim an der Bergstrasse, Germany)*, *10*(15), 3734–3740. doi:10.1002/chem.200305765 PMID:15281157

Singh, U. K., & Vannice, M. A. (2001). Kinetics of liquid-phase hydrogenation reactions over supported metal catalysts — a review. *Applied Catalysis A, General*, *213*(1), 1–24. doi:10.1016/S0926-860X(00)00885-1

Siyo, B., Schneider, M., Radnik, J., Pohl, M.-M., Langer, P., & Steinfeldt, N. (2014). Influence of support on the aerobic oxidation of HMF into FDCA over preformed Pd nanoparticle based materials. *Applied Catalysis A, General*, *478*, 107–116. doi:10.1016/j.apcata.2014.03.020

Slaa, J. C., van Ommen, J. G., & Ross, J. R. H. (1995). The synthesis of alcohols using Cu/ZnO/A12O3 + (Ce or Mn) catalysts. Topics in Catalysis, 2(1-4), 79–89. doi:10.1007/BF01491957 doi:10.1007/BF01491957

Smejkal, Q., Linke, D., & Baerns, M. (2005). Energetic and economic evaluation of the production of acetic acid via ethane oxidation. *Chemical Engineering and Processing*, *44*(4), 421–428. doi:10.1016/S0255-2701(04)00146-1

Smith, K. J., & Anderson, R. B. (1983). The higher alcohol synthesis over promoted Cu/ZnO catalysts. Canadian Journal of Chemical Engineering, 61(1), 40–45. doi:10.1002/cjce.5450610106 doi:10.1002/cjce.5450610106

Smith, K. J., & Anderson, R. B. (1984). A chain growth scheme for the higher alcohols synthesis. Journal of Catalysis, 85(2), 428–436. doi:10.1016/0021-9517(84)90232-X doi:10.1016/0021-9517(84)90232-X

Smitha, B., Suhanya, D., Sridhar, S., & Ramakrishna, M. (2004). Separation of organic-organic mixtures by pervaporation – a review. *Journal of Membrane Science*, *241*(1), 1–21. doi:10.1016/j.memsci.2004.03.042

Smith, J. A., Brzezinska, K. R., Valenti, D. J., & Wagener, K. B. (2000). Precisely Controlled Methyl Branching in Polyethylene via Acyclic Diene Metathesis (ADMET) Polymerization. *Macromolecules*, *33*(10), 3781–3794. doi:10.1021/ma9920792

Smith, R. (2005). *Chemical Process Design and Integration.* New York: Wiley.

Soliman, M., Al-Zeghayer, Y., Al-Awadi, A. S., & Al-Mayman, S. (2012). Economics of Acetic Acid Production by Partial Oxidation of Ethane. *APCBEE Procedia*, *3*, 200–208. doi:10.1016/j.apcbee.2012.06.070

Solomons, T. W. G. (1996). *Organic Chemistry* (6th ed., pp. 260–268, 283–316, 940–941). New York, USA: John Wiley & Sons, Inc.

Solsona, B., Concepción, P., Hernández, S., Demicol, B., & Nieto, J.M.L. (2012). Oxidative dehydrogenation of ethane over NiO–CeO_2 mixed oxides catalysts. *Catalysis Today*, *180*(1), 51–58. doi:10.1016/j.cattod.2011.03.056

Solsona, B., López Nieto, J. M., Concepción, P., Dejoz, A., Ivars, F., & Vázquez, M. I. (2011). Oxidative dehydrogenation of ethane over Ni–W–O mixed metal oxide catalysts. *Journal of Catalysis*, *280*(1), 28–39. doi:10.1016/j.jcat.2011.02.010

Solymosi, F., Cserenyi, J., Szoke, A., Bansagi, T., & Oszko, A. (1997). Aromatization of methane over supported and unsupported Mo-based catalysts. *Journal of Catalysis*, *165*(2), 150–161. doi:10.1006/jcat.1997.1478

Solymosi, F., & Nemeth, R. (1999). The oxidative dehydrogenation of ethane with CO_2 over Mo_2C/SiO_2 catalyst. *Catalysis Letters*, *62*(2/4), 197–200. doi:10.1023/A:1019027912597

Song, W. G., Haw, J. F., Nicholas, J. B., & Heneghan, C. S. (2000). Methylbenzenes are the organic reaction centers for methanol-to-olefin catalysis on HSAPO-34. *Journal of the American Chemical Society*, *122*(43), 10726–10727. doi:10.1021/ja002195g

Spencer, M. S. (1985). Models of strong metal-support interaction (SMSI) in Pt on TiO2 catalysts. Journal of Catalysis, 93(2), 216–223. doi:10.1016/0021-9517(85)90169-1 doi:10.1016/0021-9517(85)90169-1

Spillman, R. W. (1989). Economic of gas separation by membranes. *Chemical Engineering Progress*, *85*(1), 41–49.

Spivey, J. J., & Egbebi, A. (2007). Heterogeneous catalytic synthesis of ethanol from biomass-derived syngas. Chemical Society Reviews, 36(9), 1514–1528. doi:10.1039/b414039g PubMed doi:10.1039/b414039g PMID:17660882

Spring, A. M., Estrada, L. A., Vasilyeva, S. V., Rinzler, A. G., & Reynolds, J. R. (2014). Carbon Nanotube Adsorptive Materials Derived from Acid Degradable Poly(acetals). *Macromolecules (Washington, DC, U. S.)*, *47*(8), 2556–2560. doi:10.1021/ma500009s

Sri Hari Kumar, A., Kalevaru, V. N., Qiao, A., Alshammari, A., Lingaiah, N., Sailu, C., & Martin, A. et al. (2013b). Catalytic behavior of decomposed molybdophosphoric acid supported on alumina for oxidative dehydrogenation of ethane to ethylene. *Kinetics and Catalysis*, *54*(5), 615–619. doi:10.1134/S002315841305008X

Srinivas, G., Martin, J., Gebhard, S. C., & Mundschau, M. V. (2013). Homologation of methanol for production of ethanol using homogeneous catalysis. Energy & Fuels, 58, 2.

Stacchiola, D., Calaza, F., Burkholder, L., Schwabacher, A. W., Neurock, M., & Tysoe, W. T. (2005). Elucidation of the reaction mechanism for the palladium-catalyzed synthesis of vinyl acetate. *Angewandte Chemie International Edition*, *44*(29), 4572–4574. doi:10.1002/anie.200500782 PMID:15988776

Stacchiola, D., Calaza, F., Burkholder, L., & Tysoe, W. T. (2004). Vinyl Acetate Formation by the Reaction of Ethylene with Acetate Species on Oxygen-Covered Pd(111). *Journal of the American Chemical Society*, *126*(47), 15384–15385. doi:10.1021/ja044641w PMID:15563157

Steinberg, M. (1998). Production of hydrogen and methanol from natural gas with reduced CO2 emission. *International Journal of Hydrogen Energy*, *23*(6), 419–425. doi:10.1016/S0360-3199(97)00092-X

Stephens, W. T., Mazanec, T. J., & Anderson, H. U. (2000). Influence of gas flow rate on oxygen flux measurements for dense oxygen conducting ceramic membranes. *Solid State Ionics*, *129*(1-4), 271–284. doi:10.1016/S0167-2738(99)00332-X

Storm, D. A. (1994). The Production of Higher Alcohols from Syngas using Potassium-Promoted Co/Mo/Al2O3 and Rh/Co/Mo/Al2O3. [-CATL.]. Abstracts Of Papers Of the American Chemical Society, 207, 47.

Stratton, A., Hemming, D. F., & Teper, M. (1983). *Ethylene production from oil, gas and coal-derived feedstock*. London, UK: IEA Coal Research.

Subramani, V., & Gangwal, S. K. (2008). A Review of Recent Literature to Search for an Efficient Catalytic Process for the Conversion of Syngas to Ethanol. Energy & Fuels, 22(2), 814–839. doi:10.1021/ef700411x doi:10.1021/ef700411x

Subramanian, N. D., Balaji, G., Kumar, C. S. S. R., & Spivey, J. J. (2009). Development of cobalt–copper nanoparticles as catalysts for higher alcohol synthesis from syngas. Catalysis Today, 147(2), 100–106. doi:10.1016/j.cattod.2009.02.027 doi:10.1016/j.cattod.2009.02.027

Subramanian, N. D., Gao, J., Mo, X., Goodwin, J. G., Jr., Torres, W., & Spivey, J. J. (2010). La and/or V oxide promoted Rh/SiO2 catalysts: Effect of temperature, H2/CO ratio, space velocity, and pressure on ethanol selectivity from syngas. Journal of Catalysis, 272(2), 204–209. doi:10.1016/j.jcat.2010.03.019 doi:10.1016/j.jcat.2010.03.019

Sugden, S., & Wilkins, H. (1929). CLXVII.-The parachor and chemical constitution. Part XII. Fused metals and salts. *Journal of the Chemical Society*, 1291–1298. doi:10.1039/jr9290001291

Sugiyama, H., Fischer, U., Antonijuan, E., Hoffmann, V. H., Hirao, M., & Hungerbühler, K. (2009). How do different process options and evaluation settings affect economic and environmental assessments? A case study on methyl methacrylate (MMA) production processes. *Process Safety and Environmental Protection*, *87*(6), 361–370. doi:10.1016/j.psep.2009.08.002

Suib, S. L. (2013). New and Future Developments in Catalysis: Catalysis for Remediation and Environmental Concerns. Newnes.

Sun, J., Wan, S., Wang, F., Lin, J., & Wang, Y. (2015). Selective Synthesis of Methanol and Higher Alcohols over Cs/Cu/ZnO/Al2O3 Catalysts. Industrial & Engineering Chemistry Research, 54(32), 7841–7851. doi:10.1021/acs.iecr.5b01927 doi:10.1021/acs.iecr.5b01927

Suresh, G., Radnik, J., Kalevaru, V. N., Pohl, M.-M., Schneider, M., Lücke, B., & Brückner, A. et al. (2010). Tailoring the synthesis of supported Pd catalysts towards desired structure and size of metal particles. *Physical Chemistry Chemical Physics*, *12*(18), 4833–4842. doi:10.1039/b920464d PMID:20428566

Surisetty, V. R., Dalai, A. K., & Kozinski, J. (2010a). Effect of Rh promoter on MWCNT-supported alkali-modified MoS2 catalysts for higher alcohols synthesis from CO hydrogenation. Applied Catalysis A, General, 381(1–2), 282–288. doi:10.1016/j.apcata.2010.04.036 doi:10.1016/j.apcata.2010.04.036

Surisetty, V. R., Dalai, A. K., & Kozinski, J. (2010b). Synthesis of higher alcohols from synthesis gas over Co-promoted alkali-modified MoS2 catalysts supported on MWCNTs. Applied Catalysis A, General, 385(1–2), 153–162. doi:10.1016/j.apcata.2010.07.009 doi:10.1016/j.apcata.2010.07.009

Suzuki, S., Suzuki, T., Tomita, Y., Hirano, M., Okazaki, K.-, Kuwabata, S., & Torimoto, T. (2012). Compositional control of AuPt nanoparticles synthesized in ionic liquids by the sputter deposition technique. *CrystEngComm*, *14*(15), 4922–4926. doi:10.1039/c2ce25235j

Swaan, H. M., Toebes, A., Seshan, K., van Ommen, J. G., & Ross, J. R. H. (1992). The kinetic and mechanistic aspects of the oxidative dehydrogenation of ethane over Li/Na/MgO catalysts. *Catalysis Today*, *13*(2-3), 201–208. doi:10.1016/0920-5861(92)80143-B

Sworen, J. C., Smith, J. A., Wagener, K. B., Baugh, L. S., & Rucker, S. P. (2003). Modeling Random Methyl Branching in Ethylene/ Propylene Copolymers Using Metathesis Chemistry: Synthesis and Thermal Behavior. *Journal of the American Chemical Society*, *125*(8), 2228–2240. doi:10.1021/ja020862v PMID:12590552

Sznejer, G., & Sheintuch, M. (2004). Application of a carbon membrane reactor for dehydrogenation reactions. *Chemical Engineering Science*, *59*(10), 2013–2021. doi:10.1016/j.ces.2004.01.051

Taarning, E., Madsen, A. T., Marchetti, J. M., Egeblad, K., & Christensen, C. H. (2008). Oxidation of glycerol and propanediols in methanol over heterogeneous gold catalysts. *Green Chemistry*, *10*(4), 408–414. doi:10.1039/b714292g

Takeuchi, A., & Katzer, J. R. (1982). Ethanol formation mechanism from carbon monoxide + molecular hydrogen. Journal of Physical Chemistry, 86(13), 2438–2441. doi:10.1021/j100210a038 doi:10.1021/j100210a038

Tanaka, H., Kaino, R., Okumura, K., Kizuka, T., & Tomishige, K. (2009). Catalytic performance and characterization of Rh–CeO2/MgO catalysts for the catalytic partial oxidation of methane at short contact time. Journal of Catalysis, 268(1), 1–8. doi:10.1016/j.jcat.2009.08.008 doi:10.1016/j.jcat.2009.08.008

Tang, X., Pang, Z., Gu, Z., & Liu, S. (2007). Catalytic perovskite hollow fibre membrane reactors for methane oxidative coupling. *Journal of Membrane Science*, *302*(1-2), 109–114. doi:10.1016/j.memsci.2007.06.033

Tanksale, A., Beltramini, J. N., Dumesic, J. A., & Lu, G. Q. (2008). Effect of Pt and Pd promoter on Ni supported catalysts - A TPR/TPO/TPD and microcalorimetry study. *Journal of Catalysis*, *258*(2), 366–377. doi:10.1016/j.jcat.2008.06.024

Tanksale, A., Beltramini, J. N., & Lu, G. M. (2010). A review of catalytic hydrogen production processes from biomass. *Renewable & Sustainable Energy Reviews*, *14*(1), 166–182. doi:10.1016/j.rser.2009.08.010

Tannerhill, C. C. (1999). Nitrogen Removal Requirement from Natural Gas. *Gas Research Institute Topical Report GRI-99/0080, Washington, D.C.*

Tan, X., & Li, K. (2006). Oxidative coupling of methane in a perovskite hollow-fiber membrane reactor. *Industrial & Engineering Chemistry Research*, *45*(1), 142–149. doi:10.1021/ie0506320

Tan, X., Li, K., Thursfield, A., & Metcalfe, I. S. (2008). Oxyfuel combustion using a catalytic ceramic membrane reactor. *Catalysis Today*, *131*(1-4), 292–304. doi:10.1016/j.cattod.2007.10.081

Tan, X., Liu, Y., & Li, K. (2005). Mixed conducting ceramic hollow-fiber membranes for air separation. *AIChE Journal. American Institute of Chemical Engineers*, *51*(7), 1991–2000. doi:10.1002/aic.10475

Taralas, G., & Kontominas, M. G. (2006). Pyrolysis of solid residues commencing from the olive oil food industry for potential hydrogen production. *Journal of Analytical and Applied Pyrolysis*, *76*(1-2), 109–116. doi:10.1016/j.jaap.2005.08.004

Tarasov, A. L., Kustov, L. M., Kirichenko, O. A., & Kucherov, A. V. (2014). Method for production of ethylene RU 2528830.

Tatsumi, T. (1986). Importance of sequence of impregnation in the activity development of alkali-promoted mo catalysts for alcohol synthesis from CO-H2. Journal of Catalysis, 101(2), 553–556. doi:10.1016/0021-9517(86)90287-3 doi:10.1016/0021-9517(86)90287-3

Taylor, M. N., Carley, A. F., Davies, T. E., & Taylor, S. H. (2009). The Oxidative Dehydrogenation of Propane Using Vanadium Oxide Supported on Nanocrystalline Ceria. *Topics in Catalysis*, *52*(12), 1660–1668. doi:10.1007/s11244-009-9307-0

Temel, B., Nielsen, P. E. H., & Beato, P. (2013). Process for the selective preparation of 1-propanol, iso-butanol and other C3+ alcohols from synthesis gas and methanol. Google Patents.

ten Dam, J., & Hanefeld, U. (2011). Renewable Chemicals: Dehydroxylation of Glycerol and Polyols. *ChemSusChem*, *4*(8), 1017–1034. doi:10.1002/cssc.201100162 PMID:21887771

Ten Elshof, J. E., Bouwmeester, H. J. M., & Verweij, H. (1995). Oxidative coupling of methane in a mixed-conducting perovskite membrane reactor. *Applied Catalysis A, General*, *130*(2), 195–212. doi:10.1016/0926-860X(95)00098-4

The Industrial Uses of Palladium. (2011). Manhattan Gold & Silver. Retrieved from http://www.mgsrefining.com/blog/post/2011/09/13/The-Industrial-Uses-for-Palladium.aspx

The shale revolution. (2009). Nature, 460(7255), 551-552.

Thompson, D. L., Wagener, K. B., Schulze, U., Voit, B., Jehnichen, D., & Malanin, M. (2015). Spectroscopic Examinations of Hydrogen Bonding in Hydroxy-Functionalized ADMET Chemistry. *Macromolecular Rapid Communications*, *36*(1), 60–64. doi:10.1002/marc.201400545 PMID:25393938

Thursfield, A., & Metcalfe, I. S. (2006). Methane oxidation in a mixed ionic–electronic conducting ceramic hollow fibre reactor module. *Journal of Solid State Electrochemistry*, *10*(8), 604–616. doi:10.1007/s10008-006-0139-6

Thursfield, A., & Metcalfe, I. S. (2007). Air separation using a catalytically modified mixed conducting ceramic hollow fibre membrane module. *Journal of Membrane Science*, *288*(1-2), 175–187. doi:10.1016/j.memsci.2006.11.015

Tien-Thao, N., Hassan Zahedi-Niaki, M., Alamdari, H., & Kaliaguine, S. (2007). Effect of alkali additives over nanocrystalline Co–Cu-based perovskites as catalysts for higher-alcohol synthesis. Journal of Catalysis, 245(2), 348–357. doi:10.1016/j.jcat.2006.10.026 doi:10.1016/j.jcat.2006.10.026

Tien-Thao, N., Zahedi-Niaki, M. H., Alamdari, H., & Kaliaguine, S. (2007). Conversion of syngas to higher alcohols over nanosized LaCo0.7Cu0.3O3 perovskite precursors. Applied Catalysis A, General, 326(2), 152–163. doi:10.1016/j.apcata.2007.04.009 doi:10.1016/j.apcata.2007.04.009

Tijm, P. J., & Stevens, R. R. (2014). Production of mixed alcohols from synthesis gas. Google Patents.

Tixier, J. D., Dusserre, G., Salvi, O., & Gaston, D. (2002). Review of 62 risk analysis methodologies of industrial plants. *Journal of Loss Prevention in the Process Industries*, *15*(4), 291–303. doi:10.1016/S0950-4230(02)00008-6

Topsoe, T., Hinnemann, B., Norskov, J. K., Lauritsen, J. V., Besenbacher, F., Hansen, P. L., & Knudsen, K. G. et al. (2005). The role of reaction pathways and support interactions in the development of high activity hydrotreating catalysts. *Catalysis Today*, *107–108*, 12–22. doi:10.1016/j.cattod.2005.07.165

Torimoto, T., Ohta, Y., Enokida, K., Sugioka, D., Kameyama, T., Yamamoto, T., & Kuwabata, S. et al. (2015). Ultrathin oxide shell coating of metal nanoparticles using ionic liquid/metal sputtering. *Journal of Materials Chemistry A*, *3*(11), 6177–6186. doi:10.1039/C4TA06643J

Torimoto, T., Okazaki, K.-, Kiyama, T., Hirahara, K., Tanaka, N., & Kuwabata, S. (2006). Sputter deposition onto ionic liquids: Simple and clean synthesis of highly dispersed ultrafine metal nanoparticles. *Applied Physics Letters*, *89*(24), 243117. doi:10.1063/1.2404975

Torniainen, P. M., Chu, X., & Schmidt, L. D. (1994). Comparison of monolith-supported metals for the direct oxidation of methane to syngas. *Journal of Catalysis*, *146*(1), 1–10. doi:10.1016/0021-9517(94)90002-7

Toyoda, T., Minami, T., & Qian, E. W. (2013). Mixed Alcohol Synthesis over Sulfided Molybdenum-Based Catalysts. Energy & Fuels, 27(7), 3769–3777. doi:10.1021/ef400262a doi:10.1021/ef400262a

Treviño, H., Hyeon, T., & Sachtler, W. M. H. (1997). A Novel Concept for the Mechanism of Higher Oxygenate Formation from Synthesis Gas over MnO-Promoted Rhodium Catalysts. Journal of Catalysis, 170(2), 236–243. doi:10.1006/jcat.1997.1756 doi:10.1006/jcat.1997.1756

Tsilomelekis, G., Christodoulakis, A., & Boghosian, S. (2007). Support effects on structure and activity of molybdenum oxide catalysts for the oxidative dehydrogenation of ethane. *Catalysis Today*, *127*(1-4), 139–147. doi:10.1016/j.cattod.2007.03.026

Tsuji, J. (2004). *Palladium reagents and catalysts: new perspectives for the 21st century* (p. 90). John Wiley and Sons. doi:10.1002/0470021209

Tsyganok, A., Green, R. G., Giorgi, J. B., & Sayari, A. (2007). Non-oxidative dehydrogenation of ethane to ethylene over chromium catalysts prepared from layered double hydroxide precursors. *Catalysis Communications*, *8*(12), 2186–2193. doi:10.1016/j.catcom.2007.04.031

Tu, X., Furuta, N., & Sumida, Y. (2005). *US Patent No. US7645897(B2).*

Tu, X., Sumida, Y., Takahashi, M., & Niizuma, H. (2004). *US Patent No. US7375052(B2).*

Tuba, R., Al-Hashimi, M., Bazzi, H. S., & Grubbs, R. H. (2014). One-Pot Synthesis of Poly(vinyl alcohol) (PVA) Copolymers via Ruthenium Catalyzed Equilibrium Ring-Opening Metathesis Polymerization of Hydroxyl Functionalized Cyclopentene. *Macromolecules (Washington, DC, U. S.)*, *47*(23), 8190–8195. doi:10.1021/ma501976v

Tuzun, A., Lligadas, G., Ronda, J. C., Galia, M., & Cadiz, V. (2015). Integrating plant oils into thermally curable main-chain benzoxazine polymers via ADMET polymerization. *Eur. Polym. J.* doi: 10.1016/j.eurpolymj.2014.12.023

Ueda, W. (2013). Establishment of Crystalline Complex Mo-V-Oxides as Selective Oxidation Catalysts. *Journal of the Japan Petroleum Institute*, *56*(3), 122–132. doi:10.1627/jpi.56.122

Ugo, R. (Ed.), (1980). *Palladium catalyzed oxidation of hydrocarbons*. Dordrecht, Netherlands: D. Reidel Publishing Company.

Umpierre, A. P., Machado, G., Fecher, G. H., Morais, J., & Dupont, J. (2005). Selective Hydrogenation of 1,3-Butadiene to 1-Butene by Pd(0) Nanoparticles Embedded in Imidazolium Ionic Liquids. *Advanced Synthesis & Catalysis*, *347*(10), 1404–1412. doi:10.1002/adsc.200404313

Unverferth, M., & Meier, M. A. R. (2014). Tuning the polarity of ADMET derived star-shaped polymers via thiol-ene chemistry. *Polymer*, *55*(22), 5571–5575. doi:10.1016/j.polymer.2014.06.035

Valencia, D., & Klimova, T. (2012). Kinetic study of NiMo/SBA-15 catalysts prepared with citric acid in hydrodesulfurization of dibenzothiophene. *Catalysis Communications*, *21*, 77–81. doi:10.1016/j.catcom.2012.02.003

Valente, J. S., Lopez-Nieto, J. M., Herrera, H. A., Ramirez, A. M., Barcelo, F. I., Castillo, M., . . . Flores, E. M. (2012). *US Patent No. US20140114109(A1).*

Valente, J. S., Armendáriz-Herrera, H., Quintana-Solórzano, R., del Ángel, P., Nava, N., Masso, A., & López Nieto, J. (2014). Chemical, Structural, and Morphological Changes of a MoVTeNb Catalyst during Oxidative Dehydrogenation of Ethane. *ACS Catalysis*, *4*(5), 1292–1301. doi:10.1021/cs500143j

Valente, J. S., Quintana-Solórzano, R., Armendáriz-Herrera, H., Barragán-Rodríguez, G., & López-Nieto, J. M. (2014). Kinetic study of oxidative dehydrogenation of ethane over MoVTeNb mixed-oxide catalyst. *Industrial & Engineering Chemistry Research*, *53*(5), 1775–1786. doi:10.1021/ie402447h

Valenti, D. J., & Wagener, K. B. (1998). Direct Synthesis of Well-Defined Alcohol-Functionalized Polymers via Acyclic Diene Metathesis (ADMET) Polymerization. *Macromolecules*, *31*(9), 2764–2773. doi:10.1021/ma9714833

van de Water, L. G. A., Bergwerff, J. A., Leliveld, B. G., Weckhuysen, B. M., & de Jong, K. P. (2005). Insights into the Preparation of Supported Catalysts: A Spatially Resolved Raman and UV-Vis Spectroscopic Study into the Drying Process of CoMo/γ-Al2O3 Catalyst Bodies. *J. Phys. Chem. B*, 109(30), 14513-14522. doi:10.1021/jp051037e

van den Bergh, J., Gücüyener, C., Gascon, J., & Kapteijn, F. (2011). Isobutane dehydrogenation in a DD3R zeolite membrane reactor. *Chemical Engineering Journal*, *166*(1), 368–377. doi:10.1016/j.cej.2010.11.021

Van Dillen, A. J., Terorde, R. J. A. M., Lensveld, D. J., Geus, J. W., & De Jong, K. P. (2003). Synthesis of supported catalysts by impregnation and drying using aqueous chelated metal complexes. *J. Catal.*, 216(1-2), 257-264. doi:10.1016/S0021-9517(02)00130-6

Van Doorn, R. H. E., Bouwmeester, H. J. M., & Burggraaf, A. J. (1998). Kinetic decomposition of La0.3Sr0.7CoO3-δ perovskite membranes during oxygen permeation. *Solid State Ionics*, *111*(3-4), 263–272. doi:10.1016/S0167-2738(98)00191-X

van Dyk, J. C., Keyser, M. J., & Coertzen, M. (2006). Syngas production from South African coal sources using Sasol–Lurgi gasifiers. International Journal of Coal Geology, 65(3–4), 243–253. doi:10.1016/j.coal.2005.05.007 doi:10.1016/j.coal.2005.05.007

van Dyk, L., Miachon, S., Lorenzen, L., Torres, M., Fiaty, K., & Dalmon, J. A. (2003). Comparison of microporous MFI and dense Pd membrane performances in an extractor-type CMR. *Catalysis Today*, *82*(1-4), 167–177. doi:10.1016/S0920-5861(03)00229-3

Van Hassel, B. A., Kawada, A., Sakai, T., Yokokawa, N., Dokiya, H., & Bouwmeester, H. J. M. (1993). Oxygen permeation modelling of perovskites. *Solid State Ionics*, *66*(3-4), 295–305. doi:10.1016/0167-2738(93)90419-4

van Helden, R., Kohll, C. F., Medema, D., Verberg, G., & Jonkhoff, T. (1968). Palladium-catalyzed reactions of unsaturated compounds in nonaqueous solvents. Synthesis of vinyl acetate via palladium salt/ethylene complexes. *Recueil des Travaux Chimiques des Pays-Bas*, *87*(8), 961–991. doi:10.1002/recl.19680870813

van Hensbergen, J. A., Gaines, T. W., Wagener, K. B., Burford, R. P., & Lowe, A. B. (2014). Functional α,ω-dienes via thiol-Michael chemistry: Synthesis, oxidative protection, acyclic diene metathesis (ADMET) polymerization and radical thiol-ene modification. *Polym. Chem.*, *5*(21), 6225–6235. doi:10.1039/C4PY00783B

Vanecht, E., Binnemans, K., Patskovsky, S., Meunier, M., Seo, J. W., Stappers, L., & Fransaer, J. (2012). Stability of sputter-deposited gold nanoparticles in imidazolium ionic liquids. *Physical Chemistry Chemical Physics*, *14*(16), 5662–5671. doi:10.1039/c2cp23677j PMID:22422275

Vanrysselberghe, V., LeGall, R., & Froment, G. F. (1998). Hydrodesulfurization of 4-methyldibenzothiophene and 4,6-dimethyldibenzothiophene on a CoMo/Al_2O_3 catalyst: Reaction network and kinetic. *Industrial & Engineering Chemistry Research*, *37*(4), 1235–1242. doi:10.1021/ie970533p

Vargaftik, M. N., Zagorodnikov, V. P., Stolyarov, I. P., Moiseev, I. I., Kochubey, D. I., Likholobov, V. A., & Zamaraev, K. I. et al. (1989). Giant palladium clusters as catalysts of oxidative reactions of olefins and alcohols. *Journal of Molecular Catalysis*, *53*(3), 315–348. doi:10.1016/0304-5102(89)80066-5

Vasile, C., (Ed.), (2000). *Handbook of Polyolefins (Rev. Second ed.).*

Vasiliadou, E. S., & Lemonidou, A. A. (2011). Investigating the performance and deactivation behaviour of silica-supported copper catalysts in glycerol hydrogenolysis. *Applied Catalysis A, General*, *396*(1-2), 177–185. doi:10.1016/j.apcata.2011.02.014

Vasudevan, P. T., & Fierro, J. L. G. (1996). A review of deep hydrodesulfurization catalysis. *Catalysis Reviews. Science and Engineering*, *38*(2), 161–188. doi:10.1080/01614949608006457

Vedage, G. A., Himelfarb, P. B., Simmons, G. W., & Klier, K. (1985). Alkali-Promoted Copper-Zinc Oxide Catalysts for Low Alcohol Synthesis Solid State Chemistry in Catalysis (Vol. 279, pp. 295–312). American Chemical Society.

Vedage, G.A., Himelfarb, P.B., Simmons, G., & Klier, K. (1985). Alkali-Promoted Copper-Zinc Oxide Catalysts for Low Alcohol Synthesis. Solid State Chemistry in Catalysis (Vol. 279, pp. 295-312). American Chemical Society.

Velle, O. J., Andersen, A., & Jens, K. J. (1990). Oxidative dehydrogenation of ethane by perovskite type catalysts containing oxides of strontium, cerium and ytterbium. *Catalysis Today*, *6*(4), 567–574. doi:10.1016/0920-5861(90)85053-Q

Vennestr, øm, P. N. R., Osmundsen, C. M., Christensen, C. H., & Taarning, E. (2011). Beyond Petrochemicals: The Renewable Chemicals Industry. *Angewandte Chemie (International ed. in English)*, *50*(45), 10502–10509. doi:10.1002/anie.201102117 PMID:22045621

Verkerk, K. A. N., Jaeger, B., Finkeldei, C.-H., & Keim, W. (1999). Recent developments in isobutanol synthesis from synthesis gas. Applied Catalysis A, General, 186(1–2), 407–431. doi:10.1016/S0926-860X(99)00158-1 doi:10.1016/S0926-860X(99)00158-1

Villa, A., Campione, C., & Prati, L. (2007). Bimetallic gold/palladium catalysts for the selective liquid phase oxidation of glycerol. *Catalysis Letters*, *115*(3-4), 133–136. doi:10.1007/s10562-007-9077-x

Villa, A., Gaiassi, A., Rossetti, I., Bianchi, C. L., van Benthem, K., Veith, G. M., & Prati, L. (2010). Au on MgAl2O4 spinels: The effect of support surface properties in glycerol oxidation. *Journal of Catalysis*, *275*(1), 108–116. doi:10.1016/j.jcat.2010.07.022

Villa, A., Veith, G. M., Ferri, D., Weidenkaff, A., Perry, K. A., Campisi, S., & Prati, L. (2013). NiO as a peculiar support for metal nanoparticles in polyols oxidation. *Catalysis Science & Technology*, *3*(2), 394–399. doi:10.1039/C2CY20370G

Villa, A., Veith, G. M., & Prati, L. (2010). Selective Oxidation of Glycerol under Acidic Conditions Using Gold Catalysts. *Angewandte Chemie (International ed. in English)*, *49*(26), 4499–4502. doi:10.1002/anie.201000762 PMID:20461742

Villa, A., Wang, D., Veith, G. M., & Prati, L. (2012). Bismuth as a modifier of Au-Pd catalyst: Enhancing selectivity in alcohol oxidation by suppressing parallel reaction. *Journal of Catalysis*, *292*, 73–80. doi:10.1016/j.jcat.2012.04.021

Villa, A., Wang, D., Veith, G. M., Vindigni, F., & Prati, L. (2013). Sol immobilization technique: A delicate balance between activity, selectivity and stability of gold catalysts. *Catalysis Science & Technology*, *3*(11), 3036–3041. doi:10.1039/c3cy00260h

Vinyl Acetate Monomer (VAM) Global Market to 2020. (2013). *Vertical Edge Limited.* Retrieved from http://www.companiesandmarkets.com/Market/Chemicals/Market-Research/Vinyl-Acetate-Monomer-VAM-Global-Market-to-2020/RPT1134065

Vollmer, C., & Janiak, C. (2011). Naked metal nanoparticles from metal carbonyls in ionic liquids: Easy synthesis and stabilization. *Coordination Chemistry Reviews*, *255*(17-18), 2039–2057. doi:10.1016/j.ccr.2011.03.005

Vollmer, C., Redel, E., Abu-Shandi, K., Thomann, R., Manyar, H., Hardacre, C., & Janiak, C. (2010). Microwave irradiation for the facile synthesis of transition-metal nanoparticles (NPs) in ionic liquids (ILs) from metal-carbonyl precursors and Ru-, Rh-, and Ir-NP/IL dispersions as biphasic liquid-liquid hydrogenation nanocatalysts for cyclohexene. *Chemistry (Weinheim an der Bergstrasse, Germany)*, *16*(12), 3849–3858. doi:10.1002/chem.200903214 PMID:20187043

Voskanyan, P. (2010). Effect of the amounts and proportions of active components on the activity and selectivity of a catalyst for vinyl acetate synthesis by gas-phase ethylene acetoxylation. *Catalysis in Industry*, *2*(2), 167–172. doi:10.1134/S2070050410020121

Voskanyan, P. S. (2013). Effect of the nature of a support on the catalytic activity of a palladium catalyst in the synthesis of vinyl acetate by gas-phase ethylene acetoxylation. *Catalysis in Industry*, *5*(1), 90–97. doi:10.1134/S2070050412040198

Wagener, K. B., & O'Donnell, P. M. (2002). ATRP grafting on ADMET polyolefin backbones. *Polym. Prepr. (Am. Chem. Soc., Div. Polym. Chem.)*, *43*(2), 48.

Wagener, K. B., Boncella, J. M., & Nel, J. G. (1991). Acyclic diene metathesis (ADMET) polymerization. *Macromolecules*, *24*(10), 2649–2657. doi:10.1021/ma00010a001

Walter, K. M., Schubert, M., Kleist, W., & Grunwaldt, J.-D. (2015). Effect of the Addition of Ethanol to Synthesis Gas on the Production of Higher Alcohols over Cs and Ru Modified Cu/ZnO Catalysts. Industrial & Engineering Chemistry Research, 54(5), 1452–1463. doi:10.1021/ie504066f doi:10.1021/ie504066f

Wan, G., Duan, A., Zhang, Y., Zhao, Z., Jiang, G., Zhang, D., . . . Chung, K. (2010). NiW/AMBT catalysts for the production of ultra-low sulfur diesel. *Catal. Today*, 158(3-4), 521-529. doi:10.1016/j.cattod.2010.08.021

Wang, A., J., Ruan, L., F., Teng, Y., et al. (2005). Hydrodesulfurization of dibenzothiophene over siliceous MCM-41 -supported nickel phosphide catalysts. *J. Catal., 229,* 314-321.

Wang, A., Wang, Y., Kabe, T., Chen, Y., Ishihara, A., Qian, W., & Yao, P. (2002). Hydrodesulfurization of Dibenzothiophene over Siliceous MCM-41-Supported Catalysts. *J. Catal.*, 210(2), 319-327. doi:10.1006/jcat.2002.3674

Wang, H. Y., Liu, J. P., Fu, J. K., Wan, H. L., & Tsai, K. R. (1992). Study on the Mechanism of Ethanol Synthesis from Syngas by Insitu Chemical Trapping and Isotopic Exchange Reactions. Catalysis Letters, 12(1-3), 87–96. doi:10.1007/BF00767191 doi:10.1007/BF00767191

Wang, H. Y., Liu, J. P., Fu, J. K., Zhang, H. B., & Tsai, K. R. (1992). Study on the Method of Chemical Trapping for Formyl Intermediates. Research on Chemical Intermediates, 17(3), 233–242. doi:10.1163/156856792X00155 doi:10.1163/156856792X00155

Wang, J., Chernavskii, P. A., Khodakov, A. Y., & Wang, Y. (2012). Structure and catalytic performance of alumina-supported copper–cobalt catalysts for carbon monoxide hydrogenation. Journal of Catalysis, 286(0), 51–61. doi:10.1016/j.jcat.2011.10.012 doi:10.1016/j.jcat.2011.10.012

Wang, J., Chernavskii, P. A., Wang, Y., & Khodakov, A. Y. (2013). Influence of the support and promotion on the structure and catalytic performance of copper–cobalt catalysts for carbon monoxide hydrogenation. Fuel, 103(0), 1111–1122. doi:10.1016/j.fuel.2012.07.055 doi:10.1016/j.fuel.2012.07.055

Wang, J., Zhang, Q., & Wang, Y. (2011). Rh-catalyzed syngas conversion to ethanol: Studies on the promoting effect of FeOx. Catalysis Today, 171(1), 257–265. doi:10.1016/j.cattod.2011.03.023 doi:10.1016/j.cattod.2011.03.023

Wang, J.-J., Xie, J.-R., Huang, Y.-H., Chen, B.-H., Lin, G.-D., & Zhang, H.-B. (2013). An efficient Ni–Mo–K sulfide catalyst doped with CNTs for conversion of syngas to ethanol and higher alcohols. Applied Catalysis A, General, 468(0), 44–51. doi:10.1016/j.apcata.2013.08.026 doi:10.1016/j.apcata.2013.08.026

Wang, L., & Hall, W. K. (1980). On the genesis of molybdena-alumina catalyst. *J. Catal.*, 66(1), 251-255. doi: 10.1016/0021-9517(80)90029-9

Wang, L., Zhang, Y., Zhang, Y., Jiang, Z., & Li, C. (2009). Ultra-Deep Hydrodesulfurization of Diesel Fuels on Trimetallic NiMoW Sulfide Catalysts. *Chem. - Eur. J.*, 15(46), 12571-12575. doi:10.1002/chem.200901997

Wang, N., Fang, K., Lin, M., Jiang, D., Li, D., & Sun, Y. (2010). Synthesis of Higher Alcohols from Syngas over Fe/K/β-Mo2C Catalyst. Catalysis Letters, 136(1-2), 9–13. doi:10.1007/s10562-010-0288-1 doi:10.1007/s10562-010-0288-1

Wang, Y., Luo, H., Liang, D., & Bao, X. (2000). Different Mechanisms for the Formation of Acetaldehyde and Ethanol on the Rh–Based Catalysts. Journal of Catalysis, 196(1), 46–55. doi:10.1006/jcat.2000.3026 doi:10.1006/jcat.2000.3026

Wang, B., Zydorczak, B., Poulidi, D., Metcalfe, I. S., & Li, K. (2011). A further investigation of the kinetic demixing/decomposition of La0.6Sr0.4Co0.2Fe0.8O3−δ oxygen separation membranes. *Journal of Membrane Science*, *369*(1-2), 526–535. doi:10.1016/j.memsci.2010.12.040

Wang, B., Zydorczak, B., Wu, Z., & Li, K. (2009). Stabilities of La0.6Sr0.4Co0.2Fe0.8O3-δ oxygen separation membranes. Effects of kinetic demixing/decomposition and impurity segregation. *Journal of Membrane Science*, *344*(1-2), 101–106. doi:10.1016/j.memsci.2009.07.040

Wang, D., Villa, A., Porta, F., Prati, L., & Su, D. S. (2008). Bimetallic gold/palladium catalysts: Correlation between nanostructure and synergistic effects. *The Journal of Physical Chemistry C*, *112*(23), 8617–8622. doi:10.1021/jp800805e

Wang, D., Villa, A., Porta, F., Su, D. S., & Prati, L. (2006). Single-phase bimetallic system for the selective oxidation of glycerol to glycerate. *Chemical Communications*, (18): 1956–1958. doi:10.1039/b518069d PMID:16767248

Wang, D., Villa, A., Su, D. S., Prati, L., & Schlogl, R. (2013). Carbon-Supported Gold Nanocatalysts: Shape Effect in the Selective Glycerol Oxidation. *Chemcatchem*, *5*(9), 2717–2723. doi:10.1002/cctc.201200535

Wang, H., Fu, J. L. J., Cai, J., Zhang, H., & Cai, Q. (1993).. . *Journal of Natural Gas Chemistry*, *2*, 13–18.

Wang, H., & Prins, R. (2008). Hydrodesulfurization of dibenzothiophene and its hydrogenated intermediates over sulfided Mo/γ-Al_2O_3. *Journal of Catalysis*, *258*(1), 153–164. doi:10.1016/j.jcat.2008.06.005

Wang, H., Ren, F., Liu, C., Si, R., Yu, D., Pfefferle, L. D., & Chen, Y. et al. (2013). CoSO4/SiO2 catalyst for selective synthesis of (9, 8) single-walled carbon nanotubes: Effect of catalyst calcination. *Journal of Catalysis*, *300*, 91–101. doi:10.1016/j.jcat.2012.12.032

Wang, H., Wang, X., Zhang, M., Du, X.-Y., Li, W., & Tao, K.-Y. (2007). Synthesis of bulk and supported molybdenum carbide by a single-step thermal carburization method. *Chemistry of Materials*, *19*(7), 1801–1807. doi:10.1021/cm0615471

Wang, J., Zhao, D., & Li, K. (2009). Oxidative desulfurization of dibenzothiophene catalyzed by Brönsted acid ionic liquid. *Energy & Fuels*, *23*(8), 3831–3834. doi:10.1021/ef900251a

Wang, S., Murata, K., Hayakawa, T., Hamakawa, S., & Suzuki, K. (1999). Oxidative dehydrogenation of ethane by carbon dioxide over sulfate-modified Cr2O3/SiO2 catalysts. *Catalysis Letters*, *63*(1/2), 59–64. doi:10.1023/A:1019084030344

Wang, X., Clark, P., & Oyama, S. T. (2002). Synthesis, characterization, and hydrotreating activity of several iron group transition metal phosphides. *Journal of Catalysis*, *208*(2), 321–331. doi:10.1006/jcat.2002.3604

Wang, X., Han, M., Wan, H., Yang, C., & Guan, G. (2011). Study on extraction of thiophene from model gasoline with brønsted acidic ionic liquids. *Frontiers of Chemical Science and Engineering*, *5*(1), 107–112. doi:10.1007/s11705-010-0539-0

Wasserscheid, P., & Jess, A. (2004). Deep Desulphurization of oil refinery streams by extraction with ILs. *Green Chemistry*, *6*(7), 316–322. doi:10.1039/B407028C

Wasserscheid, P., van Hal, R., & Bösmann, A. (2002). 1-n-Butyl-3-methylimidazolium ([bmim]) octylsulfate an even 'greener' ionic liquid. *Green Chemistry*, *4*(4), 400–404. doi:10.1039/b205425f

Watson, J. M., Daly, F. P., Mazanec, T. J., & Yang, B. (2005). Catalysts having catalytic material applied directly to thermally grown alumina and catalytic methods using same, improved methods of oxidative dehydrogenation (US 2005/0272965).

WCED (World Commission on Environment and Development). (1987). *"Our Common Future", (The Brundtland Report)*. New York: Oxford University Press.

Wei, Z., Sun, J., Li, Y., Datye, A. K., & Wang, Y. (2012). Bimetallic catalysts for hydrogen generation. Chemical Society Reviews, 41(24), 7994–8008. doi:10.1039/c2cs35201j PubMed doi:10.1039/c2cs35201j PMID:23011345

Weigold, H. (1983). The nature of molybdenum oxide species mounted on alumina: an oligomer model. *J. Catal.*, 83(1), 85-98. doi:10.1016/0021-9517(83)90032-5

Wei, N., Ji, S., Guan, Y., Liu, H., & Li, C. (2011). Influence of W on structure and hydrodesulfurization performance of W-Ni_2P/SBA-15 catalysts.[Petroleum Processing Section]. *Acta Petrolei Sinica*, *27*, 852–858.

Wei, N., Ji, S., Wu, P., Guo, Y., Liu, H., Zhu, J., & Li, C. (2009). Preparation of nickel phosphide/SBA-15/cordierite monolithic catalysts and catalytic activity for hydrodesulfurization of dibenzothiophene. *Catalysis Today*, *147S*, S66–S70. doi:10.1016/j.cattod.2009.07.053

Wei, N., Zeng, P., Ji, S., Zhao, P., Liu, H., & Li, C. (2011). Effect of Lanthanum promoter on structure and hydrodesulfurization performance of La-Ni_2P/SBA-15 catalysts. *Journal of Rare Earths*, *29*, 310–315.

Weingärtner, H. (2008). Understanding ionic liquids at the molecular level: Facts, problems, and controversies. *Angewandte Chemie International Edition*, *47*(4), 654–670. doi:10.1002/anie.200604951 PMID:17994652

Weissermel, K., & Arpe, H.-J. (Fourth Ed.) (2003). Industrial Organic Chemistry. (pp. 59-63). Weinheim, Germany: Wiley-VCH GmbH & Co. KGaA.

Wei, Y., Liu, H., Xue, J., Li, Z., & Wang, H. (2011). Preparation and oxygen permeation of U-shaped perovskite hollow-fiber membranes. *AIChE Journal. American Institute of Chemical Engineers*, *57*(4), 975–984. doi:10.1002/aic.12321

Weller, S. M., & Steiner, W. A. (1950). Separation of gases by fractional permeation through membranes. *Journal of Applied Physics*, *21*(4), 279. doi:10.1063/1.1699653

Welton, T. (1999). Room-temperature ILs. Solvents for synthesis and catalysis. *Chemical Reviews*, *99*(8), 2071–2084. doi:10.1021/cr980032t PMID:11849019

West, N. M., Miller, A. J. M., Labinger, J. A., & Bercaw, J. E. (2011). Homogeneous syngas conversion. Coordination Chemistry Reviews, 255(7–8), 881–898. doi:10.1016/j.ccr.2010.08.019 doi:10.1016/j.ccr.2010.08.019

Weychardt, H., & Plenio, H. (2008). Acyclic Diene Metathesis Polymerization of Divinylarenes and Divinylferrocenes with Grubbs-Type Olefin Metathesis Catalysts. *Organometallics*, *27*(7), 1479–1485. doi:10.1021/om701277p

White, E. B., Wilson, C., Berggren, M., & Zubrin, R. M. (2014). Methods of higher alcohol synthesis: Google Patents.

Whitehurst, D. D., Isoda, T., & Mochida, I. (1998). Present state of the art and future challenges in the hydrodesulfurization of polyaromatic sulfur compounds. *Advances in Catalysis*, *42*, 345–471. doi:10.1016/S0360-0564(08)60631-8

White, L. S. (2006). Development of large-scale applications in organic solvent nanofiltration and pervaporation for chemical and refining processes. *Journal of Membrane Science*, *286*(1-2), 26–35. doi:10.1016/j.memsci.2006.09.006

Whyman, R., Wright, A. P., Iggo, J. A., & Heaton, B. T. (2002). Carbon monoxide activation in homogeneously catalysed reactions: The nature and roles of catalytic promoters. Journal of the Chemical Society, Dalton Transactions: Inorganic Chemistry, (5): 771–777. doi:10.1039/b107940a doi:10.1039/b107940a

Widiasa, I. N., Susanto, A. A., & Susanto, H. (2014). Performance of an integrated membrane pilot plant for wastewater reuse: Case study of oil refinery plant in Indonesia. *Desalination and Water Treatment*, *52*(40-42), 7443–7449. doi:10.1080/19443994.2013.830759

Wilkes, J. S., & Zaworotko, M. J. (1992). Air and water stable 1-ethyl-3-methylimidazolium based ionic liquids. *Journal of the Chemical Society. Chemical Communications*, (13): 965–967. doi:10.1039/c39920000965

Woo, H. C., Park, T. Y., Kim, Y. G., Nam, I.-S., Lee, J. S., & Chung, J. S. (1993). Alkali-Promoted MoS2 Catalysts for Alcohol Synthesis: the Effect of Alkali Promotiom and Preparation Condition on Activity and Selectivity. In F. S. L. Guczi & T. P (Eds.), Studies in Surface Science and Catalysis (Vol. Volume 75, pp. 2749-2752). Elsevier.

Worrell, E., De Beer, J. G., Faaij, A. P. C., & Blok, K. (1994). Potential energy savings in the production route for plastics. *Energy Conservation and Management*, *35*(12), 1073–1085. doi:10.1016/0196-8904(94)90011-6

Wu, J. B., Gross, A., & Yang, H. (2011). Shape and Composition-Controlled Platinum Alloy Nanocrystals Using Carbon Monoxide as Reducing Agent. *Nano Letters*, *11*(2), 798–802. doi:10.1021/nl104094p PMID:21204581

Wu, P., Lan, L., Ji, S., Wei, N., Lv, Z., & Liu, H. (2012). Co-Ni2P/SBA-15/cordierite monolithic catalysts: Preparation and performance for hydrodesulfurization of dibenzothiophene. *Wuji Huaxue Xuebao*, *28*, 565–571.

Wu, P., Li, Q., Lan, L., Liu, H., Ju, N., Piao, J., & Ji, S. (2014). Preparation, structure characterization and hydrodesulfurization performance of B-Ni_2P/SBA-15/cord monolithic catalysts. *Chinese Journal of Chemical Engineering*, *22*(8), 903–908. doi:10.1016/j.cjche.2014.06.014

Wu, P., Wang, J., Lan, L., Ju, Y., Piao, J., Jin, C., & Ji, S. et al. (2014). Preparation of La-Ni_2P/SBA-15/cordierite monolithic catalysts and their catalytic performance of hydrodesulfurization. *Industrial Catalysis*, *22*, 272–276.

Wu, Z. J., Mao, Y. Z., Wang, X. X., & Zhang, M. H. (2011). Preparation of a Cu-Ru/carbon nanotube catalyst for hydrogenolysis of glycerol to 1,2-propanediol via hydrogen spillover. *Green Chemistry*, *13*(5), 1311–1316. doi:10.1039/c0gc00809e

Xiang, M., Li, D., Li, W., Zhong, B., & Sun, Y. (2007). Synthesis of higher alcohols from syngas over K/Co/β-Mo2C catalysts. Catalysis Communications, 8(3), 503–507. doi:10.1016/j.catcom.2006.07.029 doi:10.1016/j.catcom.2006.07.029

Xiang, Y., Chitry, V., Liddicoat, P., Felfer, P., Cairncy, J., Ringer, S., & Kruse, N. (2013). Long-Chain Terminal Alcohols through Catalytic CO Hydrogenation. Journal of the American Chemical Society, 135(19), 7114–7117. doi:10.1021/ja402512r PubMed doi:10.1021/ja402512r PMID:23634891

Xiao, F.-S., Fukuoka, A., Ichikawa, M., Henderson, W., & Shriver, D. F. (1992). SiO2-supported bimetallic ruthenium cluster-derived catalysts for selectivity and activity control in CO hydrogenation toward C1-C5 alcohols. Journal of Molecular Catalysis, 74(1–3), 379–390. doi:10.1016/0304-5102(92)80255-F doi:10.1016/0304-5102(92)80255-F

Xiao, K., Bao, Z., Qi, X., Wang, X., Zhong, L., Fang, K., & Sun, Y. (2013). Structural evolution of CuFe bimetallic nanoparticles for higher alcohol synthesis. Journal of Molecular Catalysis A Chemical, 378(0), 319–325. doi:10.1016/j.molcata.2013.07.006 doi:10.1016/j.molcata.2013.07.006

Xiao, K., Qi, X., Bao, Z., Wang, X., Zhong, L., Fang, K., & Sun, Y. (2013). CuFe, CuCo and CuNi nanoparticles as catalysts for higher alcohol synthesis from syngas: A comparative study. Catalysis Science & Technology, 3(6), 1591–1602. doi:10.1039/c3cy00063j doi:10.1039/c3cy00063j

Xiaoding, X., Doesburg, E. B. M., & Scholten, J. J. F. (1987). Synthesis of higher alcohols from syngas - recently patented catalysts and tentative ideas on the mechanism. Catalysis Today, 2(1), 125–170. doi:10.1016/0920-5861(87)80002-0 doi:10.1016/0920-5861(87)80002-0

Xie, F. Z., Chu, X. W., Hu, H. R., Qiao, M. H., Yan, S. R., Zhu, Y. L., & Zhang, X. X. et al. (2006). Characterization and catalytic properties of Sn-modified rapidly quenched skeletal Ni catalysts in aqueous-phase reforming of ethylene glycol. *Journal of Catalysis*, *241*(1), 211–220. doi:10.1016/j.jcat.2006.05.001

Xu, M., & Iglesia, E. (1998). Initial carbon–carbon bond formation during synthesis gas conversion to higher alcohols on K–Cu–Mg5CeOx catalysts. Catalysis Letters, 51(1/2), 47–52. doi:10.1023/A:1019016513428 doi:10.1023/A:1019016513428

Xu, M., & Iglesia, E. (1999). Carbon–Carbon Bond Formation Pathways in CO Hydrogenation to Higher Alcohols. Journal of Catalysis, 188(1), 125–131. doi:10.1006/jcat.1999.2650 doi:10.1006/jcat.1999.2650

Xu, R., Yang, C., Wei, W., Li, W.-h., Sun, Y.-h., & Hu, T.-d. (2004). Fe-modified CuMnZrO2 catalysts for higher alcohols synthesis from syngas. Journal of Molecular Catalysis A Chemical, 221(1–2), 51–58. doi:10.1016/j.molcata.2004.07.003 doi:10.1016/j.molcata.2004.07.003

Xu, R., Zhang, S., & Roberts, C. B. (2013). Mixed Alcohol Synthesis over a K Promoted Cu/ZnO/Al2O3 Catalyst in Supercritical Hexanes. Industrial & Engineering Chemistry Research, 52(41), 14514–14524. doi:10.1021/ie3024017 doi:10.1021/ie3024017

Xu, A., Wang, Y., Ge, H., Chen, S., Li, Y., & Lu, W. (2013). An outstanding Cr-doped catalyst for selective oxidation of propane to acrylic acid. *Chinese Journal of Catalysis*, *34*(12), 2183–2191. doi:10.1016/S1872-2067(12)60671-8

Xu, N., Li, S., Jin, W., Shi, J., & Lin, Y. S. (1999). Experimental and modeling study on tubular dense membranes for oxygen permeation. *AIChE Journal. American Institute of Chemical Engineers*, *45*(12), 2519–2526. doi:10.1002/aic.690451207

Xun, S., Zhu, W., Zheng, D., Zhang, L., Liu, H., Yin, S., & Li, H. et al. (2014). Synthesis of metal-based ionic liquid supported catalyst and its application in catalytic oxidative desulphurization of fuels. *Fuel*, *136*(0), 358–365. doi:10.1016/j.fuel.2014.07.029

Xu, S., & Thomson, W. (1997). Perovskite-type oxide membranes for the oxidative coupling of methane. *AIChE Journal. American Institute of Chemical Engineers*, *43*(S11), 2731–2740. doi:10.1002/aic.690431319

Xu, S., & Thomson, W. (1998). Stability of La0.6Sr0.4Co0.2Fe0.8O3-δ perovskite membranes in reducing and nonreducing environments. *Industrial & Engineering Chemistry Research*, *37*(4), 1290–1299. doi:10.1021/ie970761j

Xu, X., Li, Y., Gong, Y., Zhang, P., Li, H., & Wang, Y. (2012). Synthesis of palladium nanoparticles supported on mesoporous N-doped carbon and their catalytic ability for biofuel upgrade. *Journal of the American Chemical Society*, *134*(41), 16987–16990. doi:10.1021/ja308139s PMID:23030399

Xu, Y. D., Bao, X. H., & Lin, L. W. (2003). Direct conversion of methane under nonoxidative conditions. *Journal of Catalysis*, *216*(1-2), 386–395. doi:10.1016/S0021-9517(02)00124-0

Xu, Y., Wang, Y., Zeng, Y., Jiang, J., & Jin, Z. (2012). A Novel Thermoregulated Ionic Liquid and Organic Biphasic System With Rh Nanoparticles for Olefin Hydroformylation. *Catalysis Letters*, *142*(7), 914–919. doi:10.1007/s10562-012-0841-1

Yacou, C., Sunarso, J., Lin, C. X. C., Smart, S., Liu, S., & Diniz da Costa, J. C. (2011). Palladium surface modified La0.6Sr0.4Co0.2Fe0.8O3−δ hollow fibres for oxygen separation. *Journal of Membrane Science*, *380*(1-2), 223–231. doi:10.1016/j.memsci.2011.07.008

Yamamatsu, S., & Yamaguchi, T. (1989). *US Patent No. US5191116(A).*

Yang, X., Zhu, X., Hou, R., Zhou, L., & Su, Y. (2011). The promotion effects of Pd on Fe–Cu–Co based catalyst for higher alcohols synthesis. Fuel Processing Technology, 92(10), 1876–1880. doi:10.1016/j.fuproc.2011.05.003 doi:10.1016/j.fuproc.2011.05.003

Yang, B., Yuschak, T., Mazanec, T., Tonkovich, A. L., & Perry, S. (2008). Multi-scale modeling of microstructured reactors for the oxidative dehydrogenation of ethane to ethylene. *Chemical Engineering Journal*, *135*(1), S147–S152. doi:10.1016/j.cej.2007.07.050

Yanglong, G., Guanzhong, L., Yunsong, W., & Ren, W. (2003). Preparation and characterization of Pd–Ag/ceramic composite membrane and application to enhancement of catalytic dehydrogenation of isobutene. *Separation and Purification Technology*, *32*(1-3), 271–279.

Yang, R. T., Hernández-Maldonado, A. J., & Yang, F. H. (2003). Desulfurization of transportation fuels with zeolites under ambient conditions. *Science*, *301*(5629), 79–81. doi:10.1126/science.1085088 PMID:12843388

Yang, S., Liang, C., & Prins, R. (2006). Preparation and hydrotreating activity of unsupported nickel phosphide with high surface area. *Journal of Catalysis*, *241*(2), 465–469. doi:10.1016/j.jcat.2006.05.014

Yan, N., Xiao, C., & Kou, Y. (2010). Transition metal nanoparticle catalysis in green solvents. *Coordination Chemistry Reviews*, *254*(9-10), 1179–1218. doi:10.1016/j.ccr.2010.02.015

Yansheng, C., Changping, L., Qingzhu, J., Qingshan, L., Peifang, Y., Xiumei, L., & Welz-Biermann, U. (2011). Desulfurization by oxidation combined with extraction using acidic room-temperature ionic liquids. *Green Chemistry*, *13*(5), 1224–1229. doi:10.1039/c0gc00745e

Yaremchenko, A. A., Buysse, C., Middelkoop, V., Snijkers, F., Buekenhoudt, A., Frade, J. R., & Kovalevsky, A. V. (2013). Impact of sulphur contamination on the oxygen transport mechanism through Ba0.5Sr0.5Co0.8Fe0.2O3-δ: Relevant issues in the development of capillary and hollow fibre membrane geometry. *Journal of Membrane Science*, *428*, 123–130. doi:10.1016/j.memsci.2012.10.033

Yin, H., Ding, Y., Luo, H., Zhu, H., He, D., Xiong, J., & Lin, L. (2003). Influence of iron promoter on catalytic properties of Rh-Mn-Li/SiO2 for CO hydrogenation. Applied Catalysis A, General, 243(1), 155–164. doi:10.1016/S0926-860X(02)00560-4 doi:10.1016/S0926-860X(02)00560-4

Yin, L., & Liebscher, J. (2007). Carbon-Carbon Coupling Reactions Catalyzed by Heterogeneous Palladium Catalysts. *Chemical Reviews*, *107*(1), 133–173. doi:10.1021/cr0505674 PMID:17212474

Yoosuk, B., Song, C., Kim, J. H., Ngamcharussrivichai, C., & Prasassarakich, P. (2010). Effects of preparation conditions in hydrothermal synthesis of highly active unsupported NiMo sulfide catalysts for simultaneous hydrodesulfurization of dibenzothiophene and 4,6-dimethyldibenzothiophene. *Catal. Today*, 149(1-2), 52-61. doi:10.1016/j.cattod.2009.05.001

Yu, J., Mao, D., Han, L., Guo, Q., & Lu, G. (2013). Synthesis of C2 oxygenates from syngas over monodispersed SiO2 supported Rh-based catalysts: Effect of calcination temperature of SiO2. Fuel Processing Technology, 106(0), 344–349. doi:10.1016/j.fuproc.2012.08.020 doi:10.1016/j.fuproc.2012.08.020

Yuan, X., Sun, G., Asakura, H., Tanaka, T., Chen, X., Yuan, Y., & Yan, N. et al. (2013). Development of palladium surface-enriched heteronuclear Au-Pd nanoparticle dehalogenation catalysts in an ionic liquid. *Chemistry (Weinheim an der Bergstrasse, Germany)*, *19*(4), 1227–1234. doi:10.1002/chem.201203605 PMID:23293076

Yu, F.-L., Wang, Y.-Y., Liu, C.-Y., Xie, C.-X., & Yu, S.-T. (2014). Oxidative desulfurization of fuels catalyzed by ammonium oxidative-thermoregulated bifunctional ionic liquids. *Chemical Engineering Journal*, *255*, 372–376. doi:10.1016/j.cej.2014.06.087

Yu, G., Li, X., Liu, X., Asumana, C., & Chen, X. (2011). Deep desulfurization of fuel oils using low-viscosity 1-ethyl-3-methylimidazolium dicyanamide ionic liquid. *Industrial & Engineering Chemistry Research*, *50*(4), 2236–2244. doi:10.1021/ie1013103

Yu, G., Zhao, D., Wen, L., Yang, S., & Chen, X. (2012). Viscosity of ionic liquids: Database, observation, and quantitative structure-property relationship analysis. *AIChE Journal. American Institute of Chemical Engineers*, *58*(9), 2885–2899. doi:10.1002/aic.12786

Yu, G., Zhao, J., Song, D., Asumana, C., Zhang, X., & Chen, X. (2011). Deep oxidative desulfurization of diesel fuels by acidic ionic liquids. *Industrial & Engineering Chemistry Research*, *50*(20), 11690–11697. doi:10.1021/ie200735p

Yumoto, M., Usui, K., Watanabe, K., Idei, K., & Yamazaki, H. (1997). Development of a Cosmo deep HDS catalyst for diesel fuel. *Catal. Today*, 35(1-2), 45-50. doi:10.1016/S0920-5861(96)00133-2

Yun, S., & Oyama, S. T. (2011). Correlations in palladium membranes for hydrogen separation: A review. *Journal of Membrane Science*, *375*(1-2), 28–45. doi:10.1016/j.memsci.2011.03.057

Yurchak, S. (1988). Development of Mobil's fixed-bed methanul-to-gasoline (MTG) process. *Studies in Surface Science and Catalysis*, *36*, 251–272. doi:10.1016/S0167-2991(09)60521-8

Zahmakiran, M., & Ozkar, S. (2011). Metal nanoparticles in liquid phase catalysis; from recent advances to future goals. *Nanoscale*, *3*(9), 3462–3481. doi:10.1039/c1nr10201j PMID:21833406

Zaman, S. F., & Smith, K. J. (2009). Synthesis gas conversion over MoP catalysts. Catalysis Communications, 10(5), 468–471. doi:10.1016/j.catcom.2008.10.022 doi:10.1016/j.catcom.2008.10.022

Zaman, S. F., & Smith, K. J. (2010). A study of K-promoted MoP–SiO2 catalysts for synthesis gas conversion. Applied Catalysis A, General, 378(1), 59–68. doi:10.1016/j.apcata.2010.01.046 doi:10.1016/j.apcata.2010.01.046

Zaman, S. F., & Smith, K. J. (2011). Synthesis gas conversion over a Rh–K–MoP/SiO2 catalyst. Catalysis Today, 171(1), 266–274. doi:10.1016/j.cattod.2011.02.017 doi:10.1016/j.cattod.2011.02.017

Zaman, S., & Smith, K. J. (2012). A Review of Molybdenum Catalysts for Synthesis Gas Conversion to Alcohols: Catalysts, Mechanisms and Kinetics. Catalysis Reviews, 54(1), 41–132. doi:10.1080/01614940.2012.627224 doi:10.1080/01614940.2012.627224

Zeiss, S., Dingerdissen, U., Baerns, M., Wolf, D., & David, L. (n. d.). *US Patent No. US6906221(B2).*

Zeng, Y., Lin, Y. S., & Swartz, S. L. (1998). Perovskite-type ceramic membrane: Synthesis, oxygen permeation and membrane reactor performance for oxidative coupling of methane. *Journal of Membrane Science*, *150*(1), 87–98. doi:10.1016/S0376-7388(98)00182-3

Zeyss, S., & Dingerdissen, U. (2001). *European Patent No. EP1286942B1.*

Zhang, M., & Yu, Y. (2013). Dehydration of Ethanol to Ethylene. Industrial & Engineering Chemistry Research, 52(28), 9505–9514. doi:10.1021/ie401157c doi:10.1021/ie401157c

Zhang, B., Zhan, X., Zhao, P., & Li, Z. (2015). Ionic liquid-mediated synthesis of unique PtPd bimetallic particles with tiny subunits for efficient electrocatalytic and catalytic applications. *RSC Advances*, *5*(71), 57640–57646. doi:10.1039/C5RA09832G

Zhang, C., Pan, X., Wang, F., & Liu, X. (2012). Extraction–oxidation desulfurization by pyridinium-based task-specific ionic liquids. *Fuel*, *102*, 580–584. doi:10.1016/j.fuel.2012.07.040

Zhang, G. R., & Xu, B. Q. (2010). Surprisingly strong effect of stabilizer on the properties of Au nanoparticles and Pt boolean AND Au nanostructures in electrocatalysis. *Nanoscale*, *2*(12), 2798–2804. doi:10.1039/c0nr00295j PMID:20938521

Zhang, H., & Cui, H. (2009). Synthesis and characterization of functionalized ionic liquid-stabilized metal (gold and platinum) nanoparticles and metal nanoparticle/carbon nanotube hybrids. *Langmuir*, *25*(5), 2604–2612. doi:10.1021/la803347h PMID:19437685

Zhang, J., Wang, A., Li, X., & Ma, X. (2011). Oxidative desulfurization of dibenzothiophene and diesel over [Bmim] 3 PMo 12 O 40. *Journal of Catalysis*, *279*(2), 269–275. doi:10.1016/j.jcat.2011.01.016

Zhang, J., Zhu, W., Li, H., Jiang, W., Jiang, Y., Huang, W., & Yan, Y. (2009). Deep oxidative desulfurization of fuels by Fenton-like reagent in ionic liquids. *Green Chemistry*, *11*(11), 1801–1807. doi:10.1039/b914130h

Zhang, M., Gao, M., Li, Q., Tao, M., & Yu, Y. (2014). DFT study on effect of CO on the system of acetoxylation of ethylene to vinyl acetate. *RSC Advances*, *4*(34), 17709–17716. doi:10.1039/c3ra47794k

Zhang, Q., Zhang, S., & Deng, Y. (2011). Recent advances in ionic liquid catalysis. *Green Chemistry*, *13*(10), 2619–2637. doi:10.1039/c1gc15334j

Zhang, S., van Houten, R., Eikelboom, D. H., Doddema, H., Jiang, Z., Fan, Y., & Wang, J. (2003). Sewage treatment by a low energy membrane bioreactor. *Bioresource Technology*, *90*(2), 185–192. doi:10.1016/S0960-8524(03)00115-9 PMID:12895562

Zhang, W., Xu, K., Zhang, Q., Liu, D., Wu, S., Verpoort, F., & Song, X.-M. (2010). Oxidative desulfurization of dibenzothiophene catalyzed by ionic liquid [BMIm] HSO4. *Industrial & Engineering Chemistry Research*, *49*(22), 11760–11763. doi:10.1021/ie100957k

Zhang, X., Guo, A., Wang, F., & Duan, X. (2010). Directly producing clean and low softening point diesel using integrated hydrotreating and hydroisomerizing catalysts. *Energy & Fuels*, *24*(7), 3772–3777. doi:10.1021/ef901373w

Zhang, X., Ye, Q., Xu, B., & He, D. (2007). Oxidative dehydrogenation of ethane over Co-BaCO3 catalysts using CO2 as oxidant: Effects of Co-promoter. *Catalysis Letters*, *117*(3-4), 140–145. doi:10.1007/s10562-007-9122-9

Zhang, X., Yu, G., Gong, Y., Jiang, D., & Xie, Y. (2000). Oxidative dehydrogenation of ethane to ethylene over NiO/Al2O3 catalyst. *Studies in Surface Science and Catalysis*, *130*, 1835–1840. doi:10.1016/S0167-2991(00)80468-1

Zhao, Di-shun, Zhi-min Sun, Fa-tang Li, & Hai-dan Shan. (2009). Optimization of oxidative desulfurization of dibenzothiophene using acidic ionic liquid as catalytic solvent. *Journal of fuel chemistry and technology,* 37(2), 194-198. doi:10.1016/S18725813(09)60015-3

Zhao, D., Fei, Z., Geldbach, T. J., Scopelliti, R., & Dyson, P. J. (2004). Nitrile-functionalized pyridinium ionic liquids: Synthesis, characterization, and their application in carbon-carbon coupling reactions. *Journal of the American Chemical Society*, *126*(48), 15876–15882. doi:10.1021/ja0463482 PMID:15571412

Zhao, P., Ji, S., Wei, N., Liu, H., & Li, C. (2010). Preparation, characterization and hydrodesulphurization catalytic performance of Al-Ni_2P/SBA-15 catalysts. *Wuji Huaxue Xuebao*, *26*, 2273–2279.

Zhao, P., Ji, S., Wei, N., Ma, Q., Liu, H., & Li, C. (2011). Effect of Boron promoter on the structure and hydrodesulfurization activity of Ni_2P/SBA-15 catalysts. *Acta Physico-Chimica Sinica*, *27*, 1737–1742.

Zhao, X., & Wang, X. (2006). Oxidative dehydrogenation of ethane to ethylene by carbon dioxide over Cr/TS-1 catalysts. *Catalysis Communications*, *7*(9), 633–638. doi:10.1016/j.catcom.2006.02.005

Zhou, C. H. C., Beltramini, J. N., Fan, Y. X., & Lu, G. Q. M. (2008). Chemoselective catalytic conversion of glycerol as a biorenewable source to valuable commodity chemicals. *Chemical Society Reviews*, *37*(3), 527–549. doi:10.1039/B707343G PMID:18224262

Zhu, H., Dong, H., Laveille, P., Saih, Y., Caps, V., & Basset, J.-M. (2014). Metal oxides modified NiO catalysts for oxidative dehydrogenation of ethane to ethylene. *Catalysis Today*, *228*, 58–64.

Zhu, W., Wu, P., Chao, Y., Li, H., Zou, F., Xun, S., & Zhao, Z. et al. (2013). A Novel Reaction-Controlled Foam-Type Polyoxometalate Catalyst for Deep Oxidative Desulfurization of Fuels. *Industrial & Engineering Chemistry Research*, *52*(49), 17399–17406. doi:10.1021/ie402513x

Zhu, W., Yu, Y., Yang, H., Hua, L., Qiao, Y., Zhao, X., & Hou, Z. (2013). Cooperative effects in catalytic hydrogenation regulated by both the cation and anion of an ionic liquid. *Chemistry (Weinheim an der Bergstrasse, Germany)*, *19*(6), 2059–2066. doi:10.1002/chem.201202707 PMID:23255466

Zhu, X. F., Liu, H. Y., Cong, Y., & Yang, W. S. (2012). Permeation model and experimental investigation of mixed conducting membranes. *AIChE Journal. American Institute of Chemical Engineers*, *58*(6), 1744–1754. doi:10.1002/aic.12710

Zope, B. N., Hibbitts, D. D., Neurock, M., & Davis, R. J. (2010). Reactivity of the Gold/Water Interface During Selective Oxidation Catalysis. *Science*, *330*(6000), 74–78. doi:10.1126/science.1195055 PMID:20929807

Zuo, Z.-J., Wang, L., Yu, L.-M., Han, P.-D., & Huang, W. (2014). Experimental and Theoretical Studies of Ethanol Synthesis from Syngas over CuZnAl Catalysts without Other Promoters. The Journal of Physical Chemistry C, 118(24), 12890–12898. doi:10.1021/jp502828c doi:10.1021/jp502828c

Zuo, D., Li, D., Nie, H., Shi, Y., Lacroix, M., & Vrinat, M. (2004). Acid–base properties of NiW/Al_2O_3 sulfided catalysts: Relationship with hydrogenation, isomerization and hydrodesulfurization reactions. *Journal of Molecular Catalysis A Chemical*, *211*(1-2), 179–189. doi:10.1016/j.molcata.2003.10.018

About the Contributors

Hamid A. Al-Megren obtained his BSc in Chemistry in 1987 and an MSc in Physical Chemistry in 1992 from the King Saud University, Riyadh, Saudi Arabia, and his D.Phil. in Inorganic Chemistry in 2002 from the Inorganic Chemistry Laboratory, University of Oxford, UK. Since 2008 he has been Director of the Petrochemical Research Institute at the King Abdulaziz City for Science and Technology (KACST), Riyadh, Saudi Arabia. He has more than 25 years of research experience in the petrochemical and catalysis reaction. Prof. Al-Megren started his career as a Scientific Researcher at KACST in 1987, then a researcher at the Saudi Arabian Basic Industry Company (SABIC, R&D), moving his way up to become the Manager of Refining and Petrochemicals Division at KACST, then Director of Petrochemical Research Institute, then Coordinator of the Materials and Nanotechnology Sector, ultimately attaining his current role of Assistant Director for Scientific Affairs in Material Research Institute (Sector). His research interests have included hydrotreatment reactions, isomerisation reactions, new mesoporous materials preparation, carbide, nitride, and sulfide catalysts. Prof. Al-Megren has authored and co-authored more than 52 publications in his research field. He is Editor-in-Chief of the journal of Applied Petrochemical Research and Editor of the book "Advances in Natural Gas Technology".

Tiancun Xiao received his BSc from Norwest University China in 1987 and his Doctoral Degree of Chemistry from Lanzhou Institute of Chemical Physics, Chinese Academy of Sciences in 1993. He worked in Shandong University until 1999, reaching the position of deputy Dean of the Environmental Engineering Department. During this period he collaborated with the World Bank and United Nation Environmental Programme to introduce cleaner production to the chemical and petrochemical industry in China. He joined the University of Oxford in 1999 as the Royal Society BP-Amoco Fellow, and co-founded Oxford Catalysts, a spin-out company in 2004. The company was created based upon his research at the Inorganic Chemistry Laboratory at the University of Oxford and floated on the London AIM Market in April 2006, and now re-branded as Velocys (http://www.velocys.com" www.velocys.com). After the company operate steadily and shifts focus to USA in 2010, he moved back to Oxford University, working as a Senior Research Fellow (<http://research.chem.ox.ac.uk/ tiancun-xiao.aspx>), and together with Professor Peter P Edwards and Professor Hamid AlMegren, he setup the KACST-Oxford Petrochemical Research Centre (KOPRC) in Oxford in 2010. The 2.5 million pounds KOPRC was commissioned by the Government of Saudi Arabia. KOPRC has recently been designated as a Centre of Excellence in Petrochemicals. Dr Xiao has industrialized 8 catalyst materials and other functional materials. The Fischer Tropsch catalysts designed and developed by him formed the foundation for the creation of Oxford Catalysts Plc. These catalysts can improve liquid productivity by 10-20 times, which makes it economically viable to utilise flare gas and convert biomass into super clean diesel. His other new technologies are being used by several new startup companies in Guangzhou China, where novel mercury removal materials as well as functional food have been developed and supplied. More details can be found at: http://www.boxenergytech.com" www.boxenergytech.com and http://www.boxabio.com" www.boxabio.com

***

Tauqeer Abbas graduated in B.Sc Chemical Engineering from COMSATS Institute of Information Technology (CIIT), Lahore in 2011. After his graduation, he joined and worked in Department of Environmental Sciences, CIIT, Vehari as Research Associate till 2012 and after that he joined Department of Chemical Engineering, CIIT Lahore campus as Research Associate. Currently, he is studying Masters in Chemical Engineering from Universiti Teknologi PETRONAS (UTP) on scholarship awarded by UTP and on Ex-Pakistan study leave from CIIT, Lahore, Pakistan. His area of interest includes separation of contaminates from oil and gas using ionic liquids, adsorption, renewable energy, application of nanoparticles.

Rashid Abro is a Chemical Engineer at the Pakistan Council of Scientific and Industrial Research (PCSIR) in Pakistan. He is a currently PhD scholar (Funded by Higher Education Commission—HEC, Pakistan) at Beijing university of Chemical Technology under the supervision of Dr Xiaochun Chen. His PhD research focuses on the applications of ionic liquids for sulfur removal from real oils. He received his Bachelors and Master's Degrees in Chemical Engineering in 2007 and 2011, respectively, from Mehran University of Engineering and Technology, Jamshoro, Pakistan. He also worked as a lecturer in the Department of Chemical Engineering, Dawood University of Engineering and Technology, Karachi for three years.

Mohammed Al-Hazmi is currently the Global Technology Director for Oxygenates and Glycols at SABIC. He has a Bachelor Degree in Chemistry from King Saud University, Riyadh, MSc in Physical Chemistry from the same university and a Ph.D. in Catalysis in 2005 from Oklahoma State University, United States. He has more than 18 years' experience in research and technology related activities, dealing with design, development, synthesis and characterization of many heterogeneous and homogenous catalytic systems used in petrochemical processes such as selective oxidation, oxidative dehydrogenation, oxidative coupling, selective hydrogenation, alkylation and isomerization, aromatization, light olefins oligomerization etc. Additionally, he was involved in development of several technologies from the lab scale to a full commercial scale. Dr. Al-Hazmi received several national and international awards and he is member of several international scientific societies such as Royal Society of Chemistry and American Chemical Society, he has about 16 publications in peer-reviewed scientific journals, 20 granted and published patents, and about 10 conference and symposium talks.

Ahmad Alshammari was born in Hail, Saudi Arabia on May 16, 1980. He received the B.Sc. degree in Chemistry from King Saud University, Riyadh, Saudi Arabia, in 2001, the M.Sc. degree in Designing Chemical Solutions from the University of Newcastle-upon-Tyne, Newcastle-upon-Tyne, U.K., in 2005, and then he studied chemistry at Leibniz Institute for Catalysis at the University of Rostock, where he completed his PhD thesis in the end of 2010 in the group of Prof. Axel Schulz and PD Dr. habil. Andreas Martin. In 2011, he joined the Petrochemical Research Institute, King Abdulaziz City for

Science and Technology (KACST), Riyadh, where among other things he focused on the application of Heterogeneous Catalysis in different area. Presently he is Assistant Research Professor in the National Center for Nanotechnology, KACST. His principle research field is focused mainly on the application of nanomaterials and metal organic frameworks in different areas such as catalysis, photocatalysis, and energy. His scientific work has been published in more than 15 original publications. In addition, 6 patent applications have been granted in the last three years.

Abdulaziz Bagabas received his B.S. degree with First Honour in Chemistry from King Saud University in 1997 and his Ph.D. degree with Excellence in Inorganic Chemistry from Oklahoma State University in 2005. After completion of his undergraduate studies, he joined the Petroleum and Petrochemicals Research Institute (PAPRI), currently National Petrochemical Technology Center (NPTC), Materials Science Research Institute (MSRI), King Abdulaziz City of Science and Technology (KACST), where he is currently working Associate Professor. He was appointed an Honorary Research Associate at the department of Chemistry, University of Glasgow in 2006. He was also a visiting scholar at Chemistry Department, Northwestern University, Evanston, Illinois, USA, from 2011-2013. He has been recently got a visiting scholar position at the Department of Chemistry, University of California, Berkeley, California, USA. His research covers broad range of chemistry of materials. In particular, his research focuses on inorganic and organic syntheses, homogeneous and heterogeneous catalysis, environmental chemistry, nanotechnology, and corrosion. His work involves several joint researches with several national and international institutions. He has published more than 30 peer-reviewed articles, two book chapters, and 12 granted patents. He is a member of the editorial board of "Applied Petrochemicals Research" journal, published jointly by KACST and Springer, and a reviewer for various international journals.

Hisham S Bamufleh is an associate professor in chemical engineering department at King Abdulaziz University (KAU). He worked in Saudi Aramco as a process engineer in oil processing units and utilities including desalination and steam generation plants. Dr. Bamufleh is a licensed professional engineer (PE). Dr. Bamufleh has MS and PhD from the University of Tulsa Oklahoma and BS from KAU all in chemical engineering. He also has MS from KAU in thermal engineering Dr. Bamufleh has 15 refereed publications. His main research interests involve diesel desulfurization using activated carbon derived from dates' stones and utilization of agriculture waste to produce valuable products such as furfural. Process integration and plant simulation using HYSYS is also among his interest.

Giuseppe Barbieri, Coordinator of the LoLiPEM FCH-JU project, responsible for ITM of projects co-funded by EC (Hydrofueler, GRACE, NanoGloWa), Ministry of Foreign Affairs and International Cooperation of Italy (METT 2013, METT 2014, etc.), Ministry for Education, University and Research (e.g., FotoRiduco2) and Norwegian Research Council (OMPA). He has expertise in the modelling and experimental testing of membranes and membrane modules in gas separation, catalytic membrane reactors for high and low temperature reactions. Author of 70 publications on peer reviewed International Journals, 9 books chapters, a two books set, an Italian patent, and numerous presentations in International Conferences, some of them as invited plenary or keynote, in the field of Membrane Science and Technology. He serves as reviewer for several International Scientific Journals.

Adele Brunetti PhD in Chemical Engineering, since 2012, researcher at ITM-CNR; Since 2004 works in the framework of several projects at international and national level. Her main expertise are in modelling and experimental testing of catalytic membrane reactors for high and low temperature reactions. Testing and characterization of metallic, polymeric, zeolitic, ceramic membranes for gas separation. Design of integrated membrane systems for high purity (CO<10 ppm) hydrogen production. Author of 40 papers published on referenced international scientific journals, 7 book chapters and more than 100 international conference proceedings, she has participated with oral presentations to several international conferences. Referee of several scientific journal of international level (ISI). Winner of the EFCE-Excellence Award in Membrane Engineering 2010 for the best recognized PhD thesis in the years 2006-2009; Awarded as young talent in the research field for the "Festival dei Giovani Talenti", supported by Italian Ministry of Youth.

Abdul Waheed Bhutto is an Assistant Professor at Dawood University of Engineering & Technology, Karachi. He is currently a Ph.D. candidate at Mehran University Jamshoro. He is concentrating his efforts on improving the efficiency of processes for biofuel production from lignocellulosic material. He received his Master's Degree from NED University Karachi in 2001. He earned his Bachelor's degree in chemical engineering from Dawood Engineering College Karachi in 1998. His research work has focused on alternate energy sources for sustainable development. He is an author of more than 30 research papers.

Carine E. Chan-Thaw is an assistant Researcher at the Università degli Studi di Milano. Her PhD research, based on the in situ UV-Vis-NIR spectroscopic study of sulphated zirconia, was mainly carried out at the Fritz Haber Institute of the Max Planck Gesellschaft under the supervision of Prof. Robert Schlögl. She received a European PhD in Chemistry, 2008, from both the Université de Strasbourg and the Technische Universität Berlin. To broad her knowledge, she accepted Post-doctoral fellowships at the Technische Universität Berlin where she worked on testing nitrogen rich highly porous framework for liquid phase oxidations. She is currently investigating on the valorization of biomass

Haoyi Chen Project Manager of Guangzhou Boxenergy Tech Ltd since March 2011, in charge of the R&D of catalysts.

Xiaochun Chen is a professor and dean of College Chemical Engineering, Beijing University of Chemical Technology. He obtained his PhD degree in Chemical Engineering in 1998. He has an extensive research and teaching background in chemical and environmental engineering. Dr Chen is an author of more than 60 research papers and patents. His current research focuses on the design, synthesis and applications of ionic liquids in separation technology, industrial catalysis, membrane separation technology, process simulation and optimization.

Youquan Deng is a Research Professor of Physics Chemistry and head of the Centre for Green Chemistry and Catalysis, Lanzhou Institute of Chemical Physics, Chinese Academy of Sciences. He is also the professor of the Graduate School of the Chinese Academy of Sciences. He received a BS degree at Lanzhou University and got his PhD at the University of Portsmouth (UK, with Dr T. G. Nevell). He has been engaging in the R&D of ionic liquids, green chemistry, non-phosgene processes and nano-catalysis for many years. He has authored 40 patents and more than 150 scientific papers in international journals.

Nikolaos Dimitratos studied chemistry at the University of Bath and received a PhD at the University of Liverpool in 2003. He was a postdoctoral researcher at the University of Milan (Professor Laura Prati and Professor Michele Rossi) and University of Cardiff (Professor Graham J. Hutchings). His current research field is catalysis, covering areas of synthesis, characterisation and reactivity in oxidation and hydrogenation reactions. He is currently University Research Fellow and Independent Research Group Leader in the area of catalysis and nanoparticle synthesis at the school of Chemistry, Cardiff University.

Enrico Drioli Distinguished Adjunct Professor, CEDT-KAU Jeddah Saudi Arabia (2012) WCU-Distinguished Professor, Department of Energy Engineering, Hanyang University, Seoul (2009) Emeritus Professor, School of Engineering of the University of Calabria (2012) Founding Director of the Institute on Membrane Technology (ITM) of CNR (from 2011) Director of the Institute on Membrane Technology (ITM) of CNR (1993 - 2008) Author of more of 700 scientific papers, 18 patents, and 18 books. His main skills are: Membranes in Artificial Organs, Integrated Membrane Processes, Membrane Preparation Transport Phenomena, Membrane Distillation and Membrane Contactors, Catalytic Membrane and Catalytic Membrane Reactors.

Mahmoud M. El-Halwagi is a professor and holder of the McFerrin Professorship in chemical engineering at Texas A&M University. He is also a distinguished adjunct professor of chemical and materials Engineering at King Abdulaziz University. His main research interests involve sustainable process design, simulation, operation, integration, and optimization. Dr. El-Halwagi is the co-author/co-editor of 8 books and more than 250 refereed papers and book chapters. He has served as a consultant to a wide variety of chemical, petrochemical, petroleum, pharmaceutical and metal finishing industries. Dr. El-Halwagi has a PhD from the University of California, Los Angeles and an MS and BS from Cairo University all in chemical engineering.

Hamdy Farag Professor and Research Scientist at Kyushu University, Mansoura University

Taylor W. Gaines is a senior materials engineer employed by Intel Corporation in Chandler, AZ and a 2015 graduate from Professor Kenneth B. Wagener's group. Taylor's graduate research focus included sulfur-functionalized polyolefins synthesized via ADMET. For his research excellence, he was awarded the University of Florida's 2014 Butler Polymer Research Award. He graduated from the University of South Carolina in 2011 with his B.S. under the advisement of Professor Brian C. Benicewicz. During his undergraduate studies, he performed proton-exchange membrane fuel cell research.

Srikant Gopal is currently a Lead Scientist at SABIC Technology Center, Riyadh. He obtained his Ph.D. in Chemical Engineering in 2003 from the University of Cincinnati, USA. He has 12 years of experience in heterogeneous catalysis and process research working at the technology centers of SABIC, Dow Chemical and Shell. In his career Dr. Srikant Gopal has performed research in the areas of zeolite catalysis for production of fuels and chemicals, catalysts for epoxidation of ethylene to ethylene oxide, decomposition of N_2O, and biomass conversion to hydrocarbon fuels. A key aspect of his work is taking forward the catalyst inventions through scale-up to commercial deployment. Currently, selective oxidation of lower alkanes is one of his major areas of research.

Ceri Hammond received his PhD in 2012 from Cardiff University, under the guidance of Prof. G. J. Hutchings FRS. Following several years at ETH Zürich (post-doc, 2011-2014), he accepted a position at the Cardiff Catalysis Institute as a University Research Fellow and Research Group Leader (tenure-track). Since October 2015, his research is fully supported by The Royal Society via the University Research Fellowship program. Ceri also holds dual affiliation as a Royal Society of Chemistry Research Fellow at Stanford University, where he spends approximately 30% of his time. His research team specializes in several aspects of applied catalysis, inorganic materials, and in situ spectroscopy. He has authored over 35 publications since 2011, holds 2 international patents, has provided 15 international presentations (7 invited), and has received a number of international awards including the "Best European PhD Thesis" award of the EU Federation of Catalysis Societies, 2011-2013.

Yu Huang, from 2011.4-present, was at Guangzhou Boxenergy Tech Ltd., and was a Project Manager 2008.9-2011.3, at East China University of Science and Technology, Master.

Shengfu Ji was born in 1957. He is the Professor and the PhD supervisor at Beijing University of Chemical Technology. He received his PhD from the Lanzhou Institute of Chemical Physics, Chinese Academy of Sciences, in 1998. He had worked at the Lanzhou Institute of Chemical Physics from 1998 to 2002. He did his research as a senior visitor in the Department of Chemistry of Oxford University from 2000 to 2002 and did cooperative research at Oxford University funded by the Great Britain Royal Society. Now, Professor JI is working at the State Key Laboratory of Chemical Resource Engineering, Beijing University of Chemical Technology. He has finished and participated in many research projects funded by the national, the provincial or ministerial, as well as the international cooperation. He has published over 150 papers and has been awarded 16 patents. His research fields are the catalytic chemistry, energy and environmental sciences. The research focuses include the catalytic conversion of light alkanes, the novel nano-catalytic materials, the green chemistry and the environmental catalysis, as well as the chemical process intensification and energy saving technology.

Venkata Narayana Kalevaru was born in Seethampeta (Khammam), Andhra Pradesh State, India. He carried out his doctoral research at Indian Institute of Chemical Technology, India on the ammoxidation of hetero aromatics and received his Ph.D. degree in Chemistry (Heterogeneous Catalysis) from Osmania University, India in 1998. Then he joined Institute for Applied Chemistry, Berlin in 2000 as a scientist dealing with various basic and industrial projects. Presently, he is a group leader in the Department of Heterogeneous Catalytic processes, Leibniz institute for Catalysis, Rostock, Germany. His research interests include catalyst synthesis, characterisation and catalyst evaluation for various reactions, e.g. selective oxidations, ammoxidation aromatics/heteroaromatics, oxy-chlorinations, acetoxylation of aromatics, natural gas and CO2 utilisation etc. are some of his key research areas. He holds ~30 patents and ca. 90 publications in the field of catalysis.

Masahiro Kishida Professor of chemical engineering at Kyushu University.

Jingdong Lin Department of Chemistry, College of Chemistry and Chemical Engineering, Xiamen University, Xiamen. Lin Jingdong was born in 1971 in Fujian, China. He graduated from Fuzhou University in 1993, obtained his Ph.D. degree in 2010 from Xiamen University.In 2004 he was appointed as an Assistant Professor of College of Chemistry and Chemical Engineering, Xiamen University. His research interests are related to catalytic conversion of syngas and photocatalysis.

Shimin Liu was born in 1984 in Shandong, China. He received his BS degree in chemistry from Shandong Normal University in 2006. He received his PhD degree in 2012 in Physical Chemistry from the Lanzhou Institute of Chemical Physics (LICP), Chinese Academy of Sciences (China, with Prof. Y. Deng). He now works at the Centre for Green Chemistry and Catalysis in LICP as a research assistant. His current research interests are focused on ionic liquids and nano-catalysis.

Helen H. Lou is the Director of Sustainable Technology Laboratory, and a professor in the Dan F. Smith Department of Chemical Engineering. His work focuses on sustainable engineering, i.e., the development of technologies for reducing the environmental impacts and improving the safety, efficiency and profitability of industrial activities. She has been applying process modeling, design, simulation, control and optimization methods in pollution prevention, waste management, and process integration. Dr. Lou has extensive leadership experience in American Institute of Chemical Engineers (AIChE). Currently, she is a Director of AIChE Fuel and Petrochemicals Division, Representative of East Asia in AIChE International Committee. In AIChE Process Development Division, she served as the Chair (2003 – 2005) of Manufacturing Area. In AIChE Sustainable Engineering Forum (SEF), she served as the Chair (2010 – 2011). In addition to her service in AIChE, Dr. Lou is a Board Member of the Gulf Coast Regional Center of Innovation and Commercialization (2009 – 2015). She is also an editorial member of Journal of Environmental Progress and Sustainable Energy.

Andreas Martin was born in Wilkau-Haßlau, Germany. He got his Ph.D. degree in Chemistry (Heterogeneous Catalysis) from the Academy of Sciences, Berlin in 1986. After his Ph.D. he was engaged in C1-Chemistry, at Central Institute of Organic Chemistry (Berlin) in MTG/MTO catalysis and pyrolysis of hydrocarbons on zeolites. He also worked for some time at University Bremen as a postdoctoral research fellow (Zeolite chemistry and catalysis). Then he moved again to Berlin and worked at the Centre for Heterogeneous Catalysis, Berlin as Head of project group "Zeolite catalysis". In 1994, he joined the Institute of Applied Chemistry Berlin-Adlershof (ACA) as a scientist dealing with oxidation and ammoxidation of methyl aromatics. In 1997, he became Head of project group "Catalysis in multi-phases and supercritical fluids (SCF)". His research interests include heterogeneously catalysed oxidation reactions in SCF, catalytic wet oxidation, high-pressure reactions, bio-mass conversion, partial oxidations and ammoxidation of methyl aromatics. Currently, he is the Head of Heterogeneous Catalytic processes department, Leibniz Institute for Catalysis at the University of Rostock, Rostock, Germany. He holds over 30 patents and ca. 250 publications for his credit in various fields of catalysis.

Mohamed M.B. Noureldin completed his PhD at Texas A&M University under the supervision of Dr. Mahmoud El-Halwagi and Dr. Nimir Elbashir. His research interests are in process integration, optimization and process design. Mohamed received his B.E. in biochemical engineering from University of Waikato, New Zealand. Mohamed's current research focuses on developing a systematic framework for simultaneous energy and chemicals production through the combination of fossil fuels (e.g. shale gas) and renewables (e.g. biomass). The research aims to design integrated eco-industrial parks capable of converting various feedstock into a range of products including fuels, chemicals and polymers. Mohamed is the recipient of several awards including the Eastman Graduate Fellowship, Texas A&M Energy Institute ConocoPhillips Fellowship and the Texas A&M Institute for Advanced Study (TIAS) Fellowship.

Laura Prati after Specialization in ''Analytical technique applied to fine chemicals" at Politecnico of Milan in 1985, was awarded a PhD in Industrial Chemistry in 1988. She became Associate Professor of Inorganic Chemistry at the Universita` degli Studi di Milano in 2001. Her scientific interests focus principally on catalytic hydrogenation and oxidation reactions and specific design of heterogeneous catalysts. She has been involved in catalytic applications for gold since 1986.

Jörg Radnik 1996 PhD at the University of Bonn, Germany 1996 - 1997 Postdoc at the CEA, Saclay 1998 Scientist at the ACA, Berlin since 2006 Scientist at LIKAT, Rostock.

Serbia M. Rodulfo-Baechler obtained the PhD degree in Chemical Engineering & Advanced material from Newcastle University in 2012 under the supervision of Professor Ian Metcalfe. She was an academic visitor in the Wolfson catalysis centre at Oxford University for 3 years (2002-2005). During this period she was working with Professor Malcolm L. H. Green FRS. She obtained a Master degree in Applied Chemistry focus on Material Studies and the degree of Licentiate in Chemistry at the University of Los Andes, Venezuela.

Miriam Sellaro Graduated in Chemistry at the University of Calabria, Rende, Italy in December 2011. Since January 2014 Research Fellow at ITM-CNR. Main expertise in catalytic membranes preparation and tests for low temperature reactors.

Ravinder Singh Chemical Engineer with extensive experience in process safety.

Jie Sun received his B.S. degree in applied chemistry at Chongqing University of China in 2010. He is now working on his doctoral degree under the supervision of Prof. Yong Wang at Xiamen University. His research mainly focuses on the highly efficient synthesis of mixed alcohols from syngas.

Alberto Villa is an Assistant Professor at Universita` degli Studi di Milano. He received his PhD in Industrial Chemistry in 2007 from the Universita` degli Studi di Milano, where his thesis focused on the development of gold based catalysts for liquid phase reactions. He worked at the Fritz Haber Institute of Max Planck Gesellschaft of Berlin, with Prof. Robert Schlögl. His current research focuses on the development of heterogeneous catalysts for biomass transformation.

Kenneth B. Wagener, the George Bergen Butler Professor of Polymer Chemistry, Department of Chemistry, University of Florida received the 2013 Herman F. Mark Award. Professor Wagener was recognized as an outstanding polymer chemist who is highly regarded for his passion for education at the undergraduate and graduate level. His research group, which totals over 145 students, pioneered the ADMET reaction, which is used worldwide, leading to numerous tangible collaborations in Germany, Japan, and several laboratories in the USA. Prior to joining Florida he served in several industrial managerial positions for eleven years within Akzo Nobel. Other prestigious awards in recent times include 2012 Visiting Professorships at ETH/Zürich, MPI/Polymers Mainz, & Kyoto University, Japan; being inducted as a 2011 Fellow of the American Chemical Society; the 2010 American Chemical Society Herty Award and the 2009 Max Planck Institute for Polymer Research Award. Soon he will receive the 2016 Paul Flory Education Award, given to one polymer chemist in the USA every two years. The award will be given at the 2016 ACS spring meeting in the USA.

Yong Wang joined PNNL in 1994 and was promoted to Laboratory Fellow in 2005. He led the Catalysis and Reaction Engineering Team from 2000 to 2007, and has served as the Associate Director of the Institute for Integrated Catalysis (http://iic.pnl.gov) since 2008. In 2009, Dr. Wang assumed a joint position at Washington State University and PNNL. In this unique position, he continues to be a Laboratory Fellow and associate director of IIC at PNNL and is the Voiland Distinguished Professor in Chemical Engineering at WSU, a full professorship with tenure. Dr. Wang is best known for his leadership in the development of novel catalytic materials and reaction engineering for the conversion of fossil and biomass feedstocks to fuels and chemicals. Dr. Wang has authored 190 peer reviewed publications with H index=44, co-edited 2 books and 5 special journal issues, and given more than 120 invited presentations since 2001. He is the inventor on 243 issued patents including 90 issued U.S. patents (>90% of his patents are licensed to industries). His discoveries in microchannel reaction technologies led to the formation of Velocys, trading under the London Stock Exchange (VLS). He is a fellow of 4 major professional societies: AIChE (American Institute of Chemical Engineers), ACS (American Society of Chemistry), RSC (Royal Society of Chemistry), and AAAS (American Association of the Advancement of Science). He has won numerous awards including 2006 Asian American Engineer of the Year Award, Presidential Green Chemistry Award, 3 R&D 100 Awards, Distinguished Alumni Achievement Award from Chemical Engineering at WSU, 2 PNNL Inventor of the Year Awards, Battelle Distinguished Inventor Award, and the first recipient of PNNL Laboratory Director's Award for Exceptional Scientific Achievement Award. He is the past chair of the Energy & Fuel Division of the American Chemical Society, and currently serves editorial board of 8 catalysis and energy related journals including ACS Catalysis andCatalysis Today.

Shaolong Wan Department of Chemistry, College of Chemistry and Chemical Engineering, Xiamen University, Xiamen.

Kathryn R. Williams has over 40 years' experience teaching university-level chemistry and writing instructional materials. After completing her PhD at the Florida State University in 1972, she joined the faculty at Florida International University, where she designed the lecture and laboratory programs in inorganic and analytical chemistry. In 1979, Dr. Williams moved to the University of Florida to become the coordinator of laboratories in analytical, physical, and radiochemistry. During her teaching career, Dr. Williams wrote seven student laboratory manuals and coauthored over 50 technical publications, many of them describing new student laboratory experiments. She also produced over 20 video tapes to enhance laboratory instruction. In 1999, she became editor of the "From Past Issues" feature of the Journal of Chemical Education, for which she wrote non-technical articles on interesting or off-beat themes encountered in previous volumes of the journal. Now semi-retired from full-time teaching, Dr. Williams works as a free-lance editor of journal manuscripts, dissertations, and proposals, and she has served as the Wagener group editor since September 2008. Realizing in her advanced years that it was time to learn some synthetic techniques, in 2013 she plunged into active laboratory work. She is now a familiar figure in the Butler Polymer Research Laboratory.

Guangren Yu is a professor in the College of Chemical Engineering, Beijing University of Chemical Engineering (BUCT). He received his BS, MS and PhD degrees in chemical engineering from Shandong University of Industrial Technology in 2000, BUCT in 2003 and Institute of Process Engineering, Chinese Academy of Sciences in 2007, respectively. In 2012, he did one-year visiting scholar research at the University of California at Berkeley, USA under the supervision of Professor John M. Prausnitz. He is an author of more than 50 research papers and patents. He is researching on the theory and application of ionic liquids in clean fuel production, acid gas capture separation, computer simulation and QSPR.

Index

A

B

C

D

E

F

G

H

S

T

V

X

www.ingramcontent.com/pod-product-compliance
Lightning Source LLC
LaVergne TN
LVHW080620300926
849162LV00005B/1219
* 9 7 8 1 4 6 6 6 9 9 7 5 5 *